OPTICAL GUIDED WAVES AND DEVICES

OPTICAL GUIDED WAVES AND DEVICES

Richard Syms
John Cozens

Department of Electrical and Electronic Engineering
Imperial College of Science, Technology and Medicine

McGRAW-HILL BOOK COMPANY

London · New York · St Louis · San Francisco · Auckland · Bogotá · Caracas
Hamburg · Lisbon · Madrid · Mexico · Milan · Montreal · New Delhi · Panama
Paris · San Juan · São Paulo · Singapore · Sydney · Tokyo · Toronto

Published by
McGRAW-HILL Book Company Europe
Shoppenhangers Road, Maidenhead, Berkshire, SL6 2QL, England
Telephone 0628 23432
Fax 0628 770224

British Library Cataloguing in Publication Data
Syms, R. R. A.
Optical Guided Waves and Devices
I. Title II. Cozens, J. R.
621.3848

ISBN 0-07-707425-4

Library of Congress Cataloging-in-Publication Data
Syms, R. R. A.
Optical guided waves and devices / R. R. A. Syms, J. R. Cozens.
p. cm.
Includes bibliographical references and index.
ISBN 0-07-707425-4 :
1. Optical wave guides. 2. Fiber optics. 3. Integrated optics.
I. Cozens, J. R. (John Ritchie), . II. Title.
TA1800.S96 1992 92–7193
621.36'9--dc20 CIP

1234 9432

Printed and bound in Great Britain by
BPCC Hazells Ltd
Member of BPCC Ltd

CONTENTS

ACKNOWLEDGEMENTS

It is a pleasure to acknowledge the very considerable assistance given to us in the preparation of this book by our colleagues in the Optical and Semiconductor Devices Section at Imperial College. Professor M. Green, A. S. Holmes, S. Makrimichalou and Dr E. M. Yeatman all read portions of the manuscript, and we are grateful for their many valuable and constructive suggestions. We owe a considerable debt to Dr K. D. Leaver, who provided a continuing stream of advice on optoelectronic devices; his own course notes were invaluable in writing Chapters 11 and 12 of the book. Many undergraduates also took the trouble to point out numerous mathematical errors in the text; to these past students, we are also indebted.

We would also like to thank a number of people the world over, who were kind enough to supply information in their research and photographic material for use in the book: Dr R. Alferness, AT&T Bell Laboratories; Dr C. J. Armistead, STC Technology Ltd.; R. Blanchard, Bell Northern Research Ltd.; Dr P. Buchmann, IBM Research Division, Zurich; Ms D. DeRose, NSG America Inc.; Dr W. Döldissen, Heinrich-Hertz-Institut; Dr R. Dyott, Andrew Corp.; Dr P. Granestrand, Ericsson Telecom AB; Dr G. Heise, Siemens AG; D. Henderson, Biccotest Ltd.; Dr A. Himeno, NTT; Dr R. A. Lawes, SERC RAL; Prof. D. Marcuse, AT&T Bell Laboratories; A. Mills, BT&D Technologies; Dr D. Mortimore, BTRL; Dr Y. Okamura, Osaka University; Prof. D. Payne, Southampton University; Dr P. M. Rogers, BTRL; Dr T. Suhara, Osaka University; Dr H. Sundaresan, BTRL; Dr S. Takagi, Furukawa Electric Co. Ltd.; D. Tedone, GTE Products Corporation; Dr A. Tobin, Sifam Ltd.; Prof. R. Ulrich, Technische Universität Hamburg-Harburg; Dr S. Valette, LETI.

CHAPTER

ONE

OVERVIEW

1.1 GUIDED WAVE OPTICAL DEVICES

There is now little doubt that guided wave optical devices will have an increasing impact on electrical engineering in the coming years. Primarily, this is because of the sustained development of two extremely successful components: optical fibre, and the semiconductor laser. Together, these are currently in the process of revolutionizing the telecommunications industry, and making considerable advances in a variety of other new and exciting application areas.

Optical fibre acts as a propagation medium for guided optical waves. The most important fibre type is single-mode optical fibre, which can support guided waves for extremely long distances (tens or hundreds of kilometres) with low loss and low signal dispersion. Single-mode fibre is generally formed from silica glasses, which are arranged in a cylindrical geometry with a core region of high refractive index surrounded by a lower-index cladding layer (as shown in Fig. 1.1).

The overall diameter of a single-mode fibre is comparable to that of a human hair. However, the light actually travels almost entirely in the much smaller core region (which typically has a diameter of only a few microns) and is contained inside this region and guided along the fibre by a phenomenon known as total internal reflection. This can occur when light is incident on an interface between two lossless dielectric media with different refractive indices, provided the light is incident from the high-index medium at a shallow enough angle, as shown in Fig. 1.2. In this

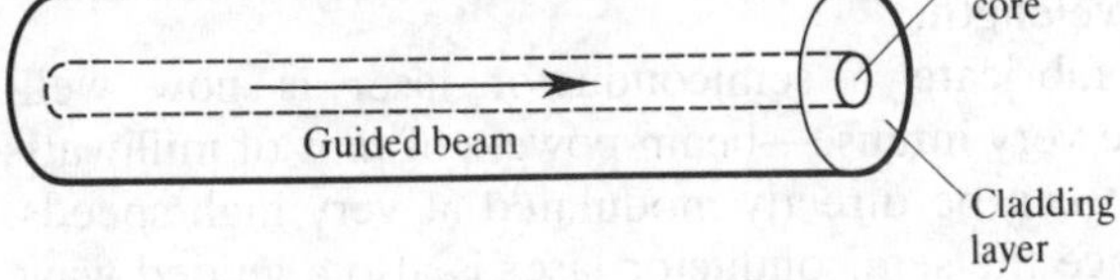

Figure 1.1 An optical fibre.

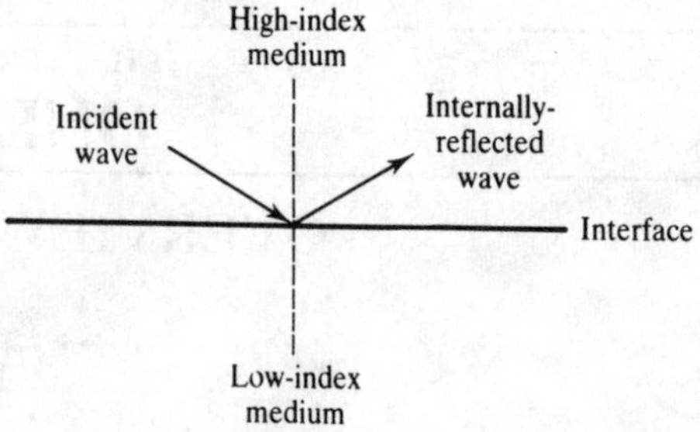

Figure 1.2 Total internal reflection

case, there is no transmission of energy across the interface, which simply acts as a highly efficient reflector.

In the last two decades the optical fibre has finally allowed the practical use of light as a carrier of information, and this application has given rise to the entirely new field of photonic engineering. Since an optical field oscillates at extremely high frequency ($\approx 10^{15}$ Hz), a light wave acts as a high-frequency carrier. In combination with the low dispersion provided by an optical fibre, this allows information to be transferred at very high data rates. In fact, the optical fibre offers an increase in channel capacity of a factor of 10^4–10^5 over its nearest competitor, the microwave guide, coupled with a considerable reduction in size, weight and cost. Because of the low cost of the raw materials used (effectively, purified sand), optical fibre will replace copper cables in many applications involving the transfer of low-power signals.

A further type of optical fibre, known as multi-mode fibre, also exists. This has a similar cylindrical geometry, but a much larger core diameter (typically 50–100 μm). It can be formed from either glass or plastic. However, as the name implies, multi-mode fibre can support more than one characteristic propagation mode, and consequently suffers from much greater dispersion than single-mode fibre. It is also extremely lossy, and is therefore suitable only for medium-bandwidth, short-haul applications.

The semiconductor laser is a small, efficient light source. It is the natural successor of the light-emitting diode (or LED), a low-coherence, low-power source which may only be modulated at slow rates (up to tens of megahertz). However, because of its relative cheapness and simplicity, the LED has been widely used in the past in multi-mode optical fibre systems. Figure 1.3 shows a commercial near-infrared buried heterostructure semiconductor laser contained in a dual in-line package. The laser itself is no bigger than a grain of sand, and is primarily formed from layers of the crystalline material indium gallium arsenide phosphide, grown in differing material compositions by sophisticated epitaxial techniques. It is designed to emit light at around 1.3 μm wavelength.

The technology required to fabricate a semiconductor laser is now well-developed. Its optical output can be very intense—beam powers of tens of milliwatts are now routinely available—and it can be directly modulated at very high speeds, typically in the gigahertz range. Since the semiconductor laser is also a guided wave

Figure 1.3 A packaged semiconductor laser with a fibre pigtail (*photograph courtesy A. Mills, BT&D Technologies*).

device, and emits light from a small stripe window which has dimensions comparable to those of the core of an optical fibre, it is immediately compatible with fibre systems. Indeed, the module in Fig. 1.3 is already fitted with a single-mode fibre pigtail. This is butted up to the emitting facet of the laser, deep inside the package, allowing very simple connection to a fibre system. Furthermore, the semiconductor laser can emit light of the wavelengths at which silica fibre shows minimum dispersion and minimum propagation loss, near $\lambda_0 = 1.3$ and $1.5\ \mu m$ respectively. With careful design, the output can also consist of very nearly a single optical frequency. This is an important advantage for communications, since it reduces the effect of dispersion.

Tuneable lasers for use in wavelength-division multiplexed communications systems, surface-emitting lasers for optical processor applications and high-power laser arrays are also the subject of an intense development effort. The latter can generate quite staggering amounts of power—many watts, continuous wave (CW)—from a tiny chip, and may well pose a challenge to large gas lasers in the future. Recent advances include the use of semiconductor lasers as in-line amplifiers and repeaters in optical fibre systems, and the development of semiconductor laser-pumped optical fibre lasers.

Together with photodiodes, which are light-detecting devices fabricated in semiconductor materials, optical fibre and semiconductor lasers form the key elements of optoelectronics. In addition, a large number of other components have been developed, which allow a complete guided wave circuit to be assembled. These ancillary components are in a more rudimentary form at present, and it is hard to predict which of them will be the most successful in the long run. As with most things, the deciding factor will almost certainly be cost rather than absolute performance.

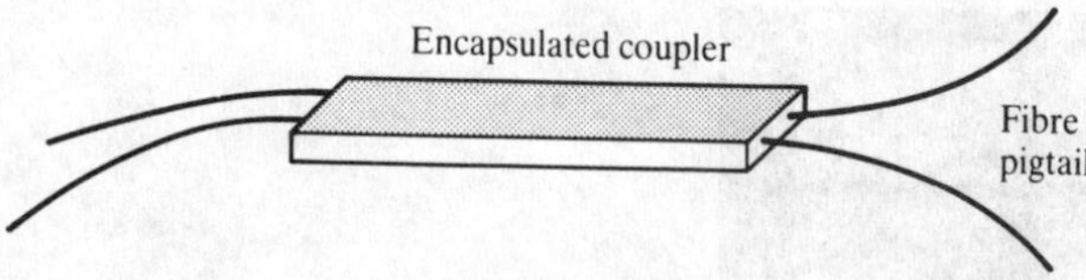

Figure 1.4 A packaged fibre coupler.

The subsidiary components have evolved in several different formats. For example, a number of passive devices are available in all-fibre form, and may readily be spliced into fibre systems. .The most widely-used device is the fibre coupler, which can be used for beam-splitting and filtering operations. Figure 1.4 shows a packaged fused tapered fibre coupler, which is available in a variety of fixed splitting ratios ranging from 50:50 to 1:99. More complicated power division functions—for example, a $1 \times N$ split—are possible using similar technology. All-fibre polarizers and polarization controllers provide ways of manipulating the polarization state of guided beams. Although the devices are fabricated using relatively simple equipment, they must generally be made individually, so the construction of a complicated optical circuit is time-consuming. Furthermore, since optical fibre is based on amorphous material, it has been found to be extremely difficult to fabricate more sophisticated active devices (such as high-speed modulators) in this way.

Another common component format, which offers a solution to some of these problems, is provided by integrated optics. As the name implies, this allows a number of guided wave devices to be combined on a common substrate. In contrast to optical fibre devices, integrated optic components are made using a mass-production technology akin to that required for very large scale integrated (VLSI) microelectronics. Complex optical circuits may therefore be constructed as easily as simple ones, the main limitation being the size of available substrates. There are two basic configurations. Historically, planar waveguide integrated optics was the first system to be developed, followed shortly by the more sophisticated and successful channel waveguide integrated optics system.

Planar waveguide integrated optics is concerned with the manipulation of sheet beams. These can propagate in any direction parallel to the surface of a high-index guiding layer, which provides optical confinement in a single direction. Figure 1.5

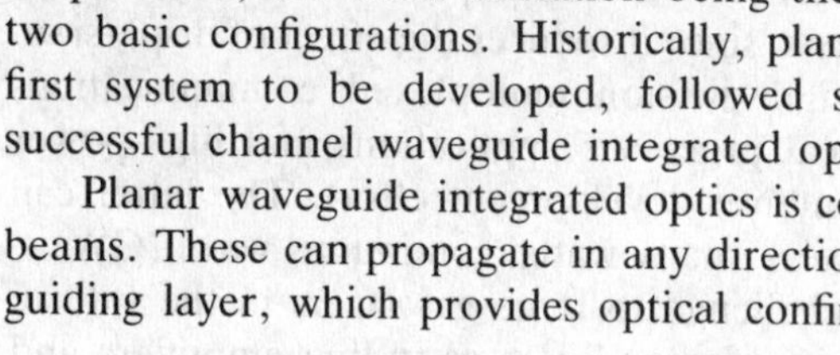

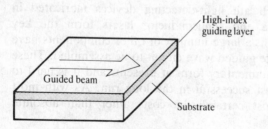

Figure 1.5 A planar waveguide.

shows a three-layer planar guide, formed by depositing a thin layer of material of high refractive index on a thicker, low-index substrate. The third layer in the system can often be air itself, or an additional low-index cover layer can be used. Once again, guidance is provided by total internal reflection at the layer interfaces. Because sheet beams allow many of the operations that are possible using free-space optics (for example, focusing by a lens, or beam deflection), planar integrated optic chips often contain circuits that are, in effect, miniaturized and ruggedized versions of bulk optic systems. Many of these are used for parallel signal processing operations, based on the Fourier transform properties of a lens.

In the second configuration, channel waveguide integrated optics, the beams propagate along high-index guiding channels. These may be formed as ridges on the surface, or as buried channels as shown in Fig. 1.6. The beams are therefore confined in two dimensions, and so can only follow predefined pathways round the chip. If the channel dimensions are chosen to correspond with those of an optical fibre core, this form of integrated optics is directly compatible with fibre optics, and fibre and channel guide components may be connected together quite simply. Integrated optic components can, of course, be formed on passive substrates such as glass or plastic. At the very least, these allow the construction of beamsplitting and combining devices. For example, Fig. 1.7 shows a number of different 1 × 2 and 1 × 4 passive splitters formed by silver–sodium ion exchange in glass. The integrated optic components themselves are contained inside the packages, and are again fitted with fibre pigtails.

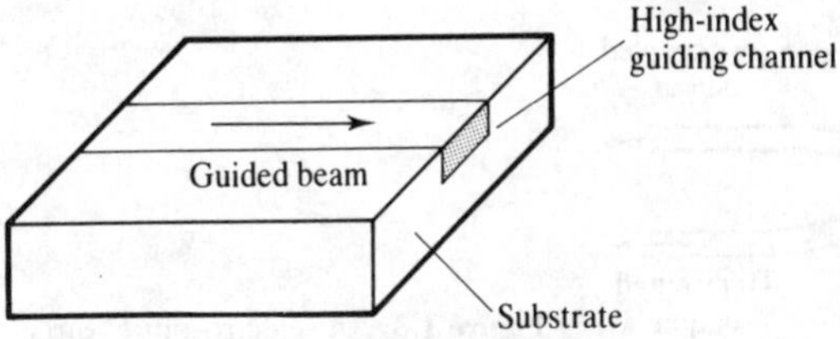

Figure 1.6 A channel waveguide

If other materials are used, more exotic functions are possible. For example, using electro-optic substrates like $LiNbO_3$, GaAs and InP, modulation and switching can be performed, often at extremely high speeds (tens of gigahertz). Figure 1.8 shows an electro-optic directional coupler, a generic device type which can be used to switch light between two adjacent parallel waveguides under electrical control, or modulate an information channel. These operations can be used to place information on the carrier wave, and then route it around a network. Similarly, filtering operations allow the channel to be multiplexed and demultiplexed by wavelength-division. These functions can even be combined with the generation and detection of light, if the waveguide is fabricated in a semiconductor substrate with a suitable bandgap (such as GaAs or InP). Integration also offers the intriguing possibility of combining optical components with their controlling electronics, in the

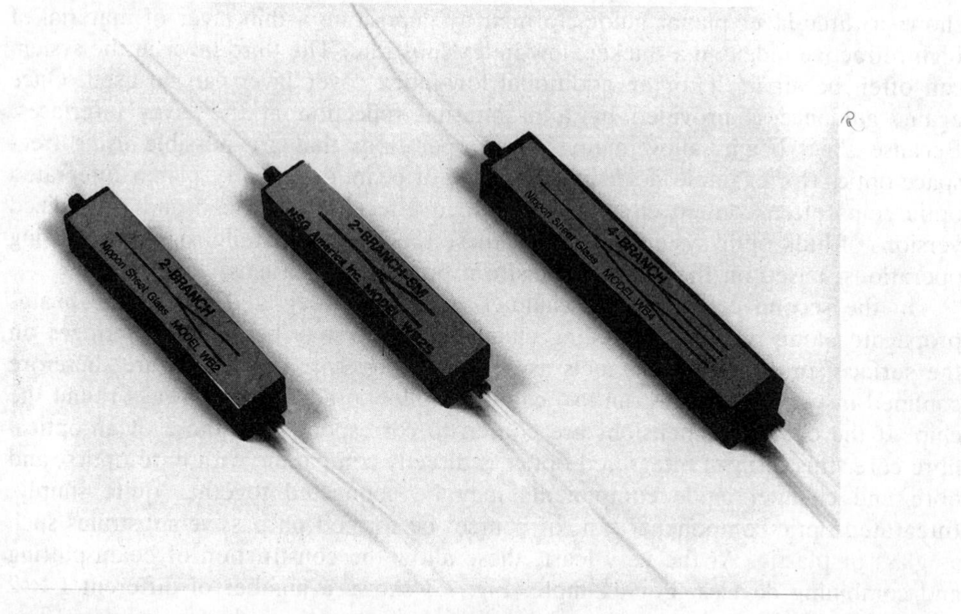

Figure 1.7 Passive channel waveguide integrated optic beamsplitting components formed by ion exchange in glass (*photograph courtesy D.DeRose, NSG America Inc.*).

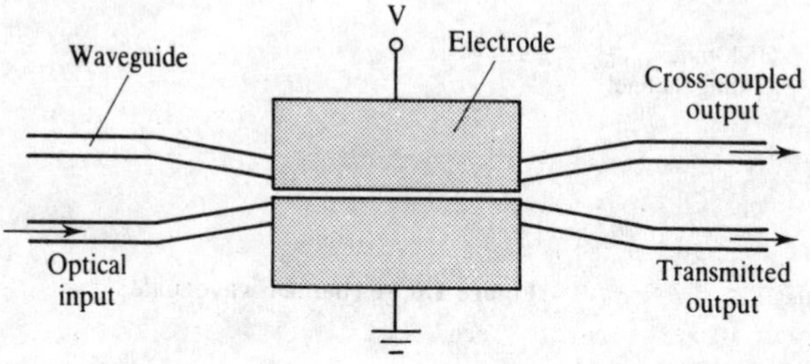

Figure 1.8 An electro-optic directional coupler switch.

form of integrated optoelectronics; or of using light as a method of communicating between very high speed electronic circuits in fast computers.

Using integrated optics, complete optical circuits can be constructed on a common substrate, often with very small dimensions. One example might be a switch matrix, capable of routeing light between N input and N output fibres in a communications system, thus forming a compact electrically-controllable optical 'telephone exchange'. Figure 1.9 shows a packaged 4×4 lithium niobate switch array, constructed from a set of directional coupler switches. Here, each directional coupler acts as a switching node, and suitable configuration of the array can allow a pathway to be set up between any input fibre and any output.

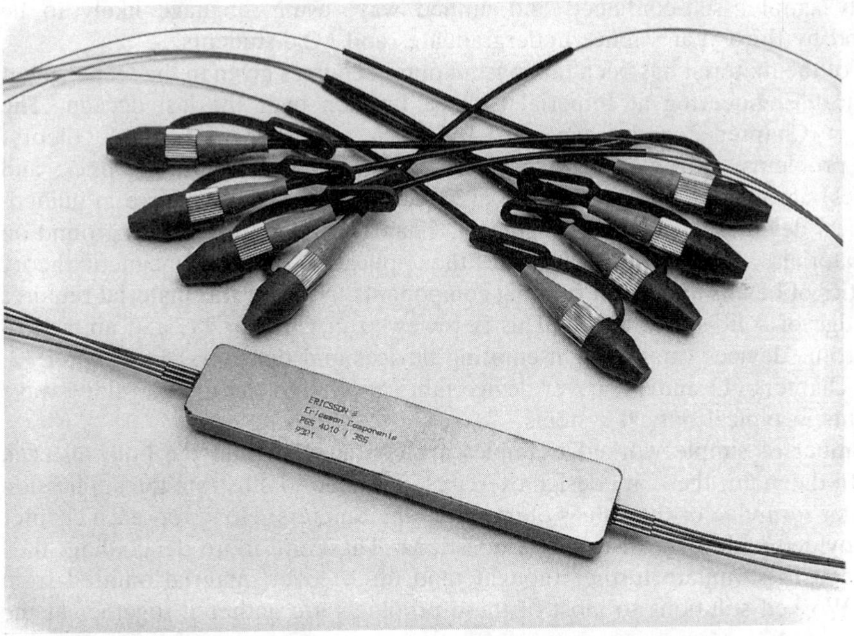

Figure 1.9 A packaged, pigtailed 4 × 4 switch array in titanium-diffused $LiNbO_3$ (*photograph courtesy P. Granestrand, Ericsson Telecom AB*).

A second application might involve the fabrication of an optical interferometer, using a coil of optical fibre (which provides the sensor element) and a single channel waveguide integrated optic chip (which carries a number of beamsplitting and signal processing components). Such devices can be arranged to sense variations in a wide range of physical parameters. For example, a Sagnac interferometer based on an optical fibre coil can act as a rotation sensor or gyroscope. In contrast to its mechanical equivalent, the optical fibre gyroscope requires no precision mechanical parts, little maintenance, and no 'run-up' time. Combined with linear accelerometers, such devices may well one day form the basis of solid-state inertial navigation systems, sensing rotational motion about three orthogonal axes.

1.2 RATIONALE

This book is intended to provide an introduction to the whole field of guided wave devices, which has proved to be one of the major technological growth areas of this decade. It is not a research monograph, and we make no claim that the topics covered are a description of the latest advances; in fact, the whole field is developing so rapidly that it is hard to identify the leading edge of technology in many cases. Instead, our aim is to describe a large range of devices and applications in a reasonably simple, self-contained and unified way, using language likely to be understood by third-year science undergraduates and MSc students.

Most of the material has been taken from option courses given in the Department of Electrical Engineering at Imperial College London over the last decade. The contents of Chapter 2 and Chapters 5–9 (which cover electromagnetic theory, interface problems, waveguides, planar and channel guide integrated optics, and fibre optics) are currently being used as a 20-lecture introductory course in guided-wave optical devices for electrical engineers. Chapter 3 adds further background on optical materials, and Chapter 4 illustrates the application of electromagnetic theory to the optics of beams and beam-forming components. None of this material requires a knowledge of solid-state theory. This is reviewed in Chapter 11, and applied to optoelectronic devices (mainly light-emitting devices and detectors) in Chapter 12. The final chapters (13 and 14) cover device fabrication, and the use of guided-wave components in typical optical systems.

A number of simple worked examples are contained within the body of each chapter. In the main, these are design exercises, intended to illustrate the application of particular formulae or the values of important parameters. However, each chapter is also provided with a set of example questions. These are more demanding; they are intended to stimulate further thought, and often cover material omitted from the text. Worked solutions to most of these problems are gathered together at the end of the book.

CHAPTER

TWO

ELECTROMAGNETIC FIELDS AND PLANE WAVES

2.1 MAXWELL'S EQUATIONS

The understanding of any field of physics or electrical engineering requires a suitable theoretical basis. In optics, we are fortunate that two highly developed and accurate theories are available. In the older theory—often described as the 'classical theory'—the behaviour of light is described in terms of electromagnetic fields and waves. This is particularly appropriate for the analysis of passive devices, where the absorption and emission of radiation is unimportant, and consequently where the interaction of a wave with matter may then be represented in a somewhat phenomenological way.

In the newer theory—the *quantum theory*—a different model is employed. Light is considered to be composed of photons, which are the elementary units or quanta of radiation. The interaction of light and matter is then understood in terms of exchanges of energy between photons and electrons. For example, the generation of a photon may be identified with the transition of a single electron between two energy levels. Quantum theory is therefore directly applicable to active optical devices. The development of this alternative model, and of its later incarnation, quantum electrodynamics, occupied the first half of the century, and involved many of the world's foremost physicists.

Quantum theory may also describe situations that do not involve matter at all, but are still not accurately represented by classical theory. One example is provided by low light levels, where the photon flux may be extremely small and the arrival of radiation in discrete units is important. However, for a high enough photon flux, the two theories are equivalent. We shall encounter both of them in this book. Since our early discussions will concentrate on passive devices, we shall begin with a classical approach, turning only to the quantum theory at a much later stage (Chapter 11).

In classical theory, the laws of electricity and magnetism are described by *Maxwell's equations*. These represent the result of a synthesis of several existing theories and experimental observations by James Clerk Maxwell (1831–1879). In effect, Maxwell's equations are a set of relations linking the values of a number of quantities that describe electric and magnetic fields. These are the *electric flux density* **D**, the *magnetic flux density* **B**, the *electric field strength* **E**, the *magnetic field strength* **H**, and the *current density* **J**. All are vector quantities, and are functions not only of the three spatial coordinates x, y and z but also of time t. Of these parameters, **D**, **B**, **E** and **H** are the most important to high frequency electromagnetic theory. In this regime they are essentially 'inventions'; they are not directly observable, but are linked by a self-consistent set of equations, which correctly predict the magnitudes of other measurable quantities (like the flow and distribution of power). Often, different representations are used for the fields; for example, it is common to separate their time- and space-variation. The electric field strength **E** may therefore be written alternatively as

$$\mathsf{E}(x,y,z,t) = \mathbf{E}(x,y,z)\,f(t) \tag{2.1}$$

Here **E** accounts for the spatial variation of the field, and $f(t)$ for temporal changes. Note the use of a bold-faced sanserif type for the complete field **E**, and a bold, serif type for the *time-independent field* **E**.

It is also common to refer to individual elements of the vectors concerned. In cartesian coordinates, we may write the time-dependent electric field **E** as

$$\mathsf{E}(x,y,z,t) = [\mathsf{E}_x(x,y,z,t), \mathsf{E}_y(x,y,z,t), \mathsf{E}_z(x,y,z,t)] \tag{2.2}$$

Alternatively, we could write for the time-independent field **E**:

$$\mathbf{E}(x,y,z) = [E_x(x,y,z), E_y(x,y,z), E_z(x,y,z)] \tag{2.3}$$

Here, E_x refers to the x-component of **E**, and so on.

Because of the nature of the fields, Maxwell's equations are written in terms of vector calculus. Though this is hard for the novice to begin with, the notation rapidly becomes a useful (not to say essential) tool-of-the-trade in electromagnetic theory. Unfortunately, for reasons of space, we must assume here some familiarity with the basic techniques involved.

The equations can be written in two different forms. From experimental observations, by and large carried out in the previous century, the common integral version was obtained. This is generally presented in the form of a series of laws. The first pair of these (Gauss' law, and its magnetic equivalent) describe relations which are most important for static fields.

Gauss' Law

Consider a closed surface, surrounding a volume containing electric charges, as shown in Fig. 2.1. *Gauss' law* (named after the German mathematician Karl Gauss, 1777–1855) states that the electric flux flowing out of the surface is equal to the charge enclosed inside the volume. This assertion can be written mathematically in terms of the vector fields involved in Maxwell's theory in the following way.

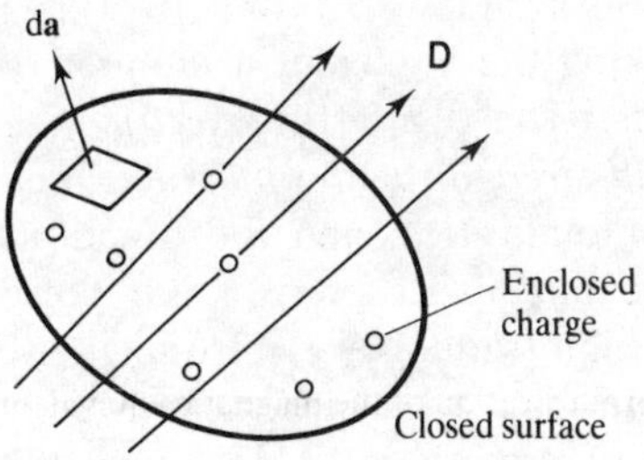

Figure 2.1 Geometry for illustration of Gauss' law.

First, we define a vector **da** as having magnitude equal to the area da of a small element of the surface, and a direction normal to that element. Second, we introduce a new operation, the *scalar* or *dot product* between two vectors **F** and **G**. This is written as **F** · **G**, and is defined in cartesian coordinates as

$$\mathbf{F} \cdot \mathbf{G} = F_x G_x + F_y G_y + F_z G_z \tag{2.4}$$

Simple trigonometry may then be used to show that Eq. (2.4) can also be written as

$$\mathbf{F} \cdot \mathbf{G} = |\mathbf{F}|\,|\mathbf{G}| \cos\theta \tag{2.5}$$

where $|\mathbf{F}|$ and $|\mathbf{G}|$ are the moduli or lengths of the two vectors, and θ is the angle between them.

Consequently, if **D** is the electric flux density, the term **D** · d**a** represents the product of the component of **D** normal to the small surface element d**a** and the area of the element. When integrated over the whole of the surface, this will give the net outward normal electric flux. Similarly, if we define the scalar term ρ as the *charge density*, the integral of ρ over the whole volume must give the charge enclosed. We may therefore write Gauss' law for vector fields as

$$\iint_A \mathbf{D} \cdot d\mathbf{a} = \iiint_V \rho \, dv \tag{2.6}$$

Here the left-hand integral is a *surface integral*, taken over the whole of the closed surface, while the right-hand one is a *volume integral*, over the volume enclosed.

Magnetic Equivalent of Gauss' Law

We can do the same thing for magnetic flux density. If we now consider the magnetic flux flowing out of a closed surface, we get a similar picture, shown in Fig. 2.2 Comparison with Fig. 2.1 shows that the resulting integral equation must have a similar form. However, no *magnetic monopoles* (the magnetic equivalent of electric charges) have ever been found experimentally, despite extensive searches. Evidence of this curious fact is provided by the simple bar magnet, which has both north and south poles. If the magnet is divided into two, each half will also have two poles,

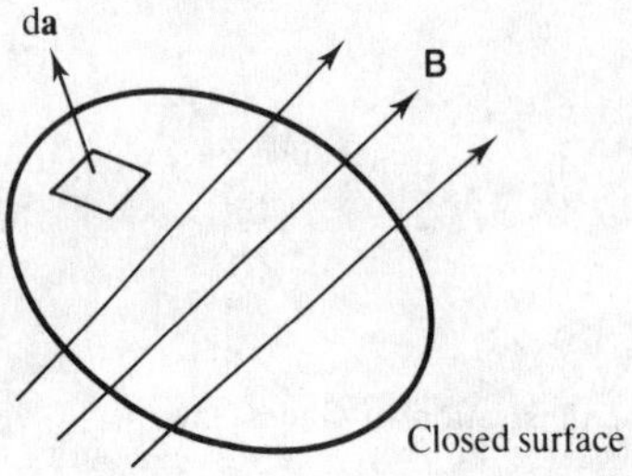

Figure 2.2 Geometry for illustration of the magnetic equivalent of Gauss' law.

and no amount of further subdivision will isolate a monopole. In this case, therefore, the right-hand side of the equation is zero, giving

$$\iint_A \mathbf{B} \cdot \mathbf{da} = 0 \tag{2.7}$$

where **B** is the magnetic flux density.

The second pair of laws (Faraday's and Ampère's laws) describe relations which are of greater significance for time-varying fields.

Faraday's Law of Magnetic Induction

Consider a time-varying magnetic flux passing through a closed loop L, defining the rim of an open surface, as shown in Fig. 2.3. The flux of magnetic induction Ψ_B through the open surface is

$$\Psi_B = \iint_A \mathbf{B} \cdot \mathbf{da} \tag{2.8}$$

The *electromotive force* (EMF) induced round the loop is therefore

$$\text{EMF} = -\partial \Psi_B / \partial t \tag{2.9}$$

We know from the integral relationship between electric potential and field strength that this can also be written as a *line integral*, in the form

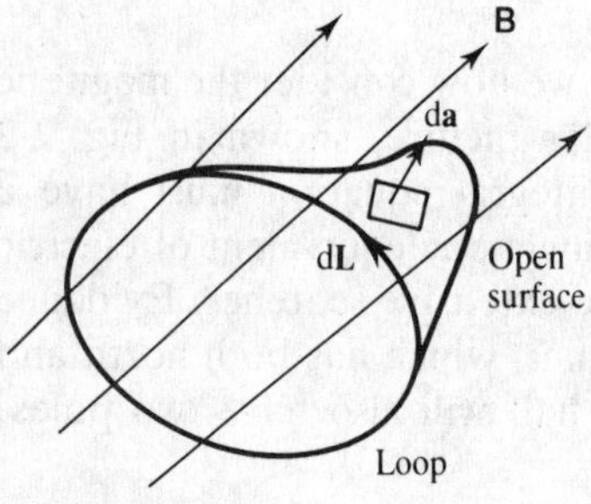

Figure 2.3 Geometry for illustration of Faraday's law.

$$\text{EMF} = \int_L \mathbf{E} \cdot d\mathbf{L} \tag{2.10}$$

Here $\mathbf{E}$ is the electric field strength, $d\mathbf{L}$ is a small element of the closed loop in Fig. 2.3, and the integral is taken round the whole loop. By comparing Eqs (2.9) and (2.10), and using Eq. (2.8) we obtain *Faraday's law*, named after Michael Faraday (1791–1867):

$$\int_L \mathbf{E} \cdot d\mathbf{L} = -\iint_A \partial\mathbf{B}/\partial t \cdot d\mathbf{a} \tag{2.11}$$

Faraday's law implies that a time-varying magnetic field must have an electric field associated with it. This feature is of great importance for electromagnetic waves, as we shall see later on.

Ampère's Law

Now consider the flow of current through a closed loop, of the same geometry, as shown in Fig. 2.4.

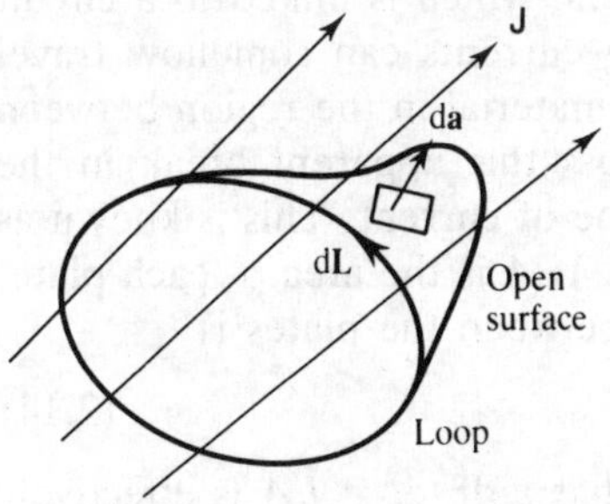

Figure 2.4 Geometry for illustration of Ampère's law.

Ampère's law (named after André Ampère, 1775–1836) states that

$$\int_L \mathbf{H} \cdot d\mathbf{L} = \iint_A \mathbf{J} \cdot d\mathbf{a} \tag{2.12}$$

where $\mathbf{H}$ is the magnetic field strength and $\mathbf{J}$ is the current density. This implies that moving charges give rise to a magnetic field, a notion in pleasing symmetry with Faraday's law.

In addition to these laws, there are a set of relationships known as the *material equations*, which link field strengths with flux densities, through a set of material coefficients that are representative of the bulk properties of matter. There are three of them, written

$$\mathbf{J} = \sigma \mathbf{E}$$

$$\mathbf{D} = \varepsilon \mathbf{E}$$

$$\mathbf{B} = \mu \mathbf{H} \tag{2.13}$$

Here σ is the *conductivity*, ε is the *permittivity* or *dielectric constant*, and μ is the *permeability*. The physical origin and significance of some of these quantities will be discussed in Chapter 3. Note that the first equation is a vectorial form of *Ohm's law*, commonly encountered in electrical engineering. For the dielectric materials used to guide high frequency electromagnetic waves (which we will mainly consider here) the conductivity is typically zero, while the permeability is that of free space. The former implies that the current density $\mathbf{J}$ is zero; there are also no free charges, so that ρ is zero.

This is not quite the end of the story, however, because experiments showed that magnetic fields can also be measured in free space—for example, between the plates of a capacitor, while it is being charged. Consequently, electromagnetic theory was modified by Maxwell (in what amounts to a stroke of genius) to cope with this observation.

Generalized Form of Ampère's Law

Consider the parallel-plate capacitor shown in Fig. 2.5, which is linked to a circuit by wires carrying a current I. Clearly, time-varying currents can somehow travel round the circuit, despite the absence of conducting material in the region between the plates. To account for the flow of current across this apparent break in the circuit, Maxwell suggested the existence of a new type of current. This is known as the *displacement current*, and is calculated as follows. If A is the area of each plate, and Q is the charge on it, then the electric field $\mathbf{E}$ between the plates is

$$\mathbf{E} = Q/\varepsilon A \tag{2.14}$$

As the charge varies, the electric field changes, so that $\varepsilon\, \mathrm{d}\mathbf{E}/\mathrm{d}t = I/A$ is effectively a current density. Maxwell therefore defined a vector displacement current density $\mathbf{J}_D$ as

$$\mathbf{J}_D = \varepsilon\, \partial\mathbf{E}/\partial t = \partial\mathbf{D}/\partial t \tag{2.15}$$

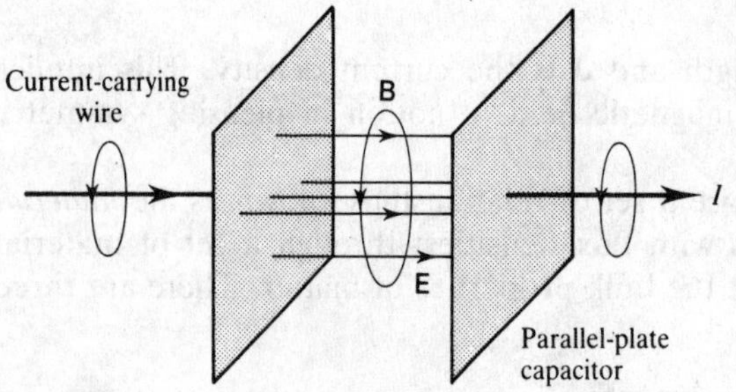

Figure 2.5 Geometry for derivation of Maxwell's displacement current.

This displacement current must be added into any calculation involving the 'normal' conduction current. The only example we have come across so far is Ampère's law. Including the displacement current, Ampère's law must be restated as

$$\int_L \mathbf{H} \cdot \mathrm{d}\mathbf{L} = \iint_A [\mathbf{J} + \partial\mathbf{D}/\partial t] \cdot \mathbf{da} \qquad (2.16)$$

This simple modification was of great historical importance, since it showed that time-varying electric fields can exist *even in the absence of normal currents* (i.e. when $\mathbf{J} = 0$). It allowed Maxwell to justify theoretically the electromagnetic waves discovered experimentally by Heinrich Hertz in 1888, thus demonstrating at a stroke that static electric and magnetic fields, and radio and light waves, are all part of the wider phenomenon of electromagnetism.

2.2 THE DIFFERENTIAL FORM OF MAXWELL'S EQUATIONS

We can use two standard vector theorems to transform the integral equations into an alternative differential form, which will be of great use in later calculations. The theorems are described in detail in many mathematics texts, and so will simply be stated here.

Gauss' Theorem

The first is *Gauss' theorem* (not to be confused with Gauss' *law*), which states that for any vector field $\mathbf{F}$ the following relation holds:

$$\iint_A \mathbf{F} \cdot \mathbf{da} = \iiint_V \operatorname{div} \mathbf{F} \, \mathrm{d}v \qquad (2.17)$$

Here the *divergence* of a vector field $\mathbf{F}$ is an important new quantity, a scalar term, which is defined in cartesian coordinates by

$$\operatorname{div} \mathbf{F} = \partial F_x/\partial x + \partial F_y/\partial y + \partial F_z/\partial z \qquad (2.18)$$

The operation div $\mathbf{F}$ is also often written as $\nabla \cdot \mathbf{F}$, where the vector operator ∇ (or 'del') is defined in cartesian coordinates as

$$\nabla = \mathbf{i}\, \partial/\partial x + \mathbf{j}\, \partial/\partial y + \mathbf{k}\, \partial/\partial z \qquad (2.19)$$

Here $\mathbf{i}$, $\mathbf{j}$ and $\mathbf{k}$ are unit vectors in the x-, y- and z-directions, respectively. Hence, Gauss' law becomes

$$\iint_A \mathbf{D} \cdot \mathbf{da} = \iiint_V \operatorname{div} \mathbf{D} \, \mathrm{d}v = \iiint_V \rho \, \mathrm{d}v \qquad (2.20)$$

By examining the latter part of Eq. (2.20), we see than an integral-free relation between $\mathbf{D}$ and ρ can be obtained, in the form:

$$\text{div } \mathbf{D} = \rho \tag{2.21}$$

We can also show, from the magnetic equivalent of Gauss' law, that div $\mathbf{B} = 0$.

Stokes' Theorem

A similar type of theorem, due to Stokes, states that for any vector field $\mathbf{F}$:

$$\int_L \mathbf{F} \cdot d\mathrm{L} = \iint_A \text{curl } \mathbf{F} \cdot d\mathbf{a} \tag{2.22}$$

Here we have introduced another new quantity, the *curl* of a vector field $\mathbf{F}$. This is itself a vector, and is defined in cartesian coordinates as

$$\text{curl } \mathbf{F} = \mathbf{i}\{\partial F_z/\partial y - \partial F_y/\partial z\} + \mathbf{j}\{\partial F_x/\partial z - \partial F_z/\partial x\} + \mathbf{k}\{\partial F_y/\partial x - \partial F_x/\partial y\} \tag{2.23}$$

Equation (2.23) is a rather long-winded expression. A more compact version (which is also rather easier to remember) can be written in the form of a determinant, as shown below:

$$\text{curl } \mathbf{F} = \begin{vmatrix} \mathbf{i} & \mathbf{j} & \mathbf{k} \\ \partial/\partial x & \partial/\partial y & \partial/\partial z \\ F_x & F_y & F_z \end{vmatrix} \tag{2.24}$$

Here $|\mathbf{M}|$ represents the determinant of $\mathbf{M}$, where $\mathbf{M}$ is a general matrix. The quantity curl $\mathbf{F}$ is also often written as $\nabla \times \mathbf{F}$, where the symbol $\times$ denotes a further new operation, the *vector product*.

The vector product $\mathbf{F} \times \mathbf{G}$ between two general fields $\mathbf{F}$ and $\mathbf{G}$ is defined as

$$\mathbf{F} \times \mathbf{G} = \mathbf{i}\{F_y G_z - F_z G_y\} + \mathbf{j}\{F_z G_x - F_x G_z\} + \mathbf{k}\{F_x G_y - F_y G_x\} \tag{2.25}$$

This can also be written in determinantal form, as

$$\mathbf{F} \times \mathbf{G} = \begin{vmatrix} \mathbf{i} & \mathbf{j} & \mathbf{k} \\ F_x & F_y & F_z \\ G_x & G_y & G_z \end{vmatrix} \tag{2.26}$$

Note that the vector product (unlike the scalar product) is not commutative, so that $\mathbf{F} \times \mathbf{G} \neq \mathbf{G} \times \mathbf{F}$. However, with the definition given above, $\nabla \times \mathbf{F}$ reduces to both Eqs (2.23) and (2.24).

Using Stokes' theorem, Faraday's law transforms to

$$\int_L \mathbf{E} \cdot d\mathrm{L} = \iint_A \text{curl } \mathbf{E} \cdot d\mathbf{a} = -\iint_A \partial \mathbf{B}/\partial t \cdot d\mathbf{a} \tag{2.27}$$

Examining the latter part of Eq. (2.27), we see that

$$\text{curl } \mathbf{E} = -\partial \mathbf{B}/\partial t \tag{2.28}$$

Similarly, we can show from Ampère's law that

$$\text{curl } \mathbf{H} = \mathbf{J} + \partial\mathbf{D}/\partial t \tag{2.29}$$

We have now derived the differential form of Maxwell's equations. Because they contain only differential operators, they are often much easier to manipulate than the integral form. For completeness, we show all the new equations grouped together below:

$$\begin{aligned} \text{div } \mathbf{D} &= \rho && (1) \\ \text{div } \mathbf{B} &= 0 && (2) \\ \text{curl } \mathbf{E} &= -\partial\mathbf{B}/\partial t && (3) \\ \text{curl } \mathbf{H} &= \mathbf{J} + \partial\mathbf{D}/\partial t && (4) \end{aligned} \tag{2.30}$$

Note that when $\rho = 0$ and $\mathbf{J} = 0$ (which is the case for electromagnetic waves in dielectric media) the equations for electric and magnetic quantities appear interchangeable.

2.3 HARMONICALLY-VARYING FIELDS AND THE WAVE EQUATION

One striking success of Maxwell's equations is to predict the existence of harmonically-varying fields, otherwise known as *electromagnetic waves*. We shall now perform a similar demonstration, with the following assumptions: (a) we restrict ourselves to non-magnetic materials, so that $\mu = \mu_0$, where μ_0 is the permeability of free space (known from electrostatic experiment to have the value $4\pi \times 10^{-7}$ m kg/C^2); (b) we assume that there are no currents flowing, and no charges present, so that $\mathbf{J}$ and ρ are both zero.

The Wave Equation

We start by deriving a suitable wave equation. If we put $\mathbf{D} = \varepsilon\, \mathbf{E}$ and $\mathbf{B} = \mu_0\, \mathbf{H}$ in Eq. (2.30), then Eqs (3) and (4) of (2.30) contain only the two variables $\mathbf{E}$ and $\mathbf{H}$. We can therefore eliminate one or other by direct manipulation. Taking the curl of Eq. (3) gives

$$\begin{aligned} \text{curl}\,[\text{curl } \mathbf{E}] &= -\,\text{curl}\,[\partial\mathbf{B}/\partial t] \\ &= -\,\mu_0\, \partial/\partial t\,[\text{curl } \mathbf{H}] \\ &= -\,\mu_0\, \partial^2\mathbf{D}/\partial t^2 \end{aligned} \tag{2.31}$$

We now simplify Eq. (2.31) using a standard vector identity:

$$\text{curl}\,[\text{curl } \mathbf{F}] = \text{grad}\,[\text{div } \mathbf{F}] - \nabla^2\mathbf{F} \tag{2.32}$$

Here we have introduced two new operators. The first is the *gradient* of a scalar function ϕ (which in the case of Eq. (2.32) is div $\mathbf{F}$). This is defined in cartesian coordinates as

$$\text{grad } \phi = \nabla\phi = \partial\phi/\partial x \, \mathbf{i} + \partial\phi/\partial y \, \mathbf{j} + \partial\phi/\partial z \, \mathbf{k} \tag{2.33}$$

The second is the operator ∇^2 (known as the *Laplacian*), defined in cartesian coordinates by

$$\nabla^2 = \partial^2/\partial x^2 + \partial^2/\partial y^2 + \partial^2/\partial z^2 \tag{2.34}$$

Using this identity, Eq. (2.31) can be reduced to:

$$\text{grad } [\text{div } \mathbf{E}] - \nabla^2\mathbf{E} = -\,\mu_0\,\partial^2(\varepsilon\mathbf{E})/\partial t^2 \tag{2.35}$$

This is now a form of wave equation. This can be simplified considerably, by making assumptions and approximations that are frequently valid:

1. Since, with no charges present, div $\mathbf{D} = 0$, it follows that div $(\varepsilon\mathbf{E}) = 0$. If the medium is homogeneous and isotropic, then ε is independent of position and direction, so div $\mathbf{E} = 0$. If ε varies only slowly with position, we can still put div $\mathbf{E} = 0$ to a reasonable approximation. Consequently, the first term in Eq. (2.35) is often zero.
2. If ε is not time-dependent, then $\partial^2(\varepsilon\mathbf{E})/\partial t^2 = \varepsilon\,\partial^2\mathbf{E}/\partial t^2$. This is normally a good assumption; even if ε varies with time, it needs to do so significantly in a period of oscillation of the field before it matters. Non-linear optics is concerned precisely with circumstances in which ε varies with t, but high fields are required for significant effects to occur.

Assuming approximations 1 and 2, Eq. (2.35) becomes

$$\nabla^2\mathbf{E} = \mu_0\varepsilon\,\partial^2\mathbf{E}/\partial t^2 \tag{2.36}$$

This is also a *vector wave equation*. In cartesian coordinates, it can be written as a set of three independent scalar equations—one for each coordinate. For the E_x component, for example, we have

$$\partial^2\mathrm{E}_x/\partial x^2 + \partial^2\mathrm{E}_x/\partial y^2 + \partial^2\mathrm{E}_x/\partial z^2 = (\mu_0\varepsilon_0)\,\varepsilon_r\,\partial^2\mathrm{E}_x/\partial t^2 \tag{2.37}$$

Here we have written the permittivity ε as a product, in the form:

$$\varepsilon = \varepsilon_0\,\varepsilon_r \tag{2.38}$$

Here ε_0 is the permittivity of free space, again known from experiments to have the value 8.85×10^{-12} s^2 C^2/m^3 kg, while ε_r is the *relative permittivity* or *relative dielectric constant* of the material concerned.

Equation (2.37) is now in the form of a simple classical scalar wave equation, which is adequate for introducing many aspects of electromagnetic wave propagation. It should be compared with the one-dimensional wave equation for waves on a string, which is often written as

$$\partial^2 y/\partial x^2 = (1/c^2)\,\partial^2 y/\partial t^2 \tag{2.39}$$

Here y is the displacement of the string, x is the distance along it, and c is the wave velocity. Both equations are essentially similar in character, involving second derivatives of some quantity with respect to space (on the left-hand side) and second derivatives with respect to time (on the right). The main difference is that electromagnetic waves are not constrained to travel in any particular direction, so that all three spatial coordinates appear in Eq. (2.37).

Note that an exactly similar equation to (2.36) can be obtained for the magnetic field $\mathbf{H}$:

$$\nabla^2\mathbf{H} = \mu_0\varepsilon_0\,\varepsilon_r\,\partial^2\mathbf{H}/\partial t^2 \tag{2.40}$$

This follows from the 'interchangeability' of the equations for electric and magnetic field quantities discussed earlier. There will be occasions when it is easier to work with Eq. (2.36) than Eq. (2.40), and vice versa.

The Time-Independent Wave Equation

The electric field $\mathbf{E}$ is, in general, a function of x, y, z and t. However, it is often the case that all the fields involved will be harmonically varying, at some single angular frequency ω. This will occur in many situations involving monochromatic light. It is then convenient to eliminate any time-dependence from the problem. One possibility is to assume cosinusoidally-varying solutions to the wave equation, but it is generally far more convenient to use the complex exponential form:

$$\mathbf{E}(x,y,z,t) = \mathrm{E}(x,y,z)\exp(\mathrm{j}\omega t) \tag{2.41}$$

Here the function $\mathrm{E}(x,y,z)$ accounts for the spatial dependence of the field, while the exponential $\exp(\mathrm{j}\omega t)$ describes the time-variation. The use of such complex notation is standard in electromagnetic theory, but it is important to note that ultimately *only the real parts of Eq. (2.41) are significant.* Generally, entire calculations are performed in books with the simple understanding that real parts are implied throughout. With this assumption, we can find time-derivative terms as

$$\partial\mathbf{E}/\partial t = \mathrm{j}\omega\,\mathbf{E}\,;\ \partial^2\mathbf{E}/\partial t^2 = -\omega^2\,\mathbf{E} \tag{2.42}$$

and so on, so that the wave equation reduces to

$$\nabla^2\mathrm{E} = -\,\omega^2\mu_0\varepsilon_0\,\varepsilon_r\,\mathrm{E} \tag{2.43}$$

Equation (2.43) is a *time-independent* vector wave equation, which is valid for field oscillating at a single angular frequency ω. We will now try to find some solutions to it.

2.4 PLANE WAVES

The simplest form of solution is a plane wave, i.e. a wave whose surfaces of constant phase are infinite planes, perpendicular to the direction of propagation. Figure 2.6 shows the geometry for a wave travelling in the $+z$ direction.

Clearly in this case no field quantities can vary with the transverse coordinates x and y; the only spatial variation is caused by changes in z. Considering the electric

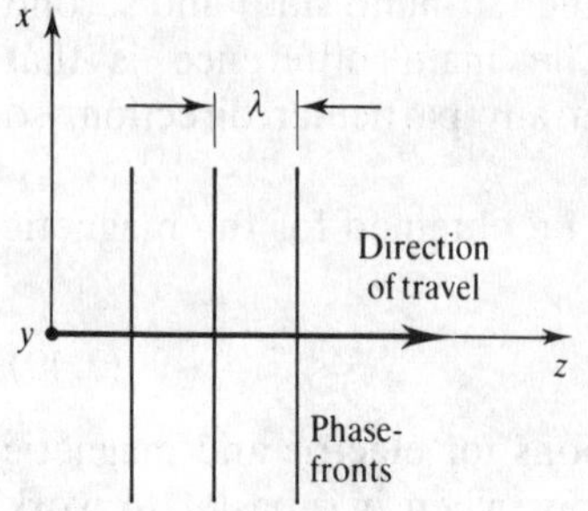

Figure 2.6 A plane wave.

field to begin with, we must therefore have

$$\partial \mathbf{E}/\partial x = \partial \mathbf{E}/\partial y = 0 \tag{2.44}$$

so that E must be a function of z only. We may therefore replace $\partial \mathbf{E}/\partial z$ by $\mathrm{d}\mathbf{E}/\mathrm{d}z$, and so on.

For simplicity we now consider the electric field to have one single component, say in the x-direction. The vectorial wave equation Eq. (2.43) then reduces to the scalar equation:

$$\mathrm{d}^2 E_x/\mathrm{d}z^2 + \omega^2 \mu_0 \varepsilon_0 \varepsilon_r E_x = 0 \tag{2.45}$$

We now guess that a possible solution has the form

$$E_x = E_{x+} \exp(-\mathrm{j}kz) \tag{2.46}$$

where E_{x+} is a constant. Direct substitution into Eq. (2.45) shows that this solution is valid, provided

$$k = \omega(\mu_0 \varepsilon_0 \varepsilon_r)^{1/2} \tag{2.47}$$

Hence the full solution, including the time variation, is

$$\mathsf{E}_x = E_{x+} \exp[\mathrm{j}(\omega t - kz)] \tag{2.48}$$

Equation (2.48) represents a *travelling wave*, moving in the $+z$-direction. E_{x+} is clearly the wave amplitude. What do the other constant terms represent? Well, $\omega = 2\pi f$ is the angular frequency of the wave; typically, the corresponding temporal frequency f is $\approx 10^{15}$ Hz for light waves. We could also write $\omega = 2\pi/T$, where T is the period of the oscillating field. By analogy, we could write $k = 2\pi/\lambda$, introducing the new quantity λ, the *spatial wavelength*. This is the distance separating planes of equal phase, as shown in Fig. 2.6. Typically, λ lies in the approximate range 0.4–0.8 μm for visible light; the lower limit corresponds to the ultra-violet or blue end of the spectrum, while the latter corresponds to deep red or near infrared wavelengths. An important new parameter, k, known as the *propagation constant*, will often be referred to later on.

Finally the velocity of the wave (known as the *phase velocity*) is given by

$$v_{ph} = \omega/k \tag{2.49}$$

This can also be written in the more familiar form:

$$v_{ph} = f\lambda \tag{2.50}$$

From Eq. (2.47), we can find the phase velocity as

$$v_{ph} = 1/\sqrt{(\mu_0 \varepsilon_0 \varepsilon_r)} \tag{2.51}$$

In a vacuum $\varepsilon_r = 1$, so $v_{ph} = 1/\sqrt{(\mu_0\varepsilon_0)}$. Putting in the numbers for μ_0 and ε_0, we find that

$$v_{ph} = 1/\sqrt{(8.85 \times 10^{-12} \times 4\pi \times 10^{-7})} \approx 3 \times 10^8 \text{ m/s} \tag{2.52}$$

This is the velocity of light, written as c, and one of the major successes of Maxwell's equations was the discovery that this directly measurable quantity could be derived so simply. In 1849, Fizeau measured c as 3.153×10^8 m/s, while in 1983 the value was fixed at 2.99792458×10^8 m/s. In other media ε_r is normally greater than unity, so that $v_{ph} < c$; light therefore generally travels slower in matter than in free space. The *refractive index* n (a useful parameter in optics) is then defined as

$$n = c/v_{ph} = \sqrt{\varepsilon_r} \tag{2.53}$$

Quantities measured in a particular material are often referred to those in free-space. For example, the spatial wavelength λ could be related to the free-space wavelength λ_0 by

$$\lambda = \lambda_0/n \tag{2.54}$$

Similarly, the propagation constant k could be related to the free-space propagation constant k_0 by

$$k = nk_0 \tag{2.55}$$

The relative permittivity ε_r and the refractive index n are generally both functions of frequency ω, as we shall see in the next chapter.

The Transverse Nature of Electromagnetic Waves

We will now consider the other field components that must accompany the solution we have found for the electric field. First, we note that with the assumptions we have made so far, Maxwell's equations can be rewritten in the following time-independent form:

$$\begin{aligned} &\text{div}\,(\varepsilon\mathbf{E}) = 0 && (1)\\ &\text{div}\,(\mu_0\mathbf{H}) = 0 && (2)\\ &\text{curl}\,\mathbf{E} = -j\omega\mu_0\,\mathbf{H} && (3)\\ &\text{curl}\,\mathbf{H} = j\omega\varepsilon\,\mathbf{E} && (4) \end{aligned} \tag{2.56}$$

Our solution has so far contained only an x-component of the electric field. In this case we get:

$$\text{curl } \mathbf{E} = \mathbf{j}\,\partial E_x/\partial z - \mathbf{k}\,\partial E_x/\partial y \tag{2.57}$$

Given that $E_x = E_{x+} \exp(-jkz)$, we find

$$\text{curl } \mathbf{E} = -jkE_{x+} \exp(-jkz)\,\mathbf{j} \tag{2.58}$$

From Eq. (3) in (2.56), we must have curl $\mathbf{E} = -j\omega\mu_0\,\mathbf{H}$. Hence, the magnetic field accompanying our solution only has a component in the y-direction. Writing this as

$$\mathbf{H} = H_{y+} \exp(-jkz)\,\mathbf{j} \tag{2.59}$$

we can obtain the following constant relation between the electric and magnetic field amplitudes:

$$H_{y+}/E_{x+} = k/\omega\mu_0 \tag{2.60}$$

The solution therefore really consists of two travelling waves—an electric and a magnetic component. Both are in-phase, but the field directions are at right-angles to each other. We can represent the complete solution at any given instant in time as in Fig. 2.7, which shows the real parts of the two components together. Both exhibit similar cosinusoidal variations with distance.

Is this solution the only one possible? It would seem reasonable to repeat the analysis, starting with the assumption that the electric field only has a component in the y-direction. In this case, we find that if $\mathbf{E} = E_{y+} \exp(-jkz)\,\mathbf{j}$, then $\mathbf{H} = H_{x+} \exp(-jkz)\,\mathbf{i}$, so the magnetic field now only has a component in the x-direction. As before, we can find a relation between the two field amplitudes. This time, we get

$$H_{x+}/E_{y+} = -k/\omega\mu_0 \tag{2.61}$$

Apart from the minus sign, the amplitude ratio is as before.

What happens if we assume instead that the electric field only has a component in the z-direction? In an isotropic medium, div $(\varepsilon\mathbf{E}) = \varepsilon$ div $\mathbf{E}$, so Eq. (1) in (2.56) must reduce to div $\mathbf{E} = 0$. Remember that we can expand this as

$$\partial E_x/\partial x + \partial E_y/\partial y + \partial E_z/\partial z = 0 \tag{2.62}$$

However, since we have already assumed that $\partial\mathbf{E}/\partial x$ and $\partial\mathbf{E}/\partial y = 0$, it follows that $\partial E_x/\partial x = \partial E_y/\partial y = 0$. Hence, $\partial E_z/\partial z$ must be zero, so E_z must be a constant

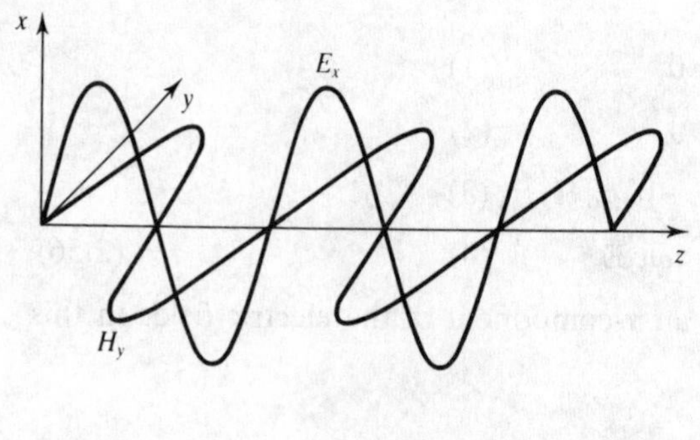

Figure 2.7 A plane electromagnetic wave.

independent of z. We therefore do not find travelling-wave solutions for E_z, and a similar argument can be used to show that there are no wave solutions for H_z. Plane electromagnetic waves are therefore strictly transverse. They are therefore often described as *TEM* (standing for *transverse electromagnetic*) waves.

Optical Polarization

We now consider some of the wider properties of the solutions found so far, beginning with the important feature of optical polarization. We start by noting that the two independent travelling wave solutions discussed above can be combined into a more general solution, in the form:

$$\mathbf{E} = E_{x+} \exp\,[j(\omega t - kz + \phi_x)]\,\mathbf{i} + E_{y+} \exp\,[j(\omega t - kz + \phi_y)]\,\mathbf{j} \qquad (2.63)$$

where ϕ_x and ϕ_y are arbitrary (but constant) phase factors. The nature of the resulting wave then depends on the values of E_{x+}, E_{y+}, ϕ_x and ϕ_y. Several combinations are particularly important.

1. If $\phi_x = \phi_y$, the solution can be written as

$$\mathbf{E} = [E_{x+}\,\mathbf{i} + E_{y+}\,\mathbf{j}] \exp\,[j(\omega t - kz + \phi)]$$
$$= \mathbf{E}_0 \exp\,[j(\omega t - kz + \phi)] \qquad (2.64)$$

In this solution the direction of the electric field vector is independent of time and space, and is defined by a new vector $\mathbf{E}_0$, which is the vectorial sum of $E_{x+}\,\mathbf{i}$ and $E_{y+}\,\mathbf{j}$ as shown in Fig. 2.8. This type of wave is known as a *linearly polarized* wave, and the direction of the electric field vector $\mathbf{E}_0$ represents the *direction of polarization*. Linearly polarized light is particularly important in engineering optics. It can be produced from natural light (which has random polarization) by passing it through a *polarizer*. More importantly, it is emitted directly by many types of laser.

2. If $E_{x+} = E_{y+}$, and $\phi_y = \phi_x \pm \pi/2$, the solution can be written as

$$\mathbf{E} = E_0 \exp\,[j(\omega t - kz + \phi)]\,\mathbf{i} + E_0 \exp\,[j(\omega t - kz + \phi \pm \pi/2)]\,\mathbf{j} \qquad (2.65)$$

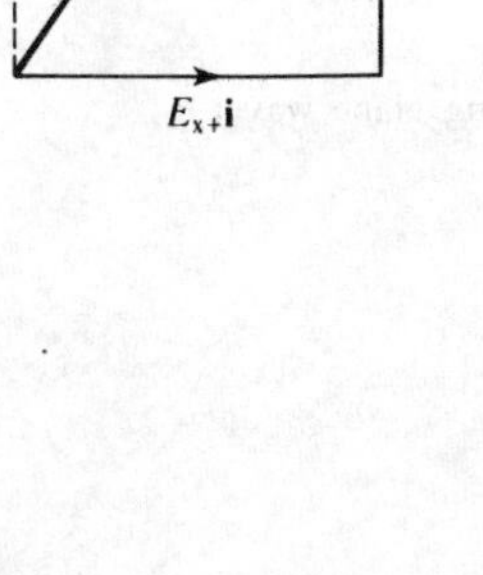

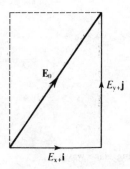

Figure 2.8 Construction of the polarization vector.

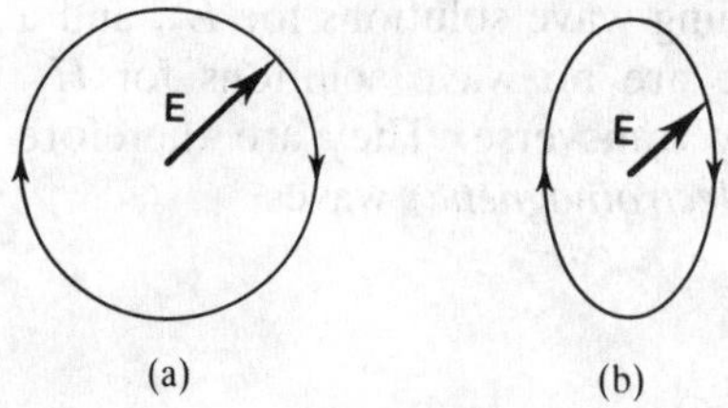

Figure 2.9 Loci of the electric field vector for (a) circular and (b) elliptic polarization.

Or alternatively as

$$\mathbf{E} = E_0 (\mathbf{i} \pm \mathrm{j}\,\mathbf{j}) \exp [\mathrm{j}(\omega t - kz + \phi)] \tag{2.66}$$

Ultimately we are interested in the real part of **E**. This is given by

$$\mathrm{Re}\,\{\mathbf{E}\} = E_0 \{\cos (\omega t - kz + \phi)\,\mathbf{i} \pm \sin (\omega t - kz + \phi)\,\mathbf{j}\} \tag{2.67}$$

In this case the amplitude of the electric field vector is still constant (and equal to E_0), but the direction of polarization is not. Instead, it rotates as a function of space and time. This solution is known as a *circularly polarized* wave, because the locus traced by the electric field vector as a function of time (at a given point) is a circle, as shown in Fig. 2.9(a). Right- and left-hand circular polarizations are both possible, depending on the sign of the $\pi/2$ phase-shift. If $E_{x+} \neq E_{y+}$, the locus becomes an ellipse, and the wave is described as being *elliptically polarized* (Fig. 2.9(b)).

Oblique Waves

Simple solutions can also be found for plane waves travelling in different directions. For example, Fig. 2.10 shows a wave, travelling in the x–z plane at an angle θ to the z-axis. If the wave is linearly polarized in the y-direction, i.e. perpendicular to the plane of the figure, the time-independent electric field is given simply by:

$$\mathbf{E} = \mathbf{j}\,E_{y+} \exp [-\mathrm{j}\,(kx \sin \theta + kz \cos \theta)] \tag{2.68}$$

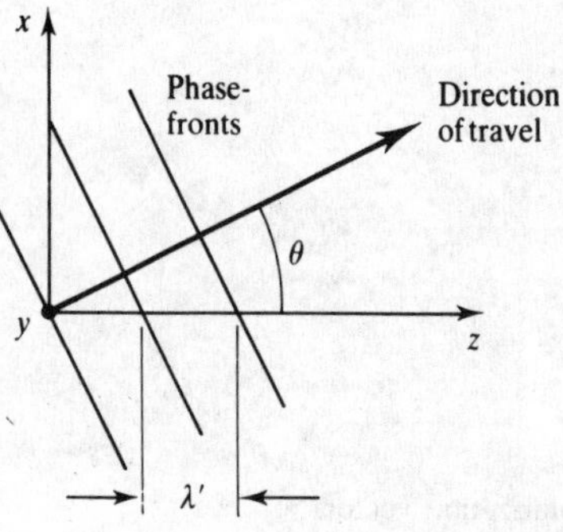

Figure 2.10 An obliquely-travelling plane wave.

This result is obtained simply by rotating the coordinate system of Fig. 2.6 about the y-axis. We can check that it is correct by putting $\theta = 0$, whereupon Eq. (2.68) reduces to our original solution, $\mathbf{E} = \mathbf{j}E_{y+} \exp(-jkz)$. Similarly, if $\theta = \pi/2$, we get $\mathbf{E} = \mathbf{j}E_{y+} \exp(-jkx)$—this is also as expected, as the wave is now travelling in the $+x$ direction.

However, we note that in the time it takes the wave to travel a distance λ in the direction of propagation, the phase-fronts have advanced *further* in the z-direction, a distance $\lambda' = \lambda/\cos\theta$. The effective phase velocity in the z-direction is therefore greater than v_{ph}, by the factor $1/\cos\theta$. Some thought is required fully to appreciate this point, since it implies faster motion of the disturbance in a direction off-axis to the direction of propagation.

Impedance

We can find the *impedance* of a medium through which a wave is propagating, as follows. Since $\omega/k = 1/\sqrt{(\mu_0\varepsilon_0\varepsilon_r)}$, the ratio E_{x+}/H_{y+} can be found as $\sqrt{(\mu_0/\varepsilon_0\varepsilon_r)}$. Since this ratio has the dimensions of Ohms, it is called the characteristic impedance Z of the medium. Thus, we may put

$$Z = \sqrt{(\mu_0/\varepsilon_0\varepsilon_r)} \tag{2.69}$$

For free space, $\varepsilon_r = 1$, so that $Z_0 = \sqrt{(\mu_0/\varepsilon_0)} = 377\ \Omega$. Anyone owning a radio tuner with an external aerial socket should note the input impedance—it will be around this value. The impedance of any other material may be related to the impedance of free space using

$$Z = Z_0/\sqrt{\varepsilon_r} = Z_0/n \tag{2.70}$$

Impedance is therefore inversely proportional to refractive index.

Waves in Lossy Dielectrics

We may also extend the analysis to describe the behaviour of plane waves propagating in slightly lossy dielectric media, as follows. We begin by including loss in a phenomenological way, by assuming that the relative dielectric constant of the material is complex-valued. We shall justify this more rigorously in Chapter 3, but for the time being we will simply take ε_r to be defined by

$$\varepsilon_r = \varepsilon_r' - j\varepsilon_r'' \tag{2.71}$$

For x-polarized waves travelling in the $+z$-direction, the scalar wave equation we must solve can be found by substituting Eq. (2.71) into Eq. (2.45). We get

$$d^2E_x/dz^2 + \omega^2(\mu_0\varepsilon_0)[\varepsilon_r' - j\varepsilon_r'']E_x = 0 \tag{2.72}$$

Assuming a solution of the form used previously, namely $E_x = E_{x+} \exp(-jkz)$, we find that the propagation constant k is now given by

$$k^2 = \omega^2 (\mu_0 \varepsilon_0) [\varepsilon_r' - j\varepsilon_r''] \tag{2.73}$$

or

$$k = \omega \sqrt{(\mu_0 \varepsilon_0 \varepsilon_r')} \sqrt{[1 - j\varepsilon_r''/\varepsilon_r']} \tag{2.74}$$

Making the additional assumption that the loss is small, so that $\varepsilon_r'' \ll \varepsilon_r'$, we may use a binomial approximation for Eq. (2.74), which gives

$$\begin{aligned} k &= \omega \sqrt{(\mu_0 \varepsilon_0 \varepsilon_r')} [1 - j\varepsilon_r''/2\varepsilon_r'] \\ &= k' - jk'' \end{aligned} \tag{2.75}$$

where the real and imaginary parts of k are given by

$$k' = \omega \sqrt{(\mu_0 \varepsilon_0 \varepsilon_r')} \quad \text{and} \quad k'' = k' \varepsilon_r''/2\varepsilon_r' \tag{2.76}$$

Since the propagation constant is complex, we can now rearrange the solution as the product of two exponentials:

$$E_x = E_{x+} \exp(-k''z) \exp(-jk'z) \tag{2.77}$$

Equation (2.77) has the form of a plane wave, whose amplitude decays exponentially with distance z. The real part of the propagation constant, k', defines the phase variation of the wave, while the imaginary part k'' defines the amplitude variation and is known as the *absorption constant*, often given the symbol α. The presence of even a small amount of loss in a material (which is often unavoidable) causes the exponential decay of a propagating wave.

We may also hypothesise the existence of media with a negative value of ε_r''. In this case, the solution corresponds to an exponentially-growing wave, of the form:

$$E_x = E_{x+} \exp(+gz) \exp(-jk'z) \tag{2.78}$$

where g, the *gain constant* of the medium, is defined by $g = -k''$. The growth of an optical wave as it passes through a medium with gain is the key to the operation of all laser devices.

Waves in Metals

Finally, we can extend the analysis to include materials with non-zero conductivity—for example, metals. All that is required is to work through Eqs (2.31) to (2.43) again, assuming to start with that curl $\mathbf{H} = \mathbf{J} + \partial\mathbf{D}/\partial t$, rather than simply $\partial\mathbf{D}/\partial t$. If this is done, a slightly revised time-independent wave equation can be obtained for monochromatic waves:

$$\nabla^2\mathbf{E} + \{\omega^2\mu_0\varepsilon - j\omega\mu_0\sigma\} \mathbf{E} = 0 \tag{2.79}$$

Note that this has exactly the same form as Eq. (2.43), but the term $\omega^2\mu_0\varepsilon$ has been replaced here by $\omega^2\mu_0\varepsilon - j\omega\mu_0\sigma$. Since the former was previously interpreted in terms of the propagation constant by putting $\omega^2\mu_0\varepsilon = k^2$, it is reasonable to write in this case:

$$k^2 = \omega^2\mu_0\varepsilon - j\omega\mu_0\sigma$$
$$= \omega^2\mu_0(\varepsilon - j\sigma/\omega) \quad (2.80)$$

If we wish, we can interpret all the terms inside the bracket in Eq. (2.80) as a modified dielectric constant, given by

$$\varepsilon = \varepsilon_\infty - j\,\sigma/\omega \quad (2.81)$$

where ε_∞ is the value as $\omega \to \infty$. If this is done, we see that the effective dielectric constant is complex once again. However, we cannot immediately equate the real and imaginary parts of ε to ε_∞ and $-\sigma/\omega$, since this implies an assumption that the conductivity σ is real. While this is justified for low values of ω, it is certainly not the case at optical frequencies. We shall see why this should be so in Chapter 3.

2.5 POWER FLOW

It is clearly important to form an expression for *power flow* in terms of **E** and **H**, since this is something we can measure. This is done by considering the work done on charges in a volume, as a combination of a change in stored energy and a flow of power into or out of the volume. We shall consider, initially, time-dependent fields.

The rate of work done by the fields per unit volume is $\mathbf{J} \cdot \mathbf{E}$, by analogy with $W = I\,V$ in electrical circuits. From Eq. (4) in (2.30) we get

$$\mathbf{J} = \text{curl}\,\mathbf{H} - \partial\mathbf{D}/\partial t \quad (2.82)$$

We can then write

$$\mathbf{J} \cdot \mathbf{E} = \mathbf{E} \cdot (\text{curl}\,\mathbf{H}) - \mathbf{E} \cdot \partial\mathbf{D}/\partial t \quad (2.83)$$

We now introduce a further vector identity, again described more fully in mathematics texts:

$$\text{div}\,(\mathbf{E} \times \mathbf{H}) = \mathbf{H} \cdot (\text{curl}\,\mathbf{E}) - \mathbf{E} \cdot (\text{curl}\,\mathbf{H}) \quad (2.84)$$

We also need another relation, from Eq. (3) in (2.30):

$$\text{curl}\,\mathbf{E} = -\partial\mathbf{B}/\partial t \quad (2.85)$$

Combining Eqs (2.83) to (2.85), we get

$$\mathbf{H} \cdot \partial\mathbf{B}/\partial t + \mathbf{E} \cdot \partial\mathbf{D}/\partial t + \text{div}(\mathbf{E} \times \mathbf{H}) + \mathbf{J} \cdot \mathbf{E} = 0 \quad (2.86)$$

We can now rewrite this in the following standard form:

$$\partial U/\partial t + \nabla \cdot \mathbf{S} + \mathbf{J} \cdot \mathbf{E} = 0 \quad (2.87)$$

where

$$\partial U/\partial t = [\mathbf{E} \cdot \partial \mathbf{D}/\partial t + \mathbf{H} \cdot \partial \mathbf{B}/\partial t] \quad (2.88)$$

and

$$\mathbf{S} = \mathbf{E} \times \mathbf{H} \quad (2.89)$$

Equation (2.87) (known as *Poynting's Theorem*, after John Poynting, 1852–1914) is effectively a power conservation relation, since it relates the rate of change of stored energy to the outward energy flow and the energy dissipated. U is the density of the energy stored in the electromagnetic fields (measured in J/m^3). This can be divided conveniently into two components: $U_E = 1/2\ \mathbf{E} \cdot \mathbf{D}$ represents the electrically-stored energy, while $U_M = 1/2\ \mathbf{H} \cdot \mathbf{B}$ is the energy in the magnetic field. Similarly the vector $\mathbf{S}$, known as the *Poynting vector*, describes the power flow (measured in W/m^2). However, it should be noted that $\mathbf{S}$ is an extremely fast-varying function. In fact, $\mathbf{S}$ contains components at 2ω (or about 10^{30} Hz for optical waves), which are clearly not measurable by any practical technique. It is therefore convenient to define an alternative quantity related to power, that is directly measurable.

Irradiance

For oscillating fields we have been using the alternative notation $\mathbf{E}(x,y,z,t) = \mathbf{E}(x,y,z) \exp(j\omega t)$, where the real part is implied. In this case we can calculate an associated *time-averaged Poynting vector* or *irradiance* S. This is the mean of $\mathbf{S}$ over many oscillations, defined as

$$\mathrm{S} = (1/T) \int_T \mathbf{S}\, \mathrm{d}t \quad (2.90)$$

where T is large compared with the period of the oscillations. Substituting harmonic solutions for $\mathbf{E}$ and $\mathbf{H}$ into Eq. (2.90), we obtain

$$\mathrm{S} = (1/T) \int_T \mathrm{Re}\,\{\mathrm{E} \exp(j\omega t)\} \times \mathrm{Re}\,(\mathrm{H} \exp(j\omega t))\}\, \mathrm{d}t \quad (2.91)$$

It is then simple to show that the irradiance is given by

$$\mathrm{S} = 1/2\ \mathrm{Re}\,[\mathrm{E} \times \mathrm{H}^*] \quad (2.92)$$

The expression for S therefore contains no time variation, as required.

Time-averaged Power

Irradiance is still not a directly measurable quantity. However, we can measure the *time-averaged power* P flowing through a given surface. This is found as the integral of the normal component of S over the surface, as shown in Fig. 2.11. Using Eq. (2.92), P may be evaluated as

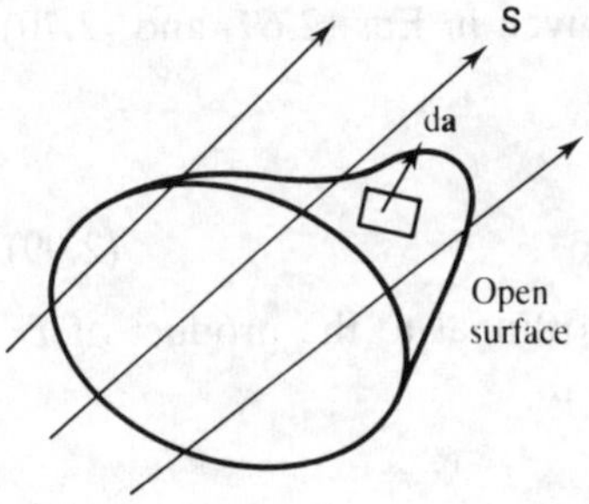

Figure 2.11 Geometry for calculation of power flow.

$$P = 1/2 \text{ Re}\left\{\iint_A [\mathbf{E} \times \mathbf{H}^*] \cdot \mathbf{da}\right\} \tag{2.93}$$

Time-averaged power is extremely important since it is one of the few parameters of high-frequency electromagnetic field that can actually be measured. Suitable detectors are the human eye, solar cells and semiconductor p–n junction photodiodes.

Design Example

As an example we shall calculate the power carried by a plane wave travelling in the $+z$-direction. Assuming that components of both polarizations are present, the electric field can be written as

$$\mathbf{E} = [\mathbf{i}\, E_{x+} + \mathbf{j}\, E_{y+}] \exp(-jkz) \tag{2.94}$$

The corresponding magnetic field can be found from Eqs (2.60) and (2.61) as

$$\mathbf{H} = [E_{y+}\,(-k/\mu_0\omega)\,\mathbf{i} + E_{x+}\,(k/\mu_0\omega)\,\mathbf{j}] \exp(-jkz) \tag{2.95}$$

The time-averaged power flow in the z-direction per unit area is therefore

$$\begin{aligned} P &= 1/2 \text{ Re}\,[\mathbf{E} \times \mathbf{H}^*] \cdot \mathbf{k} \\ &= 1/2 \text{ Re}\,[\mathbf{i}\,\{E_y H_z^* - E_z H_y^*\} + \mathbf{j}\,\{E_z H_x^* - E_x H_z^*\} \\ &\quad + \mathbf{k}\,\{E_x H_y^* - E_y H_x^*\}] \cdot \mathbf{k} \end{aligned} \tag{2.96}$$

which in this case reduces to

$$P = 1/2 \text{ Re}\,[E_x H_y^* - E_y H_x^*] \tag{2.97}$$

Substituting in the necessary values, we get

$$P = (1/2)\,\sqrt{(\varepsilon_0\varepsilon_r/\mu_0)}\,[E_{x+}^2 + E_{y+}^2] \tag{2.98}$$

This expression allows us to relate electric field strengths in V/m to power density in W/m^2. Each of the two terms above clearly corresponds to one polarization component. Writing $E_0^2 = E_{x+}^2 + E_{y+}^2$, where E_0 is the amplitude of the combined electric field, and using the definitions of impedance given in Eqs (2.69) and (2.70) we obtain

$$P = 1/2\, E_0^2/Z$$
$$= 1/2\, n\, E_0^2/Z_0 \tag{2.99}$$

The power carried by a plane wave is therefore proportional to the product of E^2 and the refractive index of the medium.

2.6 THE PROPAGATION OF GENERAL TIME-VARYING SIGNALS

We have concentrated on the behaviour of single-frequency electromagnetic fields, which in optics correspond to monochromatic light. Throughout this book, however, we will be interested in the use of optical devices in an engineering environment. One of the most obvious applications is a communications system, which might loosely be defined as an arrangement for the transmission of information between different points. Unfortunately, a perfectly monochromatic wave (which, strictly speaking, must exist for all time without changing its frequency or amplitude) cannot carry any information. Only the modulation of such a wave—for example, by switching it on and off—can do so.

In general, therefore, we will be interested in the transmission of time-varying signals. A suitable framework for their analysis is provided by *Fourier transform* theory, which states that any signal $f(t)$ may be decomposed into an infinite sum of single-frequency terms. Conventionally, this relationship is written as an integral transformation, of the form

$$f(t) = (1/2\pi) \int_{-\infty}^{\infty} F(\omega) \exp(j\omega t)\, d\omega \tag{2.100}$$

where $F(\omega)$ represents the amplitude of the component at angular frequency ω, which itself may be found from the signal using the inverse transform:

$$F(\omega) = \int_{-\infty}^{\infty} f(t) \exp(-j\omega t)\, dt \tag{2.101}$$

Design Example

As an example we may compute the frequency spectrum of a signal consisting of a short burst of a monochromatic carrier, of unity amplitude and angular frequency ω_c, as in Fig. 2.12. Assuming that the duration of the burst is ΔT, Eq. (2.101) reduces to

$$F(\omega) = \int_{-\Delta T/2}^{\Delta T/2} \exp(j\omega_c t) \exp(-j\omega t)\, dt \tag{2.102}$$

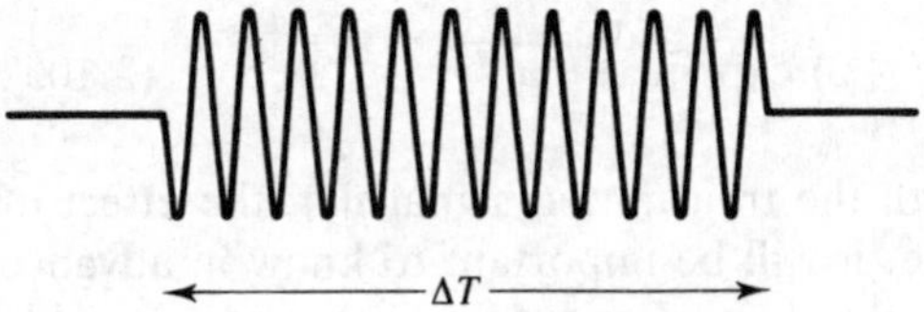

Figure 2.12 A burst of single-frequency tone.

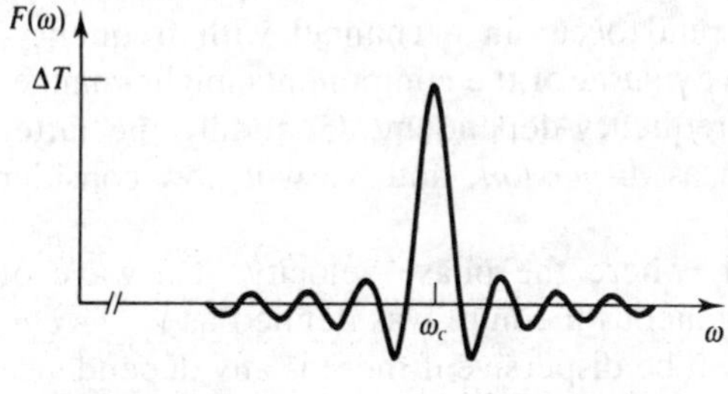

Figure 2.13 Frequency spectrum of a burst of a single tone.

Evaluation of the integral then gives

$$F(\omega') = \Delta T \text{ sinc } (\omega'\Delta T/2) \tag{2.103}$$

where $\omega' = \omega_c - \omega$ and sinc (x) = sinc $(x)/x$. Figure 2.13 shows a plot of the frequency spectrum; this peaks at $\omega = \omega_c$ and decays away on either side of this point with a typical filter envelope.

We may obtain an estimate of the width of this frequency spectrum by noting that the first zeros in Eq. (2.103) are reached when $\omega'\Delta T/2 = \pi$. This allows the definition of an approximate signal bandwidth $\Delta\omega$ (which is then about half the width of the main lobe) in the form:

$$\Delta\omega\Delta T = 2\pi \tag{2.104}$$

The signal bandwidth is therefore inversely proportional to the duration of the burst. This implies that high bit rate data transmission will involve large bandwidths.

Dispersion

The Fourier relations above may be used to analyse communication channels, as follows. Given a specified input $f(t)$, Eq. (2.101) may be used to identify the frequency components of the signal and their corresponding amplitudes $F(\omega)$. These components may then mentally be passed through the channel in turn; naturally, a component of angular frequency ω will propagate as a travelling wave of the same angular frequency. On arrival at the far end of the channel, the amplitudes of the components may well have changed, so that an amplitude $F(\omega)$ might be received as $F'(\omega)$. However, the total received signal $f'(t)$ may be reconstructed from these modified constituents by using a simple adaptation of Eq. (2.100), namely

$$f'(t) = (1/2\pi) \int_{-\infty}^{\infty} F'(\omega) \exp(j\omega t)\, d\omega \qquad (2.105)$$

By comparing the received signal $f'(t)$ with the transmitted signal $f(t)$, the effect of the channel may be assessed. For example, it will be important to know in advance what type of signals may be passed through the channel and still arrive in recognizable form. Before this can be done, however, we must identify the signal distortions that are possible. By and large, there are just two. First, the relative *amplitudes* of the frequency components may alter. This could occur in a channel with frequency-dependent attenuation. Second, the relative *phases* of the components might change, if the phase velocity of the channel is frequency-dependent. Generally the latter effect is the most significant. It is known as *dispersion*, and we will now consider some of its features.

We start by returning to Section 2.4, where the phase velocity of a wave of angular frequency ω travelling in a homogeneous medium was defined as $v_{ph} = c/n$. From this we may infer that the medium will be dispersive if there is any dependence of the refractive index on frequency. In fact, this is the case in all matter. The result is therefore normally described as *material dispersion* (to distinguish it from other effects that occur in more complicated transmission channels, particularly waveguides). We will examine the underlying reasons for this dependence in Chapter 3. However, we also obtained the alternative definition $v_{ph} = \omega/k$. Consequently, in a dispersive medium we would expect a more complicated relation between ω and k than just ω/k = constant.

One of the simplest examples of a dispersive medium is provided by an ionized medium (such as the ionosphere), for which it can be shown that

$$\omega = \sqrt{[\omega_p^2 + c^2k^2]} \qquad (2.106)$$

where ω_p is a constant, the *plasma frequency*, which will be introduced properly in Chapter 3. This type of relation can be represented as a plot of ω against k called a *dispersion diagram*, as shown in Fig. 2.14. In this case the diagram shows that ω tends to ω_p for small values of k, while for large k, ω tends to the dashed line $\omega = ck$. The phase velocity ω/k may then be found either from the dispersion diagram, or directly from Eq. (2.106). The latter process yields

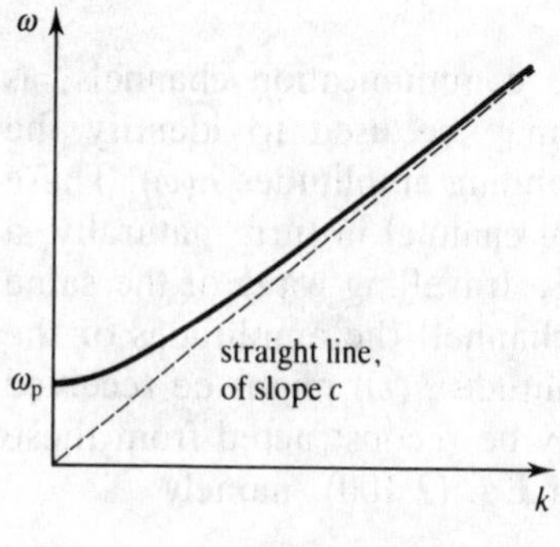

Figure 2.14 ω–k diagram for an ionized medium.

$$v_{ph} = \sqrt{[\omega_p^2/k^2 + c^2]} \tag{2.107}$$

From this it can be seen that the phase velocity is not constant; it tends to infinity as k tends to zero, and to c as k becomes large. To assess the effect of this variation we shall consider the propagation of an elementary compound signal, consisting of components at just two distinct angular frequencies $\omega + \mathrm{d}\omega$ and $\omega - \mathrm{d}\omega$, where $\mathrm{d}\omega$ is small. For simplicity we take the amplitudes of the two waves to be the same. However, we assume that the phase velocities at the two frequencies are unequal, so that the corresponding propagation constants must be written as $k + \mathrm{d}k$ and $k - \mathrm{d}k$. Assuming y polarization and z-propagation, the electric field of the signal might then be written

$$\mathrm{E}_y = E_{y+} [\exp\{\mathrm{j}((\omega + \mathrm{d}\omega)t - (k + \mathrm{d}k)z)\} + \exp\{\mathrm{j}((\omega - \mathrm{d}\omega)t - (k - \mathrm{d}k)z)\}] \tag{2.108}$$

Figure 2.15 shows a snapshot of the combined field at $t = 0$. Note that the two waves sum together to give what amounts to a carrier of constant period, modulated by an envelope (shown dashed). We may illustrate this mathematically by combining the two components slightly differently. If the common factor $\exp\{\mathrm{j}(\omega t - kz)\}$ is taken out, we get

$$\mathrm{E}_y = 2E_{y+} \exp\{\mathrm{j}(\omega t - kz)\} \cos\{\mathrm{d}\omega\, t - \mathrm{d}k\, z\} \tag{2.109}$$

This suggests that we may view propagating signals in two different ways. We may either regard them as a sum of a number of individual travelling waves (as in Eq. (2.108)) or as a single *modulated* wave (Eq. (2.109)). However, the latter viewpoint shows clearly that the information-carrying component of the signal—the modulation envelope—is also propagating as a travelling wave, defined by the term $\cos(\mathrm{d}\omega\, t - \mathrm{d}k\, z)$. This envelope must therefore also have a velocity of propagation which in general is distinct from that of the carrier wave. Since it refers to a group of waves rather than a single wave, it is known as the *group velocity* v_g, and is defined as

$$v_g = \mathrm{d}\omega/\mathrm{d}k \tag{2.110}$$

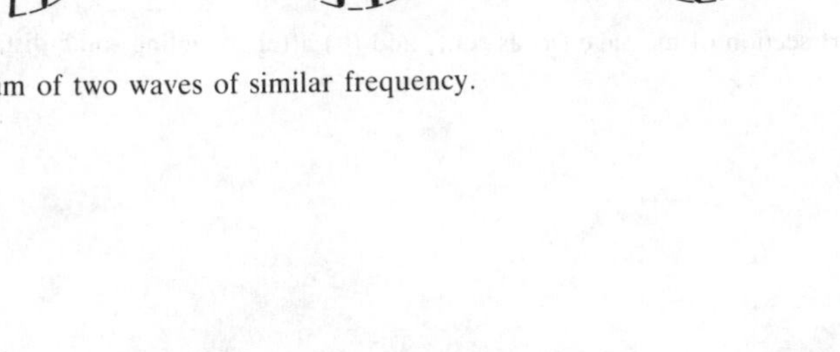

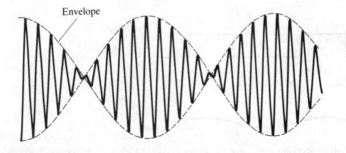

Figure 2.15 The sum of two waves of similar frequency.

For the ionosphere we may find the group velocity by differentiating Eq. (2.106), to get

$$v_g = c^2/\sqrt{[\omega_p^2/k^2 + c^2]} \tag{2.111}$$

This is clearly different from the expression derived earlier for the phase velocity. However, for large k (and thus very high frequencies), v_g tends to v_{ph}.

The foregoing analysis can be used to find the velocity of information carried by groups of waves of similar frequency. However, for groups comprising a wider range of frequencies, v_g may not be considered constant, so that different groups of a signal will travel at different speeds. This can result in a damaging effect, known as *pulse broadening*, which limits the rate of data transmissions. We can illustrate this by considering the problem of transmitting data over a distance L via successive bursts of a high-frequency carrier. Each 'one' in the data stream corresponds to a burst of duration ΔT, and the separation between successive bursts is T. Figure 2.16(a) shows the modulation envelope of a typical section of a message.

From our earlier example we may estimate the required bandwidth to be $\Delta\omega$, so that the frequencies comprising the signal range approximately from $\omega_1 = \omega_c - \Delta\omega/2$ to $\omega_2 = \omega_c + \Delta\omega/2$. At these extremes the group velocity may have different values—say, v_{g1} and v_{g2}. Consequently, different constituents of the signal must arrive at the far end of the channel at times ranging from $t_1 = L/v_{g1}$ to $t_2 = L/v_{g2}$. Assuming that $t_2 > t_1$. this spread in arrival times will amount to a lengthening of the information-carrying envelope from ΔT to $\Delta T + \Delta t$, where $\Delta t = (t_2 - t_1)$. If L is large enough we may even find that $\Delta t \approx T$. Information from successive bursts will then become mixed (as is beginning to happen in Fig. 2.16(b)) so that the message is beyond recovery by simple means.

The effective broadening of the signal may be calculated by writing

$$\Delta t = L\,\{1/v_{g2} - 1/v_{g1}\} \approx L\,\mathrm{d}(1/v_g)/\mathrm{d}\omega\,\Delta\omega \tag{2.112}$$

Since $1/v_g = \mathrm{d}k/\mathrm{d}\omega$, this yields

$$\Delta t \approx L\,\Delta\omega\,\mathrm{d}^2k/\mathrm{d}\omega^2 \tag{2.113}$$

Clearly, pulse broadening may be minimized by ensuring that the term $\mathrm{d}^2k/\mathrm{d}\omega^2$ is as small as possible, and preferably zero. We may see how this relates to material parameters by noting that $k = n\omega/c$. For a dispersive medium we can write

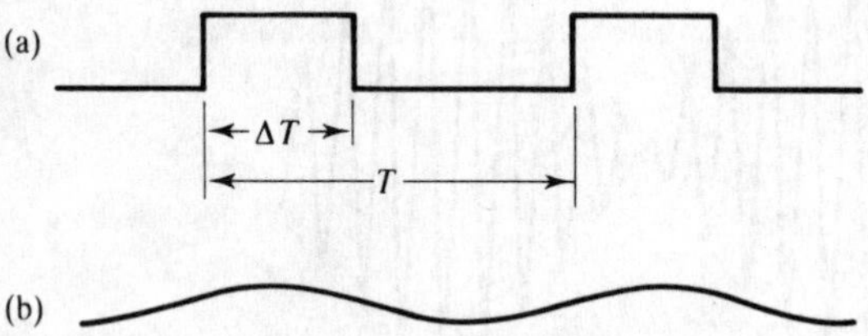

Figure 2.16 A short section of message (a) as sent, and (b) after travelling some distance in a dispersive medium.

$$dk/d\omega = 1/c\ \{n + \omega\ dn/d\omega\} \tag{2.114}$$

Similarly, since $\lambda_0 = 2\pi c/\omega$, we may put

$$dn/d\omega = (dn/d\lambda_0)\ (d\lambda_0/d\omega) = -2\pi c/\omega^2\ dn/d\lambda_0 \tag{2.115}$$

so that

$$dk/d\omega = 1/c\ \{n - \lambda_0\ dn/d\lambda_0\} \tag{2.116}$$

Differentiating a second time, we get

$$d^2k/d\omega^2 = \{d(dk/d\omega)/d\lambda_0\}\ (d\lambda_0/d\omega) = -\lambda_0/c\ (d^2n/d\lambda_0^2)\ (d\lambda_0/d\omega) \tag{2.117}$$

If we put $d\lambda_0/d\omega \approx \Delta\lambda_0/\Delta\omega$, where $\Delta\lambda_0$ is the wavelength range of the signal, we can obtain a variant of Eq. (2.113) that emphasizes the contribution of the medium to pulse broadening:

$$\Delta t = -(L\lambda_0\Delta\lambda_0/c)\ d^2n/d\lambda_0^2 \tag{2.118}$$

This suggests that dispersion can be minimized by operating near a wavelength where $d^2n/d\lambda_0^2 = 0$. In silica (the main constituent of most optical fibres) this occurs at $\lambda_0 \approx 1.27\ \mu m$. Consequently, optical communications systems almost always operate at near infrared wavelengths.

PROBLEMS

2.1 Starting from the differential form of Maxwell's equations, derive a time-dependent vector wave equation, valid for a uniform dielectric medium and containing only the magnetic field **H**.

2.2 A plane electromagnetic wave oscillates at an angular frequency of 2.978×10^{15} rad/sec. Find its wavelength in (a) free space, and (b) a material of relative dielectric constant 2.25.
[(a) 0.633 μm; (b) 0.422 μm]

2.3 A plane electromagnetic wave is defined by $\mathbf{E} = E_{y0} \exp\{-jk_0\ (ax + bz)\}\ \mathbf{j}$, where k_0 is the propagation constant of free space, and the constants a and b are given by a = 1, b = $\sqrt{3}$. Find (a) the wave amplitude, (b) the direction of polarization, (c) the direction of travel, and (d) the refractive index of the medium.
[(c) 30° to the z-axis, in the x–z plane; (d) 2]

2.4 The expression $E_y = E_{y0} \exp(-jkz)$ is a solution to the scalar equation $\nabla^2 E_y + \omega^2\mu_0\varepsilon\ E_y = 0$ that represents a plane wave travelling in the z-direction. What is k? Show that the inhomogeneous wave $E_y = E_{y0} \exp(\gamma x) \exp(-j\beta z)$ is also a solution. What relation must be satisfied by γ and β?

2.5 Sketch the loci of (a) constant phase, and (b) constant amplitude for the inhomogeneous wave in Question 2.4. What direction is it travelling in? Does it travel faster or slower than the plane wave? What do you think it represents?

2.6 A optical wave of wavelength $\lambda_0 = 0.633\ \mu m$ is travelling in a material of complex relative dielectric constant $\varepsilon_r = 2.25 - j10^{-8}$. What are the values of the propagation constant and the absorption coefficient? Find the distance the wave must travel in the medium before the power it carries decays to 1/e of its initial value.
[$14.89 \times 10^6\ m^{-1}$; $0.0331\ m^{-1}$; 15.11 m]

2.7 Show that the irradiance $\mathbf{S} = 1/2\ \mathrm{Re}\ [\mathbf{E} \times \mathbf{H}^*]$ does indeed represent the time average of the Poynting vector $\mathbf{S} = \mathbf{E} \times \mathbf{H}$, for harmonically-varying fields.

2.8 Determine the z-component of the irradiance, for the wave in Question 2.3. What is the direction of maximum irradiance?

2.9 The time-averaged power density of a linearly-polarized plane wave in free space is 100 mW/cm^2. Find the peak amplitude of the electric field.
[868 V/m]

2.10 The relationship between ω and k for electromagnetic waves in the ionosphere may be shown to have the form $\omega = \sqrt{[\omega_p^2 + c^2k^2]}$, where c is the velocity of light and ω_p is the plasma frequency. Sketch the variation of the phase and group velocity with k. Find the value of v_g when $\omega \rightarrow \omega_p$. What is the significance of this result? Find a solution for k when $\omega < \omega_p$. What form of wave do you think that this solution describes?

SUGGESTIONS FOR FURTHER READING

Born, M., and Wolf, E. "Principles of Optics", 6th edn., Chapter 1, Pergamon Press, Oxford, 1980.

Hecht, E., "Optics", 2nd edn., Chapters 1, 2, 3, and 8, Addison-Wesley Publishing Co., Reading MA., 1987.

Ramo, S., Whinnery, J. R., van Duzer, T. "Fields and Waves in Communication Electronics", 2nd edn., Chapter 3, John Wiley and Sons, New York, 1984.

CHAPTER

THREE

MATERIAL EFFECTS

3.1 INTRODUCTION

In Chapter 2 we described the basic characteristics of plane electromagnetic waves by making use of the parameters ε, σ and μ. Although these can account phenomenologically for the properties of the medium in which the wave is propagating, they treat it as if it were totally uniform and continuous. Consequently, they provide no help in visualizing the detailed interaction of the wave and the medium, and even the simplest model will allow an improvement on this rather superficial concept. We will now add such a model, basing the discussion again on a classical picture.

Materials may obviously be classified in several different ways. On the basis of their phase, we may describe them as solids, liquids or gases. Here, we will naturally be interested mainly in solids. Even these can exist in a variety of forms, however, ranging from those entirely lacking in detailed structure (amorphous materials) to those with a detailed, long-range order (crystals). In between, there are media which have local structure, known as polycrystalline materials. Furthermore, their electronic properties may be widely different. At one extreme there are insulators (which do not conduct electricity), and at the other there are conductors (which clearly do). Somewhere in the middle we find the semiconductors, which have the most complicated properties of all.

Faced with all this we might suppose that a quite separate and complex model would be required for each class of solid. For simplicity, therefore, we will begin by ignoring any detailed structure, and concentrate entirely on the electronic aspects. Our argument runs like this. Since all materials are simply collections of atoms and molecules, they can all be described as particular distributions of positive and negative charge. A model for the optical properties of a material can then be developed by considering this charge distribution and its interaction with an electromagnetic wave. Naturally, we may improve the picture later on, by inserting structural features where relevant.

The electronic model (originally due mainly to Hendrik Lorentz, 1853–1928, and Paul Drude, 1863–1906) is based on a consideration of the forces acting on the charges inside the material. For a free charge e moving with a velocity $\mathbf{v}$ through an electric field $\mathbf{E}$ and a magnetic flux density $\mathbf{B}$, the force is known from low-frequency electromagnetic experiments to be

$$\mathbf{F} = e[\mathbf{E} + \mathbf{v} \times \mathbf{B}] \tag{3.1}$$

However, all charge velocities will be small in the discussions that follow, so we may safely neglect any forces due to magnetic flux, concentrating exclusively on the interaction of the electric field with the charges. This view is reinforced by the fact that the permeability of most optically significant materials is essentially μ_0—the value for free space. We can use this approach to explain the origin of the dielectric constant in insulators and the conductivity of metals; the properties of semiconductors will be treated in a different way (using the language of solid state theory) in Chapter 11.

3.2 THE ORIGIN OF THE DIELECTRIC CONSTANT

We will start by considering an insulating, transparent, dielectric solid, whose optical properties are determined mainly by the behaviour of the outermost, bound electrons orbiting each atom. In Chapter 2 we showed that the electric field at any point will oscillate sinusoidally for a harmonic wave. From Eq. (3.1), this field must impose forces in opposite directions on any positive and negative charges located there. Consequently, we would expect the field to separate the charges in a periodic manner, producing a local distribution of *dipoles*. Figure 3.1 shows this separation for a static field.

In addition, we might expect that as the charges separate, the local charge imbalance will generate another force, which would tend to restore charge neutrality. Finally, the oscillating charges will act as microscopic radiators, giving rise to energy loss. This will also cause some disturbance to the neighbouring atoms of the structure. We can model the whole as a damping effect, which describes the transfer of energy from our particular charges into the whole structure of the solid.

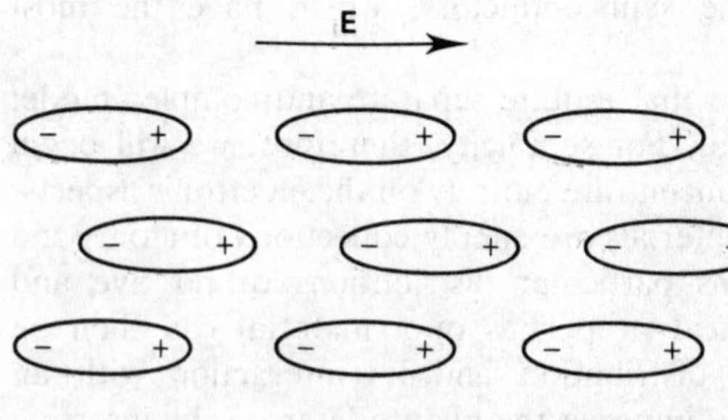

Figure 3.1 Separation of charges by an electric field.

Combining these effects together enables us to write an expression for the equation of motion of a typical charge. This will be one of the bound electrons—the positive charges, being considerably more massive (and anyhow locked into the structure, since it is solid) would tend not to move significantly at optical frequencies. For simplicity, we will also ignore the vectorial nature of the field, and consider motion in the *x*-direction only. The components of force are therefore:

The driving force due to the harmonic field Since a harmonic electric field is given by $\mathsf{E} = E \exp(j\omega t)$, it might initially be thought that this force will have the form:

$$\mathsf{F}_e = e\mathsf{E} = eE \exp(j\omega t) \tag{3.2}$$

The local field acting on an electron will be determined not only by the driving field E but also by the separation of all the other charges in its immediate neighbourhood. The net effect of this charge separation is described as an electronic *polarization* of the material, and is represented by the symbol P (not to be confused with the optical polarization of the previous chapter—it is an unfortunate coincidence that two such different phenomena should have the same name). Defining the displacement of a typical charge from its equilibrium position by x, we can write P as

$$\mathsf{P} = Ne\mathsf{x} \tag{3.3}$$

where N is the dipole density. The polarization will also vary harmonically with an angular frequency ω, since it is generated by the driving field, so we may also write

$$\mathsf{P} = P \exp(j\omega t) \tag{3.4}$$

In general, P might be a complex quantity, implying a possible phase difference between E and P.

The effect of the polarization is to modify the local electric field, increasing its value above that in free space (to, say, E_{loc}). For a non-polar material (i.e. one with no permanent dipoles in the absence of a driving field), Lorentz showed that the increase can be found very simply as

$$\mathsf{E}_{loc} - \mathsf{E} = \mathsf{P}/3\varepsilon_0 \tag{3.5}$$

Consequently the force acting on an electron is not as given in Eq. (3.2), but is instead

$$\mathsf{F}_e = e\mathsf{E}_{loc} = e(\mathsf{E} + \mathsf{P}/3\varepsilon_0) \tag{3.6}$$

Typical non-polar materials include Si, Ge, C (in diamond form) and liquid hydrogen. Polar materials, which *do* have permanent dipoles, include NaCl, TiO_2 and SiO_2; in addition, there is also a class of weakly polar materials.

The restoring force due to the charge separation Since dipoles have been produced by the driving field, we might expect a restoring force due to attraction between the separated charges. Modelling the system as a spring with extension x, this force would have the form:

$$\mathsf{F_r} = -s\mathsf{x} \tag{3.7}$$

Here s is a constant of proportionality, which depends on the exact nature of the material. For example, materials with a highly polarizable structure will have high values of s. Now, the linear dependence of $\mathsf{F_r}$ on x might be regarded as an unjustifiable assumption (indeed, we might have expected an inverse square law dependence, since this is the norm for two point charges). However, there are many interacting charges, and it is the *superposition* of large numbers of inverse square forces that produce this linear effect. Furthermore, we can argue that for the moment we will treat the linear term as but the first (and hopefully dominant) term in a general power series. We will consider the significance of higher-order terms later on.

The damping force We must also include a force which accounts for energy loss. The justification for this ultimately hangs on any success of our theory in describing the observable effects. The simplest such force is the viscous damping force of a simple harmonic oscillator, which has the form:

$$\mathsf{F_d} = -r\,\mathrm{d}\mathsf{x}/\mathrm{d}t \tag{3.8}$$

The sum of these forces must yield the product of the mass m of our electron (9.1×10^{-31} kg) and its acceleration, so combining all the terms we obtain the following typical *equation of motion*:

$$m\,\mathrm{d}^2\mathsf{x}/\mathrm{d}t^2 = -s\mathsf{x} - r\,\mathrm{d}\mathsf{x}/\mathrm{d}t + e(\mathsf{E} + \mathsf{P}/3\varepsilon_0) \tag{3.9}$$

Eliminating the displacement x using Eq. (3.3), we then obtain

$$m\,\mathrm{d}^2\mathsf{P}/\mathrm{d}t^2 + r\,\mathrm{d}\mathsf{P}/\mathrm{d}t + s\mathsf{P} = Ne^2(\mathsf{E} + \mathsf{P}/3\varepsilon_0) \tag{3.10}$$

Rearranging Eq. (3.10) slightly then gives

$$\mathrm{d}^2\mathsf{P}/\mathrm{d}t^2 + \gamma\,\mathrm{d}\mathsf{P}/\mathrm{d}t + (\omega_0^2 - Ne^2/3m\varepsilon_0)\,\mathsf{P} = Ne^2\mathsf{E}/m \tag{3.11}$$

where we have written $s/m = \omega_0^2$, and $r/m = \gamma$.

Both E and P vary harmonically. Performing the necessary differentiation, we can therefore eliminate the time-dependence to get

$$-\omega^2 P + \mathrm{j}\omega\gamma P + (\omega_0^2 - Ne^2/3m\varepsilon_0)\,P = Ne^2E/m \tag{3.12}$$

and hence find P as

$$P = E\,\{Ne^2/m\}/\{[(\omega_0^2 - \omega^2) - Ne^2/3m\varepsilon_0] + \mathrm{j}\omega\gamma\} \tag{3.13}$$

We have now found a relationship between the polarization and the driving field. For simplicity, we can combine the terms ω_0^2 and $Ne^2/3m\varepsilon_0$. Defining the *plasma frequency* ω_p using

$$\omega_p^2 = Ne^2/m\varepsilon_0 \tag{3.14}$$

we obtain the slightly modified resonant frequency ω_m, given by

$$\omega_m^2 = \omega_0^2 - \omega_p^2/3 \tag{3.15}$$

so that Eq. (3.13) becomes

$$P = E\, \varepsilon_0 \omega_p^2 / \{(\omega_m^2 - \omega^2) + j\omega\gamma\} \tag{3.16}$$

Equation (3.16) shows that the dipoles will oscillate in response to the driving field E, but generally neither the polarization nor the displacement are in phase with E. We note that the magnitude of the polarization depends on the relation of the driving frequency ω to ω_m. Close to resonance, when $\omega \approx \omega_m$, P will reach a maximum, whose magnitude is limited only by the damping term $j\omega\gamma$.

We now need to find out how the polarization P relates to the permittivity ε. From Gauss' law, we know that

$$\iint_A \mathbf{D} \cdot \mathbf{da} = \iiint_V \rho\, dv \tag{3.17}$$

This implies (among other things) that the surface charge density on the plates of a parallel-plate capacitor must be D. If the plates are spaced by vacuum, we have of course that $\mathsf{D} = \varepsilon_0 \mathsf{E}$. However, the introduction of a dielectric spacer must lead to a change in the surface charge density, since the dielectric material is polarized by the field inside the capacitor. This leads to a motion of charges towards the plates. The quantity of charge moved per unit area is $Ne\mathsf{x}$, so in general we have

$$\mathsf{D} = \varepsilon_0 \mathsf{E} + \mathsf{P} \tag{3.18}$$

A dielectric spacer therefore improves the charge storage capability of a capacitor, a discovery originally made by Faraday. Since P is proportional to E, then writing $\mathsf{P} = \varepsilon_0 \chi\, \mathsf{E}$, where χ is the *dielectric susceptibility*, leads to

$$\mathsf{D} = \varepsilon_0 (1 + \chi)\, \mathsf{E} \tag{3.19}$$

Since in Chapter 2 we defined $\mathsf{D} = \varepsilon_0 \varepsilon_r\, \mathsf{E}$, it is clear that the relative dielectric constant must be

$$\varepsilon_r = 1 + \chi = 1 + \mathsf{P}/\varepsilon_0 \mathsf{E} \tag{3.20}$$

Using Eq. (3.15), we then obtain

$$\varepsilon_r = 1 + \omega_p^2 / \{(\omega_m^2 - \omega^2) + j\omega\gamma\} \tag{3.21}$$

We have therefore found a direct connection between the dielectric constant and the displacements of dipoles. There are two interesting points to note. First, we see that the expression for ε_r is independent of the driving field, as we might expect. This has come about essentially because of our assumption that the restoring force on a general electron was linearly proportional to displacement. Second, the expression for ε_r is complex. In the next section we will discuss the implications of this, but before doing so we will briefly consider one final aspect: the relation between polarizability and the dielectric constant.

Polarizability

We have seen that the origin of the dielectric constant lies in the separation of charges by an electric field. We may therefore connect the polarization with the local field using

$$\mathsf{P} = N\alpha \mathsf{E}_{loc} \tag{3.22}$$

Here α is a constant, known as the *polarizability*. Its magnitude essentially describes the ease or otherwise with which charges may be separated by the field. Using the definition of the local field given in Eq. (3.5) we can obtain

$$\mathsf{P} = \{N\alpha/(1 - N\alpha/3\varepsilon_0)\}\, \mathsf{E} \tag{3.23}$$

However, since we have already defined $\mathsf{P} = \varepsilon_0 \chi \mathsf{E}$, it is clear that the dielectric susceptibility is related to the polarizability by

$$\chi = (N\alpha/\varepsilon_0)/(1 - N\alpha/3\varepsilon_0) \tag{3.24}$$

Equally, since $\varepsilon_r = 1 + \chi$, we must have

$$\varepsilon_r = (1 + 2N\alpha/3\varepsilon_0)/(1 - N\alpha/3\varepsilon_0) \tag{3.25}$$

Equation (3.25) shows that, in general, materials with high polarizability will have a large dielectric constant, and vice versa. Often, the equation is inverted to give α in terms of ε_r, as

$$\alpha = (3\varepsilon_0/N)\,(\varepsilon_r - 1)/(\varepsilon_r + 2) \tag{3.26}$$

This is known as the *Lorentz-Lorenz* equation. It is an interesting relation, because it enables the calculation of a microscopic quantity (α) from a macroscopic observable (ε_r).

3.3 MATERIAL DISPERSION

The dependence of the dielectric constant on frequency is known as *material dispersion*. We shall now examine the consequence of this feature, and of the complex nature of ε_r.

In the previous chapter we assumed that a lossy dielectric could be described by a complex relative permittivity of the form $\varepsilon_r = \varepsilon_r' - j\,\varepsilon_r''$. Multiplying the top and bottom of Eq. (3.21) by $\{(\omega_m^2 - \omega^2) - j\omega\gamma\}$ we obtain the following expressions for the real and imaginary parts of ε_r:

$$\varepsilon_r' = 1 + \omega_p^2\,(\omega_m^2 - \omega^2)/\{(\omega_m^2 - \omega^2)^2 + (\omega\gamma)^2\}$$
$$\varepsilon_r'' = \omega_p^2\,\omega\gamma/\{(\omega_m^2 - \omega^2)^2 + (\omega\gamma)^2\} \tag{3.27}$$

If the damping factor γ is negligibly small, then these may be approximated by

$$\varepsilon_r' \approx 1 + \omega_p^2/(\omega_m^2 - \omega^2)$$
$$\varepsilon_r'' \approx \omega_p^2\,\omega\gamma/(\omega_m^2 - \omega^2)^2 \tag{3.28}$$

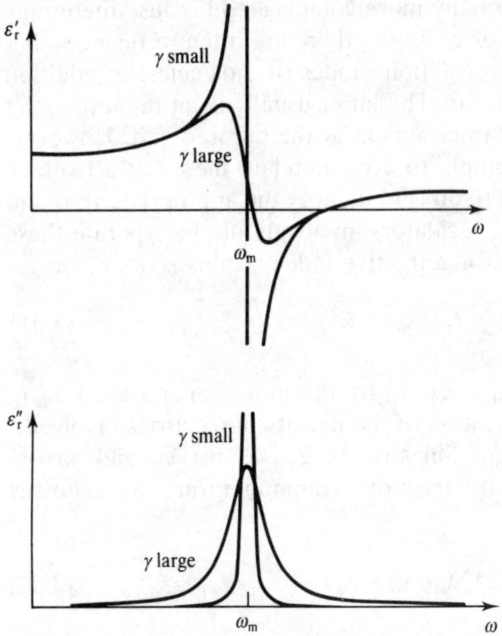

Figure 3.2 Typical variation of the complex dielectric constant with frequency.

From this, we see that the real part ε_r' will be approximately unity, unless ω is close to the resonant frequency ω_m. Exactly at $\omega = \omega_m$ it is discontinuous. Similarly, the imaginary part ε_r'' is small except near $\omega = \omega_m$, where it becomes infinite. This behaviour is plotted in Fig. 3.2. Note that it is quite possible for ε_r' to become negative, just beyond the resonant frequency.

Also shown in Fig. 3.2 are the predictions of the full expression (3.27), which does not assume that $\gamma = 0$. The introduction of damping introduces two new features. The first is that while ε_r' normally increases with ω (this behaviour is called *normal dispersion*) there is now a narrow range of frequency near ω_m where ε_r' decreases with ω. This is known as *abnormal dispersion*. The second is that ε_r'' now has a finite maximum value. The overall shape of this so-called 'absorption line' is now a function that occurs often in physics, a *Lorentzian lineshape*. For optical propagation it is often convenient to deal with the refractive index instead of ε_r. Since $n = \sqrt{\varepsilon_r}$, n must also be complex. Writing $n = n' - jn''$, we have:

$$n^2 = (n'^2 - n''^2) - 2jn'n'' = \varepsilon_r' - j\,\varepsilon_r'' \tag{3.29}$$

Making the approximation that n'' is small, we may put $n'^2 \approx \varepsilon_r'$ and $2n'n'' \approx \varepsilon_r''$. Assuming that Eq. (3.28) is also valid, we can then write

$$n'^2 \approx 1 + \omega_p^2/(\omega_m^2 - \omega^2) \tag{3.30}$$

and so on. Hence it is clear that the refractive index must depend on frequency.

The situation in real materials is normally more complicated because there may be more than one resonance involved. For example, there are often resonances that correspond to excitation of vibrational oscillation modes of molecules, in addition to the electronic resonance considered so far. The latter usually lie at the ultraviolet end of the optical spectrum, while the former appear at the infrared end. However, by slightly modifying Eq. (3.30), it is simple to accommodate these extra features in a phenomenological way. All we need to do is to identify the appropriate resonant frequencies, together with the density of oscillators involved, and incorporate these terms as a summation in the expression for refractive index. If this is done, we get

$$n'^2 \approx 1 + \omega_p^2 \sum_j f_j/(\omega_{mj}^2 - \omega^2) \tag{3.31}$$

Here we have replaced the plasma frequency ω_p^2 by the more general term $\omega_p^2 f_j$, where f_j is the *oscillator strength* (an indication of the density of electrons involved), and the resonant frequency ω_m by ω_{mj}. Substituting $2\pi c/\lambda_0$ for ω, and so on, Eq. (3.31) may then be transformed into its more common form, the *Sellmeier equation*:

$$n'^2 \approx 1 + \sum_j A_j \lambda_0^2/(\lambda_0^2 - \lambda_j^2) \tag{3.32}$$

Values of A_j and λ_j are often fitted to dispersion data to provide an estimate of refractive index over a wide spectral range.

Design Example

Table 3.1 shows values of A_j and λ_j for silica, following I. H. Maletson (1965). Note that only three resonances need to be considered. Two of these (at 0.068 μm and 0.116 μm wavelength) lie in the ultraviolet, while the third (at 9.9 μm) lies in the infrared. The corresponding dispersion curve is shown in Fig. 3.3. In this case experimental values are indistinguishable from the theoretical curve, which provides convincing support for the rather heuristic theory we have used here. Note that there is an inflexion in the curve at around $\lambda_0 = 1.27$ μm, so that $d^2n/d\lambda_0^2$ will be zero at this point. According to the discussion at the end of Chapter 2, this will prove a suitable operating point for minimum signal dispersion.

Table 3.1 Sellmeier coefficients for silica

i	A_j	λ_j (μm)
1	0.696 166 3	0.068 404 3
2	0.407 942 6	0.116 241 4
3	0.897 479 4	9.896 161 0

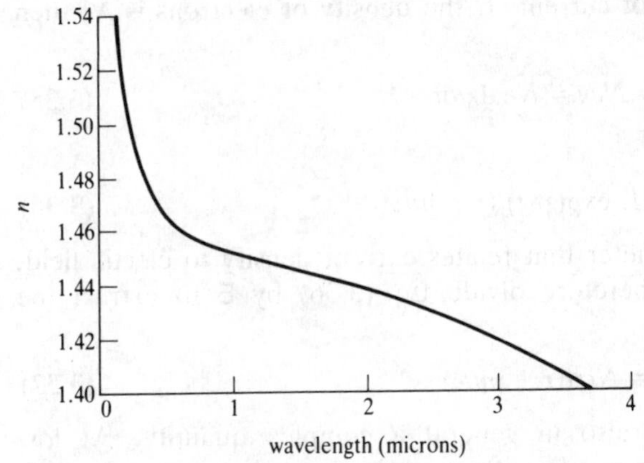

Figure 3.3 Dispersion curve for silica.

Validity of the 'Classical' Model

Figure 3.2 shows absorption loss increasing strongly in the vicinity of a resonance. While it adequately describes the *sense* of the loss, the classical treatment we have used fails completely to account for the actual transition between one bound state and another that must occur. This can only be described effectively using the ideas of quantum mechanics and solid state theory, which we will discuss in Chapter 11.

3.4 THE ORIGIN OF ELECTRICAL CONDUCTIVITY

We now move on to consider the origin of a different parameter, conductivity. In this case the dominant effects are caused by charges that are free to move, rather than bound to an atomic site as in a dielectric. For simplicity we shall restrict ourselves to the case of a single type of charge carrier; *free electrons* in a metal. An alternative title for this topic might therefore be 'the optics of metals'.

We begin by considering the response of the free electrons to the oscillating field of a propagating wave. The major difference in the behaviour of these electrons (compared with bound electrons) is that as a free electron moves under the action of the field, another one moves into the space vacated. There are, therefore, no unbalanced charges, and there can be no restoring electrostatic forces. Similarly, there is little in the way of local polarization. However, we can still expect damping, due to collisions of the electrons among themselves and with the lattice. The controlling equation of motion for each electron is therefore

$$m\, d^2\mathsf{x}/dt^2 = -\, r\, d\mathsf{x}/dt + e\mathsf{E} \tag{3.33}$$

This should be compared with Eq. (3.9); clearly the two are similar, but Eq. (3.33) lacks the spring and local polarization terms.

Assuming again that $\mathsf{E} = E \exp(\mathrm{j}\omega t)$, we may solve Eq. (3.33) to obtain

$$\mathsf{x} = eE \exp(\mathrm{j}\omega t)/(r\mathrm{j}\omega - m\omega^2) \tag{3.34}$$

Motion of charges implies a flow of current. If the density of electrons is N, then the resulting current density is

$$\mathsf{J} = Nev = Ne\,\mathrm{d}\mathsf{x}/\mathrm{d}t \tag{3.35}$$

Differentiating Eq. (3.34), we get

$$\mathsf{J} = Ne^2E\,\exp(\mathrm{j}\omega t)/(r + \mathrm{j}m\omega) \tag{3.36}$$

Conductivity is defined as a parameter that relates current density to electic field, according to $\mathsf{J} = \sigma\mathsf{E}$. We may therefore divide Eq. (3.36) by E to extract the conductivity as

$$\sigma = Ne^2/(r + \mathrm{j}m\omega) \tag{3.37}$$

Clearly, electrical conductivity is also in general a complex quantity. At low frequencies it is real (which corresponds with everyday experience) but at high enough frequencies its character changes quite dramatically, and it becomes purely imaginary.

At this point we shall revert to the notation of Chapter 2, where we showed that a material with non-zero conductivity could be described by a modified dielectric constant of the form:

$$\varepsilon = \varepsilon_\infty - \mathrm{j}\sigma/\omega \tag{3.38}$$

In a metal we have already mentioned that the contribution of bound electrons is extremely small, so ε_∞ is very close to the value obtained in free space, ε_0. A metal may therefore be described by a modified *relative* dielectric constant:

$$\varepsilon_\mathrm{r} = 1 - (Ne^2/m\varepsilon_0)/\{\omega\,(\omega - \mathrm{j}r/m)\} \tag{3.39}$$

The term $Ne^2/m\varepsilon_0$ is clearly equal to ω_p^2, where ω_p is once again the plasma frequency. Similarly, r/m corresponds to the damping factor γ of the previous sections. However, because damping in metals is essentially due to collisions, this is generally given the alternative name ω_τ, the *collision damping frequency*.

We can easily extract the real and imaginary parts of ε_r, by multiplying the top and bottom of Eq. (3.39) by $\omega + \mathrm{j}\omega_\tau$. If this is done we get

$$\varepsilon_\mathrm{r}' = 1 - \omega_\mathrm{p}^2/(\omega^2 + \omega_\tau^2)$$
$$\varepsilon_\mathrm{r}'' = -\,\omega_\mathrm{p}^2\,\omega_\tau/\{\omega\,(\omega^2 + \omega_\tau^2)\} \tag{3.40}$$

Examining Eq. (3.40) we first note the very striking result that ε_r' can actually be negative over a wide spectral range, when $\omega_\mathrm{p}^2/(\omega^2 + \omega_\tau^2) > 1$. For this to be possible, however, we require $\omega_\mathrm{p} > \omega_\tau$. Values of ω_p and ω_τ are shown in Table 3.2 for some common metals. Note that the ratio $\omega_\mathrm{p}/\omega_\tau$ is greater than about 20 for all the members of this group.

In the case of silver (Ag), which is widely used in optical surface wave devices (see Chapter 5), ε_r' will be negative for all angular frequencies up to $\omega \approx 1.2 \times 10^{16}$ rad/sec. Since this corresponds to a wavelength of $\lambda_0 \approx 150$ nm (deep in the ultraviolet), ε_r' will be negative throughout the visible and infrared

Table 3.2 Values of ω_p, ω_τ and ω_p/ω_τ for some common metals

	ω_p	ω_τ	ω_p/ω_τ
Aluminium	19.8×10^{15}	1.01×10^{15}	19.6
Copper	10.1×10^{15}	0.31×10^{15}	32.6
Gold	9.9×10^{15}	0.44×10^{15}	22.5
Silver	12.2×10^{15}	0.09×10^{15}	135.6

spectra. This result has enormous influence on the types of electromagnetic wave that may be supported by this material. Furthermore, the exceptionally high value of ω_p/ω_τ suggests that the damping of these waves may be relatively low.

Design Example

We can use Eq. (3.40) to plot the variation of ε_r' and ε_r'' with wavelength for silver through the visible spectrum. The results are as shown in Fig. 3.4.

Also shown on the figure are some experimental points (after Ordal *et al.*, 1983). The theoretical model has been matched to the data by choosing ω_p and ω_τ to give exact agreement for ε_r' at $\lambda_0 = 0.6\ \mu$m. The simple theory outlined here provides broadly the correct magnitude for ε_r', but the wavelength dependence is not quite correct. This is mainly because our model does not completely describe the mechanisms for electron energy loss through collisions. However, the agreement is almost exact for ε_r''.

Note that if damping can be neglected (as in an ionized gas, which has no lattice for the electrons to collide with), Eq. (3.40) reduces to $\varepsilon_r' = n^2 = 1 - \omega_p^2/\omega^2$. Since

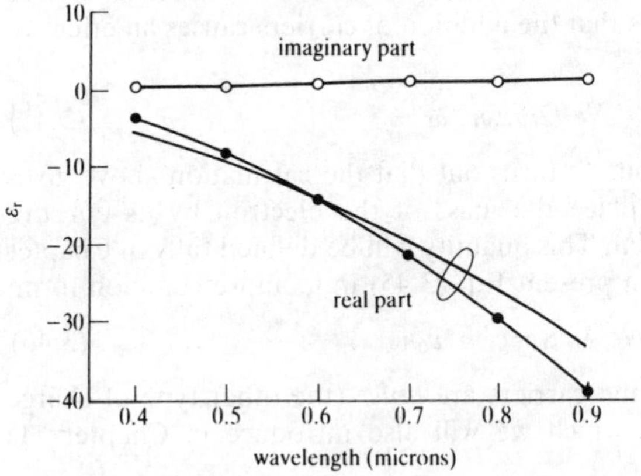

Figure 3.4 Comparison between experimental and theoretical data for ε_r' and ε_r'' in silver.

$\omega = kc/n$, we may then obtain $\omega = \sqrt{[\omega_p^2 + c^2k^2]}$—the dispersion characteristic introduced in Chapter 2 for the ionosphere.

Plasma Contribution to the Refractive Index

Of course there are materials in which the dielectric constant is determined by contributions from both bound and free electrons. The examples we have in mind are semiconductors, and though the electronic effects occuring in these media are vastly more complicated than we have encountered so far, we shall take the liberty of extending our crude model to estimate one very important quantity: the change in refractive index that follows from the introduction of significant numbers of free carriers to an otherwise poorly-conducting material.

We start by returning to Eqs (3.37) and (3.38). Assuming this time that the contribution of bound electrons is *not* negligible, we may write

$$\varepsilon_r = \varepsilon_{r\infty} - j\sigma/\varepsilon_0\omega \tag{3.41}$$

where $\varepsilon_{r\infty}$ is the relative dielectric constant in the absence of free electrons. This is no longer unity, but we take it for simplicity to be real. In the presence of free electrons σ is non-zero, but we shall assume the frequency is sufficiently high that $\omega \gg \omega_\tau$, and σ may then be approximated by

$$\sigma \approx Ne^2/jm\omega \tag{3.42}$$

In this case Eq. (3.41) becomes

$$\varepsilon_r \approx \varepsilon_{r\infty}\{1 - Ne^2/m\varepsilon_0\varepsilon_{r\infty}\omega^2\} \tag{3.43}$$

The refractive index is clearly given by the square-root of Eq. (3.43). If we now assume that the right-hand term in the brackets is small, we may use a binomial approximation to obtain

$$n \approx n_\infty - Ne^2/2m\varepsilon_0 n_\infty \omega^2 \tag{3.44}$$

where we have defined $n_\infty = \sqrt{\varepsilon_{r\infty}}$ as the refractive index in the absence of free electrons. Equation (3.44) now shows that the addition of carriers causes an effective *decrease* in index Δn, given by

$$\Delta n \approx -Ne^2/2m\varepsilon_0 n_\infty \omega^2 \tag{3.45}$$

Despite the crudity of our assumptions, it turns out that the calculation above gives the correct answer, provided we replace the mass of the electron by its *effective mass* m^* in the semiconductor material. This quantity will be defined fully in Chapter 11. Writing $\omega = 2\pi c/\lambda_0$, we may then present Eq. (3.45) in its more common form:

$$\Delta n \approx -Ne^2\lambda_0^2/8\pi^2c^2m^*\varepsilon_0 n_\infty \tag{3.46}$$

A similar result may be obtained if the carriers are *holes* (the other type of charge carrier present in a semiconductor, which we will also introduce in Chapter 11) instead of electrons.

Design Example

For n-type GaAs, $n_\infty = 3.5$ and $m^* = 0.067$ m. Consequently, $\Delta n \approx -1.9 \times 10^{-21} N \lambda_0^2$ (when N is measured in cm^{-3} and λ_0 is in μm). For $N \approx 5 \times 10^{18}\ cm^{-3}$ and $\lambda_0 = 1\ \mu m$, we then obtain $\Delta n \approx -0.01$. Although the carrier density needed to obtain such a value of Δn is very large, the reduction in index caused by the plasma contribution is extremely useful. For example, fixed changes in index may be induced by doping semiconductor material to change the carrier concentration. Alternatively, carriers may be injected to cause a dynamic change in refractive index. Both effects are employed in semiconductor integrated optics, and this topic will be further discussed in Chapter 12.

3.5 ANISOTROPIC MEDIA

We now consider the additional effect of the actual *structure* of materials, which we have ignored in favour of their electronic behaviour. Essentially, we have treated all materials as isotropic until now. This assumption implies that the result of applying a field will be independent of its direction. Furthermore, it assumes implicitly that the material polarization will also be parallel to the applied field. Although these assumptions are sometimes valid (e.g. in a truly amorphous material), in crystals we should generally allow for an *anisotropic* response. Since many of the materials used in optoelectronics are crystalline, this is clearly an important feature.

The origins of anisotropy lie in any lack of symmetry in the molecules forming the material. In general, these do not have neat spherical shapes—instead, they have configurations that are characteristic of their composition, electronic structure and chemical activity. The molecular configuration in turn determines the broad classification of any crystalline solid they may form. Generally there are seven possibilities: a crystal may be *cubic, tetragonal, hexagonal, trigonal, orthorhombic, monoclinic* or *triclinic* in structure. These systems represent the only shapes of unit cell (the basic element of a crystal lattice) that can fill all space by repetition.

Figure 3.5 shows a particularly simple example, the structure of the cubic ionic crystal NaCl. Here the Na^+ and Cl^- ions are arranged in a regular lattice, which consists of three equal *crystal axes*, all at right angles to one another. This high

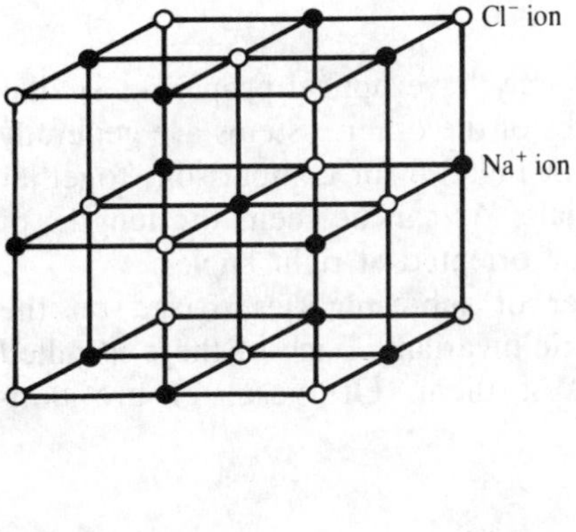

Figure 3.5 Cubic structure of NaCl.

Cubic	3 equal axes, all at right angles.	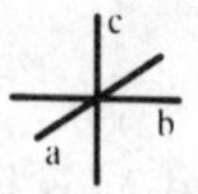	GaAs Ge InP Si ZnS (zinc blende)
Tetragonal	3 axes at right angles; the two horizontal axes of equal length, the vertical one of different length.	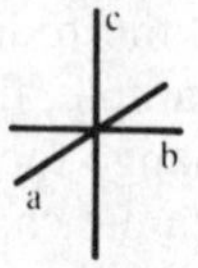	$BaTiO_3$ KH_2PO_4 (KDP) TiO_2 (rutile) $ZrSiO_4$(zircon)
Hexagonal	1 vertical and 3 horizontal axes, at angles of 120 degrees to each other. The horizontal axes of equal length.	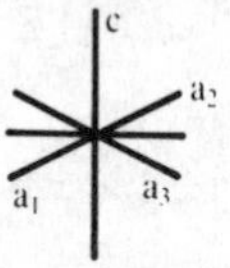	
Trigonal	Similar to hexagonal.	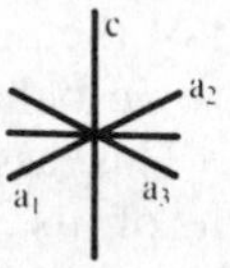	$CaCO_3$(calcite) $LiNbO_3$ $LiTaO_3$ SiO_2(quartz)
Orthorhombic	3 axes of unequal length, at right angles	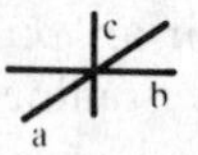	
Monoclinic	1 vertical and 1 horizontal axis at right angles, with the axis inclined towards the observer. All axes are of different length.	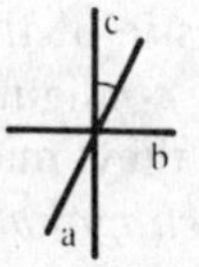	mica MNA
Triclinic	A vertical axis with 2 other axes inclined to it. These are not at right angles to each other, and are of different length.	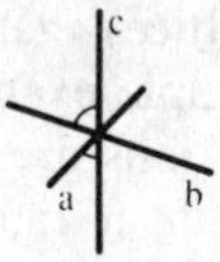	$CuSO_4.5H_2O$ (copper sulphate)

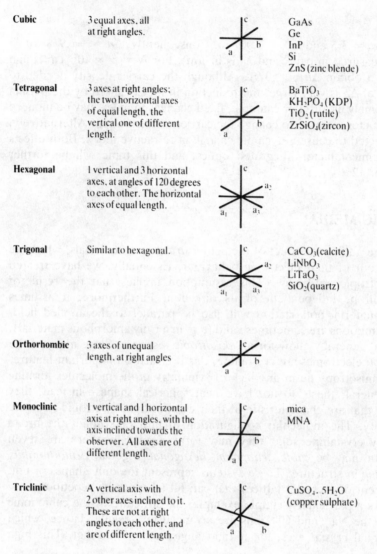

Figure 3.6 Crystal axes of the seven crystal classes.

degree of symmetry suggests that cubic crystals may have optical properties similar to those of amorphous media. However, the axes of the other systems are generally more complicated. The complete set is shown in Fig. 3.6 for comparison, together with some examples of common optical materials. As can be seen, the lengths of the axes may be different, and they need not be oriented at right angles.

Within these systems, there are a number of sub-categories, based on the possible symmetry operations that leave the lattic invariant. Each of these is called a *point group*, and there are a total of 32 of them. Of these, 21 are noncentrosymmetric (i.e. they lack a centre of symmetry). Because of these differences in symmetry, we would guess that the ease with which an electron may move in the various directions may be different in some crystals. An electronic model of optical anisotropy could therefore be developed from our previous treatment, for example by assuming several spring constants, each one for electron displacement in a particular direction. If this is done, it is found that the result of applying a field is *not* independent of direction; instead, the polarization depends on the field direction relative to the crystal axes. The relation between the polarization and the electric field must then be written (at least, for small fields) as

$$\mathbf{P} = \varepsilon_0 [\chi] \mathbf{E} \tag{3.47}$$

Although Eq. (3.47) appears more complicated than the definition $P = \varepsilon_0 \chi E$ we have used till now, it has a simple explanation. First, we have replaced the scalar fields by vector ones, to introduce some significance to field directions. Second, we have exchanged the scalar dielectric susceptibility χ for the *second rank tensor* term $[\chi]$; this allows the response of the material to be different in different directions relative to the crystal axes.

According to the rules of tensor arithmetic, Eq. (3.47) may be written out in longhand as

$$\begin{aligned} P_x &= \varepsilon_0 \{\chi_{xx} E_x + \chi_{xy} E_y + \chi_{xz} E_z\} \\ P_y &= \varepsilon_0 \{\chi_{yx} E_x + \chi_{yy} E_y + \chi_{yz} E_z\} \\ P_z &= \varepsilon_0 \{\chi_{zx} E_x + \chi_{zy} E_y + \chi_{zz} E_z\} \end{aligned} \tag{3.48}$$

Thus, Eq. (3.47) may be interpreted by saying that if the three components of **P** are written as a column vector, this may be found by multiplying a similar vector describing **E** by a 3 × 3 matrix, whose elements are those of the dielectric susceptibility tensor.

If we consider one component of polarization (say, P_x), then Eq. (3.48) shows that there will be contributions to P_x from all three of the electric field components E_x, E_y and E_z. Now, the relative magnitudes of the components of the susceptibility tensor depend in practice on our choice of coordinate system, with respect to the axes of the crystal. For the particular case in which the coordinate system is chosen to coincide with these principal axes, the off-axis components of the tensor are zero. We are then left with the simpler relations:

$$\begin{aligned} P_x &= \varepsilon_0 \chi_x E_x \\ P_y &= \varepsilon_0 \chi_y E_y \\ P_z &= \varepsilon_0 \chi_z E_z \end{aligned} \tag{3.49}$$

Here we have written χ_x for χ_{xx} in this special case, and so on; we will continue to use this reduced notation in other similar situations. Even here, **P** will only be parallel to **E** when

$$\chi_x = \chi_y = \chi_z \tag{3.50}$$

It turns out that this equality holds for amorphous materials, and also for crystals with cubic symmetry. However, for all the other crystal groups, it does not hold. We can interpret this as establishing that the dielectric constant will be different for field components along each of the crystal axes. Thus, we will also need to replace the simple relation $\mathbf{D} = \varepsilon_0 (1 + \chi) \mathbf{E}$ by

$$\mathbf{D} = \varepsilon_0 \{1 + [\chi]\} \mathbf{E} = [\varepsilon] \mathbf{E} \tag{3.51}$$

where $[\varepsilon] = \varepsilon_0 \{1 + [\chi]\}$ is a second rank tensor definition of the dielectric constant, with $\varepsilon_{xx} = \varepsilon_0 \{1 + \chi_{xx}\}$, $\varepsilon_{xy} = \varepsilon_0 \chi_{xy}$, $\varepsilon_{xz} = \varepsilon_0 \chi_{xz}$ and so on. Similarly, we may define a tensor $[\varepsilon_r]$ for the *relative* dielectric constant, with elements $\varepsilon_{rij} = \varepsilon_{ij}/\varepsilon_0$. This allows the following classification of crystals in terms of their optical properties. Materials in which two of the components ε_x, ε_y and ε_z are equal (e.g. $\varepsilon_x = \varepsilon_y \neq \varepsilon_z$) are termed *uniaxial*. This class includes tetragonal, hexagonal and trigonal crystals. Similarly, materials in which no two components are equal (e.g. $\varepsilon_x \neq \varepsilon_y \neq \varepsilon_z$) are called *biaxial*. This class contains the orthorhombic, monoclinic and triclinic crystals.

Wave Propagation in Anisotropic Media

A logical conclusion from the above is that the phase velocity of waves in anisotropic materials will also depend on the orientation of the electric field vector with respect to the crystal axes, so we would expect some complicated propagation effects to arise. We shall illustrate this with simple examples, beginning with the case of an optical wave travelling in a direction parallel to one of the crystal axes. We shall assume for simplicity that these axes are parallel to the *x*-, *y*- and *z*-directions, and that the wave is travelling in the *z*-direction. Figure 3.7 then shows one possible orientation of the electric field vector in relation to the *x*- and *y*-axes.

The component of the transverse field in the *x*-direction (E_x) will experience the dielectric constant ε_x, while E_y sees ε_y. Consequently, the wave is divided into two parts, one polarized in the *x*-direction and travelling with a phase velocity c/n_x, where $n_x = \sqrt{\varepsilon_{rx}}$, the other *y*-polarized and moving with a velocity c/n_y. If $n_x \neq n_y$, the two travel in the same direction, but at different speeds.

Design Example—Retardation Plates

This feature is exploited in *retardation plates*, which are devices that can manipulate the state of polarization of a wave. Assuming the geometry of Fig. 3.7, with a crystal thickness of *d*, the phase delays ϕ_x and ϕ_y introduced into each component in travelling through the material are

$$\phi_x = k_0 n_x d \quad \text{and} \quad \phi_y = k_0 n_y d \tag{3.52}$$

The phase difference introduced (i.e. $\phi_x - \phi_y$) will then in general convert the incident plane polarized wave into an elliptically polarized output. There are two particular cases of interest. The first occurs when there is an effective phase difference of π radians, so that

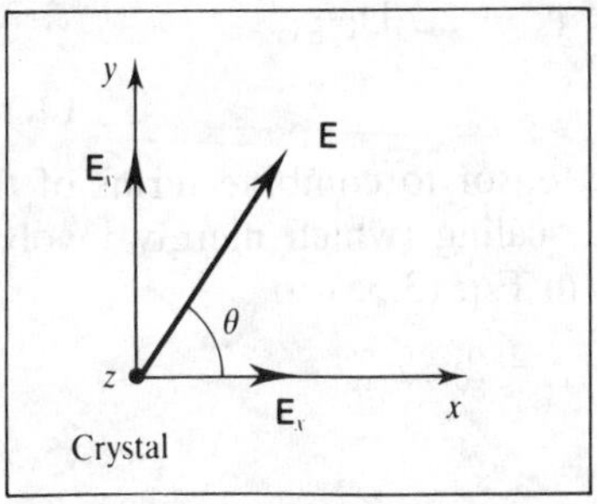

Figure 3.7 Transverse field components of a plane wave travelling in an anisotropic crystal.

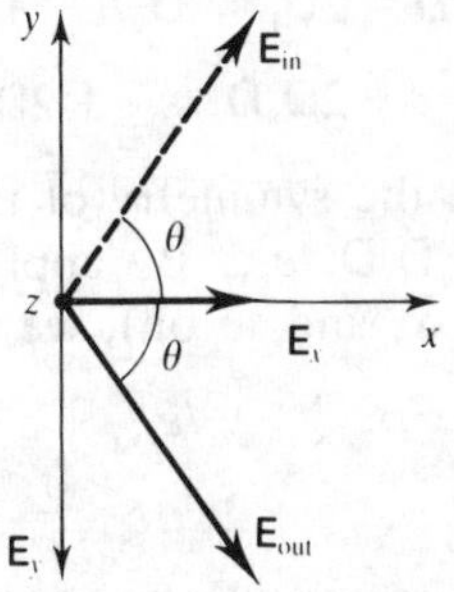

Figure 3.8 Transverse field components after a relative phase delay of π radians.

$$\phi_x - \phi_y = (2\nu + 1)\pi \tag{3.53}$$

The introduction of a π radians phase difference is equivalent to changing the sign of one of the components (say, E_y). The output electric field is therefore as shown in Fig. 3.8; as can be seen, the direction of polarization has simply rotated by 2θ, where θ is the angle of polarization of the input wave relative to the x-axis. This component (known as a *half-wave plate*) therefore allows the orientation of the polarization vector to be altered, simply by rotating the crystal. It is made simply by polishing a crystal substrate to the required thickness.

The second case of interest occurs when there is an effective phase difference of $\pi/2$ rads, so that

$$\phi_x - \phi_y = (2\nu + 1/2)\pi \tag{3.54}$$

In general the output will then be elliptically polarized. However, if we arrange for θ to equal $\pi/4$, then $E_x = E_y$ for the input wave. According to our discussion in Chapter 2 (Eq. (2.65)) the output will then be circularly polarized. The *quarter-wave plate* is thus a convenient component for converting a plane polarized input into a circularly polarized output, and vice versa.

Off-axis Wave Propagation

A much more complicated situation occurs when the direction of propagation of the input wave does not lie along a crystallographic axis, when the material polarization must be described by the full tensor expression (Eq. (3.47)). Similarly the full tensor description of the permittivity must be used. Fortunately some simplications are possible. First, it can be shown via the principle of energy conservation that the dielectric tensor must be symmetric, so that $\varepsilon_{ij} = \varepsilon_{ji}$. Second, it is possible to devise a simple geometrical interpretation of many aspects of propagation, known as the *index indicatrix* (or index ellipsoid).

The argument runs as follows. Consider the scalar product $1/2\ \mathbf{E} \cdot \mathbf{D}$; as we saw in Chapter 2, this is a measure of the stored electrostatic energy U_E. If we form this product with the relevant field components, we can show that

$$\mathbf{E} \cdot \mathbf{D} = 2\mathrm{U_E} = \mathrm{D}_x^2/\varepsilon_{xx} + \mathrm{D}_y^2/\varepsilon_{yy} + \mathrm{D}_z^2/\varepsilon_{zz} + 2\mathrm{D}_x\mathrm{D}_y/\varepsilon_{xy}$$
$$+ 2\mathrm{D}_y\mathrm{D}_z/\varepsilon_{yz} + 2\mathrm{D}_z\mathrm{D}_x/\varepsilon_{zx} \qquad (3.55)$$

Here we have used the symmetry of the dielectric tensor to combine terms of the form $\mathrm{D}_x\mathrm{D}_y/\varepsilon_{xy}$ and $\mathrm{D}_y\mathrm{D}_x/\varepsilon_{yx}$. By applying suitable scaling (which merely involves writing $\mathrm{D}_x^2/2\varepsilon_0\mathrm{U_E} = x$, and so on), we may transform Eq. (3.55) to

$$x^2/\varepsilon_{rxx} + y^2/\varepsilon_{ryy} + z^2/\varepsilon_{rzz} + 2xy/\varepsilon_{rxy} + 2yz/\varepsilon_{ryz}$$
$$+ 2zx/\varepsilon_{rzx} = 1 \qquad (3.56)$$

This is the equation of an ellipsoid, the index indicatrix mentioned earlier. In general its orientation in space is determined by the magnitudes of the off-axis elements ε_{rxy}, ε_{ryz} and ε_{rzx}. A typical view of the indicatrix is shown in Fig. 3.9.

For the special case when the off-axis elements of the dielectric tensor are all zero, Eq. (3.56) reduces to

$$x^2/\varepsilon_{rx} + y^2/\varepsilon_{ry} + z^2/\varepsilon_{rz} = 1 \qquad (3.57)$$

The index indicatrix is then oriented along the coordinate axes, as shown in Fig. 3.10, and the semi-major axes of the ellipsoid are equal to $\sqrt{\varepsilon_{rx}}$, $\sqrt{\varepsilon_{ry}}$ and $\sqrt{\varepsilon_{rz}}$ (and thus to n_x, n_y and n_z). For a uniaxial crystal (for which $n_x = n_y$, say), the indicatrix reduces to an ellipsoid of revolution, whose axis of symmetry (in this case, the z-axis) is called the *optic axis*. This particular ellipsoid can be characterized by just

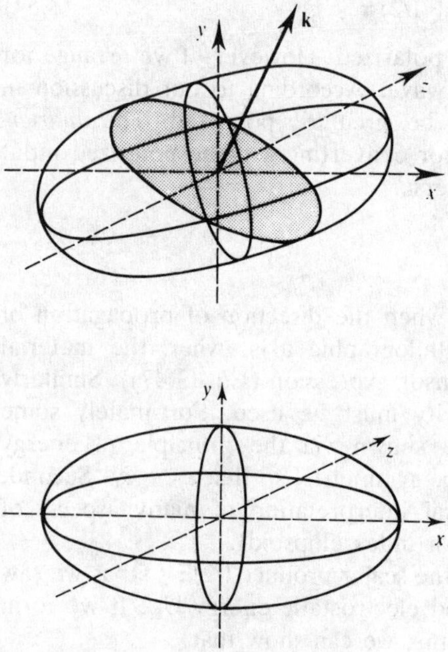

Figure 3.9 General orientation of the index indicatrix.

Figure 3.10 The index indicatrix aligned with the crystal coordinates.

Table 3.3 Indices of refraction of some uniaxial crystals

	n_o	n_e
Calcite ($CaCO_3$)	1.658	1.486
Lithium niobate ($LiNbO_3$)	2.286	2.200
Lithium tantalate ($LiTaO_3$)	2.176	2.180
Quartz (SiO_2)	1.544	1.553
Rutile (TiO_2)	2.616	2.903

two values of refractive index: n_o (the *ordinary index*), which corresponds to n_x or n_y, and n_e (the *extraordinary index*), corresponding to n_z.

Typical values of n_o and n_e are given in Table 3.3 for some common optical materials. Note that in some cases n_o may be greater than n_e, while the reverse is true in others.

We have seen that the dimensions of the ellipsoid can be usefully interpreted when it is oriented parallel to the coordinate axes. If this is not the case, or if the direction of propagation is *not* along a principal axis, the intepretation is more complicated. Let us return to the general ellipsoid, Fig. 3.9, which also shows the direction of propagation of a plane wave, marked **k**. It can be shown that such a wave can be split into two components, plane-polarized at right angles to each other. The directions of polarization are themselves parallel to the axes of the shaded ellipse, which is formed by the intersection of a plane through the origin, normal to **k**, with the index ellipsoid. The effective indices for the two component waves are then given by the semi-major axes of this ellipse.

This general phenomenon (the multi-valued nature of the refractive index) is called *birefringence*, and was first observed by Erasmus Bartolinus (1625–1692). However, from our previous electronic model, we might expect the following additional effects to occur. First, a frequency-dependence of the refractive indices, which results in dispersion of the birefringence. Second, a polarization-dependence of the absorption, known as *dichroism*, which can be explained by the introduction of a different damping factor for electronic motion in different directions. If the effect is sufficiently strong, dichroic material can be used to polarize natural light by preferential absorption of the unwanted polarization. The most famous examples of an artificial dichroic medium are the molecular 'Polaroid' sheets invented by Edwin Land in 1928.

3.6 NON-LINEAR EFFECTS

In our model of the interaction of radiation and matter, we have supposed that a dipole can be created by an electric field, and that the restoring force on the charges is proportional to displacement. This led to a form of simple harmonic motion, and to a polarization proportional to the driving field. It is reasonable to suppose that this is merely a first approximation to the true behaviour of a dipole, since all real oscillators demonstrate simple harmonic behaviour only for vanishingly small oscillations. We might therefore expect that the relation between **P** and **E** should really be written as a power series, of which Eq. (3.47) is simply the first term. If this is done, our model will describe *non-linear* optical effects.

Any of Eqs (3.48) may be written using the following notation:

$$\mathsf{P}_i = \varepsilon_0 \sum_j \chi_{ij}\, \mathsf{E}_j \tag{3.58}$$

More generally, if we allow higher-order dependence of the polarization on the electric field, we must write

$$\mathsf{P}_i = \varepsilon_0 \left\{ \sum_j \chi_{ij}\, \mathsf{E}_j + \sum_j \sum_k \chi_{ijk}\, \mathsf{E}_j\, \mathsf{E}_k + \sum_j \sum_k \sum_l \chi_{ijkl}\, \mathsf{E}_j\, \mathsf{E}_k\, \mathsf{E}_l + \ldots \right\} \tag{3.59}$$

This shows that (for example), the product of any pair of field components E_j and E_k can contribute to P_i through a third-rank susceptibility tensor whose elements are χ_{ijk}. The first point to note is that the higher-order terms normally get progressively smaller, so that the elements χ_{ij} are the dominant factors determining the dielectric constant. The importance of the higher-order terms lies in a number of new and entirely different effects they make possible.

Consider, for example, an optical field (say, in the *x*-direction) which at any point can be described as $\mathsf{E}_x = E_x \exp(\mathrm{j}\omega t)$. The third-rank susceptibility tensor now makes an additional contribution P'_x to the total polarization P_x, of the form:

$$\begin{aligned} \mathsf{P}'_x &= \varepsilon_0 \sum_j \sum_k \chi_{xjk}\, \mathsf{E}_j\, \mathsf{E}_k \\ &= \varepsilon_0\, \chi_{xxx}\, E_x^2 \exp(2\mathrm{j}\omega t) \end{aligned} \tag{3.60}$$

The interesting point here is that P'_x now has a contribution at an angular frequency 2ω, i.e. at twice the frequency of the incident wave. This effect is known as *second harmonic generation*, because this component of **P** can be arranged to generate a new wave at the doubled frequency. In fact, frequency doublers based on this principle are now widely used in optoelectronics. For example, a non-linear crystal pumped at near infrared wavelengths ($\lambda_0 \approx 0.84\ \mu$m) can be used to generate visible, blue light (at $\lambda_0 \approx 0.42\ \mu$m), a truly remarkable feat. However, we note here that merely having a significant value of the appropriate coefficient χ_{ijk} is not a sufficient condition for a strong second harmonic wave—a further requirement is that the phase velocities of the waves must be matched. We shall find this an important condition in other situations later on.

Another possibility occurs when the field components are associated with *different* sources. For example, we might have $\mathsf{E}_x = E_x \exp(\mathrm{j}\omega t)$ and

$E_y = E_s \exp(j\omega_s t)$. While the former may still be a component of an optical field, the latter might be of much lower frequency, derived from an external source such as a signal generator. In this case, the new contribution to the polarization is

$$P'_x = \varepsilon_0 \chi_{xxy} E_x \exp(j\omega t) E_s \exp(j\omega_s t) \tag{3.61}$$

We can interpret this as follows. In our earlier, first-order derivation we had $\mathbf{P} = \varepsilon_0 [\chi] \mathbf{E}$, with the dielectric tensor being given by $[\varepsilon] = \varepsilon_0 \{1 + [\chi]\}$. From Eq. (3.61), we see that χ_{xx} has effectively been replaced by $\chi_{xx} + \chi_{xxy} E_s \exp(j\omega_s t)$. Thus, the dielectric constant has now become a function of the applied low-frequency field E_s. This effect (known as the *linear electro-optic* or *Pockels effect*) has extremely important consequences, since it allows the modification of the phase velocity of a wave through the application of an external field. The magnitude of the effect in absolute terms is rather small. Typical changes in refractive index are often less than 10^{-4} (with n in the broad range 1.45 to 2.50), so that propagation distances of the order of 5×10^3 wavelengths ($\approx$ 1 mm) are required for a phase change of π radians. This implies that useful electro-optic devices will be physically large (at least, in comparison to an integrated electronic device).

3.7 THE ELECTRO-OPTIC EFFECT

We shall now look at this effect in more detail. The first point to note is that many materials show no electro-optic effect at all! The basic consideration is whether or not the point group has a symmetry which leaves the structure invariant after a change of the coordinates of all the elements from r to $-r$. This can be illustrated by considering the index change that might result from the application of fields E and $-E$. In the first case Δn will be given by $\Delta n = kE$, where the constant k is a measure of the strength of the electro-optic effect, while in the second, we would expect $\Delta n = k(-E)$. However, if the material has inversion symmetry, then changing from E to $-E$ should make no difference. This implies that k must be zero. The electro-optic effect can therefore only exist in materials lacking inversion symmetry. These are called *non-centro-symmetric*. Of the 32 point groups within the 7 crystal systems, 21 are of this type. All amorphous materials have inversion symmetry, and therefore show no linear electro-optic effect.

We note, however, that regardless of whether a material is centro-symmetric or not (or indeed of whether it is crystalline), the coefficient of the third term in the expansion for the polarization (the fourth-rank susceptibility tensor, whose elements are χ_{ijkl}) is never zero. This term gives rise to the *Kerr effect*, in which we can imagine one component of an applied field generating an asymmetry, while a second produces a form of electro-optic modification of the refractive index, which then characterizes the phase velocity of a third (optical) component. However, since the higher-order non-linear coefficients become progressively smaller, higher fields are needed to exploit the Kerr effect. Consequently, for practical devices, the electro-optic effect has been almost exclusively used. This may change as materials with higher Kerr coefficients are discovered.

The tensor describing the electro-optic effect will have a total of 27 elements χ_{ijk} (for each of the three possible directions of the polarization P_i there will be three directions for the optical field E_j and three for the low frequency field E_k). However, the argument that showed that the dielectric tensor must be symmetric (so that $\varepsilon_{ij} = \varepsilon_{ji}$ and $\chi_{ij} = \chi_{ji}$) can also be shown to lead to $\chi_{ijk} = \chi_{jik}$. Thus, the number of *different* elements needed is only 18. However, the χ_{ijk} terms themselves are not normally used to characterize the effect. Since they modify the polarization (and hence the dielectric constant), we expect that the presence of an external electric field will alter the shape of the index indicatrix. It is these changes that are traditionally used instead.

Returning to the characteristic equation of the index ellipsoid, Eq. (3.56), we note that it could be written in the following alternative form:

$$B_{xx}\, x^2 + B_{yy}\, y^2 + B_{zz}\, z^2 + 2B_{xy}\, xy + 2B_{yz}\, yz + 2B_{zx}\, zx = 1 \tag{3.62}$$

where $B_{xx} = 1/\varepsilon_{rxx}$, and so on. Changes in these B coefficients due to the external field can then be written as

$$\Delta B_{ij} = \sum_k r_{ijk}\, E_k \tag{3.63}$$

The terms r_{ijk} are called the electro-optic or *Pockels coefficients* (after Carl Pockels, 1865–1913). Clearly these new elements are simply related to the χ_{ijk} terms. Following the arguments above we would therefore expect $r_{ijk} = r_{jik}$. It is therefore customary to contract the indices i and j into a single index I, following the set of 'rules' shown in Table 3.4.

The notation 11 for xx, 12 for xy is also often employed. Hence, r_{yzx} (which can also be written r_{231}) becomes contracted to r_{41}. By symmetry this is also equal to r_{321}. Similarly r_{52} corresponds to r_{132} $(= r_{312})$. This notation may also be applied to the indices of the B_{ij} coefficients. If this is done, Eq. (3.63) may be written in matrix form as

$$\begin{bmatrix} \Delta B_1 \\ \Delta B_2 \\ \Delta B_3 \\ \Delta B_4 \\ \Delta B_5 \\ \Delta B_6 \end{bmatrix} = \begin{bmatrix} r_{11} & r_{12} & r_{13} \\ r_{21} & r_{22} & r_{23} \\ r_{31} & r_{32} & r_{33} \\ r_{41} & r_{42} & r_{43} \\ r_{51} & r_{52} & r_{53} \\ r_{61} & r_{62} & r_{63} \end{bmatrix} \begin{bmatrix} E_x \\ E_y \\ E_z \end{bmatrix} \tag{3.64}$$

Thus, for example, we have $\Delta B_1 = r_{11}E_x + r_{12}E_y + r_{13}E_z$, and so on. One important feature of the electro-optic matrix is that it is rather sparse since the possible elements that can be non-zero are governed entirely by the symmetry of the appropriate point group. Table 3.5 is a standard representation of these elements, for the 21 non-centro-symmetric groups.

A second point is that the magnitudes of the coefficients can vary widely. This limits the usability of some materials, and places restrictions on the crystal cut and

Table 3.4 Contraction rules for the Pockels coefficients

$$\underset{ij}{\begin{bmatrix} xx & xy & xz \\ yx & yy & yz \\ zx & zy & zz \end{bmatrix}} \Rightarrow \underset{I}{\begin{bmatrix} 1 & 6 & 5 \\ & 2 & 4 \\ & & 3 \end{bmatrix}}$$

on the orientation of the fields that must be used, if significant effects are to be obtained from fields of reasonable strength. Table 3.6 shows the values of the electro-optic coefficients for representative materials. Clearly $LiNbO_3$ would be a strong contender for any device applications, due to the high value of its r_{33} coefficient. New organic materials (such as the crystal 2-methyl-4-nitroaniline, or MNA) have also been found to have large coefficients. However, the relatively weak coefficients of the semiconductors GaAs and InP (which form the basis of many optoelectronic devices) should be noted.

How do we make use of all this information? The simplest way is through the index indicatrix. Let us assume that (in the absence of a field) we have chosen a coordinate system that is aligned with the principal axes of the crystal, so that the equation for the ellipsoid is of the form:

$$B_x x^2 + B_y y^2 + B_z z^2 = 1 \tag{3.65}$$

and the ellipsoid is as shown in Fig. 3.10. Let us also assume the material is $LiNbO_3$, a trigonal crystal with point group $3m$. From Tables 3.5 and 3.6, we see that there are a number of non-zero elements in the electro-optic matrix. Consequently several additional terms may be generated in the equation for the indicatrix, depending on the direction of the applied field.

Assuming that the field is applied in the x-direction, the significant elements are r_{51} and $-r_{22}$. Performing the matrix multiplication in Eq. (3.64) we can find the equation of the new index indicatrix as

$$B_{xx} x^2 + B_{yy} y^2 + B_{zz} z^2 + 2r_{51}E_x xz - 2r_{22}E_x xy = 1 \tag{3.66}$$

Note that the element $-r_{22}$ in the above is really r_{61} (which is in turn equivalent to r_{131}). However, this coefficient has the same magnitude but opposite sign to r_{22} (equivalent to r_{222}) for the $3m$ point group. These extra terms will produce a new ellipsoidal shape, which is shifted in both orientation and scale from the original one.

Alternatively, assuming the field is applied in the z-direction, we obtain

$$(B_{xx} + r_{13}E_z) x^2 + (B_{yy} + r_{13}E_z) y^2 + (B_{zz} + r_{33}E_z) z^2 = 1 \tag{3.67}$$

Once again Eq. (3.67) describes a new ellipsoid, but this time the result is different. The ellipsoid is still aligned with the coordinate system, but the lengths of the

Table 3.5 Electro-optic matrices for the seven crystal systems (after Kaminow *et al.*, 1971)

Triclinic

Group 1
$$\begin{bmatrix} r_{11} & r_{12} & r_{13} \\ r_{21} & r_{22} & r_{23} \\ r_{31} & r_{32} & r_{33} \\ r_{41} & r_{42} & r_{43} \\ r_{51} & r_{52} & r_{53} \\ r_{61} & r_{62} & r_{63} \end{bmatrix}$$

Monoclinic

Group 2
$$\begin{bmatrix} 0 & r_{21} & 0 \\ 0 & r_{22} & 0 \\ 0 & r_{23} & 0 \\ r_{41} & 0 & r_{43} \\ 0 & r_{52} & 0 \\ r_{51} & 0 & r_{63} \end{bmatrix}$$

Group *m*
$$\begin{bmatrix} r_{11} & 0 & r_{13} \\ r_{21} & 0 & r_{23} \\ r_{31} & 0 & r_{33} \\ 0 & r_{42} & 0 \\ r_{51} & 0 & r_{53} \\ 0 & r_{62} & 0 \end{bmatrix}$$

Orthorhombic

Group 222
$$\begin{bmatrix} 0 & 0 & 0 \\ 0 & 0 & 0 \\ 0 & 0 & 0 \\ r_{41} & 0 & 0 \\ 0 & r_{52} & 0 \\ 0 & 0 & r_{63} \end{bmatrix}$$

Group *mm*2
$$\begin{bmatrix} 0 & 0 & r_{13} \\ 0 & 0 & r_{23} \\ 0 & 0 & r_{33} \\ 0 & r_{42} & 0 \\ r_{51} & 0 & 0 \\ 0 & 0 & 0 \end{bmatrix}$$

Tetragonal

Group 4
$$\begin{bmatrix} 0 & 0 & r_{13} \\ 0 & 0 & r_{13} \\ 0 & 0 & r_{33} \\ r_{41} & r_{51} & 0 \\ r_{51} & -r_{41} & 0 \\ 0 & 0 & 0 \end{bmatrix}$$

Group $\bar{4}$
$$\begin{bmatrix} 0 & 0 & r_{13} \\ 0 & 0 & -r_{13} \\ 0 & 0 & 0 \\ r_{41} & -r_{51} & 0 \\ r_{51} & r_{41} & 0 \\ 0 & 0 & r_{63} \end{bmatrix}$$

Group 422
$$\begin{bmatrix} 0 & 0 & 0 \\ 0 & 0 & 0 \\ 0 & 0 & 0 \\ r_{41} & 0 & 0 \\ 0 & -r_{41} & 0 \\ 0 & 0 & 0 \end{bmatrix}$$

Group 4mm
$$\begin{bmatrix} 0 & 0 & r_{13} \\ 0 & 0 & r_{13} \\ 0 & 0 & r_{33} \\ 0 & r_{51} & 0 \\ r_{51} & 0 & 0 \\ 0 & 0 & 0 \end{bmatrix}$$

Group $\bar{4}2m$
$$\begin{bmatrix} 0 & 0 & 0 \\ 0 & 0 & 0 \\ 0 & 0 & 0 \\ r_{41} & 0 & 0 \\ 0 & r_{41} & 0 \\ 0 & 0 & r_{63} \end{bmatrix}$$

Trigonal

Group 3
$$\begin{bmatrix} r_{11} & -r_{22} & r_{13} \\ -r_{11} & r_{22} & r_{13} \\ 0 & 0 & r_{33} \\ r_{41} & r_{51} & 0 \\ r_{51} & -r_{41} & 0 \\ -r_{22} & -r_{11} & 0 \end{bmatrix}$$

Group 32
$$\begin{bmatrix} r_{11} & 0 & 0 \\ -r_{11} & 0 & 0 \\ 0 & 0 & 0 \\ r_{41} & 0 & 0 \\ 0 & -r_{41} & 0 \\ 0 & -r_{11} & 0 \end{bmatrix}$$

Group 3m
$$\begin{bmatrix} 0 & -r_{22} & r_{13} \\ 0 & r_{22} & r_{13} \\ 0 & 0 & r_{33} \\ 0 & r_{51} & 0 \\ r_{51} & 0 & 0 \\ -r_{22} & 0 & 0 \end{bmatrix}$$

Table 3.5 Continued

Hexagonal

Group 6 $$\begin{bmatrix} 0 & 0 & r_{13} \\ 0 & 0 & r_{13} \\ 0 & 0 & r_{33} \\ r_{41} & r_{51} & 0 \\ r_{51} & -r_{41} & 0 \\ 0 & 0 & 0 \end{bmatrix}$$

Group $\bar{6}$ $$\begin{bmatrix} r_{11} & -r_{22} & 0 \\ -r_{11} & r_{22} & 0 \\ 0 & 0 & 0 \\ 0 & 0 & 0 \\ 0 & 0 & 0 \\ -r_{22} & -r_{11} & 0 \end{bmatrix}$$

Group 622 $$\begin{bmatrix} 0 & 0 & 0 \\ 0 & 0 & 0 \\ 0 & 0 & 0 \\ r_{41} & 0 & 0 \\ 0 & -r_{41} & 0 \\ 0 & 0 & 0 \end{bmatrix}$$

Group 6*mm* $$\begin{bmatrix} 0 & 0 & r_{13} \\ 0 & 0 & r_{13} \\ 0 & 0 & r_{33} \\ 0 & r_{51} & 0 \\ r_{51} & 0 & 0 \\ 0 & 0 & 0 \end{bmatrix}$$

Group $\bar{6}m2$ $$\begin{bmatrix} 0 & -r_{22} & 0 \\ 0 & r_{22} & 0 \\ 0 & 0 & 0 \\ 0 & 0 & 0 \\ 0 & 0 & 0 \\ -r_{22} & 0 & 0 \end{bmatrix}$$

Cubic

Group 432 $$\begin{bmatrix} 0 & 0 & 0 \\ 0 & 0 & 0 \\ 0 & 0 & 0 \\ 0 & 0 & 0 \\ 0 & 0 & 0 \\ 0 & 0 & 0 \end{bmatrix}$$

Group $\bar{4}3m$
Group 23 $$\begin{bmatrix} 0 & 0 & 0 \\ 0 & 0 & 0 \\ 0 & 0 & 0 \\ r_{41} & 0 & 0 \\ 0 & r_{41} & 0 \\ 0 & 0 & r_{41} \end{bmatrix}$$

Table 3.6 Electro-optic coefficients for representative materials

Material	Point group	r, 10^{-12} m/V	
GaAs	$\bar{4}3m$	$r_{41} = -1.5$	
InP	$\bar{4}3m$	$r_{41} = -1.4$	
KD_2PO_4 (KDP)	$\bar{4}2m$	$r_{41} = 8$	$r_{63} = 11$
$LiNbO_3$	$3m$	$r_{33} = 30.8$	$r_{51} = 28$
		$r_{13} = 8.6$	$r_{22} = 3.4$
MNA	m	$r_{11} = 67$	
SiO_2 (Quartz)	32	$r_{41} = 1.4$	$r_{11} = 0.59$

principal axes have altered slightly. We can interpret these changes as follows. Recalling that $B_{xx} = 1/\varepsilon_{rxx} = 1/n_x^2$, to a first approximation we can write

$$\Delta B_{xx} = -2\Delta n_x / n_x^3 \tag{3.68}$$

However, from Eq. (3.67) we know that $\Delta B_{xx} = r_{13}E_z$. Equating these two definitions, we can show that the effect of the field E_z is to change the index n_x by an amount

$$\Delta n_x = -n_x^3 r_{13} E_z/2 \tag{3.69}$$

The result of the applied field is therefore an index change proportional to the field strength. Note that the important feature is $n_x^3 r_{13}$, rather than simply r_{13}. A relatively weak electro-optic coefficient can therefore be compensated for to some extent, if the material has a high refractive index. Similar changes can be found for n_y and n_z, as

$$\Delta n_y = -n_y^3 r_{13} E_z/2 \qquad \Delta n_z = -n_z^3 r_{33} E_z/2 \tag{3.70}$$

We can use these index changes to construct phase modulators (devices which can alter the phase of an optical wave under electrical control), as described below.

Longitudinal Phase Modulators

Figure 3.11 shows a *longitudinal phase modulator*. Transparent electrodes have been added to the input and output faces of a slab of electro-optic crystal. This has length L in the z-direction, so the application of a voltage V to the electrodes will lead to a longitudinal field $E_z = V/L$. Light is passed through the crystal in the z-direction, with the optical field polarized in either the x- or y-directions. For $LiNbO_3$ we can see from Eqs (3.69) and (3.70) that the static electric field will cause an equal refractive index change for both components. In length L this will induce phase shifts

$$\begin{aligned}\phi_x = \phi_y &= k_0 \Delta n_x L \\ &= -\pi n_x^3 r_{13} V/\lambda_0\end{aligned} \tag{3.71}$$

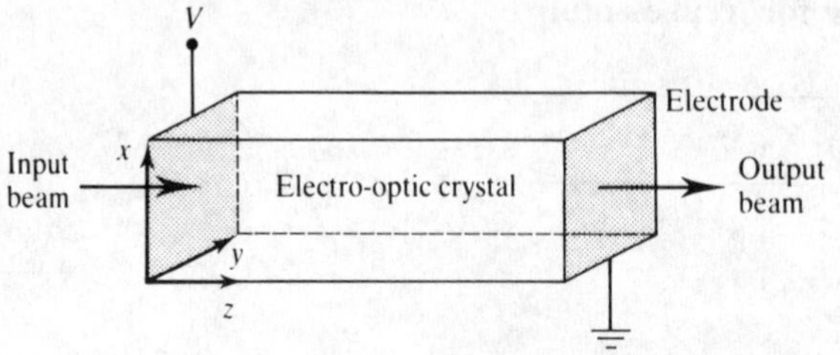

Figure 3.11 A longitudinal phase modulator.

Note that L has disappeared from Eq. (3.71) because a longer crystal results in a greater phase shift for a given index change, but a smaller electric field for a given voltage. Consequently the drive voltage must be independent of the modulator dimensions.

Design Example

We may calculate the magnitude of the voltage V_π required for a phase shift of π radians from Eq. (3.71) as

$$V_\pi = \lambda_0/(n_x^3 r_{13}) \tag{3.72}$$

For $LiNbO_3$, $n_x = 2.286$ and $r_{13} = 8.6 \times 10^{-12}$ m/V. Hence, for $\lambda_0 \approx 1\ \mu$m, $V_\pi \approx 9.7$ kV. This is a very high value, presenting an inherent problem for longitudinal phase modulators.

Transverse Phase Modulators

Improvements are offered by an alternative modulator design, which we now describe. Figure 3.12 shows a *transverse phase modulator*. Electrodes have been deposited on the top and bottom faces of the crystal, which has been rotated so that the static electric field is still in the z-direction. The electrode separation is taken to be g, so that the field is now $E_z = V/g$, but the length of the crystal in the direction of optical propagation remains L. This time the input optical wave may be polarized in the x- or z-directions. Equations (3.69) and (3.70) show that the resulting phase shifts are now different, being given by

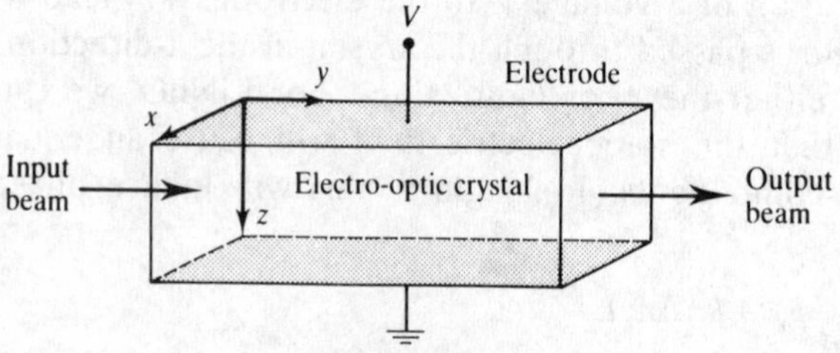

Figure 3.12 A transverse phase modulator.

$$\phi_x = k_0 \Delta n_x d = -\pi n_x^3 r_{13} VL/(\lambda_0 g)$$

$$\phi_z = k_0 \Delta n_z d = -\pi n_z^3 r_{33} VL/(\lambda_0 g) \quad (3.73)$$

Since $r_{33} > r_{13}$ in $LiNbO_3$, the greatest effect will be experienced by the z-polarized wave. The magnitude of the voltage required for π radians phase shift of this component is now

$$V_\pi = \lambda_0 g/(n_z^3 r_{33} L) \quad (3.74)$$

Equation (3.74) is no longer independent of the crystal dimensions, since it contains g and L. This suggests that if the ratio g/L can be arranged to be $\approx 10^{-3}$ (say, by using an electrode gap of a few microns and a length of several millimetres), V_π will reduce from kilovolts to volts. Since this can be done in guided wave optics, the door has been opened to the widespread use of electro-optic devices.

3.8 STRESS-RELATED EFFECTS

In this section we consider a number of stress-related effects, all of which have been exploited for optical device applications. We begin with the *photoelastic effect*, whereby a transparent, isotropic material may become optically anisotropic under the influence of an applied stress. This phenomenon, also known as *stress-induced birefringence*, was first investigated by David Brewster in 1816.

The stress to which a solid body is subjected may be characterized by a second-rank stress tensor $[T]$, whose elements T_{ij} represent the force applied along the i direction per unit area of a plane normal to the j direction. The diagonal components of $[T]$ therefore represent normal components of stress, while the off-diagonal components are shear stresses. By considering moments, it is simple to show that $[T]$ must be symmetric, to avoid any net torque on the body. Similarly the elastic strain of the body due to applied stresses may be described by a symmetric, second-rank tensor $[S]$, whose diagonal elements S_{ii} (defined as $S_{ii} = \partial U_i/\partial x_i$, where U_i is the local displacement of the body in the i direction, and x_i is the coordinate in the same direction) represent normal strains, and whose off-diagonal elements S_{ij} (defined as $S_{ij} = \{\partial U_i/\partial x_j + \partial U_j/\partial x_i\}$) are shear strains (note that we have avoided the use of the notation σ and ε for stress and strain, to avoid confusion with other optical parameters).

Because of the symmetry of $[T]$ and $[S]$ a contracted notation is often used to describe the elements T_{ij} and S_{ij}, as was done previously for the Pockels coefficients. Since the rules for this have already been given in Table 3.4, we shall not repeat them. Suffice it to say that if this notation is employed, $[T]$ and $[S]$ may be represented by two six-element vectors $\mathbf{T} = (T_1, T_2, ..., T_6)$ and $\mathbf{S} = (S_1, S_2, ..., S_6)$, which may be used to characterize the photoelastic effect in terms of the changes in the B coefficients of the index ellipsoid induced by stresses or strains. Two alternative definitions are used. First, in terms of stresses, we may write in longhand:

$$\begin{bmatrix} \Delta B_1 \\ \Delta B_2 \\ \Delta B_3 \\ \Delta B_4 \\ \Delta B_5 \\ \Delta B_6 \end{bmatrix} = \begin{bmatrix} q_{11} & q_{12} & q_{13} & q_{14} & q_{15} & q_{16} \\ q_{21} & q_{22} & q_{23} & q_{24} & q_{25} & q_{26} \\ q_{31} & q_{32} & q_{33} & q_{34} & q_{35} & q_{36} \\ q_{41} & q_{42} & q_{43} & q_{44} & q_{45} & q_{46} \\ q_{51} & q_{52} & q_{53} & q_{54} & q_{55} & q_{56} \\ q_{61} & q_{62} & q_{63} & q_{64} & q_{65} & q_{66} \end{bmatrix} \begin{bmatrix} T_1 \\ T_2 \\ T_3 \\ T_4 \\ T_5 \\ T_6 \end{bmatrix} \tag{3.75}$$

where the terms q_{ij} are elements of a second-rank *stress-optical tensor* $[q]$, which have units of m^2/N. Second, in terms of strains, and using a more compact notation, we may put

$$\mathbf{\Delta B} = [p]\,\mathbf{S} \tag{3.76}$$

where $\mathbf{\Delta B} = (\Delta B_1, \Delta B_2, \ldots, \Delta B_6)$ and $[p]$ is a corresponding *strain-optical tensor*, whose elements p_{ij} are dimensionless. For elastic deformation there is a linear relation between stress and strain. With the contracted notation used above, this may be written as

$$\mathbf{T} = [c]\,\mathbf{S} \tag{3.77}$$

where $[c]$ is a second-rank *stiffness tensor* with elements c_{ij}. Clearly the stress- and strain-optical tensors are related, since $[p] = [q][c]$. The two representations given above are therefore entirely complementary. With knowledge of the coefficients of $[p]$ or $[q]$, the effect of any stress or strain distribution may easily be computed, as we have done before for the electro-optic effect.

It can be shown that the non-zero elements in the photoelastic tensors are generally in different positions to those of the electro-optic tensor, but in the same positions as those of the Kerr effect, so the photoelastic effect is observable in centro-symmetric crystals, liquids (e.g. water) and amorphous solids (e.g. plastics, silica glass). Table 3.7 shows the stress-optical tensor for an isotropic medium; this is particularly simple because it is described entirely by two coefficients, q_{11} and q_{12}.

Table 3.7 Stress-optical tensor for an isotropic medium

$$[q] = \begin{bmatrix} q_{11} & q_{12} & q_{12} & 0 & 0 & 0 \\ q_{12} & q_{11} & q_{12} & 0 & 0 & 0 \\ q_{12} & q_{12} & q_{11} & 0 & 0 & 0 \\ 0 & 0 & 0 & q_{11}-q_{12} & 0 & 0 \\ 0 & 0 & 0 & 0 & q_{11}-q_{12} & 0 \\ 0 & 0 & 0 & 0 & 0 & q_{11}-q_{12} \end{bmatrix}$$

We may use this example to evaluate the effect of various applied stresses. From Eq. (3.75) it is apparent that a uniaxial stress (say, a non-zero value of T_1) will result in the following changes to the B coefficients: $\Delta B_1 = q_{11}T_1$, $\Delta B_2 = \Delta B_3 = q_{12}T_1$ and $\Delta B_4 = \Delta B_5 = \Delta B_6 = 0$. An initially isotropic medium will have a spherical index ellipsoid, independent of the choice of coordinate axes, and a uniform refractive index n. After application of the uniaxial stress, the refractive index in the x-, y- and z-directions will therefore have altered by

$$\Delta n_x = -n^3 q_{11} T_1/2, \qquad \Delta n_y = \Delta n_z = -n^3 q_{12} T_1/2 \tag{3.78}$$

Clearly if q_{11} and q_{12} are unequal (as they often are), the material will now be birefringent, with principal axes located parallel and perpendicular to the direction of the applied stress. In each case the index change is of the form $\Delta n = -n^3 qT/2$, where q is the relevant stress-optical coefficient. By analogy, we may obtain the corresponding expression in terms of strain, as $\Delta n = -n^3 pS/2$, where p is the corresponding strain-optical coefficient.

Similarly if the stress is now applied hydrostatically (so that $T_1 = T_2 = T_3 = T$, say), the index changes become

$$\Delta n_x = \Delta n_y = \Delta n_z = -n^3(q_{11} + 2q_{12})T/2 \tag{3.79}$$

This time, the refractive index change is isotropic, so the material does not become birefringent. By convention, compressive stresses are defined as being negative in sign (since they cause negative strain), while tensile stresses are positive. An increase in hydrostatic pressure will therefore result in a uniform increase in refractive index that is linearly proportional to the pressure.

The Acousto-optic Effect

An *acoustic wave* is a form of travelling wave, consisting of moving variations in stress. From the foregoing it is clear that the passage of an acoustic wave through a photoelastic medium should be accompanied by a travelling refractive index variation, with peaks and troughs in index correlated with points of high and low stress in the wave. The variation must therefore be periodic, with the same periodicity as the acoustic wave ($\Lambda = v/f$, where v is the acoustic velocity and f is the frequency). In 1922 Brillouin suggested that this variation could be used to diffract an optical beam, a prediction that was verified almost simultaneously in 1932 by Debye and Sears in America and Lucas and Biquard in France. In this form the photoelastic effect is commonly known as the *acousto-optic effect*. We shall describe the diffraction process itself in Chapter 7; in the meantime we shall simply calculate the peak index change expected from an acoustic wave of a given power.

Acoustic waves may take several different forms, depending on whether they propagate in solids or liquids, and in the bulk of a medium or near its surface. In solids bulk waves can exist as either longitudinal or transverse waves (which have mechanical displacements parallel and normal to the direction of propagation). Figure 3.13 shows a longitudinal wave, which has a strain variation only in the direction of propagation. Only one component of strain (say, S_1) is non-zero, so we

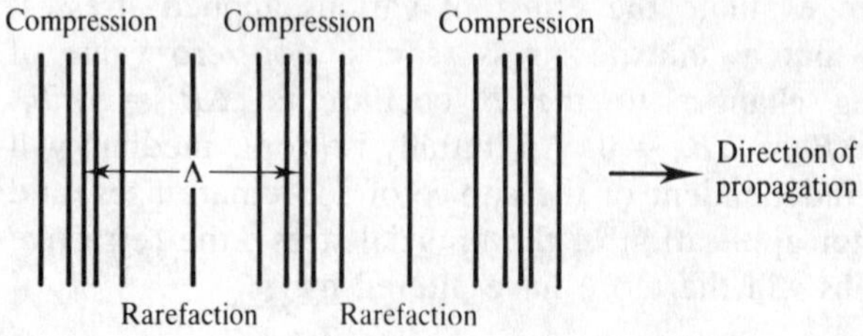

Figure 3.13 Propagation of an acoustic wave.

can describe the wave by the travelling variation $S_1 = S_{1\ max} \cos(\omega t - kx)$, where $\omega = 2\pi f$ and $k = 2\pi/\Lambda$.

Note that it is also possible for *surface acoustic waves* to exist; these are propagation modes that have both longitudinal and transverse components, and are constrained to propagate with finite amplitude only close to a surface boundary. They are of considerable importance in guided wave optics because they can interact efficiently with optical waves propagating in a guide near a surface. It can be shown that the power per unit area carried by the longitudinal wave above is given by $S_{1\ max}^2 \rho v^3/2$, where ρ is the density of the medium supporting the acoustic wave. Since this equals P/A, where P is the total power in the acoustic beam and A is its cross-section, we must have

$$S_{1\ max} = \surd\{2P/A\rho v^3\} \tag{3.80}$$

We may then obtain the peak index change in a given direction as

$$\Delta n_{max} = n^3 p S_{1\ max}/2 = \surd\{Pp^2 n^6/2A\rho v^3\} \tag{3.81}$$

where p is the relevant strain-optical coefficient. This result is often written as $\Delta n_{max} = \surd\{PM_2/2A\}$, where $M_2 = p^2 n^6/\rho v^3$ is one of a number of *acousto-optic figures of merit* (actually, the most important for many practical purposes), useful for comparing different acousto-optic materials.

Design Example

We may estimate M_2 for a typical material, fused silica, as follows. Fused silica is isotropic, with non-zero strain-optical coefficients $p_{11} = 0.121$ and $p_{12} = 0.27$. Its refractive index, density and acoustic velocity are 1.46, 2.2×10^3 kg/m^3 and 5960 m/s, respectively. The largest value of M_2 is obtained from p_{12}, as $M_2 = 0.27^2 \times 1.46^6/(2.2 \times 10^3 \times 5960^3) = 1.51 \times 10^{-15}$ s^3/kg. An acoustic beam of power 1 W and cross-sectional area 1 mm^2 would then yield a peak index change of $\Delta n_{max} = \surd\{1 \times 1.51 \times 10^{-15}/(2 \times 10^{-6})\} = 2.75 \times 10^{-5}$. Table 3.8 shows values of M_2 for some other acousto-optic materials. One of the most widely used is $LiNbO_3$ since it is also piezoelectric. This allows simple excitation of an acoustic wave by an electical drive signal, as we shall describe below.

Table 3.8 Parameters of typical acousto-optic materials (from Chang, 1976)

Material	v (m/s)	ρ (g/cm^3)	M_2 (10^{-15} s^3/kg)
Fused silica	5960	2.2	1.51
$LiNbO_3$	6570	4.64	6.95
$PbMoO_4$	3630	6.95	35.8
TeO_2	4200	6.0	34.6
GaAs	5150	5.34	104.19

The Piezoelectric Effect

All crystals that exhibit the linear electro-optic effect possess one further stress-related characteristic: they are *piezoelectric*. The piezoelectric effect was first noted by the brothers Pierre and Jacques Curie in 1880, and is strong in materials such as $BaTiO_3$, $LiNbO_3$, PLZT, quartz and ZnO. It has two manifestations. In the first, the application of an electric field to a piezoelectric crystal may give rise to an internal strain. If the field is a d.c. one, the crystal dimensions will therefore alter slightly.

This particular effect has found widespread application in guided wave optics since it allows the construction of devices capable of moving small components through short distances (say, 10 μm), with a resolution of the order of 1 nm. Using a three-axis *piezoelectric micropositioner* it is possible to position one component (e.g. a single-mode optical fibre) relative to another (perhaps a semiconductor laser) to high accuracy, so that transmission between the two is optimized. Often this is done under closed-loop control. Piezoelectric devices are therefore essential to the construction of low-loss jointed systems.

For an a.c. electric field, on the other hand, the strain distribution inside the crystal will be time-varying. If the crystal is arranged as a parallel-sided slab†, it may act as an acoustic resonator, so that the overall pattern is equivalent to a standing acoustic wave. This can be used to excite travelling acoustic waves in a non-piezoelectric material in close mechanical contact with the crystal. A piezoelectric crystal may therefore be used as an extremely convenient transducer between an electrical drive signal and an acoustic wave, for example in an acousto-optic device†.

In the second manifestation of the piezoelectric effect, the application of a stress may result in an internal electric field, so that a voltage is developed across the crystal. Here the major consequence is that the effect of an acoustic wave is more complicated than we have described so far. For example, the travelling strain

† Transducers for surface acoustic waves (SAW) are rather different, and will be described in Chapter 7.

distribution can produce a change in refractive index directly, through the photoelastic effect. However, it can also generate an electric field (via the piezoelectric effect), which in turn produces a change in index (via the electro-optic effect). Depending on the material, the index change produced by this second, indirect mechanism may either add to or subtract from that produced directly.

PROBLEMS

3.1 Using the Sellmeier coefficients given in Table 3.1, compute the refractive index of silica at wavelengths of 1.3 and 1.55 μm.
[1.447; 1.444]

3.2 Sellmeier's equation $n'^2 = 1 + \sum_j A_j \lambda_0^2/(\lambda_0^2 - \lambda_j^2)$ can be used to model material dispersion over a wide spectral range. Show that, for the case of a single resonance located in the ultraviolet, dispersion over the visible spectrum can be approximated by *Cauchy's equation* $n' \approx A + (B/\lambda_0^2 + (C/\lambda_0^4)$, and find A, B and C in terms of the Sellmeier coefficients. What phenomenon would you expect Cauchy's equation to be unable to describe?

3.3 Using the data given in Table 3.2, estimate the density of free electrons in silver. Over what spectral range would the real part of the dielectric constant (ε_r') be negative?
[$4.68 \times 10^{28}/\mathrm{m}^3$; $\lambda_0 > 0.155$ μm]

3.4 By constructing a simple model for the behaviour of electrons in a high frequency electromagnetic field, show that a reasonable approximation to the dispersion relation for a low-density, ionised gas (e.g. the ionosphere) is given by the expression $\omega^2 = \sqrt{\{\omega_p^2 + c^2k^2\}}$.

3.5 The dielectric constant of a semiconductor may be modified by the introduction of free carriers, through the plasma contribution. Find an approximate expression for the change in *complex* relative dielectric constant, assuming that collision damping is not negligible. What consequence will follow from the resulting finite value of ε_r''?

3.6 The figure below shows an optical experiment involving a birefringent wave plate and a polarizer. The former is constructed from a lithium niobate slab of thickness 3.684 mm ($n_o = 2.286$, $n_e = 2.2$ for $LiNbO_3$). You may assume that the latter is perfect, i.e. that it blocks the unwanted polarization absolutely while allowing the desired polarization through without attenuation. A plane wave (of wavelength 0.633 μm) is passed through the two components, which are initially rotated about the optical axis to maximise the output power. What will be the percentage transmission if the birefringent plate is now rotated by (a) 22.5°, and (b) 45°?

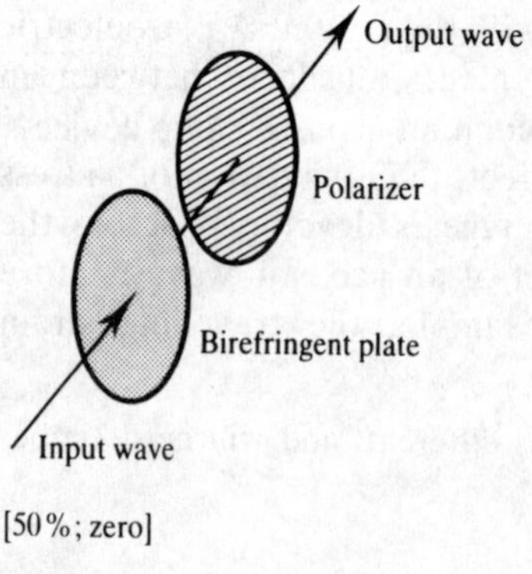

[50%; zero]

3.7 The general equation for the index indicatrix in anisotropic media may be written as $B_{xx}\,x^2 + B_{yy}\,y^2 + B_{zz}\,z^2 + 2B_{xy}\,xy + 2B_{yz}\,yz + 2B_{zx}\,zx = 1$, where $B_{xx} = 1/\varepsilon_{rxx}$ and so on.

Show that, if the coordinates are represented by a radius vector $\mathbf{r}$, and the B coefficients are assembled into a symmetric, second-rank tensor $[B]$, this equation may be rewritten in the alternative *quadratic form* $\mathbf{r}^{\mathrm{T}}\,[B]\,\mathbf{r} = 1$, where $\mathbf{r}^{\mathrm{T}}$ represents the transpose of $\mathbf{r}$.

3.8 The 3×3 matrix $[B]$ possesses a set of eigenvalues λ_i and eigenvectors $\mathbf{v}_i$, defined using $[B]\,\mathbf{v}_i = \lambda_i\,\mathbf{v}_i$. Since $[B]$ is symmetric, the eigenvectors are orthogonal, so $\mathbf{v}_i^{\mathrm{T}}\,\mathbf{v}_j = 0$. It may also be assumed that the eigenvectors are normalized so that $\mathbf{v}_i^{\mathrm{T}}\,\mathbf{v}_i = 1$.

Show that, if the eigenvalues and eigenvectors are assembled into two new matrices $[\Lambda]$ (whose diagonal elements are λ_i) and $[V]$ (which contains the vectors $\mathbf{v}_i$ arranged in columns), $[B]$ may be written as $[B] = [V]\,[\Lambda]\,[V]^{\mathrm{T}}$. Hence, find an expression for a new coordinate system $\mathbf{r}'$, which coincides with the axes of the index ellipsoid. What are the lengths of the semi-major axes of this ellipsoid?

3.9 A KH_2PO_4 (KDP) crystal is oriented with its crystal axes parallel to the x-, y- and z-directions. From the information given in Tables 3.5 and 3.6, determine the equation of the index ellipsoid when an electric field is applied in the z-direction.

3.10 Using the methods of Questions 3.7 and 3.8, show that the lengths of the semi-major axes of the index ellipsoid for the KDP crystal in Question 3.9 are unaltered by the application of the electric field.

SUGGESTIONS FOR FURTHER READING

Born, M., and Wolf, E. "Principles of Optics", 6th ed., Chapters 2, 13 and 14, Pergamon, Oxford, 1980.

Chang, I. C. Acousto-optic devices and applications, IEEE Trans. Sonics and Ultrasonics **SU-23**, 2–21, 1976.

Fleming, J. W. Material dispersion in lightguide glasses, Elect. Lett. **14**, 326–328, 1978.

Guenther, R. "Modern Optics", Chapters 7, 13, 14 & 15, John Wiley and Sons, New York, 1990.

Kaminow, I. P. "An Introduction to Electrooptic Devices", Academic Press, London, 1974.

Lipscomb, G. F., Garito, A. F., and Narang, R. S. A large linear electro-optic effect in a polar crystal 2-methyl-4-nitroaniline, Appl. Phys. Lett. **38**, 663–665, 1981.

Maletson, I. H. Interspecimen comparison of the refractive index of fused silica, JOSA **55**, 1205–1209, 1965.

Miller, R. C. Optical second harmonic generation in piezoelectric crystals, Appl. Phys. Lett. **5**, 17–19, 1964.

Ordal, M. A., Long, L. L., Bell, R. J., Bell, S. E., Bell, R. R., Alexander, R. W., and Ward, C. A. Optical properties of the metals Al, Co, Cu, Au, Fe, Pb, Ni, Pd, Pt, Ag, Ti and W in the infrared and far infrared, Appl. Opt. **22**, 1099–1119, 1983.

Ramo, S., Whinnery, J. R., and Van Duzer, T. "Fields and Waves in Communication Electronics", 2nd ed., Chapter 13, John Wiley and Sons, New York, 1984.

Soref, R. A. and Bennett, B. R. Electrooptical effects in silicon, IEEE J. Quant. Elect. **QE-23**, 123–129, 1987.

Turner, E. H. High frequency electro-optic coefficients of lithium niobate. Appl. Phys. Lett. **8**, 303–304, 1966.

Yariv, A. "Optical Electronics", Chapters 1, 8 and 9, Holt, Rinehart and Winston, New York, 1965.

Yariv, A., and Yeh, P. "Optical Waves in Crystals", John Wiley and Sons, New York, 1983.

Zernike, F., and Midwinter, J. E. "Applied Nonlinear Optics", John Wiley and Sons, New York, 1973.

CHAPTER

FOUR

THE OPTICS OF BEAMS

4.1 SCALAR THEORY

The analysis of Chapter 2 was based on the vectorial form of Maxwell's equations, which we used to demonstrate the existence of plane waves. Clearly other types of wave must exist—an infinite plane wave is only an idealization, and real waves have the form of *bounded beams*. Unfortunately vector theory rapidly becomes unmanageable when applied to more complicated geometries. We therefore introduce an alternative scalar theory, which can be used to illustrate the optics of beams.

First, we impose the condition that the entire field is harmonically-varying, at a single angular frequency ω. Second, we assume that the electric field vector always points in the same direction. This is reasonable for linearly polarized, paraxial fields, i.e. beams with amplitudes that are only non-zero close to some axis of propagation. Finally we take the medium to be free-space. With these assumptions, the time-independent vector wave equation Eq. (2.43) reduces to the simple scalar equation:

$$\nabla^2 E + \omega^2 \mu_0 \varepsilon_0 E = 0 \tag{4.1}$$

It is simple to repeat our previous analysis and find plane wave solutions to Eq. (4.1). For a wave travelling in the $+z$-direction, for example, we would obtain $E = E_0 \exp(-jk_0 z)$, where E_0 is some fixed amplitude and k_0 (the propagation constant) is again given by $k_0^2 = \omega^2 \mu_0 \varepsilon_0$. Now we shall find solutions for non-planar waves. We shall begin with another wave of simple geometrical form, the *spherical wave*.

4.2 SPHERICAL WAVES

Until now, we have worked in cartesian coordinates, for which the Laplacian in Eq. (4.1) is given by $\nabla^2 = \partial^2/\partial x^2 + \partial^2/\partial y^2 + \partial^2/\partial z^2$. Unfortunately the evaluation of ∇^2 in other coordinate systems is (in general) rather difficult. This makes the wave equation hard to set up, let alone solve. In this case, however, there is an observation we can use to simplify matters. It is that spherical waves are, by definition, spherically symmetric, so our field solution must depend only on a radial coordinate r. This implies that $E(x,y,z) = E(r)$, so the first derivative of E with respect to x can be found as

$$\partial E/\partial x = \mathrm{d}E/\mathrm{d}r \; \partial r/\partial x \tag{4.2}$$

Similarly we may obtain the corresponding second derivative as:

$$\partial^2 E/\partial x^2 = \mathrm{d}^2 E/\mathrm{d}r^2 \; (\partial r/\partial x)^2 + \mathrm{d}E/\mathrm{d}r \; \partial^2 r/\partial x^2 \tag{4.3}$$

Since $r^2 = x^2 + y^2 + z^2$, we must have

$$2r \; \mathrm{d}r = 2x \; \mathrm{d}x \quad \text{or} \quad \partial r/\partial x = x/r \tag{4.4}$$

and

$$\begin{aligned} \partial^2 r/\partial x^2 &= (1/r) + x \; \partial(1/r)/\partial x \\ &= (1/r)\,(1 - x^2/r^2) \end{aligned} \tag{4.5}$$

so that

$$\partial^2 E/\partial x^2 = (x^2/r^2) \; \mathrm{d}^2 E/\mathrm{d}r^2 + (1/r)\,(1 - x^2/r^2) \; \mathrm{d}E/\mathrm{d}r \tag{4.6}$$

We can of course find $\partial^2 E/\partial y^2$ and $\partial^2 E/\partial z^2$ in a similar way. By adding all three second derivative terms together, we can obtain an expression for the Laplacian that involves r only:

$$\nabla^2 E = \mathrm{d}^2 E/\mathrm{d}r^2 + (2/r) \; \mathrm{d}E/\mathrm{d}r \tag{4.7}$$

For spherically-symmetric fields the scalar wave equation therefore reduces to

$$\mathrm{d}^2 E/\mathrm{d}r^2 + (2/r) \; \mathrm{d}E/\mathrm{d}r + \omega^2 \mu_0 \varepsilon_0 \; E = 0 \tag{4.8}$$

This is still quite a hard equation to solve, so we now proceed as follows. Let us define a new variable $F(r)$, such that

$$F(r) = r \; E(r) \tag{4.9}$$

Derivatives of F may then be evaluated as:

$$\mathrm{d}F/\mathrm{d}r = r \; \mathrm{d}E/\mathrm{d}r + E \tag{4.10}$$

and

$$\begin{aligned} \mathrm{d}^2 F/\mathrm{d}r^2 &= r \; \mathrm{d}^2 E/\mathrm{d}r^2 + \mathrm{d}E/\mathrm{d}r + \mathrm{d}E/\mathrm{d}r \\ &= r \; \mathrm{d}^2 E/\mathrm{d}r^2 + 2 \; \mathrm{d}E/\mathrm{d}r \end{aligned} \tag{4.11}$$

Using these, Eq. (4.8) may be rewritten as

$$(1/r) \; \mathrm{d}^2 F/\mathrm{d}r^2 + \omega^2 \mu_0 \varepsilon_0 \; (1/r) \; F = 0 \tag{4.12}$$

Apart from the common factor of $1/r$ (which we can remove by cancellation) Eq. (4.12) looks just like a one-dimensional wave equation, with the radial coordinate

r now being the only variable. By analogy with our previous results for plane waves, we can therefore guess that solutions travelling in the $+r$ direction (i.e. outward-propagating waves) have the form

$$F(r) = E_0 \exp(-jk_0 r) \tag{4.13}$$

where E_0 is again a constant factor. Complete solutions for $E(r)$ are therefore given by

$$E(r) = E_0/r \exp(-jk_0 r) \tag{4.14}$$

Equation (4.14) is the spherical wave solution we have been seeking. It has two components. The first is the term $\exp(-jk_0 r)$, which describes the phase of the wave, and is similar to the complex exponential variation we have previously found in plane waves. However, the surfaces of constant phase (known as *phase-fronts*) are now spheres instead of planes, as shown in Fig. 4.1 opposite. (Figure 4.1 illustrates another important concept, the *ray*. This is the local normal to a phase-front, and it effectively defines the direction the wave is moving in at any particular point. This idea is often used to deduce many of the features of an optical system, without invoking waves at all.)

The second term has a $1/r$ dependence, which implies that the amplitude falls off with distance from the origin. The explanation for this is power conservation. We found for plane waves that the time-averaged power per unit area is proportional to the electric field amplitude squared, and we might expect the same to be true for a spherical wave. However, at any radius r, a spherical wave must cross a hypothetical surface of area $4\pi r^2$. In order for the total power passing through this surface to be independent of radius, the wave amplitude must vary inversely with r. The solution unfortunately predicts an infinite amplitude at the origin. However, we shall ignore this for the time being, since it only occurs with a perfect spherical wave. In practice, an additional effect, known as *diffraction*, places an inherent limit on the amplitude achieveable with bounded beams, as we shall see shortly.

Paraxial Approximation

We shall now briefly introduce a *paraxial approximation* to the spherical wave, which will prove useful later on. Consider the wave amplitude in an x–y plane, a distance z from the origin, as shown in Fig. 4.2. The phase variation of the wave is defined by the exponential $\exp(-jk_0 r)$, where $r^2 = z^2 + R^2$ and $R^2 = x^2 + y^2$. For small R, we can put

$$r = z\sqrt{(1 + R^2/z^2)} \approx z + R^2/2z \tag{4.15}$$

so we can approximate the phase variation by

$$\exp(-jk_0 r) \approx \exp(-jk_0 z) \exp(-jk_0 R^2/2z) \tag{4.16}$$

We could make a similar approximation for the $1/r$ dependence of the wave amplitude. However, because $1/r$ is such a slowly-varying function, we can make

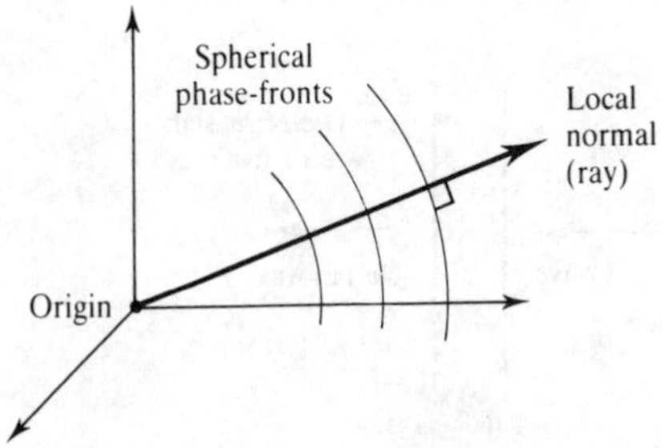

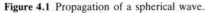

Figure 4.1 Propagation of a spherical wave.

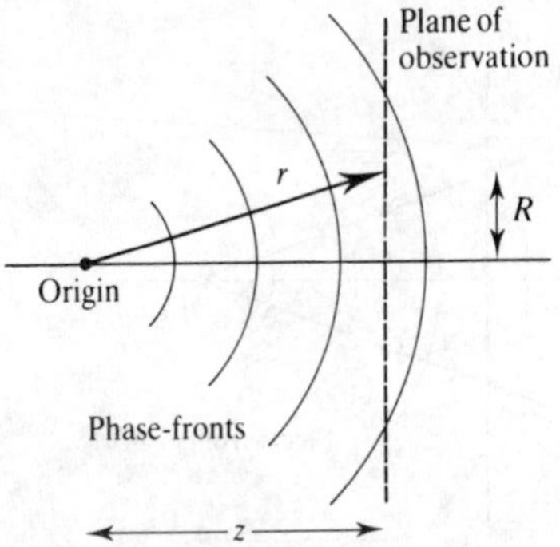

Figure 4.2 Geometry for calculation of the paraxial approximation to a spherical wave.

the approximation a cruder one in this case, and put $r \approx z$. The complete form of the wave is then

$$E = E_0/z \exp(-jk_0R^2/2z) \exp(-jk_0z) \tag{4.17}$$

We can now see clearly the significant features of our paraxial spherical wave. It travels mainly in the z-direction because of the $\exp(-jk_0z)$ term in Eq. (4.17). However, in any plane z = constant (i.e. in a plane normal to the main axis of propagation) the amplitude is roughly constant, while the phase varies parabolically with distance R from the axis. Thus we may write

$$E(R,z) \approx A(z) \exp(-jk_0R^2/2z) \tag{4.18}$$

It is easy to obtain a similar approximation for a converging wave, rather than a diverging one. The key information is contained in the sign of the exponential in Eq. (4.18), which describes the curvature of the phase-front. With the negative sign above, we have seen that the wave diverges from an origin a distance z behind the plane of observation. If the sign is made positive instead, the phase curvature is reversed. A converging wave may therefore be defined to vary as

$$E'(R,z) \approx A(z)' \exp(+jk_0R^2/2z) \tag{4.19}$$

This wave will converge on a different origin, a distance z in front of the plane of observation, as shown in Fig. 4.3.

The astute reader will have noticed that the arguments used above are somewhat self-contradictory. We cannot have a uniform amplitude (which implies an unbounded beam) while maintaining the validity of a paraxial approximation. Some form of 'envelope' must be introduced to limit the beam cross-section in a truly paraxial wave, which then cannot be spherically symmetric ...! We will show how to find a more consistent solution later on. In the meantime we shall continue with our approximate description of a spherical wave, and examine the way in which it can be manipulated by a lens.

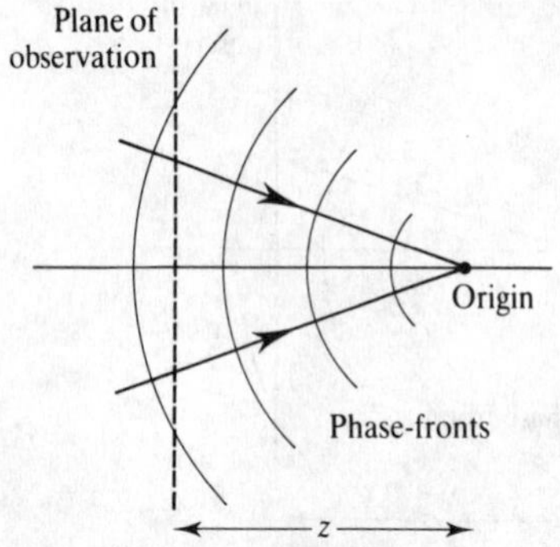

Figure 4.3 A converging, paraxial spherical wave.

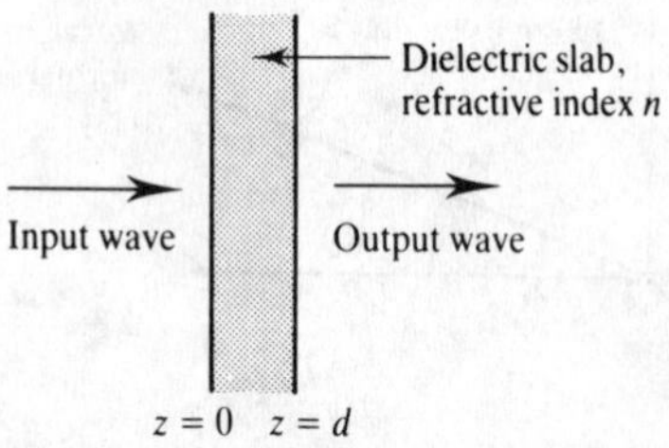

Figure 4.4 A plane wave passing through a slab of dielectric.

4.3 LENSES

Lenses are transmissive optical components, which operate by altering the phase of an optical wave as it passes through. In the simplest approximation, their action can be described by a complex transmissivity or *transparency function*. To illustrate this we shall consider the propagation of a plane wave at normal incidence through a dielectric slab. We assume for simplicity that the surround is air. The slab itself has refractive index n and thickness d, as in Fig. 4.4.

When the wave strikes the first boundary, at $z = 0$, it will actually be reflected and refracted. The first term implies that not all of the wave will penetrate the slab, and the second that the direction of the transmitted component may change. Both will be explained fully in Chapter 5, when we learn how to solve boundary-matching problems exactly. In the meantime we will simply borrow a result from the full solution: at normal incidence, there is little reflection and no refraction. We may therefore consider, to a good approximation, that the wave enters the slab unaltered.

Wave solutions in a medium of refractive index n have the form $E = E_0 \exp(-jk_0nz)$. By the time the wave emerges from the slab at $z = d$, it will therefore have accumulated additional phase, defined by the exponential term $\exp(-jk_0nd)$. We can therefore relate the input and output wave amplitudes by

$$E_{out} = \tau_s E_{in} \tag{4.20}$$

where

$$\tau_s = \exp(-jk_0nd) \tag{4.21}$$

The complex transparency function τ_s now describes the effect of the slab: it is to impart a phase shift, which depends on the *optical thickness nd*. Because the wave travels slower in a medium of refractive index greater than unity, this amounts to a retardation of the wave by the material.

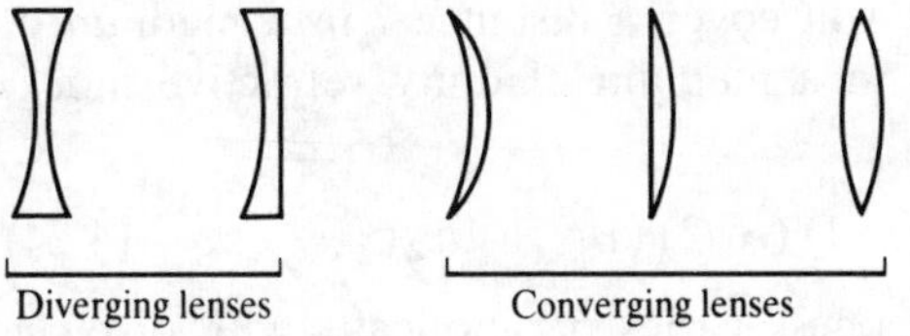

Figure 4.5 Lens types.

The *thin lens* is a more complicated optical element, which works by the same principle. It consists of a lump of dielectric (often glass), of refractive index n and maximum thickness d, with two highly polished, spherical surfaces of radii r_1 and r_2. Figure 4.5 shows some of the possible cross-sectional shapes a thin lens may take, depending on the magnitude and sign of r_1 and r_2. Again the effect of the dielectric is to retard a wave. Clearly the retardation is greater in thicker regions of the lens. In a diverging lens the dielectric is thicker in the outer regions; this retards the outermost parts of the wave, causing it to diverge. The reverse is true in a converging lens, which has thicker glass in its central region. If one of the lens surfaces is plane, the lens is described as plano-convex (for a converging lens) or plano-concave (for a diverging lens). In either case, one of the radii r_1 or r_2 must be infinite.

We can analyse a thin lens by first finding the relevant transparency function. This can be done by evaluating the phase shift incurred by passing between two parallel planes, say at $z = 0$ and $z = d$. Figure 4.6 shows the more detailed diagram needed for this calculation, for a converging lens.

First, we work out the thickness t of the glass at any radial distance R from the axis as

$$t = d - [r_1 (1 - \cos \theta_1) + r_2 (1 - \cos \theta_2)] \tag{4.22}$$

where θ_1 and θ_2 are as shown in Fig. 4.6. Using small-angle approximations, we may put

$$\begin{aligned} t &\approx d - (r_1 \theta_1^2/2 + r_2 \theta_2^2/2) \\ &\approx d - (R^2/2)(1/r_1 + 1/r_2) \end{aligned} \tag{4.23}$$

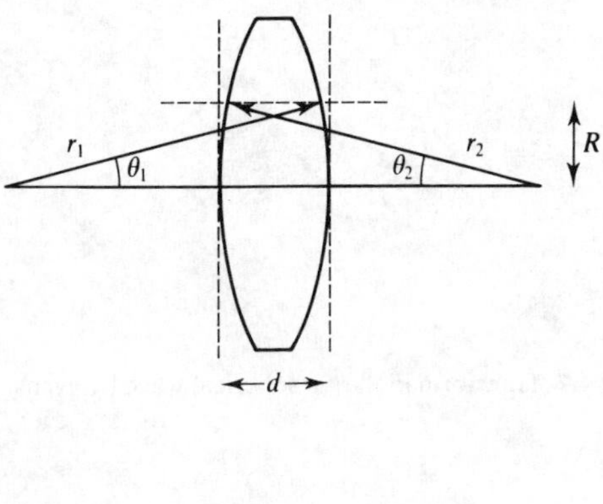

Figure 4.6 Detailed geometry of a converging lens.

In travelling from $z = 0$ to $z = d$, a wave will cover a distance t in a medium of refractive index n, and a distance $(d - t)$ in a medium of unity refractive index. This yields a total optical thickness δ of

$$\delta = nt + (d - t) \approx nd - (n - 1)(R^2/2)(1/r_1 + 1/r_2) \tag{4.24}$$

Equation (4.24) shows that the optical thickness varies parabolically with distance R from the axis. This variation has two components. The first is a constant, which would arise if the lens were to be replaced by a uniform slab of thickness d; this is relatively unimportant. The second is a term proportional to R^2—this is far more interesting since it accounts for the focusing property of the lens.

The lens transfer function τ_L can now be found (by analogy with the slab) as

$$\begin{aligned} \tau_L &= \exp(-jk_0\delta) \\ &= \tau_s \exp[+jk_0(n - 1)(R^2/2)(1/r_1 + 1/r_2)] \end{aligned} \tag{4.25}$$

Here τ_s is the slab transfer function previously found in Eq. (4.21). Note that we can write Eq. 4.25 in the alternative form

$$\tau_L = \tau_s \exp(+jk_0R^2/2f) \tag{4.26}$$

Here we have introduced an important new parameter, the *focal length* f of the lens, given by

$$1/f = (n - 1)(1/r_1 + 1/r_2) \tag{4.27}$$

Equation (4.27) is known as the *lens maker's equation*, since it offers a useful formula for choosing the surface curvatures required in a lens of a given focal length. Using the transfer function, we can now examine the effect of the lens on a spherical wave. We shall see that a lens can be used to transform an incoming wave whose origin is located at one particular point, into an outgoing wave with a different origin. This process is called *imaging*.

Imaging

Figure 4.7 shows a thin lens, illuminated on-axis by a paraxial spherical wave originating from an *object point* u away from the lens. We shall now show that the

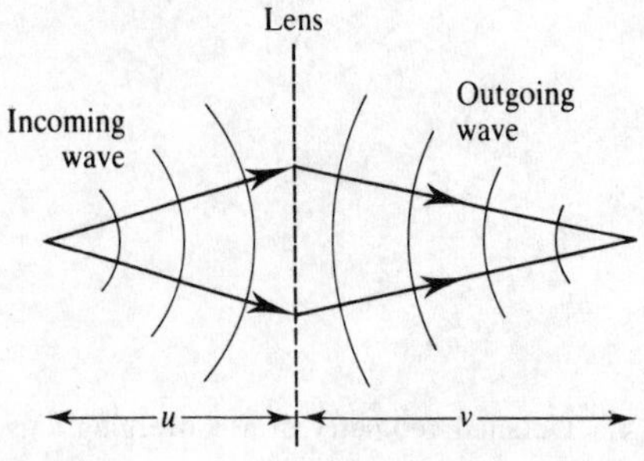

Figure 4.7 Transformation of a spherical wave by a lens.

transmitted output is a wave converging on a different origin (or *image point*) a distance v away from the lens.

The input electric field E_{in} can be found in the plane of the lens from Eq. (4.17) as

$$E_{in} = A(u) \exp(-jk_0R^2/2u) \tag{4.28}$$

The output field E_{out} is then found by multiplying the input field by τ_L, to give

$$E_{out} = \tau_L E_{in} = A'(u) \exp(+jk_0R^2/2f) \exp(-jk_0R^2/2u) \tag{4.29}$$

where we have written $A'(u)$ for $A(u)\ \tau_s$. The important feature of this new field is that the phase still varies parabolically with the distance R from the axis. This implies that we could write it as

$$E_{out} = A'(u) \exp(+jk_0R^2/2v) \tag{4.30}$$

where v is given by

$$1/u + 1/v = 1/f \tag{4.31}$$

The exponential in Eq. (4.30) has the positive sign associated with a converging wave, which will travel towards an origin or *focus* at a distance v from the lens. This shows that a lens can be used to convert one paraxial spherical wave into another, and Eq. (4.31) (known as the *thin lens imaging equation*) provides a convenient relation between the object and image points.

Design Examples

We can illustrate lens imaging with a number of examples. If $u = \infty$ (so that the phase-front of the input wave is effectively plane), Eq. (4.31) shows that $v = f$. A lens can therefore focus a plane wave to a point a distance f away from the lens, as shown in Fig. 4.8.

Similarly if $u = 2f$ we find that $v = 2f$. It is therefore possible to design a repetitive sequence of identical lenses, each spaced a distance $4f$ apart, which will continually focus and refocus a wave to points equally spaced between the lenses,

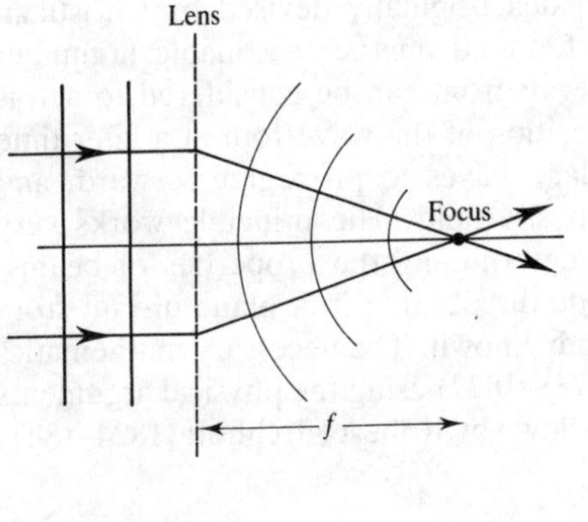

Figure 4.8 Focusing of a plane wave.

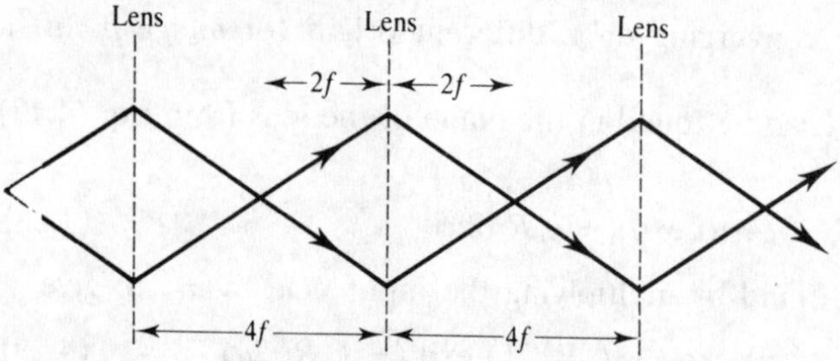

Figure 4.9 An iterative sequence of lenses.

as in Fig. 4.9. We will return to this structure later on because it represents one of the earliest attempts to guide an optical wave (known, with startling originality, as the *lens waveguide*).

Image Types

Note that v is positive in the examples above since f is positive and $u > f$. The images formed are therefore *real* ones, formed on the opposite side of the lens to the object. If v is negative, however, the image is created on the same side as the object. It is then known as a *virtual image*. This occurs when $u < f$. However, a virtual image is always formed when f is negative, i.e. for a diverging lens.

It is possible to extend the argument above, and show that a lens will also image a spherical wave originating from an off-axis point source onto an off-axis image point. It can be shown that Eq. (4.31) is still valid in this case, which implies that the entire plane passing through u is imaged as a plane passing through v. This simple theory (known as the *Gaussian theory of image formation*) accounts for the formation of extended images by a lens.

4.4 DIFFRACTION THEORY

We now consider the effect of the bounded nature of real beams. We base the argument on *Huygens' principle*, a powerful idea originally devised by Christiaan Huygens (1629–1695). Huygens' principle is founded on the reasonable argument that every point on a (possibly complicated) wavefront can be considered to act as a source of secondary spherical waves. The position of the wavefront at a later time can then be found by allowing these secondary waves to propagate forward, and combining their contributions in a new composite field. The principle works very well for many diffraction problems, and for determining the properties of beams. It is generally used to determine the amplitude distribution in a plane distant from another one, for which the distribution is already known. The necessary mathematics was initially developed by Augustin Fresnel (1788–1827) using the physical arguments outlined above. However, it was subsequently shown by Robert Kirchhoff (1824–1887)

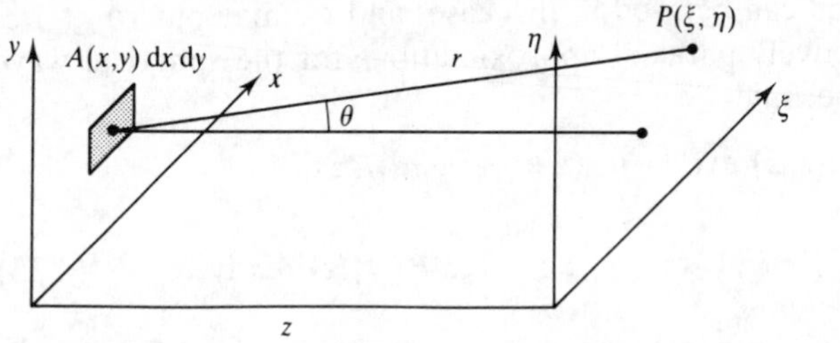

Figure 4.10 Geometry for the calculation of diffraction integrals.

to have a valid basis, as an approximate solution to the scalar wave equation. Since Kirchhoff's procedure is rather abstract, we shall ignore it here and follow Huygens and Fresnel instead.

Suppose the amplitude distribution $A(x,y) = A(x,y,0)$ is known in the x–y plane and we wish to determine the new distribution $A(\xi,\eta) = A(\xi,\eta,z)$ in a plane a distance z away, as in Fig. 4.10. We can find the answer by integrating the contributions made by all the elementary areas dxdy to the field at P. The input weight attached to each such contribution is $A(x,y)$ dxdy. However, as the contribution travels forward, it propagates as a spherical wave. At a distance r from the origin it will have changed both its phase and its amplitude, becoming $A(x,y)$ $(1/r) \exp(-jk_0r)$ dxdy.

In addition, Fresnel found it was necessary to multiply each elementary contribution by a further scaling factor $f(\theta)/j\lambda_0$, where $f(\theta)$ is a slowly-varying function of θ, defined by

$$f(\theta) = 1 \text{ for } \theta = 0; \qquad f(\theta) = 0 \text{ for } \theta = \pi \tag{4.32}$$

The function $f(\theta)$, which was later justified rigorously by Kirchhoff, has the effect of removing all backward-propagating waves from the analysis. Integrating all the contributions, we then get

$$A(\xi,\eta) = \iint_A [f(\theta)/j\lambda_0 r]\, A(x,y) \exp(-jk_0 r)\, dx dy \tag{4.33}$$

where the integration is taken over the whole area of the x–y plane of interest—typically, x and y will each range from minus infinity to plus infinity.

For small θ, we can put $f(\theta) \approx 1$. In addition, we assume that the effective range of $A(x,y)$ is much less than z, so we have a fairly narrow beam. In this case $z \gg x$, y, ξ or η so we can put

$$\begin{aligned} r &= \sqrt{[z^2 + (\xi - x)^2 + (\eta - y)^2]} \\ &\approx z + (\xi^2 + \eta^2)/2z + (x^2 + y^2)/2z - (\xi x + \eta y)/z \end{aligned} \tag{4.34}$$

This approximation can be used in the exponential contained in Eq. (4.33). There is also a $1/r$ amplitude term in Eq. (4.33); however, since this function is so slowly-varying, a cruder approximation can be used in this case, and we may put $1/r \approx 1/z$. These simplifications are effectively paraxial approximations for the spherical wave terms. Equation (4.33) then becomes

$$A(\xi,\eta) \approx (1/j\lambda_0 z) \exp[-jk_0(z + \{\xi^2+\eta^2\}/2z)] \cdot$$

$$\iint_A A(x,y) \exp[-jk_0(\{x^2+y^2\}/2z - \{\xi x+\eta y\}/z)] \, dxdy \qquad (4.35)$$

Equation (4.35) is known as the *diffraction integral*, and forms the basis of scalar diffraction theory. Once $A(x,y)$ is known, the integration can be carried out and $A(\xi,\eta)$ can be found. To evaluate such integrals we distinguish two important cases known as Fraunhofer and Fresnel diffraction.

4.5 FRAUNHOFER DIFFRACTION

When the distance z is much larger than the size of the object field, the term $k_0(x^2 + y^2)/2z$ is always much less than one, so that the exponential $\exp[-j\, k_0(x^2 + y^2)/2z]$ is always approximately unity. When this condition is satisfied, we talk of *Fraunhofer diffraction*. The consequences are not only that the integral is simplified, but that $A(\xi,\eta)$ looks very like the Fourier transform of $A(x,y)$. This distant field is called the *far-field*. We can estimate the minimum distance z that satisfies the criterion above, by setting the maximum allowable phase angle to about $\pi/8$. We then have

$$k_0(x^2 + y^2)/2z \leq \pi/8 \qquad (4.36)$$

Bearing in mind that $k_0 = 2\pi/\lambda_0$, we may rearrange this condition as

$$z \geq 8(x^2 + y^2)/\lambda_0 \qquad (4.37)$$

Putting in some numbers, we find that if the input beam is about 1 millimetre in diameter ($x = y = 10^{-3}$ m), and the wavelength is 0.5 μm (corresponding to green light), then z must be greater than 32 m. This is clearly a large distance! However, if the approximation is valid, we can write

$$A(\xi,\eta) \approx (1/j\lambda_0 z) \exp[-jk_0(z + \{\xi^2 + \eta^2\}/2z)] \cdot$$

$$\iint_A A(x,y) \exp[+jk_0(\xi x + \eta y)/z] \, dxdy \qquad (4.38)$$

Given that the term outside the integral is mainly a phase factor, we have

$$|A(\xi,\eta)|^2 = (1/\lambda_0^2 z^2) \left| \iint_A A(x,y) \exp[+j(\omega_x x + \omega_y y)] \, dxdy \right|^2 \qquad (4.39)$$

where ω_x and ω_y are given by

$$\omega_x = k_0\xi/z \quad \text{and} \quad \omega_y = k_0\eta/z \qquad (4.40)$$

If ω_x and ω_y are interpreted as *spatial frequencies*, Eq. (4.39) looks very like a Fourier transform relationship. A useful two-dimensional signal processing operation can therefore be carried out on an image simply by allowing an optical beam to propagate through the required distance, and allowing diffraction to do all the work. This is the origin of the much-quoted parallel processing capability of light. The large propagation distance involved is, unfortunately, a major drawback, which renders the whole process impractical without modification.

Fourier Transformation Using a Lens

A useful Fourier transform can, however, be carried out if a lens is introduced into the scheme, which effectively moves the far-field to a convenient nearby point. Figure 4.11 shows the setup. We can find the effect of the lens using the transfer function approach previously described. It is first assumed that an input amplitude distribution $A'(x,y)$ is prepared just behind the lens. This might be done by passing a plane wave through a suitable transparency. After passing through the lens, this distribution is modified to $A(x,y)$, given by

$$A(x,y) = A'(x,y)\, \tau_s \exp\,[+jk_0(x^2 + y^2)/2f] \tag{4.41}$$

If Eq. (4.41) is now substituted into Eq. (4.39) we obtain

$$|A(\xi,\eta)|^2 = (1/\lambda_0^2 z^2) \left| \iint_A A'(x,y) \exp\,[-jk_0(\{x^2 + y^2\}\{1/2f - 1/2z\} - \{\xi x + \eta y\}/z)]\, dx\, dy \right|^2 \tag{4.42}$$

When $z = f$ the first part of the exponent cancels and we are left with

$$|A(\xi,\eta)|^2 = (1/\lambda_0^2 z^2) \left| \iint_A A'(x,y) \exp\,[+jk_0(\xi x + \eta y)/z]\, dx\, dy \right|^2 \tag{4.43}$$

Equation (4.43) shows that we can obtain the far-field distribution at a distance f from the lens, in a plane known as the *Fourier transform plane*. This arrangement

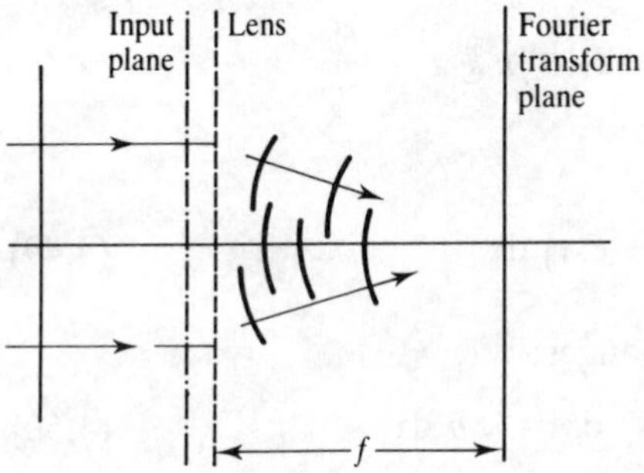

Figure 4.11 Optical Fourier transform using a lens.

is much more convenient, and is often used in signal processing systems. Note, however, that the result is not an exact transform because of the presence of extra phase terms outside the integral in Eq. (4.38). This is often unimportant (especially if we are only interested in intensity variations), but if an exact transform is essential it can be obtained by a further simple rearrangement of the optical setup.

4.6 FRESNEL DIFFRACTION AND GAUSSIAN BEAMS

When the far-field condition does not hold, we speak instead of *Fresnel diffraction*. As an example application of scalar diffraction theory, we shall consider what happens if we have a *Gaussian amplitude distribution* in the x–y plane, defined by

$$A(x,y) = A_0 \exp\left[-(x^2 + y^2)/a^2\right] \tag{4.44}$$

The amplitude distribution in a plane a distance z away is found by substituting this function into Eq. (4.35), a process which gives

$$\begin{aligned} A(\xi,\eta) = (1/\mathrm{j}\lambda_0 z) \exp\left[-\mathrm{j}k_0(z + \{\xi^2 + \eta^2\}/2z)\right] \iint_A A_0 \exp\left[-(x^2 + y^2)/a^2\right] \\ \times \exp\left[-\mathrm{j}k_0(\{x^2 + y^2\}/2z - \{\xi x + \eta y\}/z)\right] \mathrm{d}x\, \mathrm{d}y \end{aligned} \tag{4.45}$$

Equation (4.45) looks extremely complicated. However, because the variables are separated, it is actually easier to evaluate than it appears. We first note that $A(\xi,\eta)$ can be written in the form:

$$A(\xi,\eta) = (1/\mathrm{j}\lambda_0 z) \exp\left[-\mathrm{j}k_0(z + \{\xi^2 + \eta^2\}/2z)\right] I_1 I_2 \tag{4.46}$$

where I_1 and I_2 are two new integrals defined by

$$\begin{aligned} I_1 &= \int_{-\infty}^{\infty} \exp\left[-(\mathrm{j}k_0/2z + 1/a^2)x^2\right] \exp\left[\mathrm{j}k_0\xi x/z\right] \mathrm{d}x \\ I_2 &= \int_{-\infty}^{\infty} \exp\left[-(\mathrm{j}k_0/2z + 1/a^2)y^2\right] \exp\left[\mathrm{j}k_0\eta y/z\right] \mathrm{d}y \end{aligned} \tag{4.47}$$

Because of the similar appearance of I_1 and I_2, we need only evaluate one of them, say I_1. We now make the substitutions

$$b = \mathrm{j}k_0/2z + 1/a^2 \quad \text{and} \quad c_1 = \mathrm{j}k_0\xi/z \tag{4.48}$$

which reduce I_1 to

$$I_1 = \int_{-\infty}^{\infty} \exp\left[-bx^2 + c_1 x\right] \mathrm{d}x \tag{4.49}$$

We then follow with the additional variable changes:

$$p = x\sqrt{b} - c_1/2\sqrt{b} \quad \text{and} \quad \mathrm{d}p = \sqrt{b}\, \mathrm{d}x \tag{4.50}$$

which transform I_1 to

$$I_1 = (1/\sqrt{b}) \exp(c_1^2/4b) \int_{-\infty}^{\infty} \exp(-p^2)\, dp \qquad (4.51)$$

Since:

$$\int_{-\infty}^{\infty} \exp(-p^2)\, dp = \sqrt{\pi} \qquad (4.52)$$

we then obtain

$$I_1 = \sqrt{(\pi/b)} \exp(c_1^2/4b) \qquad (4.53)$$

The y-integral I_2 may be evaluated in a similar way. If we define

$$c_2 = jk_0\eta/z \qquad (4.54)$$

we get

$$I_2 = \sqrt{(\pi/b)} \exp(c_2^2/4b) \qquad (4.55)$$

In this way we can obtain an expression for the field distribution $A(\xi,\eta)$ in the form:

$$A(\xi,\eta) = (A_0/j\lambda_0 z)\,[2\pi z a^2/(jk_0a^2 + 2z)] \exp(-jk_0z)$$
$$\times \exp[-(k_0/2z)(j + k_0a^2/\{jk_0a^2 + 2z\})(\xi^2 + \eta^2)] \qquad (4.56)$$

Equation (4.56) still looks rather complicated. However, we note that, for any output plane, z is a constant, so the first part of the expression must also be constant. Hence we can write

$$A(\xi,\eta) = A(z) \exp[-(k_0/2z)(j + k_0a^2/\{jk_0a^2 + 2z\})(\xi^2 + \eta^2)] \qquad (4.57)$$

Here $A(z)$ describes the z-variation of the amplitude, while the information about the amplitude distribution in the ξ–η plane is contained in a much simpler exponential. We now notice that the coordinates ξ and η only appear together, in the form $f(\xi,\eta) = (\xi^2 + \eta^2)$, so the amplitude distribution must be circularly symmetric. If we put $(\xi^2 + \eta^2) = R^2$ we get

$$A(R) = A(z) \exp[-(k_0/2z)\,\{j4z^2/(4z^2 + k_0^2a^4)$$
$$+ 2k_0za^2/(4z^2 + k_0^2a^4)\}\, R^2] \qquad (4.58)$$

We shall use this expression to examine the basic properties of Gaussian beams.

Amplitude Variations

The real part of the exponential in Eq. (4.58) must describe the amplitude distribution of $A(R)$, while the imaginary part gives its phase variation. For the real part we have

$$|A(R)| = |A(z)| \exp[-k_0^2a^2R^2/(4z^2 + k_0^2a^4)] \qquad (4.59)$$

Since this still has the form of a Gaussian function, the overall shape of the beam must still be Gaussian, even after it has travelled the distance z from the x–y plane to the ξ–η plane. If we define the *beam width* w as the radius at which the amplitude falls to 1/e of its peak value, we get

$$w^2 = (4z^2 + k_0^2a^4)/k_0^2a^2 \tag{4.60}$$

Equation (4.60) shows that the beamwidth varies parabolically with z. We can examine this behaviour in more detail in two important regions:

1. *In the near field*, when $4z^2 \ll k_0^2a^4$, we have $w^2 \approx a^2$. This implies that the beam has constant width, independent of z. We can therefore usefully define $w_0 = a$ as the radius of the beam waist.
2. *In the far field*, when $4z^2 \gg k_0^2a^4$, we have $w^2 \approx 4z^2/k_0^2a^2$. Here the beamwidth varies linearly with z. In this region we can define a *half-angle of divergence* θ, such that

$$\theta = w/z = 2/k_0w_0 \tag{4.61}$$

We can also define a *whole-angle of divergence* Θ *as*:

$$\Theta = 2\theta = 4\lambda_0/\pi D \tag{4.62}$$

where D (the beam diameter at its waist) equals $2w_0$.

We can summarize the results above in normalized units. Suppose we define the transition between one regime and the other as the point $z = z_0$. This occurs when

$$4z_0^2 = k_0^2a^4 \quad \text{or} \quad z_0 = k_0w_0^2/2 \tag{4.63}$$

Equation (4.60) can then be rewritten as

$$(w/w_0)^2 = 1 + (z/z_o)^2 \tag{4.64}$$

The normalized beamwidth w/w_0 therefore depends parabolically on the normalized distance z/z_0. This is shown in Fig. 4.12, where w/w_0 is plotted as a function of z/z_0, for both positive and negative z/z_0. In the far-field the variation tends to the linear dashed-line asymptote shown. Note that, as w_0 increases, z_0 increases and θ decreases. A wide beam (measured in wavelengths) therefore expands slowly, and travels a longer distance before starting to expand.

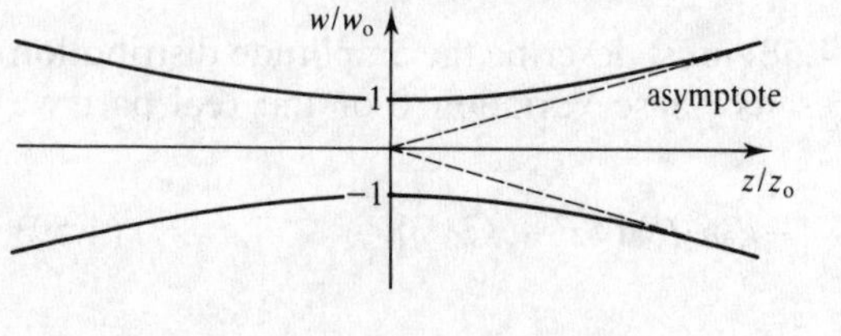

Figure 4.12 Parabolic variation of beamwidth with distance.

Design Examples

We can illustrate this with some examples. A typical visible laser has the parameters $\lambda_0 = 0.633\ \mu$m and $w_0 \approx 0.3$ mm; this corresponds to a helium-neon laser emitting red light. In this case, the beam is roughly 1000 wavelengths wide, and we find that $z_0 = 0.45$ m ($\approx$ 700 000 wavelengths) and $\theta = 0.67 \times 10^{-3}$ rad. Similarly a microwave dish might have $\lambda_0 = 1$ cm and $w_0 = 10$ cm (so the beam is 20 wavelengths wide), which gives $z_0 = 3.14$ m ($\approx$ 300 wavelengths) and $\theta = 31.8 \times 10^{-3}$ rad.

Phase Variations

Now we return to the complex part of the exponential in Eq. (4.58), which determines the phase of the Gaussian beam. The relevant part is the term:

$$A'(R) = \exp\,[-\mathrm{j}2k_0 zR^2/(4z^2 + k_0^2 a^4)] \tag{4.65}$$

This can be rewritten using our normalized variables as

$$A'(R) = \exp\,[-\mathrm{j}k_0 R^2/2(z^2 + z_0^2)] \tag{4.66}$$

In any plane z = constant, the phase therefore varies parabolically with the distance R from the axis. This is very similar to the behaviour previously demonstrated by a paraxial spherical wave. In fact, Eq. (4.66) can be written in the alternative form:

$$A'(R) = \exp\,[-\mathrm{j}k_0 R^2/2r(z)] \tag{4.67}$$

Equation (4.67) should be compared with Eq. (4.18). It shows that the effective radius of curvature $r(z)$ of the phase-front is given by

$$r(z) = (z^2 + z_0^2)/z \tag{4.68}$$

We can now work out the local radius of curvature at a number of important points along the beam:

1. In the *near field* (when $z \approx 0$) we find that $r \approx \infty$. The phase-front is therefore approximately plane, as might be expected from symmetry.
2. At the *transition point* (when $z = z_0$) we find that $r = 2z_0$.
3. In the *far-field* (when $z \to \infty$) we see that $r \to z$. The phase-front is therefore roughly spherical, and the field is equivalent to a spherical wave emanating from a point source at z.

These results are summarized in Fig. 4.13 on page 88. At the left, the wavefront is spherical and the beam is converging. As the beam passes through its waist, the phase-front slowly becomes plane. At the right, the wavefront gradually becomes spherical once more, but now the beam is diverging.

Diffraction of Arbitrary Field Distributions

We have seen how the width of a Gaussian beam spreads as we move away from the beam waist. This illustrates an important general point: an arbitrary bounded

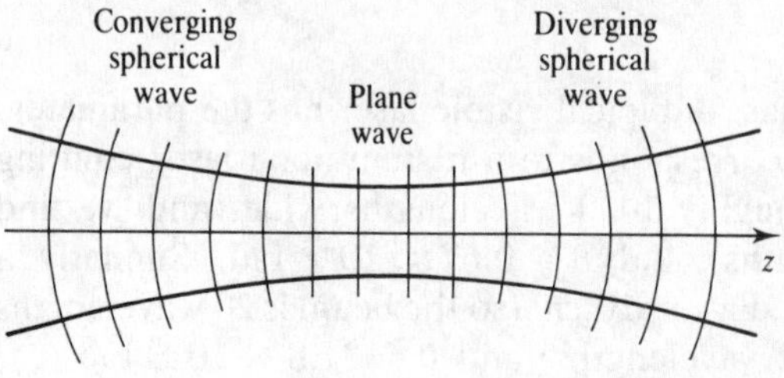

Figure 4.13 Variation of the phase-front curvature of a Gaussian beam with distance.

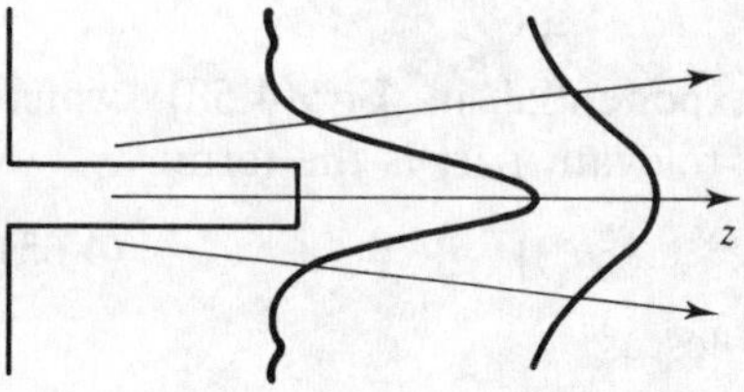

Figure 4.14 The spread of a bounded beam.

beam will, in general, spread through diffraction. Furthermore, the more the beam is confined, the worse this effect becomes. For example, Fig. 4.14 shows a 'top-hat' intensity distribution which rapidly broadens and loses its well-defined shape as it propagates, with a consequent reduction in power density. Efficient line-of-sight optical communication is therefore not practical over long distances.

An exception to this general rule is the recently discovered set of *diffraction-free beams*. These have rather more complicated beam shapes, but can (in theory) propagate without any spreading at all. At the moment it is too early to say if any practical use can be made of this feature.

4.7 TRANSFORMATION OF GAUSSIAN BEAMS BY LENSES

From the previous section we know that the main features of a Gaussian beam are a Gaussian amplitude distribution and a spherical phase-front. We shall now see what happens when such a beam is passed through a lens. Clearly the amplitude distribution will be unaltered. Equally, the emerging wave will still be spherical because the lens can only introduce a quadratic phase factor. However, the phase curvature will be modified because the spherical wave will have different parameters. A lens can therefore only transform a Gaussian input beam into a Gaussian output. We shall now see how this feature can be exploited to control the parameters of a beam.

Transformation by a Single Lens

First we shall consider the transformation of a Gaussian beam by a single lens. Figure 4.15 shows the geometry. The input beam has a waist of width $2w_{10}$ located at a distance u behind the lens.

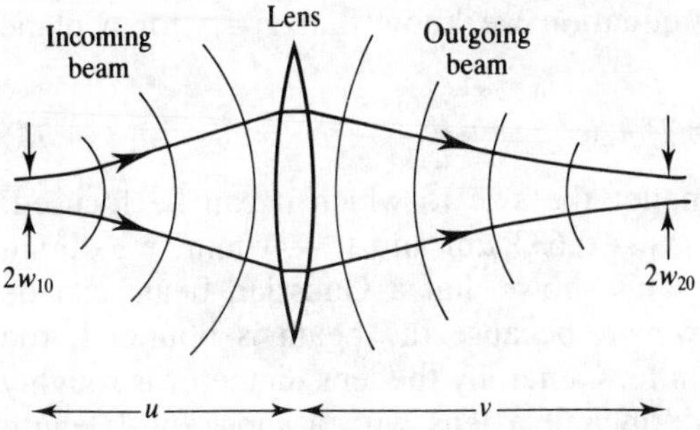

Figure 4.15 Transformation of a Gaussian beam by a lens.

It is simplest to assume initially that $u \gg w_{10}$ so that the lens is in the far-field of the beam. If this is the case we know from Eq. (4.68) that the radius of curvature of the phase-front at the lens is roughly u. Similarly from Eq. (4.60), we know that the beamwidth at the lens is $2w$, where

$$w \approx 2u/k_0 w_{10} \tag{4.69}$$

From the thin-lens imaging equation, Eq. (4.31), we can find the radius of curvature of the emerging beam as v. If we adopt a far-field approximation once again, we can assume that the new beamwaist will be located at a distance v in front of the lens. If the new waist has width $2w_{20}$ we know that w_{20} is defined (by analogy with Eq. (4.69)) by

$$w \approx 2v/k_0 w_{20} \tag{4.70}$$

Combining Eqs (4.69) and (4.70), we can relate w_{20} to w_{10} using

$$w_{20}/w_{10} = v/u \tag{4.71}$$

If $v < u$, therefore, the new waist will be smaller than the old one. We can illustrate this with examples.

Focusing of a Gaussian beam First we consider the focusing of a plane beam with a Gaussian envelope, as shown in Fig. 4.16. In this case we assume that the input beam is approximately parallel, with a Gaussian amplitude distribution of width $2w$

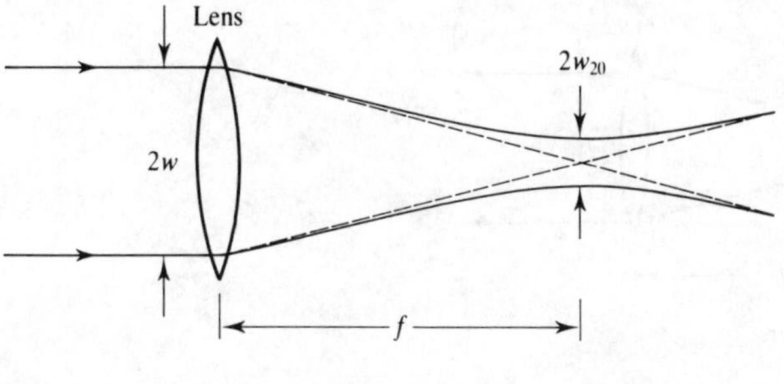

Figure 4.16 Focusing of a Gaussian beam.

at the lens. From the thin-lens imaging equation we know that $v = f$ for a plane input. In this case we obtain

$$w_{20} = 2f/k_0w \tag{4.72}$$

Thus, the wider the input beam, the smaller the size to which it can be focused. For the typical parameters of $f = 1$ cm, $\lambda_0 = 0.633$ μm and $w = 1$ mm, we obtain $w_{20} = 2$ μm, a respectably small figure. This shows that a Gaussian beam can be focused down to small dimensions. However, because the beam is bounded, the energy density at the focal spot remains finite. Generally the lens diameter is roughly matched to the expected input beam width, and a lens with a short focal length compared with its diameter is then said to have a high *numerical aperture* or *NA*. For a medium of refractive index n_e external to the lens, this quantity is defined as

$$\text{NA} = n_e \sin \theta_{max} \tag{4.73}$$

where θ_{max} is the maximum angle of the converging cone of rays (e.g. the largest value of θ_2 in Fig. 4.6).

Naturally there is a trade-off between parameters. For example, the wider the beam, the larger the angle of convergence, and hence the smaller the extent of the focused region. This has important consequences for the design of electro-optic modulators (see Chapter 3). The voltage needed to drive a transverse modulator depends on the electrode gap, and if the beam has a small diameter this gap can be small. As we have seen, a lens can be used to obtain a small beam diameter; however, if the depth of focus is very small the usable interaction length is very short, and this in turn increases the drive voltage. To make a more useful modulator, more effective confinement of the beam is needed.

Telescope beam expansion We now consider an application that involves the use of two lenses. When arranged as a *telescope*, these can be used to carry out beam expansion. Figure 4.17 shows the geometry. The input beam is again assumed to be approximately parallel, and of width $2w$ at the lens. If the focal length of the first lens is f_1 then the focal spot is defined as before by $w_{20} = 2f_1/k_0w$. However, if the second lens has focal length f_2 then $w_{20} = 2f_2/k_0w'$. Equating the two we find the output beam width w' as

$$w' = w\,(f_2/f_1) \tag{4.74}$$

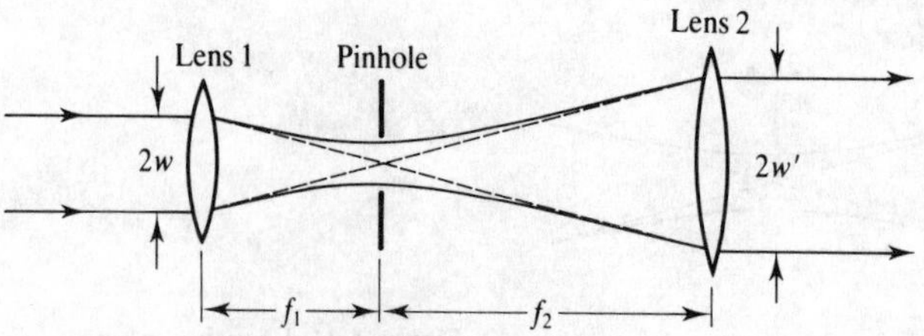

Figure 4.17 A telescope beam expander.

The output beam is therefore an expanded version of the input, enlarged by the ratio of the lens focal lengths. This reduces any residual beam divergence by the same factor. In many cases the input beam is not quite Gaussian but suffers from imperfections caused by noise of high spatial frequency. These can be removed by placing a pinhole at the focus between the lenses. Since the field distribution in this plane is closely related to a Fourier transform of the input distribution, blocking the outermost regions removes the high frequency components and can greatly improve the beam quality. This process is known as *spatial filtering*.

The lens waveguide One solution to the problem of confining a beam within a narrow cross-section over a long distance is provided by the iterative sequence of lenses, which we discussed earlier in Fig. 4.9 in terms of simple imaging theory. If we now assume that the waveform propagating between each lens is a Gaussian beam, it is clear that the beamwidth must vary through the structure much as shown in Fig. 4.18. Note that the separation between lenses is now taken as $4s$, rather than $4f$.

We shall calculate the parameters of the Gaussian beam propagating through the guide, assuming this time that the focal spot is not small, so that a far-field approximation cannot be used. In this case, the thin-lens imaging equation, Eq. (4.31), is invalid. However, a suitable modification can be obtained simply by replacing u and v with the radii of curvature of two Gaussian beams, one originating from a waist at u, the other converging to a waist at v. If these are denoted u' and v', we get

$$1/u' + 1/v' = 1/f \tag{4.75}$$

For the geometry shown in Fig. 4.18, $u' = v'$, so that $u' = 2f$. However, Eq. (4.68) implies that the radius of curvature can also be found as

$$u' = (u^2 + z_0^2)/u \tag{4.76}$$

where $u = 2s$ and z_0 is the transition distance of the beam. Hence z_0 is given by:

$$z_0 = \sqrt{(u'u - u^2)} = 2\sqrt{(fs - s^2)} \tag{4.77}$$

For z_0 to be real, we require $s < f$. If this is the case we may invert Eq. (4.63) to give

$$w_0 = \sqrt{(2z_0/k_0)} \tag{4.78}$$

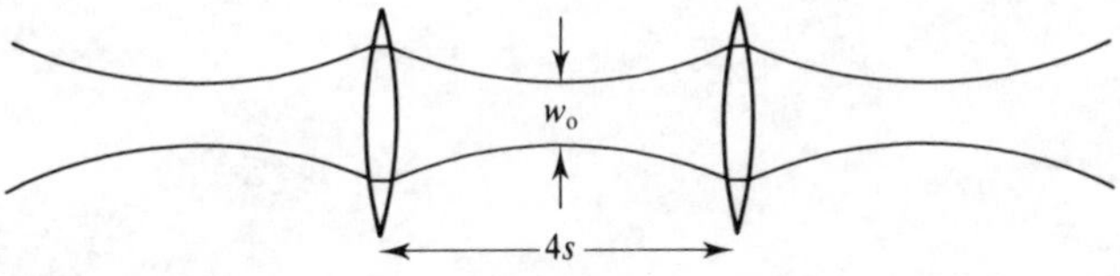

Figure 4.18 The lens waveguide.

Substituting from Eq. (4.77) we then get

$$w_0 = [16\,(fs - s^2)/k_0^2]^{1/4} = [4\lambda_0^2\,(fs - s^2)/\pi^2]^{1/4} \tag{4.79}$$

Equations (4.77) and (4.79) therefore imply that the parameters z_0 and w_0 both depend strongly on the separation of the lenses.

Design Example

For a lens waveguide formed by a sequence of lenses of focal length 100 mm, and light of wavelength 0.633 μm, we obtain $w_0 = 0.14$ mm for $s = 50$ mm and $w_0 = 53.3$ μm for $s = 99.5$ mm. Note that as s tends to f, w_0 and z_0 both tend to zero, and the solution reverts to the geometrical optics approximation discussed previously.

The lens waveguide can support a number of other modes of propagation besides the simple Gaussian beam described above. Unfortunately it is hard to make such a guide work well in practice. For example, the positioning of the lenses is very critical since any lateral misalignment causes the beam to wander off-axis; equally, the air/glass interfaces cause considerable scatter and loss. Luckily there are considerably better types of guide, as we shall see later. However, some of these (parabolic index fibres) can also support a Gaussian transverse field distribution.

4.8 MIRRORS AND RESONATORS

We now consider the use of an alternative optical component, the mirror. This often consists of a smooth, polished surface coated with metal. It is common experience that light does not penetrate such a surface to any significant extent, but is entirely reflected instead. It is also well known that the angle of reflection is equal and opposite to the angle of incidence, as shown in Fig. 4.19 (both these assumptions will be fully justified in Chapter 5).

Mirrors clearly have many applications, but for the time being we shall consider just two of them. The first example is the *Fabry-Perot cavity*, which is a resonant structure formed from two plane, parallel mirrors. When filled with an optical gain medium the device can oscillate, and thus forms the basis of the laser. The second is the *spherical mirror*. This can be used to transform one spherical wavefront into

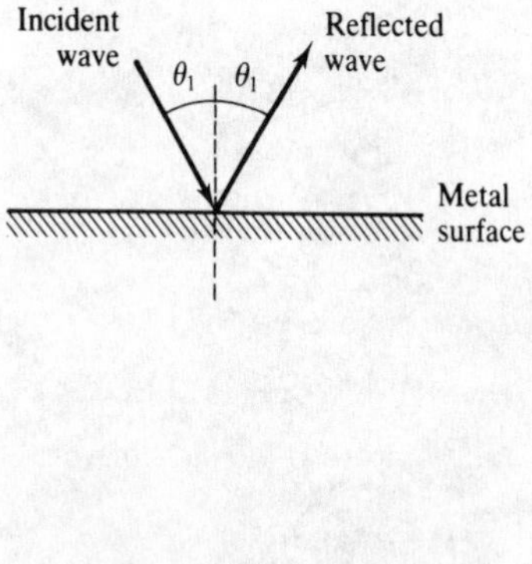

Figure 4.19 Light incident on a polished, metallized surface.

another and so (just like a lens) it can perform imaging operations. It turns out that a laser cavity can be greatly improved by the use of at least one spherical end-mirror, since this confines the beam inside the cavity more effectively. When this is done, the laser output is generally a Gaussian beam, which is one reason why such beams are so common in the first place.

The Fabry-Perot Cavity

Figure 4.20 shows a Fabry-Perot cavity consisting of two plane, parallel mirrors with reflectivities R_1 and R_2, separated by a distance L.

We start by considering the behaviour of the device when the two mirrors are perfect (so that $R_1 = R_2 = 1$) and the cavity is filled with a dielectric of refractive index n. We first assume that there is a plane wave propagating up and down the cavity, following the zig-zag path shown. We now consider the resulting field amplitude at a particular point A. Suppose the wave is initially travelling in the $+z$ direction, and that its amplitude is given by E_A. By the time it has completed one round-trip of the cavity the new field E_A' is found as

$$E_A' = E_A \exp(-j2k_0nL) \tag{4.80}$$

Clearly the combined field travelling in the $+z$ direction at A is found as a summation of many such contributions. These will add up in-phase if

$$2k_0nL = 2\nu\pi \tag{4.81}$$

where ν is an integer. If Eq. (4.81) does not hold, on the other hand, the field will consist of many contributions with different phase, which will add up to zero on average. Rearranging Eq. (4.81) slightly we find that the condition for a non-zero field inside the cavity is

$$L = \nu(\lambda_0/2n) \tag{4.82}$$

This is known as the *longitudinal resonance condition*, and it implies that only a certain set of field patterns can exist inside the cavity. Another way to calculate it is to assume that the total field travelling in both directions must everywhere be given by the sum of one combined field travelling in the $+z$ direction and a similar field travelling in the $-z$ direction. Thus we could write

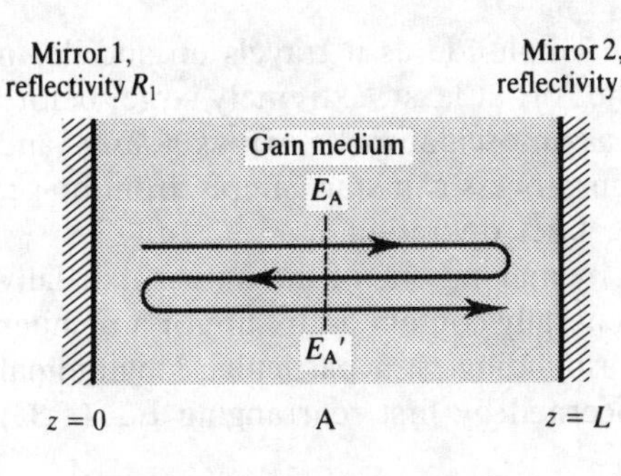

Figure 4.20 A Fabry-Perot cavity formed by two mirrors.

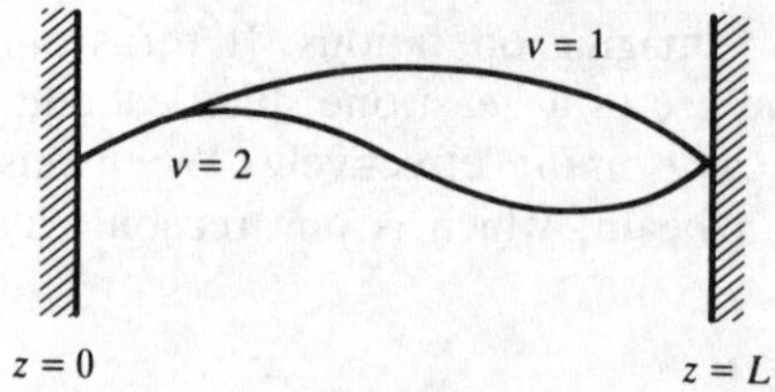

Figure 4.21 Standing-wave patterns inside the cavity.

$$E(z) = E_+ \exp(-jk_0nz) + E_- \exp(+jk_0nz) \tag{4.83}$$

If the mirrors are perfect, light cannot penetrate them, and the boundary conditions to be satisfied are $E = 0$ at $z = 0$ and at $z = L$ (again, these assumptions will be reviewed in Chapter 5). We can satisfy the former condition by taking $E_- = -E_+$, which allows us to write Eq. (4.83) as

$$E(z) = E_0 \sin(k_0nz) \tag{4.84}$$

The latter condition is satisfied when $k_0nL = \nu\pi$, which is effectively the same as the longitudinal resonance condition previously found in Eq. (4.82). Note, however, that this time we have also calculated the actual field distribution inside the cavity, which must be a standing-wave pattern as shown in Fig. 4.21. Each pattern of this type is called a *longitudinal mode*.

We now consider the case when there are losses, so that $R_1, R_2 < 1$. This could arise if power is simply dissipated in the mirrors, or if we have arranged one of the mirrors to be only partially reflecting. In this case the amplitude of the wave trapped inside the cavity will swiftly fall to zero as it suffers repeated reflections. If the cavity is filled with an *optical gain medium*, however, these losses may be made up, and the wave can actually be amplified as it passes back and forth.

If the gain medium provides a gain g, then a forward-travelling wave will grow as defined by the exponential term $\exp(+gz)$. Including this gain and the mirror loss, we can now find the field E'_A in terms of the field E_A as

$$E'_A = E_A R_1 R_2 \exp(+2gL) \exp(-j2k_0nL) \tag{4.85}$$

Clearly the longitudinal resonance condition is exactly as before. However, we note that $|E'_A| > |E_A|$, provided

$$R_1 R_2 \exp(+2gL) > 1 \tag{4.86}$$

In this case the wave will continually grow in amplitude as it travels up and down the cavity, so the total field will become infinite (or, at least, extremely large, before gain saturation sets in). The cavity then acts as an oscillator, known as a *laser*, and (as with nearly all such devices) the oscillation process can start simply from noise. Equation (4.86) is therefore the condition for laser operation.

We can tap off some of the stored energy by making one of the mirrors partially reflecting. In this case the output beam will generally contain a mixture of a number of different wavelengths, or lines, each corresponding to a particular longitudinal mode. We can calculate the frequencies concerned by first rearranging Eq. (4.82) to give the wavelength of the ν th mode as

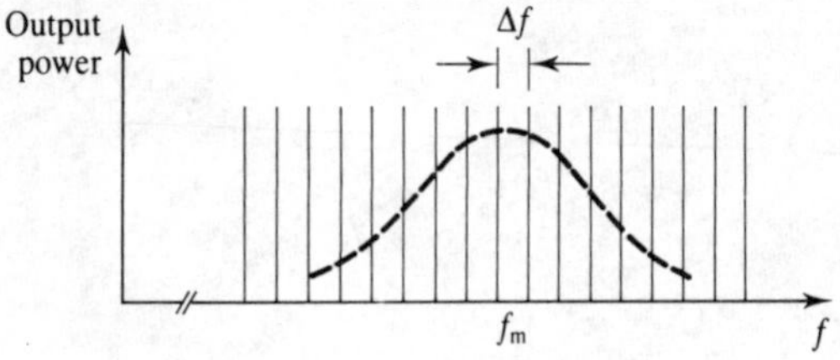

Figure 4.22 Emission spectrum of a Fabry-Perot laser.

$$\lambda_\nu/n = 2L/\nu \qquad (4.87)$$

Since frequency and wavelength are related to the velocity of light by $c = f_\nu \lambda_\nu$, we can find the frequency of the ν th mode as

$$f_\nu = c\nu/2Ln \qquad (4.88)$$

so the frequency separation between adjacent modes is given by

$$\Delta f = c/2Ln \qquad (4.89)$$

We can now plot the frequency content of the laser output (known as the *emission spectrum*) as shown in Fig. 4.22. For a 1 metre long cavity with $n \approx 1$, the mode spacing is $\Delta f \approx 150$ MHz. In practice, the exact number of modes that are observed is limited by the existence of optical gain only over a finite frequency range, or *gain bandwidth*. This is indicated by the dashed-line envelope in Fig. 4.22. However, the number of longitudinal modes experiencing sufficient gain to lase may still be large. The Fabry-Perot laser is therefore a multi-mode device, unless special efforts are made to select a particular line using additional optical components.

The Spherical Mirror Resonator

We have previously found the field distribution inside the cavity as a standing-wave pattern. Of course this only represents a longitudinal variation, and we now consider the field distribution in a transverse plane and its relation to the construction and operation of the laser. The optical gain will generally be provided by electrical pumping. The power needed for this will be proportional to the volume of active material, which can be reduced by making the cavity cross-section small. To achieve adequate round-trip gain, however, the cavity length L must be quite large. These specifications are conflicting, since they require a narrow, bounded beam to travel for long distances. Clearly the beam will spread laterally en route, as we have previously shown, so energy will leak sideways out of the cavity. Once more, therefore, we need a way of confining a bounded beam.

One answer is provided by the *spherical mirror*. If correctly located, this can exactly retroreflect a spherical wave, as shown in Fig. 4.23. Here the radius of curvature of the mirror is r, and the origin of the spherical wave is located on-axis, at a distance r from the mirror. Remember that the phase-front of a Gaussian beam is spherical. This type of beam can therefore also be reflected back on itself using a spherical mirror with the correct curvature. If two mirrors are used, an optical

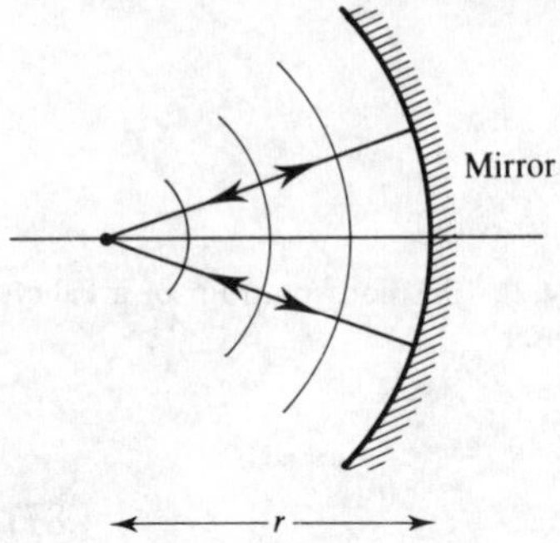

Figure 4.23 Retroreflection of a spherical wave by a spherical mirror.

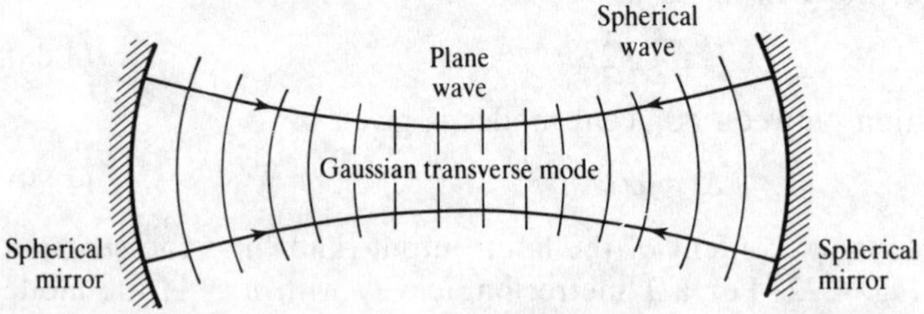

Figure 4.24 The symmetric spherical mirror resonator.

cavity is formed. This allows the beam to travel back and forth between the mirrors as shown in Fig. 4.24, losing very little energy out of the sides of the cavity.

This arrangement is termed a *symmetric spherical mirror resonator*, and the Gaussian beam is one of its possible *transverse oscillation modes*. It turns out that there are other transverse modes, but the Gaussian is the lowest-order one, with most of its energy concentrated near the axis. We can calculate the parameters of such a beam using analysis very similar to that previously employed for the lens waveguide. Assuming the cavity length is L, the radius of curvature of the beam at each mirror can be found from Eq. (4.68) as

$$r = [(L/2)^2 + z_0^2]/(L/2) \tag{4.90}$$

where (once again) z_0 is the transition length of the beam. For the beam to be correctly retroreflected, the radius of curvature of the beam and the mirror must match, so that z_0 is given by

$$z_0 = \sqrt{[(L/2)(r - L/2)]} \tag{4.91}$$

As with the lens waveguide, only a certain combination of parameters will give a real value of z_0. In this case we require $L/2 < r$. If this is the case the width w_0 of the beam can be found as

$$w_0 = \{4[r L/2 - (L/2)^2]/k_0^2\}^{1/4} = \{\lambda_0^2 [r L/2 - (L/2)^2]/\pi^2\}^{1/4} \tag{4.92}$$

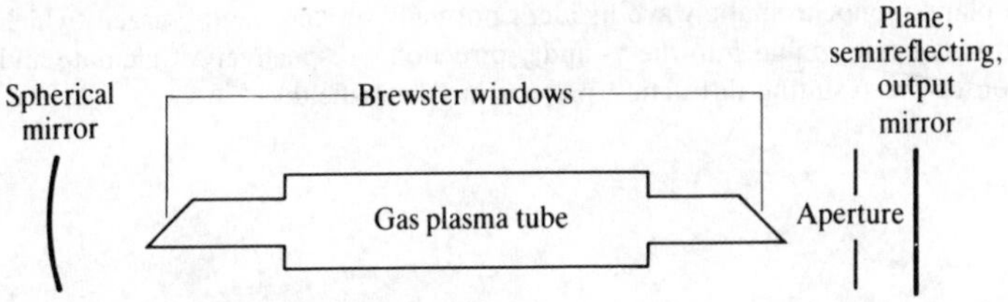

Figure 4.25 Schematic of a typical He-Ne laser.

The beamwidth is therefore dependent on the cavity length and on the mirror curvature. For a cavity of length $L = 0.5$ m, fitted with end mirrors of radius 0.5 m, and a wavelength of 0.633 μm, we obtain $w_0 = 0.225$ mm. This corresponds to a beam diameter of 0.45 mm.

There are other possible ways of arranging a resonator of this type. One common method is to use one spherical and one plane mirror, in an asymmetric configuration. The beam waist now occurs at the plane mirror; if this is made partially reflecting, an output can be obtained from the cavity. This arrangement is used in many gas laser types, including the ubiquitous *helium-neon laser*, shown in Fig. 4.25. The optical gain is provided by a gas plasma tube which contains an electrically-excited gas mixture. The ends of the tube are sloped at the *Brewster angle*; as we shall see in Chapter 5, this allows one polarization to have greatly reduced reflection loss, through the *Brewster effect*. Consequently it has a lower lasing threshold and the ouput of the laser is linearly polarized—this provides a convenient way of generating polarized light. An aperture is also introduced into the cavity. This allows the desired Gaussian beam to pass with little loss, but heavily attenuates higher-order transverse modes which have more energy in off-axis regions of the beam.

Note that an entirely different solution to the problem of confining a narrow beam inside a region of optical gain is provided by a waveguide structure. This solution is adopted in all semiconductor lasers, and will be discussed in detail in Chapter 12.

PROBLEMS

4.1 Find the focal length of a converging glass lens, having (a) two spherical surfaces, each of radius 200 mm, and (b) one plane surface, and one spherical surface of radius 400 mm.
[(a) 200 mm; (b) 800 mm]

4.2 The lenses in Question 4.1 are used in turn to form an image of a source positioned at a point 400 mm away. Find the image position in each case. What is the difference between the image types?
[(a) 400 mm away, on the opposite side of the lens; (b) 800 mm away, on the same side]

4.3 A source is positioned at point S, a distance δ off axis, u away from a converging lens of focal length f. Find the position of the image. Hence show that the lens will form a plane image of a plane, two-dimensional object, and find the magnification involved.

4.4 A sequence of identical lenses, each of focal length f, are separated by alternating distances d_1 and d_2. Find a relation between d_1 and d_2 that must be satisfied if the sequence is to act as a lens waveguide.

4.5 The figure below shows a plane monochromatic wave incident normally on an opaque screen, which has a central rectangular aperture of sides a and b in the x- and y-directions, respectively. Calculate and sketch the intensity distribution of the resulting diffraction pattern in the far-field.

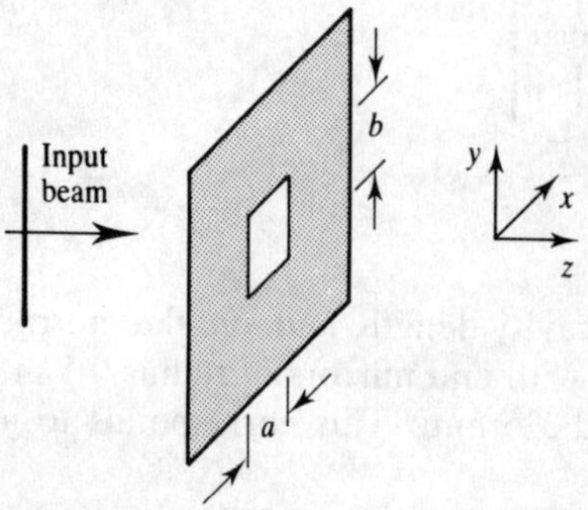

4.6 The figure below shows a plane monochromatic wave incident normally on a periodic structure known as an *amplitude grating.* This is a screen of locally-varying transmission, defined by the transparency function $\tau = \tau_0 + \Delta\tau \cos(Ky)$, where $K = 2\pi/\Lambda$ is the spatial frequency of the periodic modulation in the transmission. What is the output of the grating in the far-field?

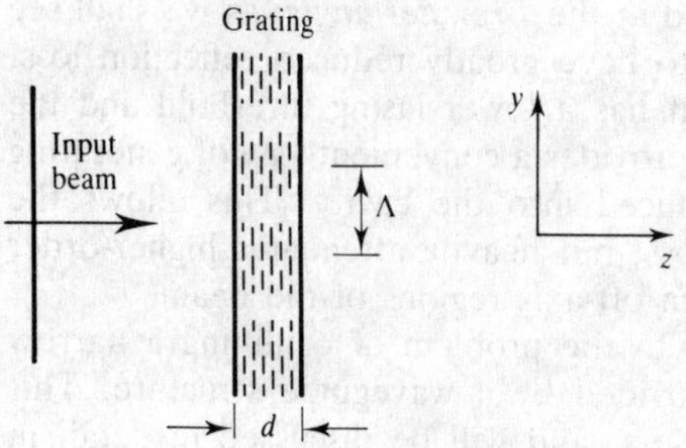

4.7 Show that an exact two-dimensional Fourier transform can be carried out between the input distribution A_{in} and the output field A_{out} using the arrangement below.

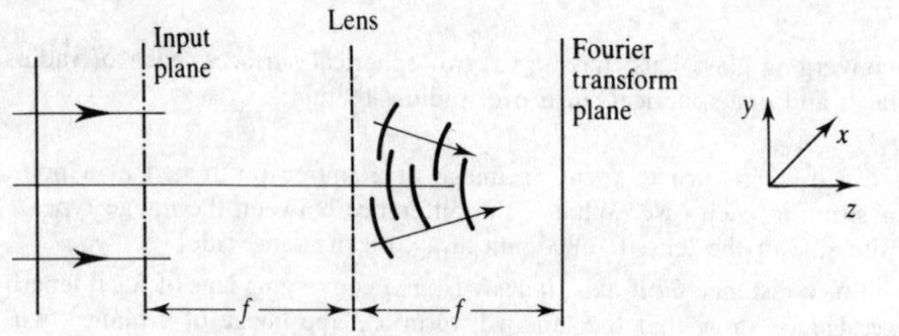

4.8 (a) Find the half-angle of divergence of a Gaussian beam of wavelength 0.633 μm and diameter 1 mm.

(b) How far from the beam waist would you expect to find the transition point between plane and spherical wave behaviour?
[(a) 4×10^{-4} rad; (b) 1.24 m]

4.9 A gas laser is to be constructed with an optical cavity of length 250 mm and end-mirrors of amplitude reflectivity $R_1 = 1.0$ and $R_2 = 0.9$. Determine the threshold gain g_{th} required from the lasing medium. Estimate the frequency difference between adjacent longitudinal modes.
[0.211 m^{-1}; 600 MHz]

4.10 The optical cavity of a helium-neon laser (for which $\lambda = 0.633$ μm) is to be constructed from one spherical mirror of radius 2 m and one plane, partially-reflecting mirror. Describe steps that could be taken to ensure the output is a Gaussian beam. If the two mirrors are 1 m apart, what will be the diameter of the beam?
[0.9 mm]

SUGGESTIONS FOR FURTHER READING

Durnin, J. Exact solutions for nondiffracting beams. I. The scalar theory, *J.O.S.A.*, **4**, 651–654, 1987.
Fox, A. G., and Li, T. Resonant modes in a maser interferometer, *Bell Syst. Tech. J.* **40**, 453–488, 1961.
Goubau, G., and Schwering, F. On the guided propagation of electromagnetic wave beams, *IRE Trans. on Antennas and Propagation*, **AP-9**, 248–256, 1961.
Hecht, E. "Optics", 2nd ed., Chapters 5, 10 & 11, Addison-Wesley Publishing Co., Reading MA., 1987.
Kogelnik, H., and Li, T. Laser beams and resonators, *Appl. Opt.*, **5**, 1550–1567, 1966.
Marcuse, D. "Light Transmission Optics", Chapters 4, 5 & 6, Van Nostrand Reinhold Co., New York, 1972.
O'Neill, E. L. Spatial filtering in optics, *IRE Trans. Inform. Theory*, **IT-2**, 56–65, 1956.
Pierce, J. R. Modes in sequences of lenses, *Proc. Natl Acad. Sci.*, **47**, 1808–1813, 1961.
Ramo, S., Whinnery, J. R., and van Duzer, T. "Fields and Waves in Communication Electronics", 2nd ed., Chapter 14, John Wiley and Sons, New York, 1984.
Steward, E. G. "Fourier Optics", Ellis Horwood Ltd., Chichester, 1983.
Yariv, A. "Optoelectronics", Chapters 2 and 4, Holt, Rinehart and Winston, New York, 1985.

CHAPTER

FIVE

REFLECTION AND REFRACTION AT A SINGLE INTERFACE

5.1 THE BEHAVIOUR OF LIGHT AT A DIELECTRIC INTERFACE

The previous chapters have been concerned with the propagation of waves in empty space or in uniform, homogeneous media. It is now time to introduce more complicated geometries, building towards those appropriate to optical waveguides. We begin with a qualitative discussion of the phenomena that occur when light strikes a *dielectric interface*. Figure 5.1 shows a plane wave incident at an angle θ_1 on two semi-infinite media with refractive indices n_1 and n_2, respectively. These are chosen such that $n_1 > n_2$.

Experience suggests that for near-normal incidence (small θ_1) the incident beam will give rise to a plane reflected beam in medium 1, together with a transmitted beam in medium 2. The reflected beam travels at an angle equal and opposite to that made by the incident wave with the interface normal. The angle of the transmitted beam is defined by *Snell's law* (named after Willebrord Snell, 1591–1626). This was originally an experimental observation, and requires that

$$n_1 \sin \theta_1 = n_2 \sin \theta_2 \tag{5.1}$$

Figure 5.1 Light incident on an interface between two media.

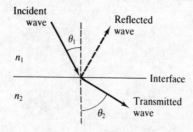

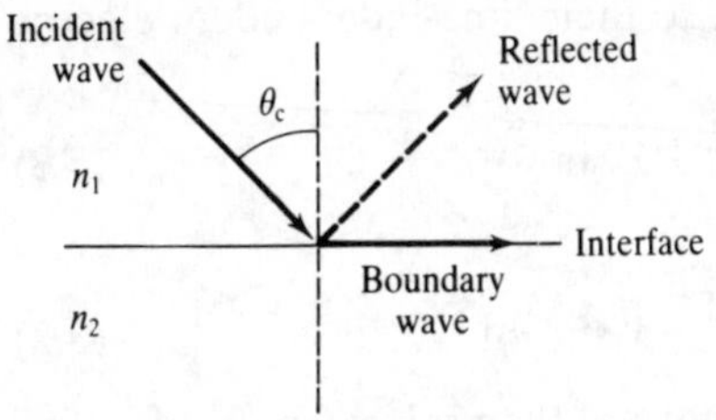

Figure 5.2 Light incident on an interface at the critical angle.

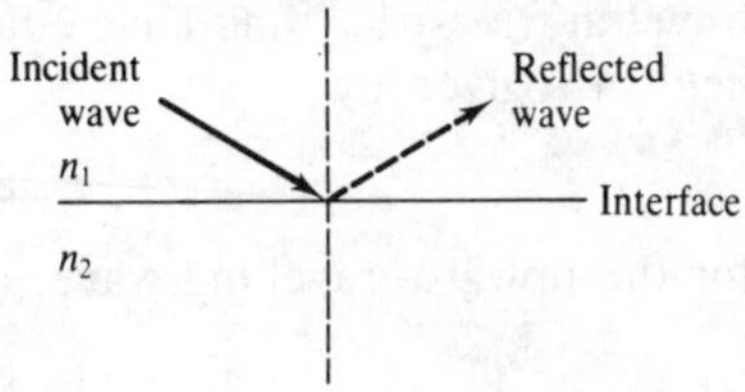

Figure 5.3 Total internal reflection.

However, we note that for $\theta_1 = \sin^{-1}(n_2/n_1)$ we get $\theta_2 = \pi/2$. In this case the transmitted wave travels parallel to the interface, as shown in Fig. 5.2. This angle of incidence is called the *critical angle* θ_c. For an interface between glass and air (when $n_1 = 1.5$ and $n_2 = 1.0$) the critical angle is $\theta_c = 41.8°$. However, θ_2 can never reach $\pi/2$ if we exchange the values of n_1 and n_2, so there is no critical angle when the wave is incident from the low-index side. For $\theta_1 > \theta_c$, there is no real solution for θ_2, which implies that a propagating transmitted wave cannot arise at all. This is known as *total internal reflection*, and is illustrated in Fig. 5.3.

It is easy to show that when total internal reflection occurs, no power crosses the interface. Total internal reflection therefore presents a mechanism for confining a field in one region of space. We shall now proceed to a more rigorous analysis of the problem. The first point to note is that the presence of two waves (one reflected, one transmitted) in medium 1 implies that there must be a more complicated field distribution in this region. We will therefore start by examining the way plane wave solutions to Maxwell's equations can be combined. Since the equations are linear, this is very simple.

5.2 SUPERPOSITION OF FIELDS

Perfectly monochromatic fields can produce a range of strikingly beautiful effects, which fall under the general heading of *interference* phenomena. The simplest of these occurs when two plane waves of the same frequency cross at an angle, as shown in Fig. 5.4. We assume that both waves are polarized in the y-direction,

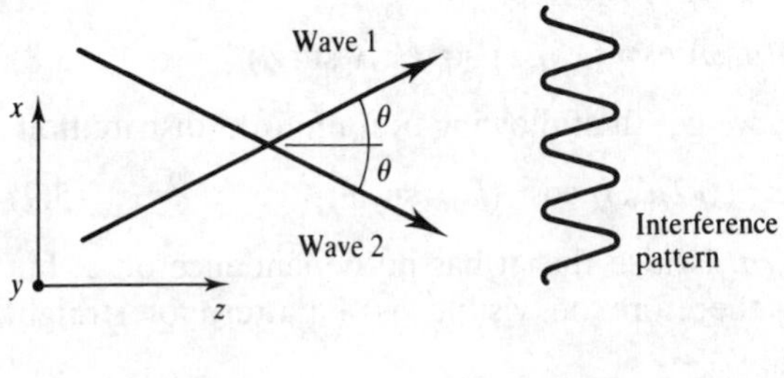

Figure 5.4 Interference between two plane waves.

travel in free space, and have equal amplitude E_y, so their time-independent electric fields are given by

$$\mathbf{E}_1 = \mathbf{j}\, E_y \exp\left[-\mathrm{j}k_0(z \cos\theta + x \sin\theta)\right] \tag{5.2}$$

for the upward-travelling wave, and

$$\mathbf{E}_2 = \mathbf{j}\, E_y \exp\left[-\mathrm{j}k_0(z \cos\theta - x \sin\theta)\right] \tag{5.3}$$

for the downward one. Because Maxwell's equations are themselves linear, the total field can be written as a linear superposition of these two fields, in the form:

$$\begin{aligned}\mathbf{E} &= \mathbf{j}\, E_y \{\exp\left[-\mathrm{j}k_0(z \cos\theta + x \sin\theta)\right] + \exp\left[-\mathrm{j}k_0(z \cos\theta - x \sin\theta)\right]\} \\ &= \mathbf{j}\, 2E_y \exp(-\mathrm{j}k_0 z \cos\theta) \cos(k_0 x \sin\theta)\end{aligned} \tag{5.4}$$

Here we have combined the x-dependent parts of the two fields to yield a real, cosinusoidal amplitude distribution. We can write Eq. (5.4) in the alternative form:

$$\mathbf{E} = \mathbf{j}\, E(x) \exp\left[-\mathrm{j}\beta z\right] \tag{5.5}$$

where

$$E(x) = 2E_y \cos(k_0 x \sin\theta) \quad \text{and} \quad \beta = k_0 \cos\theta \tag{5.6}$$

The result can therefore be regarded as a field with non-uniform amplitude $E(x)$, travelling in the z-direction. However, its propagation constant is reduced from the normal value by the factor $\cos\theta$.

Design Example

As an exercise we shall now calculate the time average of the power carried by the combined field above, per unit area in the z-direction. This is representative of the way the field would be detected by the eye, or by photographic film oriented in a plane normal to the z-axis. Once again we need to evaluate the quantity $P_z = 1/2 \operatorname{Re}[\mathbf{E} \times \mathbf{H}^*] \cdot \mathbf{k}$. The time-independent electric field is given by Eq. (5.4). The time-independent magnetic field, on the other hand, is best calculated from the curl relationship between $\mathbf{E}$ and $\mathbf{H}$:

$$\nabla \times \mathbf{E} = -\mathrm{j}\omega\mu_0 \mathbf{H} \tag{5.7}$$

Since $\mathbf{E}$ has just a y-component, it is easy to show that $\mathbf{H}$ has only x- and z-components, given by

$$\begin{aligned}H_x &= 2E_y(-\beta/\mu_0\omega) \exp(-\mathrm{j}\beta z) \cos(k_0 x \sin\theta) \\ H_z &= 2E_y(-\mathrm{j}k_0 \sin(\theta)/\mu_0\omega) \exp(-\mathrm{j}\beta z) \sin(k_0 x \sin\theta)\end{aligned} \tag{5.8}$$

Substituting into the expression for P_z we get the following non-uniform distribution:

$$P_z = -E_y H_x^* = 4E_y^2(\beta/2\mu_0\omega) \cos^2(k_0 x \sin\theta) \tag{5.9}$$

This is known as an *interference pattern*. Notice that it has no dependence on z, but varies periodically with x. It would therefore be visible as a pattern of straight

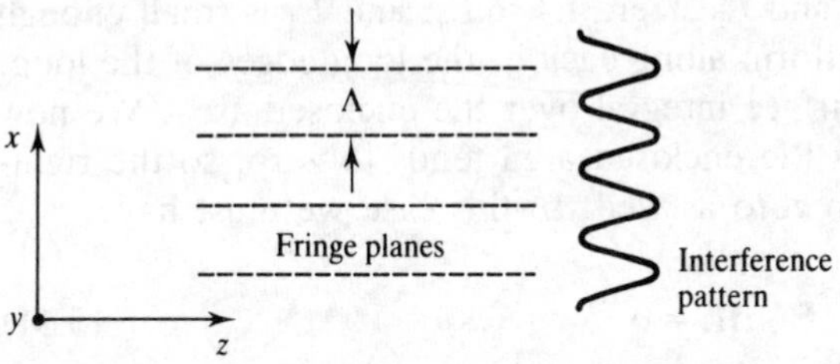

Figure 5.5 Fringe planes and the interference pattern.

fringes, oriented parallel to the z-axis, as shown in Fig. 5.5. The *spatial frequency* of the pattern is $K = k_0 \sin \theta$, so the separation between fringes is $\Lambda = 2\pi/K$. Notice that Λ decreases as the angle between the two beams increases, and is of the same order as the wavelength of the light. Interference patterns of this type are particularly important in many branches of optics. However, they do not arise if the two beams are orthogonally polarized, or if the light used is not coherent.

5.3 BOUNDARY MATCHING

There is one further complication to the interface problem. When electromagnetic waves meet changes in dielectric constant, *boundary conditions* must be satisfied. We can find the necessary conditions by applying Maxwell's equations at a boundary discontinuity. Figure 5.6 shows a suitable junction between two different semi-infinite media, medium 1 (where the time-dependent electric field is $\mathbf{E}_1$), and medium 2 (where it is $\mathbf{E}_2$).

First, recall from Faraday's law that

$$\int_L \mathbf{E} \cdot \mathrm{d}\mathbf{L} = -\iint_A \partial\mathbf{B}/\partial t \cdot \mathrm{d}\mathbf{a} \tag{5.10}$$

We shall now evaluate the left-hand side of Eq. (5.10) by performing a line integration round the rectangular dashed loop shown in Fig. 5.6. This crosses the

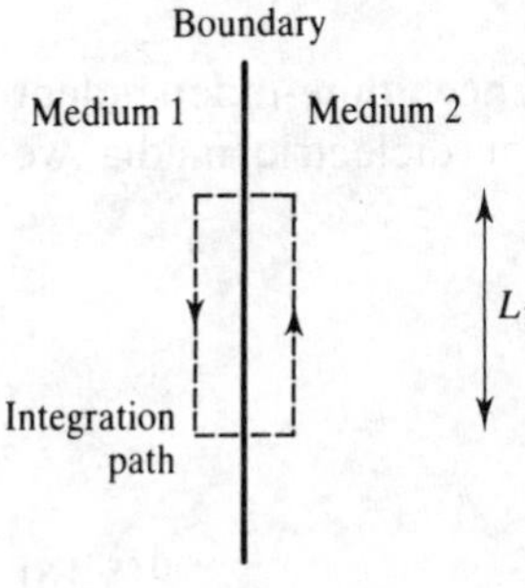

Figure 5.6 Line integral performed at a boundary.

boundary and has sides of length L_1 and L_2. Here $L_2 \ll L_1$, and L_1 is small enough for the fields to be approximately uniform along each of the long edges of the loop. The right-hand side is of course a surface integral over the enclosed area. We now let L_2 tend to zero. As this happens the enclosed area tends to zero, so the right-hand side of Eq. (5.10) must tend to zero as well. In this case we must have

$$\int_L \mathbf{E} \cdot \mathrm{d}\mathbf{L} = 0 \tag{5.11}$$

At this point the integral in Eq. (5.11) is well approximated by

$$\int_L \mathbf{E} \cdot \mathrm{d}\mathbf{L} = (\mathrm{E}_{t1} - \mathrm{E}_{t2})\, L_1 \tag{5.12}$$

where E_{t1} and E_{t2} are the components of $\mathbf{E}_1$ and $\mathbf{E}_2$ tangential to the boundary. This implies that

$$\mathrm{E}_{t1} - \mathrm{E}_{t2} = 0 \tag{5.13}$$

In other words, the tangential components of $\mathbf{E}$ must match across the boundary. This is the first of our boundary conditions. We can also write this without explicitly mentioning E_{t1} and E_{t2}, as

$$\mathbf{n} \times (\mathbf{E}_2 - \mathbf{E}_1) = 0 \tag{5.14}$$

where the vector $\mathbf{n}$ is a new vector, the local normal to the boundary.

We can do the same sort of thing for $\mathbf{H}$, $\mathbf{D}$ and $\mathbf{B}$. The complete set of boundary conditions is then

$$\begin{aligned} \mathbf{n} \times (\mathbf{E}_2 - \mathbf{E}_1) &= 0 \\ \mathbf{n} \times (\mathbf{H}_2 - \mathbf{H}_1) &= \mathbf{K} \\ \mathbf{n} \cdot (\mathbf{D}_2 - \mathbf{D}_1) &= \rho_s \\ \mathbf{n} \cdot (\mathbf{B}_2 - \mathbf{B}_1) &= 0 \end{aligned} \tag{5.15}$$

Here $\mathbf{K}$ is the *surface current* (which can be ignored in dielectrics) and ρ_s is the *surface charge density* (again, zero in dielectrics). Generally not all the Eqs (5.15) are needed to solve any particular problem; in fact, picking the most useful subset of conditions is something of an art in itself.

Similar equations can be obtained for the single-frequency, time-independent representation, simply by replacing $\mathbf{E}$ by E and so on. For dielectric media we therefore obtain

$$\begin{aligned} \mathbf{n} \times (\mathbf{E}_2 - \mathbf{E}_1) &= 0 \\ \mathbf{n} \times (\mathbf{H}_2 - \mathbf{H}_1) &= 0 \\ \mathbf{n} \cdot (\mathbf{D}_2 - \mathbf{D}_1) &= 0 \\ \mathbf{n} \cdot (\mathbf{B}_2 - \mathbf{B}_1) &= 0 \end{aligned} \tag{5.16}$$

5.4 ELECTROMAGNETIC TREATMENT OF THE INTERFACE PROBLEM

We must now calculate what happens at the interface more exactly. We assume that the governing equation is the time-independent vector wave equation. For a uniform dielectric this takes the form:

$$\nabla^2 \mathbf{E} + n^2 k_0^2 \mathbf{E} = 0 \tag{5.17}$$

We can use Eq. (5.17) to solve the interface problem by first finding solutions for each region as if it were infinite, and then matching them at the boundary. The geometry is shown in Fig. 5.7. The junction is assumed to lie between two infinite sheets of dielectric with indices n_1 and n_2. These are lying in the x–z plane, with the interface at $x = 0$, and the aim is to calculate the field distribution with a plane wave incident from medium 1 at an angle θ_1.

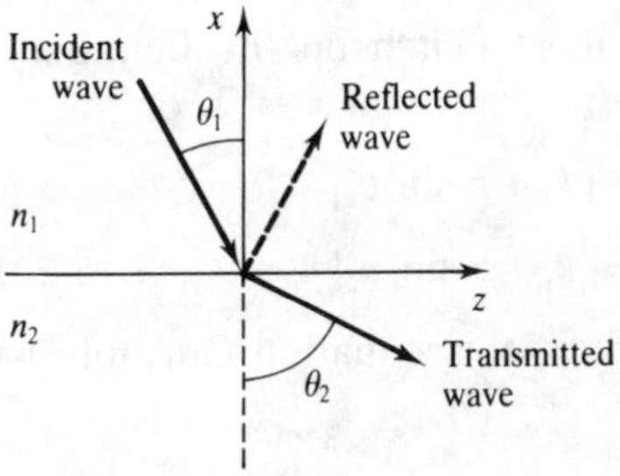

Figure 5.7 Geometry for the interface problem.

TE incidence If we assume as before that the incident wave is polarized in the y-direction, the electric field is entirely parallel to the interface and only has a y-component. This is known as *transverse electric* or *TE* incidence. Without justification (apart from common sense) we assume that the transmitted and reflected waves are similarly polarized, so we can work with E_y throughout. The vector wave equation therefore reduces to a single scalar equation. Furthermore, because everything is uniform in the y-direction, $\partial/\partial y$ must be zero for all field quantities. The relevant equation is therefore

$$\partial^2 E_y/\partial x^2 + \partial^2 E_y/\partial z^2 + n^2 k_0^2 E_y = 0 \tag{5.18}$$

Reflection and Refraction

We start by assuming that total internal reflection does not occur, so we need only consider the case of reflection and refraction. The equations we must solve in each medium are

$$\partial^2 E_{y1}/\partial x^2 + \partial^2 E_{y1}/\partial z^2 + n_1^2 k_0^2 E_{y1} = 0 \quad \text{in medium 1}$$
$$\partial^2 E_{y2}/\partial x^2 + \partial^2 E_{y2}/\partial z^2 + n_2^2 k_0^2 E_{y2} = 0 \quad \text{in medium 2} \tag{5.19}$$

We now assume that the field in each region is a combination of plane waves, the solutions we would find if the media were infinite and separate. A suitable guess is to choose incident and reflected waves in medium 1, and just a transmitted wave in medium 2, so that

$$E_{y1} = E_I \exp[-j\, k_0 n_1 (z \sin\theta_1 - x \cos\theta_1)] + E_R \exp[-j\, k_0 n_1 (z \sin\theta_1' + x \cos\theta_1')]$$

$$E_{y2} = E_T \exp[-j\, k_0 n_2 (z \sin\theta_2 - x \cos\theta_2)] \qquad (5.20)$$

Here we have assumed for generality that the reflected wave travels upwards at an angle θ_1', which may be different from θ_1. There are a number of unknowns in Eq. (5.20). Given that we know θ_1 and E_I, we must find θ_1', θ_2, E_R and E_T. We can find them from the boundary conditions in Eq. (5.16). In principle we could use any of these conditions, but here we choose to start with

$$\mathbf{n} \times (\mathbf{E}_2 - \mathbf{E}_1) = 0 \qquad \text{on } x = 0 \qquad (5.21)$$

This requires that the tangential components of **E** must match on the boundary. Since **E** is wholly tangential anyway, we must have $E_{y1} = E_{y2}$ on $x = 0$, or

$$E_I \exp[-j\, k_0 n_1 z \sin\theta_1] + E_R \exp[-j\, k_0 n_1 z \sin\theta_1'] = E_T \exp[-j\, k_0 n_2 z \sin\theta_2] \qquad (5.22)$$

Because it contains exponential terms that vary with z, this equation can only be satisfied for all z if

$$n_1 \sin\theta_1 = n_1 \sin\theta_1' = n_2 \sin\theta_2 \qquad (5.23)$$

If this is the case, the z-variation is removed and Eq. (5.22) can be satisfied by a correct choice of the constants E_R and E_T. Equation (5.23) requires first that $\theta_1' = \theta_1$, so the angle of the reflected beam is indeed equal to that of the incident beam (verifying our initial intuitive guess). It also shows that Snell's law appears automatically in the electromagnetic analysis, which is highly satisfactory. Equation (5.22) then reduces to

$$E_I + E_R = E_T \qquad (5.24)$$

Performing the boundary matching for the electric field, we therefore find that the wave amplitudes must be continuous at the interface, which is physically reasonable.

The second boundary condition we choose is that

$$\mathbf{n} \times (\mathbf{H}_2 - \mathbf{H}_1) = 0 \qquad \text{on } x = 0 \qquad (5.25)$$

This implies that the tangential components of **H** must also match at $x = 0$. We can find the necessary components of **H** by using the curl relation between **E** and **H** given previously in Eq. (5.7). Since $E_x = E_z = 0$, we have

$$H_x = (-j/\omega\mu_0)\, \partial E_y/\partial z\,; \qquad H_y = 0\,; \qquad H_z = (+j/\omega\mu_0)\, \partial E_y/\partial x \qquad (5.26)$$

The only tangential component of **H** is H_z, so the second boundary condition will be satisfied if

$$\partial E_{y1}/\partial x = \partial E_{y2}/\partial x \qquad \text{on } x = 0 \tag{5.27}$$

Using this condition, we find that

$$\begin{aligned} &\mathrm{j}\,k_0 n_1 \cos(\theta_1)\, E_\mathrm{I} \exp[-\mathrm{j}\,k_0 n_1 z \sin\theta_1] \\ &- \mathrm{j}k_0 n_1 \cos(\theta_1)\, E_\mathrm{R} \exp[-\mathrm{j}\,k_0 n_1 z \sin\theta_1] \\ &= \mathrm{j}\,k_0 n_2 \cos(\theta_2)\, E_\mathrm{T} \exp[-\mathrm{j}\,k_0 n_2 z \sin\theta_2] \end{aligned} \tag{5.28}$$

Following the same argument as before we can remove all the exponentials to get

$$n_1 \cos(\theta_1)\, E_\mathrm{I} - n_1 \cos(\theta_1)\, E_\mathrm{R} = n_2 \cos(\theta_2)\, E_\mathrm{T} \tag{5.29}$$

Equations (5.24) and (5.29) now represent two simultaneous equations with two unknowns. By a fairly simple rearrangement we can therefore find E_R in terms of E_I. Normally the result is expressed in terms of a *reflection coefficient* Γ_E, which can be shown to reduce to

$$\Gamma_\mathrm{E} = E_\mathrm{R}/E_\mathrm{I} = [n_1 \cos\theta_1 - n_2 \cos\theta_2]/[n_1 \cos\theta_1 + n_2 \cos\theta_2] \tag{5.30}$$

We can also find the *transmission coefficient* T_E, which is given by

$$\mathrm{T}_\mathrm{E} = E_\mathrm{T}/E_\mathrm{I} = 2n_1 \cos\theta_1/[n_1 \cos\theta_1 + n_2 \cos\theta_2] \tag{5.31}$$

These expressions satisfy physical intuition. For example, if the two media have the same refractive index, the reflection coefficient is zero and the transmission coefficient is unity. Surprisingly, the reflection coefficient can be large, even at normal incidence. For a glass/air interface (where $n_1 = 1.5$ and $n_2 = 1$) we find that $\Gamma_\mathrm{E} = (1.5 - 1)/(1.5 + 1) = 0.2$.

The other boundary conditions At this point we pause and consider the fact that there were two more boundary conditions we could have used. What about them? The third condition is that

$$\mathbf{n} \cdot (\mathbf{D}_2 - \mathbf{D}_1) = 0, \quad \text{or} \quad \mathbf{n} \cdot (\sqrt{n_2}\,\mathbf{E}_2 - \sqrt{n_1}\,\mathbf{E}_1) = 0 \qquad \text{on } x = 0 \tag{5.32}$$

Since **E** has no component normal to the boundary, all this equation says is 0 = 0. The third condition is therefore not too helpful and we can safely ignore it. The fourth is that

$$\mathbf{n} \cdot (\mathbf{B}_2 - \mathbf{B}_1) = 0, \quad \text{or} \quad \mathbf{n} \cdot (\mathbf{H}_2 - \mathbf{H}_1) = 0 \qquad \text{on } x = 0 \tag{5.33}$$

Since the only normal component of **H** is H_x, this implies that

$$H_{x2} = H_{x1} \qquad \text{on } x = 0 \tag{5.34}$$

Bearing in mind the definition of **H** given in Eq. (5.26), we find that

$$\partial E_{y2}/\partial z = \partial E_{y1}/\partial z \qquad \text{on } x = 0 \tag{5.35}$$

Doing the necessary differentiation, we get

$$-\mathrm{j}\, k_0 n_1 \sin(\theta_1)\, E_I \exp[-\mathrm{j}\, k_0 n_1 z \sin\theta_1]$$
$$-\mathrm{j}\, k_0 n_1 \sin(\theta_1)\, E_R \exp[-\mathrm{j}\, k_0 n_1 z \sin\theta_1]$$
$$= -\mathrm{j}\, k_0 n_2 \sin(\theta_2)\, E_T \exp[-\mathrm{j}\, k_0 n_2 z \sin\theta_2] \qquad (5.36)$$

If we now use Snell's law again to remove the exponentials, this reduces to Eq. (5.24). The two other boundary conditions therefore introduce no new features, so our original choice was a good one.

TM incidence The analysis can easily be repeated for the case when the magnetic field is oriented in the y-direction. This is known as *transverse magnetic* or *TM* incidence. We shall not work through all the mathematics, which is somewhat repetitive, but merely quote the results. These are slightly different; for the reflection coefficient we get

$$\Gamma_H = E_R/E_I = [n_2 \cos\theta_1 - n_1 \cos\theta_2]/[n_2 \cos\theta_1 + n_1 \cos\theta_2] \qquad (5.37)$$

and for the transmission coefficient we find that

$$\mathrm{T}_H = E_T/E_I = 2n_1 \cos\theta_1/[n_2 \cos\theta_1 + n_1 \cos\theta_2] \qquad (5.38)$$

Collected together, Γ_E, T_E, Γ_H and T_H are often known as the *Fresnel coefficients*. Most interestingly Γ_H becomes zero at a particular angle of incidence θ_1. This occurs when

$$n_2 \cos\theta_1 = n_1 \cos\theta_2 \qquad (5.39)$$

We can find a suitable angle θ_1 for Eq. (5.39) to hold, as follows. Squaring both sides of the equation and using Snell's Law, we obtain

$$n_2^2 \cos^2\theta_1 = n_1^2 [1 - (n_1^2/n_2^2) \sin^2\theta_1] \qquad (5.40)$$

or

$$n_2^2 \cos^2\theta_1 + (n_1^4/n_2^2) \sin^2\theta_1 = n_1^2 \qquad (5.41)$$

The $\cos^2$ term above can now be eliminated by combining it with the $\sin^2$ term. If this is done we get

$$n_2^2 [\cos^2\theta_1 + \sin^2\theta_1] + [(n_1^4 - n_2^4)/n_2^2] \sin^2\theta_1 = n_1^2 \qquad (5.42)$$

so that

$$[(n_1^4 - n_2^4)/n_2^2] \sin^2\theta_1 = n_1^2 - n_2^2 \qquad (5.43)$$

We therefore find that

$$\sin^2\theta_1 = n_2^2/(n_1^2 + n_2^2) \qquad (5.44)$$

so the relevant angle is given by

$$\theta_1 = \sin^{-1}[n_2/\surd(n_1^2 + n_2^2)] \qquad (5.45)$$

This is often written in the alternative form:

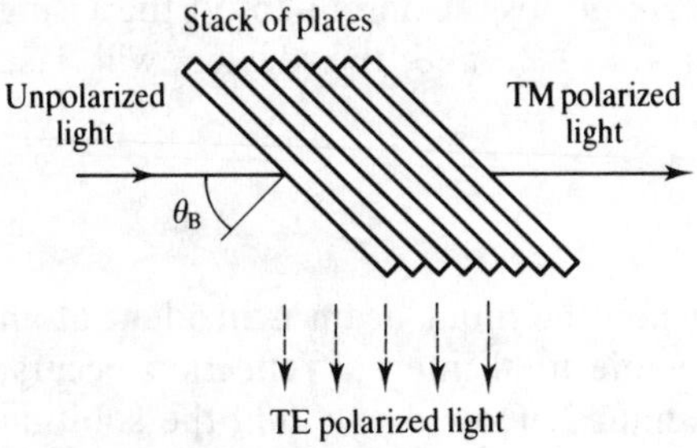

Figure 5.8 The pile-of-plates polarizer.

$$\theta_1 = \tan^{-1}(n_2/n_1) \tag{5.46}$$

The reduction of the reflection coefficient to zero for TM incidence is called the *Brewster effect*, after David Brewster (1781–1868), and is often exploited in the design of polarizing components. Similarly the angle defined in Eq. (5.46) is known as the *Brewster angle* θ_B. This can be calculated for a glass/air interface (for which $n_1 = 1.5$ and $n_2 = 1.0$) as $\theta_B \approx 33.7°$. Note that the Brewster effect does not occur for TE incidence.

Figure 5.8 shows one way in which the Brewster effect may be used to make a polarizer. A number of identical plates are piled up and oriented so that an incident, unpolarized beam strikes the stack at the Brewster angle. Any TE polarization component in the beam will be diminished by reflection at each interface, while the TM component will pass through unaffected. After many such reflections the output beam will therefore be largely TM polarized. This is known as the *pile-of-plates polarizer*. Though simple, it has been entirely superseded by dichroic polarizers.

Figure 5.9 shows a much more effective use of the Brewster effect, inside the resonant cavity of a He-Ne laser. We have previously described the required optical layout in Chapter 4 and concentrate here on the provision of optical gain by an electrically-excited gas plasma which is contained inside a glass tube. In lasers that emit unpolarized light the plasma tube is simply capped with end-plates that are orthogonal to the bore axis. In polarized lasers the tube is sealed with windows slanted at the Brewster angle (known, unsurprisingly, as *Brewster windows*). TE polarized light must suffer some reflection loss at these windows. However, because

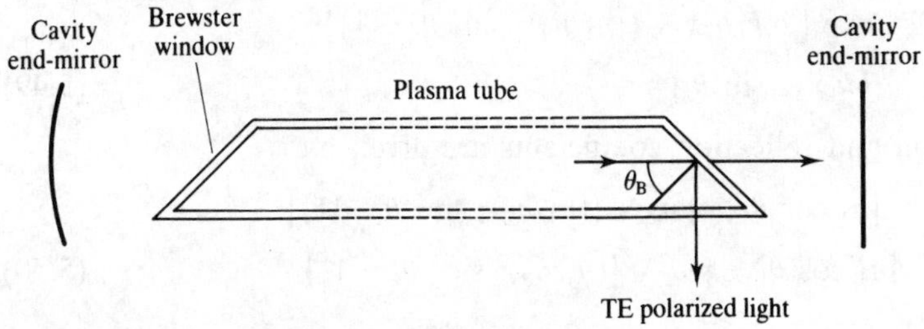

Figure 5.9 Brewster windows on a laser plasma tube.

of the Brewster effect, TM polarized light suffers no loss. Consequently, the lasing threshold for TM light is much lower than for TE, and the device will lase preferentially in a TM-polarized mode.

Total Internal Reflection

We now consider the case when $\theta_1 > \theta_c$, i.e. when the input beam is incident at an angle greater than the critical angle. In this regime total internal reflection occurs, so that a propagating wave cannot arise in medium 2. How do we find the solution now? There is nothing wrong with the analysis we have used so far, but we cannot satisfy the Snell's law equation, $n_1 \sin \theta_1 = n_2 \sin \theta_2$. Or rather, we cannot satisfy it with a real value of θ_2. It turns out that we can satisfy it with a complex value instead. This might seem a bizarre idea, but there is no need to dwell on the physical implications involved; it is simplest just to consider this as a mathematical trick. The important thing is that we can now calculate $\cos \theta_2$ for $\theta_1 > \theta_c$ as

$$\cos \theta_2 = \pm \mathrm{j} \sqrt{\{(n_1/n_2)^2 \sin^2 \theta_1 - 1\}} \qquad (5.47)$$

This identity will allow us to repeat the previous calculation under the new conditions.

TE incidence Once again we will begin with the case when the electric field vector is parallel to the interface. Returning to the solution in medium 2, $E_{y2} = E_T \exp[-\mathrm{j}\, k_0 n_2 (z \sin \theta_2 - x \cos \theta_2)]$, and inserting the value for the cosine given in Eq. (5.47), we find

$$E_{y2} = E_T \exp[\pm\, k_0 n_2 x \sqrt{\{(n_1/n_2)^2 \sin^2 \theta_1 - 1\}}] \exp[-\mathrm{j}\, k_0 n_1 z \sin \theta_1] \qquad (5.48)$$

The solution no longer has the form of a plane wave propagating at an angle θ_2 because its x-dependence is not a complex exponential. Instead, it is a wave whose amplitude distribution varies exponentially in the x-direction, and that propagates along the interface in the z-direction. This is called an *evanescent wave*. Clearly the amplitude must decay to zero for large negative x, so we will keep only the positive sign in Eq. (5.48), which corresponds to the negative sign in Eq. (5.47).

What does the complete field distribution look like now? We now know that

$$E_{y1} = E_I \{\exp[+\mathrm{j} k_0 n_1 x \cos \theta_1] + \Gamma_E \exp[-\mathrm{j}\, k_0 n_1 x \cos \theta_1]\} \cdot \exp[-\mathrm{j}\, k_0 n_1 z \sin \theta_1]$$

$$E_{y2} = E_I \mathrm{T}_E \exp[+\, k_0 n_2 x \sqrt{\{(n_1/n_2)^2 \sin^2 \theta_1 - 1\}}] \cdot \exp[-\mathrm{j}\, k_0 n_1 z \sin \theta_1] \qquad (5.49)$$

where the transmission and reflection coefficients are given by

$$\Gamma_E = [n_1 \cos \theta_1 + \mathrm{j}\, n_2 \sqrt{\{(n_1/n_2)^2 \sin^2 \theta_1 - 1\}}] / [n_1 \cos \theta_1 - \mathrm{j}\, n_2 \sqrt{\{(n_1/n_2)^2 \sin^2 \theta_1 - 1\}}] \qquad (5.50)$$

and

$$\mathrm{T_E} = 2n_1 \cos\theta_1/[n_1 \cos\theta_1 - \mathrm{j}\, n_2 \sqrt{\{(n_1/n_2)^2 \sin^2\theta_1 - 1\}}] \tag{5.51}$$

These expressions are rather long-winded. However, we can simplify them by defining three new parameters β, κ and γ as

$$\begin{aligned} \beta &= k_0 n_1 \sin\theta_1 \\ \kappa &= \sqrt{(n_1^2 k_0^2 - \beta^2)} \\ \gamma &= \sqrt{(\beta^2 - n_2^2 k_0^2)} \end{aligned} \tag{5.52}$$

in which case Eq. (5.49) becomes

$$\begin{aligned} E_{y1} &= E_\mathrm{I}\{\exp[+\mathrm{j}\kappa x] + \Gamma_\mathrm{E} \exp[-\mathrm{j}\kappa x]\} \exp[-\mathrm{j}\beta z]\} \\ E_{y2} &= E_\mathrm{I}\,\mathrm{T_E} \exp[+\gamma x] \exp[-\mathrm{j}\beta z] \end{aligned} \tag{5.53}$$

and Eqs (5.50) and (5.51) simplify to

$$\begin{aligned} \Gamma_\mathrm{E} &= [\kappa + \mathrm{j}\,\gamma]/[\kappa - \mathrm{j}\,\gamma] \\ \mathrm{T_E} &= 2\kappa/[\kappa - \mathrm{j}\gamma] \end{aligned} \tag{5.54}$$

The reflection coefficient Γ_E is therefore now a complex number in the form $\Gamma_\mathrm{E} = z/z^*$, where z is also a complex number and * denotes the operation of complex conjugation. It is easy to show that the *power reflection coefficient* $|\Gamma_\mathrm{E}|^2$ is unity, since

$$|\Gamma_\mathrm{E}|^2 = \Gamma_\mathrm{E}\Gamma_\mathrm{E}^* = (z/z^*)(z^*/z) = 1 \tag{5.55}$$

This implies that all of the incident power is reflected, which is consistent with our earlier qualitative discussion of total internal reflection. It might be argued that a non-zero transmission coefficient is at odds with a unity power reflection coefficient. The evanescent wave will then have a finite amplitude, and must therefore carry some power. However, it is easy to show that it does not carry any power normal to the interface, but only parallel to it. Power conservation is therefore not violated after all.

Since Γ_E is a complex number with modulus unity, it can be written in complex exponential form as

$$\Gamma_\mathrm{E} = \exp(\mathrm{j}2\phi) \tag{5.56}$$

where the phase term ϕ is given by

$$\tan\phi = \gamma/\kappa \tag{5.57}$$

Similarly, the transmission coefficient $\mathrm{T_E}$ can be written as

$$\mathrm{T_E} = 2\cos(\phi)\exp(\mathrm{j}\phi) \tag{5.58}$$

We can obtain entirely analogous results for the TM transmission and reflection coefficients when total internal reflection occurs. The phase shift is different, however; this time we get $\tan\phi = (n_1^2/n_2^2)\,\gamma/\kappa$.

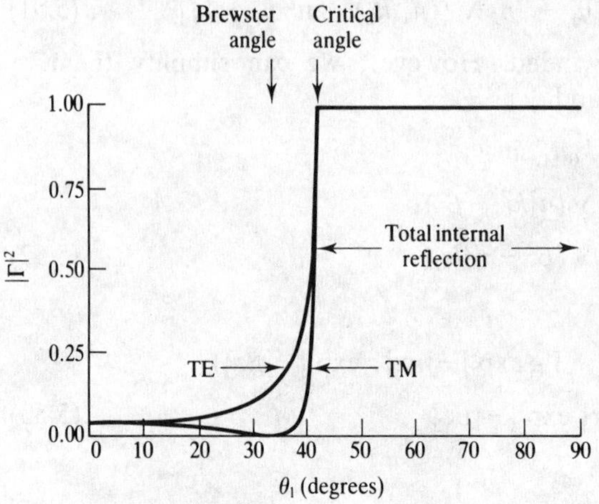

Figure 5.10 Power reflection coefficients for a glass/air interface.

Design Example

We can combine the solutions obtained so far and plot the power reflectivity $|\Gamma|^2$ versus angle over the complete range of incidence. Figure 5.10 shows the curves for TE and TM incidence, and the particular example of a glass/air interface (where $n_1 = 1.5$ and $n_2 = 1$). The reflectivities are the same for normal incidence, when $|\Gamma|^2 = 0.2^2 = 0.04$. In each case the reflectivity is unity after the critical angle, $\theta_1 = 41.8°$. Notice, however, that it falls to zero for TM incidence alone, at the Brewster angle (33.7°).

The Transverse Field Distribution Near the Interface

Knowing the transmission and reflection coefficients, we can now rewrite the field distributions in the two media as

$$\begin{aligned} E_{y1} &= 2E_1 \exp(\mathrm{j}\phi)\cos(\kappa x - \phi)\exp(-\mathrm{j}\beta z) \\ &= E\cos(\kappa x - \phi)\exp(-\mathrm{j}\beta z) \end{aligned} \tag{5.59}$$

and

$$\begin{aligned} E_{y2} &= 2E_1 \exp(\mathrm{j}\phi)\cos(\phi)\exp(\gamma x)\exp(-\mathrm{j}\beta z) \\ &= E\cos(\phi)\exp(\gamma x)\exp(-\mathrm{j}\beta z) \end{aligned} \tag{5.60}$$

Both fields again have the form $E_y = E(x)\exp(-\mathrm{j}\beta z)$, so the pattern corresponds to a non-uniform wave travelling in the z-direction, parallel to the interface. More specifically, it is a *standing-wave pattern* in medium 1, and a decaying or evanescent wave in medium 2. Near the interface the transverse field function $E(x)$ is therefore as shown in Fig. 5.11. Notice that the transverse field is continuous at the boundary, as is its first derivative. This follows from application of the boundary matching conditions. It is also worth noting that as β tends to n_2k_0, γ tends to zero, implying

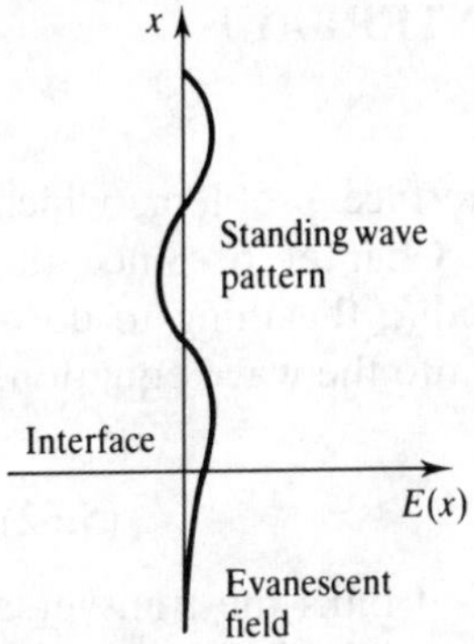

Figure 5.11 The transverse field distribution near an interface.

that, as the angle of incidence approaches the critical angle, the decay of the evanescent field gets slower. The confinement of the field is therefore less good and it extends further and further into medium 2. If we wish to avoid this, the incidence angle should be chosen to lie suitably far away from the critical angle.

The results are qualitatively similar for TM incidence; we still get standing waves in medium 1 and an evanescent wave in medium 2. However, the details at the boundary are slightly different. The main change is that the transverse electric field distribution is no longer continuous.

The Reflectivity of Metals

We conclude with a brief discussion of the reflectivity of metals, which we assumed without much justification to be high in Chapter 4. The analysis developed above will still be valid provided merely that we insert the correct expression for refractive index. In Chapter 3 we found that (neglecting collision damping losses) a metal may be described at optical frequencies as a medium with a large, negative value of ε_r'. Consequently the refractive index of a metal will be almost purely imaginary, given by $n \approx jn'$, where $n' = \sqrt{|\varepsilon_r'|}$.

At normal incidence, for example, the reflection coefficient at the interface between a dielectric (of refractive index n_1) and a metal (of index jn_2') may then be found from Eq. (5.30) as

$$\Gamma_E \approx [n_1 - jn_2']/[n_1 + jn_2'] \tag{5.61}$$

Once again Γ_E is a complex number, of the form $\Gamma_E = z/z^*$. We therefore conclude without further ado that the power reflectivity will be 100 per cent, justifying our previous assumptions. In practice, collision damping losses reduce this figure somewhat, but (as we know from our bathrooms) even the most basic metal layer will act as a good reflector. Furthermore, if $n_2 \gg n_1$ (as is the case for good conductors), the reflection coefficient Γ_E is ≈ -1. This means the total electric field at the interface (i.e. the sum of the incident and reflected waves) must be close to zero, providing a simple justification for the heuristic boundary condition adopted in Chapter 4.

5.5 MODAL TREATMENT OF THE DIELECTRIC INTERFACE PROBLEM

We now consider an easier route to the solution to the interface problem, which will prove a useful tool for the analysis of waveguides in Chapter 6. Since the solution has the form $E_y(x,z) = E(x) \exp[-\mathrm{j}\beta z]$ in both media, the thing to do is to assume this at the outset. Substituting this solution directly into the wave equation, Eq. (5.18), we get

$$\mathrm{d}^2E/\mathrm{d}x^2 + [n^2k_0^2 - \beta^2]\,E = 0 \tag{5.62}$$

This type of equation is known as a *waveguide equation*, and it links the transverse field $E(x)$ with the propagation constant β. In this case it is a standard second-order differential equation, of the form:

$$\mathrm{d}^2E/\mathrm{d}x^2 + C^2\,E = 0 \tag{5.63}$$

Note that its solutions are sines and cosines if $C^2 > 0$, and exponentials if $C^2 < 0$.

For our particular geometry, the waveguide equations we must solve are:

$$\begin{aligned} \mathrm{d}^2E_1/\mathrm{d}x^2 + [n_1^2k_0^2 - \beta^2]\,E_1 &= 0 \qquad \text{in medium 1} \\ \mathrm{d}^2E_2/\mathrm{d}x^2 + [n_2^2k_0^2 - \beta^2]\,E_2 &= 0 \qquad \text{in medium 2} \end{aligned} \tag{5.64}$$

where E_1 and E_2 are the transverse field distributions in the two media. Knowing (as we do) the exact solution, the sensible thing is to choose a trial solution in the same form. We therefore take

$$\begin{aligned} E_1 &= E\cos(\kappa x - \phi) \\ E_2 &= E'\exp(\gamma x) \end{aligned} \tag{5.65}$$

where κ and γ are as in Eq. (5.52) and E, E' and ϕ are unknown constants. Once again we can determine their values by boundary matching. The first of our original boundary conditions was that $E_{y1} = E_{y2}$ at the interface; for the transverse fields this implies that $E_1 = E_2$ on $x = 0$, so

$$E\cos\phi = E' \tag{5.66}$$

The second condition was that $\partial E_{y1}/\partial x = \partial E_{y2}/\partial x$ at the interface; for the transverse fields this implies that $\mathrm{d}E_1/\mathrm{d}x = \mathrm{d}E_2/\mathrm{d}x$ on $x = 0$. Performing the necessary differentiation we get

$$-\kappa E\sin(-\phi) = \gamma E' \tag{5.67}$$

Dividing Eq. (5.67) by Eq. (5.66) we obtain

$$\tan\phi = \gamma/\kappa \tag{5.68}$$

Since this is identical to Eq. (5.57), we have obtained exactly the same results as before without invoking the Fresnel coefficients at all. A solution of this type is called a *modal solution*.

5.6 SURFACE PLASMA WAVES

We have become accustomed to solutions to interface problems that involve either propagating waves on both sides of the interface, or propagating waves on one side and evanescent ones on the other. It is now reasonable to ask whether other possibilities exist. Can we, for example, have a solution that is evanescent on both sides of an interface? It turns out that this is indeed possible, and is of engineering significance.

The geometry involved is the interface between a metal (most often, silver or gold) and a dielectric, as shown in Fig. 5.12. Because the wave is evanescent on both sides, it is effectively confined to the interface, and is therefore a surface wave. Its correct title is a *surface plasma wave* or *surface plasmon* since propagation in the metal layer involves plasma oscillations of the free electrons inside the metal, but this is not important to our electromagnetic analysis. What is important, however, is that surface plasma waves can be shown to exist only for TM polarization.

We shall tackle the problem using the modal approach above. First we recall that at optical frequencies a metal can be adequately modelled as a dielectric, but with a relative permittivity that is large and negative. The methods we have used so far will therefore still prove suitable. It is sensible to perform the analysis in terms of the magnetic field since this is always oriented in the same direction. In this case the relevant time-independent wave equation is

$$\nabla^2 \mathbf{H} + \varepsilon_r k_0^2 \mathbf{H} = 0 \qquad (5.69)$$

Assuming that the magnetic field is parallel to the y-axis, a suitable modal solution has the form:

$$\mathbf{H} = H(x) \exp(-\mathrm{j}\beta z)\,\mathbf{j} \qquad (5.70)$$

where $H(x)$ is the transverse variation of the magnetic field and β is the propagation constant. Substituting this solution into Eq. (5.69) we obtain the following waveguide equation:

$$\mathrm{d}^2H/\mathrm{d}x^2 + [\varepsilon_r k_0^2 - \beta^2]\,H = 0 \qquad (5.71)$$

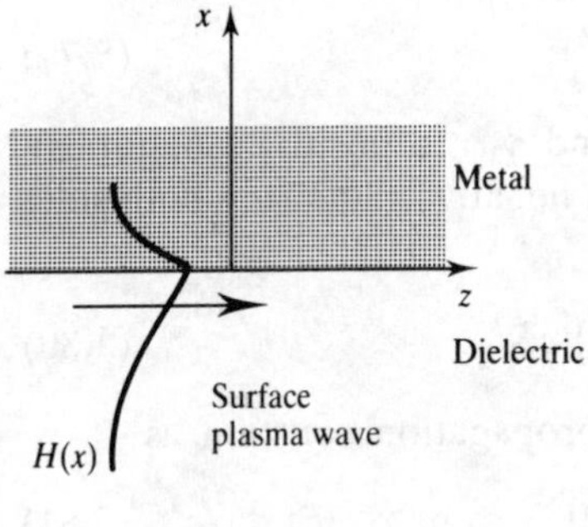

Figure 5.12 Geometry for surface plasma waves.

To solve the plasmon problem we simply assume a relative dielectric constant of ε_{r1} in medium 1 (the metal layer), and ε_{r2} in medium 2 (the dielectric). The equations we must solve are therefore

$$\mathrm{d}^2H_1/\mathrm{d}x^2 + [\varepsilon_{r1}k_0^2 - \beta^2]\,H_1 = 0 \qquad \text{in medium 1}$$
$$\mathrm{d}^2H_2/\mathrm{d}x^2 + [\varepsilon_{r2}k_0^2 - \beta^2]\,H_2 = 0 \qquad \text{in medium 2} \tag{5.72}$$

Assuming exponential decay of the transverse magnetic field on either side of the interface, the solutions in each region can then be specified as

$$H_1(x) = H\exp(-\gamma_1 x)$$
$$H_2(x) = H'\exp(\gamma_2 x) \tag{5.73}$$

where H and H' are arbitrary constants, and γ_1 and γ_2 are given by

$$\gamma_1 = \sqrt{(\beta^2 - \varepsilon_{r1}k_0^2)}$$
$$\gamma_2 = \sqrt{(\beta^2 - \varepsilon_{r2}k_0^2)} \tag{5.74}$$

As before, the boundary conditions are that the tangential components of the magnetic and electric fields must match at the interface, which lies at $x = 0$. Matching the magnetic field (which is wholly tangential) we see simply that $H' = H$. To match the electric fields we must first evaluate **E**. This is best done using the curl relation between **H** and **E**:

$$\nabla \times \mathbf{H} = \mathrm{j}\omega\varepsilon\,\mathbf{E} \tag{5.75}$$

Given that $H_x = H_z = 0$, we can then find the individual components of **E** as

$$E_x = (+\mathrm{j}/\omega\varepsilon)\,\partial H_y/\partial z\,; \qquad E_y = 0\,; \qquad E_z = (-\mathrm{j}/\omega\varepsilon)\,\partial H_y/\partial x \tag{5.76}$$

Since the only tangential component of the electric field is E_z, the second condition will be satisfied if

$$(1/\varepsilon_{r1})\,\partial H_{y1}/\partial x = (1/\varepsilon_{r2})\,\partial H_{y2}/\partial x \qquad \text{on } x = 0 \tag{5.77}$$

For the modal solution we have assumed, this condition requires

$$(1/\varepsilon_{r1})\,\mathrm{d}H_1/\mathrm{d}x = (1/\varepsilon_{r2})\,\mathrm{d}H_2/\mathrm{d}x \qquad \text{on } x = 0 \tag{5.78}$$

Performing the necessary differentiation we then find that

$$-\gamma_1/\varepsilon_{r1} = \gamma_2/\varepsilon_{r2} \tag{5.79}$$

Although Eq. (5.79) looks insoluble (given that γ_1 and γ_2 are positive quantities), it can be satisfied in this particular case, because ε_{r1} is negative. Squaring both sides we obtain

$$[\beta^2 - \varepsilon_{r1}k_0^2]/\varepsilon_{r1}^2 = [\beta^2 - \varepsilon_{r2}k_0^2]/\varepsilon_{r2}^2 \tag{5.80}$$

After some rearrangement we may then extract the propagation constant as

$$\beta = k_0\sqrt{[\varepsilon_{r1}\varepsilon_{r2}/(\varepsilon_{r1} + \varepsilon_{r2})]} \tag{5.81}$$

Knowing ε_{r1} and ε_{r2}, we may therefore calculate β. The transverse magnetic field distribution is then much as shown in Fig. 5.12. The field decays exponentially on both sides of the interface, but the decay constant is different in each medium. Because of the sign of ε_{r1}, Eq. (5.74) shows that the rate of decay in the metal is much larger than that in the dielectric. As a result, the light hardly penetrates the metal, and the major part of the field extends into the dielectric.

The analysis above is an extremely simple demonstration that a single interface may actually 'guide' an optical wave. However, to be more realistic, one major modification must be made to our model. As we saw in Chapter 3, the dielectric constant of a metal is actually complex (because of collision damping), so ε_{r1} really has the form $\varepsilon_{r1} = \varepsilon'_{r1} - \mathrm{j}\,\varepsilon''_{r1}$. If this expression is used in Eq. (5.81) it is found that the propagation constant is also complex, so the plasmon mode decays as it propagates. In real materials this is very significant, and plasmon modes have an extremely short range.

Design Example

We shall now calculate the range of a surface plasmon in a typical material, silver. In this case the data in Fig. 3.4 shows that $\varepsilon'_{r1} \approx -16.9$ and $\varepsilon''_{r1} \approx 0.55$ at $\lambda_0 = 0.633\ \mu$m. Assuming that the surrounding dielectric is air, for which $\varepsilon_r = 1, \beta$ is given by

$$\beta = (2\pi/0.633 \times 10^{-6}) \times \surd[(-16.9 - \mathrm{j}\,0.55)/(-15.9 - \mathrm{j}\,0.55)]$$

$$= 10.2 \times 10^{6} - \mathrm{j}\,1.05 \times 10^{4} \tag{5.82}$$

The actual form of the plasmon wave is given by Eq. (5.70). Splitting the propagation constant into real and imaginary parts we can write

$$\beta = \beta_r - \mathrm{j}\beta_i \tag{5.83}$$

With this notation Eq. (5.70) then becomes

$$\mathbf{H} = H(x)\exp(-\mathrm{j}\beta_r z)\exp(-\beta_i z)\,\mathbf{j} \tag{5.84}$$

This shows that the mode amplitude will decay to 1/e of its initial value in a distance z_e, given by

$$z_e = 1/\beta_i \tag{5.85}$$

We may take z_e to be representative of the range of the plasmon mode. Inserting the relevant value from Eq. (5.82), we then obtain a range of $z_e \approx 100\ \mu$m, a very small figure. In practice, other effects reduce the range still further, making a surface plasmon wave useless for communication. Nonetheless, one highly important application has been found for optical plasmons: sensing.

Sensing with Surface Plasma Waves

We have already mentioned that because of the imbalance between γ_1 and γ_2, the vast proportion of the field extends into the dielectric. As a result, it is possible to show that the value of β_r is highly sensitive to the exact nature of the dielectric. If this medium is made in the form of a sensing layer with a dielectric constant that can be altered by some parameter of interest, a variation in that parameter will then result in a detectable change in the plasmon propagation constant.

The experimental apparatus needed to measure this change is extremely simple. It is known as the *prism coupler* and will be described in detail in Chapter 7. For the present we merely note that many sensor materials have already been tested for suitability. The most promising application area is biosensing, where it is hoped that surface plasmon devices will provide a range of extremely cheap disposable medical sensors (for example, for measuring enzyme levels).

PROBLEMS

5.1 (a) Find the critical angle for an interface between a high-index glass ($n_1 = 1.7$) and air ($n_2 = 1$). (b) What is the corresponding Brewster angle, if the incident beam travels in the glass?
[(a) 36.03°; (b) 30.46°]

5.2 The critical angle for an interface between two media is 30°. Find the transmission coefficient, for normal incidence from the high-index side of the interface.
[1.333]

5.3 Find the power reflectivity Γ^2 and transmissivity T^2, for TE incidence at $\theta_1 = 30°$ on an interface between air ($n_1 = 1$) and glass ($n_2 = 1.5$).
[0.0577; 0.577]

5.4 The figures below show the definitions of the reflection and transmission coefficients at a dielectric interface. In the left-hand figure, incidence is from medium 1, and the coefficients are Γ and T, respectively. In the right-hand figure, incidence is from medium 2, and the corresponding values are Γ' and T'. Using only the fact that light rays are reversible, show that (a) $\Gamma = -\Gamma'$, and (b) $\Gamma^2 + TT' = 1$.

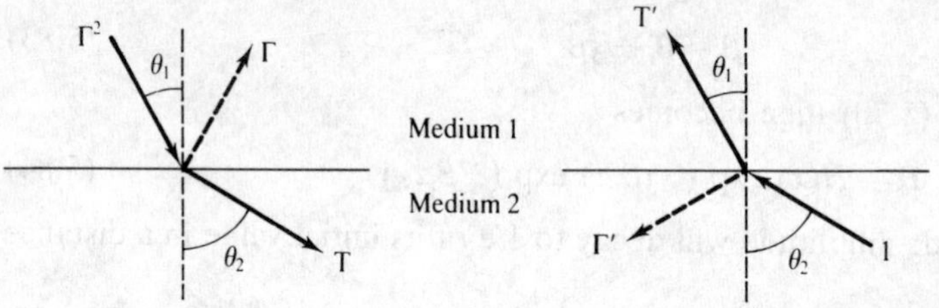

5.5 Verify the results of Question 5.4 analytically, using the Fresnel coefficients.

5.6 The figure below shows a three-layer dielectric stack. Media 1 and 2 are semi-infinite, with refractive indices n_1 and n_2, respectively; medium 3 has thickness d, and index n_3. The effect of the stack on a plane optical wave may be analysed by summing the contributions from all possible paths between the input and the output, using the transmission and reflection coefficients to calculate the path amplitudes and taking account of any optical phase changes. For example, the reflectivity may be found by summing all paths starting and ending in medium 1. Path 1 simply involves reflection at the first interface. The amplitude for this is Γ_{13}, where Γ_{13} is the reflection coefficient at the boundary between media 1 and 3. Path 2 involves transmission through the first interface, reflection at the second, and transmission through the first interface again; the amplitude for this is $T_{13}\Gamma_{32}T_{31} \exp(-j2k_0n_3d)$, where T_{13} and T_{31} are transmission coefficients at the first interface, and Γ_{32} is the reflection coefficient at the second.

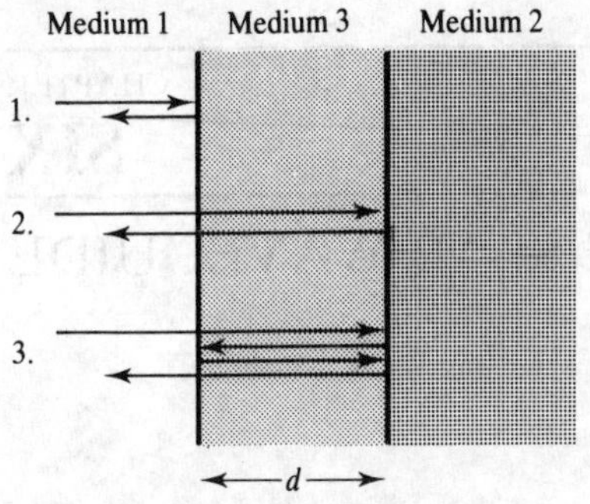

Write down the amplitude for path 3. Generalize your results to form an infinite series for the reflectivity of the stack, and sum the series. Hence show that there will be no reflection at all, if the refractive index of medium 3 is $n_3 = \sqrt{(n_1 n_2)}$ and if its thickness d represents a quarter of a wavelength in the medium. What application might this have?

5.7 Derive the Fresnel coefficients for TM incidence on a dielectric interface.

5.8 The figure below shows a plane wave incident from air on a glass slab of finite thickness. Show that there is no reflection from *either* interface, for TM incidence at the Brewster angle θ_B.

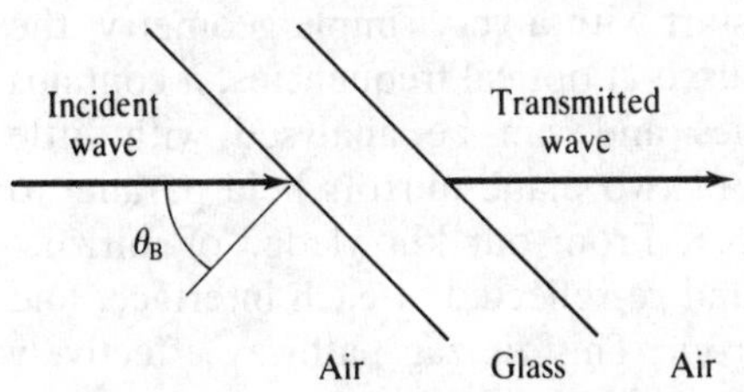

5.9 Show (a) analytically, and (b) numerically, using data from Question 5.3, that the Fresnel coefficients for TE incidence satisfy the relation: $\Gamma^2 + T^2 (n_2/n_1) [\cos \theta_2/\cos \theta_1] = 1$.

5.10 Interpret the result of Question 5.9 in terms of power conservation.

SUGGESTIONS FOR FURTHER READING

Flanagan, M. T., and Pantell R. H. Surface plasmon resonance and immunosensors, *Elect. Lett.*, **20**, 968–970, 1984.

Hecht, E., "Optics", 2nd ed., Chapter 4, Addison-Wesley Publishing Co., Reading MA., 1987.

Kretschmann, E. The determination of the optical constants of metals by the excitation of surface plasmons, *Z. Phys.*, **214**, 313–324, 1971.

Liedberg, B., Nylander, C., and Lundström, I. Surface plasmon resonance for gas detection and biosensing, *Sensors and Actuators*, **4**, 299–304, 1983.

Otto, A. Excitation of nonradiative surface plasma waves by the method of frustrated total reflection, *Z. Phys.*, **216**, 398–410, 1968.

Ramo, S., Whinnery, J. R., and van Duzer, T. "Fields and Waves in Communication Electronics", 2nd ed., Chapter 6, John Wiley and Sons, New York, 1984.

CHAPTER

SIX

THE SLAB WAVEGUIDE

6.1 GUIDED WAVES IN A METAL GUIDE

It is now time to consider more effective guiding structures for electromagnetic waves. To introduce the basic ideas, we shall start with a very simple geometry, the *metal-walled guide*. Though this is not actually used at optical frequencies, it contains nearly all the features of more useful guides and can be analysed with little mathematics. The metal-walled guide consists of two plane mirrors held parallel to one another in free-space, as shown in Fig. 6.1. From our knowledge of mirrors, we would expect plane waves to be reflected and re-reflected at each interface, and hence bounce up and down between the mirrors. This zig-zag pathway effectively results in travel only in the z-direction, so the field is guided down the z-axis.

We must begin by working out what kind of field can exist in the guide. First, we recall that the time-independent field due to a y-polarized plane wave travelling at an angle θ to the z-axis (either upwards or downwards) is

$$E_y = E \exp\left(-jk_0\{z \sin\theta \pm x \cos\theta\}\right) \tag{6.1}$$

Since we can guess that the field inside the guide is a combination of an upward- and a downward-travelling wave, we will assume the solution:

$$E_y = E_+ \exp\left(-jk_0\{z \sin\theta + x \cos\theta\}\right) + E_- \exp\left(-jk_0\{z \sin\theta - x \cos\theta\}\right) \tag{6.2}$$

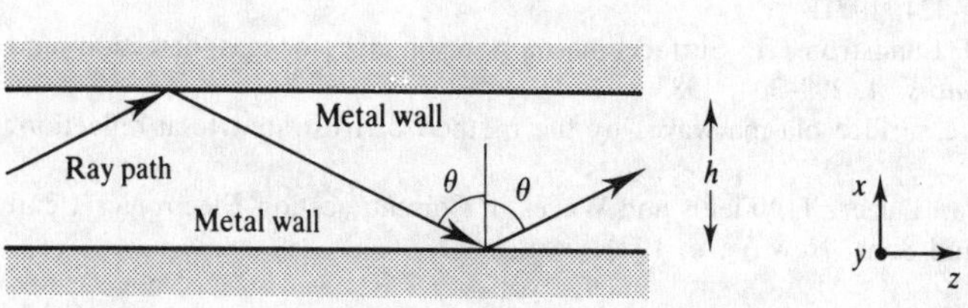

Figure 6.1 A metal-walled waveguide.

where E_+ and E_- are unknown constants. We must satisfy the usual set of boundary conditions at any interfaces. If we adopt the approximations used in Chapters 3, 4 and 5 for a good conductor, these reduce to a requirement that the electric field vanishes at the metal walls (so that $E_y = 0$ at $x = 0$ and $x = h$). We can satisfy the first condition if $E_- = -E_+$, i.e. if the solution has the form:

$$E_y = E \sin(k_0 x \cos\theta) \exp(-jk_0 z \sin\theta) \tag{6.3}$$

The solution is therefore a wave with a sinusoidal amplitude envelope, and which travels in the z-direction. It is called a *guided mode* because it has its energy confined within the guide walls. Note that the field can once again be written in the standard form:

$$E_y = E(x) \exp(-j\beta z) \tag{6.4}$$

where $E(x)$ is the *transverse field distribution* and β is the *propagation constant*. Clearly in this case β is given by

$$\beta = k_0 \sin\theta \tag{6.5}$$

so the smaller the ray angle inside the guide, the smaller the propagation constant. The exact value of θ (and thus of β) is fixed by the second boundary condition, namely

$$\sin(k_0 h \cos\theta) = 0 \tag{6.6}$$

Equation (6.6) is satisfied whenever

$$2k_0 h \cos\theta = 2\nu\pi \qquad (\text{where } \nu = 1,2,\ldots) \tag{6.7}$$

This is called the *eigenvalue equation* of the guide. It may be viewed as a condition for *transverse resonance* since it implies that the round-trip phase accumulated by bouncing up and down between the walls must be a whole number of multiples of 2π.

Each solution corresponds to a particular *guided mode*, defined by the *mode index* ν. However, in general, only a fixed number of modes can be supported by the guide. For example, if $\nu\pi/k_0 h > 1$, there is no solution, so the condition for a guided mode of order ν to exist is

$$h > \nu\pi/k_0 \tag{6.8}$$

Consequently if the guide width h is too small, no modes are supported all. This occurs when

$$h < \pi/k_0 \qquad \text{or} \qquad h < \lambda_0/2 \tag{6.9}$$

This puts a lower limit on the usable value of h of half the optical wavelength. If h is slightly bigger, just one mode is supported and the guide is then called *single-moded*. If h is bigger still, a finite number of modes can propagate, so the guide is *multi-moded*. Any mode which cannot propagate is described as being *cut off*. The higher the order of the mode, the smaller the value of θ, and, exactly at cut-off,

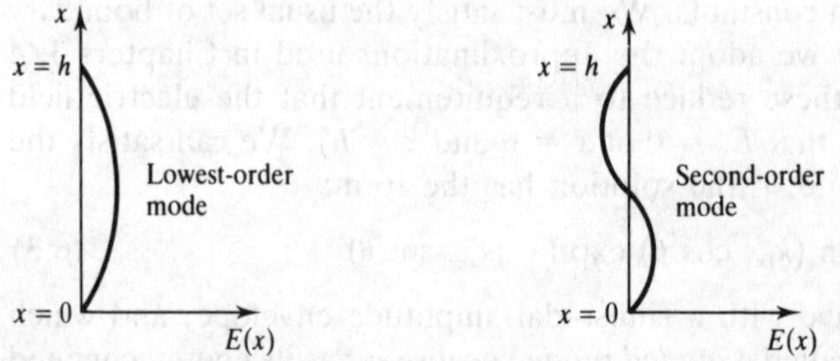

Figure 6.2 Transverse field distributions of the two lowest-order modes in a metal-walled guide.

the ray angle is zero, so at this point the rays just bounce up and down between the metal walls, making no progress down the guide.

Each guided mode has its own particular transverse field distribution. The field patterns of the first two modes are shown in Fig. 6.2. Note that the lowest-order mode has no sign reversals, while the second-order mode has one, and so on.

The metal-walled guide is very effective at microwave frequencies when the conductivity of the walls is high and the reflectivity is correspondingly good. However, at optical frequencies, σ is lower, causing unacceptable propagation loss. Fortunately an alternative guiding mechanism is available, based on total internal reflection at a dielectric interface.

6.2 GUIDED WAVES IN A SLAB DIELECTRIC WAVEGUIDE

Figure 6.3 shows a more useful structure, a *dielectric waveguide*. This consists of three layers of dielectric: layer 1 (which has thickness h, and refractive index n_1), and layers 2 and 3 (which are both semi-infinite, and which have indices n_2 and n_3 respectively). We shall assume that $n_1 > n_2$ and $n_1 > n_3$, so that total internal reflection can occur at each interface. Often layer 1 is referred to as the *guiding layer*, while layers 2 and 3 are the *substrate* and the *cover layer*. A guide of this type can be formed simply by depositing a high-index guiding layer onto a polished substrate—the cover can then often be air. Because of this common geometry, it is usual to describe the structure as an *asymmetric slab guide*, and take $n_1 > n_2 > n_3$.

We can represent the refractive index variation using a discontinuous function $n(x)$, as shown in Fig. 6.4(a). Other smoother profiles can result from using different

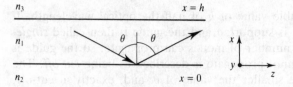

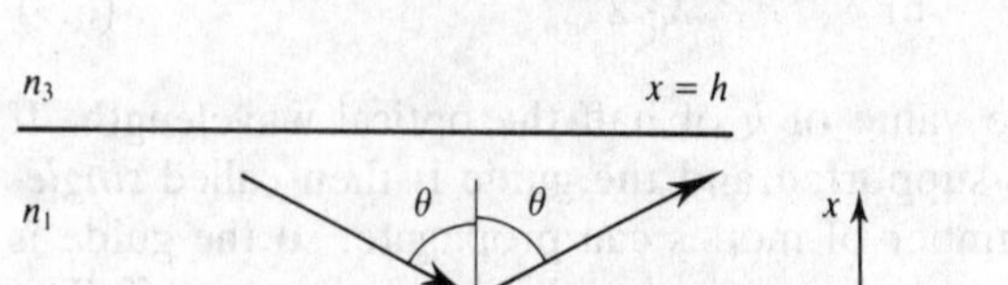

Figure 6.3 A slab dielectric waveguide.

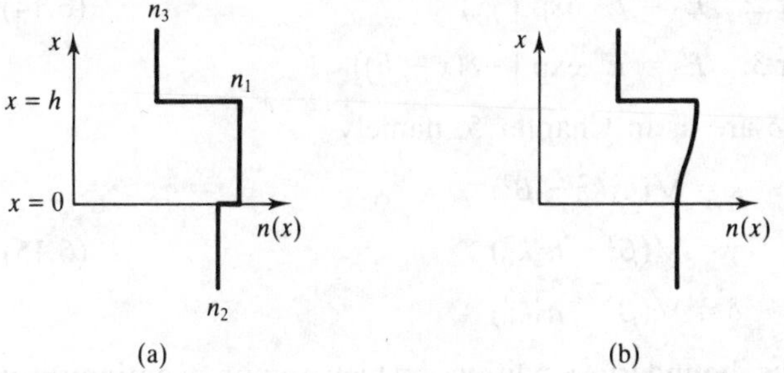

Figure 6.4 Refractive index profiles for asymmetric 1-D waveguides.

fabrication methods; for example, Fig. 6.4(b) shows the graded-index profile obtained when a guide is made by diffusion of metal atoms into a crystalline substrate. In this case the index distribution below the crystal surface is a Gaussian function (as will be shown in Chapter 13), given by

$$n(x) = n_2 + (n_1 - n_2) \exp [-(x - h)^2/L_D^2] \quad \text{for } x < h \qquad (6.10)$$

where L_D is a constant related to the diffusion conditions, the *diffusion length*. However, because many of these methods are also used to fabricate more complicated channel guide structures, we shall postpone their detailed discussion and stay with the simple slab guide for the time being.

Through total internal reflection, waves may bounce to and fro between the guide walls much as before. However, the form of solution used above—combinations of plane waves—rapidly becomes too complicated when used with dielectric guides. Instead, we shall base the analysis on the modal solution introduced in Chapter 5, and merely look for solutions moving at constant speed in the z-direction. For simplicity we assume y-polarization once again, so that the mode is *transverse electric*, or *TE*. In each layer the scalar wave equation we must solve is therefore given by

$$\nabla^2 E_{yi}(x,z) + n_i^2 k_0^2 E_{yi}(x,z) = 0 \quad (i = 1,2,3) \qquad (6.11)$$

while the solutions can be written in the form:

$$E_{yi}(x,z) = E_i(x) \exp(-\mathrm{j}\beta z) \quad (i = 1,2,3) \qquad (6.12)$$

The waveguide equation (which links the transverse field E_i with the propagation constant β) is then given in each layer by

$$\mathrm{d}^2E_i/\mathrm{d}x^2 + [n_i^2k_0^2 - \beta^2]\, E_i = 0 \qquad (6.13)$$

Since we are interested in fields that are confined within the guide (which will be standing waves inside the guiding layer, and evanescent fields outside) we will assume the following trial solutions:

$$\text{in layer 1:} \quad E_1 = E \cos(\kappa x - \phi)$$

$$\text{in layer 2:} \quad E_2 = E' \exp(\gamma x) \tag{6.14}$$
$$\text{in layer 3:} \quad E_3 = E'' \exp[-\delta(x - h)]$$

where the constants κ, γ and δ are as in Chapter 5, namely

$$\kappa = \sqrt{(n_1^2 k_0^2 - \beta^2)}$$
$$\gamma = \sqrt{(\beta^2 - n_2^2 k_0^2)} \tag{6.15}$$
$$\delta = \sqrt{(\beta^2 - n_3^2 k_0^2)}$$

Once more we must satisfy the boundary conditions, which require continuity of $E_i(x)$ and its gradient dE_i/dx at each interface. Matching fields at $x = 0$ gives the answer we found in Chapter 5, namely

$$E' = E \cos \phi \tag{6.16}$$

while matching the field gradients gives:

$$\gamma E' = -\kappa E \sin(-\phi) \tag{6.17}$$

Dividing Eq. (6.17) by Eq. (6.16) we may eliminate the field amplitudes from the problem and obtain a closed-form expression for ϕ:

$$\tan \phi = \gamma/\kappa, \quad \text{or} \quad \phi = \tan^{-1}(\gamma/\kappa) \tag{6.18}$$

Similarly, matching fields at $x = h$ gives

$$E'' = E \cos(\kappa h - \phi) \tag{6.19}$$

while matching field gradients gives

$$-\delta E'' = -\kappa E \sin(\kappa h - \phi) \tag{6.20}$$

Dividing Eq. (6.20) by Eq. (6.19) we obtain

$$\tan(\kappa h - \phi) = \delta/\kappa \tag{6.21}$$

Self-consistent solutions are only possible when Eqs (6.16)–(6.21) are all satisfied simultaneously. The most important point is that the trigonometric relations in Eqs (6.18) and (6.21) are both satisfied. Once this is done the field amplitudes E' and E'' can be found in terms of E, simply by substituting values for ϕ and κ into Eqs (6.16) and (6.19).

We can reduce the two main equations to a single one, as follows. Using the standard trigonometrical identity

$$\tan(A - B) = (\tan A - \tan B)/(1 + \tan A \tan B) \tag{6.22}$$

we can convert the left-hand side of Eq. (6.21) to

$$\tan(\kappa h - \phi) = [\tan(\kappa h) - \tan \phi]/[1 + \tan(\kappa h) \tan \phi] \tag{6.23}$$

Substituting the value of $\tan \phi$ from Eq. (6.18) we then obtain

$$[\tan(\kappa h) - \gamma/\kappa]/[1 + \tan(\kappa h)\,\gamma/\kappa] = \delta/\kappa \tag{6.24}$$

Finally, after some rearrangement we get

$$\tan(\kappa h) = \kappa[\gamma + \delta]/[\kappa^2 - \gamma\delta] \tag{6.25}$$

This is the eigenvalue equation for the dielectric guide. Once again it can be shown that only certain values of β can satisfy it, so this guide will also only support a discrete set of guided modes. Since the parameters κ, γ and δ are all functions of the propagation constant β, the eigenvalue equation is obviously a function of β as well. However, it is a transcendental equation (which means the solution cannot be written in closed form), so the β-values must be found numerically. This is not difficult, but before we examine the solutions themselves we will examine the connection between the eigenvalue equation and the transverse resonance condition mentioned previously.

The Transverse Resonance Condition

We first recall the definition of the parameter ϕ from our discussion of total internal reflection in Chapter 5. Parameter ϕ first appeared when the reflection coefficient Γ_E was written in the form:

$$\Gamma_E = \exp(j2\phi) \tag{6.26}$$

Since this is a complex exponential, -2ϕ can be interpeted as a phase shift between the incident and reflected waves (note the negative sign here). We can therefore define two new parameters, $-2\phi_{12}$ and $-2\phi_{13}$, for our three-layer guide. These are the shifts that occur on reflection from the interfaces between layers 1 and 2, and layers 1 and 3, respectively. Because ϕ_{12} can be identified with our previous parameter ϕ, we may write $\tan(\phi_{12}) = \gamma/\kappa$. By analogy, we can also put $\tan(\phi_{13}) = \delta/\kappa$. Equation (6.21) can therefore be written as

$$\tan(\kappa h - \phi_{12}) = \tan(\phi_{13}) \tag{6.27}$$

Taking the inverse tangent of both sides of Eq. (6.27), allowing for the periodic nature of the tan function (which repeats every π radians), and multiplying both sides by a factor of two, we get

$$2\kappa h - 2\phi_{12} - 2\phi_{13} = 2\nu\pi \tag{6.28}$$

Equation (6.28) is now a transverse resonance condition, much like Eq. (6.7); however, there are some differences. First, the term κh has replaced $k_0 h \sin\theta$. This is not a real change, since it is easy to show that $\kappa = k_0 n_1 \sin\theta$; the new term therefore merely reflects the fact that the guiding layer is a medium of refractive index n_1. Second, the phase terms $-2\phi_{12}$ and $-2\phi_{13}$ have been introduced. The overall interpretation of Eq. (6.28) is therefore that the total phase accumulated in bouncing between the guide walls, including the phase changes experienced on reflection, must be a whole number of multiples of 2π. The astute reader will be able to explain why this extra feature (the phase changes due to reflection) did not apparently show up in our analysis of the metal-walled guide.

The Transverse Field Patterns

Once the values of β are found, the transverse field patterns can be drawn. It is easiest to do this for a symmetric guide, which can only support modes with symmetric or anti-symmetric field patterns. In this case it can be shown that the eigenvalue equation Eq. (6.25) reduces to

$$\tan(\kappa h/2) = \gamma/\kappa \tag{6.29}$$

for all modes with symmetric fields, and

$$\tan(\kappa h/2) = -\kappa/\gamma \tag{6.30}$$

for all anti-symmetric modes. If Eqs (6.29) and (6.30) are solved numerically, the following results are obtained. At low optical frequencies (when λ_0 is large and k_0 is low) the guide is *single-moded* and only symmetric patterns of the type shown in Fig. 6.5 can be supported. Note that the field varies cosinusoidally inside the guide and exponentially outside, as expected. At extremely low frequencies a significant amount of power propagates in the evanescent field. A mode of this type is described as poorly-confined. As the frequency rises, however, the field concentrates more towards the centre of the guide. We shall see later that this improved confinement of the mode is advantageous since it reduces the propagation loss that occurs when the waveguide is formed into a bend. As the frequency rises further, the guide becomes two-moded and a second, anti-symmetric field solution becomes possible. This is shown in Fig. 6.6.

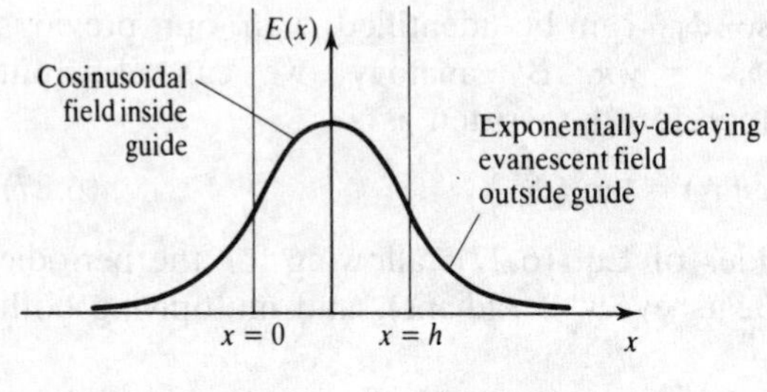

Figure 6.5 The lowest-order symmetric mode of a slab waveguide.

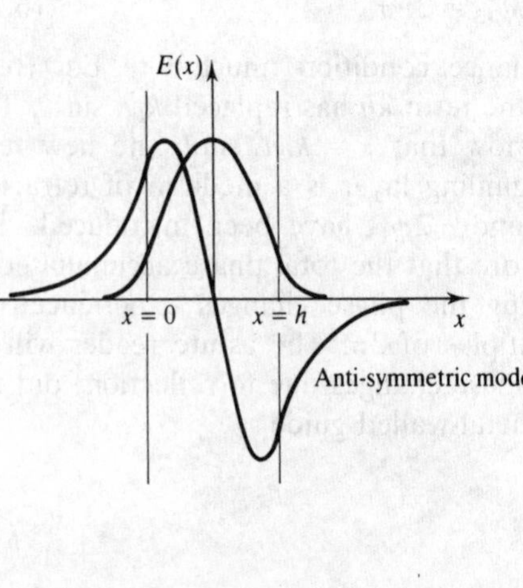

Figure 6.6 Mode patterns for a two-moded slab guide.

Cut-off Conditions

At higher frequencies still, even more patterns are possible. We can work out how many modes can propagate at any given frequency as follows. Particular modes cease to be guided when the ray angle inside the guide tends to the critical angle (note that this is different from the metal-walled guide, where $\theta_{\text{cut-off}} = 0$). Since $\beta = n_1 k_0 \sin\theta$, and Snell's law requires $n_1 \sin\theta = n_2$ at $\theta = \theta_c$, this implies that $\beta \to n_2 k_0$ at cut-off. At the same time we can show that $\gamma \Rightarrow 0$, so the confinement of the field tends to zero at this point. For symmetric modes the cut-off condition then reduces to

$$\tan(\kappa h/2)_{\text{c.o.}} = 0 \tag{6.31}$$

and hence

$$(\kappa h/2)_{\text{c.o.}} = 0, \pi, 2\pi, \ldots \tag{6.32}$$

Similarly, for anti-symmetric modes, the cut-off condition is that

$$(\kappa h/2)_{\text{c.o.}} = \pi/2, 3\pi/2, \ldots \tag{6.33}$$

So in general we must have

$$(\kappa h/2)_{\text{c.o.}} = \nu\pi/2 \tag{6.34}$$

where ν is again the mode number. This is the cut-off condition for all the modes in a symmetric slab guide. However, it can be expressed more conveniently with a little further manipulation, as we now show. Since $\kappa = \surd(n_1^2 k_0^2 - \beta^2)$, then at cut-off we must have

$$\kappa_{\text{c.o.}} = k_0 \surd(n_1^2 - n_2^2) \tag{6.35}$$

so that

$$(\kappa h/2)_{\text{c.o.}} = (k_0 h/2) \surd(n_1^2 - n_2^2) \tag{6.36}$$

This parameter is often called the *'V' value* of the guide; it is a dimensionless number, which can be used to characterize a guide—generally a guide will be heavily multi-moded if $V \gg 1$. Combining Eqs (6.34) and (6.36) we get

$$(k_0 h/2) \surd(n_1^2 - n_2^2) = \nu\pi/2 \tag{6.37}$$

This represents the cut-off condition for all possible guided modes in a guide of width h, constructed from dielectric layers of refractive indices n_1 and n_2. We can use it as follows: for the lowest order mode (corresponding to $\nu = 0$), the cut-off condition is satisfied when

$$(k_0 h/2) \surd[n_1^2 - n_2^2] = 0 \tag{6.38}$$

Since h, n_1 and n_2 are all finite, it follows that k_0 must be zero to satisfy Eq. (6.38). Effectively therefore, there is no cut-off for the lowest-order mode; this is a particular property of the symmetric slab guide. Similarly, for the second lowest-order mode ($\nu = 1$) cut-off occurs when

$$(k_0 h/2) \surd[n_1^2 - n_2^2] = \pi/2 \tag{6.39}$$

and so on. We can also use the cut-off condition to work out the dimensions needed for a guide to be single-moded, given the guide indices and the optical wavelength. In this case we require the second-order mode to be just cut off. Rearranging Eq. (6.39) slightly we then find that

$$h < \lambda_0/\{2\sqrt{(n_1^2 - n_2^2)}\} \tag{6.40}$$

There is therefore a strict upper limit on the allowable width of the guide.

Design Example

We can now put in some example numbers to illustrate typical dimensions of single-mode guides. We choose $\lambda_0 = 0.633\ \mu m$, corresponding to the He-Ne laser wavelength, and guide indices of $n_1 \approx n_2 \approx 1.5$; these are typical for different types of glass. With the relatively large index difference of $n_1 - n_2 = 0.01$, we then get $h < 1.8\ \mu m$. For the same parameters, but with the smaller index difference of $n_1 - n_2 = 0.001$, we get $h < 5.8\ \mu m$. Single-mode guides are clearly small in practical materials!

Phase Velocity, Group Velocity and Waveguide Dispersion

Naturally enough, we will be concerned with the information-carrying capacity of waveguides. This will prove especially important for optical fibres, which we will discuss in Chapter 8. Our previous analysis of plane waves (in Chapter 2) showed that dispersion occurs whenever v_{ph} (or ω/k) varies with frequency. The analogous quantity can be found for the ν th guided mode, simply by inserting the relevant value for the propagation constant. In this case we get

$$v_{ph\nu} = \omega/\beta_\nu \tag{6.41}$$

It is also often convenient to define an *effective index* $n_{eff\nu}$ for the mode, such that the phase velocity is given by

$$v_{ph\nu} = c/n_{eff\nu} \tag{6.42}$$

A moment's thought then shows that the effective index and the propagation constant are related by

$$\beta_\nu = 2\pi n_{eff\nu}/\lambda_0 \tag{6.43}$$

For a given mode, the effective index may therefore be interpreted as the refractive index of an equivalent bulk medium which would give identical values for the phase velocity and propagation constant to those obtained in the guide.

Once more the dispersion characteristics of modes can be represented by a graph of ω against β. For the symmetric slab guide it is as shown in Fig. 6.7. Here the lowest-order mode has been labelled the TE_0 mode, the second lowest the TE_1 mode, and so on.

Note that each characteristic lies between two lines which have slopes of c/n_1 and c/n_2, respectively, so that the phase velocity of any mode cannot be less than

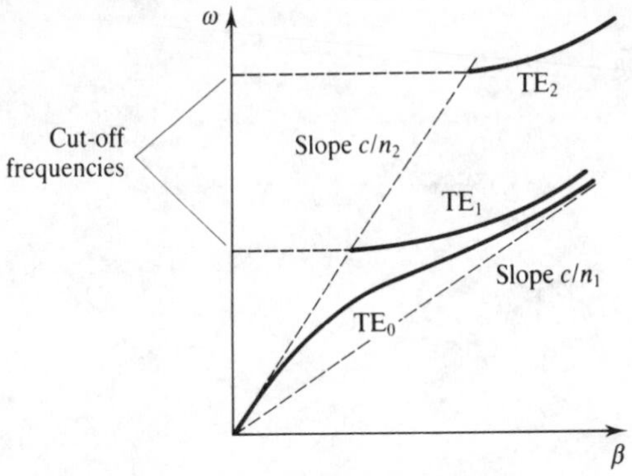

Figure 6.7 Dispersion diagram for the symmetric slab guide.

c/n_1 or greater than c/n_2. The former value is approached if the mode is well-confined, when most of the field is travelling inside the guide (layer 1, which has refractive index n_1). The latter is approached as the mode tends towards cut-off, when most of the field is travelling outside the guide (in a medium of index n_2). Consequently the effective index of all guided modes must lie between n_1 and n_2.

Since the phase velocity clearly depends on frequency (and also on the mode number), the guide must be dispersive. For each mode we may define a group velocity $v_{g\nu}$ as

$$v_{g\nu} = d\omega/d\beta_\nu \tag{6.44}$$

In general the group velocity will also depend on frequency and mode number. Note that this effect (known as *waveguide dispersion*) is distinct from the material dispersion discussed in Chapters 2 and 3, which for simplicity was ignored in Fig. 6.7; it forms an extra contribution to the overall dispersion, and one which may be extremely large. Often the two components of waveguide dispersion are referred to separately, as *inter-* and *intra-modal* dispersion, respectively. The former refers to the variation in v_g between the different modes, the latter to changes in a particular mode's group velocity with frequency. We shall return to both aspects in Chapter 8, which is concerned with optical fibres.

6.3 OTHER TYPES OF MODE

The modal solutions found so far are, of course, descriptive of light confined inside the guide. However, further solutions must exist which account for light propagating *outside* the guide. These are known as *radiation modes*, and we now describe the way they are calculated. Remember that we solved Eq. (6.13) to the find the TE modes of a slab guide. This is, of course, a general second-order differential equation, which can be written in the form:

$$d^2E_i/dx^2 + C_i^2 E_i = 0 \tag{6.45}$$

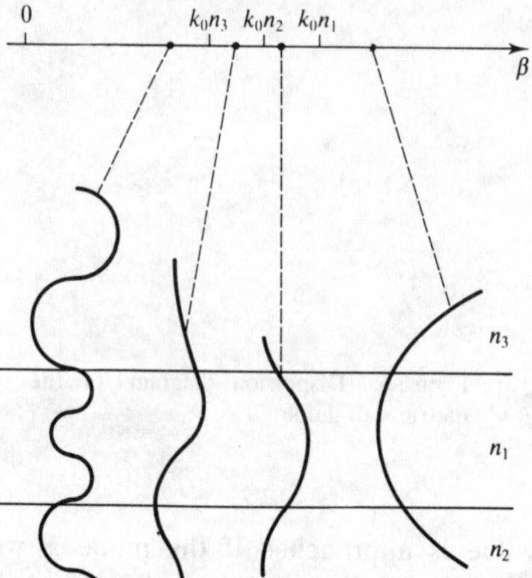

Figure 6.8 Modal solutions represented in β-space.

In our previous analysis the form of the solution was exponential or sinusoidal, depending on the sign of the term $C_i^2 = (n_i^2k_0^2 - \beta^2)$. If we consider all possible values of β, it turns out that a wider range of solutions can be found. If we again take $n_1 > n_2 > n_3$, which is often the case in real guides, the complete set can be represented as a diagram in β-space as shown in Fig. 6.8. The essential features of the diagram are

1. For $\beta > k_0n_1$, the solutions are exponential in all three layers. Since this implies infinite field amplitudes at large distances from the guide (which are physically unrealistic) we will ignore these solutions for the time being.
2. For $k_0n_1 > \beta > k_0n_2$, there are a discrete number of bound or guided modes, which are the solutions already found. These vary cosinusoidally inside the guide core and decay exponentially outside the guide.
3. For $k_0n_2 > \beta > k_0n_3$ the solutions vary exponentially in the cover (layer 3) and cosinusoidally in both the guide (layer 1) and substrate (layer 2). Since these fully penetrate the substrate region, they are called *substrate modes*. Any value of β is allowed (between the two limits given above) so the set forms a continuum.
4. For $k_0n_3 > \beta$, solutions vary cosinusoidally in all three layers. These particular field patterns are known as *radiation modes*. Once again any value of β is allowed in the range above, so the set forms another continuum.

The introduction of radiation modes suggests that we should alter our previous dispersion diagram to account for them. Equally, we should allow for the possibility of *backward-travelling modes* (which have negative values of β). Both modifications are simple to carry out, and the complete set of modes for a symmetric slab guide

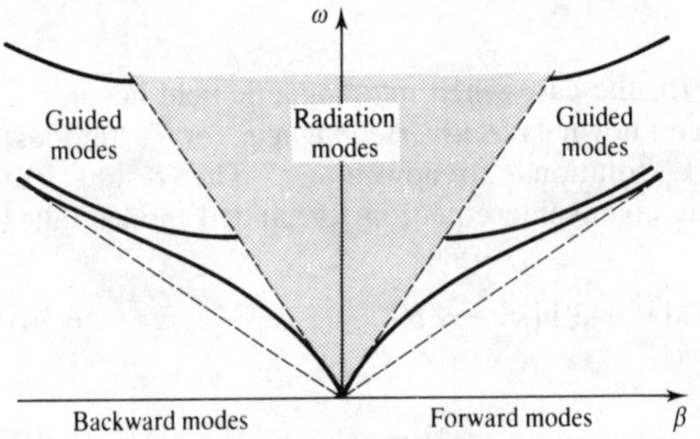

Figure 6.9 Full dispersion diagram for the symmetric slab guide.

is then as shown in the modified ω–β diagram of Fig. 6.9. We will find this representation useful in the solution of coupled mode problems in Chapter 10.

The Magnetic Field Patterns

Having found the complete set of all possible solutions for the electric field, it is simple to obtain the corresponding magnetic field patterns. This gives a self-consistent, vectorial solution for the entire guided electromagnetic field. First, recall that the TE solution for **E** is a y-polarized guided mode, written as $\mathbf{E} = \mathbf{j}E(x) \exp(-\mathrm{j}\beta z)$. From the curl relation between **E** and **H**, we can show that the associated magnetic field has two components, given by

$$H_x = (1/\mathrm{j}\omega\mu_0)\, \partial E_y/\partial z = (-\beta/\omega\mu_0)\, E(x) \exp(-\mathrm{j}\beta z) \tag{6.46}$$

and

$$H_z = (-1/\mathrm{j}\omega\mu_0)\, \partial E_y/\partial x = (-1/\mathrm{j}\omega\mu_0)\, \mathrm{d}E/\mathrm{d}x \exp(-\mathrm{j}\beta z) \tag{6.47}$$

Unlike the electric field, the magnetic field has a longitudinal component (i.e. a component in the direction of propagation) as well as a transverse one.

The Vectorial Representation of Modes

It is often convenient to represent both the fields in terms of vectorial transverse fields. Thus for the ν th mode we would write

$$\mathbf{E} = \mathbf{E}_\nu \exp(-\mathrm{j}\beta_\nu z) \text{ and } \mathbf{H} = \mathbf{H}_\nu \exp(-\mathrm{j}\beta_\nu z) \tag{6.48}$$

where $\mathbf{E}_\nu$ and $\mathbf{H}_\nu$ are vectorial descriptions of the transverse fields. This notation will prove useful in some of the later proofs in this section. We may find expressions for $\mathbf{E}_\nu$ and $\mathbf{H}_\nu$ for TE modes by comparing Eq. (6.48) with the results above. If this is done we get:

$$\mathbf{E}_\nu = \mathbf{j}E_\nu \quad \text{and} \quad \mathbf{H}_\nu = \mathbf{i}\,(-\beta/\omega\mu_0)\, E_\nu + \mathbf{k}\,(-1/\mathrm{j}\omega\mu_0)\, \mathrm{d}E_\nu/\mathrm{d}x \tag{6.49}$$

We can see from Eq. (6.49) that (unlike the transverse electric field, which is real) the transverse magnetic field is complex, since $H_{\nu z}$ is imaginary.

TM Modes

A similar analysis can be performed for the case when the magnetic field lies in the y-direction. The field patterns are then known as *transverse magnetic*, or *TM* modes, and are complementary to the TE solutions already found. The results are qualitatively similar, but this time the eigenvalue equation for guided modes takes the form:

$$\tan(\kappa h) = \kappa[\gamma' + \delta']/[\kappa^2 - \gamma'\delta'] \tag{6.50}$$

where

$$\gamma' = \gamma[n_1^2/n_2^2] \quad \text{and} \quad \delta' = \delta[n_1^2/n_3^2] \tag{6.51}$$

Broadly speaking, the behaviour of TM modes is similar to that of TE modes, however, detailed analysis shows that the field profiles no longer have the smooth, continuous form found earlier. Similarly, the detailed shape of the dispersion diagram is slightly different because of the refractive index ratios that appear in Eq. (6.51). However, if the index differences are small (so that the ratios approach unity) the two are practically identical. This implies that a useful simplification can be made in this case, known as the *weak-guidance approximation*.

6.4 THE WEAK-GUIDANCE APPROXIMATION

The vectorial analysis used so far is very complicated, and a simpler approach is often desirable. This is especially true for the two-dimensional refractive index distributions that occur in channel guides or optical fibres. Luckily it turns out that if the index difference Δn forming the guide is small, a scalar approximation can be used instead. Since this is often the case in practical guides, the approximation is a very useful one. It is based on the observation that, for a small value of Δn, the critical angle θ_c at a single interface is very large. Hence for a guided mode, the ray directions inside the guide are almost parallel to the axis of propagation, as shown in Fig. 6.10.

Since the total field inside the guide may be considered as a summation of similar bouncing waves, the combined E- and H-vectors must be almost exactly orthogonal to each other, and to the axis of propagation. This will be the case anyway for our TE mode, assuming that dE/dx is small in Eq. (6.47) (the physical justification being that the field inside a weak guide must be a fringe pattern of

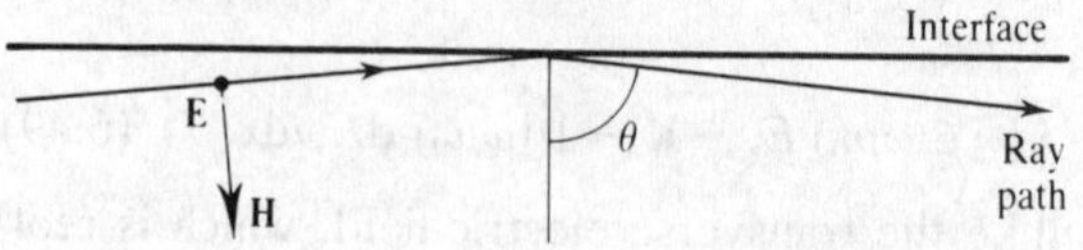

Figure 6.10 Field directions for total internal reflection at a low-Δn interface.

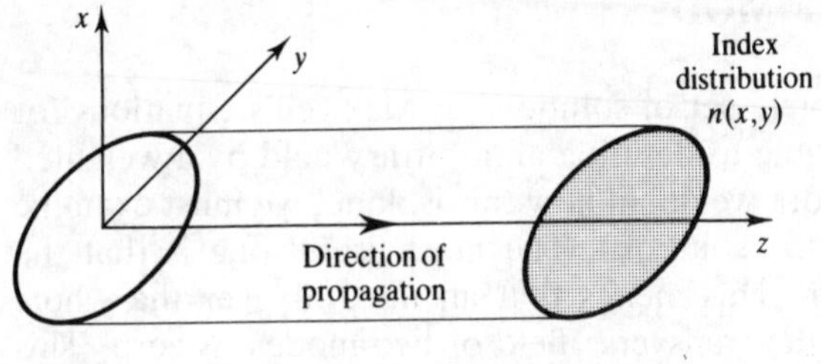

Figure 6.11 A two-dimensional guide.

large periodicity). The resulting modes are known as *TEM modes* (implying that both the electric and the magnetic field components are approximately transverse).

As it turns out, the following analysis can also be used to treat two-dimensional guides of arbitrary cross-section, provided the index changes involved are small. They may also be gradual changes rather than the discrete steps we have encountered so far. To illustrate this, let us consider a guide formed by a 2-D index distribution $n(x,y)$ lying in the z-direction as shown in Fig. 6.11.

If we pick one field component to work with (say **E**), then this can point in either the x- or the y-direction. For example, let us suppose that it points in the y-direction. The scalar amplitude $E_y(x,y,z)$ then satisfies (at least, approximately) the three-dimensional scalar wave equation:

$$\nabla^2 E_y(x,y,z) + n^2(x,y)k_0^2\, E_y(x,y,z) = 0 \tag{6.52}$$

We now assume a modal solution in the usual form, namely as a product of a transverse electric field distribution and an exponential propagation term. This time the transverse field must also be two-dimensional, so we take

$$E_y(x,y,z) = E(x,y)\ \exp\,(-\mathrm{j}\beta z) \tag{6.53}$$

where $E(x,y)$ is the transverse electric field. Substituting into Eq. (6.52) we obtain the following waveguide equation linking E with β:

$$\nabla^2_{x,y}\, E(x,y) + [n^2(x,y)k_0^2 - \beta^2]\, E(x,y) = 0 \tag{6.54}$$

Here the subscripts on the Laplacian indicate that differentiation is to be performed with respect to x and y only. Though Eq. (6.54) is very similar to our previous waveguide equation, it is also valid for 2-D guides. Although several standard techniques exist for solving this type of equation, we will postpone any attempt at a solution until Chapter 8. For now we simply note that a complete set of two-dimensional modes—both guided and radiation modes—can exist. However, we can assume directly that the transverse *magnetic* field points in the x-direction, and that its amplitude can be found from that of the transverse electric field as:

$$H(x,y) = -(\beta/\omega\mu_0)\, E(x,y) \tag{6.55}$$

The two transverse fields are therefore related by a simple scaling factor.

6.5 THE ORTHOGONALITY OF MODES

Since the modes above form a complete set of solutions to Maxwell's equations for a waveguide geometry, we must be able to describe an arbitrary field by a weighted summation of all types of mode. Before we detail how this is done, we must examine some further general properties of the solutions. The most useful one is that the modes are *orthogonal* to each other. This means that an integral over the whole guide cross-section, of a product of the transverse field of two modes, is zero. This can be proved using the power conservation theorem, as we now show.

The power propagating in a guide can be found by integrating the normal component of the time-averaged Poynting vector over the guide cross-section. For a guide lying in z-direction, we obtain

$$P_z = 1/2 \text{ Re}\left[\iint_A (\mathbf{E} \times \mathbf{H}^*) \cdot \mathbf{k} \, dx \, dy\right] \tag{6.56}$$

Since power is conserved in a lossless system, this implies that $dP_z/dz = 0$, or

$$d/dz\left[1/4 \iint_A (\mathbf{E} \times \mathbf{H}^* + \mathbf{E}^* \times \mathbf{H}) \cdot \mathbf{k} \, dx \, dy\right] = 0 \tag{6.57}$$

where we have used the identity Re $(z) = 1/2[z + z^*]$. Let us now assume that any field can be written as a sum of modes. For simplicity we restrict the expansion to forward-travelling guided modes (although generally we should include backward-travelling and radiation modes as well). Suppose that the μ th mode has transverse fields $\mathbf{E}_\mu$ and $\mathbf{H}_\mu$, and propagation constant β_μ. We can then write

$$\mathbf{E} = \sum_\mu a_\mu \mathbf{E}_\mu \exp(-j\beta_\mu z) \quad \text{and} \quad \mathbf{H} = \sum_\mu a_\mu \mathbf{H}_\mu \exp(-j\beta_\mu z) \tag{6.58}$$

Here the terms a_μ are *expansion coefficients*, which describe the amplitude of each mode contributing to the total field. Substituting these expressions into Equation 6.52, we get

$$d/dz\left[1/4 \iint_A \sum_{\mu,\nu} \{a_\mu a_\nu^* \exp[j(\beta_\nu - \beta_\mu)z]\mathbf{E}_\mu \times \mathbf{H}_\nu^* + a_\nu^* a_\mu \exp[j(\beta_\nu - \beta_\mu)z]\mathbf{E}_\nu^* \times \mathbf{H}_\mu\} \cdot \mathbf{k} \, dx \, dy\right] = 0 \tag{6.59}$$

or:

$$\sum_{\mu,\nu} a_\mu a_\nu^* \, j(\beta_\nu - \beta_\mu) \exp[j(\beta_\nu - \beta_\mu)z] \iint_A [\mathbf{E}_\mu \times \mathbf{H}_\nu^* + \mathbf{E}_\nu^* \times \mathbf{H}_\mu] \cdot \mathbf{k} \, dx \, dy = 0 \tag{6.60}$$

The coefficients a_μ and a_ν^* are arbitrary, and also the propagation constants of different modes are unequal, so that $\beta_\mu \neq \beta_\nu$. The only way Eq. (6.60) can be satisfied for all z is if

$$\iint_A (\mathbf{E}_\mu \times \mathbf{H}_\nu^* + \mathbf{E}_\nu^* \times \mathbf{H}_\mu) \cdot \mathbf{k}\, dx\, dy = 0 \tag{6.61}$$

This is called the *orthogonality condition*. It is also possible to show (by including backward-travelling modes as well) that each of the two parts of the integral are themselves zero, so that

$$\iint_A (\mathbf{E}_\mu \times \mathbf{H}_\nu^*) \cdot \mathbf{k}\, dx\, dy = 0 \tag{6.62}$$

We will now show how this condition is used in practice, using a number of examples.

Design Example 1

We begin with the case of TE modes in a slab guide, for which there is clearly no need to integrate in the y-direction. The orthogonality condition therefore reduces to

$$\int_{-\infty}^{\infty} (\mathbf{E}_\mu \times \mathbf{H}_\nu^*) \cdot \mathbf{k}\, dx = 0 \tag{6.63}$$

or, in terms of the relevant vectorial components:

$$\int_{-\infty}^{\infty} -E_{\mu y} H_{\nu x}^*\, dx = 0 \tag{6.64}$$

Substituting for $H_{\nu x}$, and assuming that $E_{\mu y}$ is real, we then obtain

$$\int_{-\infty}^{\infty} (E_{\mu y} E_{\nu y})\, dx = 0 \tag{6.65}$$

In other words, the integral of the product of two different modes, when taken over the cross-section of the guide, is zero. This is confirmed by Fig. 6.6, which shows the field distributions for the lowest-order symmetric and antisymmetric modes in a symmetric guide. Clearly the integral of the product of these two functions must be zero.

Design Example 2

We can obtain a similar result for TEM modes in a weak two-dimensional graded-index guide by using the relation between the transverse electric and magnetic fields given in Eq. (6.55). Substituting into Eq. (6.62) and making no assumption that the fields are real, we get

$$\iint_A (E_{\mu y} E_{\nu y}^*)\, dx\, dy = 0 \tag{6.66}$$

In both cases described above, the orthogonality condition is often written in shorthand as

$$\langle E_{\mu y}, E_{\nu y} \rangle = 0 \tag{6.67}$$

The bracket notation employed here is often used in quantum mechanics. The operation of multiplying two functions together and integrating over a range is called an *inner product*.

6.6 THE POWER CARRIED BY A MODE

We can use the orthogonality relation to define the time-averaged power carried down the guide by each mode. From the previous section we know that the total power flow P_z in the z-direction is given by

$$P_z = 1/4 \iint_A \sum_{\mu,\nu} \{a_\mu a_\nu^* \exp[\mathrm{j}(\beta_\nu - \beta_\mu)z]\, \mathbf{E}_\mu \times \mathbf{H}_\nu^* \\ + a_\nu^* a_\mu \exp[\mathrm{j}(\beta_\nu - \beta_\mu)z]\, \mathbf{E}_\nu^* \times \mathbf{H}_\mu\} \cdot \mathbf{k}\, \mathrm{d}x\, \mathrm{d}y \tag{6.68}$$

However, from the orthogonality relation we also know that many of the terms in Eq. (6.68) must be zero. Removing these gives

$$P_z = \sum_\mu a_\mu a_\mu^* \left[1/4 \iint_A \{\mathbf{E}_\mu \times \mathbf{H}_\mu^* + \mathbf{E}_\mu^* \times \mathbf{H}_\mu\} \cdot \mathbf{k}\, \mathrm{d}x\, \mathrm{d}y \right] \tag{6.69}$$

We can now simplify this expression by defining a *modal power coefficient* P_μ, such that

$$P_\mu = 1/4 \iint_A [\mathbf{E}_\mu \times \mathbf{H}_\mu^* + \mathbf{E}_\mu^* \times \mathbf{H}_\mu] \cdot \mathbf{k}\, \mathrm{d}x\, \mathrm{d}y \tag{6.70}$$

in which case we obtain

$$P_z = \sum_\mu a_\mu a_\mu^* P_\mu \tag{6.71}$$

This a very simple result. It implies that the power carried by forward-travelling modes is found as the sum of the modulus-squares of the mode amplitudes, weighted by the coefficients P_μ.

Design Example

Calculation of the modal power coefficients is quite simple. For TE modes in a slab guide, for example, we can define P_μ per unit width in the y-direction as

$$P_\mu = 1/4 \int_{-\infty}^{\infty} (\mathbf{E}_\mu \times \mathbf{H}_\mu^* + \mathbf{E}_\mu^* \times \mathbf{H}_\mu) \cdot \mathbf{k}\, \mathrm{d}x \tag{6.72}$$

Inserting the relevant vectorial components, and substituting for $H_{\mu x}$ using Eq. (6.49) we get

$$P_\mu = (\beta/4\omega\mu_0) \int_{-\infty}^{\infty} (E_{\mu y}E_{\mu y}^* + E_{\mu y}^* E_{\mu y})\, \mathrm{d}x \tag{6.73}$$

It is obvious that each component in the integral above contributes an equal amount to the total. Furthermore, the transverse fields $E_{\mu y}$ are real in this case. We therefore obtain

$$\mathrm{P}_\mu = (\beta/2\omega\mu_0) \int_{-\infty}^{\infty} E_{\mu y}^2\, \mathrm{d}x \tag{6.74}$$

Note that this can of course be written in inner product notation as $\mathrm{P}_\mu = (\beta/2\omega\mu_0)\, \langle E_{\mu y}, E_{\mu y}\rangle$.

6.7 THE EXPANSION OF ARBITRARY FIELDS IN TERMS OF MODES

The orthogonality condition can be also used to show how arbitrary transverse fields can be written in terms of the modal solutions by finding the expansion coefficients themselves. Suppose we wish to expand a forward-travelling field $\mathbf{E}(x,y,0)$ at $z = 0$ in this way. We first put

$$\mathbf{E}(x,y,0) = \sum_\mu a_\mu \mathbf{E}_\mu(x,y) \tag{6.75}$$

If we now take vector products of both sides of Eq. (6.75) with the complex conjugate of the transverse magnetic field of one mode (say, the ν th), and integrate over the cross-section, we get

$$\iint_A [\mathbf{E} \times \mathbf{H}_\nu^*] \cdot \mathbf{k}\, \mathrm{d}x\, \mathrm{d}y = \iint_A \sum_\mu a_\mu [\mathbf{E}_\mu \times \mathbf{H}_\nu^*] \cdot \mathbf{k}\, \mathrm{d}x\, \mathrm{d}y \tag{6.76}$$

Since the field is forward-travelling, we could also have put

$$\mathbf{H} = \sum_\mu a_\mu \mathbf{H}_\mu \tag{6.77}$$

Performing a similar operation, but this time using the complex conjugate of the transverse *electric* field of the ν th mode, we get

$$\iint_A [\mathbf{E}_\nu^* \times \mathbf{H}] \cdot \mathbf{k}\, \mathrm{d}x\, \mathrm{d}y = \iint_A \sum_\mu a_\mu [\mathbf{E}_\nu^* \times \mathbf{H}_\mu] \cdot \mathrm{d}x\, \mathrm{d}y \tag{6.78}$$

Adding Eqs (6.76) and (6.78) together then gives

$$\iint_A [\mathbf{E} \times \mathbf{H}_\nu^* + \mathbf{E}_\nu^* \times \mathbf{H}] \cdot \mathbf{k}\, \mathrm{d}x\, \mathrm{d}y = \iint_A \sum_\mu a_\mu [\mathbf{E}_\mu \times \mathbf{H}_\nu^* + \mathbf{E}_\nu^* \times \mathbf{H}_\mu] \cdot \mathbf{k}\, \mathrm{d}x\, \mathrm{d}y \tag{6.79}$$

From the orthogonality condition we know that the right-hand side of Eq. (6.79) must be zero unless $\mu = \nu$. Hence we can rearrange it to extract the modal coefficient a_ν as

$$a_\nu = \iint_A [\mathbf{E} \times \mathbf{H}_\nu^* + \mathbf{E}_\nu^* \times \mathbf{H}] \cdot \mathbf{k} \, dx \, dy \Big/ \iint_A [\mathbf{E}_\nu \times \mathbf{H}_\nu^* + \mathbf{E}_\nu^* \times \mathbf{H}_\nu] \cdot \mathbf{k} \, dx \, dy \qquad (6.80)$$

The coefficients can therefore be found by evaluating Eq. (6.80) for each mode in turn. Once more we shall illustrate this with an example.

Design Example

To expand a TE field $E_y(x)$ in terms of the TE modes of a slab guide we follow the usual procedure of dropping the integration with respect to the y-variable. Substituting the relevant vectorial components into Eq. (6.80) we get

$$a_\nu = \int_{-\infty}^{\infty} E_y \, E_{\nu y} \, dx \Big/ \int_{-\infty}^{\infty} E_{\nu y}^2 \, dx \qquad (6.81)$$

Note that the integral in the denominator of Eq. (6.81) really just represents a normalization factor. The integral in the numerator is more important; it is often called an *overlap integral* because it represents the 'overlap' of the two fields E_y and $E_{\nu y}$. This is really a measure of how similar the total field is to the ν th mode, rather like a correlation. Clearly, if the two are actually identical, we obtain $a_\nu = 1$ (and $a_\mu = 0$ for $\nu \neq \mu$). This is common sense. Finally we note that Eq. (6.81) can be written in inner product notation as $a_\nu = \langle E_y, E_{\nu y} \rangle / \langle E_{\nu y}, E_{\nu y} \rangle$.

6.8 APPLICATION OF THE OVERLAP PRINCIPLE

Overlap integrals are particularly useful in calculating the coupling efficiency in *end-fire couplers* when an external field is used to excite a guided mode. They can also be used to explain what happens when a mode passes through a waveguide discontinuity or a taper. We shall postpone a description of end-fire coupling until Chapter 8, concentrating for now on discontinuities and tapers.

Discontinuities

Figure 6.12 shows a number of different waveguide discontinuities. In each case the substrate is continuous, but there is some difference in the guiding layer on either side of the junction. This could involve a change in layer thickness, or the addition of an overlay of a different material. Both can be achieved easily using conventional planar processing techniques. Alternatively, the orientation of the guide can change. This requires machining of the substrate surface to the desired topology before deposition of the guiding layer. A junction can also be formed between two entirely

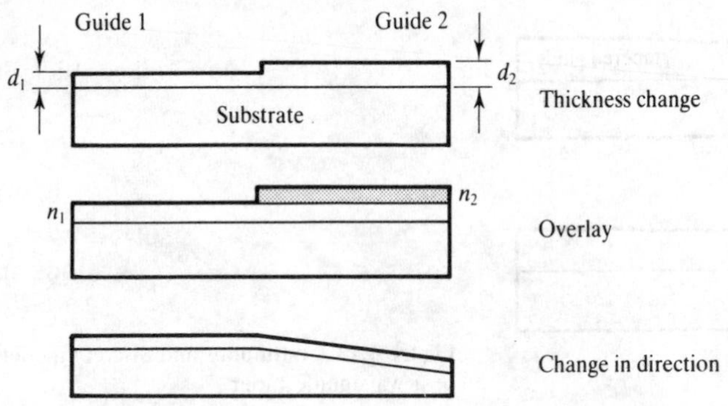

Figure 6.12 Examples of waveguide discontinuities.

different guides, with no common substrate. In this case the process is called *butt-coupling*. In both cases, Eq. (6.80) can be used to calculate the effect of the discontinuity. For example, let us assume in a TEM model that the set of modes on the input side of the discontinuity have transverse field distributions $E_{\nu y}$. On the other side, the corresponding fields are $E'_{\nu y}$. We will now show how to find the efficiency with which an input mode (for example, the lowest-order mode E_{0y}) is coupled across the junction. We may take the input field at the junction as $a_0 E_{0y}$, where a_0 is the input mode amplitude. The power P_{in} carried by this field is then

$$P_{in} = (\beta/2\omega\mu_0)|a_0|^2\langle E_{0y}, E_{0y}\rangle \tag{6.82}$$

After the junction, the portion of the total field carried by the lowest-order mode is now $a'_0 E'_{0y}$, where a'_0 is the new mode amplitude. This can be found as

$$a'_0 = \langle a_0 E_{0y}, E'_{0y}\rangle / \langle E'_{0y}, E'_{0y}\rangle \tag{6.83}$$

The power P_{out} carried by this field is (by analogy with Eq. (6.82)):

$$P_{out} = (\beta/2\omega\mu_0)|a'_0|^2 \langle E'_{0y}, E'_{0y}\rangle \tag{6.84}$$

Substituting for a'_0, this reduces to

$$P_{out} = (\beta/2\omega\mu_0)|a_0|^2|\langle E_{0y}, E'_{0y}\rangle|^2/\langle E'_{0y}, E'_{0y}\rangle \tag{6.85}$$

If we now define the *coupling efficiency* η as the ratio P_{out}/P_{in}, we obtain

$$\eta = |\langle E_{0y}, E'_{0y}\rangle|^2/\{\langle E_{0y}, E_{0y}\rangle\langle E'_{0y}, E'_{0y}\rangle\} \tag{6.86}$$

Clearly η is a dimensionless number which tends to unity as E_{0y} tends to E'_{0y}. Although this is what one would intuitively expect, the implication is that the input mode will cross a discontinuity without much in the way of conversion to other modes, merely provided it is small enough. We can also deduce that the most effective waveguide joint will be one between two guides with matched fields. Though we have not included radiation modes in the calculation, this would be required in a rigorous model; it would then be found that some power is lost to radiation at any discontinuity.

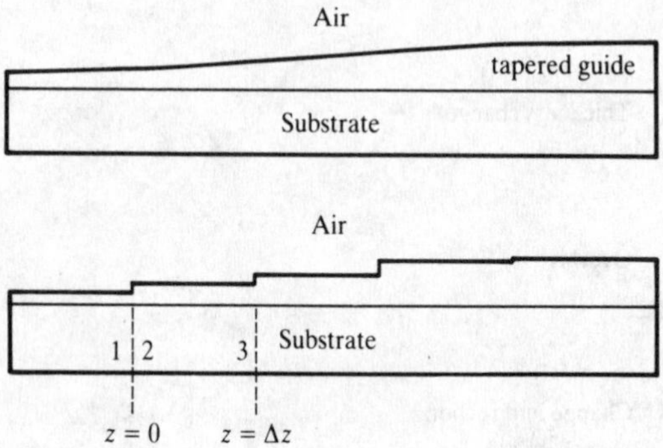

Figure 6.13 Continuous and discrete models of a waveguide taper.

Tapers

We can extend this principle to discover what happens in a tapered waveguide. The upper diagram in Fig. 6.13 shows a typical slow taper, which involves a gradual change in the guide cross-section. The lower one shows a discrete approximation to this shape, consisting of a series of steps. Once again we might wish to discover how a mode of a particular order travels through the taper. If we can avoid any mode conversion, we might then be able to design the structure to enlarge the cross-section of a particular mode in a slow and controlled manner. The taper would then act as a beam expander, for use as a matching section between two different guides.

The theory we need to analyse the problem is called the method of *local normal modes*. Using a TEM model the strategy is to proceed as follows. We start by arranging that the steps in our discrete approximation are at regular intervals Δz apart. We then consider the field at three different points. The first is point 1, just to the left of the discontinuity at $z = 0$ in Fig. 6.13. Here it is assumed that the local normal modes have transverse field functions $E_{\mu y}$ and propagation constants β_μ. If we expand the field E_y in terms of these modes, we can write

$$E_y = \sum_\mu a_\mu E_{\mu y} \tag{6.87}$$

The second place of interest is point 2, just to the right of the discontinuity at $z = 0$. In this region, the local transverse fields and propagation constants have the slightly different values of $E'_{\mu y}$ and β'_μ. If we expand the same field in terms of these modes instead, we can put

$$E_y = \sum_\mu b_\mu E'_{\mu y} \tag{6.88}$$

Here the terms b_μ are an entirely different set of mode amplitudes, but we can relate them to our original a-values using the overlap principle. If this is done we get

$$b_\mu = \sum_\nu a_\nu \langle E_{\nu y}, E'_{\mu y}\rangle / \langle E'_{\mu y}, E'_{\mu y}\rangle \tag{6.89}$$

This relation between the b-values and the a-values effectively tells us how the mode amplitudes change simply in crossing the step at $z = 0$. The next place of interest is point 3. This is just to the left of the discontinuity at $z = \Delta z$, and we call the field here E'_y. This can be found by propagating the field at point 2 over the distance Δz between points 2 and 3. All that happens is that the phase of each mode changes slightly en route, so we must have

$$E'_y = \sum_{\mu} b_\mu E'_{\mu y} \exp(-j\beta'_\mu \Delta z) \tag{6.90}$$

Our aim is to see how the mode amplitudes change with distance. We should, therefore, have expressed the field at point 3 in a form comparable with that used at point 1, i.e. as

$$E'_y = \sum_{\mu} a'_\mu E'_{\mu y} \tag{6.91}$$

Here the terms a'_μ represent the amplitudes of the local normal modes at the start of the next discontinuity at $z = \Delta z$. These new amplitudes now can be extracted as follows. First, we equate Eqs (6.91) and (6.90), which gives

$$a'_\mu = b_\mu \exp(-j\beta'_\mu \Delta z) \tag{6.92}$$

Then we substitute for b_μ using Eq. (6.89), to get

$$a'_\mu = \left\{ \sum_{\nu} a_\nu \langle E_{\nu y}, E'_{\mu y} \rangle / \langle E'_{\mu y}, E'_{\mu y} \rangle \right\} \exp(-j\beta'_\mu \Delta z) \tag{6.93}$$

At this point we note that the quantity $a'_\mu - a_\mu$ is actually the *change* in a_μ that has occurred over the distance Δz. We might call this Δa_μ. Using Eq. (6.93), we may write it as

$$\Delta a_\mu = \{ a_\mu \langle E_{\mu y}, E'_{\mu y} \rangle / \langle E'_{\mu y}, E'_{\mu y} \rangle \} \{ \exp(-j\beta'_\mu \Delta z) - 1 \} + \left\{ \sum_{\nu, \nu \neq \mu} a_\nu \langle E_{\nu y}, E'_{\mu y} \rangle / \langle E'_{\mu y}, E'_{\mu y} \rangle \right\} \exp(-j\beta'_\mu \Delta z) \tag{6.94}$$

We now make a number of approximations, which will be valid provided the distance Δz is short enough and the discontinuity is small. We start by assuming that the transverse modal field changes only a small amount in the distance Δz, so we can put $E'_{\mu y} = E_{\mu y} + \Delta E_{\mu y}$. To reasonable accuracy, we can then say

$$\langle E_{\mu y}, E'_{\mu y} \rangle / \langle E'_{\mu y}, E'_{\mu y} \rangle \approx 1 \tag{6.95}$$

Using the orthogonality principle we may also show

$$\langle E_{\nu y}, E'_{\mu y} \rangle / \langle E'_{\mu y}, E'_{\mu y} \rangle \approx \langle E_{\nu y}, \Delta E_{\mu y} \rangle / \langle E_{\mu y}, E_{\mu y} \rangle \tag{6.96}$$

We now assume that the propagation constant also changes only a small amount in the distance Δz, so we can put $\beta'_\mu = \beta_\mu + \Delta \beta_\mu$. To first-order approximation we can then write

$$\exp(-j\beta'_\mu \Delta z) \approx (1 - j\beta_\mu \Delta z) \approx 1 \tag{6.97}$$

However, we note that a slightly different approximation must be used for the other exponential term in Eq. (6.94) because it is the *difference* between two similar quantities. We therefore write:

$$\exp(-j\beta'_\mu \Delta z) - 1 \approx -j\beta_\mu \Delta z \tag{6.98}$$

When all these values are substituted into Eq. (6.94), it reduces to

$$\Delta a_\mu = -j\beta_\mu \Delta z\, a_\mu + \sum_{\nu,\nu\neq\mu} a_\nu \langle E_{\nu y}, \Delta E_{\mu y}\rangle / \langle E_{\mu y}, E_{\mu y}\rangle \tag{6.99}$$

Dividing both sides by Δz and letting Δz tend to zero we then get

$$\mathrm{d}a_\mu/\mathrm{d}z = -j\beta_\mu a_\mu + \sum_{\nu,\nu\neq\mu} a_\nu \langle E_{\nu y}, \partial E_{\mu y}/\partial z\rangle / \langle E_{\mu y}, E_{\mu y}\rangle \tag{6.100}$$

where a_μ and β_μ are functions of z, and $E_{\mu y}$ is a function of x, y and z. Equation (6.100) is now a differential equation which describes the way the amplitudes of the local normal modes change with distance. In fact there is one such equation for each value of μ (i.e. for each mode of interest) and all the equations must be solved simultaneously for a given set of boundary conditions. Because each mode amplitude appears in every equation, they are known as *coupled mode equations*.

In principle, the equations can be solved as follows. The local normal modes must first be found at all points along the taper so that the functions $E_{\mu y}(x,y,z)$ and $\beta_\mu(z)$ are known. The equations are then integrated numerically. However, this process is exceptionally tedious, so we will not perform it here; instead, we will simply draw some broad conclusions. First, we note that the solution for a particular mode must be $a_\mu = A_\mu \exp(-j\beta_\mu z)$ (where A_μ is a constant) if the summation in Eq. (6.100) is zero. Under these conditions the mode amplitude does not change, only its phase. To design a conversion-free taper we must simply arrange for it to be slow enough, so that

$$\langle E_{\nu y}, \partial E_{\mu y}/\partial z\rangle \Rightarrow 0 \tag{6.101}$$

Even if the taper does not satisfy these conditions we can extract a significant result from the mathematics. We have already seen that the transverse fields of a symmetric slab guide are either symmetric or antisymmetric patterns. If we taper such a guide symmetrically, the same must be true for the local normal modes. More importantly, we can also say that $\partial E_{\mu y}/\partial z$ must be symmetric for a symmetric mode, and antisymmetric otherwise. It therefore follows that $\langle E_{\nu y}, \partial E_{\mu y}/\partial z\rangle = 0$ if one of the modes is symmetric and the other antisymmetric. Consequently such a taper cannot convert a symmetric mode to an antisymmetric one, or vice versa. We will find this rule useful in Chapter 9.

PROBLEMS

6.1 Sketch the dispersion diagram for guided modes in a metal-walled waveguide.

6.2 Find the cut-off frequency of a planar waveguide formed from a slab of dielectric with metallized walls. The slab has thickness 0.5 μm and refractive index 1.5.
[2×10^{14} Hz].

6.3 Show that the eigenvalue equation for symmetric modes in a symmetric slab dielectric waveguide is given by: $\tan(\kappa h/2) = \gamma/\kappa$, where γ, κ and h have their usual meanings.

6.4 A symmetric slab guide is to be used at 1.5 μm wavelength. What is the V-value of the guide if its thickness is 6 μm, the refractive indices of the layers are defined by $n_1 \approx n_2 \approx 1.5$, and (a) $n_1 - n_2 = 0.05$, and (b) $n_1 - n_2 = 0.005$? How many modes are supported in each case?
[(a) 4.866; (b) 1.539; 4 modes; 1 mode].

6.5 Show that the eigenvalue equation for TM modes in an asymmetric slab dielectric waveguide is given by: $\tan(\kappa h) = \kappa(\gamma' + \delta')/\kappa^2 - \gamma'\delta')$, where γ, κ and h have their usual meanings, and γ' and δ' are given by: $\gamma' = \gamma\,(n_1^2/n_2^2)$ and $\delta' = \delta\,(n_1^2/n_3^2)$.

6.6 Sketch the variation of the transverse magnetic field for the lowest-order TM mode of a symmetric slab dielectric waveguide (a) far from cut-off, and (b) near to cut-off.

6.7 A symmetric slab guide has refractive indices $n_1 = 1.505$, $n_2 = 1.5$. If the guide thickness is 1 μm, estimate the effective index of the lowest-order TE mode of the guide at $\lambda_0 = 0.633$ μm.
[1.50125].

6.8 Show by direct integration that two symmetric TE modes of a symmetric slab dielectric waveguide are orthogonal to one another, if the modes are of different order.

6.9 Two identical planar waveguides are butt-coupled together, as shown below. Assuming that the transverse variation of the electric field in the left-hand guide is $E_y(x) = E_0 \exp(-\alpha x^2)$, calculate and sketch the dependence of the coupling efficiency on any accidental lateral misalignment δ.

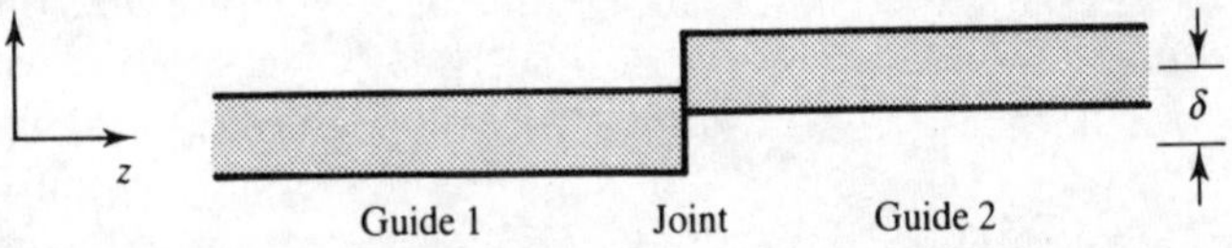

6.10 (a) Figure (i) below shows a discontinuity between two different symmetric slab dielectric waveguides. Describe qualitatively what you expect to occur when the lowest-order guided mode is incident from the left, assuming that the discontinuity is (a) small and (b) large. (b) The two guides are now joined by a taper section, as shown in Figure (ii). Describe what will occur under similar circumstances, if the taper is (a) fast and (b) slow.

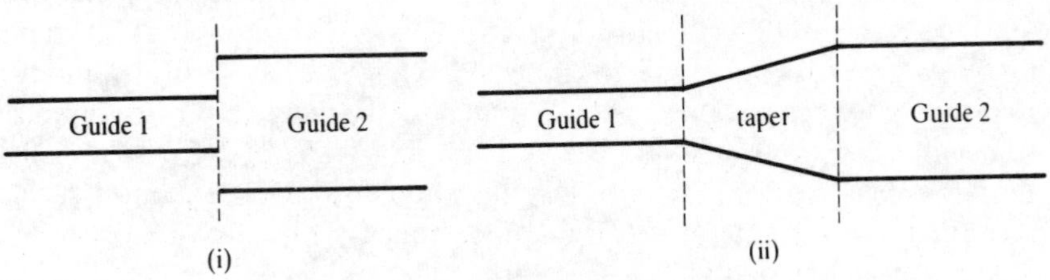

SUGGESTIONS FOR FURTHER READING

Kogelnik, H. Theory of dielectric waveguides. In "Integrated Optics", T. Tamir (ed.), Springer-Verlag, Berlin, 1979.

Lee, D. L. "Electromagnetic Principles of Integrated Optics", Chapter 4, John Wiley and Sons, New York, 1986.

Ramo, S., Whinnery, J. R. and van Duzer, T. "Fields and Waves in Communication Electronics", Chapter 14, John Wiley and Sons, New York, 1984.

Synder, A. W. and Love, J. D. "Optical Waveguide Theory", Chapters 1, 11, 12 and 13, Chapman and Hall, London, 1983.

Taylor, H. F. and Yariv, A. Guided wave optics, *Proc. IEEE*, **62**, 1044-1060, 1974.

Tien, P. K. Light waves in thin films and integrated optics, *Appl. Opt.*, **10**, 2395–2412, 1971.

Yariv, A. "Optical Electronics", Chapter 13, Holt, Rinehart and Winston, New York, 1985.

CHAPTER

SEVEN

PLANAR WAVEGUIDE INTEGRATED OPTICS

7.1 OVERVIEW OF PLANAR WAVEGUIDE COMPONENTS

We will now consider the range of devices that comprise the system of planar waveguide integrated optics. Since these operate on sheet beams (i.e. beams that are confined in one direction only), they are analogous to the devices used to manipulate free-space beams. Consequently the major building blocks are refractive components (generally lenses) and diffractive components (known as *gratings*). However, we must also consider the important question of input and output couplers, needed to connect the guided wave system to the outside world.

Before proceeding further we must first ask what information is needed to understand the behaviour of an optical device. The general answer is that we will be happy if we can predict the directions and amplitudes of the output beams, for any given input. Some geometries are relatively simple and yield this information quite readily—for example, in Chapter 5 we found the complete solution to the single interface problem. Others are less tractable and force us to accept a partial solution. Many integrated optic devices fall into this 'difficult' category. The beam directions are easy to predict, but often the amplitudes can only be found after considerable mathematics. We will therefore concentrate on a simple method of finding the beam directions, based on the principle of *phase matching*.

7.2 PHASE MATCHING AT A SINGLE INTERFACE

We begin by revisiting the interface problem mentioned above. Figure 7.1 shows reflection and refraction at the interface between two semi-infinite media. Here a beam of wavelength λ_0 is incident at an angle θ_1, and the two media have refractive indices n_1 and n_2 respectively. We assume that n_1 is greater than n_2, but that θ_1 is less than the critical angle so that a propagating transmitted wave arises.

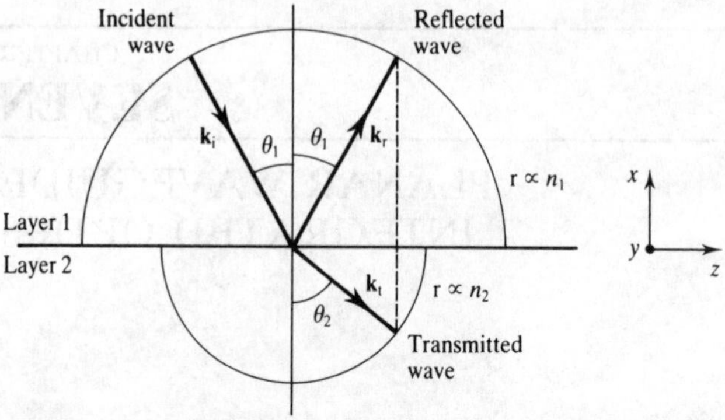

Figure 7.1 Phase matching at a single interface—transmission and reflection.

The procedure in any phase-matching problem is the same. We start by identifying all the propagation constants of interest. Here there are only two, with values k_1 in layer 1 and k_2 in layer 2, given by

$$k_1 = 2\pi n_1/\lambda_0 = k_0 n_1 \qquad \text{and} \qquad k_2 = k_0 n_2 \tag{7.1}$$

We then consider the input wave. Assuming TE incidence, the electric field of the input beam has only a y-component. We know from past experience that this can be written in the form:

$$E_{yi} = E_i \exp\left[-jk_0 n_1(z \sin\theta_1 - x\cos\theta_1)\right] \tag{7.2}$$

We note, however, that it could be written in shorthand form:

$$E_{yi} = E_i \exp(-j\mathbf{k}_i \cdot \mathbf{r}) \tag{7.3}$$

Here $\mathbf{r}$ is the radius vector, while $\mathbf{k}_i$ (known as the *propagation vector*) has components:

$$\mathbf{k}_i = -k_1 \cos\theta_1 \mathbf{i} + k_1 \sin\theta_1 \mathbf{k} \tag{7.4}$$

It is easy to demonstrate equivalence between the two forms, simply by evaluating the scalar product term in Eq. (7.3).

Examining Eq. (7.4), we see that the propagation vector is oriented parallel to the direction of travel of the wave, and has modulus:

$$|\mathbf{k}_i| = \surd(k_{ix}^2 + k_{iz}^2) = k_1 \tag{7.5}$$

Since we are only really interested in beam directions, this vector contains all the information we need—in fact it corresponds roughly to the ray introduced in our earlier discussion of bulk optical devices. What properties does it have? Well, if one end of the vector is fixed, the locus of all possible positions of the other end in the x–z plane must be a circle of radius k_1. Figure 7.2 shows two equally valid representations of this circle. In three dimensions the locus is a sphere. For the range of possible input beams in our interface problem the locus is adequately

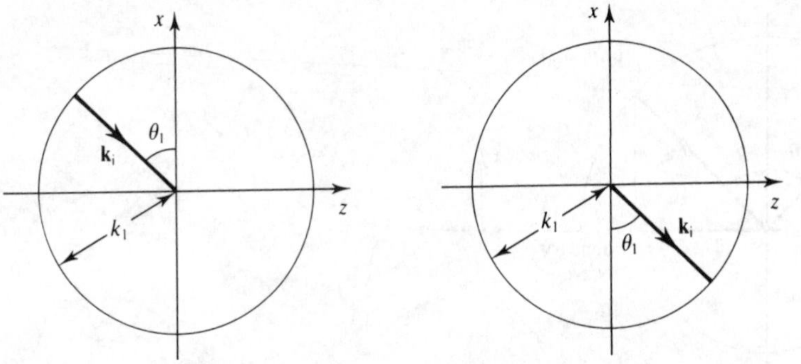

Figure 7.2 Locus of propagation vectors in the *x–z* plan—two alternative representations.

represented by the upper half-circle in Fig. 7.1. This can be drawn with a radius proportional to k_1 (or to n_1).

In the same way, we can represent the reflected beam by the vector $\mathbf{k}_r$, and the transmitted beam by $\mathbf{k}_t$. The moduli of these two vectors are $|\mathbf{k}_r| = k_1$ and $|\mathbf{k}_t| = k_2$, so the locus of the free end of $\mathbf{k}_r$ is also defined by the upper half-circle in Fig. 7.1, while that of $\mathbf{k}_t$ may be taken as the lower half-circle. Note that the radius of this second circle is proportional to k_2 (and thus to n_2), and is therefore smaller than the upper one, since $n_2 < n_1$. We will now use these vectors to interpret our previous results in a new way. We know that the incident and reflected beams make equal angles with the interface normal, so the *z*-components of their wave vectors must be the same. We may therefore put

$$k_{rz} = k_{iz} \tag{7.6}$$

We also know that the angle θ_2 of the transmitted wave is governed by Snell's law, $n_1 \sin \theta_1 = n_2 \sin \theta_2$. This requires that the *z*-component of the transmitted wave vector is also the same, so

$$k_{tz} = k_{iz} \tag{7.7}$$

Together these equations imply that the *z*-components of *all three* vectors are identical. This is consistent with our earlier result that all parts of the field travel at the same speed in the *z*-direction. It has, however, given us a new way to find the beam directions. We first draw circles of the correct relative radii, together with the propagation vector of the incident beam. We then add in the reflected beam vector, and compute the transmitted beam vector from the intersection with the lower circle of the interface normal through the end of the reflected wave vector.

Of course θ_1 may be greater than the critical angle. In this case total internal reflection occurs and the dashed-line construction yields no intersection at all with the lower circle (as in Fig. 7.3). Our interpretation of this event is that the field in layer 2 cannot now consist of a propagating wave; instead, it must be a boundary wave of some kind, travelling parallel to the interface.

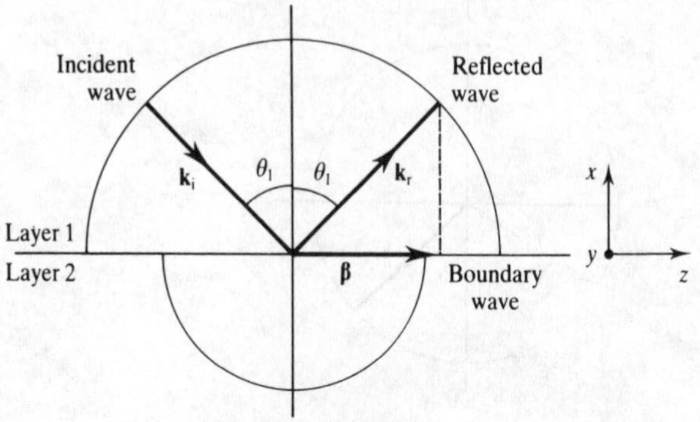

Figure 7.3 Phase matching at a single interface—total internal reflection.

From previous experience we recognize it to be an evanescent wave with propagation vector $\boldsymbol{\beta}$. The direction of $\boldsymbol{\beta}$ must be parallel to the interface, and its magnitude is given by:

$$|\boldsymbol{\beta}| = k_{iz} \tag{7.8}$$

The propagation vector $\boldsymbol{\beta}$ is therefore defined by the revised dashed-line construction in Fig. (7.3).

Figures 7.1 and 7.3 are both illustrations of the general principle of phase matching, which requires that the components of all propagation vectors parallel to an interface must match at that interface. We shall now see how this can be used to analyse more complicated device geometries.

7.3 THE FTIR BEAMSPLITTER

We begin with the cube beamsplitter, which is also often known as the *frustrated total-internal-reflection* (or FTIR) *beamsplitter*. This normally consists of two symmetric 90° glass prisms, separated by a small, accurately-defined air gap. Figure 7.4 shows the device in use. Here a beam input through the left-hand face of the

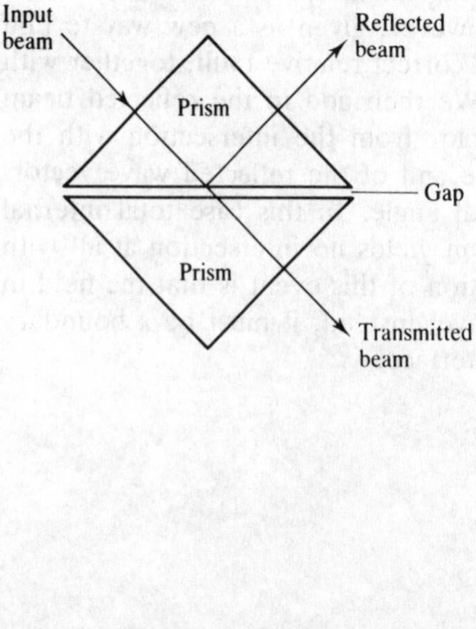

Figure 7.4 The FTIR beamsplitter.

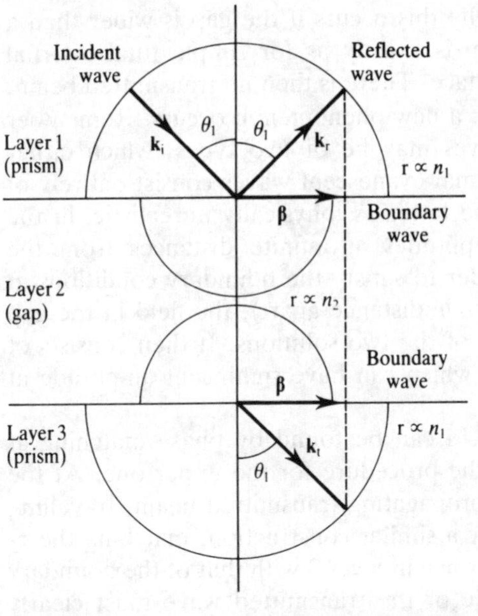

Figure 7.5 Phase matching in the FTIR beamsplitter.

upper prism is divided into two components: a reflected wave emerging from the right-hand face of the upper prism, and a transmitted wave from the corresponding face of the lower prism. It turns out (as we shall see later) that the power splitting ratio between the two beams can be set by varying the gap width. Why should this behaviour occur?

The central region of the device can be considered as a stacked structure, consisting of three separate layers. The two outer layers correspond to the prisms and have refractive index n_1, while the inner layer (the gap) has refractive index n_2. Since the device is a multi-layer, it would seem reasonable that its operation can be explained in terms of the phase matching principle. In fact Fig. 7.5 shows the required phase matching diagram, and we will now explain its construction.

As before, we start by considering the fate of the incident wave. If we take $n_1 \approx 1.5$ and $n_2 = 1$ (i.e. glass prisms, separated by an air gap) then the critical angle at the upper prism/gap interface is $\theta_c = \sin^{-1}(1/1.5) = 41.8°$. If the input beam travels at the typical angle of $\theta_1 = 45°$, it will suffer total internal reflection. The upper part of the phase matching diagram is therefore constructed exactly as for Fig. 7.3; the incident and reflected wave-vectors are first drawn in, and the dashed line construction is used to show that an evanescent wave exists beneath the surface of the first prism.

To construct the lower part of the diagram we must consider the field in the gap more carefully. If the gap is large it is reasonable to expect that the evanescent field below the upper prism will decay quickly enough for it to be unaffected by the presence of the lower interface. Typically this occurs if the gap is wider than a few μm. In this case the physical situation is exactly as for simple total internal reflection, and no power crosses either interface. There is then no transmitted beam.

If the gap is sufficiently small, however, a new phenomenon occurs. Remember that the field solutions for evanescent waves may be of two types, which either decay or grow away from an interface. Normal evanescent waves consist entirely of the former type, and we have discarded the latter as 'physically unrealistic' in the past because they imply infinite field amplitudes at infinite distances from the interface. It turns out, however, that in order to satisfy the boundary conditions at the two interfaces here (which are only a *finite* distance apart), the field in the gap region must be constructed as a linear sum of the two solutions. It then consists of a rather more complicated boundary wave which can have significant amplitude at the lower interface.

The waves generated by this hybrid field can be found by phase matching at both interfaces. We have already covered the procedure for the upper one. At the lower one there is only one possibility—a propagating transmitted beam, travelling downwards at the angle θ_1. This is found by a similar construction, matching the z-component of the propagation vector of the wave in layer 3 with that of the boundary wave at the lower interface. The amplitude of the transmitted wave must clearly tend to unity as the gap tends to zero because both interfaces will then disappear. In fact it can be set to any desired value between unity and zero by choosing the gap correctly. Since the device is lossless, the power in the reflected beam must decrease as the transmitted beam increases. Generally the gap is adjusted to equalize the two, whereupon the device acts as a 50:50 beamsplitter.

Rather surprisingly, therefore, power can be transferred across a gap region by a non-propagating, evanescent field, provided the gap is small enough. The process is often known as *optical tunnelling*. It explains the name of the FTIR splitter—the tunnelling effectively frustrates the total internal reflection which would normally occur at the upper interface. The gap required for significant tunnelling depends primarily on the decay rate of the evanescent fields. If θ_1 is much larger than the critical angle the fields decay very rapidly and a very small gap is needed. As θ_1 approaches θ_c the fields decay more slowly, and the gap may be increased. We will now show how optical tunnelling can be exploited for the excitation of the guided modes of a planar waveguide.

7.4 THE PRISM COUPLER

Planar guided modes can be excited by an external free-space beam using a *prism coupler*, shown in Fig. 7.6. This consists of a prism of suitable refractive index, which is separated from the waveguide by a small gap. Typically the prism index should be somewhat higher than that of the substrate, and the gap should be of the order of 1 μm. This can be achieved by clamping the prism and guide together—dust particles then serve as spacers.

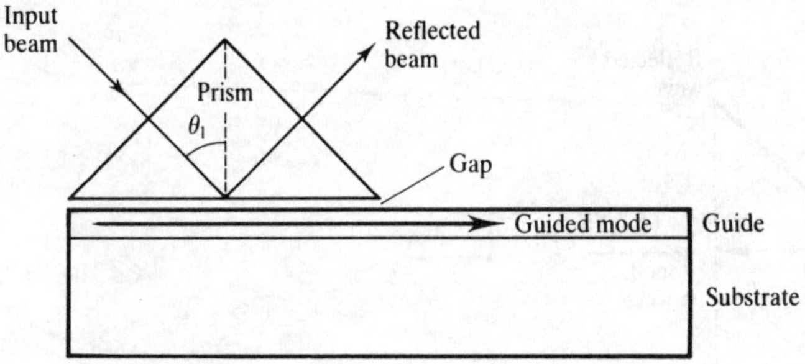

Figure 7.6 The prism coupler.

The input beam is again assumed to be a plane wave, which enters the prism from free-space at an angle θ_1. If θ_1 is large enough this beam is total-internally-reflected at the prism/air-gap interface. Once again, therefore, the field beneath the prism is an exponentially-decaying evanescent wave, travelling parallel to the interface. The propagation constant of this field is given by

$$\beta = k_1 \sin \theta_1 \tag{7.8}$$

This field can be used to excite a guided mode in the waveguide underneath, if β is matched to the propagation constant of the mode, as we now demonstrate.

The basic mechanism is illustrated by the phase matching diagram of Fig. 7.7. Here the upper part of the diagram shows the excitation of a boundary wave beneath the prism. If the gap is very large, nothing further occurs. However, if it is small enough, the field solution in the gap must be modified as described for the FTIR splitter. This implies that there is again the possibility of power transfer across the two interfaces by optical tunnelling. If conditions are correct, this can give rise to a travelling wave in the waveguide layer. The propagation vector $\mathbf{k}_g$ of this wave is found using a similar construction—the intersection of the dashed line with a circle of radius proportional to n_3 gives $\mathbf{k}_g$.

By choosing the angle of incidence correctly, the angle θ_2 can be made to correspond exactly with the direction of a downward-travelling, propagating wave in a zig-zag ray model of a guided mode. This wave will therefore be total-internally-reflected at the lower interface since there is no intersection of the dashed line with the lowest circle (which has radius proportional to n_4). Naturally such a reflection must give rise to an upward-travelling wave in the guide layer (with propagation vector $\mathbf{k}_g'$), and an evanescent boundary wave in the substrate.

We have now identified all the waves in Fig. 7.7. Note, however, that the conditions for the constructions above to be valid are quite critical. The prism index must be higher than that of the gap material so that the total internal reflection does indeed occur at the upper interface. However, it must also be higher than that of the guide itself so that the ray angles θ_2 and θ_2' (which are close to 90° in a weak-index guide) are obtained at a more convenient external angle θ_1 (say, $30° < \theta_1 < 60°$).

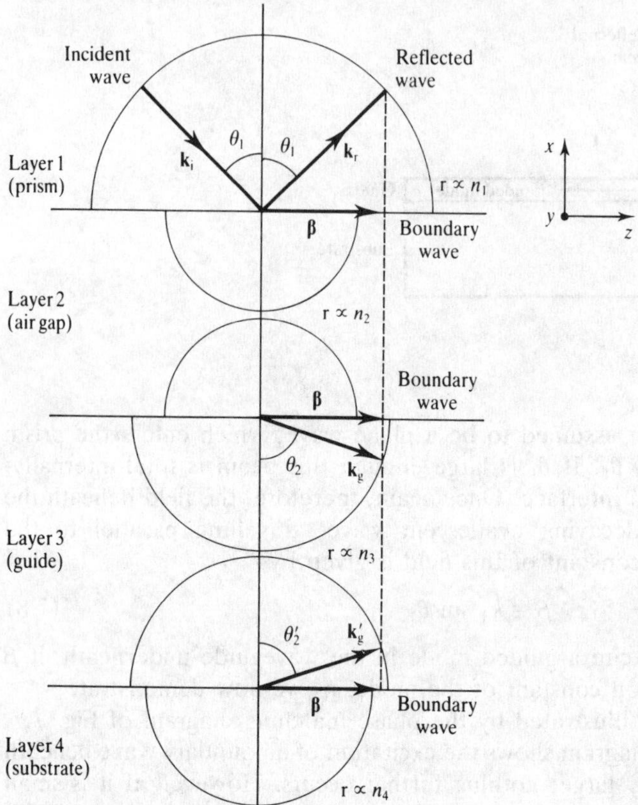

Figure 7.7 Phase matching in the prism coupler.

Figure 7.8 shows how the whole of the transverse field pattern might look at this point. Near the waveguide the pattern looks very similar to that of an ordinary guided mode. In the substrate it must be evanescent, while in the guide layer itself it must be a standing-wave pattern, composed of the sum of two zig-zag waves. Certainly there are differences in the gap region, where the field is a hybrid boundary wave, and in the prism itself, where it is again a standing wave. However, if the gap is not too close to zero, it is not inconceivable that the 'desired' portion of the field could be made to detach itself from the rest and propagate forward as a guided mode. We can achieve this by removing the external excitation at a suitable value of z. Could we then be said to have launched a guided mode? The short answer is yes, but there are several provisos. First, to make a useful input coupler the prism should be relatively small compared with the substrate length, so that the guided beam propagates out from under the prism to other devices. In this case the input beam can no longer be a plane wave, but must be a bounded beam. The amplitude of the guided mode must therefore build up from zero at the left-hand end of the

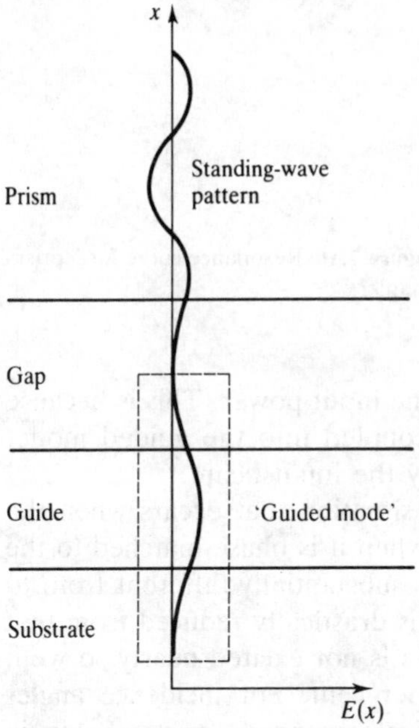

Figure 7.8 The transverse field distribution in the prism coupler.

coupler, to some value at the right. Second, the power transfer process is bi-directional—a prism coupler can just as easily be used in reverse, to couple a guided mode out into free-space. If we are not careful, therefore, the guided mode will start coupling back out of the coupler at its right-hand end.

Fortunately it has been found that a halved prism coupler can convert a bounded beam into a guided mode with good efficiency. This provides a sharp termination to the coupling region at a point when the guided mode has built up to a reasonable amplitude (Fig. 7.9). If the beam reflected from the prism is now measured, it will

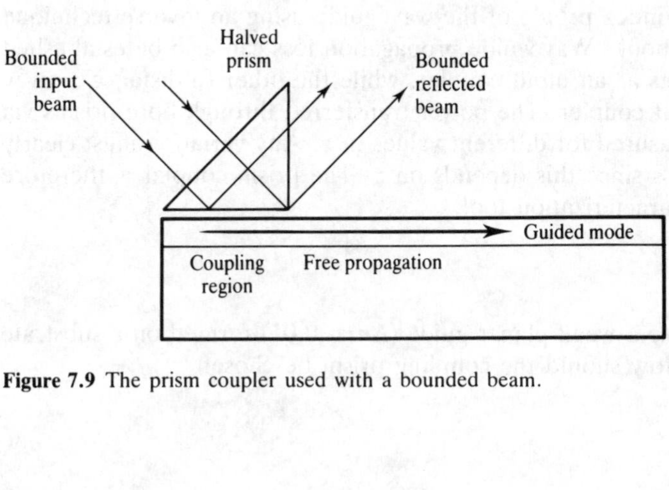

Figure 7.9 The prism coupler used with a bounded beam.

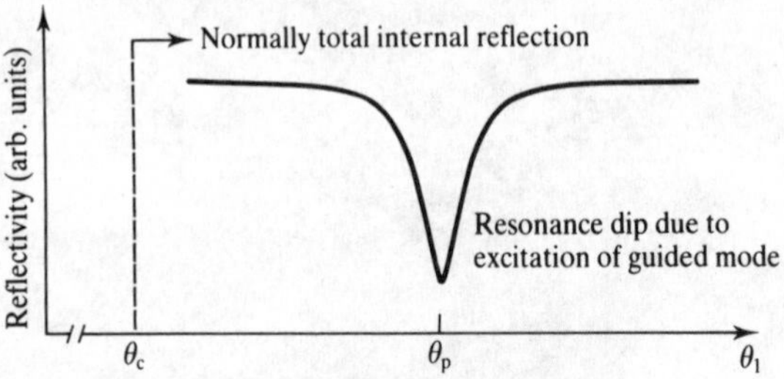

Figure 7.10 Resonance curve for a prism coupler.

be found to contain rather less than 100 per cent of the input power. This is because a substantial fraction of the power has now been coupled into the guided mode, and the required energy must clearly be provided by the input beam.

Our discussion to date has concentrated on the situation that occurs when the input beam is incident at the correct angle $\theta_1 = \theta_p$, when it is phase matched to the guided mode. What happens at other angles? If θ_1 is substantially different from θ_p it turns out that the field amplitude inside the guide is drastically reduced from that shown in Fig. 7.8. This implies that the guided mode is not excited nearly so well, and the prism reflectivity returns to close to 100 per cent. For incidence angles near to θ_p the reflectivity falls, following the typical *resonance curve* shown in Fig. 7.10.

Both the width and the depth of the resonance depend on several factors—the precise variation of the air-gap, the length of the prism, the shape of the input beam, and absorption losses within the system. However, measurement of the reflectivity minimum can be used to determine the phase matching angle θ_p, and hence the propagation constant β of the guided mode (from Eq. (7.8)). Note that angles are generally measured external to the prism, and therefore correction must be made for refraction at its input face. Consequently the prism angles and refractive index must be known accurately.

A prism coupler can also provide other information. For example, the number of minima in the reflectivity curve yields the number of modes supported by the guide. Knowledge of the effective indices of all the modes may even be used to reconstruct the refractive index profile of the waveguide using an inverse technique beyond the scope of this book. Waveguide propagation loss can also be established using two prisms. One acts as an input coupler, while the other (a distance z away from the first) is an output coupler. The power transferred through both prisms via the waveguide is then measured for different values of z. Any variation must clearly be due to propagation loss since this depends on z. The prism coupler is therefore an extremely versatile characterization tool.

Design Example

It is desired to characterize a weak planar guide ($\Delta n \approx 0.01$) formed on a substrate of refractive index 1.5. How should the coupling prism be chosen?

The effective index of the guide must lie in the range $1.5 < n_{\text{eff}} < 1.51$. The phase matching condition (Eq. (7.8)) can be written in the form:

$$n_{\text{eff}} = n_1 \sin \theta_1 \tag{7.9}$$

For $\theta_1 \approx 60°$ we obtain $n_1 \approx 1.7$. A prism of high-index glass would therefore prove suitable. It would be convenient to make this as a right-angle prism, with hypotenuse angles of 60° and 30°.

7.5 PHASE MATCHING FOR GUIDED MODES

Until now we have considered only one-dimensional propagation of a mode, down the z-axis. However, a planar guided mode can travel in any direction in the y–z plane. To proceed further we need a more general description. This is provided by a vectorial notation similar to that introduced at the beginning of this chapter. Assuming that the mode has a transverse field distribution $E(x)$ and propagation constant β, a suitable scalar solution might have the form:

$$E(x,y,z) = E(x) \exp(-j\boldsymbol{\beta} \cdot \mathbf{r}) \tag{7.10}$$

where

$$|\boldsymbol{\beta}| = 2\pi n_{\text{eff}}/\lambda_0 \tag{7.11}$$

Equation (7.10) still describes a mode with an x-dependent transverse field. However, its phase variation is defined by the vector $\boldsymbol{\beta}$, which is assumed to lie in the y–z plane. The orientation of $\boldsymbol{\beta}$ gives the wave direction, while its magnitude describes the effective index of the mode. The locus of all possible propagation vectors in the y–z plane is, once again, a circle, this time of radius β. If the guide supports more than one bound mode (say n of them), the loci of all possible modes of guided propagation are a set of concentric circles of radii β_1, β_2 ,.., β_n. Figure 7.11 shows the circles for a three-moded guide, drawn with radii proportional to n_{eff1}, n_{eff2} and n_{eff3}.

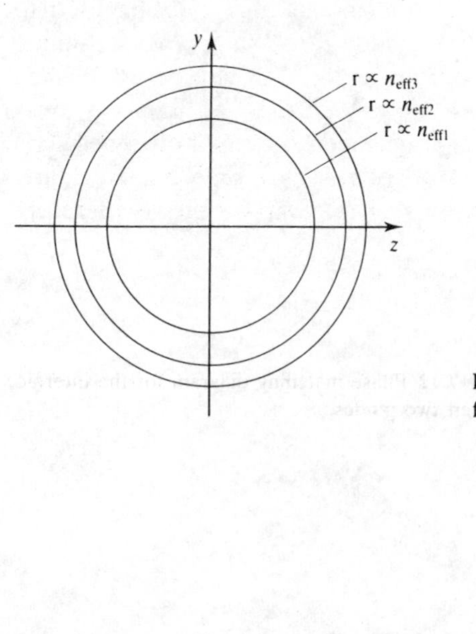

Figure 7.11 Loci of propagation vectors in the y–z plane for a three-moded guide.

Just as in bulk optics, a guided mode can be reflected and refracted at a discontinuity. However, instead of the discontinuity being between two homogeneous media, it will generally be between two different types of waveguide. As discussed in Chapter 6, these could have different guide layer thicknesses, or the same thickness but different refractive indices, or different overlays, or be inclined at an angle to each other. Consequently the propagation vectors in each guide may have different magnitudes, or lie in different planes.

Solution of the general boundary matching problem at an interface is very complicated. To find a full solution it is often necessary to assume the presence of the full set of bound and radiation modes on either side of the interface, including both polarizations. This implies that mode conversion, polarization conversion and scattering into radiation modes can all occur at interfaces. The possible propagation directions are then governed by the phase matching principle.

Here we will consider a simplified example involving a coplanar discontinuity in a three-moded guide. The interface is parallel to the y-axis, and we will restrict our attention to the bound modes. Figure 7.12 shows the phase matching diagram for the case when the lowest-order mode (which has effective index n_{eff1}) is incident obliquely from the left. The diagram is constructed by first drawing circles of radii proportional to n_{eff1}, n_{eff2} and n_{eff3} (using the relevant values on either side of the discontinuity). The propagation vector of the input wave (labelled ***I***) is then drawn in at the correct angle, and with length proportional to n_{eff1}. The component of this mode that is reflected must travel at an equal and opposite angle to the interface normal, allowing the vector ***R*** to be added. All other possible propagation vectors can then be found from the intersection of the dashed line—the normal to the interface passing through the end of ***R***—with the two sets of circles. In this example there are six possible vectors, corresponding to three transmitted waves (collectively labelled T) and three reflected waves (labelled R). The input beam therefore suffers reflection, refraction and mode conversion.

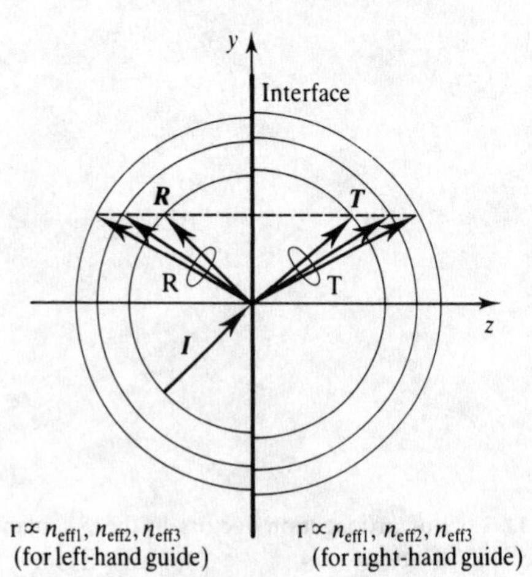

Figure 7.12 Phase matching diagram for the interface between two guides.

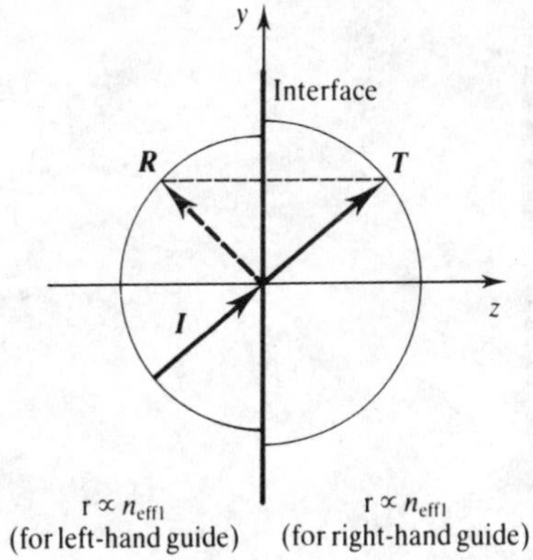

Figure 7.13 Approximate phase matching diagram for the interface between two guides.

If the discontinuity is small most of these effects can be neglected. From Chapter 6 we know that little mode conversion occurs at weak interfaces. The most important of the waves described above will therefore be the reflected component **R** and the transmitted component **T** that are of the same order as the input mode. A reasonable approximation to Fig. 7.12 is therefore provided by the reduced diagram of Fig. 7.13, which retains only the waves **I**, **R** and **T**. Furthermore, unless the angle of incidence is very oblique, we might expect the amplitude of **T** to be much larger than that of **R**, so that the main effect occurring is one of refraction of the input mode. We might guess that this will be governed by a modified Snell's law equation of the form:

$$n_{1\mathrm{eff1}} \sin \theta_1 = n_{2\mathrm{eff1}} \sin \theta_2 \qquad (7.12)$$

where $n_{1\mathrm{eff1}}$ is the effective index of the lowest-order mode in waveguide 1, θ_1 is its angle of travel, and $n_{2\mathrm{eff1}}$ and θ_2 are the corresponding values in a waveguide 2. To illustrate this, Fig. 7.14 overleaf shows the refraction of a real guided beam at an interface between two guides of different thickness.

To summarize, we may assume that if the interface is sufficiently weak, the input field will travel across it without losing too much power. Its transverse field distribution will not alter much, but its direction of propagation will change through refraction. This means that refractive optical components may operate by slight local modification of an otherwise uniform planar guide.

7.6 REFRACTIVE OPTICAL COMPONENTS

Refractive guided wave components can be subdivided into a number of classes. The simplest work through the sudden change in direction that takes place at an abrupt discontinuity, as described above. This allows guided wave lenses to be constructed; these are analogous to bulk optical lenses (except that in general they may only convert one cylindrical guided beam into another). For example, a simple

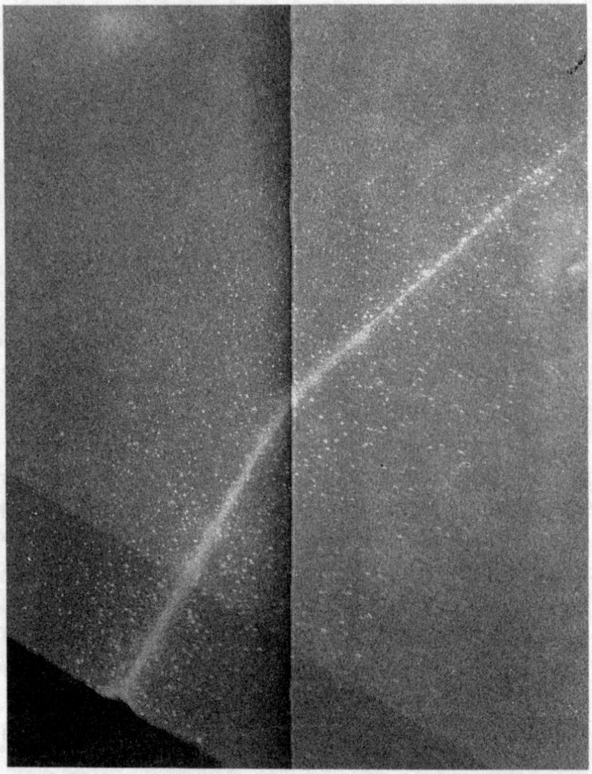

Figure 7.14 Refraction of guided beam at boundary between guide layers of thickness 800 and 2400 Å (*photograph courtesy R. Ulrich, Technische Universität Hamburg-Harburg*).

plano-convex lens might be made by the deposition of high-index overlay material in a region bounded by a parabola and a straight line, as shown in Fig. 7.15.

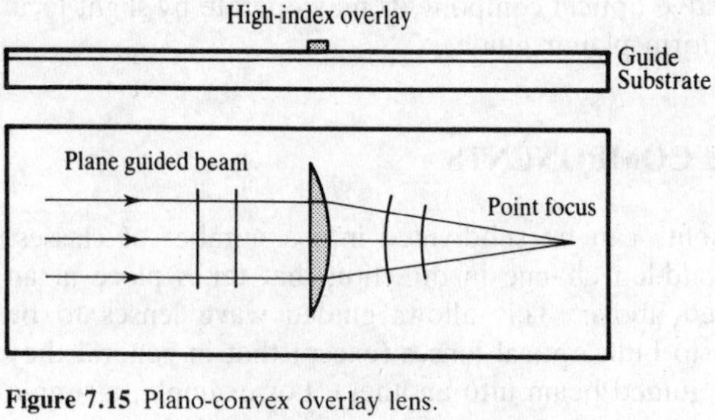

Figure 7.15 Plano-convex overlay lens.

Design Example

We will now try to estimate the factors governing the performance of overlay lenses. Previously we obtained the following formula for the focal length of a bulk-optic lens made of material of refractive index n_L and with two spherical surface of radii r_1 and r_2:

$$1/f = (n_L - 1)(1/r_1 + 1/r_2) \tag{7.13}$$

In this case the surrounding medium is air (which has unity refractive index). More generally we might expect that for a surround of index n_M the formula above modifies to

$$1/f = (n_L/n_M - 1)(1/r_1 + 1/r_2) \tag{7.14}$$

If a guided wave lens also works by refraction, we would expect its operation to be controlled by the values of the effective index inside and outside the lens. If these are n_{Leff} and n_{Meff}, the corresponding formula is

$$1/f = (n_{Leff}/n_{Meff} - 1)(1/r_1 + 1/r_2) \tag{7.15}$$

where r_1 and r_2 are now the radii of curvature of the two circular boundaries of the high-index overlay.

We can now directly compare the lens shapes required to perform the same task in bulk and guided wave optics. Suppose we wish to design a plano-convex lens with a given focal length f. In this case, we may take r_1 to be infinite, so that r_2 is given by

$$r_2 = (n_L/n_M - 1)f \quad \text{(in bulk optics)}$$
$$r_2 = (n_{Leff}/n_{Meff} - 1)f \quad \text{(in guided wave optics)} \tag{7.16}$$

Typically $n_L/n_M - 1$ will be around $1.5 - 1 = 0.5$ in the bulk device. What is the corresponding value of $n_{Leff}/n_{Meff} - 1$? With a substrate of refractive index $n_s = 1.5$ we might expect to be able to form guides with effective indices $n_{Meff} \approx 1.6$ using a high-index guiding layer. The discontinuities forming the lens must be weak to prevent scattering to radiation modes. This implies that $n_{Leff} - n_{Meff}$ must be much smaller than $n_{Meff} - n_s$. An upper limit might be $n_{Leff} - n_{Meff} = 0.02$. We then find that $n_{Leff}/n_{Meff} - 1$ is $0.02/1.6 = 0.0125$. Substituting these values into Eq. (7.16), we conclude that the guided wave lens has a radius of curvature $0.5/0.0125 = 40$ times smaller than the bulk device.

If we now take the focal length of the guided wave lens to be 5 cm (the maximum that will fit on a chip of reasonable dimensions) we find that $r_2 = 0.63$ mm. This implies that the largest possible aperture of the lens ($2r_2$) is only 1.26 mm, and to achieve this the lens must differ in shape quite considerably from the ideal thin lens assumed in the calculation—in fact, it must be semicircular. Consequently the lens will have a very low numerical aperture, and will also produce a very aberrated focus.

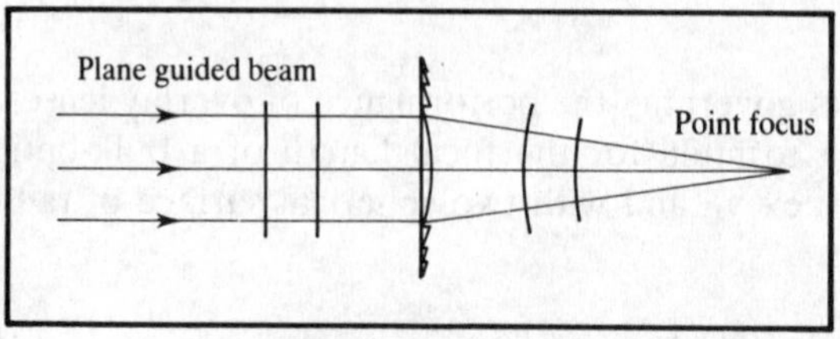

Figure 7.16 Fresnel overlay lens.

Fresnel, Luneberg and Geodesic Lenses

There are several solutions to this problem. Remember that the transfer function of a lens is given by

$$\tau_L = \tau_s \exp(jk_0 R^2/2f) \tag{7.17}$$

where τ_s is a constant, and R is the distance from the axis. The focusing property of the lens can thus be ascribed to the parabolic variation of phase shift it imparts. In guided wave optics τ_L must have a similar form. Any phase shift of $2\nu\pi$ radians (where ν is an integer) can be disregarded since all these are equivalent to no shift at all. We can therefore safely remove any parts of the overlay that simply contribute shifts of $2\nu\pi$ and still obtain a parabolic phase variation. The modified overlay pattern shown in Fig. 7.16, where each step in thickness is a step of 2π in phase, is therefore equivalent to a plano-convex lens; however, it is much thinner. This is known as a *Fresnel lens*.

Though the Fresnel lens is useful, it does not provide an answer to the underlying problem: the weak change in effective index possible at an abrupt interface leads to a requirement for small radii of curvature of the lens boundaries. The answer is to use larger changes in effective index, which are applied gradually to prevent mode conversion. We cannot analyse such lenses here, but merely state that there are two common types. Both are circularly symmetric, and can therefore nominally focus a plane guided wave with aberrations that are independent of the angle of incidence of the beam. In fact, they are two-dimensional forms of the *Maxwell fish-eye* or *Luneburg* lens in bulk optics.

The first is the *overlay Luneberg lens*, shown in Fig. 7.17. As before, the lens is formed from a high-index overlay, but this time with a graded, circularly-symmetric thickness profile. For some particular overlay profiles focusing is achieved through continuous refraction of the input wave, and the ray paths inside the lens are therefore curved. The desired overlay profile may be approximated by sputtering material through a circular aperture.

An equivalent lens can also be formed without using an overlay, as follows. Before fabrication of the guide a circularly-symmetric depression (generally spherical) is precision-turned in the substrate using a diamond lathe. Usually the edges of the depression are also rounded to provide a gradual input transition that reduces scattering. The guide is then fabricated, with uniform thickness, to follow the topology of the depression. The slow change in local orientation of the guide then causes continuous refraction so that rays follow a geodesic path (the shortest path,

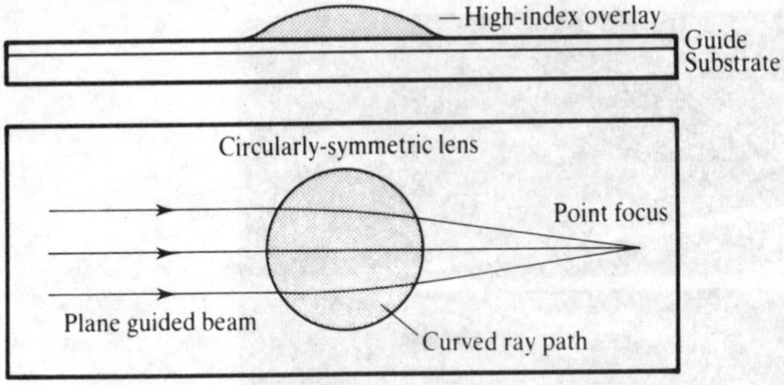

Figure 7.17 Overlay Luneberg lens.

in accordance with Fermat's principle) through the lens, which is consequently known as a *geodesic lens*.

Figure 7.18 shows a geodesic lens. Note that the optical path through the centre of the lens is larger than that at the edges because of the increased distance that must be travelled in crossing the deepest parts of the depression. This retards the centre of the wavefront so that a converging output is obtained. Figure 7.19 on page 162 shows ray paths through a real geodesic lens; the rays are made visible through the use of a fluorescing dye.

Clearly it is impossible to manufacture a diverging lens using the geodesic technique. However, we note that two waveguide lenses with differing focal lengths can be used as a telescope beam expander (previously described in Chapter 4). This arrangement is often used to expand a beam coupled directly into the guide from a laser to a more useful cross-section.

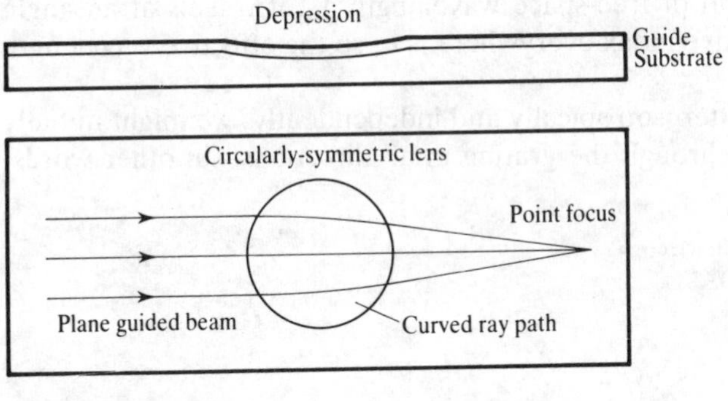

Figure 7.18 Geodesic lens.

7.7 GRATINGS

A further important class of components operate by diffraction rather than refraction. These are periodic structures, known as *gratings*. In many ways they are ideal guided

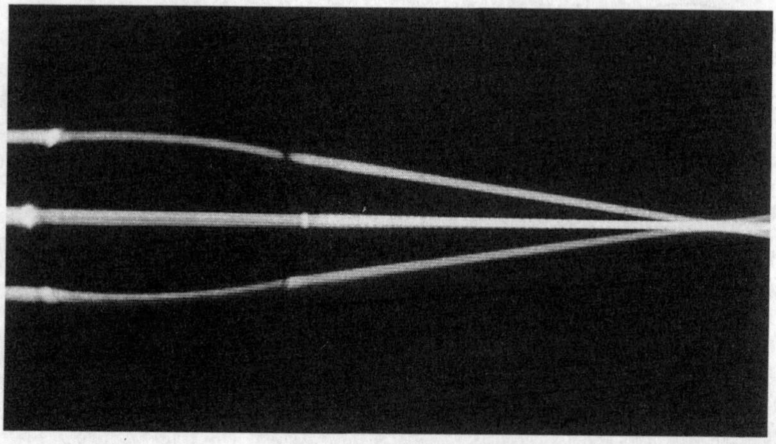

Figure 7.19 Ray paths in a geodesic lens (*photograph courtesy Y. Okamura, Osaka University*).

wave components, and they may be analysed using an extension of the phase-matching principle. We will show how this is done first in bulk optics, and then in guided wave optics.

Optically Thin Gratings

To investigate the operation of a grating we need a model. Figure 7.20 shows the simplest possible representation of a periodic structure, which is made from an infinite array of point scatterers spaced at equal distances Λ_y along the y-axis. We can work out what happens when a light wave strikes it using qualitative arguments. The wave is infinite, plane, and of free-space wavelength λ_0; it travels at an angle θ_0 to the z-axis, and the refractive index everywhere is n, so the effective wavelength is λ_0/n.

Allowing each point to scatter isotropically and independently, we might initially expect light to be transmitted through the grating in all directions—in other words,

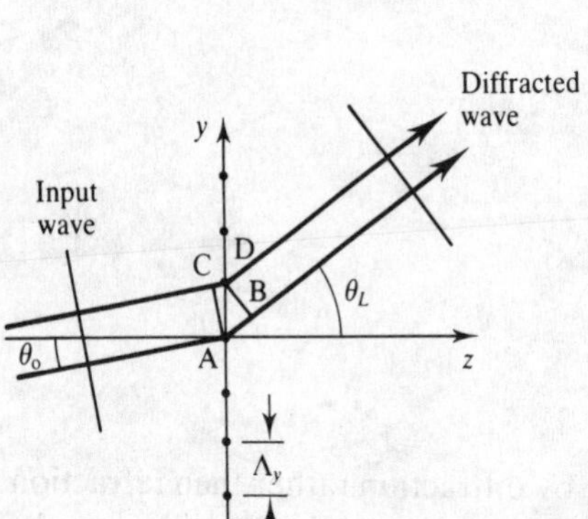

Figure 7.20 Simplified model of scattering by a periodic array.

as a spectrum of all possible plane waves. However, this argument completely ignores the periodicity of the structure. Further thought reveals that a plane wave output is actually possible only when the scattered components from any two adjacent points (e.g. A and D) add up exactly in-phase. If this is not the case, the net contribution from all the scatterers will average to zero in the far-field of the grating (see, for example, Question 4.6). This type of constructive interference can only occur when the difference between the path lengths AB and CD in Fig. 7.20 is a whole number of wavelengths. We can calculate the relevant path difference for an output angle θ_L as

$$AB - CD = \Lambda_y [\sin \theta_L - \sin \theta_0] \tag{7.18}$$

For constructive interference we require AB − CD = $L \lambda_0/n$, where L is an integer, so that

$$\sin \theta_L = \sin \theta_0 + L\lambda_0/n\Lambda_y \tag{7.19}$$

Equation (7.19) implies that constructive interference only occurs at a set of discrete angles θ_L. The action of the grating is thus to split the input beam up into a number of plane waves, travelling in different directions. These are known as *diffraction orders*; associated with each is an index L, and the solution of Eq. (7.19) gives the direction of the L th order. Since the equation contains λ_0, the angles depend on the wavelength of the optical beam, so gratings are intrinsically dispersive and will produce a series of output spectra from a polychromatic input. Equation (7.19) also shows that the diffraction orders will be more widely separated the closer the spacing of the scatters, i.e. as Λ_y decreases. To get a useful separation and dispersive power Λ_y should be of the same order as λ_0.

We cannot estimate the intensities of the individual diffraction orders from this argument, but we would expect to see many of them, of roughly equal intensity. This is often a nuisance as we may be interested in just one order. Equally we would not expect the intensities to depend much on the wavelength or the angle. This type of grating is therefore unselective, favouring no particular incidence condition or diffraction order. It is usually known as an *optically thin grating*, and the multi-wave diffraction regime is called the *Raman-Nath* regime, after its original investigators in the field of ultrasonics (Sir C. V. Raman and N. S. N. Nath).

Before proceeding further we note that the results above may be interpreted geometrically using the modified phase matching diagram shown in Fig. 7.21. This is rather different to the diagrams we have used so far and deserves a reasonably thorough explanation. First, an important new parameter, known as the *grating vector*, is defined. For historical reasons this is labelled **K**. It is oriented in the direction of the periodicity of the grating (i.e. in the y-direction in this case), and has modulus:

$$|\mathbf{K}| = 2\pi/\Lambda_y \tag{7.20}$$

Next, the usual circular locus for the propagation vectors is drawn with radius proportional to $k = nk_0$, and the propagation vector of the input wave is added. To avoid confusion with the grating vector, this is labelled $\boldsymbol{\rho}_0$. The propagation

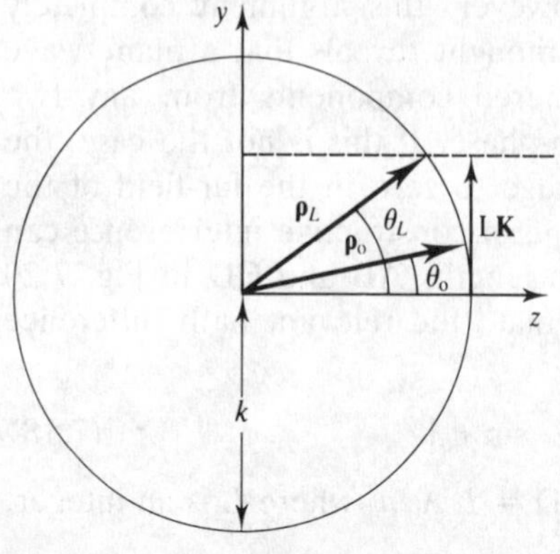

Figure 7.21 Phase matching diagram for diffraction orders in a grating.

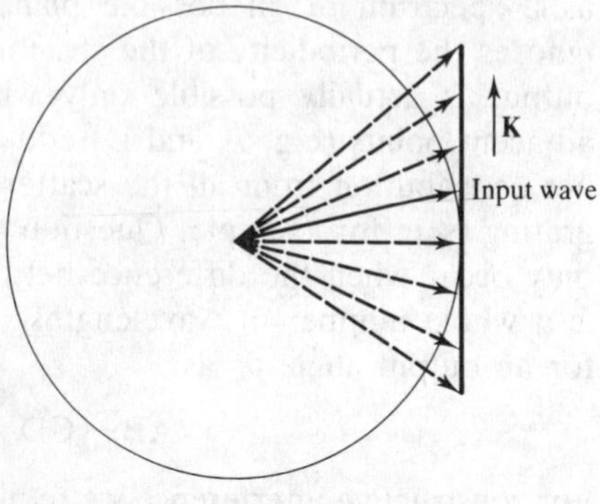

Figure 7.22 Reduced phase matching diagram for higher diffraction orders in the Raman-Nath diffraction regime.

vector $\boldsymbol{\rho}_L$ of the L th diffraction order is then found by first adding L times the grating vector to $\boldsymbol{\rho}_0$, and then finding the intersection with the circular locus of the dashed line, drawn normal to the grating boundary through the tip of this new vector. Simple geometry then shows that:

$$\rho_{Ly} = \rho_{0y} + L|\mathbf{K}| \tag{7.21}$$

so that

$$nk_0 \sin\theta_L = nk_0 \sin\theta_0 + L|\mathbf{K}| \tag{7.22}$$

or

$$(2\pi n/\lambda_0)\sin\theta_L = (2\pi n/\lambda_0)\sin\theta_0 + 2\pi L/\Lambda_y \tag{7.23}$$

Apart from a slight rearrangement, it should be clear that Eq. (7.23) is identical to Eq. (7.19). The modified phase matching diagram, which is known as a *K-vector diagram*, is therefore a useful way to predict the direction of diffraction orders geometrically. Often the dashed-line construction is omitted for clarity, being implicitly understood, and the wave vectors are taken simply as

$$\boldsymbol{\rho}_L' = \boldsymbol{\rho}_0 + L\mathbf{K} \tag{7.24}$$

This relation is termed *K-vector closure*. Figure 7.22 shows how this modified construction can be used to predict the wave directions for an optically thin grating. In this case Λ_y is large (so $|\mathbf{K}|$ is small) and the output consists of many diffraction orders, all travelling roughly parallel to the z-axis. The correction in wave direction that results from completing the construction is then only a small one.

Optically Thick Gratings

To remove the unwanted diffraction orders we need to introduce some kind of inherent selectivity into the grating. Figure 7.23 shows a suitable structure (known

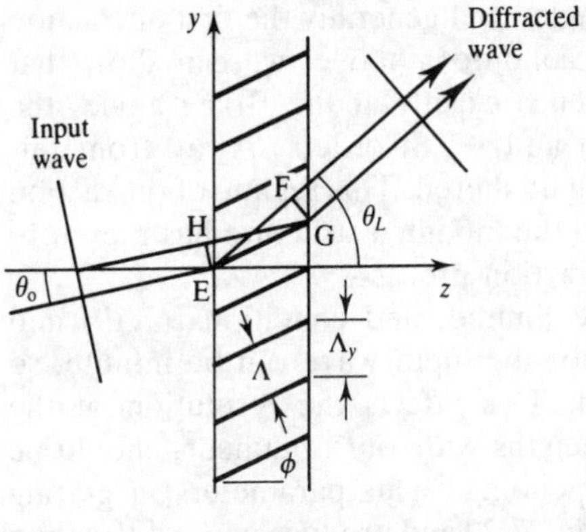

Figure 7.23 An optically thick grating.

as an *optically thick grating*) which now has a finite thickness, and consists of fringes or extended scatterers instead of points. These are slanted at an angle ϕ to the z-axis and are spaced at distances Λ apart, but the y-component of their spacing is still $\Lambda_y = \Lambda/\cos\phi$.

Again we ask what happens when the grating is illuminated by the same wave as before. Equation (7.19) must still be valid because the y-period of the grating is the same. However, the scattering is now distributed in the z-direction. This constrains the allowed diffraction angles even more. Ignoring multiple scattering, we argue that constructive interference only occurs when all scattered components add in-phase. This must be true not only for components scattered by different fringes, but also for contributions from all points along the same fringe. Path lengths like EF and HG in Fig. 7.23 must therefore be equal. This restricts the diffraction angles additionally to

$$\theta_L = \theta_0 \quad \text{or} \quad \theta_L = 2\phi - \theta_0 \tag{7.25}$$

The fringes then act as partial mirrors, allowing only transmission or reflection.

If $\theta_L = \theta_0$ the only solution to Eq. (7.19) is $L = 0$, so the only allowed wave is the 0 th diffraction order (the input wave). However, if $\theta_L = 2\phi - \theta_0$ we get

$$\begin{aligned} \sin(2\phi - \theta_0) - \sin\theta_0 &= 2\sin(\phi - \theta_0)\cos\phi \\ &= L\lambda_0/n\Lambda_y \\ &= L\lambda_0\cos\phi/n\Lambda \end{aligned} \tag{7.26}$$

or

$$2\Lambda\sin(\phi - \theta_0) = L\lambda_0/n \tag{7.27}$$

This implies that other diffraction orders can exist, but only at certain specific angles. The -1 th order, for example, is allowed at the first *Bragg angle*, which is defined by *Bragg's law*:

$$2\Lambda\sin(\theta_0 - \phi) = \lambda_0/n \tag{7.28}$$

Other waves are allowed at higher Bragg angles, but generally the first diffraction order is the most useful. Combining the two conditions above, we can show that up to two waves can exist at once. For incidence exactly at the Bragg angle, the two permitted waves are the 0 th (the input) and −1 th orders. Away from this angle, the input wave travels through the grating unaltered. This regime of diffraction is known as the two-wave or *Bragg regime*, and the modified structure has therefore introduced the desired selectivity into the diffraction process.

In practice the size of the grating will be limited, and conditions (7.19) and (7.28) relax accordingly. Neither the grating nor the input wave can be infinite, so the wave directions will be less well defined. This affects the resolution of the grating, but provided it is many optical wavelengths wide our arguments should be valid. Finite thickness is more important. Depending on its parameters, a grating may either act like the optically thin grating in Fig. 7.20 and produce many diffraction orders, or more like the optically thick grating in Fig. 7.23 and give only two. A reasonable guide is that two-wave behaviour occurs whenever the input wave has to cross many fringe planes before emerging from the grating.

Design Example

We can work out the value of Λ needed to satisfy Bragg's Law (Eq. (7.28)) for some typical values. Assuming that $\lambda_0 = 0.5145$ μm (a green argon-ion laser line) and $n = 1.6$, that the grating is unslanted (so $\phi = 0$) and that the Bragg angle is $\theta_0 = 45°$, we get $\Lambda \approx 0.23$ μm. This is an extremely small figure, which implies that a grating with a large deflection angle will be hard to make.

The Bandwidth of Optically Thick Gratings

Even in the two-wave regime restricted thickness will result in a finite angular and wavelength range over which significant diffraction occurs. The *bandwidth* can be estimated by assuming that the efficiency of the first diffraction order ($L = -1$) will be zero when there is a whole wavelength difference between contributions scattered from either end of a fringe. We can work out the change in optical wavelength needed for this to occur as follows, assuming an unslanted grating of thickness d.

The path difference between the two components is

$$\Phi = d\,\{\cos\theta_0 - \cos\theta_{-1}\} \tag{7.29}$$

When the Bragg condition is satisfied we must have $\theta_{-1} = -\theta_0$, so the path difference is zero. The change in path difference $\Delta\Phi$ accompanying a change in wavelength $\Delta\lambda_0$ is given by

$$\Delta\Phi \approx (\mathrm{d}\Phi/\mathrm{d}\lambda_0)\,\Delta\lambda_0 \tag{7.30}$$

We may evaluate Eq. (7.30) by first expanding it using the chain rule for differentiation, as

$$\Delta\Phi \approx (\mathrm{d}\Phi/\mathrm{d}\theta_{-1})\,(\mathrm{d}\theta_{-1}/\mathrm{d}\lambda_0)\,\Delta\lambda_0 \qquad (7.31)$$

$\mathrm{d}\Phi/\mathrm{d}\theta_{-1}$ and $\mathrm{d}\theta_{-1}/\mathrm{d}\lambda_0$ may then be found by differentiating Eqs (7.29) and (7.19) in turn, to get

$$\mathrm{d}\Phi/\mathrm{d}\theta_{-1} = d \sin \theta_{-1} \quad \text{and} \quad \mathrm{d}\theta_{-1}/\mathrm{d}\lambda_0 = -1/\{n\Lambda \cos \theta_{-1}\} \qquad (7.32)$$

Substituting these expressions into Eq. (7.31), and putting $\theta_{-1} \approx -\theta_0$, we then find that

$$\Delta\Phi \approx d\,\Delta\lambda_0 \sin \theta_0/\{n\Lambda \cos \theta_0\} \qquad (7.33)$$

According to our criterion the efficiency is zero when $\Delta\Phi = \lambda_0/n$, so that

$$\Delta\lambda_0/\lambda_0 \approx (\Lambda/d) \cot \theta_0 \qquad (7.34)$$

This important result implies that the bandwidth of an optically thick grating is inversely proportional to its thickness, so that the thicker the grating, the smaller the range of wavelengths over which it will diffract light efficiently.

Design Example

For the example used before, namely $\lambda_0 = 0.5145\ \mu$m, $\theta_0 = 45°$ and $\Lambda = 0.23\ \mu$m, and a moderate thickness of $d = 10\ \mu$m, we get $\Delta\lambda \approx 12$ nm. For a thicker grating with $d = 1$ mm, this reduces to $\Delta\lambda \approx 1.2$ Å. An optically thick grating can therefore act as an extremely narrow-band *wavelength filter*, and it is equally simple to show that its angular selectivity must be correspondingly high.

Optically Thick Phase Gratings

In addition to high selectivity, we will generally also require high diffraction efficiency. It turns out that an optically thick grating can be very efficient indeed, provided the structure is lossless and the fringe planes are formed by periodic modulation of the refractive index of the medium. The desired structure is therefore an optically thick *phase grating*. We cannot calculate the exact response characteristic for such a grating using the information above (though we will see how this is done for one particular geometry in Chapter 10). However, using *coupled wave theory*, it can be shown that the response characteristic is as in Fig. 7.24. This figure shows how the transmitted and diffracted beam intensities vary as either the angle of incidence or the wavelength of the input beam is changed. At Bragg incidence when $\theta_0 = \theta_B$ (or $\lambda_0 = \lambda_B$), the diffracted beam is excited efficiently. There is then a strong depletion of the transmitted beam, which may reach 100 per cent if the grating is correctly designed (as in the example above). As θ_0 (or λ_0) is changed the diffracted beam intensity falls. The transmitted intensity shows a corresponding rise, and both curves have a typical 'filter' characteristic. It turns out that the expression for the bandwidth calculated above gives surprisingly good agreement with the predictions of coupled wave theory.

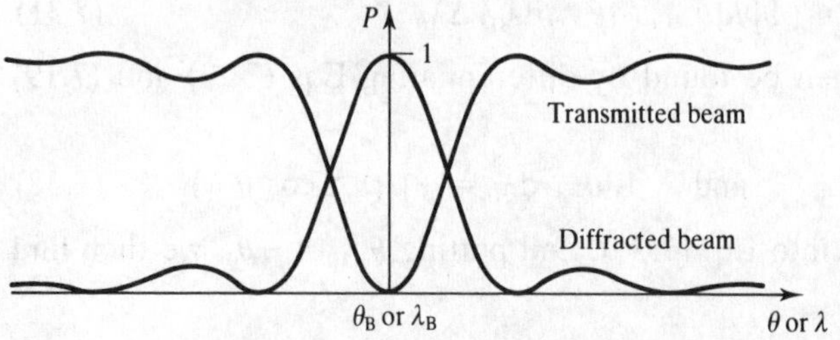

Figure 7.24 Filter response of a high efficiency, optically thick grating.

K-vector Closure in an Optically Thick Grating

The behaviour of a slanted, optically thick grating can also be predicted using a modified K-vector closure diagram. To cope with slanted fringes we simply introduce a more general grating vector **K**. This is oriented normal to the plane of the fringes, and its magnitude is

$$|\mathbf{K}| = 2\pi/\Lambda \tag{7.35}$$

The diagram is then constructed as before. For example, Fig. 7.25 shows the diagram for a slanted grating, replayed at the Bragg angle. The input wave vector $\boldsymbol{\rho}_0$ is first drawn in. The corresponding vectors $\boldsymbol{\rho}'_L$ for the diffraction orders are then computed from Eq. (7.19). When this is done it is found that the end of one such vector, $\boldsymbol{\rho}'_{-1}$, lies exactly on the circle, so that $\boldsymbol{\rho}'_{-1} = \boldsymbol{\rho}_{-1}$. This particular circumstance only arises exactly at Bragg incidence, and the -1 th order is then said to be phase-matched by the grating. This order can then be excited efficiently, because all the components scattered in the direction concerned will add in-phase. The ends of other propagation vectors (e.g. $\boldsymbol{\rho}'_{-2}$) lie much further away from the circle. This suggests that the -2 th order is far from being phase-matched and will not be excited efficiently. For other diffraction orders (e.g. the $+1$ th) the dashed line construction provides no intersection with the circle at all. These waves cannot exist as propagating waves,

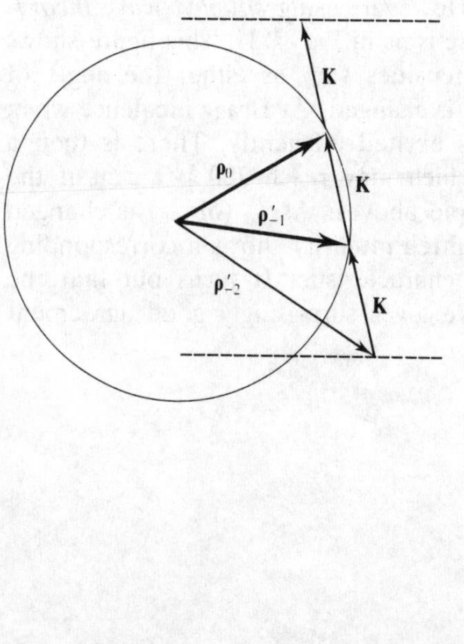

Figure 7.25 Reduced phase matching diagram for a slanted, thick grating in the Bragg diffraction regime.

and are therefore *cut off*. Thus, according to the diagram, only the two waves defined by the bold vectors are important, and we may reduce the K-vector construction accordingly. We will do so in the discussion below, which is devoted to the use of gratings in guided wave optics.

7.8 GRATINGS IN GUIDED WAVE OPTICS

A wide variety of methods can be used to construct a phase grating in guided wave optics. All that is required is that the effective index be varied periodically. Figure 7.26 shows one example, where the surface of a planar guide has been corrugated by etching through a patterned mask. The resulting changes in thickness of the guide layer then give rise to the necessary variation in effective index.

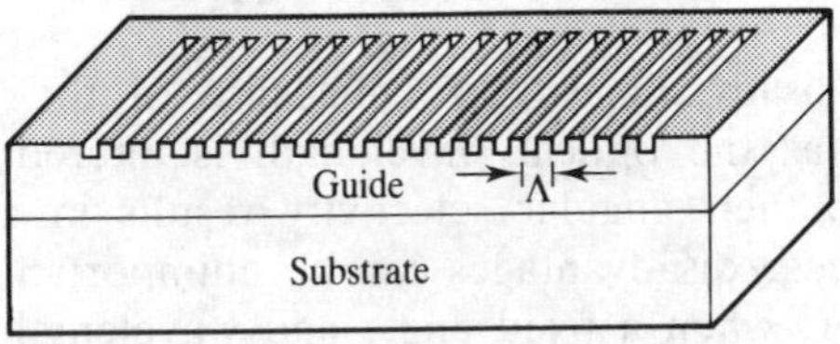

Figure 7.26 A corrugated grating.

Transmission Gratings

Figure 7.27 shows how such a corrugation might be used as a *transmission grating*. This is assumed to be optically thick, and therefore operating in the two-wave diffraction regime. Figure 7.27(a) shows a plan view of the waveguide; the input beam is incident from the left, and gives rise to two transmitted waves, the 0 th and −1 th diffraction orders, on the right. Figure 7.27(b) shows the K-vector diagram for incidence at the Bragg angle. Note that only two propagation vectors are included in the construction, and that the grating vector (which is normal to the horizontal fringes in Fig. 7.27(a)) is oriented vertically. This demonstrates an important point.

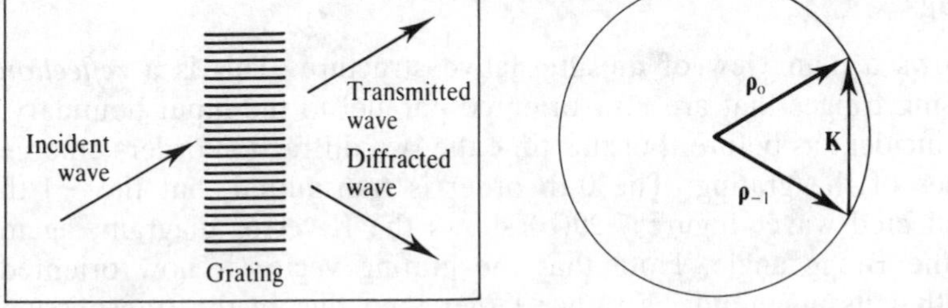

Figure 7.27 Transmission grating: (a) schematic, and (b) K-vector diagram.

A large angular deflection of the input beam has been achieved using a relatively small perturbation to the effective index of the guide, because the device operates by diffraction rather than refraction. Since small index changes are the norm in guided wave optics, gratings are ideal waveguide components in this respect.

Grating Lenses

Diffractive devices can act as more effective waveguide lenses than the overlay components discussed earlier because larger beam deflection allows a higher numerical aperture. Figure 7.28 shows one example, the *chirped grating lens*. Here a slow variation in the grating period produces a corresponding change in the deflection angle over the lens aperture, which results in an approximate focusing action. A linear chirp (i.e. a linear variation in period) is suitable for focusing an off-axis plane input beam, as shown. More generally the grating fringes must be curved, and the local fringe orientation and spacing is found by local application of the Bragg condition.

Grating lenses suffer from two main disadvantages. First, because of their inherent dispersion, the focal spot moves as the optical wavelength is altered, and can become highly aberrated. Second, their angular selectivity results in a greatly restricted field of view. However, these disadvantages can be unimportant (for example, in a telescope beam expander, when a fixed angle and wavelength are used).

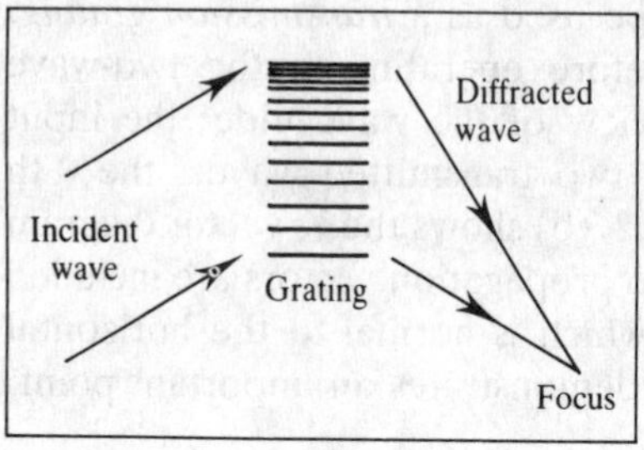

Figure 7.28 The chirped grating lens.

Reflection Gratings

Figure 7.29(a) shows a plan view of an alternative structure. This is a *reflection grating*, formed using fringes that are now oriented parallel to the input boundary. The input wave is incident as before, but this time the two diffraction orders emerge from different sides of the grating. The 0 th order is transmitted, but the −1 th order is now a reflected wave. Figure 7.29(b) shows the K-vector diagram, again for incidence at the Bragg angle. Note that the grating vector is now oriented horizontally, and that its magnitude is rather larger than that in the transmission grating above. Reflection gratings therefore typically require much smaller grating periods.

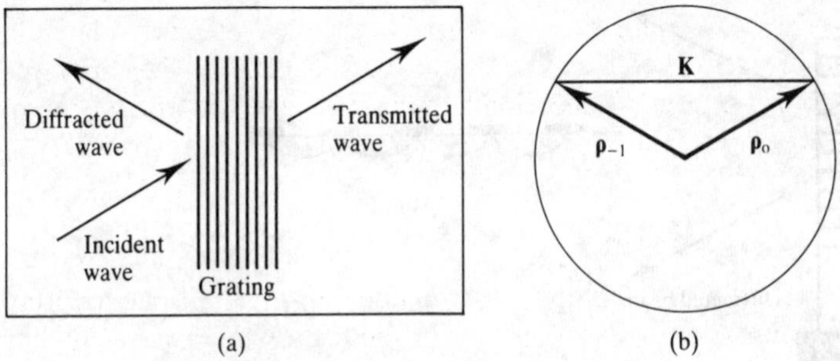

Figure 7.29 Reflection grating: (a) schematic, and (b) K-vector diagram.

'Two-dimensional' Gratings

It is of course unnecessary to consider grating boundary shapes to be fixed in the form of the simple slab geometries considered so far. Figure 7.30 shows a grating whose fringes are oriented at 45° to the edges of a boundary, which has the form of a long, thin rectangle. Here a relatively narrow input beam can be diffracted as a much wider output beam. The device therefore functions as a compact *beam expander*, which could replace an expansion telescope. However, the grating strength must be carefully tailored along its length to ensure a uniform amplitude distribution in the output beam.

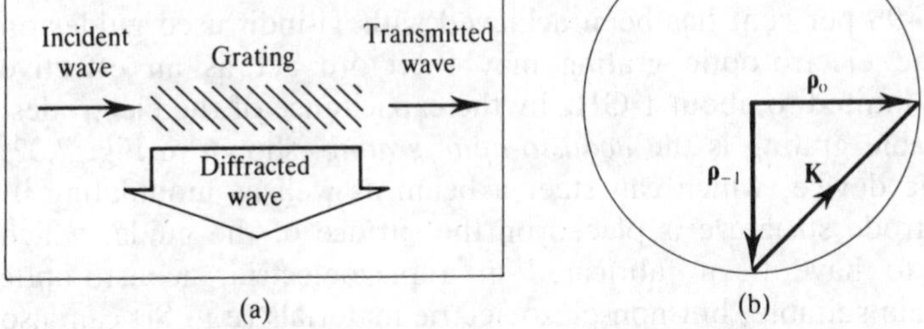

Figure 7.30 Beam expander grating: (a) schematic, and (b) K-vector diagram.

Electro- and Acousto-optic Gratings

It is not necessary to consider the grating to be a static entity. There are two common forms of electrically-switchable grating, which operate using combinations of the material effects discussed in Chapter 3. The first is the *electro-optic grating*, shown in Fig. 7.31(a). Here a periodic metal structure, known as an *interdigital electrode*, is placed on the surface of the guide, which is assumed to have been

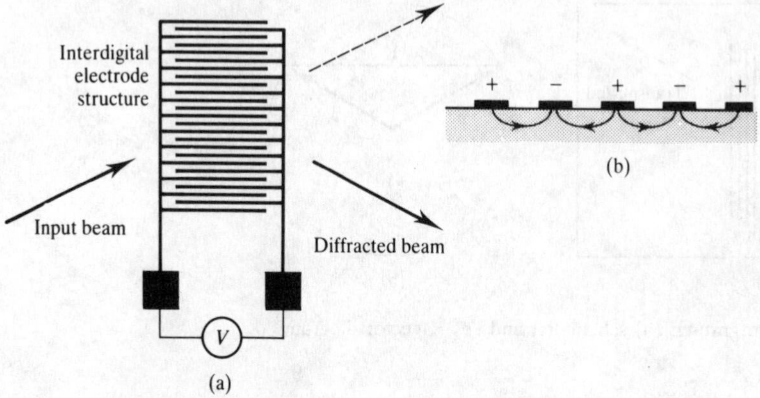

Figure 7.31 The electro-optic grating: (a) basic principle, and (b) electric field distribution.

fabricated in an electro-optic material (e.g. $LiNbO_3$ or $LiTaO_3$). With no voltage applied to the electrodes there is no grating. However, when a static voltage is applied, the 'finger' voltages alternate in sign. This results in a periodic variation in electric field beneath the electrodes (Fig. 7.31(b)), which induces a corresponding variation in index in the guide through the electro-optic effect. The grating may therefore be switched on and off at will; however, the beam deflection angle (which is determined by the electrode pitch) is fixed. Usually it is rather small since the lithographic process used to define the electrodes is limited to a linewidth of ≈ 0.5–$1\ \mu m$, which yields a period of $\Lambda \approx 2$–$4\ \mu m$. Despite this, the electrodes may be sufficiently long that the device acts as a volume grating, so the diffraction efficiency can be high—98 per cent has been achieved with Ti-indiffused guides on $LiTaO_3$ substrates. The electro-optic grating may therefore act as an effective modulator. Its speed is limited to about 1 GHz by the capacitance of the electrodes.

The second switchable grating is the *acousto-optic grating*, shown in Fig. 7.32. This is a more versatile device, which can steer a beam as well as modulating it. Again, a periodic electrode structure is placed on the surface of the guide, which is assumed this time to have been fabricated in a piezoelectric, acousto-optic material. $LiNbO_3$ is again suitable, but non-piezoelectric materials (e.g. Si) can also be used if a layer of piezoelectric material (typically ZnO) is deposited between the electrodes and the substrate.

This electrode structure, which is known as an *interdigital transducer*, does not itself act as the grating. However, when it is excited by an a.c. radio-frequency voltage it creates a time-varying, spatially-periodic electric field in the material beneath, which in turn creates a similar pattern of stress through the piezoelectric effect. This type of stress distribution effectively corresponds to a pattern of standing acoustic waves, which can of course be decomposed into the sum of two travelling waves. The net result is the excitation of two acoustic waves which emerge from

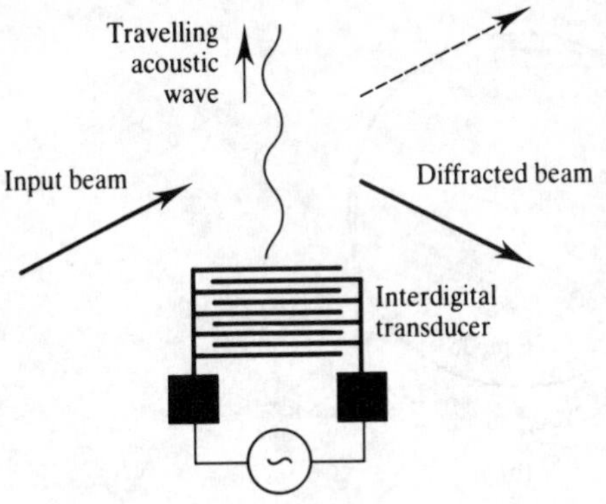

Figure 7.32 The acousto-optic grating.

beneath the transducer travelling in opposite directions. Since the waves propagate near the surface of the material they are known as *surface acoustic waves* (or SAWs). Typically only one wave is required, and the other is removed using a surface absorber. The remaining wave creates a moving variation in effective index through the acousto-optic effect, and it is this which acts as the grating. Once again the grating may be switched on and off, so the device can act as a modulator. This time, speed is mainly limited by excessive acoustic propagation losses at frequencies above $\approx$ 1 GHz. Most importantly, the grating period may be varied over a limited range (depending on the transducer bandwidth) by changing the RF frequency. This alters the diffraction angle, so acousto-optic gratings may also be used as beam deflectors. Diffraction may again take place in the Bragg regime at high frequencies, but the Raman-Nath regime is often used to minimize the efficiency variations that occur when the frequency is altered. One further important difference from the electro-optic grating is that the diffraction orders are all frequency-shifted (by integer multiples of the acoustic frequency) through the *Doppler effect*. The acousto-optic grating may therefore also be used as a frequency modulator.

Design Example

We may estimate the parameters of a guided wave acousto-optic beam deflector formed on a lithium niobate substrate as follows. For $LiNbO_3$, the acoustic velocity is 6.57×10^3 m/s and the refractive index is $n_e = 2.2$. Hence at a typical operating frequency of 500 MHz, the acoustic wavelength is $6.57 \times 10^3/500 \times 10^6 = 1.31 \times 10^{-5}$ m, or 13.1 μm. The period Λ of the interdigital transducer must have the same value, so assuming a mark-to-space ratio of unity, each finger electrode must be $13.1/4 = 3.275$ μm wide, a feature size well within the capacity of photolithography. Assuming an optical wavelength of 0.633 μm (red light from a He-Ne laser) and normal incidence, the angle of deflection of the first diffraction order will be $\theta_1 = \sin^{-1}\{\lambda_0/n\Lambda\} = \sin^{-1}\{0.633/2.2 \times 13.1\} \approx 1.25°$.

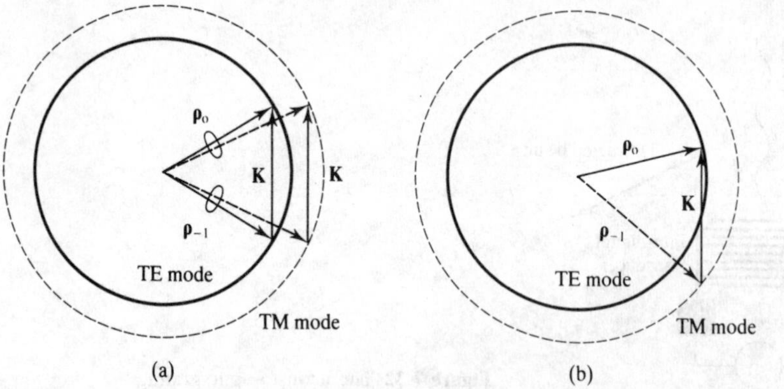

Figure 7.33 K-vector diagrams illustrating (a) TE-TE and TM-TM diffraction, and (b) TE-TM mode conversion.

Mode Conversion by a Grating

Gratings are not restricted to interactions involving the phase matching of two modes of the same type. Consider, for example, a single-moded guide. We have seen that this can support two polarization modes; consequently, even in this case there are two possible loci for propagation vectors. Figure 7.33(a) shows the K-vector diagram illustrating the conversion of a TE-moded input into a TE-moded diffracted beam. Here both $\boldsymbol{\rho}_0$ and $\boldsymbol{\rho}_{-1}$ lie on the full-line circle, the TE-mode locus. However, an equally valid possibility is the diffraction of a TM-moded input into a TM-moded output. This is shown by the alternative construction, where $\boldsymbol{\rho}_0$ and $\boldsymbol{\rho}_{-1}$ lie on the dashed-line TM-mode locus. Because of the difference in radii of the circles, the Bragg angles for TE-TE and TM-TM diffraction are slightly different. One further possibility exists: a TE-moded input may be diffracted as a TM-moded output. The phase matching condition for this case is shown in Fig. 7.33(b); once again the Bragg angle involved is different. In this case *mode conversion* accompanies diffraction. In designing a grating device, therefore, care is needed to ensure that only the desired interaction occurs.

Grating Couplers

A grating may even be used to phase match guided and radiation modes, forming an input/output coupler similar to the prism coupler discussed earlier. This device is known as the *grating coupler*, and its principle of operation is shown in Fig. 7.34.

Here a plane input beam is incident from free-space on a corrugated grating, which is fabricated on the surface of a planar waveguide. Figure 7.35 shows the corresponding phase matching diagram. If the angle of incidence θ_1 is correct, the z-component of the first diffraction order of the grating will equal the z-component of the propagation vector of the guided mode. This occurs when

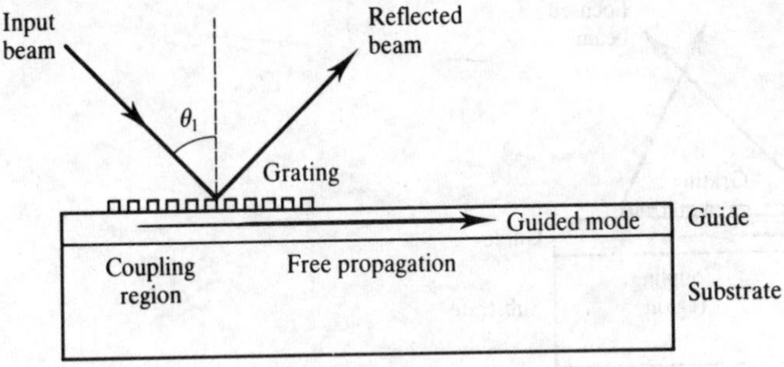

Figure 7.34 The grating coupler.

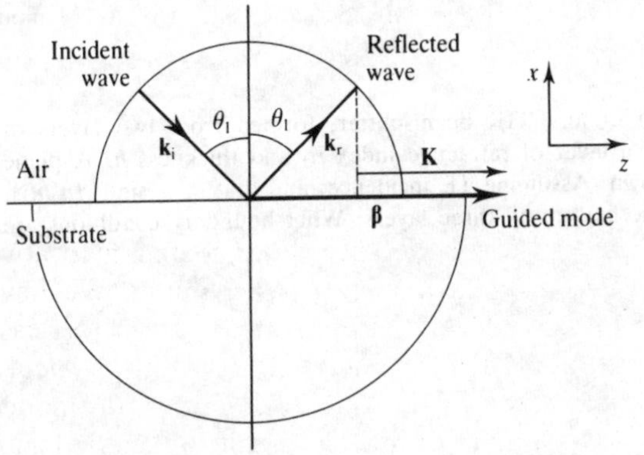

Figure 7.35 Phase matching in the grating coupler.

$$k_{iz} + |\mathbf{K}| = |\boldsymbol{\beta}| \tag{7.36}$$

If this is the case power will be transferred to the guided beam, much as in the prism coupler.

Although prism and grating couplers are functionally similar, the latter have several important advantages. First, they are flat, rugged, and fully integrated with the waveguide. Second, it is possible to design grating couplers with a varying periodicity, which can perform multiple functions—for example, they can couple a guided beam into free-space, and focus it at the same time. Figure 7.36 shows the *chirped grating coupler*, which contains a grating with linearly-varying period. A guided mode incident on such a grating will be diffracted into a free-space beam whose direction of propagation depends on the local grating period. In this case the output will be a cylindrical wave converging on a line focus above the guide. However, by simultaneously chirping the grating period and curving the grating fringes, an output can be generated that converges onto a point focus. This is important in one of the applications discussed in Chapter 14, the integrated optic disc pickup.

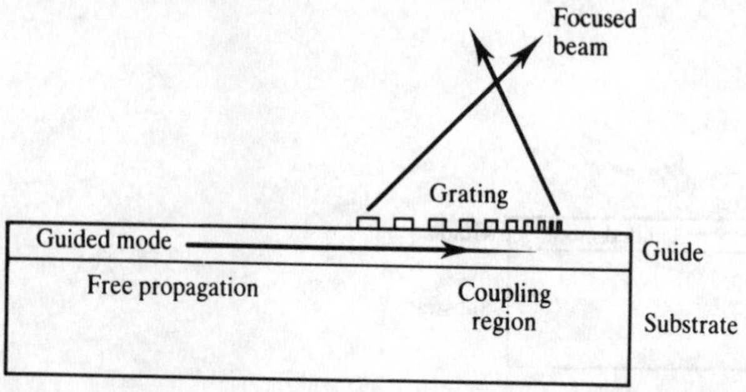

Figure 7.36 The chirped grating coupler.

PROBLEMS

7.1 The figure below shows the cross-section of an FTIR beamsplitter, formed from two layers of dielectric of refractive index n_1, separated by a layer of refractive index n_2 and thickness h. A plane wave is incident on the upper interface as shown. Assuming TE incidence, and that $\theta_1 > \sin^{-1}(n_2/n_1)$, write down general expressions for the electric field in all three layers. What boundary conditions are satisfied by these expressions?

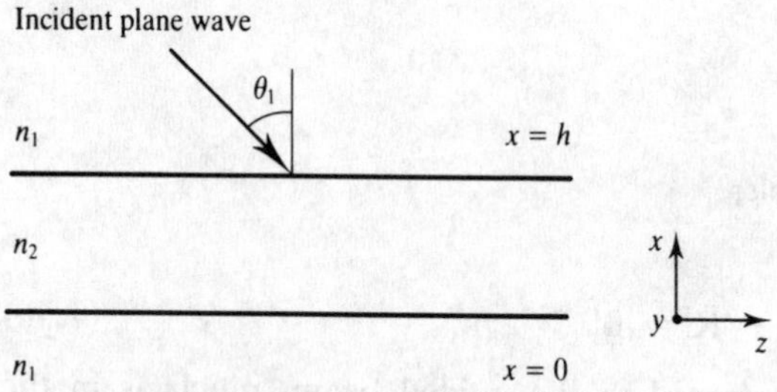

7.2 Repeat the analysis of Question 7.1, assuming a solution in modal form. Show that, provided h is large enough, the amplitude of the transmitted wave decreases exponentially with h.

7.3 A high-index glass prism ($n = 1.7$) is used at 1.5 μm wavelength to characterize a planar guide formed by ion exchange in glass. Assuming that the guide supports three modes, with effective indices of 1.515, 1.530 and 1.545, respectively, sketch the variation with launch angle that you would expect in prism reflectivity. What are the propagation constants of the three modes?

7.4 A step-change in thickness is etched into a planar guide. On the left of the step the guide is two-moded (with effective indices for the two modes of 1.504 and 1.505, respectively). On the right it is single-moded, with effective index 1.5025. The lowest-order mode is incident from the left at an angle θ_1. Sketch the phase matching diagrams you would expect for (a) $\theta_1 \approx 0°$, and (b) $\theta_1 = 87.5°$, and interpret your results.

7.5 Find the most compact arrangement you can for a $\times 5$ telescope beam expander, assuming that it is to be fabricated on a planar waveguide using step-index overlay lenses. The input beam width is 0.5 mm, the effective index of the guide is 1.6, and the maximum change in effective index that can be obtained using the overlay without inducing mode conversion is 0.02.

7.6 Calculate the Bragg angle for an unslanted transmission grating of period $\Lambda = 1.0\ \mu m$, in a material of refractive index 1.6. The optical wavelength is 1.5 μm. Estimate the spectral bandwidth of the grating, assuming that it is optically thick, with a physical thickness of 50 μm. How many fringe planes are crossed by the input beam before it emerges from the grating?
[27.95°; 56.5 nm; ≈ 27]

7.7 Show that the angular bandwidth of an unslanted transmission grating of period Λ and thickness d is given by $\Delta\theta_0 \approx \Lambda/d$. Hence calculate the bandwidth of a grating of thickness 20 μm in a medium of refractive index 1.6, assuming that the Bragg angle is 30° at $\lambda_0 = 0.633\ \mu m$.
[1.13°]

7.8 Show that Bragg's law may be written in terms of the grating vector and the propagation constant as $2k \sin(\theta_0 - \phi) = |\mathbf{K}|$. Interpret this geometrically using a phase-matching diagram.

7.9 Sketch the K-vector diagrams you would expect for the following geometries: (a) an unslanted transmission grating of thickness 5 μm and period 5 μm, for normal incidence, and (b) an unslanted reflection grating of thickness 5 μm and period 0.5 μm, for incidence at the Bragg angle. The refractive index of the medium is 1.6, and the optical wavelength is 0.633 μm.

7.10 Discuss the advantages and disadvantages of Bragg grating lenses, as compared with other types of waveguide lens. You should consider size, bandwidth, and ease of fabrication.

SUGGESTIONS FOR FURTHER READING

Chen, B., and Ramer, O. G. Diffraction limited geodesic lenses for integrated optic circuits, *IEEE J. Quant. Elect.*, **QE-15**, 853–860, 1979.

Dakss, M. L., Kuhn, L., Heidrich, P. F., and Scott, B. A. Grating coupler for efficient excitation of optical guided waves on thin films, *Appl. Phys. Lett.*, **16**, 523–525, 1970.

Dalgoutte, D. G., and Wilkinson, C. D. W. Thin grating couplers for integrated optics: an experimental and theoretical study, *Appl. Opt.*, **14**, 2983–2998, 1975.

Flanders, D. C., Kogelnik, H., Schmidt, R. V., and Shank, C. V. Grating filters for thin-film optical waveguides, *Appl. Phys. Lett.*, **24**, 194–196, 1974.

Forouhar, S., Lu. R.-X., Chang, W. S. C., Davis, R. L., and Yao, S. K. Chirped grating lenses on Nb_2O_5 transition waveguides, *Appl. Opt.* **22**, 3128–3132, 1983.

Gerber, J., and Kowarschik, R. Photoresist Bragg gratings as beam splitters and beam reflectors for integrated optical systems, *Opt. Quant. Elect.*, **19**, 49–58, 1987.

Hammer, J. M. Modulation and switching of light in dielectric waveguides. In "Integrated Optics", T. Tamir (ed.), Springer-Verlag, Berlin, 1979.

Heitmann, D., and Ortiz, C. Calculation and experimental verification of two-dimensional focusing grating couplers, *IEEE J. Quant. Elect.*, **QE-17**, 1257–1263, 1981.

Lean, E. G. H., White, J. M., and Wilkinson, C. D. W. Thin-film acousto-optic devices, *Proc. IEEE*, **64**, 779–788, 1976.

Ohmachi, Y. Acousto-optic TE_0-TM_0 mode conversion in a thin film of amorphous tellurium dioxide, *Elect. Lett.*, **9**, 539–541, 1973.

Schmidt, R. V. Acoustooptic interactions between guided optical waves and acoustic surface waves, *IEEE Trans. Sonics & Ultrasonics*, **SU-23**, 22–33, 1976.

Suhara, T., and Nishihara, H. Integrated optics components and devices using periodic structures, *IEEE J. Quant. Elect.*, **QE-22**, 845–867, 1986.

Tien, P. K., Ulrich, R., and Martin, R. J. Modes of propagating light waves in thin deposited semiconductor films, *Appl. Phys. Lett.*, **14**, 291–294, 1969.

Tsai, T. S. Guided-wave acousto-optic Bragg modulators for wide-band integrated optic communications and signal processing, *IEEE Trans. Circuits & Systems*, **CAS-26**, 1072–1098, 1979.

Ulrich, R., and Martin, R. J. Geometrical optics in thin-film light guides, *Appl. Opt.*, **10**, 2077–2085, 1971.

Valette, S., Morque, A., and Mottier, P. High performance integrated Fresnel lenses on oxidised silicon substrates, *Elect. Lett.*, **18**, 13–15, 1982.

Walpita, L. M., and Pitt, C. W. Fabrication and evaluation of linear constant period beam expander gratings, *Opt. Comm.*, **52**, 241–246, 1984.

Yao, S. K., and Anderson, D. B. Shadow sputtered diffraction limited waveguide Luneberg lenses, *Appl. Phys. Lett.*, **33**, 307–309, 1978.

Yariv, A., and Nakamura, N. Periodic structures for integrated optics, *IEEE J. Quant. Elect.*, **QE-13**, 233–253, 1977.

CHAPTER

EIGHT

OPTICAL FIBRES AND FIBRE DEVICES

8.1 OPTICAL FIBRE TYPES

In this chapter we shall consider the properties of *optical fibre*, by far the most important type of waveguide for the transmission of information at optical frequencies. Optical fibre usually (though not always) has a circular symmetry, and is most often fabricated from very pure silica, or SiO_2. This has a refractive index of $n \approx 1.458$ at $\lambda_0 = 850$ nm. Figure 8.1 shows the typical geometry of a silica fibre; by selective doping, the centre of the fibre (which is known as the *core*) is arranged to have a higher refractive index than the surrounding region (the *cladding*), thus forming a waveguide. Useful dopants include germania (GeO_2) and phosphorus pentoxide (P_2O_5), both of which increase the refractive index of silica, and boric oxide (B_2O_3) and fluorine (F_2), which reduce it. A typical fibre might therefore consist of a GeO_2:SiO_2 core, with a SiO_2 cladding. Alternatively, a pure SiO_2 core could be used, with a B_2O_3:SiO_2 cladding.

Other low-loss fibre materials that have been investigated include low melting-point silicate glasses (soda-lime silicates, germanosilicates and borosilicates) and halide crystals (TlBrI, TlBr, KCl, CsI and AgBr). Most recently, attention has been transferred to the fluoride glasses (mixtures of GdF_3, BaF_2, ZrF_4 and AlF_3) for reasons which will be discussed in the following section. However, whatever the

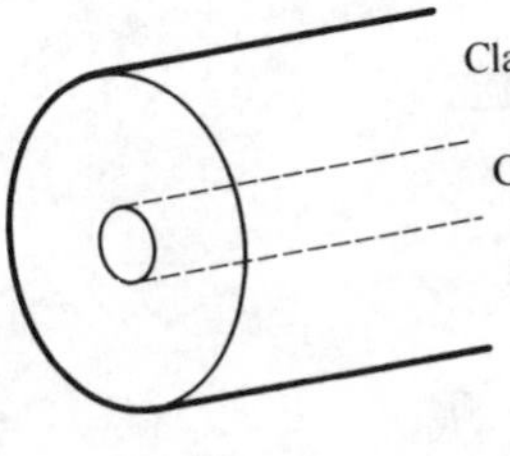

Figure 8.1 An optical fibre.

exact composition of the fibre, the cladding is almost always surrounded by a protective plastic jacket, which acts as a strain relief and prevents the ingress of water.

Alternatively, *plastic-coated silica* (PCS) *fibres* which might consist of a natural quartz core clad in silicone resin) and all-plastic fibres (for example, consisting of a polystyrene core with a methyl methacrylate cladding) are often used. Although these are cheap, they suffer from considerably higher propagation loss than silica fibre ($\approx$ 10 dB/km for PCS fibre and $\approx$ 500 dB/km for all-plastic fibre), and are therefore suitable only for transmission over short distances. For all-plastic fibres the limit is normally tens of metres.

There are many variations on the basic geometry. Some are illustrated in Fig. 8.2, which shows a number of different cylindrically-symmetric fibre cross-sections, and the radial refractive index distribution defining them. The most important distinctions are as follows. First, optical fibre is generally either single-moded or very multi-moded—the number of modes supported by a multi-moded fibre is normally very large (hundreds). Second, the core may be defined either by an abrupt increase in the refractive index, or by a more gradual variation in index (which is typically parabolic). The former is known as step-index fibre, the latter as graded-index fibre.

The dimensions of these fibre types differ quite considerably. Single-mode fibre has roughly the dimensions of a human hair, with a cladding diameter of the order of 125 μm, and a core diameter of 8–12 μm (depending on the operating wavelength). Multi-mode fibre is rather thicker: cladding diameters range from 125 to 400 μm, while the core diameters range from 50 to 200 μm. For field use fibres are supplied in cable form. Cable styles vary widely, but a typical construction would consist of a plastic tube, strengthened with steel wires or Kevlar to resist tensile strain, which is filled with the fibre in a loose helical lay. A recent development has been the availability of fibre in the form of *ribbon cable*. This consists of a number (e.g. 8) of single-mode fibres held parallel in a flat ribbon by a plastic matrix. The individual fibres are often colour-coded, and may be separated out into individual strands, allowing multiple-fibre links to be constructed very simply.

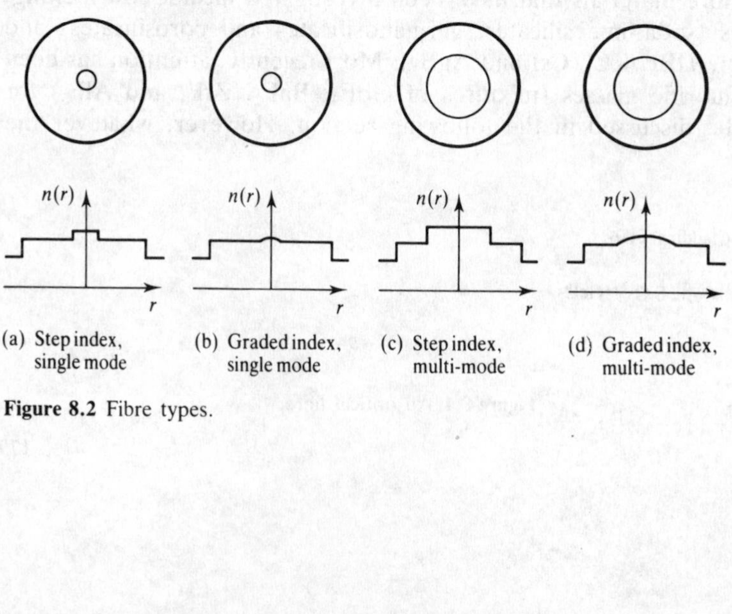

Figure 8.2 Fibre types.

8.2 LOSS IN SILICA AND FLUORIDE GLASS FIBRE

One of the most important features of the optical fibre is its exceptionally low propagation loss, which is normally quoted in units of *decibels* (dB), according to

$$\text{loss} = -10 \log_{10} (P_{out}/P_{in}) \text{ dB} \tag{8.1}$$

Thus, reduction of the power by 50 per cent between input and output corresponds to 3 dB loss. Expressed in this way, loss varies linearly with distance, so fibre may be characterized by attenuation in dB/km. Current fibre performance has been achieved through a considerable research and development effort, which began in the mid-1960s. In fact, loss in silica-based fibre has been reduced quite staggeringly, from more than 1000 dB/km in 1966 to the present value of ≈ 0.2 dB/km (at $\lambda_0 = 1.55\ \mu$m).

There are a number of fundamental mechanisms which combine to set a lower limit on transmission loss. These are illustrated by Fig. 8.3, which shows a typical plot of attenuation versus optical wavelength for silica fibre. At short wavelengths ($\lambda_0 < 1.6\ \mu$m) attenuation is mainly dominated by scattering loss (rather than by absorption due to electronic transitions at ultraviolet wavelengths, as might be expected). The effect itself is known as *Rayleigh scattering*; its origins lie in any small inhomogeneities and imperfections in the structure of the glass forming the fibre—for example, compositional fluctuations on solidification, trapped gas bubbles, dopants and so on. It can be shown that this type of scattering varies as $1/\lambda_0^4$, so it is responsible for the sharp rise in attenuation in Fig. 8.3 at short wavelengths.

At longer wavelengths the attentuation is mainly caused by *intrinsic absorption*, arising from the excitation of lattice transitions at near-infrared wavelengths. Minimum attenuation is therefore obtained when the Rayleigh scattering and infrared absorption curves cross, which occurs at around $\lambda_0 = 1.55\ \mu$m. If low loss is the main criterion, this is the most desirable operating point. Modern fibres approach the two envelopes of Rayleigh scattering and infrared absorption very closely. However, there may also be significant attenuation near a number of discrete wavelengths, as can be seen in Fig. 8.3. This type of *extrinsic absorption* can be due

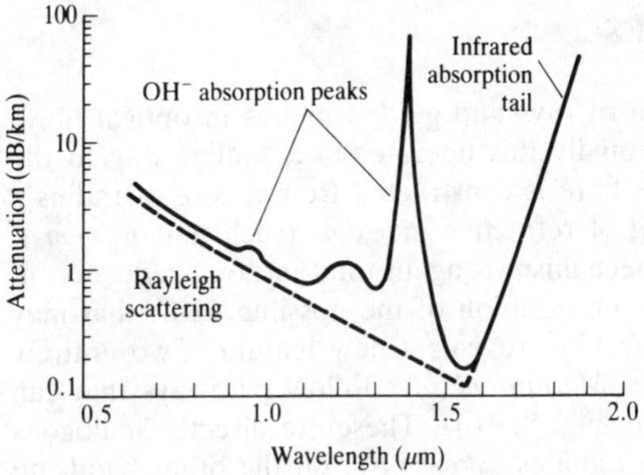

Figure 8.3 Typical attenuation characteristics for silica-based optical fibre.

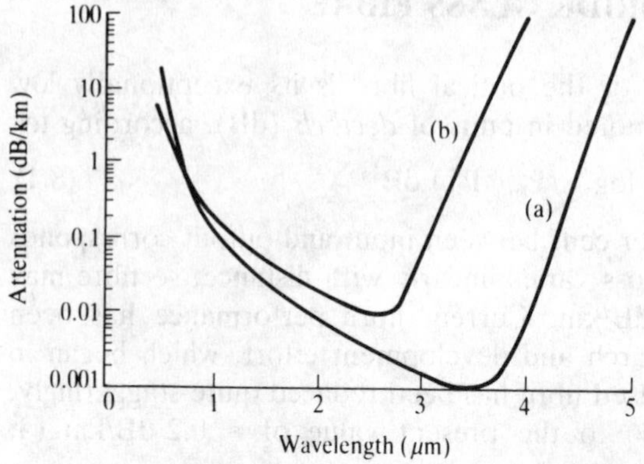

Figure 8.4 Predicted loss spectra for two fluoride glass fibres (*based on data by Shibata et al 1981*).

either to the dopants themselves, or to additional impurities. The most significant absorption band lies at $\lambda_0 = 1.39\ \mu m$, and is caused by the presence of residual *hydroxyl* (OH^-) *ions*, which also give rise to a number of smaller absorption peaks. These ions originate as water contamination, and must be removed by careful dehydration. If this is done, the ultimate limit of low-loss can be achieved almost exactly.

If different materials could be found that have their infrared absorption lines at longer wavelengths, the operating wavelength could be shifted in the same direction. The effect of Rayleigh scattering would then be reduced, and lower minimum loss obtained. This is the thinking behind current research on *fluoride glass fibres*. It is believed that such fibres could have an optimum attenuation as low as 0.001 dB/km, an incredibly small figure. Figure 8.4 shows theoretically-predicted loss spectra for two different fluoride glass fibres based on (a) BaF_2-GdF_4-ZrF_4 glass and (b) CaF_2-BaF_2-YF_3-$A1F_4$ glass. These suggest that improved performance should be available at slightly longer wavelengths, in the range $\lambda_0 = 2.5$–$3.5\ \mu m$. Although this encouraging prediction has yet to be fulfilled (the current "best" is only 0.7 dB/km), research is actively being continued.

8.3 STEP-INDEX OPTICAL FIBRES

We shall now consider the propagation of rays and guided modes in optical fibre, beginning with step-index fibre. Functionally this operates in a similar way to the slab guide introduced in Chapter 6; the fibre is constructed from a core of radius a and refractive index n_1, and a cladding of refractive index n_2 (such that $n_1 > n_2$), as shown in Fig. 8.5(a). The guiding mechanism is again total internal reflection at the core/cladding interface. However, consideration of the possible paths that may be followed by rays travelling down the fibre reveals a new feature. Two entirely different types of ray are now possible. *Meridional rays* follow pathways that can be drawn on a single plane (as shown in Fig. 8.5(b)). These are directly analogous to the rays that propagate in planar slab guides. *Skew rays*, on the other hand, do

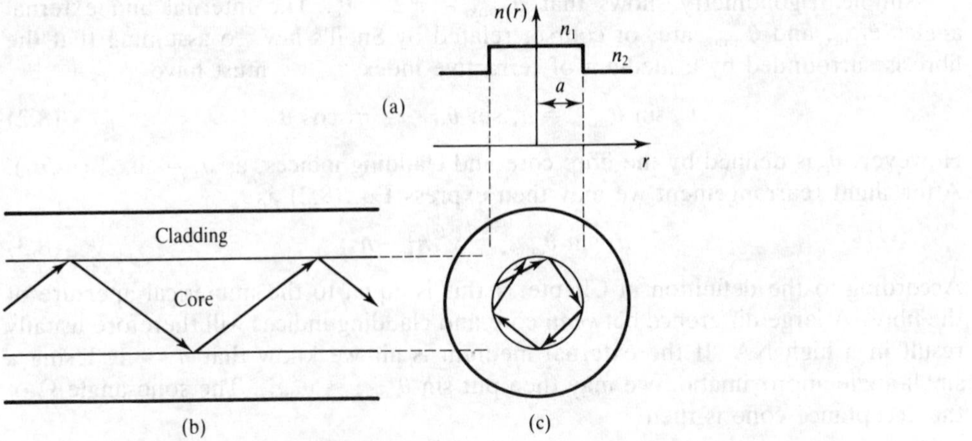

Figure 8.5 (a) The refractive index distribution of step-index fibre; (b) and (c) propagation of meridional and skew rays.

not travel in the exact centre of the fibre, but propagate in an annular region near the outer edge of the core, following off-axis helical paths that effectively circle the fibre axis after a finite number of reflections. These have no such analogy in a slab model and are best visualised in terms of their projection on the fibre cross-section (Fig. 8.5(c)). The existence of such rays implies that the analysis of an optical fibre will be more difficult than a simple extension to cylindrical geometry would suggest.

Numerical Aperture

One of the most important parameters of a fibre is its *numerical aperture*, or NA. This determines the acceptance cone of the fibre and the angular spread of radiation it may emit. Despite the distinction between meridional and skew rays outlined above, a reasonable approximation to the NA may be obtained by considering only the former, as shown in Fig. 8.6. The extent of the acceptance cone will be determined by the maximum angle θ_{emax} that an external ray entering the end-face of the fibre may have, while still giving rise to totally-reflected rays within. This is limited by the onset of cut-off, so the largest angle θ_{1max} allowed inside the fibre must be set by the critical angle, θ_c.

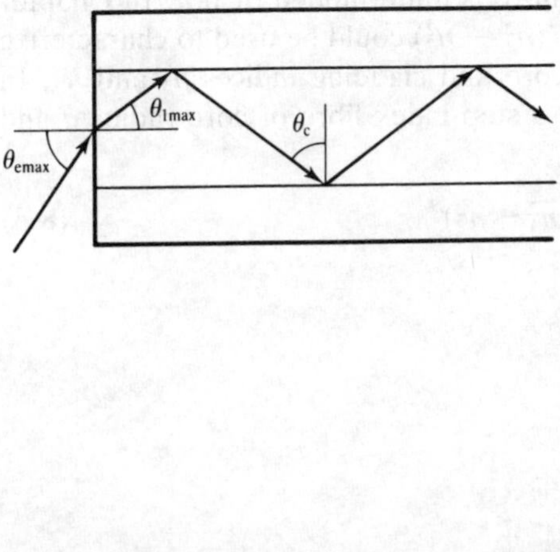

Figure 8.6 Geometry for calculation of the numerical aperture of a step-index fibre.

Simple trigonometry shows that $\theta_{1\max} = \pi/2 - \theta_c$. The internal and external angles $\theta_{1\max}$ and $\theta_{c\max}$ are, of course, related by Snell's law, so assuming that the fibre is surrounded by a medium of refractive index n_c, we must have

$$n_c \sin \theta_{c\max} = n_1 \sin \theta_{1\max} = n_1 \cos \theta_c \tag{8.2}$$

However, θ_c is defined by the fibre core and cladding indices, as $\theta_c = \sin^{-1}(n_2/n_1)$. After slight rearrangement we may then express Eq. (8.2) as

$$n_c \sin \theta_{c\max} = \sqrt{(n_1^2 - n_2^2)} \tag{8.3}$$

According to the definition in Chapter 4 this is equal to the numerical aperture of the fibre. A large difference between core and cladding indices will therefore usually result in a high NA. If the external medium is air we know that $n_c = 1$. Using a small-angle approximation we may then put $\sin \theta_{c\max} \approx \theta_{c\max}$. The solid-angle Ω of the acceptance cone is then

$$\Omega = \pi\, \theta_{c\max}^2 \approx \pi(n_1^2 - n_2^2) \tag{8.4}$$

If skew rays are considered as well it can be shown that both the numerical aperture and the value of Ω derived above are underestimates, but they remain reasonable approximations.

Exact Modes of a Step-index Fibre

Full analysis of a step-index fibre requires the solution of Maxwell's equations for a cylindrical geometry. Consequently the fields are constructed from more abstract mathematical functions than were required for the slab guide (actually, Bessel functions), and the modes themselves are more complicated. For many of them the axial field components E_z and H_z are both non-zero, so these modes are known as *hybrid* rather than TE or TM modes. Furthermore, due to the circular cross-section, the transverse fields have an angular variation as well as a radial one, so they require two identifying parameters instead of a single mode number. Because of these complications we will omit the detailed analysis (to be found in many other texts) and present only the broad conclusions.

The modes of the step-index optical fibre are, in general, classified as $TE_{0,\nu}$, $TM_{0,\nu}$, $EH_{\mu,\nu}$ or $HE_{\mu,\nu}$ types. The last two are hybrid fields, which have E_z and H_z as the dominant axial component, respectively. The most important mode is the $HE_{1,1}$ mode (roughly equivalent to the TE_0 and TM_0 modes of the slab guide) which cuts off at zero frequency. However, the $TE_{0,1}$ and $TM_{0,1}$ modes are also significant. These have the next lowest cut-off frequencies, and so their presence or absence may be used to define whether the fibre is multi-moded or not. In Chapter 6 we showed that a parameter $V = (k_0 h/2)\sqrt{(n_1^2 - n_2^2)}$ could be used to characterize a symmetric slab guide of thickness h and core and cladding indices n_1 and n_2. In the same way a V-value may be defined for a step-index fibre of core radius a and similar indices, in the form:

$$V = (k_0 a)\sqrt{(n_1^2 - n_2^2)} \tag{8.5}$$

It turns out that the $TE_{0,1}$ and $TM_{0,1}$ modes cut off when $V = 2.405$. Thus, if the frequency is such that $V < 2.405$, only one mode will propagate and the fibre is single-moded. Assuming that $n_1 \approx n_2 \approx n$, and $n_1 - n_2 = \Delta n$ (which will be the case in weakly-guiding fibres) the minimum wavelength for single-mode operation is therefore:

$$\lambda_0 = (\pi d/2.405)\sqrt{(2n\,\Delta n)} \qquad (8.6)$$

where d is the fibre diameter. For $d = 8\ \mu m$, $n \approx 1.5$, and $\Delta n = 0.001$, say, we require $\lambda_0 > 0.57\ \mu m$.

LP Modes

The hybrid modes have a complicated polarization, although the $HE_{1,1}$ mode is almost plane-polarized. However, for weak guides, it can be shown that all the other modes can be grouped together in combinations which are *degenerate* (which means they have the same propagation constant or phase velocity). A linear combination of these modes can then be used to construct a new set of nearly plane-polarized fields which represent a considerable simplification over the exact ones. These modes are called 'linearly polarized' and are given new labels of the form $LP_{\mu,\nu}$. Table 8.1 shows the constituents of some low-order LP modes. The particular case of the $HE_{1,1}$ mode is the $LP_{0,1}$ mode; as a rule, the $LP_{1,\nu}$ mode is constructed from the $HE_{2,\nu}$, $TM_{0,\nu}$ and $TE_{0,\nu}$ modes, and the $LP_{\mu,\nu}$ mode from the $HE_{\mu+1,\nu}$ and $EH_{\mu-1,\nu}$ modes (for $\mu \neq 0$ or 1).

Figure 8.7 shows approximate representations of the intensity distributions of three low-order LP modes. The $LP_{0,1}$ mode is circularly-symmetric with a field distribution containing a single peak at the centre of the fibre core. The higher-order modes, on the other hand, have multi-lobed patterns with 2μ maxima as measured in the circumferential direction, and ν maxima in the radial direction.

The dispersion diagram for the step-index fibre is of the general form shown in Fig. 8.8. The main difference in relation to the slab guide is that higher-order modes start to propagate in much greater numbers as soon as the V-value exceeds the limit for single-mode operation. In fact, it can be shown that the number of modes propagating is given approximately by

$$N \approx V^2/2 \qquad (8.7)$$

provided N is large. A fibre with a V-value of 10 would therefore support approximately 50 modes.

Table 8.1 Constituents of some low-order LP modes

LP mode	Exact mode	LP mode	Exact mode constituents
$LP_{0,1}$	$HE_{1,1}$	$LP_{0,2}$	$HE_{1,2}$
$LP_{1,1}$	$HE_{2,1}$, $TM_{0,1}$ & $TE_{0,1}$	$LP_{1,2}$	$HE_{2,2}$, $TM_{0,2}$ & $TE_{0,2}$
$LP_{2,1}$	$HE_{3,1}$ & $EH_{1,1}$	$LP_{2,2}$	$HE_{3,2}$ & $EH_{1,2}$

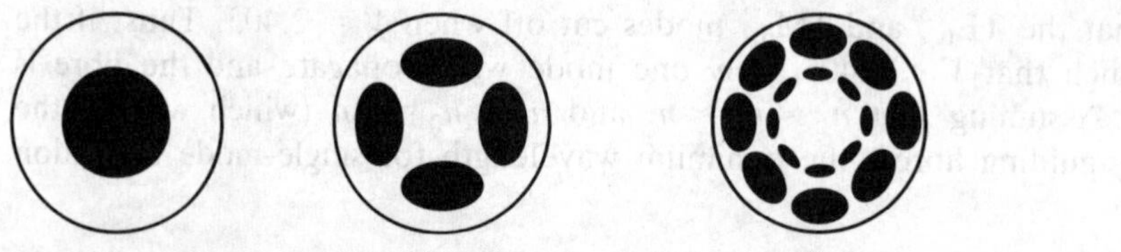

Figure 8.7 Approximate intensity distributions for the $LP_{0,1}$, $LP_{2,1}$ and $LP_{4,2}$ modes.

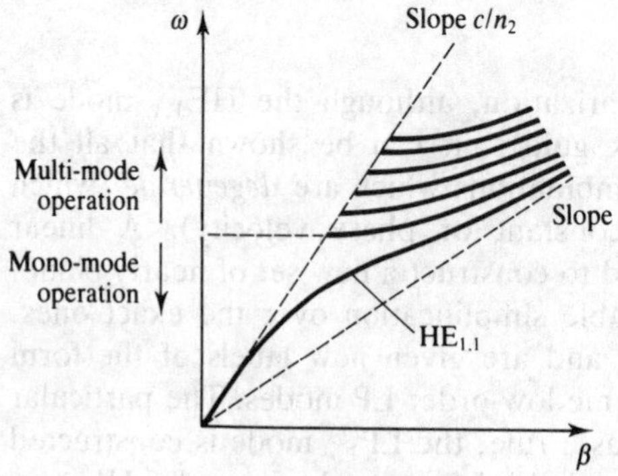

Figure 8.8 General dispersion diagram for step-index optical fibre.

Design Example

Using the approximate guidelines presented so far we may compare the main features of single- and multi-mode fibres. With the typical values of $d = 8\ \mu m$, $n_1 = 1.5$ and $n_2 = 1.499$ for the former, and $d = 50\ \mu m$, $n_1 = 1.5$ and $n_2 = 1.49$ for the latter, we obtain the parameters shown in Table 8.2 for a wavelength of $\lambda_0 = 0.85\ \mu m$. The high numerical aperture of the multi-mode fibre should be noted, together with the large number of modes that it supports.

Other Fibre Modes

Just as in the slab guide, we may expect the guided modes of a fibre to be complemented by radiation modes, and that the full range of modes will exist in

Table 8.2 Characteristic parameters of typical single- and multi-mode fibres

Fibre	$d(\mu m)$	n_1	n_2	NA	V-value	No. of modes
#1	8	1.5	1.499	0.055	1.62	1
#2	50	1.5	1.49	0.17	31.95	510

two sets, each corresponding to one of two possible polarizations. While this is true, there are two important new features. First, the radiation modes are normally subdivided into fields that propagate inside and outside the cladding. The former are known as *cladding modes*, and are often a nuisance. They may be removed using a device known as a *mode-stripper*. In this, the plastic jacket is removed from a short section of the fibre, which is placed on a glass plate in a pool of index-matching liquid. Reflection at the edge of the cladding is then greatly reduced, and most of the unwanted light radiates into the glass plate. Second, since the fibre is cylindrically-symmetric, the two sets of polarization modes are degenerate. A single-moded fibre therefore actually supports two modes of orthogonal polarization, which propagate at the same speed. As we will show later, this normally results in a random polarization fluctuation.

8.4 PARABOLIC-INDEX OPTICAL FIBRES

We now consider parabolic-index fibre. Here the possible ray trajectories are still more complicated. Because the refractive index distribution is now a continuous function of radius, rays no longer propagate as straight lines with total internal reflection causing abrupt changes in direction. Instead, their paths are determined by a process of continual refraction. This has the effect of gradually bending the direction of any ray travelling away from the axis, until it turns towards the axis once again. For meridional rays the result is a sinusoidal trajectory, oscillating on either side of the axis. Figure 8.9(a) shows two different ray paths in the same fibre. As we shall shortly demonstrate, each has almost exactly the same periodicity in the axial direction, although the turning points (e.g. z_a and z_b) lie at different radii. This has important consequences for signal transmission.

For skew rays a similar process of continual refraction occurs. Consequently the projection of a skew ray trajectory on the fibre cross-section has the appearance of a smooth, rotating orbit, as shown in Fig. 8.9(b). In fact, the ray path can again be shown to lie entirely between two coaxial cylinders. These are known as *caustic surfaces*, and represent the loci of all possible turning points.

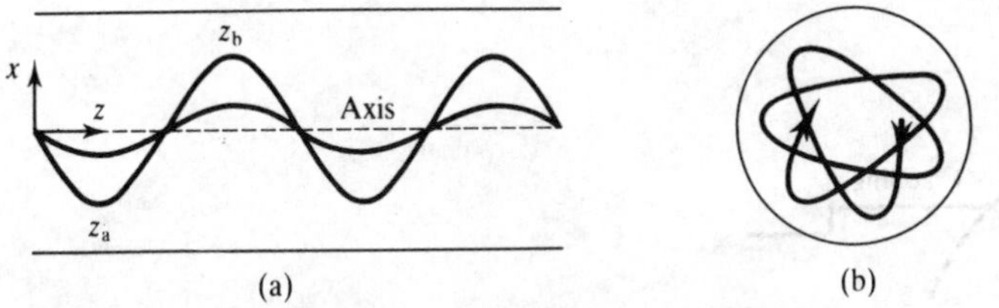

Figure 8.9 Trajectories of (a) meridional and (b) skew rays in a parabolic-index fibre.

Ray Trajectories in Parabolic-index Media

Using some fairly simple scalar analysis it is possible to calculate the trajectory of a ray in a graded-index guide, as we now show. We start by noting that the exact refractive index profile of the fibre is given by the full line in Fig. 8.10. However, on the assumption that the light is confined mainly to the region of parabolic variation, we can disregard the index steps at the edge of the core and cladding, and model the index distribution by a continuous parabolic function of the form:

$$n(r) = n_0 \sqrt{\{1 - (r/r_0)^2\}} \tag{8.8}$$

where n_0 is the refractive index at the centre of the fibre, and r_0 is a characteristic radius. This profile is shown dashed in Fig. 8.10; while it implies a negative refractive index when $r > r_0$, we can assume that r_0 is large enough for this to be unimportant for all practical purposes.

At this point we shall adopt a one-dimensional model, ignoring the y-variation of the refractive index and constraining the trajectory to lie in the x–z plane. This will allow us to deduce many of the features of a two-dimensional guide in a qualitative way. Later on we will check for agreement with the more complicated geometry. With this approximation we may put $r = x$ and $r_0 = x_0$ in Eq. (8.8). The direction of the ray at any point (x, z) may be then defined by a vector $\mathbf{k}$, such that $|\mathbf{k}|$ is equal to the local value of the propagation constant, $k_0 n(x)$. Similarly the local slope of the trajectory may be found as

$$\mathrm{d}x/\mathrm{d}z = k_x/k_z \tag{8.9}$$

where k_x and k_z are the x- and z-components of $\mathbf{k}$. However, if the entire trajectory corresponds to a guided mode, we may also set $k_z = \beta$, where β is the modal propagation constant. Since k_x and k_z are related by $k_x^2 + k_z^2 = |\mathbf{k}|^2$ we obtain

$$k_x = \sqrt{(k_0^2 n^2 - \beta^2)} \tag{8.10}$$

Substituting from Eq. (8.8) we then get

$$k_x = \sqrt{\{(k_0^2 n_0^2 - \beta^2) - k_0^2 n_0^2 (x/x_0)^2\}} \tag{8.11}$$

Equation (8.9) may therefore be written as

$$\beta^2 (\mathrm{d}x/\mathrm{d}z)^2 = (k_0^2 n_0^2 - \beta^2) - k_0^2 n_0^2 (x/x_0)^2 \tag{8.12}$$

We might guess that the solution of Eq. (8.12) is indeed a sinusoidal trajectory of the general form $x = A \sin(Bz + \Psi)$, where Ψ is an arbitrary phase factor. We

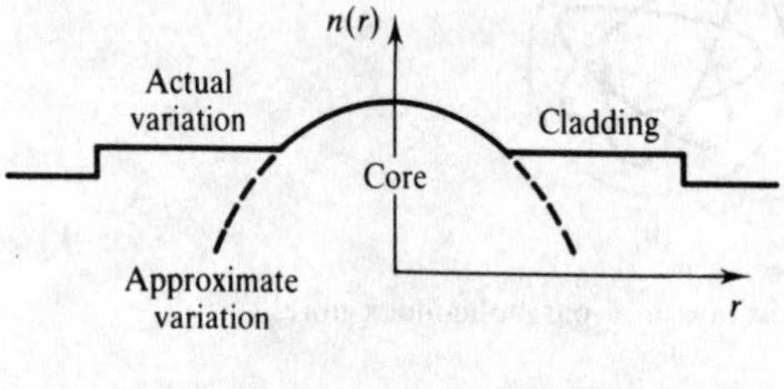

Figure 8.10 Exact and approximate refractive index distribution of a parabolic-index fibre.

shall now try to verify this hypothesis and simultaneously determine the unknown constants A and B. Peforming the necessary differentiation, substituting into Eq. (8.12), and rearranging the terms, we get

$$(k_0^2n_0^2 - \beta^2) - (k_0^2n_0^2A^2/x_0^2)\sin^2(Bz + \Psi) - \beta^2A^2B^2\cos^2(Bz + \Psi) = 0 \quad (8.13)$$

The only way Eq. (8.13) can be satisfied for all z is if the $\sin^2$ and $\cos^2$ terms can somehow be removed. This can be done if their coefficients are equal (since $\sin^2\theta + \cos^2\theta = 1$), which requires

$$k_0^2n_0^2A^2/x_0^2 = \beta^2A^2B^2 \quad \text{or} \quad B = k_0n_0/\beta x_0 \quad (8.14)$$

This places an immediate restriction on the value of B, the spatial frequency for the periodic oscillation of the trajectory. Assuming that Eq. (8.14) is indeed satisfied, Eq. (8.13) reduces to

$$(k_0^2n_0^2 - \beta^2) - (k_0^2n_0^2A^2/x_0^2) = 0 \quad (8.15)$$

so that the amplitude A of the oscillation is given by

$$A = x_0\sqrt{(1 - \beta^2/k_0^2n_0^2)} \quad (8.16)$$

Equation (8.16) shows that the smaller the value of β (i.e. the higher the order of the mode) the larger the amplitude. Our solution for the ray trajectory is therefore

$$x = x_0\sqrt{(1 - \beta^2/k_0^2n_0^2)}\sin\{(k_0n_0/\beta x_0)z + \Psi\} \quad (8.17)$$

For weak guides (when $\beta \approx k_0n_0$ for all modes), $B \approx 1/x_0$. With this approximation we can put

$$x \approx x_0\sqrt{(1 - \beta^2/k_0^2n_0^2)}\sin(z/x_0 + \Psi) \quad (8.18)$$

Equation (8.10) implies that the spatial frequency of the trajectory is virtually independent of β, i.e. the same for all modes. It turns out that the above analysis is modified somewhat in the case of a two-dimensional parabolic-index fibre, but only qualitatively so, providing comforting evidence for the shape of the ray paths in Fig. 8.9. More importantly it allows the use of a parabolic-index medium as a type of imaging device, known as a *GRIN-rod lens* (where the GRIN acronym stands for gradient-index).

GRIN-rod Lenses

A GRIN-rod lens is a short length of glass rod of moderately large diameter (1–3 mm), which has the radial index variation of Eq. (8.8). Formally, it is a section of highly multi-moded fibre, but in practice no cladding is used. It is fabricated by ion exchange of thallium- or caesium-doped silica glass in a bath of a molten salt. Typically, KNO_3 is used, at a temperature of 500°C. During the exchange, sodium and either thallium or caesium ions diffuse out of the glass and are replaced by potassium ions from the melt. The difference in polarizability of K^+ and Tl^+ ions then causes a reduction in the refractive index. The local concentration of Tl^+ at the end of the process is determined by standard diffusion equations. Due to the cylindrically-symmetric geometry, the concentration is higher at the centre of the rod than at the periphery, following a parabolic variation in between. The concentration of K^+ ions, on the other hand, is approximately uniform. The net result is a refractive index profile corresponding well to the ideal variation. GRIN-rod lenses are produced commercially by the Nippon Sheet Glass Company of Japan, under the trade name 'SELFOC micro-lens'.

We shall again base the following argument on one-dimensional analysis. First, we can guess that rays in a GRIN-rod lens will also follow periodic trajectories. The period P of the oscillation is defined as the *pitch length*, given by

$$P = 2\pi x_0 \tag{8.19}$$

It is normal for the lens length to be chosen as some fraction of a pitch, close to either $P/2$ (a half-pitch lens) or $P/4$ (quarter-pitch). If this is done, a GRIN-rod lens may perform imaging operations, as we now demonstrate. Figure 8.11(a) shows a half-pitch lens used to form an image of a point source of light located on-axis at A. Naturally this source will emit rays in all directions, but each will begin at the same point. Assuming that this is defined as $x = 0$, $z = 0$, the trajectories of all rays inside the lens have the form of Eq. (8.18), with the phase factor $\Psi = 0$. Each sinusoidal trajectory will have a different amplitude, so that the slope of the trajectory at $z = 0$ matches the direction of an input ray. However, after travelling a distance $z = P/2$ through the lens, all rays will have returned to the axis once again, at the point A′. Consequently, the lens may be considered to form a real image of A at A′. Since an off-axis point object above A can also be shown to form a real point image displaced the same distance below A′, the lens forms an inverted image of the entire input plane at the output end-face. Similar reasoning shows that a quarter-pitch lens forms an image of a point source at infinity, thus collimating a point emitter (Fig. 8.11(b)).

For practical reasons, GRIN-rod lenses are normally made in lengths somewhat shorter than those above. For example, Fig. 8.11(c) shows a 0.23P lens which will

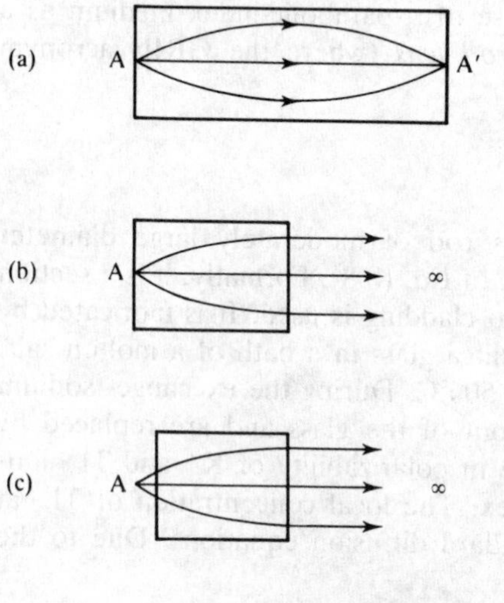

Figure 8.11 (a) A 0.5P GRIN-rod lens, imaging a point source at A at A′, (b) a 0.25P lens, used to image a point source at A at infinity, and (c) a 0.23P lens, with working tolerance.

collimate light from a point source slightly displaced from the lens, thus allowing a working tolerance for alignment and so on. This is particularly important since dispersion of the lens material generally results in a wavelength-dependent value of P. Without such a tolerance, operation at a wavelength for which P is shorter than the design value would require the source to be placed inside the lens.

GRIN-rod lenses have several advantages over conventional lenses. Because they are small they are well-suited to use in micro-optic systems, and are often used to connect optical fibre components. Most importantly, their imaging properties do not stem from refraction at their input and output surfaces (which are planar); instead, redirection of rays is performed by the medium itself. This allows components to be fixed directly to the end faces with index-matching cement, reducing reflection losses and eliminating scratchable or movable interfaces. Consequently, the assembly can be highly rugged. Figure 8.12 shows a collection of GRIN-rod lenses.

The ray picture is just one explanation for the imaging behaviour of a GRIN-rod lens, and an alternative view is provided by the transfer function model introduced in Chapter 4. Remember that the properties of a conventional lens stemmed from the parabolic variation of its optical thickness. The required variation was effected by controlling the thickness of a material with constant refractive index. In a GRIN-rod lens much the same result is achieved by control of the radial distribution of refractive index, keeping the physical thickness constant. We shall provide one further explanation later on, in terms of guided modes.

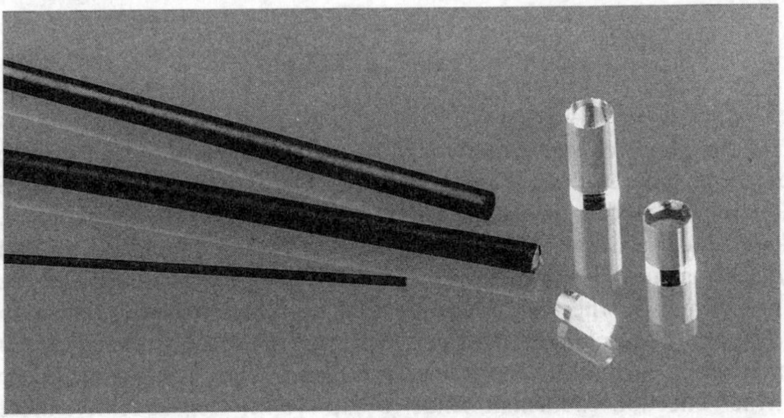

Figure 8.12 GRIN-rod lenses (*photograph courtesy D. DeRose, NSG America Inc.*).

The Eigenvalue Equation for a Parabolic-index Guide

The solution for ray paths given in Eq. (8.17) represents almost the entire story. However, to obtain a closed form solution we need to know the allowed values of β. This requires the solution of an eigenvalue equation. To set up a suitable equation we shall make use of the transverse resonance condition previously introduced for slab guides in Chapter 6, which required the total phase change accumulated in a round-trip between the guide walls to a whole number of multiples of 2π. Remember that this consisted of two terms: the phase change incurred simply in propagating between the walls, and the phase shift resulting from reflection at the walls.

In the one-dimensional parabolic-index guide propagation between any two consecutive turning points (e.g. z_a and z_b) clearly corresponds to half of one round-trip. The first term may therefore be found by integrating the x-component of the propagation constant over the transverse distance between these points:

$$2\int_{x_a}^{x_b} k_x \, dx = 2\int_{z_a}^{z_b} k_x \, (dx/dz) \, dz \tag{8.20}$$

Using Eq. (8.9) this may be transformed to

$$2\int_{x_a}^{x_b} k_x \, dx = 2\beta\int_{z_a}^{z_b} (dx/dz)^2 \, dz \tag{8.21}$$

Substituting our solution for the trajectory (Eq. (8.17)) we then get

$$2\int_{x_a}^{x_b} k_x \, dx = (2/\beta)\,(k_0^2 n_0^2 - \beta^2)\int_{z_a}^{z_b} \cos^2 \{(k_0 n_0/\beta x_0)z + \Psi\} \, dz \tag{8.22}$$

Since z_a and z_b are separated by half a cycle, we must have $(k_0 n_0/\beta x_0)\,(z_b - z_a) = \pi$. Integrating over this distance we get:

$$2\int_{x_a}^{x_b} k_x \, dx = x_0\pi\,(k_0 n_0 - \beta^2/k_0 n_0) \tag{8.23}$$

The second term—the phase change experienced on reflection—is more difficult to deal with. After all, if the ray path is determined by continuous refraction, total internal reflection is not needed to change the direction of the ray. Nonetheless, the sign of k_x does actually change at each of the turning points, which implies that some form of reflection must be occurring there. To assess the consequences we need to return to Chapter 5, where we found general expressions for the reflection coefficients for TE- and TM-polarized light at a dielectric interface. In each case it can be shown that when the index step tends to zero and the angle of incidence approaches 90°, the reflection coefficient tends to $\exp(j\pi/2)$. Consequently, there is an additional phase shift of $\phi_a = -\pi/2$ at each of the turning points.

Including all contributions, the transverse resonance condition can therefore be written

$$2\int_{x_a}^{x_b} k_x \, dx + 2\phi_a = 2\nu\pi \qquad (\nu = 1,2,\ldots) \tag{8.24}$$

Inserting the values found above we then get

$$x_0\pi\,(k_0 n_0 - \beta^2/k_0 n_0) - \pi = 2\nu\pi \tag{8.25}$$

which yields the following allowed values for the propagation constant:

$$\beta_\nu = k_0 n_0 \sqrt{\{1 - (2\nu + 1)/(k_0 n_0 x_0)\}} \tag{8.26}$$

If the second term in the square root above is small (which implies weak guidance, and therefore paraxial ray paths) we may approximate Eq. (8.26) as

$$\beta_\nu \approx k_0 n_0 - (2\nu + 1)/2x_0 \tag{8.27}$$

Since $k_0 = \omega/c$, the dispersion characteristics of a parabolic-index guide consist of a set of parallel straight lines, as shown in Fig. (8.13). This has extremely important consequences for signal transmission in fibres, which we shall discuss later on. Notice that we have deliberately obscured details of the cut-off conditions, when ray paths must stray out to the edge of the core, since these are not modelled realistically using the approximate index profile we have assumed.

In the mean time we simply note the following result, which is relevant to image transmission through a GRIN-rod lens. As we saw in Chapter 6, a general one-dimensional, forward-travelling field $E_{in}(x)$ specified at an input plane $z = 0$ may be expanded in terms of guided modes as

$$E_{in}(x) = \sum_\nu a_\nu E_\nu(x) \tag{8.28}$$

where a_ν is the amplitude of the ν th mode, and $E_\nu(x)$ is its transverse field distribution. After propagating a distance z, this field will be modified to

$$E_{out}(x) = \sum_\nu a_\nu E_\nu(x) \exp(-j\beta_\nu z) \tag{8.29}$$

Inserting the propagation constants specified by Eq. (8.27) into the above we get

$$E_{out}(x) = \exp(-j\psi) \sum_\nu a_\nu E_\nu(x) \exp(j\nu z/x_0) \tag{8.30}$$

where $\psi = (k_0 n_0 - 1/2x_0)z$. We now note that if $z = 2\pi x_0$ (i.e. if $z = P$), Eq. (8.30) reduces to

$$E_{out}(x) = \exp(-j\psi) \sum_\nu a_\nu E_\nu(x) \tag{8.31}$$

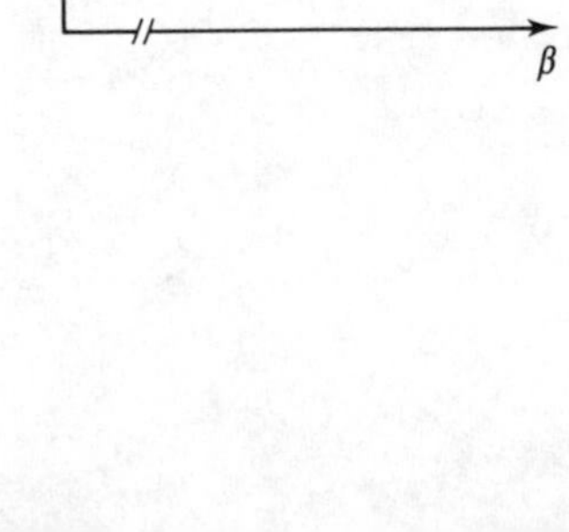

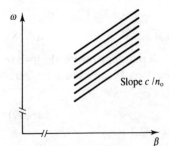

Figure 8.13 Dispersion characteristics of a one-dimensional parabolic-index guide.

At this point E_{out} is equal to E_{in} multiplied by an unimportant phase factor. Propagation through one pitch length will therefore leave an image effectively unchanged, as we would expect from the ray model of a GRIN-rod lens described earlier. We can interpret the formation of an inverted real image by a half-pitch lens in a similar way.

Modal Analysis of Parabolic-index Fibre

We shall now show how scalar theory may be used to determine the actual guided modes in a two-dimensional parabolic-index fibre. We base the analysis on the TEM model introduced in Chapter 6, which assumes that the index changes forming the guide are weak. For the refractive index distribution of Eq. (8.8) the scalar waveguide equation we must solve is

$$\nabla^2_{xy}E(x,y) + \{n_0^2k_0^2[1 - (r/r_0)^2] - \beta^2\}\,E(x,y) = 0 \qquad (8.32)$$

where $E(x,y)$ is the transverse field and β is the propagation constant. Because the fibre is cylindrical, we shall begin by looking for a solution that varies only in the radial direction. In this case we may put $E(x,y) = E(r)$, and evaluate its drivative with respect to x as

$$\partial E/\partial x = \mathrm{d}E/\mathrm{d}r\ \partial r/\partial x = (x/r)\ \mathrm{d}E/\mathrm{d}r \qquad (8.33)$$

Similarly its second derivative with respect to x is given by

$$\partial^2 E/\partial x^2 = [1/r - x^2/r^3]\ \mathrm{d}E/\mathrm{d}r + (x^2/r^2)\ \mathrm{d}^2E/\mathrm{d}r^2 \qquad (8.34)$$

$\partial^2 E/\partial y^2$ may be evaluated in the same way, so we obtain

$$\nabla^2_{xy}E = (1/r)\ \mathrm{d}E/\mathrm{d}r + \mathrm{d}^2E/\mathrm{d}r^2 \qquad (8.35)$$

Equation (8.35) may now be substituted into Eq. (8.32) to give

$$\mathrm{d}^2E/\mathrm{d}r^2 + (1/r)\ \mathrm{d}E/\mathrm{d}r + \{n_0^2k_0^2[1 - (r/r_0)^2] - \beta^2\}\,E = 0 \qquad (8.36)$$

At this point we shall make a guess at the solution. Somewhat arbitrarily, we assume that the transverse field is given by a Gaussian distribution of the form:

$$E(r) = E_0 \exp(-r^2/a^2) \qquad (8.37)$$

where E_0 and a are constants. Substituting into Equation (8.36) and removing the common terms we get

$$r^2\{4/a^4 - n_0^2k_0^2/r_0^2\} + \{n_0^2k_0^2 - 4/a^2 - \beta^2\} = 0 \qquad (8.38)$$

where we have grouped the terms into two blocks, one dependent on r^2, the other a constant. In order to satisfy Eq. (8.38) for all r, the two blocks must be zero independently. The former condition leads to

$$a = \sqrt{(2r_0/n_0k_0)} \qquad (8.39)$$

while the latter requires

$$\beta^2 = n_0^2 k_0^2 - 4/a^2 \quad \text{or} \quad \beta = n_0 k_0 \surd\{1 - 2/(n_0 k_0 r_0)\} \qquad (8.40)$$

This shows that the Gaussian transverse field we have assumed is indeed a possible mode of the parabolic-index fibre. Since its phase-front is plane, this mode will radiate into free-space from an end-face of the fibre as a Gaussian beam; the beam waist (which has radius a) will be located at the fibre end, and behaviour in the far-field will be as described earlier in Chapter 4.

Using similar methods, but this time adopting rectangular coordinates, it is possible to show that the higher-order guided modal solutions are all *Hermite-Gaussian modes* with transverse field variations of the form:

$$E_{\mu,\nu}(x,y) = H_\mu(\surd 2x/a)\ \mathrm{H}_\nu(\surd 2y/a)\ \exp\{-(x^2 + y^2)/a^2\} \qquad (8.41)$$

where μ and ν are two mode numbers, and $\mathrm{H}_\mu(\zeta)$ is a Hermite polynomial of order μ, defined as satisfying the differential equation:

$$\mathrm{d}^2\mathrm{H}_\mu(\zeta)/\mathrm{d}\zeta^2 - 2\zeta\ \mathrm{dH}_\mu(\zeta)/\mathrm{d}\zeta + 2\mu\ \mathrm{H}_\mu(\zeta) = 0 \qquad (8.42)$$

while the propagation constants $\beta_{\mu,\nu}$ are given by

$$\beta_{\mu,\nu} = n_0 k_0 \surd\{1 - 2(\mu + \nu + 1)/(n_0 k_0 r_0)\} \qquad (8.43)$$

The first few Hermite polynomials are as given in Table 8.3. From this it should be clear that the Gaussian solution above (for which $\mathrm{H}_\mu(\surd 2x/a) = \mathrm{H}_\nu(\surd 2y/a) = 1$) is the lowest-order mode, with $\mu = \nu = 0$. This is confirmed by the agreement of Eqs (8.40) and (8.43). Although the latter differs slightly from Eq. (8.26), found using our earlier ray model, it still has the same general form. We may thus deduce that the general characteristics of a parabolic-index fibre will indeed be qualitatively similar to those for a one-dimensional parabolic-index guide.

8.5 SIGNAL DISPERSION

We shall now consider a number of aspects of signal propagation in optical fibre, beginning with dispersion. In Chapter 6 we showed that there are generally two contributions to dispersion in waveguides: intermodal and intramodal dispersion. The latter can be further subdivided into contributions from material and waveguide dispersion. We shall consider them separately, and examine the dominant effects in both step- and graded-index fibre. First, we reiterate that all forms of dispersion

Table 8.3 Some low-order Hermite polynomials

$\mathrm{H}_0(\zeta) = 1$	$\mathrm{H}_3(\zeta) = -12\zeta + 8\zeta^3$
$\mathrm{H}_1(\zeta) = 2\zeta$	$\mathrm{H}_4(\zeta) = 12 - 48\zeta^2 + 16\zeta^4$
$\mathrm{H}_2(\zeta) = -2 + 4\zeta^2$	$\mathrm{H}_5(\zeta) = 120\zeta - 160\zeta^3 + 32\zeta^5$

arise from variations in group velocity over the signal bandwidth. *Intermodal dispersion* results from differences in v_g between different modes, and is therefore important only in multi-mode fibre. However, if it is present, it normally dominates all other contributions.

Clearly, to evaluate the group velocities of the relevant modes, we must have already made an accurate calculation of the dispersion diagram. Since we have not done this for step-index fibre, we must use an approximation. Our argument runs like this. Modes that are far from cut-off will have most of their energy contained inside the core, so their group velocity will be close to that of the core material. Ignoring all other forms of dispersion, this will be roughly equal to the phase velocity in a medium of refractive index n_1, namely c/n_1. Similarly, modes that are close to cut-off will have their energy mostly propagating in the cladding so that their group velocity will be roughly c/n_2. In a distance L, the time-spread of a signal due to intermodal dispersion is then

$$\Delta t = L\{1/v_{\text{gmin}} - 1/v_{\text{gmax}}\} \approx (L/c)(n_1 - n_2) \tag{8.44}$$

Assuming the typical value of $n_1 - n_2 = 0.01$, we obtain $\Delta t/L \approx 33$ ns/km. In a digital communications system pulses of duration T are normally sent separated by gaps of T. Intersymbol interference will occur when consecutive pulses start to overlap, i.e. when $\Delta t \approx T$. Intermodal dispersion therefore limits the maximum bit rate B to:

$$B \approx 1/(2\Delta t) \tag{8.45}$$

This is normally expressed in terms of a bit-rate:length product BL, given by:

$$BL \approx c/\{2(n_1 - n_2)\} \tag{8.46}$$

For the fibre above, we obtain $BL \approx 15$ Mbit/s km, implying that data may be sent at 15 MBit/s over a distance of 1 km, or at 1.5 Mbit/s over 10 km. The small value of BL renders this fibre impractical for long-distance, high bit-rate communications.

A considerable improvement is offered by parabolic index fibre, for which we have already calculated the dispersion characteristics. Assuming weak guidance, Eq. (8.43) may be differentiated to give $d\omega/d\beta_{\mu,\nu} \approx c/n_0$. This shows that the group velocities of all modes are approximately equal, a highly advantageous feature. The explanation is found in the ray paths of Fig. (8.9). Trajectories with small amplitudes of oscillation obviously travel through shorter physical distances than do those with a large amplitude. However, due to the parabolic index variation, paraxial rays spend more time in regions of high index, and therefore travel more slowly. The two effects—decreased physical distance, and increased refractive index—balance out, so that the transit times of all rays are roughly equalized. Consequently, a drastic reduction in intermodal dispersion is obtained, allowing a considerable increase in the bit-rate:length product, typically to 1 Gbit/s km. For this reason, multi-mode parabolic-index fibre has been very successful, allowing transmission at repectable data rates over short distances. However, for higher rates and longer distances, it is essential to eliminate intermodal effects entirely by using single-mode fibre. The major limitations on bandwidth then have different origins.

In Chapter 2 we showed that the broadening of a signal due to variations in the propagation constant with frequency could be written as $\Delta t \approx L\,\Delta\omega\, d^2k/d\omega^2$, where $\Delta\omega$ is the signal bandwidth. Although this was originally derived for plane waves, we would expect that the following analogous expression would describe similar effects caused by variations in the β-value of a single guided mode:

$$\Delta t \approx L\,\Delta\omega\, d^2\beta/d\omega^2 \qquad (8.47)$$

This phenomenon is known as *intramodal dispersion*. In general it consists of contributions from material and waveguide dispersion, as we now show. First, we recall from Chapter 2 that dispersion in plane waves could be related to variations in the refractive index with wavelength, according to $\Delta t \approx -(L\lambda_0\Delta\lambda_0/c)\, d^2n/d\lambda_0^2$, where $\Delta\lambda_0$ is the wavelength range of the signal. For a guided mode we simply replace n by n_{eff} to obtain

$$\Delta t \approx -(L\lambda_0\Delta\lambda_0/c)\, d^2n_{eff}/d\lambda_0^2 \qquad (8.48)$$

Intramodal dispersion therefore follows from variations in the effective index with wavelength. These will arise if the constituents of the core or the cladding display any wavelength-dependence, i.e. through material dispersion. We have already plotted the dispersion characteristic of pure silica in Chapter 3, and noted that dispersion goes to zero at $\lambda_0 \approx 1.27\ \mu m$. This would be the ideal operating point if signal distortion were the only criterion, but (as we showed earlier) the optimum wavelength for transmission is around $\lambda_0 = 1.55\ \mu m$. Several methods may be used to move the point of zero dispersion to longer wavelengths. First, the addition of some dopants (e.g. GeO_2) to silica shifts the point in the required direction. Second, there is a further contribution to dispersion that we have not yet included: variations in effective index arise simply from the changes in angle between any guided ray and the fibre axis that must occur with changes in wavelength. This is known as *waveguide dispersion*, and is independent of material effects.

In step-index fibre particular parameter choices can lead to waveguide dispersion of opposite sign to the material dispersion, resulting in cancellation of the total dispersion at the required wavelength. Unfortunately this type of *dispersion-shifted fibre* normally has a rather small core, leading to problems with jointing. However, by using a more complicated refractive index profile (e.g. a W-shaped profile) it is possible to achieve similar cancellation using cores of larger cross-section.

8.6 MODE CONVERSION IN FIBRES

In our analysis we have concentrated on infinite, perfect guides. The exception has been a discussion of tapers in Chapter 6, where we showed that variations in guide shape can induce mode conversion. Although fibre is nominally made with a uniform cross-section, there will inevitably be small irregularities (e.g. at the interface between core and cladding, as in Fig. 8.14(a)). Any change in ray direction resulting from scattering by such an irregularity then corresponds to mode conversion. Slight changes in direction (Fig. 8.14(b)) lead to similar effects. Over the large distances

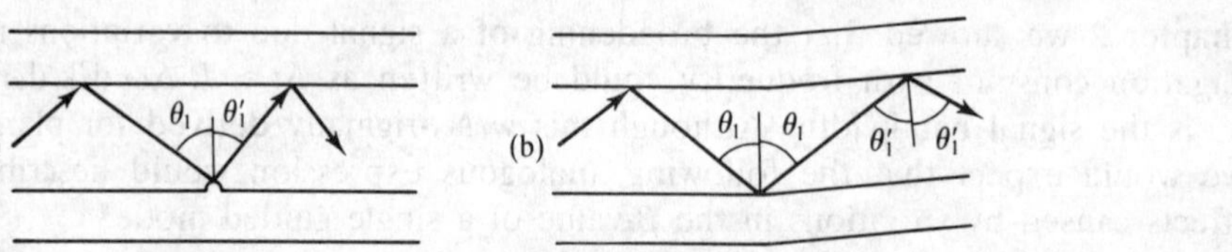

Figure 8.14 Mode conversion, at (a) a small irregularity, and (b) a slight change in direction.

involved in optical fibre transmission, even a small amount of *intermodal scattering* can lead to a significant transfer of power between modes. This leads to a range of consequences which assume differing importance in the various fibre types. We shall begin by discussing the effects in multi-mode fibre.

It might initially be thought that intermodal dispersion can be avoided by launching only a single mode. However, since the processes above lead to a gradual transfer of energy to higher-order modes, multi-mode operation must occur after a finite distance. Intermodal dispersion is therefore inevitable. It is also likely that scattering will lead to some rays striking the core/cladding interface at angles lower than the critical angle. These will not be reflected, leading to energy loss through the excitation of radiation modes. This is known as *microbending loss* if it is caused by small changes in guide direction (or as *bend loss*, if the changes are large). High-order modes, which strike the interface at angles closer to θ_c, are more susceptible to bend loss. It is likely that intermodal scattering will sooner or later lead to the excitation of modes near to cut-off, which are very sensitive to radiation loss. Consequently, multi-mode guides show an increase in bend loss with distance. One way to alleviate this is to decrease the critical angle, which requires a large index difference between core and cladding. However, in step-index fibre, this increases intermodal dispersion.

As we have mentioned, intermodal dispersion is not a problem in single-mode guides. Nonetheless, intermodal scattering still leads to undesirable effects. Bend loss remains, although this can be held to acceptable levels by maximizing the index change forming the guide, and ensuring that the bend radius always exceeds a specified minimum value. We shall discuss this point more fully in Chapter 9. Here we will concentrate on an alternative phenomenon: *polarization fluctuation*. This occurs in circularly-symmetric fibre because of random perturbations acting along the length of the fibre.

It can be shown that energy may be coupled most effectively between any two modes by an external perturbation if it is a periodic one with a spatial frequency that phase matches the modes. We have already met one example of this in Chapter 7 (the phase matching of beams travelling in different directions by an optically thick grating) and we will encounter some others in Chapter 10. The general idea is that two codirectional modes with propagation constants β_ν and β_μ can be phase matched by a perturbation with a spatial frequency K such that

$$K = 2\pi/\Lambda = |\beta_\nu - \beta_\mu| \tag{8.49}$$

Any random perturbation along the length of a fibre may be resolved into a set of Fourier components, each corresponding to a given spatial frequency of perturbation. Although the exact frequency spectrum cannot be specified, we can make some broad conclusions as to its general form: (i) it will be difficult to avoid a component at zero spatial frequency, since this represents the 'average' level of perturbation; (ii) the spectrum will fall off rapidly at higher frequencies because anything else would require rapid variations in the fibre with distance; (iii) the spectrum will not be constant because the most common disturbances (temperature changes and vibration) are inherently time-varying.

As we have mentioned before, even a single-moded fibre actually supports two modes. These have orthogonal polarizations, and are degenerate in circularly-symmetric fibre. Consequently, the zero-spatial-frequency component above can phase match them. Random coupling between the two then leads to a rotation of the polarization. This is a problem in interferometric systems, where the polarization of the interfering beams must be held parallel. Clearly, these require a different type of fibre, which holds the plane of polarization. This is known as *polarization-preserving fibre*. Its improved performance is obtained by adopting a non-circularly-symmetric core geometry, which eliminates the degeneracy between the two polarization modes. Since these now have different propagation constants (say, β_x and β_y) they are no longer phase matched by the zero-spatial-frequency component of any external perturbation. In fact, the larger the difference between β_x and β_y the better, since this will result in a larger phase matching frequency (which will have a smaller component in the spectrum of a typical perturbation). The effectiveness of polarization-preserving fibre is normally measured in terms of the *beat length* Λ between the two modes, defined as

$$\Lambda = 2\pi/|\beta_x - \beta_y| \tag{8.50}$$

Values of Λ in available fibre are generally of the order of millimetres.

It will be appreciated that a difference in propagation constant between two polarization modes is just another manifestation of the familiar phenomenon of birefringence. Polarization-preserving fibre is therefore also known as *high-birefringence* or hi-bi fibre. There are two basic methods of inducing the required effect; the first operates by modifying the physical symmetry of the fibre cross-section, while the second works by removing the isotropy of the fibre material. In each case preferred axes are imposed on the overall geometry, which become the polarization axes of the modified fibre.

To obtain *form-induced birefringence* the core is made non-circular, usually by adopting an elliptical cross-section. The two preferred polarizations are then aligned parallel to the semi-major axes of the ellipse. So that these may readily be identified, a flat is often ground on the cladding, parallel to one axis (Fig. 8.15(a)). This also improves the accessibility of the core, which is often advantageous (for example, in the directional coupler structures that will be discussed in Chapter 10). For obvious reasons this type of fibre is known as *D-fibre*.

Alternatively, in *stress-induced birefringence*, the circular shape of the core is retained, but an anisotropic stress is applied to it. The most effective method uses

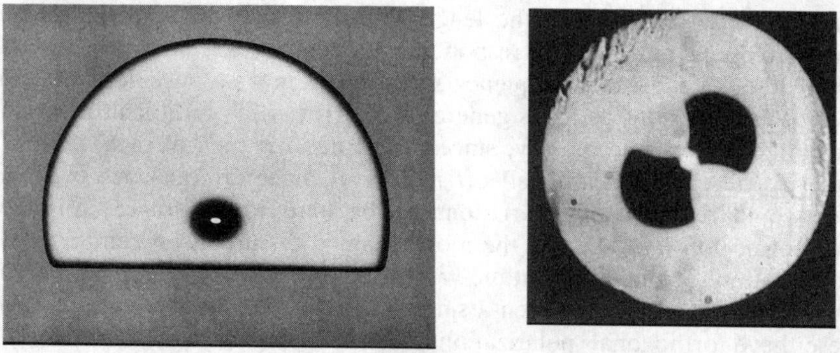

(a)

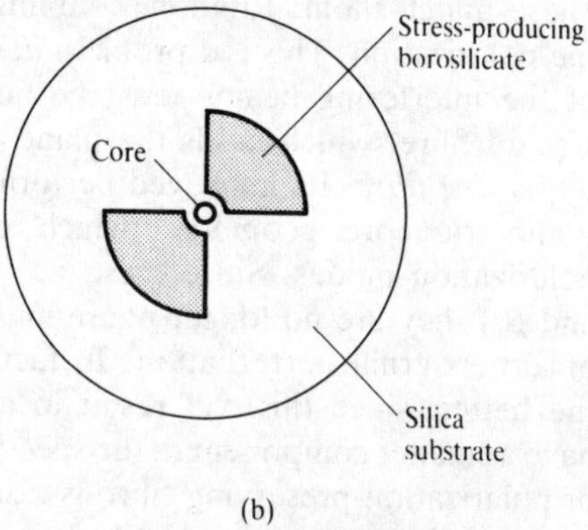

(b)

Figure 8.15 Polarization-preserving fibres: (a) D-fibre (*photograph courtesy R. Dyott, Andrew Corporation*), and (b) bow-tie fibre (*photograph courtesy D. Payne, Southampton University*).

the segmented construction shown in Fig. 8.15(b). Here an annular region of the cladding is divided into quadrants with alternating segments made of glasses with different thermal expansion coefficients. Silica (with expansion coefficient α_1) is used for most of the cladding, while the shaded areas consist of B_2O_3-doped silica (which has a higher coefficient α_2). During the cooling stages of the fibre fabrication process, differential contraction causes a large uniaxial stress to be applied to the core, which in turn induces birefringence in the core material (GeO_2-doped silica) through the *photoelastic effect*. This type of fibre is known as *bow-tie fibre* (from the shape of the stress-producing sectors) and offers beat lengths as short as 0.5 mm at $\lambda_0 = 0.633\ \mu$m.

8.7 COUPLING TO FIBRES

We will now examine some of the technology required to connect optical fibre and other guided wave components in complete systems. This is important because it is only the availability of low-loss, low-cost joints and splices that has allowed the full potential of optical fibre to be realized outside the laboratory. Two aspects are involved: the coupling of free-space beams into guided-wave components, and the direct connection of waveguides. Both may be tackled using a general method known as *transverse coupling*, and analysed using similar mathematical apparatus. We shall concentrate on the former in this section, leaving the latter to the next.

The technique of coupling free-space beams into guided modes by transverse excitation is known as *end-fire coupling*. For multi-mode fibres the process is easy to visualize. In a ray model each guided mode is defined by a characteristic angle of propagation relative to the fibre axis. Ignoring reflections, a field comprising an arbitrary collection of modes will then exit through the end-face of the fibres as a cone of light whose apex angle is set by the numerical aperture defined earlier. Reversing this argument, we would expect that a cone of external radiation impinging on a fibre end-face will excite a random set of guided modes which will then propagate down the fibre. The required external field can be generated very simply by a lens, leading to the typical arrangment shown in Fig. 8.16.

Notice that this argument contains no details of the *proportions* in which the modes are excited. However, we can deduce that it will be unproductive to use a beam whose numerical aperture is much greater than that of the fibre because a considerable fraction of the input power must then be wasted on the excitation of cladding modes. Equally we can guess that a low-NA input field will launch a greater proportion of low-order modes, which have shallower ray angles. A simple design approach would therefore be to match the numerical aperture of the input to that of the fibre. For example, fibre #2 in Table 8.2 has an NA of 0.17. Assuming that the parallel beam has a diameter of 1 mm, the required focal length of the lens is approximately $0.5/0.17 \approx 3$ mm.

Another feature ignored by the model above is the relative positions of the components. For example, if the fibre is placed a considerable distance from the focus of the lens, we would expect low coupling efficiency. To include this aspect we must consider the excitation of individual modes rather more carefully. This can be done by examining the fields at the boundary between the free-space and waveguide systems in more detail, as in Fig. 8.17. On the left we have the input wave. Near the focus this can be specified by a transverse field E_{in} which travels almost entirely in the axial direction. Exactly at the focal plane, E_{in} will be a Gaussian function if the input beam is itself Gaussian. The particular mode of interest (say, the ν th mode, whose transverse field is E_ν) is on the right. According to the analysis of Chapter 6 the excitation efficiency of the ν th mode may then be found in terms of overlap integrals as

$$\eta = |\langle E_{in}, E_\nu \rangle|^2 / \{\langle E_{in}, E_{in} \rangle \langle E_\nu, E_\nu \rangle\} \quad (8.51)$$

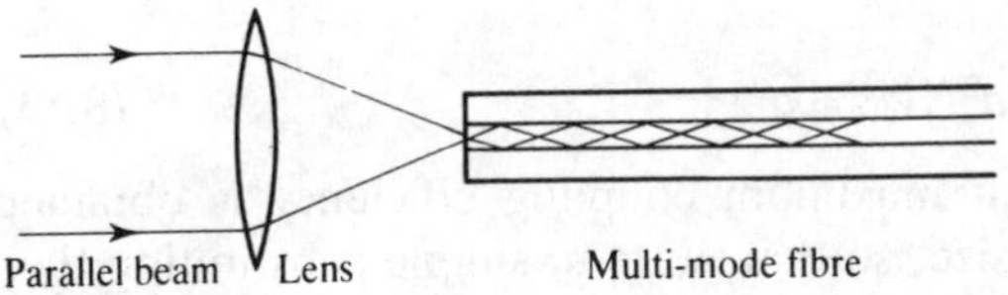

Figure 8.16 End-fire coupling into a multi-mode fibre.

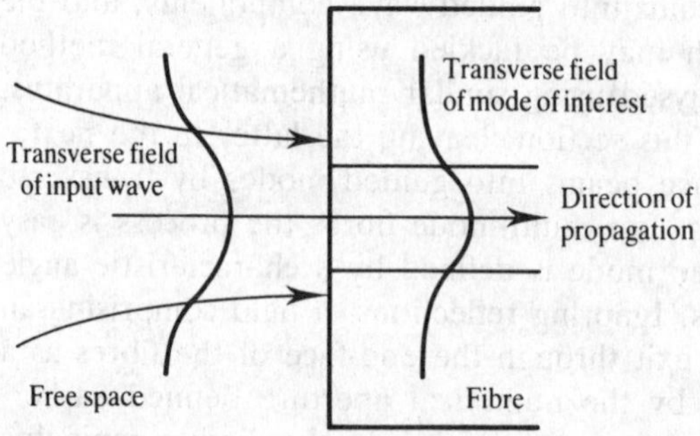

Figure 8.17 Geometry for modal analysis of end-fire coupling.

The main feature determining η is the overlap between E_{in} and E_ν. For high efficiency the two fields should coincide spatially as far as possible. Given that the field diameter will be of the order of μm in single-mode fibre, this implies stringent alignment tolerance. Furthermore, the fields should match each other closely in shape and size. We shall investigate this using an example.

Design Example

A typical end-fire coupling operation might involve the launching of light from a He-Ne laser into a single-moded parabolic-index fibre using a microscope objective. We can estimate the dependence of coupling efficiency on the focal spot size, as follows. First, from the results of Chapter 4 we can assume that since this type of laser emits a Gaussian beam, the phase-front at the focus of the lens will be planar with the diffraction-limited amplitude distribution $E_L(r) = A_L \exp(-r^2/w_0^2)$. Similarly, we may take the guided mode to have a Gaussian transverse field given by $E_1(r) = A_1 \exp(-r^2/a^2)$. The fibre is assumed to be aligned exactly on-axis with its end-face at the focal plane. In this case the overlap integrals are best performed in radial coordinates. If this is done we obtain

$$\langle E_L, E_L \rangle = \int_0^\infty 2\pi r\, E_L^2(r)\, dr = (w_0^2\pi/2)\, A_L^2$$

$$\langle E_1, E_1 \rangle = \int_0^\infty 2\pi r\, E_1^2(r)\, dr = (a^2\pi/2)\, A_1^2$$

$$\langle E_L, E_1 \rangle = \int_0^\infty 2\pi r\, E_L(r)\, E_1(r)\, dr = \{w_0^2 a^2 \pi/(w_0^2 + a^2)\}\, A_L A_1 \qquad (8.52)$$

Substitution into Eq. (8.51) then yields

$$\eta = 4w_0^2 a^2/\{w_0^2 + a^2\}^2 \qquad (8.53)$$

By differentiation it can be shown that maximum coupling efficiency is obtained when the fields are exactly matched in size, so that $w_0 = a$; simple substitution then shows that $\eta_{max} = 100$ per cent. However, away from this condition η falls rapidly; for example, when $w_0 = 2a$, we obtain $\eta = 64$ per cent. Although this figure might still seem reasonable, it actually corresponds to an insertion loss of ≈ 2 dB, equivalent to the attenuation incurred in propagating through about 10 km of low-loss fibre!

Coupling Between Laser Diodes and Optical Fibre

End-fire coupling via a lens may be used with other types of laser. For example, laser diodes (to be covered in Chapter 12) emit radiation from a narrow stripe, so their output is highly divergent with a numerical aperture that is normally much larger than that of single-mode fibre. However, a lens may be used to collect the output and redirect it towards the fibre in a converging beam of reduced NA. This process is equivalent to the formation of a magnified image of the lasing stripe on the fibre end. In practice, highly compact optical arrangements are often used. Figure 8.18 shows three examples. In Fig. 8.18(a) a separate GRIN-rod lens is used between the laser diode (LD) and the fibre. In Fig. 8.18(b) a hemispherical glass microlens is attached to the end of the fibre, which has been tapered down to a reduced diameter. Finally in Fig. 8.18(c) the end of a tapered fibre has been melted to form a lens directly in the fibre material.

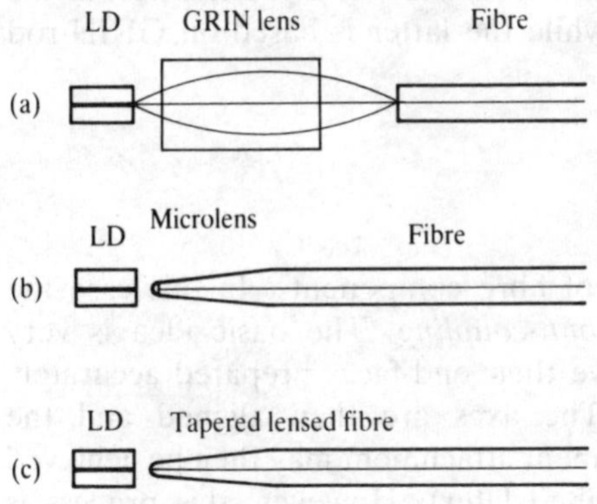

Figure 8.18 Laser diode-to-fibre connection schemes: (a) using a GRIN-rod lens, (b) using a microlens attached directly to the fibre, and (c) using a tapered lensed fibre.

Lensed Fibre Connectors

The end-fire coupling principle can be used to construct a demountable connector known as a *lensed fibre connector*. Figure 8.19(a) shows the basic idea. Each half of the connector consists of one prepared fibre end and one lens, which has a numerical aperture matched to that of the fibre. The left-hand lens is used to collimate the output from fibre #1, while the right-hand lens focuses this beam down onto the input of fibre #2. Because the expansion-contraction process results in a near-plane beam of relatively large diameter passing between the two connector halves, errors on longitudinal alignment have virtually no effect, and the significance

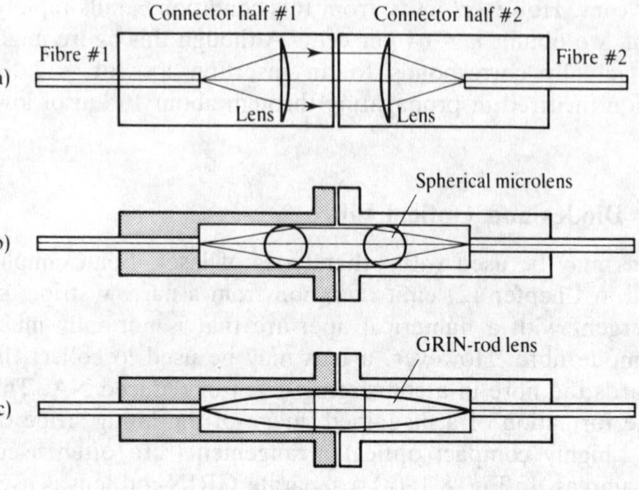

Figure 8.19 (a) Principle of lensed fibre connector; (b) and (c) connectors based on spherical microlenses and GRIN-rod lenses (*after A. Nicia, 1978*).

of errors in lateral position is drastically reduced. Naturally extremely small and cheap lenses are required. Figures 8.19(b) and 8.19(c) show two possibilities; in the former, small spherical glass balls are used, while the latter is based on GRIN-rod lenses.

8.8 FIBRE INTERCONNECTS

We now consider the direct interconnection of fibre components. In this case the transverse coupling procedure is known as *butt-coupling*. The basic idea is very simple; the two components to be joined have their end-faces prepared accurately orthogonal to their axes of propagation. The axes are then aligned and the components are simply butted together. Permanent attachment may then be achieved by gluing or by fusion splicing (to be described later). However, this process is naturally subject to error. Figure 8.20 shows examples of the difficulties that may arise in connecting two fibres. The fibres could be displaced from their ideal relative positions, by (a) a short axial distance Δz, (b) a small transverse distance Δx, or (c) a rotational misalignment $\Delta\theta$. All such errors result in reduced coupling efficiency. The effects may be calculated in terms of the overlap between the field on the left-hand side of the joint and that on the right using an expression similar to Eq. (8.51). Alternatively, they may be measured directly. For example, Fig. 8.21 shows experimental measurements for the variation in coupling efficiency with off-axis misalignment Δx, for joints between identical single-mode fibres. Two sets of data are shown: fibre #1 has a diameter of 8.6 μm, a V-value of 2.41 and a numerical aperture of 0.056, while fibre #2 has d = 4.3 μm, V = 2.18 and NA = 0.102. In

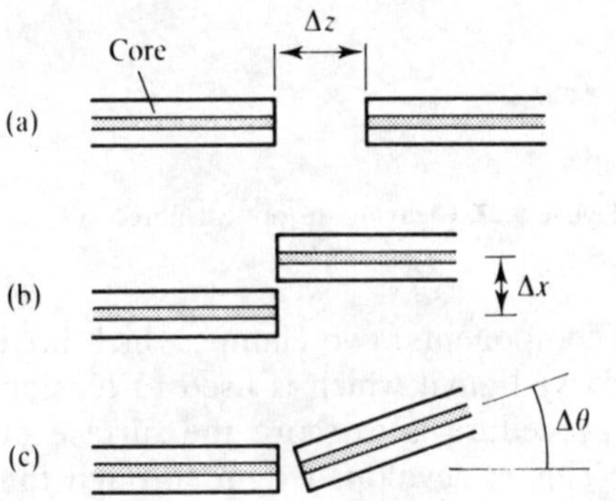

Figure 8.20 Misalignment of a fibre joint in (a) axial, (b) lateral and (c) rotational orientation.

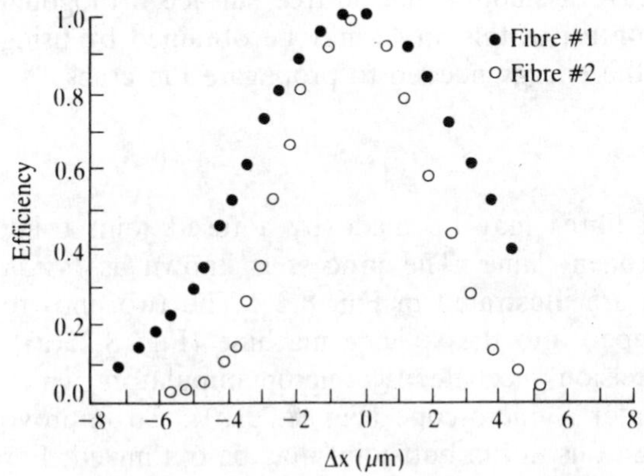

Figure 8.21 Experimental variation of coupling efficiency with transverse displacement in butt-coupled joints (*after W.A. Gambling* et. al., *1978*).

each case efficiency falls drastically for offsets measured in μm, although the larger-diameter fibre is more tolerant. The only answer is stringent alignment accuracy.

Fibre End Preparation

The problems involved in achieving the necessary alignment accuracy are enormous. To begin with, the ends of the components to be joined must be smooth, flat and accurately orthogonal to their axis of propagation. Two basic methods are used to achieve this: *polishing* and *cleaving*. In the former the fibre is mounted in a chuck, allowing one end to be held perpendicular to a rotating lap. The end-face is first ground flat, using an abrasive such as carborundum or aluminium oxide. Several grinding stages are needed, using abrasive of successively finer quality. Finally, the end is polished using diamond grit to remove as far as possible the scratches induced by grinding. Alternatively, a chemical polish such as syton (colloidal silica) may be used. Although the procedure appears laborious, automatic machines exist that may polish a number of fibres simultaneously.

In the case of cleaving, a free end is first stripped of its protective jacket for a few inches using a chemical stripper. It is then placed in a cleaving tool as shown

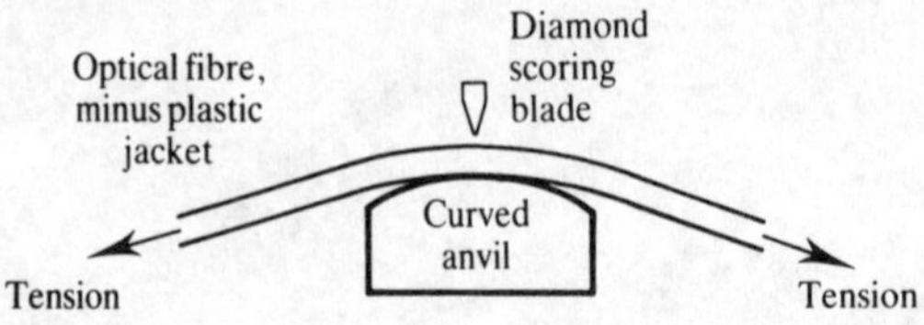

Figure 8.22 Cleaving an optical fibre.

in Fig. 8.22. This usually consists of four main components: two clamps which hold the fibre at points a few centimetres apart, a curved anvil which is used to tension the fibre, and a diamond scoring blade. The procedure is to score the surface of the fibre with the diamond, inducing a crack. This is developed right through the cross-section by moving the anvil so that the tension in the fibre is increased past a critical point. The tip of the crack travels at the speed of sound, and, provided the tool is correctly adjusted, can leave a smooth, hackle-free surface orthogonal to the axis to within 0.5°. Greater control of this angle may be obtained by using a focused ultrasonic wave to supply the energy needed to propagate the crack.

Fusion Splicing

Permanent splices between identical fibres may be made by a fused joint using either an electric arc or an oxyhydrogen flame. The process is known as *fusion splicing* and the basic steps involved are illustrated in Fig. 8.23. The two ends to be spliced are first cleaved and clamped into the splicing machine (Fig. 8.23(a)). This normally contains a set of precision piezoelectric micromanipulators which allow the two ends to be aligned under a microscope (Fig. 8.23(b)). To improve the alignment the transmission through this iterim butt-joint may be optimized. For example, light may be launched into fibre #1 by injection through a bend, and detected by radiation from a bend in fibre #2.

The position of the stages may then be adjusted under the control of a microprocessor, which is driven by a hill-climbing algorithm programmed to maximize the transmitted power. When the alignment is satisfactory the two fibres are separated axially by a short distance (Fig. 8.23(c)) and the electric arc is used to pre-fuse their tips. Surface tension causes each one to melt into an accurate hemisphere (Fig. 8.23(d)) removing any irregularities caused by cleaving. The two fibres are then fed together and the electric arc is used to perform the final fusion (Fig. 8.23(e)). In this stage surface tension forces once again perform a useful function, pulling the fibres into improved transverse alignment (Fig. 8.23(f)). Modern fusion splicing machines are semi-automatic and routinely achieve splices with extremely low loss (≈ 0.1 dB); Fig. 8.23(g) shows a typical commercial splicing machine. Splicers have also been developed for jointing fibre ribbon cable. This is a highly demanding task requiring the fabrication of joints between closely-spaced fibres. Besides the constraints of space, each new fusion operation must not disturb the previous one.

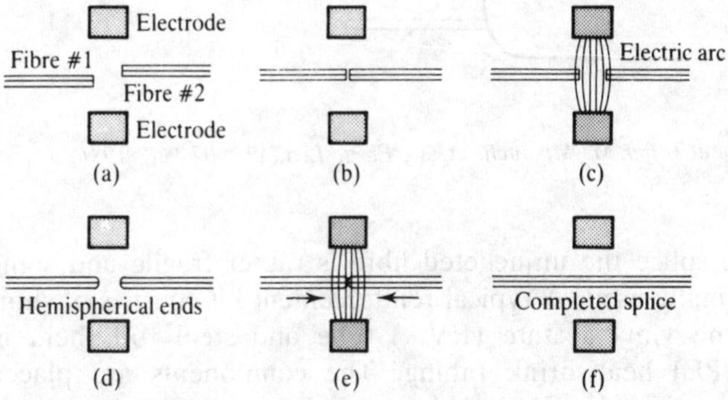

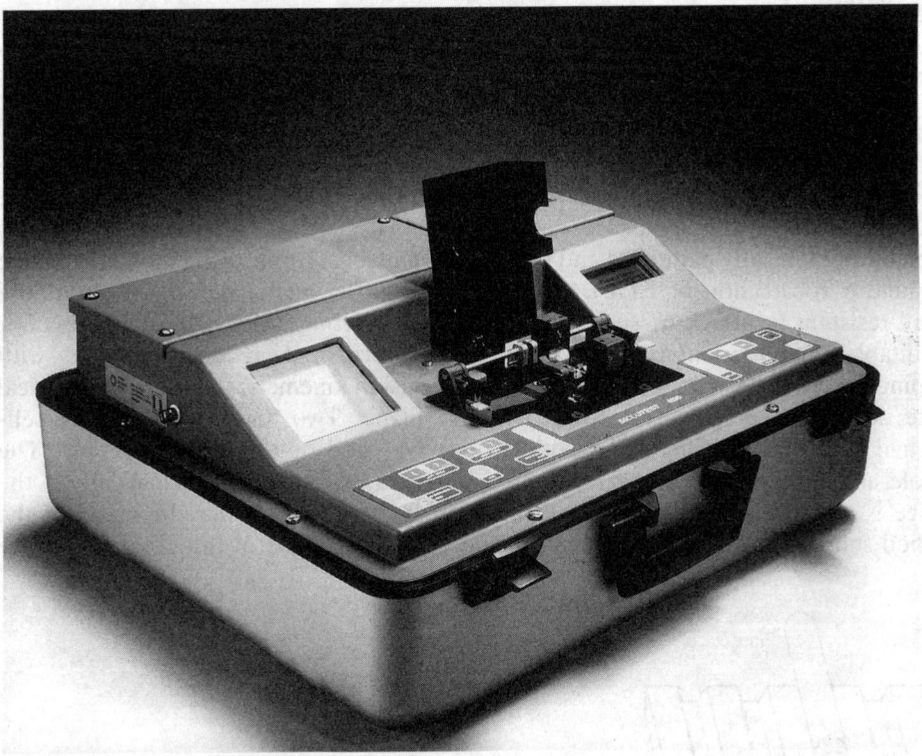

(g)

Figure 8.23 (a)–(f) Steps involved in fusion splicing; (g) an automatic fusion splicer, the BICCOTEST 4100 (*photograph courtesy D. Henderson, Biccotest Ltd.*)

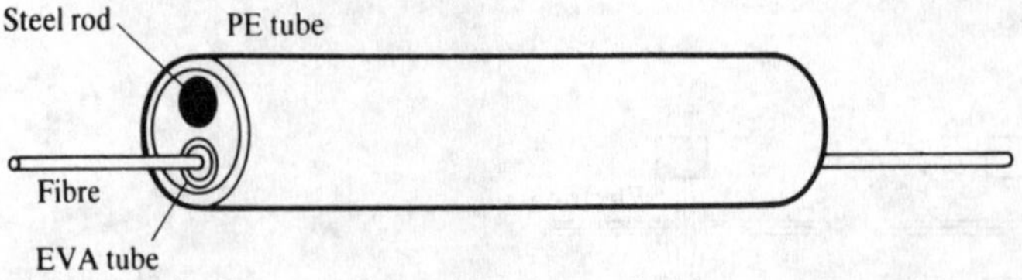

Figure 8.24 Fibre splice reinforcement (*after M. Miyauchi* et al., *Elect. Lett*, **17**, *907–908, 1981*).

After completion of the splice the unjacketed fibre is rather fragile and some kind of reinforcement is normally used. A typical reinforcement kit consists of short sections of meltable ethylene vinyl acetate (EVA) tube and steel rod, held in a length of polyethylene (PE) heat-shrink tubing. The components are placed over the splice and heated so that the EVA melts onto the fibre. Simultaneously the PE tubing is collapsed to hold the steel rod firmly against the EVA-coated splice (Fig. 8.24).

Mechanical Splices

For less-demanding applications mechanical splices are often used. In this case the two fibres are placed in a precision assembly which is secured with epoxy. Figure 8.25 shows one example, based on a preferentially-etched Si substrate. This has been lithographically patterned and processed in an anisotropic etch (such as ethylene diamine pyrocatechol, or EDP). Crystal planes with a higher density of atoms are more resistant to attack by such a mixture, so the etching follows these planes. For a suitable crystal orientation the result is a pattern of deep etched grooves with sloping side walls (*V-grooves*) which act as a precision kinematic mount for cylindrical fibre, locating it accurately in the transverse plane. Two fibres are therefore self-aligned simply by placing them both in a groove and butting them together. The whole assembly is then glued together under pressure, or held mechanically if the splice is to be reusable. For mass-production, plastic replicas may be made of the etched substrates. In the first stage a negative copy is made of the grooved surface

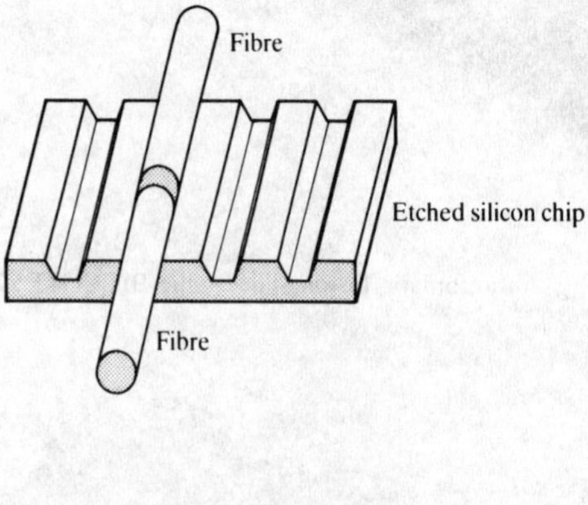

Figure 8.25 A self-aligning mechanical splice based on preferentially-etched silicon.

by nickel electroforming. This 'master' is then used to produce large numbers of positive copies by precision moulding.

Demountable Fibre Connectors

Reusable splices have only limited lifetime, and proper demountable connectors are often required. We have already described lensed connectors; *butt-jointed connectors* form the other main type. In single-fibre connectors each fibre end is normally encased in a precision ferrule which increases the effective diameter and allows accurate coaxial alignment of the two ends by a mating sleeve. The ferrules may be cylindrical (in which case a cylindrical sleeve is used), but some misalignment inevitably follows from the clearance between the ferrules and the sleeve. Errors also follow from any lack of concentricity between each ferrule and its fibre, but these may be reduced by using a core-centering lathe for assembly. Connectors of this type are suitable for multi-mode fibres. For higher precision, conical ferrules are used. Insertion of the ferrule into a tapered sleeve then allows the virtual elimination of machining tolerance, so that dust particles set the final limit on the transverse alignment. This is the principle underlying the operation of the *biconical taper connector*, which is often used for single-mode fibres. Figure 8.26 on page 210 shows a number of different fibre connector plugs.

Any approach involving discrete ferrules is impractical for the multiple joints required in *ribbon fibre connectors*. Most often these are based on etched silicon substrates, using an extension of the principle previously described for mechanical splicing. For example, the two ribbon fibre ends may be mounted on precision-etched baseplates so that each fibre in the array is accurately fixed in a V-groove. On the male half of the connector two large grooves are used to locate a pair of precision steel pins which mate with corresponding grooves on the female half. A joint may then be made simply by aligning the pins and sliding the two connector halves together. Figure 8.27 (page 211) shows a typical multi-fibre connector based on similar principles.

8.9 FIBRE-BASED COMPONENTS

We now consider a number of simple passive components constructed from optical fibre, for use in multi-mode and single-mode systems. In the former, devices are often based around GRIN-rod lenses and bulk-optic beamsplitters. Figure 8.28 shows two examples. The first is a *3 dB power splitter*, which divides an input beam equally between two outputs (Fig. 8.28(a)). The beam enters from the left-hand fibre and is collimated by a 1/4-pitch GRIN-rod lens. The resulting parallel beam is then passed into a frustrated-total-internal-reflection beamsplitter of the type described in Chapter 7. This divides the beam equally into two components which are end-fire coupled into output fibres using further 1/4-pitch GRIN-rod lenses. With the addition of a fourth unused port the device is symmetric, and may act as a bidirectional 2×2 splitter.

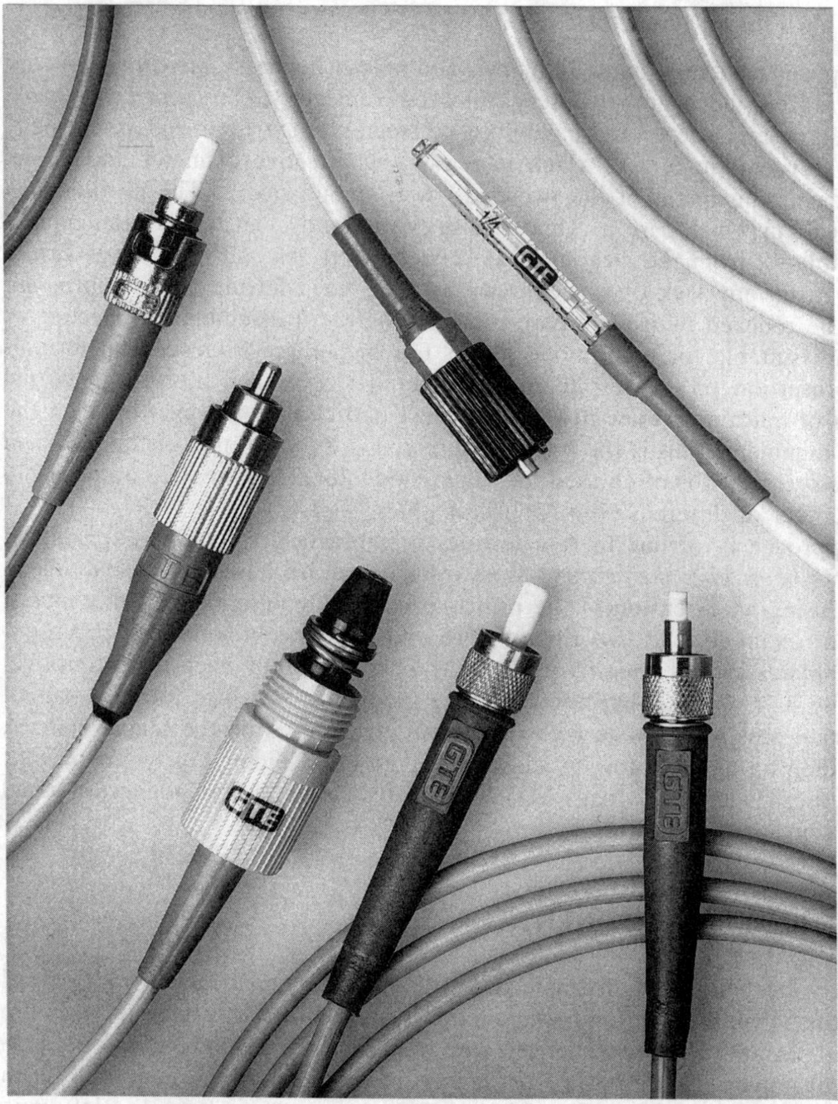

Figure 8.26 Fibre ends fitted with a variety of different connector plugs. A biconical taper plug is shown at the bottom left (*photograph courtesy D. Tedone, GTE Products Corporation*).

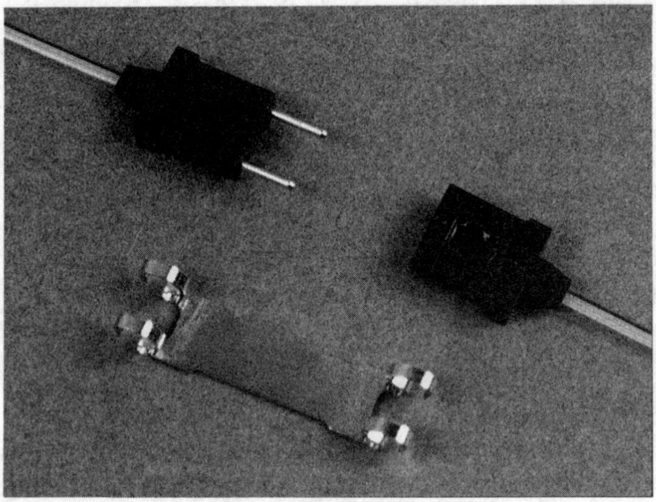

Figure 8.27 A multifibre connector (*photograph courtesy S. Takagi, Furukawa Electric Co. Ltd.*).

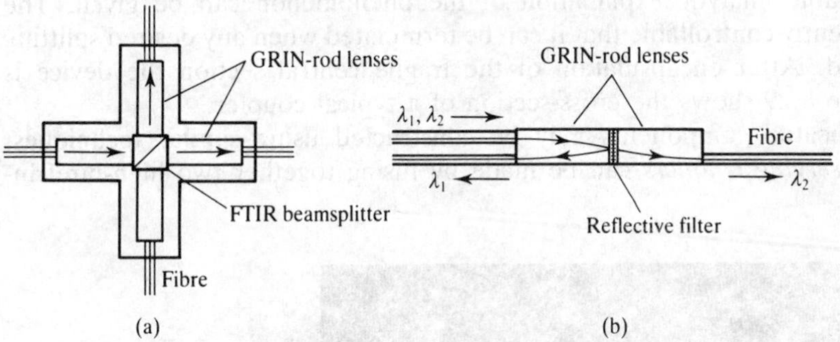

Figure 8.28 Multimode fibre components based on GRIN-rod lenses: (a) beamsplitter, and (b) wavelength demultiplexer (*after S. Sugimoto* et al., *1978*).

The second component (Fig. 8.28(b)) is a *demultiplexer*, used to separate or combine light of two different wavelengths in a wavelength-division-multiplexed communications system. Just as in the previous example, a bulk-optic component (in this case, a multi-layer filter, similar to the Bragg reflection gratings discussed in Chapter 7) is sandwiched between 1/4-pitch GRIN-rod lenses. However, only two lenses are used, and the input and output fibres are not located on-axis. Light of wavelength λ_1 is therefore collimated by the left-hand lens as an off-axis parallel beam. This particular wavelength is reflected by the filter, so the parallel beam is returned off-axis by an equal and opposite angle, and is then end-fire coupled into the left-hand output fibre. Light of a wavelength outside the reflective band of the filter (say, at λ_2) passes straight through and is end-fire coupled into the right-hand output. The operation of the device could be reversed so that two inputs at λ_1 and λ_2 are combined into a single output.

Single-mode fibre components are rather different. Because of the stringent alignment tolerances, end-fire coupling operations are inherently lossy; in fact, the best results are normally obtained by sticking to single-mode operation throughout. A component that is functionally equivalent to the 3 dB splitter above is known as a *fused tapered coupler*. To fabricate the device, two sections of single-mode fibre are first stripped of their protective jackets, placed parallel, twisted together and clamped by their free ends. Flame heating is then applied to the twisted section so that the fibres gradually melt into one another. At the same time the clamped ends are gradually drawn apart so that the central section is tapered. As each fibre is thinned it approaches cut-off, and light propagating in either core is spread further into the cladding. Since the fibres are melting together, the cores approach each other, so the modal fields in the two fibres start to overlap. There is then a possibility for power to be transferred from one fibre to the other. In this case the mechanism is complicated, and we will simply assume that power exchange takes place. However, in Chapter 10 we will examine a similar device—the *directional coupler*—where a reasonable analytic explanation of the phenomenon can be given. The process is sufficiently controllable that it can be terminated when any desired splitting ratio is achieved. After encapsulation of the fragile central section the device is complete. Figure 8.29 shows the cross-section of a typical coupler.

More-complicated components may be constructed using similar techniques. *Polarization-preserving couplers* can be made by fusing together two high-birefrin-

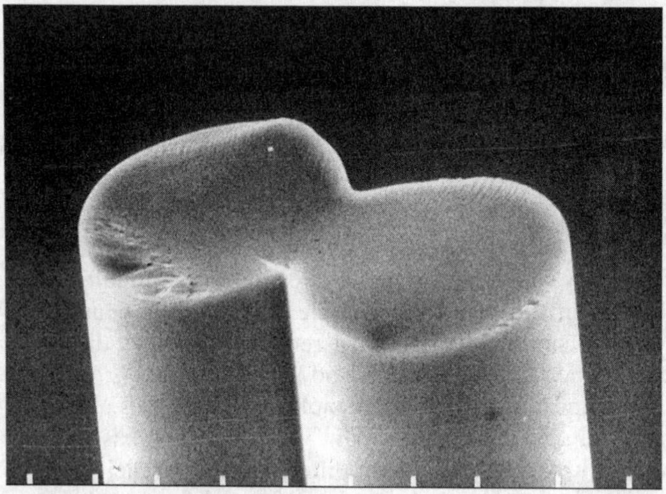

Figure 8.29 Cross-section of a fused tapered coupler (*photograph courtesy D. Mortimore, British Telecom Research Laboratories*).

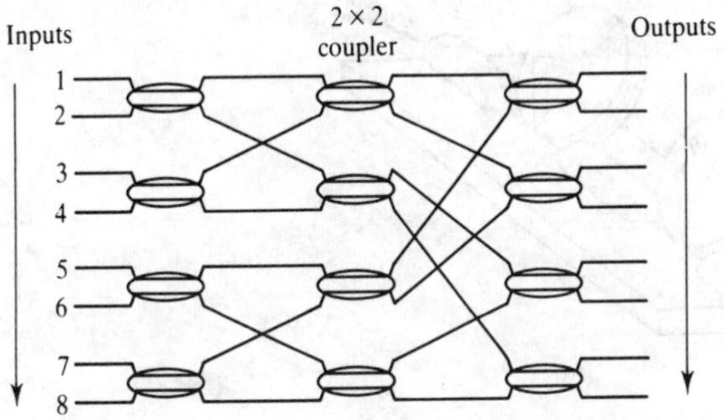

Figure 8.30 An 8 × 8 star coupler, built using fused tapered couplers.

gence fibres—for example, two D-fibres arranged with their flat surfaces in contact. Similarly, large numbers of fibres may be fused together to make 1 × N or $N \times N$ power splitters (where N might lie in the range 3–7). When N becomes much larger the monolithic approach is impractical and it is simpler to synthesize the required structure by splicing together a number of smaller devices in an array. For example, Fig. 8.30 shows an 8 × 8 *star coupler*, built up from twelve 2 × 2 fused tapered couplers. Note that there is exactly one pathway from each input to each output. Furthermore, all paths pass through the same number of 2 × 2 couplers (three) and have the same transmission through each (50 per cent). Consequently, light incident on any of the eight input ports is equally divided among all the outputs.

We now introduce an entirely different single-mode fibre device: a *polarization controller*. We have already noted the phenomenon of polarization fluctuation in circularly-symmetric fibre. In many situations it is necessary to adjust the output polarization from a fibre before coupling into another component (which may be polarization-sensitive).

In bulk optics it can be shown that a plane wave of arbitrary polarization may be transformed into any desired output polarization state using two quarter-wave plates. We recall from Chapter 3 that these are slabs of birefringent material with a thickness chosen to give a $\pi/2$ phase difference between the two orthogonal polarization components. An in-line single-mode polarization controller can perform a similar trick using the fibre-optic equivalent of a quarter-wave plate. Figure 8.31(a) shows the principle. The fibre is formed into a coil of radius R and length L. If R is sufficiently small, bending the fibre in this way results in a uniaxial stress in the core, acting towards the centre of the loop. As we saw in the bow-tie fibre, this causes stress-induced birefringence, and L is simply chosen so that a $\pi/2$ phase shift is obtained between the two polarization modes in travelling round the loop. It is also necessary to be able to achieve an effect analogous to the rotation of a wave plate. This is done by fixing the input and output of the coil (points A′ and B′). The entire coil is then rotated about the line A′–B′, twisting the sections A′–A and B–B′. This can induce the necessary rotation of the plane of polarization with respect to the principal axes of the coil. Two coils are needed in a complete polarization controller, which is then as shown in Fig. 8.31(b).

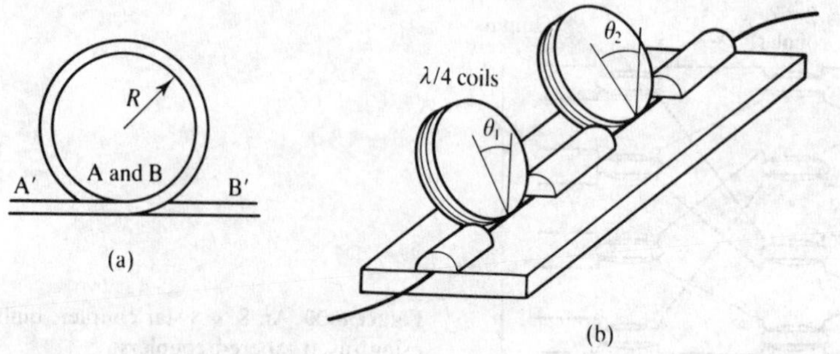

Figure 8.31 Single-mode fibre fractional wave devices: (a) basic principle, and (b) polarization controller.

PROBLEMS

8.1 Attenuation in silica glass is approximately 4 dB/km at $\lambda_0 = 0.65$ μm. What is the fraction of power transmitted through 10 km of silica-based fibre? Assuming that the main contribution to loss arises from Rayleigh scattering, estimate the attenuation at $\lambda_0 = 1.3$ μm. What is the fraction of power transmitted through the same length of fibre at this wavelength?
[0.01 %; 0.25 dB/km; 56.2 %]

8.2 A multi-mode fibre with a numerical aperture of 0.2 supports approximately 1000 modes at a wavelength of 0.85 μm. Estimate its diameter.
[60.5 μm]

8.3 Using (i) a one-dimensional ray-optic argument, and (ii) a modal argument, explain the formation of an inverted real image by a half-pitch GRIN-rod lens.

8.4 Show that the reflection coefficients for TE- and TM-polarized light at a dielectric interface both tend to exp $(\mathrm{j}\pi/2)$ when the refractive index step tends to zero and the angle of incidence approaches 90°.

8.5 Calculate the half-width of the lowest-order mode of a parabolic-index fibre assuming that the peak refractive index of the fibre is $n_0 = 1.5$, the characteristic radius is $r_0 = 25$ μm and the wavelength is $\lambda_0 = 0.85$ μm. What will be the half-angle of divergence of the far-field radiation pattern of this mode? (Hint: refer to Eq. (4.61).)
[2.1 μm; 7.4°]

8.6 The waveguide equation for an idealized model of a parabolic-index fibre may be taken as

$$\nabla^2_{xy} E_{\mu,\nu}(x,y) + \{n_0^2 k_0^2 [1 - (r/r_0)^2] - \beta^2_{\mu,\nu}\} E_{\mu,\nu}(x,y) = 0$$

Show by direct substitution that permissable solutions for the transverse fields of higher-order guided modes are given by

$$E_{\mu,\nu}(x,y) = H_\mu(\sqrt{2}x/a)\, H_\nu(\sqrt{2}y/a) \exp\{-(x^2 + y^2)/a^2\}$$

where $H_\mu(\zeta)$ is a Hermite polynomial of order μ and $a = \sqrt{(2r_0/n_0k_0)}$; and that the corresponding propagation constants $\beta_{\mu,\nu}$ are given by

$$\beta_{\mu,\nu} = n_0 k_0 \sqrt{\{1 - 2(\mu + \nu + 1)/(n_0 k_0 r_0)\}}\ .$$

8.7 Show that the first few low-order Hermite polynomials given in Table 8.3 do indeed satisfy the governing differential equation

$$d^2H_\mu(\zeta)/d\zeta^2 - 2\zeta\, dH_\mu(\zeta)/d\zeta + 2\mu\, H_\mu(\zeta) = 0$$

Sketch the variation of the transverse fields

$$E_{\mu,\nu}(x,y) = H_\mu(\sqrt{2}x/a)\, H_\nu(\sqrt{2}y/a)\, \exp\{-(x^2 + y^2)/a^2\}$$

along the line $y = 0$. Under what conditions will the solution prove inaccurate?

8.8 What properties are the most desirable in optical fibre that is to be used in a long-distance telecommunications system? You should discuss loss, dispersion, ease of jointing, and the availability of other systems components such as sources and detectors.

8.9 The theoretical expression for the end-fire coupling efficiency between a Gaussian beam of characteristic radius w_0 and the lowest-order mode of a parabolic-index fibre (which also has a Gaussian transverse field, of radius a) is given by $\eta = 4w_0^2a^2/\{w_0^2 + a^2\}^2$. Sketch the variation of η with the parameter $\alpha = \log(w_0/a)$. Prove that the variation is symmetric about the point $\alpha = 0$, when maximum efficiency of 100 per cent is obtained.

8.10 Sketch the layout of a 16 × 16 star coupler based on an array of 2 × 2 fused tapered couplers.

SUGGESTIONS FOR FURTHER READING

Adams, M. J. "An Introduction to Optical Waveguides", Ch. 7, John Wiley and Sons, New York, 1981.

Adams, M. J., Payne, D. N., and Ragdale, C. M. Birefringence in optical fibres with elliptical cross-section, *Elect. Lett.*, **15**, 298–299, 1979.

Ainslie, B. J., and Day, C. R. A review of single mode fibres with modified dispersion characteristics, *IEEE J. Lightwave Tech.*, **LT-4**, 967–979, 1986.

Birch, R. D., Payne, D. N., and Varnham, M. P. Fabrication of polarization-maintaining fibres using gas-phase etching, *Elect. Lett.*, **18**, 1036–1038, 1982.

Dyott, R. D., and Schrank, P. F. Self-locating elliptically cored fibre with an accessible guiding region, *Elect. Lett.*, **18**, 980–981, 1982.

Fleming, J. W. Material dispersion in lightguide glasses, *Elect. Lett.*, **14**, 326–328, 1978.

Gambling, W. A., Matsumura, H., and Cowley, A. G. Jointing loss in single-mode fibres, *Elect. Lett.*, **14**, 54–55, 1978.

Gambling, W. A., Payne, D. N., and Matsumura, H. Radiation from curved single-mode fibres, *Elect. Lett.*, **12**, 567–569, 1976.

Gloge, D. Weakly guiding fibers, *Appl. Opt.*, **10**, 2252–2258, 1971.

Kao, K. C., and Hockham, G. A. Dielectric fiber surface waveguides for optical frequencies, *Proc. IEE*, **113**, 1151–1158, 1966.

Keiser, G. "Optical Fiber Communications", McGraw-Hill International, London, 1986.

Khoe, G.-K., Kock, H. G., Kuppers, D., Poulissen, J. H. F. M., and de Vrieze, H. M. Progress in monomode optical fibre interconnection devices, *IEEE J. Lightwave Tech.*, **LT-2**, 217–227, 1984.

Koehler, B. G., and Bowers, J. E. In-line single-mode fiber polarization controllers at 1.55, 1.3 and 0.63 μm, *Appl. Opt.*, **24**, 349–353, 1985.

Marshall, I. W. Low-loss coupling between semiconductor lasers and single-mode fibre using tapered lensed fibres, *Br. Telecom. J.*, **4**, 114–121, 1986.

Midwinter, J. E. "Optical Fibers for Transmission", John Wiley and Sons, New York, 1978.

Miller, C. M. Fibre-optic array splicing with etched silicon chips, *Bell Syst. Tech. J.*, **57**, 75–90, 1978.

Miller, T. J., Hart, A. C., Vroom, W. J., and Bowden, M. J. Silicone- and ethylene-vinyl-acetate-coated laser-drawn silica fibres with tensile strengths > 3.5 GN/m^2 (500 kp.s.i.) in > 3 km lengths, *Elect. Lett.*, **14**, 603–605, 1978.

Miya, T., Terunuma, Y., Hosaka, T., and Miyashita, T. Ultimate low-loss single-mode fibre at 1.55 μm, *Elect. Lett.*, **15**, 106–108, 1979.

Miyauchi, M., Matsumoto, M., Toda, Y., Matsuno, K., and Tokumaru, Y. New reinforcement for arc-fusion spliced fibre, *Elect. Lett.*, **17**, 907–908, 1981.

Mortimore, D. B. Low-loss 8 × 8 single-mode star coupler, *Elect. Lett.*, **21**, 502–504, 1985.

Nicia, A. Practical low-loss lens connector for optical fibres, *Elect. Lett.*, **14**, 511–512, 1978.

Payne, D. B., McCartney, D. J., and Healey, P. Fusion splicing of a 31.6 km monomode optical fibre system, *Elect. Lett.*, **18**, 82–84, 1982.

Pinnow, D. A., Gentile, A. L., Standler, A. G., Timper, A. T., and Holbrook, L. M. Polycrystalline fiber optical waveguides for infrared transmission, *Appl. Phys. Lett.*, **33**, 28–29, 1978.

Senior, J. M. "Optical Fiber Communications: Principles and Practice", Prentice-Hall International, London, 1985.

Shibata, S., Horiguchi, M., Jinguji, K., Mitachi, S., Kanamori, T., and Manabe, T. Prediction of loss minima in infra-red optical fibres, *Elect. Lett.*, **17**, 775–777, 1981.

Tomlinson, W. J. Applications of GRIN-rod lenses in optical fiber communication systems, *Appl. Opt.*, **19**, 1127–1138, 1980.

Villaruel, C. A., and Moeller, R. P. Fused single mode fibre access couplers, *Elect. Lett.*, **17**, 243–244, 1981.

White, K. I., and Nelson, B. P. Zero total dispersion in step index monomode fibers at 1.30 and 1.55 μm, *Elect. Lett.*, **15**, 396–397, 1979.

CHAPTER

NINE

CHANNEL WAVEGUIDE INTEGRATED OPTICS

9.1 CHANNEL WAVEGUIDE TYPES

The second important system of integrated optics arises when the waveguides take the form of *channel guides*. These still lie on a planar substrate, but are now arranged to confine the optical field in two directions, x and y. Figure 9.1 shows cross-sections through the most common geometries. We shall now describe these in some detail, together with a summary of typical fabrication methods; this information was omitted from our previous discussion of planar guides in Chapter 6, on the grounds that many of the structures used are common to each waveguide system. Further processing information can be found in Chapter 13.

Buried channel guides Figure 9.1(a) shows a *buried channel guide*, which is made by modifying the properties of the substrate material so that a higher refractive index is obtained locally. Most fabrication processes result in a weak, graded-index guide buried just below the surface, and, although channel guides normally have one axis of symmetry, they usually lack a well-defined cross-sectional shape.

Diffusion is often used to fabricate guides of this type. For example, titanium metal can be diffused into lithium niobate or lithium tantalate substrates, by first depositing the metal in patterned strips of about 1000 Å thickness and then carrying

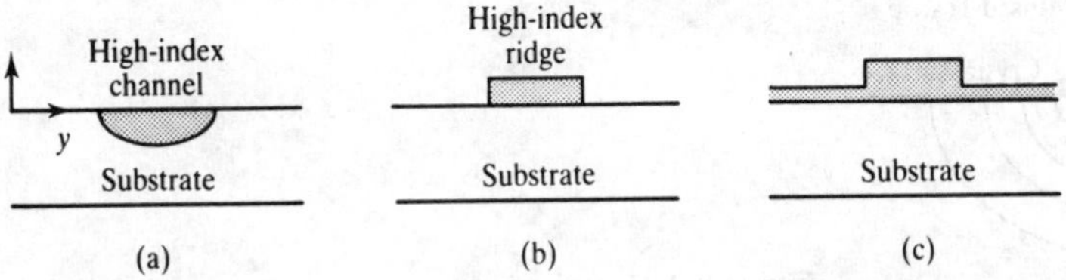

Figure 9.1 Channel waveguide types: (a) buried, (b) ridge, and (c) strip-loaded.

out an in-diffusion at a higher temperature (≈ 1000°C) for three to nine hours. This is known as the *Ti:LiNbO₃ process*. The additional impurities cause a change in refractive index that is approximately proportional to their concentration with a typical maximum value of $\Delta n \approx 0.01$. Figure 9.2 shows contours of constant refractive index that are obtained by diffusion of Ti metal into $LiNbO_3$.

Alternatively, material can be *exchanged* with the substrate. For example, protons (H^+ ions) can be exchanged with Li^+ ions in $LiNbO_3$. This is done by first covering the substrate with a masking layer, which is patterned so that narrow stripe openings are made. The crystal is then placed in a suitable hot melt (in this case, benzoic acid, heated to around 200°C), and exchange of material takes place through the crystal surface in regions exposed by the mask openings. Proton exchange yields an index change of $\Delta n \approx +0.12$ for one polarization mode (the TM mode), but the index change is actually *negative* for the other one: a waveguide is formed only in the former case. Silver ions can be exchanged with Na^+ ions in soda-lime glass in a similar way by immersing the substrate in molten $AgNO_3$, at temperatures of 200–350°C. In this case the time required to form a guide is large (several hours), but if a d.c. electric field is used to assist the ion migration, the process can be speeded up quite considerably. The index change obtained is $\Delta n \approx 0.1$.

Finally, material can be *implanted* using an ion implanter. This is a large, expensive item of vacuum equipment which can be used to select a particular species of ions and then accelerate the desired ions electrostatically, so that they strike the substrate surface at high speed and penetrate some distance. Because this process is ballistic, it causes damage to the lattice as well as a change in refractive index. Ion implantation is therefore normally followed by an annealing step, which shakes out the lattice damage. This allows properties that rely on a regular, undamaged crystal lattice (e.g. electro-optic behaviour) to be preserved, and greatly reduces waveguide propagation losses. Following annealing, He^+ implantation has been found to be reasonably successful in $LiNbO_3$.

Ridge waveguides Figure 9.1(b) shows a *ridge waveguide*. This is a step-index structure, usually fabricated from a planar guide that has been patterned and etched to leave a ridge. Total internal reflection then takes place at the sides of the guide, as well as the top and bottom faces. Often the etching is designed to follow specific crystal planes, so the edges of the ridge may be sloped. Similarly, material is often (especially in laser structures) regrown around the sides of the guide. Much higher

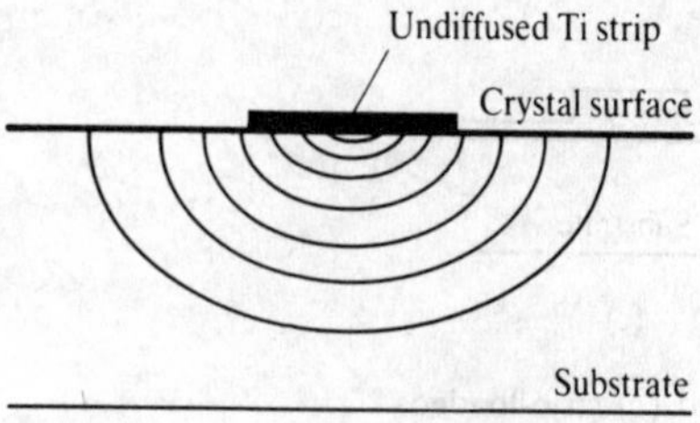

Figure 9.2 Contours of constant refractive index in a Ti-diffused channel waveguide.

field confinement is possible when using a ridge rather than a buried guide. As mentioned above, the index change that can be obtained by material modification is very small, so the evanescent field of any guided mode must extend for some distance outside the core. In a ridge guide, on the other hand, there can be a large index difference between the ridge and its surround. This is an important advantage; for example, bends can be much tighter in ridge-guide systems, as we shall see later.

Waveguides are often made as ridges in semiconductor materials because of the ease with which a mesa structure may be made by etching. Although the actual characteristics of semiconductors will not be covered until Chapter 11, we shall now describe the basic semiconductor waveguide types. There are two: *heterostructure guides*, which operate by the refractive index differences obtained between different materials, and *homostructure guides*, which use the index reduction following from an increase in the majority carrier density.

The two most common semiconductor materials used in optoelectronics are based on the ternary alloy *gallium aluminium arsenide* and the quaternary alloy *indium gallium arsenide phosphide*, respectively. To fabricate heterostructure waveguides in the GaAs/GaAlAs system, successive layers of material are first grown on a GaAs substrate in differing compositions of the general form $Ga_{1-x}Al_xAs$ (where x is the mole fraction of the constituent). Because GaAs and GaAlAs have similar lattice parameters, these additional layers are lattice-matched to the substrate and the composite crystal can grow without strain. In $Ga_{1-x}Al_xAs$ the refractive index ranges roughly linearly from $n \approx 3.57$ for $x = 0$ to $n \approx 3.36$ for $x = 0.35$ (at near infrared wavelengths). The variation of the alloy content between the layers then yields the refractive index difference necessary for a waveguide. For example, Fig. 9.3 shows how a planar guide might be fabricated using a layer of high-index GaAs grown epitaxially on top of a thick layer of $Ga_{1-x}Al_xAs$ (which acts as a low-index isolation layer, separating the guide from the high-index GaAs substrate). The guiding layer may then be etched down to a narrow rib to form a ridge guide.

Waveguide fabrication is similar in the InP/InGaAsP system, in which layers of the quaternary alloy $In_{1-x}Ga_xAs_{1-y}P_y$ are grown on an InP substrate. However, lattice-matching to the substrate is only obtained for particular combinations of x and y, which are detailed in Chapter 13. More complicated layered structures are also often used. For example, Fig. 9.4 shows a *double heterostructure* waveguide, which has a low-index GaAlAs confining layer on both sides of the guide. This arrangement is used extensively in semiconductor lasers since the difference in

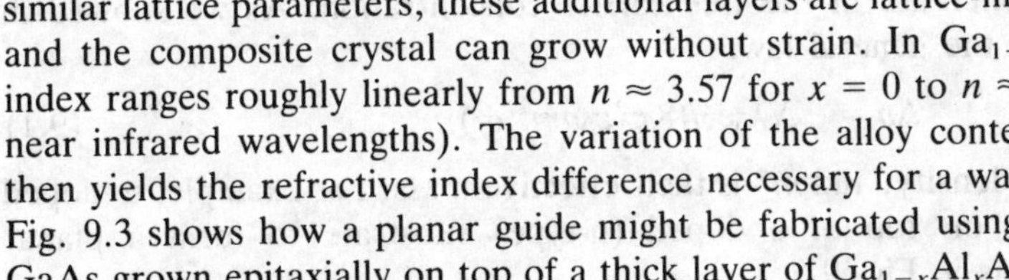

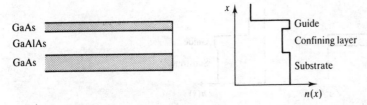

Figure 9.3 A GaAs/GaA1As planar waveguide.

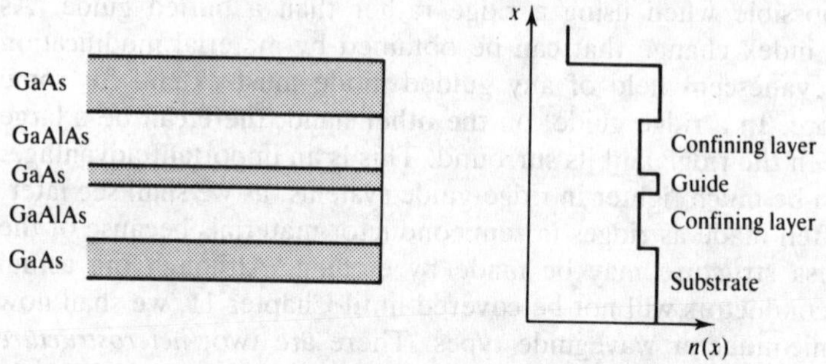

Figure 9.4 A double heterostructure GaAs/GaA1As planar waveguide.

bandgap between the GaAs and GaAlAs layers can provide carrier confinement in addition to confinement of the optical field. It will be discussed in more detail in Chapter 12.

Instead of using compositional variations to form a layered guiding structure, waveguides can also be formed by making use of the reduction in refractive index that follows from an increase in the free carrier concentration. As we showed in Chapter 3 this follows the general law:

$$\Delta n = -N\lambda_0^2 e^2/(8\pi^2\varepsilon_0 n m^* c^2) \tag{9.1}$$

where N is the carrier density, and m^* is their effective mass. For example, undoped GaAs may be grown on top of a doped (n-type) substrate to form a planar homostructure guide, as in Fig. 9.5.

Alternatively, a similar structure may be created by proton bombardment of an n-type substrate. This creates damage sites near the surface, which trap free carriers. This is known as *free carrier compensation*, and produces an increase in refractive index. We showed in Chapter 3 that the index change obtained by doping may be found for n-type GaAs as $\Delta n = -0.01$, when $N = 5 \times 10^{18}\ \text{cm}^{-3}$ and $\lambda_0 = 1\ \mu\text{m}$. This implies that the index changes possible in homostructures are much lower than in heterostructures, giving reduced confinement in any guide structure. In fact, since $m_h^* \gg m_e^*$ in InP, the hole-induced index difference is much smaller than the equivalent electron-induced value, so only n-type material is usable for waveguiding.

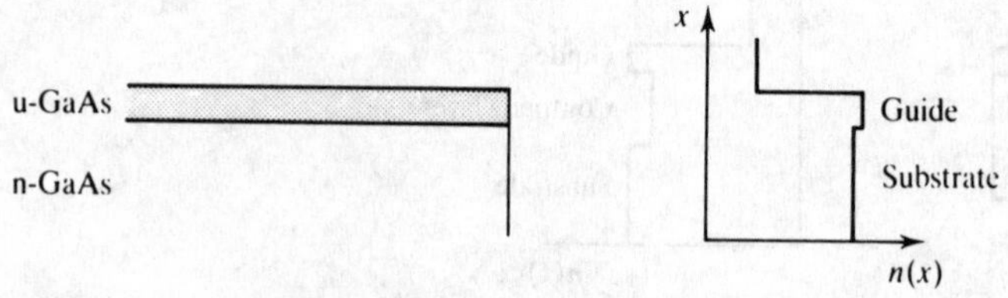

Figure 9.5 A planar waveguide in GaAs formed by changes in free carrier concentration.

A further problem arising in homostructures is that the increased carrier concentration in the substrate causes a rise in absorption which can result in unacceptable propagation loss. Heterostructures therefore offer several advantages.

Ridge guides can also be made in amorphous material on semiconducting substrates. One common system involves silicon oxynitride guides constructed on a silicon dioxide buffer layer which in turn lies on a silicon substrate. Silicon oxynitride has a relatively high refractive index, but one which is lower than that of the silicon, for which $n \approx 3.5$. The two are therefore separated by a thick layer of SiO_2, which acts as a low-index ($n \approx 1.47$) spacer. Other systems use doped silica (e.g. an SiO_2/TiO_2 mixture) on top of the silica buffer layer. These are important since they allow the fabrication of waveguides in a form compatible with VLSI electronics.

Strip-loaded waveguides Two-dimensional confinement may also be less obvious. Figure 9.1(c) shows a *strip-loaded guide*. A three-layer structure has been used, which has a substrate, a planar layer, and then a ridge. The planar layer is arranged as a guide, but one which is just cut off. However, the addition of a further high-index overlay is sufficient to induce guiding in a localized region near the ridge. Confinement of the optical field in the y-direction is obtained because the refractive index, when averaged in the x-direction, is higher in the region of the ridge. The additional ridge overlay may have a different refractive index to that of the planar guide, or (as in Fig. 9.1(c)) the same index.

Analysis of Channel Waveguides

The analysis of a channel guide is generally a complicated problem, and different approaches are used for ridge and channel guides. Since the index changes concerned in the former are so large, a TEM model cannot be used, and a rigorous solution is normally required. Figure 9.6 shows one possible model of a ridge guide, which consists of a region of dielectric of rectangular cross-section surrounded by regions of differing refractive indices.

It has been shown that modes of such a guide can be found (at least approximately) by assuming solutions for the transverse variation of the electric field in the form of a product, i.e. as $E_T(x,y) = E_1(x)\ E_2(y)$. Since the layers are individually uniform, we may guess that $E_1(x)$ and $E_2(y)$ must have the form either of cosinusoidal or of exponential functions, by analogy with the solution for a slab guide. These solutions are then built up in a piecewise fashion, as follows:

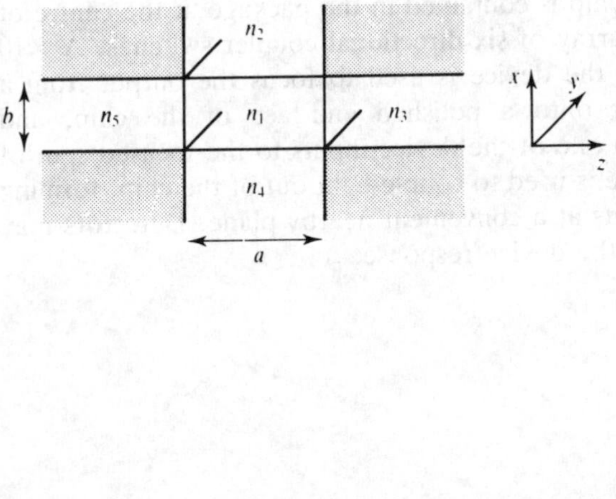

Figure 9.6 Model for a rectangular dielectric waveguide.

In region 1: the field is assumed to vary cosinusoidally with both x and y.
In region 2: the field is assumed to decay exponentially with x, and vary cosinusoidally with y.
In region 3: the field is assumed to vary cosinusoidally with x, and decay exponentially with y.

And so on. These fields are matched rigorously at the interfaces between regions 1 and 2, 1 and 3, 1 and 4, and 1 and 5, but only approximately in the shaded corner regions. The result is a set of two-dimensional mode patterns which are neither truly TE nor truly TM. Often the confinement provided by the layering is much stronger than that in the lateral direction, so this type of mode may be non-circular, with a high aspect ratio (e.g. 3:1).

Unfortunately, the index distribution obtained by other fabrication methods can be less well-defined. For example, we have already seen in Fig. 9.2 the refractive index distribution arising when a strip of Ti metal is diffused into a $LiNbO_3$ crystal. Since the index change is weaker in this case, the waveguide modes are TEM-like and the approximate model (Eq. 6.54) developed in Chapter 6 may be used to find them. We shall not attempt to solve this equation here, but simply present the results. Figure 9.7 shows contours of constant intensity for the transverse field $E(x,y)$ in a single-mode guide with the index distribution of Fig. 9.2. Notice that the mode is roughly circular, but there is increased confinement at the surface due to the large index change at the substrate/air interface.

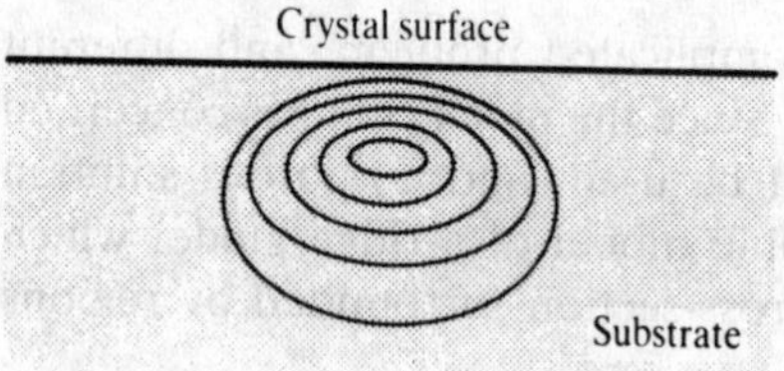

Figure 9.7 Isointensity contours for the modal field in a Ti-diffused channel waveguide.

9.2 INPUT AND OUTPUT COUPLING

As with optical fibres, power is generally coupled into a channel guide by transverse excitation using end-fire or butt-coupling. For example, Fig. 9.8 shows the experimental set-up for end-fire coupling into and out of a Ti:$LiNbO_3$ channel waveguide device. The Ti:$LiNbO_3$ chip is contained in the package at the centre of the photograph and consists of an array of six directional coupler switches. A ×10 microscope objective to the left of the device is used to focus the output from a 1.523 μm wavelength He-Ne laser onto a polished end-face of the chip, and micromanipulators are used to align one of the device inputs to the focused spot of light. A second microscope objective is used to couple light out of the chip, forming real images of the two device outputs at a convenient nearby plane. Detectors may be placed at this plane to measure the device response.

Figure 9.8 Apparatus for end-fire coupling into a $Ti:LiNbO_3$ channel waveguide device.

A channel guide device can be joined, more usefully, to a different type of guide (e.g. a fibre) simply by polishing an end of each component, aligning the field-carrying cores together, and forming a permanent cemented joint. There are two requirements for high efficiency in butt-coupling, as we saw in Chapters 6 and 8. (1) The modal field in the channel guide must be matched (in overall shape and dimensions) as far as possible with that in the fibre. Perfect *mode-matching* is never possible, but considerable success has been achieved in many cases by careful choice of the fabrication conditions. For $Ti:LiNbO_3$ devices, for example, this may be done by adjusting the width and thickness of the Ti metal strip, and the diffusion temperature and time. (2) The field-carrying cores must be accurately aligned with respect to one another in relative spatial position and angular orientation. Typical tolerances for the former are $< 1\ \mu m$. For single joints this can be achieved by micromanipulation using piezoelectric positioners. If conditions (1) and (2) are satisfied the *fibre-device-fibre coupling loss* can be as low as 1.5 dB in $Ti:LiNbO_3$ devices at near-infrared wavelengths.

Micromanipulation is, unfortunately, not a feasible method of adjusting the relative positions of closely-spaced arrays of fibres and channel guides. An alternative self-alignment technique is therefore preferred. Figure 9.9 shows one method used with $Ti:LiNbO_3$ devices. This is based on the V-grooved silicon substrates used in the fibre connectors and splices discussed in Chapter 8. The fibre array is simply sandwiched between two grooved substrates and the end of the entire assembly is

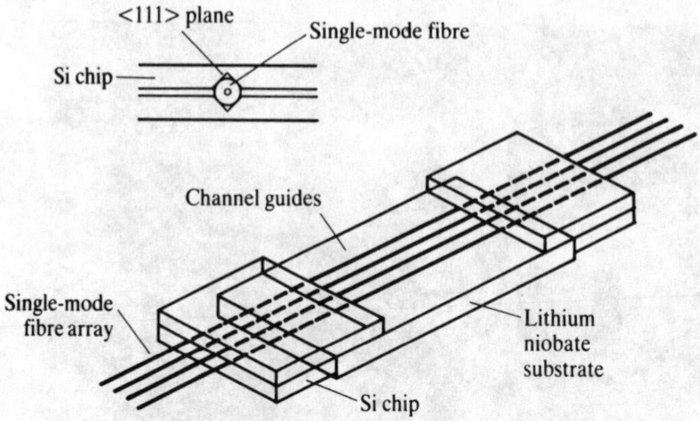

Figure 9.9 Precision alignment structure for multiple fibre end-fire coupling to an integrated optic channel waveguide chip (after *E. J. Murphy* et al., © *IEEE 1986*).

polished. It is then only necessary to position each array correctly, instead of each individual fibre. In the example above, one half of each fibre-carrying sandwich is extended as a ledge, which mates with the integrated optic chip. Only one adjustment (parallel to the substrate surface) is then needed to align each fibre array with the channel guides on the chip.

9.3 SOURCES OF PROPAGATION LOSS

Fundamental Loss Mechanisms

The main causes of propagation loss in straight waveguides are absorption, and scattering from material inhomogeneities and boundary imperfections. Note that the latter is greatly reduced in fibre manufacture by the process of pulling a preform down to the final fibre. This 'smoothing' does not occur in the fabrication of an integrated device, which is produced as a 1:1 copy of a master mask set. Consequently, scattering losses are much higher in integrated optics. However, all forms of loss vary considerably, depending on the precise nature of the guide. We shall illustrate this with examples.

For $Ti:LiNbO_3$ buried channel guides, losses are typically 1.0 dB/cm at $\lambda = 0.633\ \mu m$, 0.5 dB/cm at $\lambda = 1.15\ \mu m$ and 0.2 dB/cm at $\lambda = 1.52\ \mu m$. In this case the major contribution to loss is the *photorefractive effect*. This is a form of optical damage, which results in the optically-induced formation of refractive index changes. These scatter light out of the guide causing increased loss. Photorefractive effects are very dependent on the level of optical power in the guide. They are extremely significant at visible wavelengths, but reduce enormously in the near-infrared.

For GaAs ridge guides, losses were originally extremely high (about 4.0 dB/cm), but these have been reduced considerably in recent years, to around 0.2 dB/cm. Much of the loss in homostructures arises from the presence of free carriers in the low-index confining layers outside the guide, which gives rise to loss through collision damping. For low loss a mode must therefore be well-confined inside the more resistive guiding layer. However, further loss arises from scattering from the guide walls. If these are etched down crystal planes they may well adopt a 'staircase' profile, causing significant scattering.

Bend loss A common requirement in a guided wave system is to be able to route light round a bend, for example to establish links between different devices on an integrated optic chip. However, as mentioned in Chapter 8, any curvature of a waveguide results in a loss additional to the normal propagation loss. Figure 9.10 illustrates the effect, which is caused by continuous conversion of the guided light into radiation modes. We shall now give an alternative physical explanation of the process.

One way to understand the underlying cause of the radiation is to consider the profile of the transverse field as the guided mode travels round the bend. To keep in step this pattern must rotate around the centre of curvature of the guide, like a spoke in a rotating wheel. Consequently, the further from the centre, the faster the field must move. However, the outermost evanescent part of the field extends to infinite radius, so at some distance from the core the field would have to exceed the speed of light in the substrate material to keep up. The portion of the field outside this point must be radiated, reducing the power in the guided mode. Bend loss is therefore very dependent on the confinement offered by the guide. For indiffused guides the index change forming the guide is small, so the confinement is poor. Consequently, if low loss is required, bends must be shallow, with typical radii of curvature greater than 1 cm for Ti:$LiNbO_3$ devices. This limits the device packing density considerably. In ridge guide systems the index change is larger, so bends can be considerably tighter.

A detailed analysis of the conversion of guided light into radiation modes in a bend can be performed using the method of local normal modes, which we previously

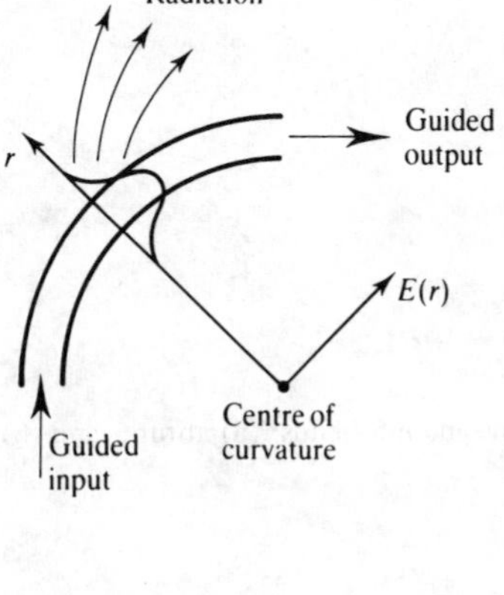

Figure 9.10 Radiation loss at a waveguide bend.

used to investigate slow tapers. This time, however, the axis of propagation of the guide must change with distance around the bend, so the analysis is rather more complicated. An approximation to the coupled mode solution is therefore often used. Typically it is found that the guided power falls off as

$$P_z = P_0 \exp(-\alpha z) \tag{9.2}$$

where P_z is the power after travelling a distance z and P_0 is the input power. The attenuation coefficient α has the semi-empirical form:

$$\alpha = c_1 \exp(-c_2 r) \tag{9.3}$$

where r is the radius of curvature and c_1 and c_2 are constants. Equation (9.2) can be interpreted as saying that the rate of power loss is proportional to the local guided power. This is reasonable, provided that no radiated power is subsequently coupled back into the guide. Since this process can and does occur, Eq. (9.2) should be used with some caution.

Note that, for a bend of a given angle θ, the transmitted power is $P_\theta = P_0 \exp\{-\alpha r\theta\}$. For bends of large radius, P_θ tends to unity since the loss coefficient α tends to zero. However, for extremely small radius bends, P_θ tends to unity once again. This is because the distance travelled round the bend now becomes very small, which counters the effect of an increasingly large value of α. In between, there is a minimum in P_θ. By differentiation, this can be shown to occur at a critical radius $r = 1/c_2$.

Bends are often used to provide offset transitions at the input and output of devices, or to separate guides at the edge of a chip for ease of coupling to fibres. In either case an *S-bend* is used. This can take the form of an abrupt transition, as shown in Fig. 9.11(a), or a smooth curve (Fig. 9.11(b)). Generally the former works better if the offset is small, and the latter otherwise.

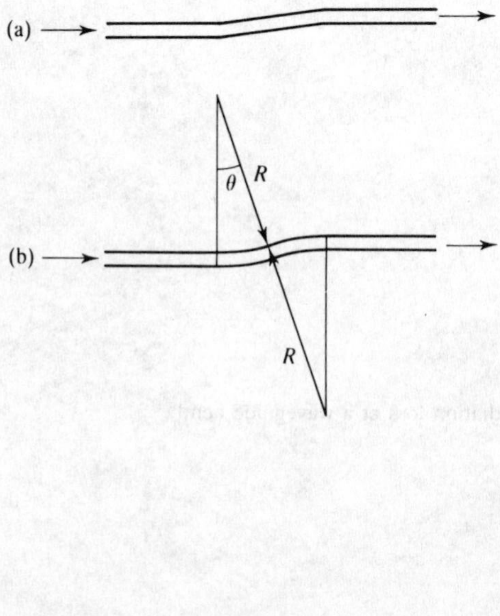

Figure 9.11 Waveguide S-bends: (a) abrupt, and (b) gradual.

9.4 POLARIZERS

We now consider a range of passive channel waveguide devices which can be used as building blocks in a more complicated guided wave optical circuit. We begin with *polarizers*. In bulk optics these are devices which will pass one polarization component of a plane wave, while attenuating the other (preferably blocking it entirely). An integrated optic polarizer is a similar component, which will pass only one polarization mode. There are a number of ways to achieve this. The major strategies are:

1. To radiate one polarization mode preferentially. This can be done by adding a birefringent overlayer to the guide so that one polarization mode is cut off while the other still propagates. For example, Fig. 9.12 shows an early polarizer for a waveguide formed in a glass substrate. This consisted of an overlayer of calcite, for which the ordinary and extraordinary indices are 1.658 and 1.486, respectively. Since the guide index is approximately 1.5, only the TE mode is guided with the crystal arranged in the orientation shown, and the TM mode is radiated. Although this approach produces a functionally adequate polarizer, it is not suitable for integration. More recently, fully integrated processes have been developed. One example combines short sections of proton-exchanged waveguide with guides formed by the Ti:$LiNbO_3$ process. This produces a TM-pass polarizer because proton exchange only forms a waveguide for the TM mode.
2. To absorb one polarization mode preferentially. This can be done using a metal overlay in conjunction with a dielectric. For Ti:$LiNbO_3$ guides an Si_3N_4/Al combination is suitable. Strong absorption of one mode can then occur if it is coupled into a surface plasmon (previously encountered in Chapter 5) propagating in the metal layer.
3. To couple one polarization mode preferentially into another guide. This can be performed using a polarization-selective *directional coupler*.

Both 2 and 3 involve the phenomenon of codirectional coupling, which will be covered in Chapter 10.

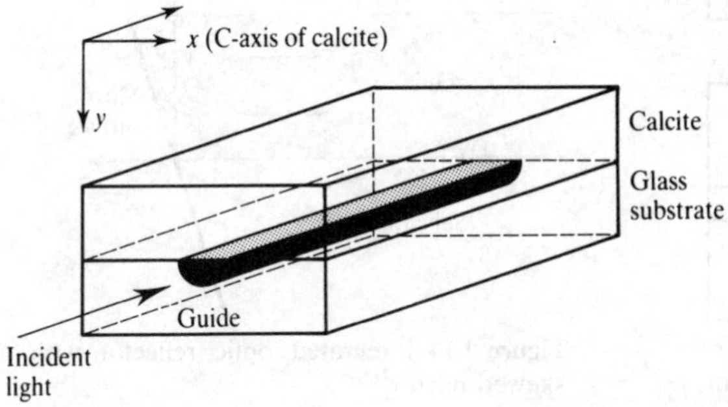

Figure 9.12 Waveguide polarizer using a calcite overlay (*after S. Uehara* et al., *1974*).

9.5 MIRRORS

We now consider the integrated optic equivalent of a *mirror*, which offers a simple method of changing the direction of a guided beam. It can occupy considerably less space than a bend and can theoretically be more efficient. Such a device can operate by the reflection of guided modes from a flat, highly reflective interface, arranged exactly orthogonal to the guide. A suitable interface can be made by cutting a trench across the guide, as shown in Fig. 9.13. This can be done using a highly anisotropic etch process—*reactive ion etching* (RIE) and *focussed ion beam micromachining* are both suitable. The interface must then be metallized to increase its reflectivity.

While the integrated mirror is conceptually a simple device, it is hard to make in practice because any reduction in surface flatness, orthogonality or reflectivity will lower the overall reflection efficiency, as we now show. This can be calculated by modelling the local mirror reflectivity as the product of a real function $R(x,y)$, which accounts for the amplitude reflectivity, and a phase function $\phi(x,y)$, which accounts for mirror tilt and flatness. An incident mode with transverse field distribution $E_\nu(x,y)$ will therefore be reflected as a new field with transverse field distribution $E'_\nu(x,y)$ given by

$$E'_\nu(x,y) = E_\nu(x,y)\, R(x,y) \exp\{-\mathrm{j}\phi(x,y)\} \tag{9.4}$$

The problem can then be treated as one of end-fire coupling, where we simply wish to know the efficiency η_r with which the new field E'_ν will excite the old field E_ν. From Chapter 6 we know that this will be given by

$$\eta_r = |\langle E'_\nu, E_\nu\rangle|^2 / \{\langle E'_\nu, E'_\nu\rangle \langle E_\nu, E_\nu\rangle\} \tag{9.5}$$

The important feature in Eq. (9.5) is (as usual) the overlap integral between E'_ν and E_ν. The reflection efficiency is clearly 100 per cent if $E'_\nu = E_\nu$. However, it is reduced if there are significant differences between E'_ν and E_ν.

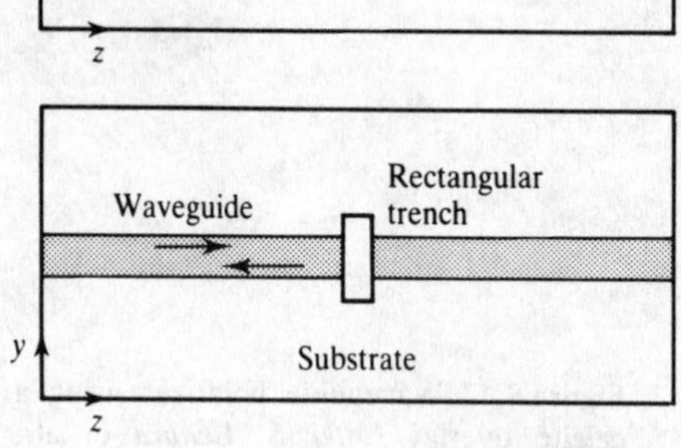

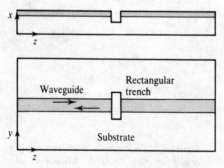

Figure 9.13 Integrated optic normal reflector.

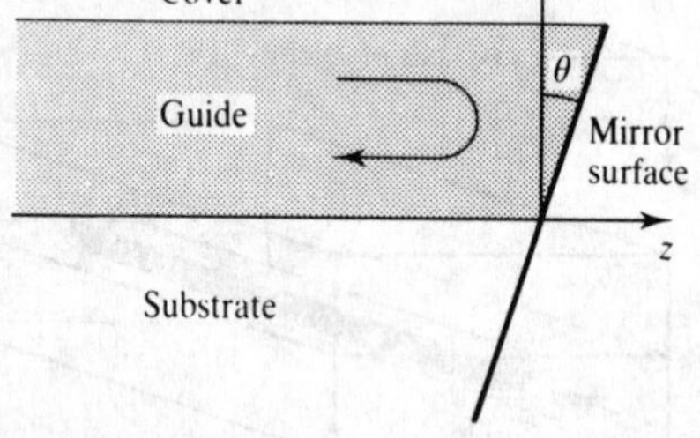

Figure 9.14 Integrated optic reflector with a skewed mirror.

Design Example

We may use this approach to estimate the tolerance on mirror tilt angle, assuming that the mirror is tilted about the y-axis through an angle θ from the vertical plane as shown in Fig. 9.14. We shall evaluate the reflection efficiency η_r using fields defined on the plane $z = 0$. We start by assuming that the phase-front of the incident mode is flat, so that $E_\nu(x,y)$ is real at $z = 0$. To calculate the reflected field E'_ν we simply assume that each part of the mode must travel a different optical distance (which depends on the tranverse coordinate x) in moving from $z = 0$ to the mirror and back again. The reflected field is therefore given approximately by

$$E'_\nu(x,y) \approx E_\nu(x,y) \exp\{-j2\beta x \tan\theta\} \tag{9.6}$$

where the exponential term accounts for the additional phase shift accumulated in travelling the extra path. The phase function ϕ in Eq. (9.4) therefore has the form

$$\phi(x,y) = 2\beta x \tan\theta \tag{9.7}$$

so the functional description of the tilted mirror is particularly simple: it imparts a linearly-varying phase shift across the guided beam. We may now estimate that η_r will fall to zero when there is a phase shift of π radians across the beam, as the overlap integral in the numerator of Eq. (9.5) will then average to zero. For a waveguide of depth h this implies that

$$2\beta h \tan\theta \approx 2\beta h\theta \approx \pi \tag{9.8}$$

or

$$\theta \approx \lambda_0/4n_{\mathrm{eff}}h \tag{9.9}$$

Using the typical parameters of $\lambda_0 = 1\ \mu$m, $h = 5\ \mu$m and $n_{\mathrm{eff}} = 2.2$, we get $\theta \approx 1.3°$. Since θ must be considerably lower than this for good reflectivity, the allowable tilt angle is extremely small.

Reflective Bends and Branches

Mirrors can also be used to turn a beam through 90°, for example by making an interface at 45° at an intersection between two orthogonal ridge guides, as shown in Fig. 9.15(a). The device is similar to the direct reflector, but there is no need for coating to increase reflectivity—the interface itself is normally totally-reflecting due to the large index difference between the ridge guide material and its surround (air). This particular technique is important because it allows the construction of bendless integrated optical circuits. Figures 9.15(b) and 9.15(c) show examples of reflective waveguide bends formed using GaAs and high-silica waveguides, respectively.

Figure 9.15(d) shows a related device formed from two 90° reflectors separated by a small air-gap. This is made using a narrow etched trench, so that the light can cross the air-gap by optical tunnelling. The device then acts as a frustrated-total-

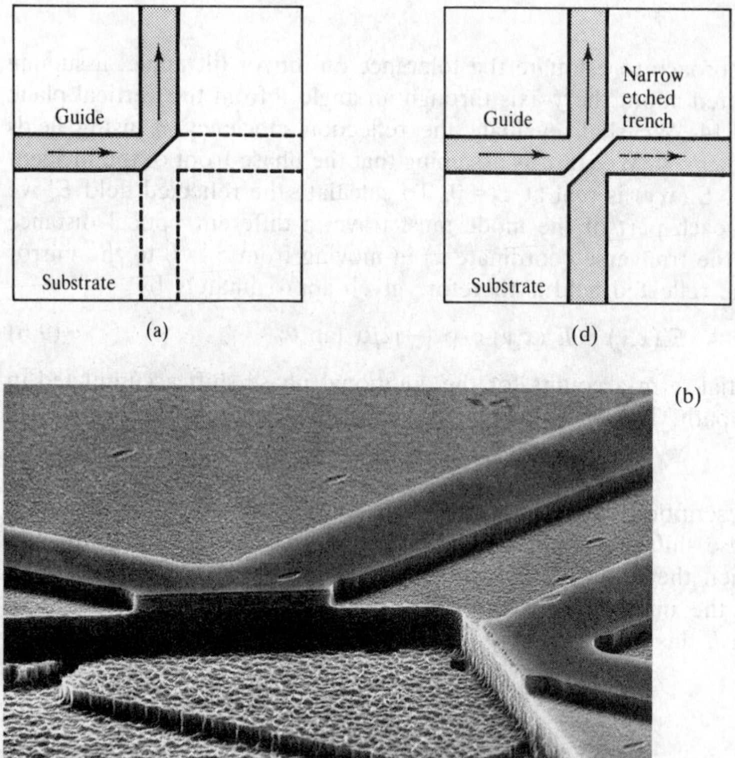

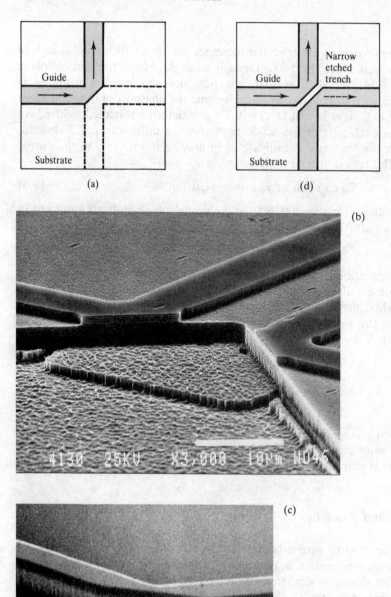

Figure 9.15 (a) Total-internal reflection mirror; (b) and (c) reflective bends in GaAs ridge guides and high silica ridge guides, respectively (*photographs courtesy P. Buchmann, IBM Research Division, Zurich, and A. Himeno, NTT*); (d) frustrated total-internal-reflection beamsplitter.

internal-reflection beamsplitter, similar to the bulk-optic FTIR device described in Chapter 7, and the splitting ratio is controlled by adjusting the gap width.

9.6 TAPERS AND Y-JUNCTIONS

We now consider devices which can change the cross-section of a guided beam by a gradual tapering of the guide parameters. The first such device is the waveguide *horn*, shown in Fig. 9.16. This was important in the early days of integrated optics because it was used as a conversion element between single-moded guides and much wider, multi-mode guides. Since the development of all-single-mode integrated optics, it has been rather less in demand.

A horn operates by a slow expansion of the guide cross-section. As we saw in Chapter 6 the theory of local normal modes implies that, if this process is slow enough, the input mode will gradually evolve into the lowest-order, symmetric mode of the multi-moded output guide, without mode conversion. Generally the allowed rate of expansion at any point along the taper depends on the number of guided modes supported at that point. At the left-hand end the number of modes is small, so the expansion rate can be relatively fast. At the right hand end the number is large, so the rate must be much slower. This leads to the characteristic parabolic variation in cross-section shown in Fig. 9.16.

The second such device, the *Y-junction*, is the simplest kind of branching element. It consists of a single-mode input guide which is gradually separated by a forked transition region into two similar single-mode output guides as shown in Fig. 9.17. The operation of the device is slightly more complicated than it appears and one useful viewpoint is provided by considering the characteristic modes of the entire structure at each end. This is particularly simple at the left-hand end. Here the structure consists of one single-moded guide which supports a characteristic mode with a transverse field distribution $E_L(x,y)$ given by (say)

$$E_L(x,y) = E(x,y) \tag{9.10}$$

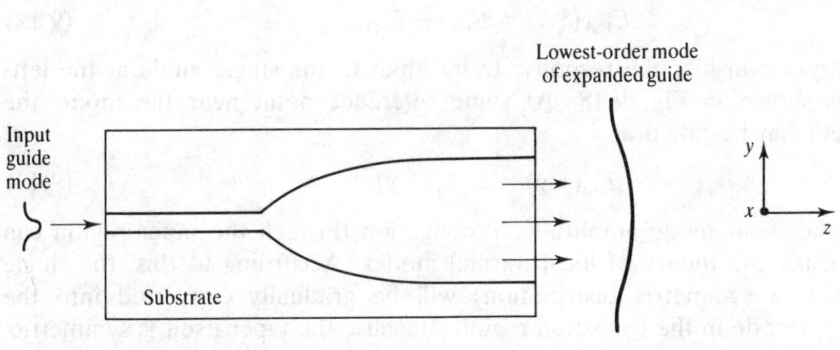

Figure 9.16 A channel waveguide horn.

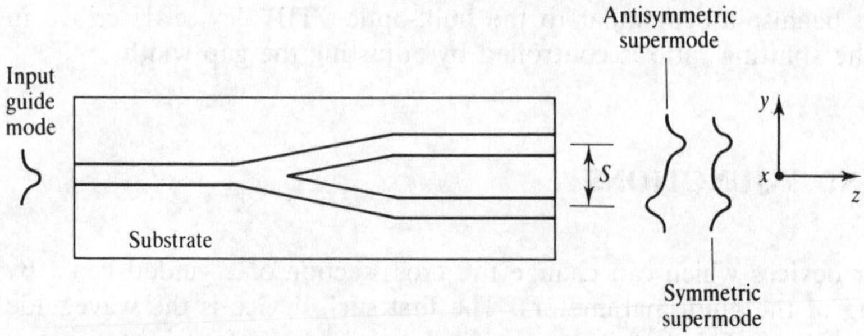

Figure 9.17 Characteristic modes of a symmetric Y-junction.

At the right-hand end the device consists of two similar single-moded guides separated by a distance S. If we define the transverse field patterns of the upper and lower guides to be E_{RU} and E_{RL}, respectively, we may write

$$E_{RU}(x,y) = E(x,y - S/2) \quad \text{and} \quad E_{RL}(x,y) = E(x,y + S/2) \tag{9.11}$$

The combined structure at the right-hand end can also be thought of as a composite guide which supports its own set of characteristic modes, known as *supermodes*. We will now guess the form this set might take. First, we shall assume that there are only two guided supermodes. This is reasonable since the structure has twice the guiding cross-section of a single guide. Second, we guess that each supermode can be written as a linear combination of the modes of each individual guide. This is also realistic; because the wave equation in the vicinity of each guide must look very like that for a single guide, the solutions must be similar. Finally, since the structure is inherently symmetric, we shall assume that the supermodes are symmetric and anti-symmetric field patterns. We shall label these patterns E_{RS} and E_{RA}, respectively.

At the right-hand end of the device, therefore, we assume that the combined structure can support two guided supermodes of the form:

$$E_{RS}(x,y) = E_{RU} + E_{RL}$$
$$E_{RA}(x,y) = E_{RU} - E_{RL} \tag{9.12}$$

We shall now consider the response to an input to the single guide at the left-hand end, as shown in Fig. 9.18. At some reference point near the input, the transverse field can be taken as

$$E_{in}(x,y) = a_{in}E_L(x,y) \tag{9.13}$$

where a_{in} is the input mode amplitude. Propagation through the taper region can be analysed using the theory of local normal modes. According to this, the single input (which is a symmetric distribution) will be gradually converted into the symmetric supermode in the transition region. Because the taper itself is symmetric, no anti-symmetric modes—especially the anti-symmetric supermode—can be excited.

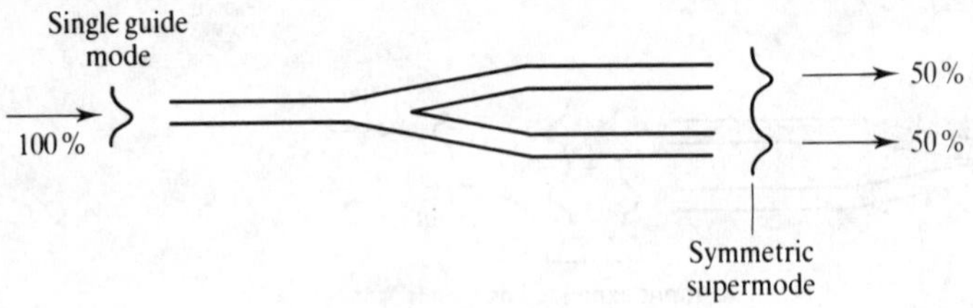

Figure 9.18 Excitation of a Y-junction at the left-hand end.

Similarly, provided the taper rate is low enough, no symmetric radiation modes will be generated. In a carefully-constructed Y-junction, conversion into the symmetric supermode will therefore be 100 per cent efficient. At some reference point near the output the transverse field can therefore be written as

$$E_{\text{out}}(x,y) = a_{\text{out}}\, E_{\text{RS}}(x,y) \tag{9.14}$$

Assuming 100 per cent efficiency we can calculate the field amplitudes. The input power is given by

$$P_{\text{in}} = (\beta/2\omega\mu_0)\, a_{\text{in}}^2 \langle E,E \rangle \tag{9.15}$$

While the output power is

$$P_{\text{out}} = (\beta/2\omega\mu_0)\, a_{\text{out}}^2 \langle E_{\text{RS}},E_{\text{RS}} \rangle \tag{9.16}$$

If the two output guides are sufficiently far apart Eq. (9.16) can be approximated as

$$\begin{aligned} P_{\text{out}} &= (\beta/2\omega\mu_0)\, a_{\text{out}}^2 \{\langle E_{\text{RU}},E_{\text{RU}} \rangle + \langle E_{\text{RL}},E_{\text{RL}} \rangle\} \\ &= 2\,(\beta/2\omega\mu_0)\, a_{\text{out}}^2 \langle E,E \rangle \end{aligned} \tag{9.17}$$

where we have assumed that $\langle E_{\text{RU}},E_{\text{RL}} \rangle \approx 0$. Equating Eqs (9.15) and (9.17) we see that

$$a_{\text{out}} = a_{\text{in}}/\sqrt{2} \tag{9.18}$$

We now change our viewpoint and consider the device output in terms of the modes travelling in each of the two right-hand guides individually. Equation (9.14) can be rewritten as

$$E_{\text{out}}(x,y) = (a_{\text{in}}/\sqrt{2})\, E_{\text{RU}}(x,y) + (a_{\text{in}}/\sqrt{2})\, E_{\text{RL}}(x,y) \tag{9.19}$$

Equation (9.19) implies that the Y-junction divides the input equally between the two output guides. Each has a mode amplitude of $1/\sqrt{2}$ times the input amplitude, and therefore carries half the power.

We now consider operation in the reverse direction, when light is incident from just one of the two channels on the right (as shown in Fig. 9.19).

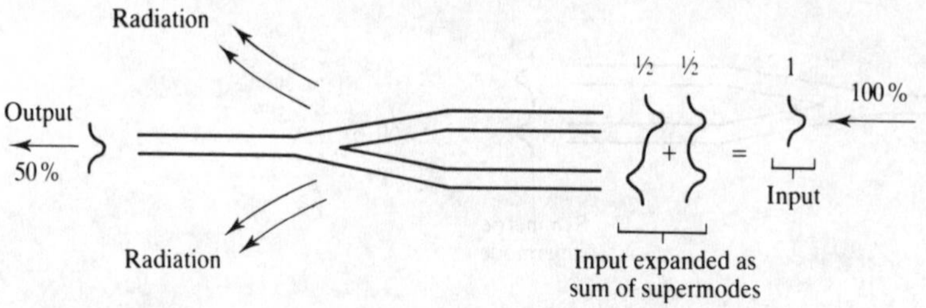

Figure 9.19 Excitation of a Y-junction at the right-hand end.

This time the incident beam can be written as

$$E'_{in}(x,y) = a'_{in} E_{RU}(x,y) \tag{9.20}$$

Combining Eqs (9.12) we see that this field can be expanded in terms of the supermodes, as

$$E'_{in}(x,y) = (a'_{in}/2)\,[E_{RS}(x,y) + E_{RA}(x,y)] \tag{9.21}$$

This is physically reasonable; the supermodes add together to give a finite field amplitude in the upper guide and cancel to give nothing in the lower one.

What happens to these two components at the junction? For the symmetric supermode the situation is effectively that of Fig. 9.18 in reverse, so it must be gradually converted into the single guide mode. This contribution can be written as

$$E'_{out}(x,y) = a'_{out} E(x,y) \tag{9.22}$$

By analogy with Eq. (9.18) we can say that

$$a'_{out} = \sqrt{2}\,(a'_{in}/2) \tag{9.23}$$

The amplitude of the emerging mode is $1/\sqrt{2}$ times that of the input mode and hence the power carried by this contribution is half the input power. We now consider the fate of the component carried by the anti-symmetric supermode. This cannot be converted into a symmetric distribution by a symmetrically-tapered Y-junction and hence cannot emerge as a guided mode at all. Instead, this component must be converted into antisymmetric radiation modes. Excitation of the Y-junction from the right-hand side with a single input therefore results in an automatic loss of 50 per cent of the power to radiation. This feature is not immediately apparent and should therefore be considered carefully.

Y-junctions can be combined together in binary *tree structures*. These allow a single input to be split into 2^N outputs, all carrying equal power, using N levels of two-way splitting. Figure 9.20 shows an example with three levels, which produces eight output beams from a single input. Y-junction trees are often used as input devices for parallel arrays of other components. They can also be used in reverse, to combine 2^N guided inputs in a single guided output. However, for the reasons just given, 100 per cent efficiency is only achieved if all of the inputs have equal amplitude and phase.

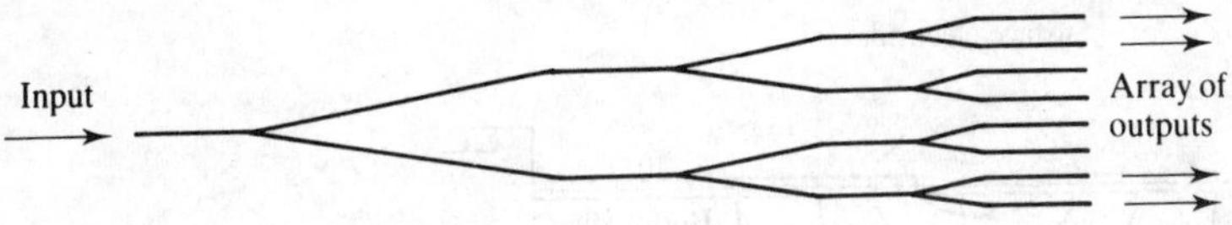

Figure 9.20 A Y-junction tree structure.

9.7 PHASE MODULATORS

We now move on to discuss a range of electrically-controllable devices which require the fabrication of a waveguide in an electro-optic substrate. The Ti:$LiNbO_3$ process is an extremely simple and versatile method of constructing such devices in lithium niobate. However, many other materials (especially the semiconductor systems based on GaAs/GaAlAs and InP/InGaAsP) are also electro-optic. The first such device is a *phase modulator*. As we saw in Chapter 3 the most desirable configuration is a transverse modulator, where the electric field is applied in a direction orthogonal to the axis of propagation. The drive voltage required is proportional to g/L where g is the electrode gap and L is the device length. In bulk optics it is difficult to make g/L small, which results in a high drive voltage. However, integrated optic devices require much lower voltages.

Integrated Optic Phase Modulators in Insulating Crystals

Figure 9.21 shows typical geometries used with an insulating crystal such as $LiNbO_3$, which illustrate many of the differences between bulk and guided-wave electro-optic devices. First, the electric field must now be applied through a pair of electrodes deposited on the same surface of the crystal, as shown in Fig. 9.21(a). Second, these must be arranged to lie somewhere above a waveguide, as in Fig. 9.21(b) and (c). (Note that in $LiNbO_3$ devices, a thin dielectric *buffer layer*—usually around 2000 Å of SiO_2—is inserted to space the electrodes away from the waveguides to avoid any undesirable polarization effects.) The resulting static electric field may then be used to alter the refractive index in the region of the guide, which in turn alters the phase of a guided optical wave. Due to the strong confinement of the mode, the electrode gap can now be very small (of the order of a waveguide width), so that a high field can be created in the gap region using a greatly reduced voltage (around 10 V).

Calculation of the field distribution $E_s(x,y)$ produced by the electrodes is a difficult electrostatic problem, which requires the solution of Laplace's equation for a relatively complicated set of boundary conditions. Although this can be done, it requires knowledge of an analytical technique known as conformal mapping, or knowledge of numerical methods for the solution of partial differential equations. Since both topics lie some way from our main theme, we shall simply assume the known solution directly. The field distribution can be shown to be highly non-uniform, as indicated by the field lines in Fig. 9.21(a). Their direction defines the

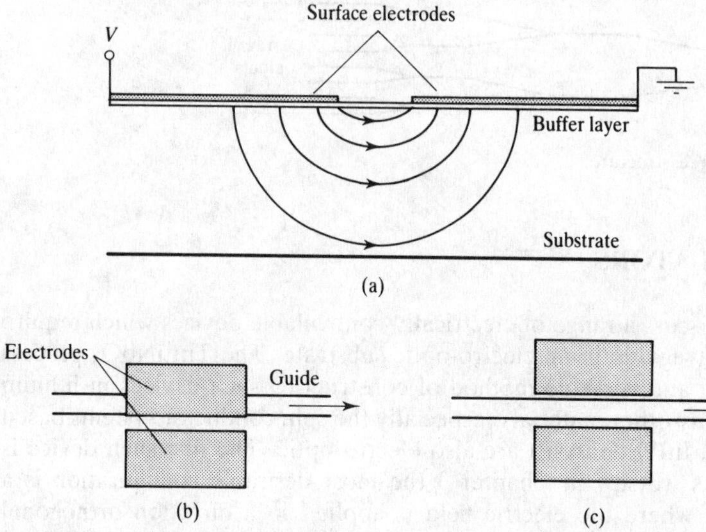

Figure 9.21 Surface electrode phase shifters: (a) electric field lines; (b) and (c) guide placement for exploitation of vertical and horizontal electric field component, respectively.

local field direction, and their spacing its strength. As can be seen, they run down from one electrode and up into the other; in between, they are roughly horizontal. The vertical component of the field therefore peaks just below the edge of each electrode, while the horizontal field peaks in the gap region. Since the optical field is significant only near the waveguide core, we can exploit either the vertical or horizontal parts of the electric field by placing the guide in different positions beneath the electrodes. Figure 9.21(b) shows how the guide should be placed if the vertical field component is to be used. For Z-cut $LiNbO_3$ with the optical wave propagating in the Y-direction, the TM mode is then affected by the strong r_{33} electro-optic coefficient, while the TE mode is controlled by r_{13}. Figure 9.21(c) shows the placement for the horizontal field, which requires a different crystal cut.

Using semiconductors, the construction of an electro-optic device is more complicated and special precautions are required to obtain a high field in the guide core. These will be discussed in Chapter 12.

Calculation of the Induced Phase Shift

The two effects described above—non-uniformity of electrical and optical fields—make it harder to work out the induced phase shift in integrated optic devices. However, this can be done approximately using a perturbation method, as we shall now show. We suppose that the guide is single-moded, and oriented in the z-direction. If it is formed by a two-dimensional index distribution $n(x,y)$, the modal field must satisfy the scalar waveguide equation

$$\nabla_{xy}^2 E(x,y) + [n^2(x,y)k_0^2 - \beta^2]E(x,y) = 0 \tag{9.24}$$

We now suppose that the effect of the static field $E_s(x,y)$ is to induce a two-dimensional refractive index perturbation $\Delta n(x,y)$ in the region from $z = 0$ to $z = L$, so that the new refractive index in this region is defined by $n' = n + \Delta n$. Next we assume that the effect on the guided mode will be to change its propagation constant to $\beta' = \beta + \Delta\beta$ without altering its transverse field distribution. This is reasonable if the mode is initially well-confined. The waveguide equation for the modified guide is therefore as Eq. (9.24), reading n' for n and β' for β. If Δn and $\Delta\beta$ are small we may put

$$n'^2 = (n + \Delta n)^2 \approx n^2 + 2n\Delta n \qquad \text{and} \qquad \beta'^2 \approx \beta^2 + 2\beta\Delta\beta \tag{9.25}$$

so that the waveguide equation for the perturbed mode is given by

$$\nabla_{xy}^2\, E(x,y) + \{[n(x,y)^2 + 2n(x,y)\Delta n(x,y)]\, k_0^2 - [\beta^2 + 2\beta\Delta\beta]\}E(x,y) = 0 \tag{9.26}$$

This looks rather formidable. However, we note that many of the terms can be cancelled out using the unperturbed waveguide equation, Eq. (9.24). Removing these terms leaves

$$\{2n(x,y)\, \Delta n(x,y)\, k_0^2 - 2\beta\Delta\beta\}E(x,y) = 0 \tag{9.27}$$

Unfortunately Eq. (9.27) is a two-dimensional equation since it contains a number of spatially-varying quantities; really, we wish only to extract the scalar term $\Delta\beta$. This can be done multiplying both sides of the equation by $E^*(x,y)$ and integrating over all space. If this is done we get

$$k_0^2\, \langle n\Delta nE,E\rangle - \beta\Delta\beta\, \langle E,E\rangle = 0 \tag{9.28}$$

so the change in propagation constant is given by

$$\Delta\beta = (k_0^2/\beta)\, \langle n\Delta nE,E\rangle \,/\, \langle E,E\rangle \tag{9.29}$$

In Eq. (9.29) the numerator is the most important term since the denominator just represents a normalization factor (which can be ignored if we have arranged that $\langle E,E\rangle = 1$). The change in propagation constant is then found as an overlap between the electrically-induced index change $\Delta n(x,y)$ and the modal field $E(x,y)$. This reflects the fact that the electric and optical fields are both localized; we must therefore arrange them to overlap spatially or the index perturbation will have no effect on the guided mode. Assuming that $\langle E,E\rangle$ is indeed unity and that the guide is weak, we may find the change in effective index as

$$\Delta n_{\text{eff}} = (k_0/\beta)\, \langle n\Delta n\, E,E\rangle \approx \langle \Delta n, |E|^2\rangle \tag{9.30}$$

From the results of Chapter 3 we know that the index change Δn will be linearly proportional to one particular component of the electric field (which we define as E_s'), following

$$\Delta n(x,y) = -n^3 r\, E_s'(x,y)/2 \tag{9.31}$$

Here n and r are the relevant refractive index and electro-optic coefficient, respectively. Thus

$$\Delta n_{\text{eff}} = -n^3 r/2 \langle E'_s, |E|^2 \rangle \tag{9.32}$$

The field E' will in turn be proportional to the drive voltage V, and inversely proportional to the electrode gap g. It is therefore often useful to write Eq. (9.32) in the alternative form:

$$\Delta n_{\text{eff}} \approx -n^3 rV\Gamma/2g \tag{9.33}$$

where we have introduced an *overlap factor* Γ defined as

$$\Gamma = g/V \langle E'_s, |E|^2 \rangle \tag{9.34}$$

The overlap factor is normally calculated by numerical integration and lies in the range $0 \leq \Gamma \leq 1$. Incomplete overlap between the electrical and optical fields therefore causes a reduction in the effective index change over the value expected for uniform fields. For well-designed devices Γ is 0.6–0.7. The phase shift $\Phi = \Delta\beta L$ corresponding to the change in effective index above is then

$$\Phi = 2\pi \Delta n_{\text{eff}} L/\lambda_0 = -\pi n^3 rVL\Gamma/g\lambda_0 \tag{9.35}$$

In a device with a given electrode gap, the phase shift is linearly proportional to both drive voltage and device length. It is then convenient to characterize transverse modulators in terms of the voltage–length product $(VL)_\pi$ required for a phase shift of $|\Phi| = \pi$ radians, given by

$$(VL)_\pi = g\lambda_0/n^3 r\Gamma \tag{9.36}$$

Note that this implies that larger drive voltages are required at longer optical wavelengths.

Design Example

We may estimate $(VL)_\pi$ for Ti:$LiNbO_3$ devices as follows. For Z-cut lithium niobate, for example, with propagation in the Y-direction and TM polarization, the refractive index is altered by the Z-component of the electric field via the r_{33} coefficient (note the use of upper-case letters to describe the crystal axes, which are slightly different from the normal x, y, z coordinates. In this case we may write $(VL)_\pi = g\lambda_0/n_e^3 r_{33}\Gamma$. Assuming that $\lambda_0 = 1.5\ \mu$m, that the electrode gap is 7 μm, and that the overlap factor Γ is 0.6, we obtain $(VL)_\pi = 7 \times 10^{-6} \times 1.5 \times 10^{-6} / (2.2^3 \times 30.8 \times 10^{-12} \times 0.6) = 0.053$ V m, or roughly 50 V mm. A 5 mm-long device would then require a 10 V drive.

Cut-off Modulators

If the field is poorly confined, our assumption that the transverse field distribution of the mode will remain unaltered when the index change is applied becomes invalid.

For example, the guide index may be depressed sufficiently far that the mode becomes cut off and radiates into the substrate. This type of device is known as a *cut-off modulator*, and can be used to provide intensity modulation directly, rather than a simple phase change. Figure 9.22(a) shows a schematic of the device, which consists of a pair of phase-modulator electrodes operating in a region of weak optical confinement.

With no applied voltage the guided beam is transmitted through the active section with little attenuation (Fig. 9.22(b)). If a voltage of the correct sign is applied this section approaches cut-off, and the guided beam suffers heavy attenuation through conversion to radiation modes (Fig. 9.22(c)).

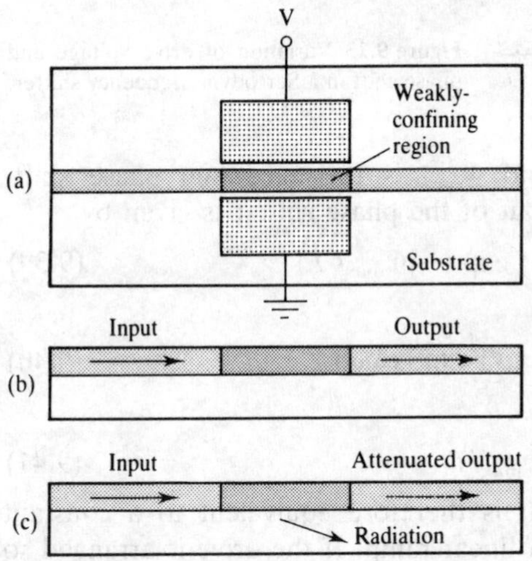

Figure 9.22 A cut-off modulator: (a) schematic, (b) operation with no applied voltage, and (c) operation with a voltage sufficient to induce cut-off.

9.8 FREQUENCY SHIFTING AND HIGH SPEED OPERATION

A phase modulator can be used to obtain *frequency modulation* instead of phase shifting, by using a time-varying electrical drive. The simplest method, known as the *Serrodyne*, uses a sawtooth drive voltage as shown in Fig. 9.23. In the regions of linear ramp, the voltage varies linearly with time so that $V(t)$ is given by

$$V(t) = V_{\max} t/T \tag{9.37}$$

where T is the period of the sawtooth and $V_{\max}$ is the maximum voltage. This results in a sawtooth variation in phase shift which (again in the regions of linear ramp) varies as

$$\Phi(t) = \alpha V(t) = \Phi_{\max} t/T \tag{9.38}$$

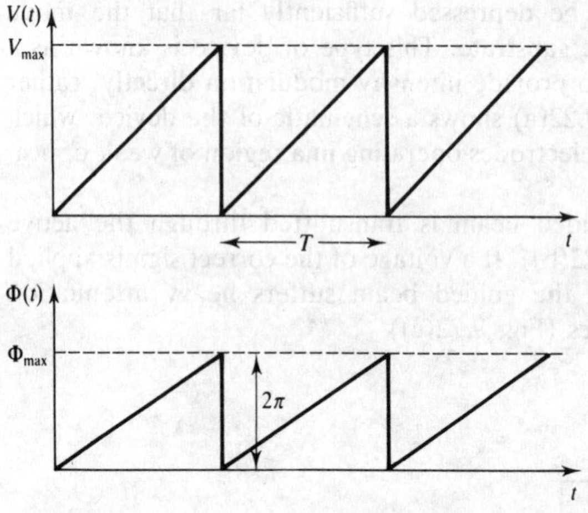

Figure 9.23 Variation of drive voltage and phase shift in a Serrodyne frequency shifter.

where α is a proportionality constant and Φ_{max} is the maximum phase shift. Consequently, the electric field at the output of the phase shifter is given by

$$E_{out}(x,y,t) = E_{out}(x,y) \exp(-j\Phi_{max}t/T) \tag{9.39}$$

This can be written as

$$E_{out}(x,y,t) = E_{out}(x,y) \exp(j\Delta\omega t) \tag{9.40}$$

where

$$\Delta\omega = -\Phi_{max}/T \tag{9.41}$$

The effect of the time-varying phase shift is therefore equivalent to a constant downward frequency shift in the regions of linear ramp. If the drive is arranged so that the voltage steps between the linear regions correspond to discontinuities of $2\nu\pi$ in Φ, these jumps have no noticeable effect and the output is always a frequency-shifted wave. The Serrodyne method is very simple and effective, but the frequency shift is limited by the difficulty of generating a sawtooth wave at high frequencies.

Design Example

We may use the analysis above to calculate the frequency shift produced by a Serrodyne phase modulator of 5 mm length, driven by a 20 V peak-to-peak sawtooth drive voltage with a period of 1 μs, given that the static voltage–length product required for π radians phase shift is $(VL)_\pi = 50$ V mm. In this case $\Phi_{max} = \pi \times 20 \times 5/50 = 2\pi$ and $T = 1\ \mu$s, so $\Delta f = 1$ MHz.

Modulation with Sinusoidal Voltages

Another common form of drive voltage (which is considerably easier to generate at high frequencies) is a sinusoid. If the modulation frequency is ω_m the phase shift has the form:

$$\Phi(t) = \Phi_{max} \sin(\omega_m t) \tag{9.42}$$

The wave emerging from the phase shifter is then given by

$$E_{out}(x,y,t) = E_{out}(x,y) \exp\{-j\Phi_{max} \sin(\omega_m t)\} \tag{9.43}$$

We can then use a well-known mathematical identity to expand the exponential in Eq. (9.43) as a sum of frequency-shifted waves, giving

$$E_{out}(x,y,t) = E_{out}(x,y) \sum_n (-1)^n J_n(\Phi_{max}) \exp(jn\omega_m t) \tag{9.44}$$

where n is an integer and J_n is the n th order Bessel function. As can be seen, the use of a sinusoidal drive does not yield a single frequency-shifted component. Instead, a whole series of upper and lower sidebands are generated. The individual components have angular frequencies of $\omega + n\omega_m$, so they lie at regular spacings of ω_m in frequency space and have Bessel function amplitude coefficients. This frequency spectrum is very similar to that of an FM wave in communications theory. Consequently, the effect of raising Φ_{max} by increasing the drive voltage is not an increase in amplitude of the desired frequency-shifted component—rather, it is that more and more sidebands are generated. This is often undesirable since it increases the overall signal bandwidth.

Figure 9.24 shows the variation of the three lowest-order Bessel functions $J_0(\Phi)$, $J_1(\Phi)$ and $J_2(\Phi)$ with Φ. These are slowly-varying functions whose values are found tabulated in many standard mathematical texts. The fraction of the optical power P_n contained in the n th sideband is given by

$$P_n = J_n^2(\Phi_{max}) \tag{9.45}$$

Thus, as an example, for a maximum phase shift of π radians, the power contained in the first (upper) sideband is $P_1 = J_1^2(\pi) = 0.0813$, slightly more than 8 per cent of the input power.

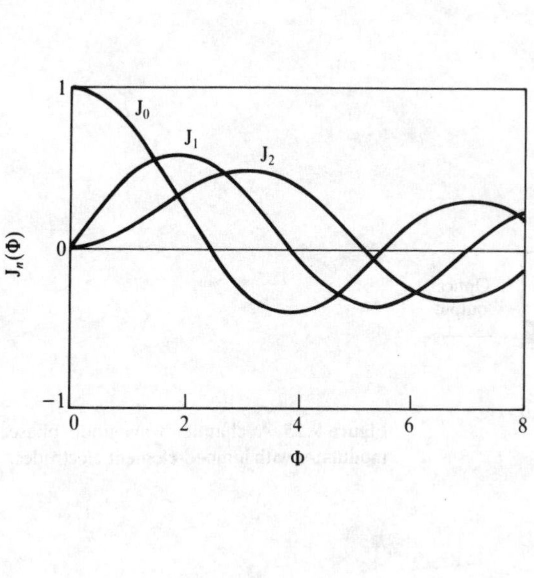

Figure 9.24 The variation of the three lowest-order Bessel functions.

High-speed Modulation

The speed at which a phase modulator can be driven is of course limited, and the high-frequency cut-off depends strongly on the electrode design. Figure 9.25 shows a typical layout for a Ti:$LiNbO_3$ device, where the electrodes can be considered as *lumped elements*. These act as capacitors, which must be charged and discharged by the voltage source. Typically, this has a fixed 50 Ω output impedance, so the high-frequency cut-off is determined by an RC time constant, where R = 50 Ω and C is the electrode capacitance. Capacitance C can be reduced either by decreasing the electrode length or by increasing the electrode gap. Unfortunately, both these strategies also increase the drive voltage; the first because the voltage–length product $(VL)_\pi$ of a given design is fixed, and the second because it results in a weaker electric field. One possible solution is to match the electrodes to the generator using a matching network, but broad-band matching is virtually impossible. Broad-band lumped-element devices therefore have a high-frequency cut-off at around 1 GHz.

One way of improving high-frequency performance is to replace the lumped-element electrodes with electrodes that act as a waveguide at microwave frequencies, known as *travelling-wave electrodes*. Typically, the required structure is an asymmetric coplanar stripline, constructed from a pair of electrodes with greatly increased metal thickness (around 15 μm), as shown in Fig. 9.26. This can be arranged to present a constant, real impedance at microwave frequencies, which can be matched to the output impedance of the generator quite simply. The electrical signal is then launched as a travelling microwave at one end of the electrodes, travels the length of the device, and is absorbed in a matched termination at the far end. This time the high-frequency cut-off is determined by a mismatch in the velocity of the optical and electrical waves due to the difference in dielectric constant at the two frequencies concerned (see, for example, Question 9.8). This is highly significant for $LiNbO_3$ devices. The result is reduced modulation at high frequencies, but operation at 10–20 GHz is quite possible. Figure 9.27 shows a fully packaged Ti:$LiNbO_3$ microwave phase modulator fitted with fibre pigtails.

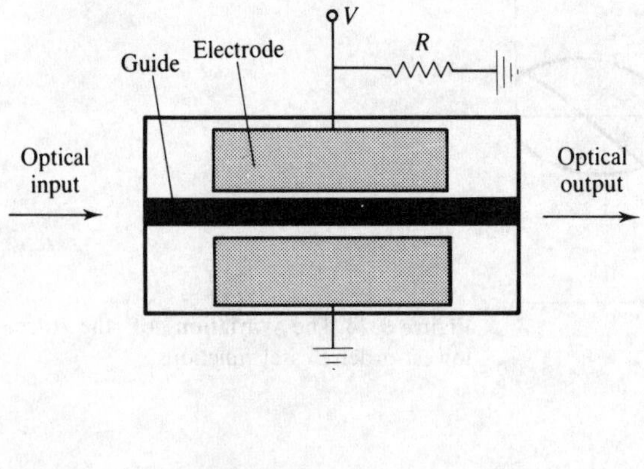

Figure 9.25 A channel waveguide phase modulator with lumped-element electrodes.

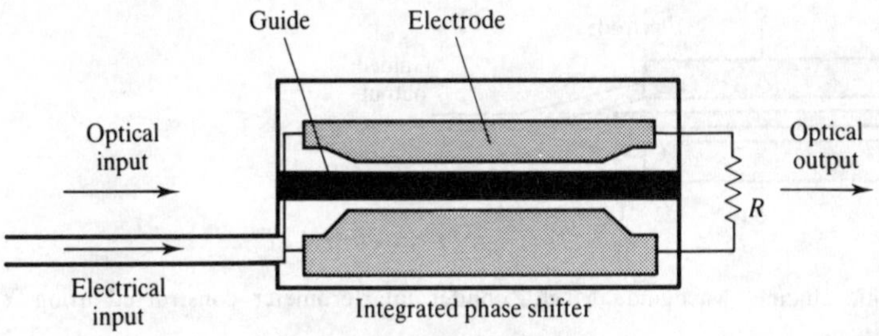

Figure 9.26 A channel waveguide phase shifter with travelling-wave electrodes.

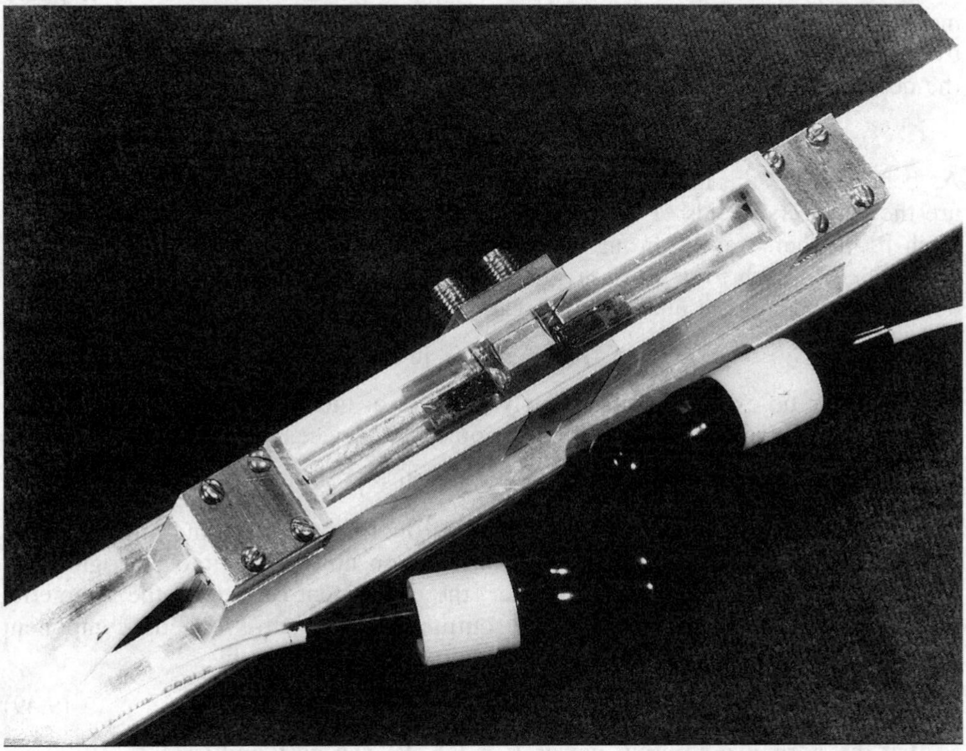

Figure 9.27 A packaged Ti:$LiNbO_3$ phase modulator (*photograph courtesy R.C. Alferness, AT&T Bell Laboratories*).

9.9 INTERFEROMETERS

Phase modulation can be converted into amplitude modulation by inserting a phase shifter in one arm of an interferometer. The simplest is the *Mach-Zehnder*

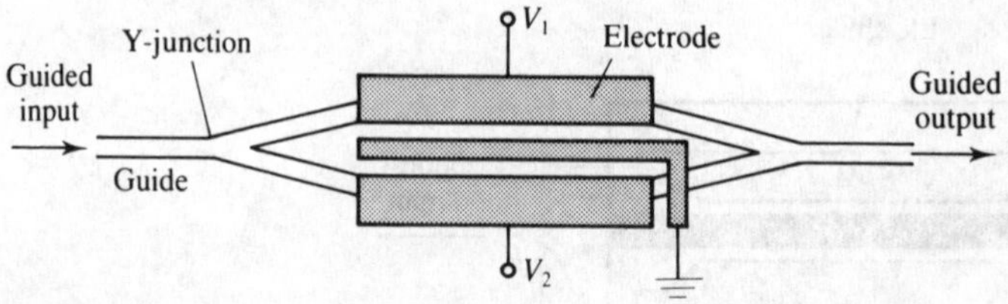

Figure 9.28 An electro-optic channel waveguide Mach-Zehnder interferometer constructed using Y-junctions.

interferometer, shown in Fig. 9.28. The structure consists of two back-to-back Y-junctions separated by a region containing a pair of identical electro-optic phase modulators. It operates as follows. The input wave is first split into two components by the left-hand Y-junction. Using our previous results, the transverse field across the device just before the phase shifter region can be written as

$$E(x,y) = (a_{in}/\sqrt{2})\,[E_{RU}(x,y) + E_{RL}(x,y)] \tag{9.46}$$

As before, a_{in} is the modal amplitude of the single-guide input and E_{RU} and E_{RL} are the transverse fields of the modes in the upper and lower guides, respectively. Each beam then travels through a phase-shifter section, and opposite shifts of Φ are applied to each to get a push–pull effect. The transverse field across the whole device just beyond the phase shifters can then be written as

$$E(x,y) = (a_{in}/\sqrt{2})\,[E_{RU}(x,y)\exp(-j\Phi) + E_{RL}(x,y)\exp(+j\Phi)] \tag{9.47}$$

This field can be expressed alternatively in terms of the characteristic modes of the two-guide system forming the central region of the device. If this is done we obtain

$$E(x,y) = (a_{in}/\sqrt{2})\,[E_{RS}(x,y)\cos\Phi - jE_{RA}(x,y)\sin\Phi] \tag{9.48}$$

where E_{RS} and E_{RA} are the transverse fields of the symmetric and antisymmetric supermodes in the central region. This distribution now passes into the right-hand Y-junction. For the reasons given earlier, the component carried by the symmetric supermode emerges from the single-guide output, while the anti-symmetric component is radiated. The amplitude of the guided output is thus given by

$$a_{out} = a_{in}\cos\Phi \tag{9.49}$$

so that the normalized output power $P = P_{out}/P_{in}$ can be found as

$$P = \cos^2\Phi = [1 + \cos(2\Phi)]/2 \tag{9.50}$$

The normalized output therefore varies cosinusoidally with Φ, as shown in Fig. 9.29. The output power is unity when no phase shift is applied, falling to zero when $\Phi = \pi/2$, $3\pi/2$, etc. In each of these cases the electrodes induce a relative phase shift of π radians between the components travelling in the two arms. On/off intensity modulation may therefore be achieved very simply by arranging the induced phase shift to jump between zero and $\pi/2$ radians.

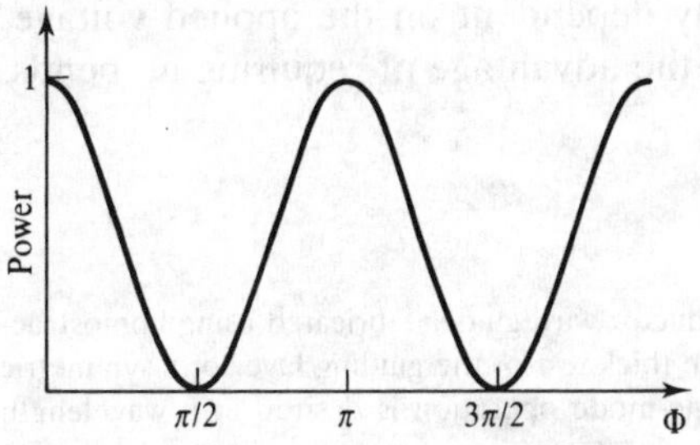

Figure 9.29 Variation of guided output power with phase imbalance for a Mach-Zehnder interferometer.

An interferometer of a rather different kind can be made by allowing two waveguides to cross at a shallow angle, as shown in Fig. 9.30. This is known as an *X-switch*. In general, such a device has two possible modes of operation (one interferometric, the other operating via switchable total-internal-reflection). Exactly which occurs depends on the design parameters, but here we will assume the interferometric mode is operative. If the device is fabricated in an electro-optic material, it can perform intensity modulation in a similar manner to the Mach-Zehnder interferometer. However, because it has two guided output ports, it can also function as a two-way switch.

In the central region of the switch the guide is arranged to support two modes which are as usual symmetric and anti-symmetric in character. This can be done by making the index change forming the guide in this region double that elsewhere. An input to one of the switch arms will excite both the characteristic modes equally. These travel at different speeds to the two output arms where they are recombined. Depending on their relative phase, a high output is obtained from either of the arms. Electrodes with a narrow central gap are now added to the device. When a voltage is applied, the static electric field is high down the device centreline, and weak elsewhere. The consequence is an induced index change in the gap region. The overlap between the symmetric mode and this perturbation is strong, so its phase is altered according to the phase modulation principle described earlier. However, the overlap with the anti-symmetric mode is much smaller, since this must have zero amplitude at its centre. A relative phase difference is then induced between the two modes so that interferometric addition or cancellation occurs, and

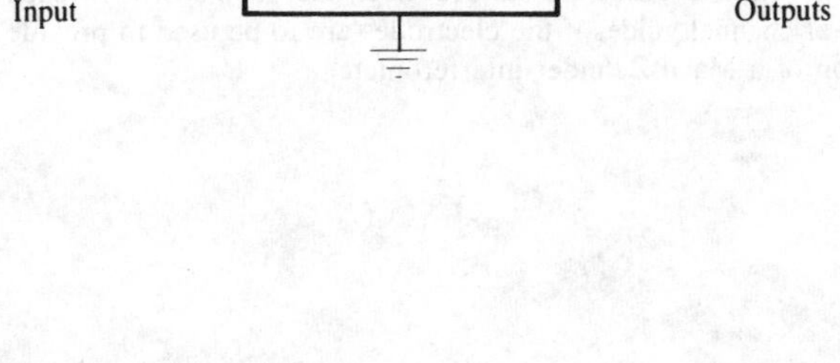

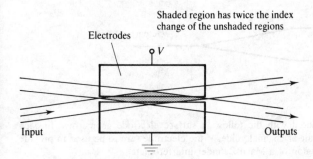

Figure 9.30 The X-switch.

the power output from each arm is sinusoidally dependent on the applied voltage. Although the device is difficult to make, it has the advantage of requiring no bends.

PROBLEMS

9.1 Describe briefly the difference in structure of semiconductor waveguides fabricated using homostructures and heterostructures. Calculate the maximum possible thickness of the guiding layer of a symmetric $GaAs/Ga_{0.85}Al_{0.15}As$ double heterostructure guide, if single-mode operation is desired at a wavelength of 0.9 μm.
[0.55 μm]

9.2 For a particular channel guide system, experimental measurements of the radiation loss incurred in 90° waveguide bends showed a 10 dB loss for a 1 mm bend radius, and 3 dB loss for 1 cm radius. Estimate the loss incurred in a 45° bend of 5 mm radius.
[5.27 dB]

9.3 A 90° bend is formed in a single-mode semiconductor ridge waveguide system using a total-internal-reflection mirror. Unfortunately, due to an error, the reflecting interface is offset by δ from its optimum position, as shown below. Explain how you would estimate the transmission through the bend. Can the transmission reach 100 per cent if $\delta = 0$?

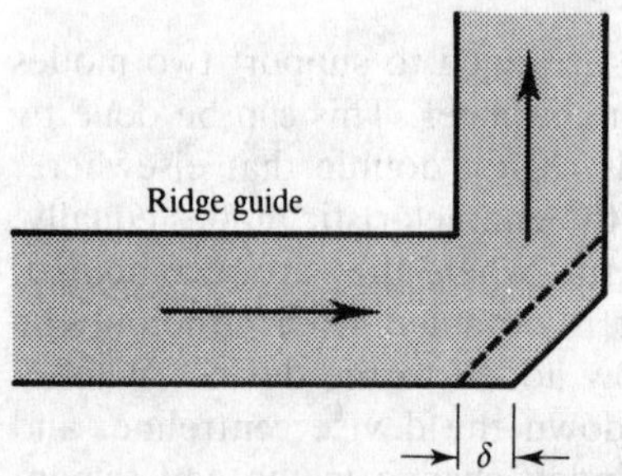

9.4 The figure below shows a Y-junction tree used as a beam combiner for eight single-mode inputs. If the input mode amplitudes in guides 1,. . .,8 are 1, −2, 3, 0, −1, 2, 4 and −1, respectively, calculate the efficiency of the beam-combining process. How does the efficiency improve if corrections are made to the input phases so that the contributions sum in-phase?

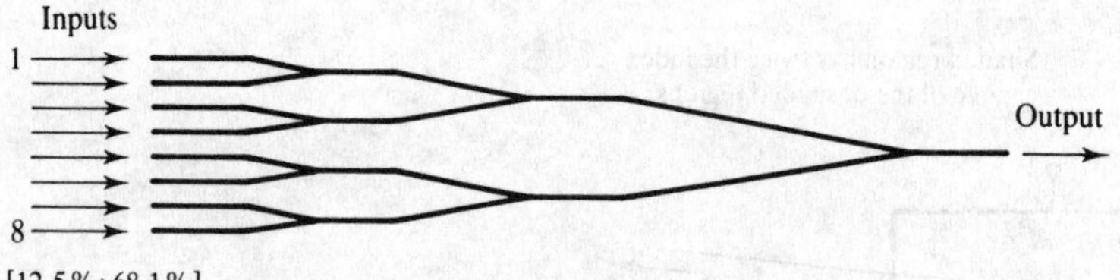

[12.5%; 68.1%]

9.5 Sketch the electric field lines associated with the following surface electrode geometries. In each case indicate where you would place a pair of channel guides if the electrodes are to be used to provide push-pull phase shifting in the central region of a Mach-Zehnder interferometer.

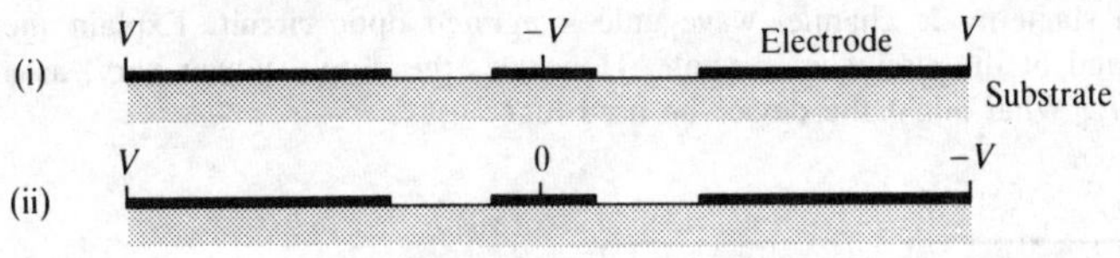

9.6 A cut-off modulator is based on a single-moded $Ti:LiNbO_3$ channel waveguide with an effective index of $n_{eff} = 2.201$. Assuming that the substrate index is $n_s = 2.2$, that the device will operate via the r_{33} electro-optic coefficient (30.8×10^{-12} m/V) using surface electrodes with a 7 μm gap width, and that overlap between the electrical and optical fields is $\Gamma = 0.6$, estimate the voltage required to induce cut-off. Why might your estimate prove inaccurate?
[71.1 V]

9.7 A 10 mm-long electro-optic phase modulator is driven with a 10 V peak-to-peak sinusoidal voltage at a frequency of 10 MHz. Assuming that the voltage–length product required for π radians phase shift is $(VL)_\pi = 50$ V mm, find the percentage of the output power contained in a bandwidth ranging from 20 MHz beow the optical carrier frequency to 20 MHz above it.
[72.5%]

9.8 The figure below shows a travelling-wave electro-optic phase modulator. Assuming that the angular frequencies of the electrical drive and the optical wave are ω_m and ω_o, respectively, and that the effective indices of the stripline waveguide and the optical guide are n_m and n_o, calculate and sketch the dependence of the induced optical phase shift Φ on ω_m. At what frequency does Φ first fall to zero?

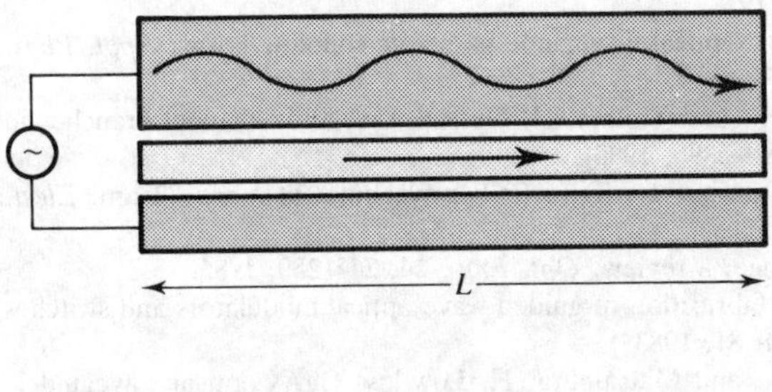

9.9 A Mach-Zehnder interferometer is constructed using two identical phase shifters, which provide push–pull modulation. Each is 1 cm in length and has a voltage–length product for π radians phase shift of $(VL)_\pi = 50$ V mm. Sketch accurately the variation of guided output power with time for a constant optical input and the d.c.-biased triangle-wave drive shown below.

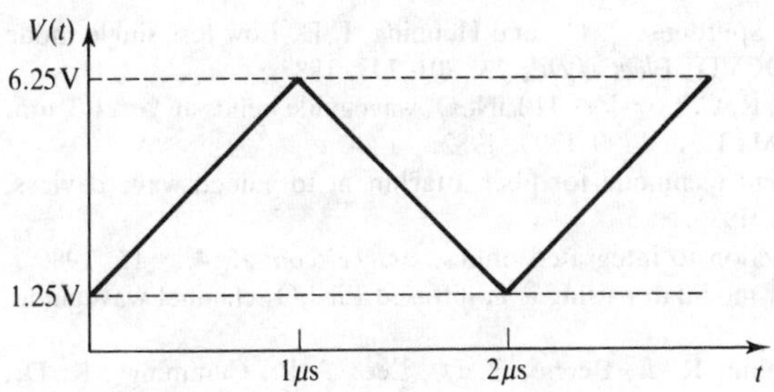

9.10 The figure below shows a single-mode channel waveguide integrated optic circuit. Explain the operation of each component, and of the circuit as a whole. How does the output power vary, as a function of the path imbalance *L*? What might the device be used for?

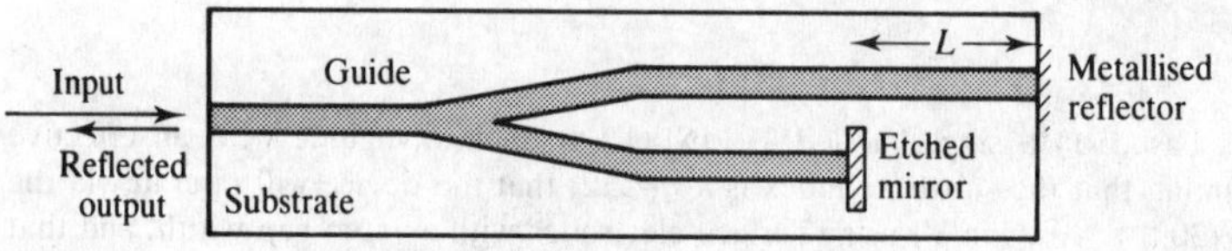

SUGGESTIONS FOR FURTHER READING

Alferness, R. C. Waveguide electro-optic modulators, *IEEE Trans. Micr. Theory & Tech.*, **MTT-30**, 1121–1137, 1982.

Betts, G. E., and Chang, W. S. C. Crossing-channel waveguide electrooptic modulators, *IEEE J. Quant. Elect.*, **QE-22**, 1027–1038, 1984.

Boyd, J. T., Wu, R. W., Zelmon, D. E., and Naulmaan A., Timlin, H. A., Jackson, H. E. Guided wave optical structures utilizing silicon, *Opt. Engng.*, **24**, 230–234, 1985.

Buchmann, P., Kaufmann, H., Melchior, H., and Guebos, H. Totally reflecting mirrors: fabrication and application in GaAs rib devices, *Proc. 3rd European Conf. on Integrated Optics*, Berlin, Germany, pp. 135–139, 6–8 May, 1985.

Burns, W. K., and Hocker, G. B. End-fire coupling between optical fibres and diffused channel waveguides, *Appl. Opt.* **16**, 2048–2050, 1977.

Burns, W. K., Milton, A. F., and Lee, A. B. Optical waveguide parabolic coupling horns, *Appl. Phys. Lett.*, **30**, 28–30, 1977.

Burns, W. K., Moeller, R. P., Bulmer, C. H., and Yajima, H. Optical waveguide channel branches in Ti-diffused $LiNbO_3$, *Appl. Opt.*, **19**, 2890–2896, 1980.

Ctyroky, J., and Henning, H.-J. Thin-film polarizer for Ti:$LiNbO_3$ waveguides at $\lambda = 1.32\ \mu m$, *Elect. Lett.*, **22**, 756–758, 1986.

Findakly, T. Glass waveguides by ion exchange: a review, *Opt. Eng.*, **24**, 244–250, 1985.

Houghton, A. J. N., and Rogers, P. M. The fabrication of guided wave optical modulators and switches in semiconductors, *Br. Telecom J.*, **1**, 78–81, 1983.

Inoue, H., Hiruma, K., Ishida, K., Asai, T., and Matsamura, H. Low-loss GaAs optical waveguides, *IEEE J. Lightwave Tech.*, **LT-3**, 1270–1276, 1985.

Jackel, J. L., Rice, C. E., and Veselka, J. J. Proton exchange for high-index waveguides in $LiNbO_3$, *Appl. Phys. Lett.*, **41**, 607–608, 1982.

Kaminow, I. P., Stulz, L. W., and Turner, E. H. Efficient strip-waveguide modulator, *Appl. Phys. Lett.*, **27**, 555–557, 1975.

Martin, W. E. A new waveguide switch/modulator for integrated optics, *Appl. Phys. Lett.*, **26**, 562–564, 1975.

McIlroy, P. W., Rogers, P. M., Singh, J. S., Spurdens, P. C., and Henning, I. D. Low loss single mode InP/GaInAsP waveguides grown by MOCVD, *Elect. Lett.*, **23**, 701–703, 1987.

Minford, W. J, Korotky, S. K., and Alferness, R. C. Low-loss Ti:$LiNbO_3$ waveguide bends at $\lambda = 1.3\ \mu m$, *IEEE Trans. Micr. Theory and Tech.*, **MTT-30**, 1790–1794, 1982.

Murphy, E. J., and Rice, T. C. Self-alignment technique for fiber attachment to guided wave devices, *IEEE J. Quant. Elect.*, **QE-22**, 928–932, 1986.

Nayar, B. K., and Booth, R. C. An introduction to integrated optics, *Br. Telecom J.*, **4**, 5–15, 1986.

Neyer, A., and Sohler, W. High-speed cutoff modulator using a Ti-diffused $LiNbO_3$ channel waveguide, *Appl. Phys. Lett.*, **35**, 256–258, 1979.

Osinski, J. S., Zah, C. E., Bhat, R., Contolini, R. J., Beebe, E. D., Lee, T. P., Cummings, K. D., and Harriot, L. R. Miniature integrated optical beamsplitter in AlGaAs/GaAs ridge waveguides, *Elect. Lett.*, **23**, 1156–1158, 1987.

Stulz, L. W. Titanium indiffused $LiNbO_3$ optical waveguide fabrication, *Appl. Opt.*, **18**, 2041–2044, 1979.

Takato, N., Yasu, M., and Kawachi, M. Low-loss high-silica single-mode channel waveguides, *Elect. Lett.*, **22**, 321–322, 1986.

Uehara, S., Izawa, T., and Nakagome, H. Optical waveguiding polarizer, *Appl. Opt.*, **13**, 1753–1754, 1974.

Walker, R. G., Bennion, I., and Carter, A. C. Low voltage, 50 Ω, GaAs/AlGaAs travelling-wave modulator with bandwidth exceeding 25 GHz, *Elect. Lett.*, **25**, 1549–1550, 1989.

Wilson, R. G., Betts, D. A., Sadana, D. K., Zavada, J. M., and Hunsperger, R. G. Proton deuteron and helium implantation into GaAs and $LiNbO_3$ for waveguide fabrication, *J. Appl. Phys.*, **57**, 5006–5010, 1985.

Wong, K. K., and Wright, S. An optical serrodyne frequency translator, *Proc. 1st European Conf. on Integrated Optics,* London, pp. 63–65, 14–15 Sept. 1981.

CHAPTER

TEN

COUPLED MODE DEVICES

10.1 INTRODUCTION

Until now we have been concerned mainly with the properties of perfect, uniform guides. We have shown them all to support a set of characteristic modes, which are (by definition) field structures that propagate down the guide with only a change in phase. We have encountered some exceptions to this rule—most notably, waveguide tapers, bends, gratings and beam couplers, where the mode amplitudes also change due to power redistribution between guided modes, or conversion from guided to radiation modes—but we have managed to avoid any quantitative calculations of the amplitude changes concerned. It is now time to rectify this omission.

It turns out that a common physical principle underpins the behaviour of nearly all perturbed waveguide structures: namely, that the introduction of a perturbation to an otherwise perfect guide can cause an interchange of energy amongst the modes of that guide. This interchange can be highly significant if a number of conditions are satisfied. In fact, one individual mode can be converted into another with close to 100 per cent efficiency in many cases. This principle, known as *mode coupling*, is ubiquitous, and occurs in many branches of physics other than guided wave optics (most notably, in quantum mechanics). As far as we are concerned, it is important to note that besides explaining the performance of waveguides suffering from some imperfection or distortion, the principle can also be exploited for use in new guided-wave devices. We shall illustrate this point by considering some of the wide range of devices that operate through coupling between co- and contra-directional modes.

10.2 THE DIRECTIONAL COUPLER—BASIC PRINCIPLES

We will start with one of the most successful and versatile devices in both integrated and fibre optics, the *directional coupler*, which works by coupling together two modes travelling in the same direction. In its simplest form this acts as a beamsplitter, but more complicated devices can be used as two-way switches or modulators; further variants can be used as filters or polarizers. In broad outline, the device works as follows.

From previous experience, we know that there is an evanescent field extending outside any dielectric waveguide. If two parallel guides are placed sufficiently close together, these parts of the field must overlap spatially. Usually the interwaveguide gap required for this overlap to be significant is of the order of the guide width. For example, Fig. 10.1 shows the situation for the lowest-order modes in two identical, adjacent symmetric slab waveguides. The refractive index distribution of the twin-guide system is also shown. The indices of the two guide cores are n_{g1} and n_{g2}, respectively, and the index in the substrate material outside the guide has the uniform value n_s.

Because of the intermodal overlap, it is hard to decide which guide any light in the overlap region actually belongs to. This implies that there must be a mechanism for light in one guide to be transferred to the other. It turns out (as we shall see later) that, if the two guides satisfy a particular set of conditions and run parallel for sufficient distance, this interchange of power can be highly significant, reaching almost 100 per cent in many cases. The light then starts coupling back, so the power transfer process must be periodic with distance. We can represent this as shown in Fig. 10.2. Here the input is to guide 1, at $z = 0$. We shall assume that complete coupling occurs at $z = L$ when all the light has been transferred to guide 2. At $z = 2L$, therefore, the light will have been completely coupled back to guide 1, and

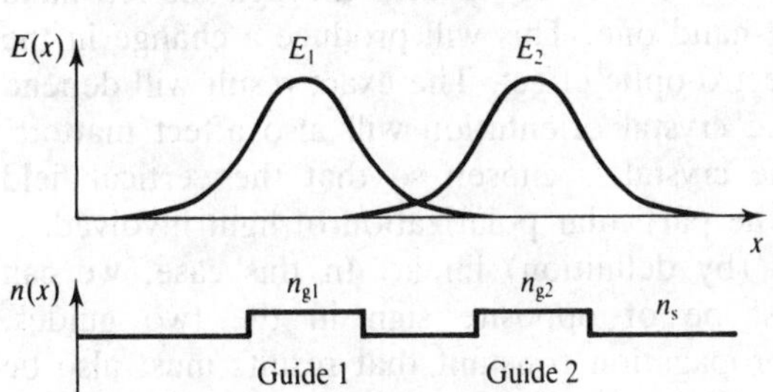

Figure 10.1 The overlap of the transverse fields of two adjacent, parallel symmetric slab guides.

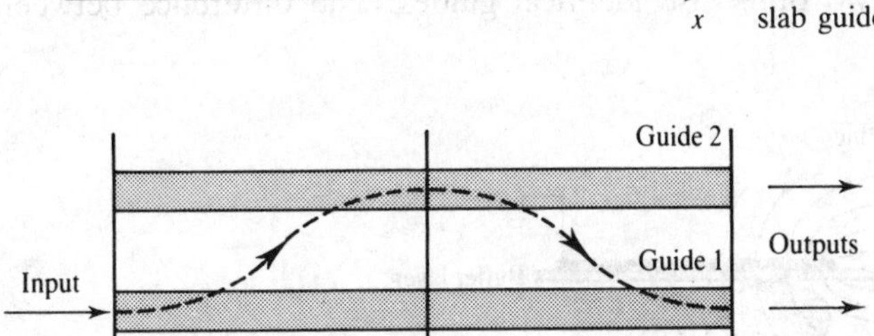

Figure 10.2 Conceptual representation of the coupling process.

so on. In fact, the process is a continuous, bidirectional one, of a type known as *multiple scattering*. The condition the guides must satisfy for 100 per cent power transfer is therefore that they are *synchronous* (i.e. have the same propagation constants) so that all the scattered contributions add up in-phase. This usually requires them to be identical.

We can use this effect to make a device that can switch light between two waveguides, as follows. We start with two identical single-moded guides, and choose the length of the coupling region so that the device runs from $z = 0$ to $z = L$. Under normal circumstances an input to guide 1 will then be completely coupled across and emerge as an output from guide 2. If we can now arrange to desynchronize the guides under electrical control, we can prevent the power transfer process occurring. In this case the light emerges from guide 1, without any cross-coupling. Since the choice of output guide is now electrically-alterable, the device functions as a two-way switch.

Integrated Optic Directional Couplers in Insulating Crystals

Figure 10.3 shows the cross-section of a structure suited to realization in integrated optics. This is made from two channel waveguides buried in an insulating, electro-optic substrate, with a pair of surface electrodes. The Ti:$LiNbO_3$ process is now well developed for fabrication of such devices. We have already discussed the effect of this type of electrode pair in Chapter 9. When a d.c. voltage is applied between the two electrodes, the result is a static electric field, represented by the field lines in Fig. 10.3. Note that the field is directly mainly downwards through the left-hand guide, and upwards through the right-hand one. This will produce a change in the refractive index through the linear electro-optic effect. The exact result will depend on the precise field variation, but the crystal orientation will also affect matters. For simplicity let us assume that the crystal is chosen so that the vertical field component is the important one for the particular polarization of light involved.

The linear electro-optic effect is (by definition) linear. In this case, we can deduce that the index change must be of opposite sign in the two guides. Consequently, any change in guide propagation constant that results must also be of opposite sign. The application of a d.c. voltage to the electrodes can therefore be used to desynchronize two otherwise identical guides. The difference between

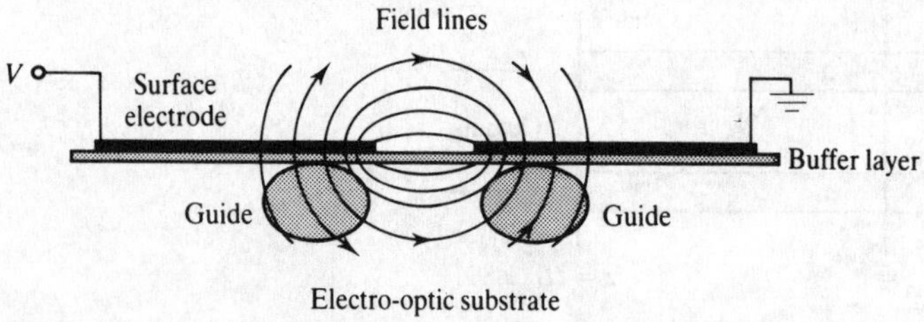

Figure 10.3 Cross-section of an electro-optic directional coupler.

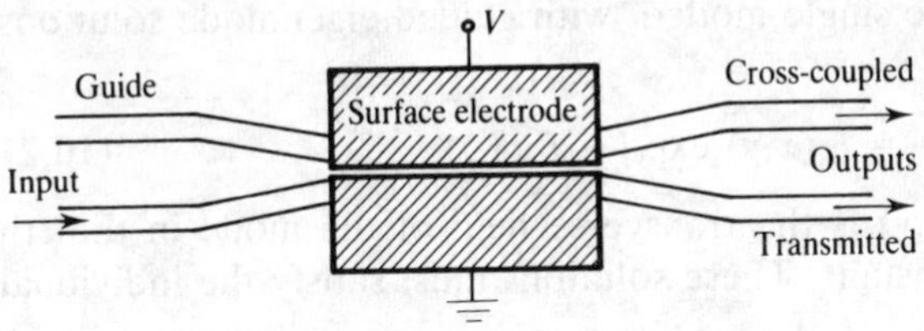

Figure 10.4 Plan view of a directional coupler switch.

the propagation constants is often written as $\Delta\beta = \beta_2 - \beta_1$, and is proportional to voltage. The electrode structure shown is consequently often known as a pair of *$\Delta\beta$ electrodes*. An electro-optic device using such electrodes is called a *directional coupler switch*, shown in plan view in Fig. 10.4. Here additional bend sections have been included to separate the waveguides at the input and output, for example for coupling to pairs of fibres (which have a large core-to-core separation due to their cladding). Integrated optic directional couplers in semiconducting crystals require a rather different electrode construction, which will again be covered in Chapter 12.

10.3 THE DIRECTIONAL COUPLER—THEORETICAL ANALYSIS

We will now form a theoretical model of the directional coupler. The derivation of the relevant equations is, unfortunately, rather lengthy, so for simplicity we base the analysis on scalar theory, and concentrate on the central region of the device where the guides run parallel. We start by assuming that the two guides are oriented in the z-direction, and are described in isolation by the refractive index distributions $n_1(x,y)$ and $n_2(x,y)$, respectively. Figure 10.5 shows how these distributions might look for a pair of symmetric slab waveguides. Here the refractive index outside the guides is everywhere n_s, while the indices in the two guiding layers are n_{g1} and n_{g2} respectively. Assuming that the electric field is polarized in the y-direction, the scalar wave equation governing the variation of the electric field $E_{yi}(x,y,z)$ for each guide in isolation is then given by

$$\nabla^2 E_{yi}(x,y,z) + n_i^2(x,y)k_0^2 E_{yi}(x,y,z) = 0 \qquad (10.1)$$

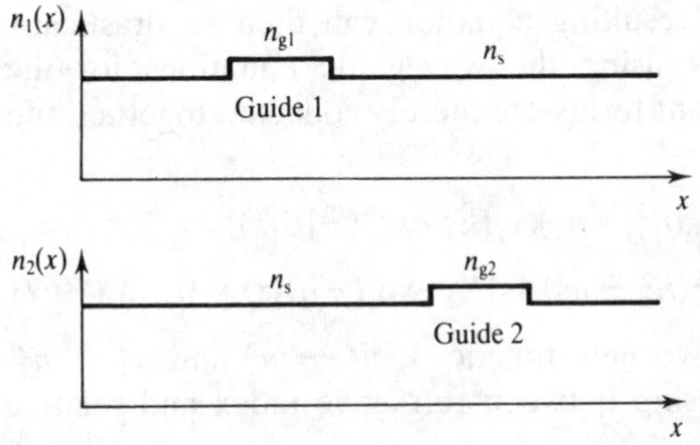

Figure 10.5 The refractive index distributions $n_1(x)$ and $n_2(x)$ for two isolated slab guides.

Both guides are now assumed to be single-moded, with guided eigenmode solutions defined by:

$$E_{yi}(x,y,z) = E_i(x,y) \exp(-j\beta_i z) \tag{10.2}$$

where $i = 1$ or 2, and where $E_i(x,y)$ is the transverse field of the mode in the ith guide, and β_i is its propagation constant. These solutions must satisfy the individual waveguide equations:

$$\nabla^2_{xy}E_i(x,y) + [n_i^2(x,y)k_0^2 - \beta_i^2]E_i(x,y) = 0 \tag{10.3}$$

Here the subscripts on the Laplacian imply that the differentiation is to be performed with respect to x and y only. We take it that these equations have previously been solved by some means, so that we know both of the transverse field distributions $E_i(x,y)$ and both propagation constants β_i.

We assume that the complete coupler is described by a refractive index distribution $n_T(x,y)$, which corresponds to both guides together. We have already shown in Fig. 10.1 how this distribution might look for the slab waveguide model used above. The scalar wave equation for the complete coupler is then

$$\nabla^2 E_y(x,y,z) + n_T^2(x,y)k_0^2 E_y(x,y,z) = 0 \tag{10.4}$$

Because we expect power to be coupled slowly from one guide to the other, we assume that we can describe the solution for $E_y(x,y,z)$ as a superposition of the modes in the two isolated guides, i.e. as a linear combination of the two modal solutions previously found. We therefore put

$$E(x,y,z) = A_1(z)E_1(x,y) \exp(-j\beta_1 z) + A_2(z)E_2(x,y) \exp(-j\beta_2 z) \tag{10.5}$$

Here the functions $A_1(z)$ and $A_2(z)$ are the local amplitudes of the two modes. Clearly, these must vary with distance to describe the slow changes in power associated with the coupling process.

Carrying out the necessary differentiation, we can show that

$$\begin{aligned}\nabla^2 E_y(x,y,z) = {} & [A_1 \nabla^2_{xy}E_1 + (d^2A_1/dz^2 - 2j\beta_1 \, dA_1/dz - \beta_1^2 A_1)E_1] \exp(-j\beta_{1z}) \\ & + [A_2 \nabla^2_{xy}E_2 + (d^2A_2/dz^2 - 2j\beta_2 \, dA_2/dz - \beta_2^2 A_2)E_2] \exp(-j\beta_2 z)\end{aligned} \tag{10.6}$$

We now substitute this expression and the assumed solution of Eq. (10.5) into the scalar wave equation (Eq. (10.4)). The resulting equation can then be drastically simplified by eliminating the ∇^2_{xy} terms using the waveguide equations for the individual guides (Eq. (10.3)). The relevant terms are merely collected together and set equal to zero. The result is

$$\begin{aligned}&[d^2A_1/dz^2 - 2j\beta_1 \, dA_1/dz + k_0^2(n_T^2 - n_1^2)A_1]E_1 \exp(-j\beta_1 z) \\ &+ [d^2A_2/dz^2 - 2j\beta_2 \, dA_2/dz + k_0^2(n_T^2 - n_2^2)A_2]E_2 \exp(-j\beta_2 z) = 0\end{aligned} \tag{10.7}$$

Note the appearance in Eq. (10.7) of two new functions, $n_T^2 - n_1^2$ and $n_T^2 - n_2^2$. Bearing in mind the square-law relationship between refractive index and relative dielectric constant, $n^2 = \varepsilon_r$, we can interpret these as follows. The first represents a perturbation in dielectric constant $\Delta\varepsilon_{r1}$ to guide 1, caused by the introduction of a neighbouring waveguide, guide 2. This will clearly be non-zero only in the neighbourhood of guide 2. The second is a similar perturbation $\Delta\varepsilon_{r2}$ caused to guide 2 by guide 1, and will in turn be significant only near guide 1.

Of course we can make further simplifications. For example, we can neglect second derivatives of the amplitudes d^2A_i/dz^2 on the grounds that these represent modal envelopes which will probably vary only slowly with distance. We can also divide through by $\exp(-j\beta_1 z)$. This yields

$$\begin{aligned}&[-2j\beta_1\, dA_1/dz + k_0^2(n_T^2 - n_1^2)A_1]E_1 \\ &+ [-2j\beta_2\, dA_2/dz + k_0^2(n_T^2 - n_2^2)A_2]E_2 \exp(-j\Delta\beta z) = 0\end{aligned} \quad (10.8)$$

where $\Delta\beta = \beta_2 - \beta_1$ represents the mismatch in propagation constant between the guides. We are now nearly there; Eq. (10.8) is almost (but not quite) usable. The problem is that we really wish to find an equation governing the changes in the wave amplitudes A_i with distance. Since these are one-dimensional functions, it is not sensible to solve a three-dimensional equation. Somehow we must remove the x- and y-variations contained in the quantities E_i and $n_T^2 - n_i^2$. We do this in two steps. First, we multiply the whole of Eq. (10.8) by the complex conjugate of the transverse field in guide 1 (E_1^*), and integrate over the device cross-section. This yields the following new equation:

$$\begin{aligned}&[-2j\beta_1 \langle E_1, E_1\rangle\, dA_1/dz + \langle k_0^2(n_T^2 - n_1^2)E_1, E_1\rangle A_1] \\ &+ [-2j\beta_2 \langle E_2, E_1\rangle\, dA_2/dz + \langle k_0^2(n_T^2 - n_2^2)E_2, E_1\rangle A_2] \exp(-j\Delta\beta z) = 0\end{aligned} \quad (10.9)$$

Here we have used the inner product notation introduced in Chapter 6 to represent all the integrals that arise. These are of course independent of x and y, so the operation has had the desired effect.

We now make a number of assumptions concerning the relative sizes of the remaining terms.

Term (1) We assume that the *overlap term* $\langle E_2, E_1\rangle$ is negligible, i.e. that the spatial overlap of the two modes is small. This is reasonable if the guides are not too close together.

Term (2) We also assume that the *self-coupling term* $\langle k_0^2(n_T^2 - n_1^2)E_1, E_1\rangle$ is negligible. This is physically reasonable since the integral is only non-zero in the region of the perturbation $\Delta\varepsilon_{r1}$, which lies in the evanescent region where the field function E_1 is itself small.

Term (3) We assume on the other hand that the *self-overlap term* $\langle E_1, E_1\rangle$ is non-negligible since E_1 must of course overlap completely with itself.

Term (4) We also assume that the *coupling term* $\langle k_0^2(n_T^2 - n_2^2)E_2, E_1\rangle$ is non-negligible since E_1 is significant in the region where $\Delta\varepsilon_{r2}$ is non-zero, the core of guide 1.

Neglecting terms (1) and (2), and retaining terms (3) and (4), leaves Eq. (10.9) in the form:

$$-2j\beta_1 \langle E_1, E_1\rangle \, dA_1/dz + \langle k_0^2(n_T^2 - n_2^2)E_2, E_1\rangle A_2 \exp(-j\Delta\beta z) = 0 \tag{10.10}$$

We could of course repeat the process above, this time multiplying Eq. (10.8) by E_2^*. We can guess that the result will be the slightly different equation:

$$-2j\beta_2 \langle E_2, E_2\rangle \, dA_2/dz + \langle k_0^2(n_T^2 - n_1^2)E_1, E_2\rangle A_1 \exp(+j\Delta\beta z) = 0 \tag{10.11}$$

We now make two more simplifications. First, since the propagation constants will not change by very much during the switching process, we assume that $\beta_1 \approx \beta_2 \approx \beta_0$ for the purpose of division. Second, we assume that the symmetry of the device will ensure that

$$\langle k_0^2(n_T^2 - n_2^2)E_2, E_1\rangle / \langle E_1, E_1\rangle \approx \langle k_0^2(n_T^2 - n_1^2)E_1, E_2\rangle / \langle E_2, E_2\rangle \tag{10.12}$$

With these assumptions, Eqs (10.10) and (10.11) become

$$\begin{aligned} dA_1/dz + j\kappa A_2 \exp(-j\Delta\beta z) &= 0 \\ dA_2/dz + j\kappa A_1 \exp(+j\Delta\beta z) &= 0 \end{aligned} \tag{10.13}$$

Here we have introduced a new quantity κ which is known as the *coupling coefficient*. This is arguably the most important parameter of the coupler and is defined as

$$\kappa = (k_0^2/2\beta_0)\, \langle (n_T^2 - n_2^2)E_2, E_1\rangle / \langle E_1, E_1\rangle \tag{10.14}$$

Equations (10.13) are known as *coupled mode equations*, and represent the final stage in the derivation. Essentially they imply that variations in the amplitude of the mode in guide 1 are linked to the amplitude of the mode in guide 2 through the coupling coefficient κ, and vice versa. In other words, the mode amplitudes are coupled together. We shall see how to solve the equations very shortly; first, we will briefly consider factors affecting the value of the coupling coefficient.

Factors Affecting the Coupling Coefficient

The expression for κ in Eq. (10.14) is relatively complicated. However, we note that the first term, $k_0^2/2\beta_0$, is really a constant since it depends mainly on the optical wavelength. Similarly the last term, $\langle E_1, E_1\rangle$, is a normalization factor. The salient features are all contained in the central term, $\langle (n_T^2 - n_2^2)E_2, E_1\rangle$. This is, in effect, an overlap integral between $\Delta\varepsilon_{r2}$ (the dielectric constant perturbation seen by guide 2 due to guide 1) and the two fields E_1 and E_2.

In our slab guide model $\Delta\varepsilon_{r2}$ is non-zero only within the core of guide 1, so this region will give the only contribution to the integral. The main requirement for κ to be large is therefore that the evanescent tail of the field E_2 penetrates guide 1 to a significant extent. We can then say that κ is most strongly affected by the interwaveguide gap g, which should be small (but not too small, or our equations will be invalid). Because evanescent fields fall off roughly exponentially, it turns out that κ also depends exponentially on g, to good approximation. Typically g should be of the order of the guide width (usually, a few μm).

The confinement of the modes also affects κ. For example, a poorly-confined mode will have an evanescent field that extends a long distance from the guide core, which will result in a strong coupling coefficient. However, weak confinement may be undesirable because it increases bend loss, so the optimum confinement is something of a compromise. Finally, κ is affected by polarization. Even though we have assumed a scalar model here, the use of a birefringent electro-optic substrate generally causes different coupling rates for TE and TM modes. This is clearly unfortunate since the device will no longer be polarization-independent. Fortunately careful choice of the fabrication parameters can equalize κ_{TE} and κ_{TM} in some cases (for example, in the Ti:$LiNbO_3$ process).

10.4 SOLUTION OF THE EQUATIONS AT SYNCHRONISM

We shall see first what happens when the two guides are synchronous, i.e. when there is no propagation constant mismatch, so that $\Delta\beta = 0$. The two coupled mode equations then reduce to

$$dA_1/dz + j\kappa A_2 = 0; \quad dA_2/dz + j\kappa A_1 = 0 \tag{10.15}$$

We can solve these equations as follows. Differentiating the left-hand one we get

$$d^2A_1/dz^2 + j\kappa\, dA_2/dz = 0 \tag{10.16}$$

Substituting for dA_2/dz using the right-hand one then gives

$$d^2A_1/dz^2 + \kappa^2 A_1 = 0 \tag{10.17}$$

This is a standard second-order differential equation with the general solution:

$$A_1 = C_1 \cos(\kappa z) + C_2 \sin(\kappa z) \tag{10.18}$$

Here C_1 and C_2 are constants that must be chosen to satisfy the boundary conditions. If we initially launch light into guide 1 only, with unit modal amplitude, these must be

$$A_1 = 1 \text{ and } A_2 = 0 \text{ on } z = 0 \tag{10.19}$$

From Eq. (10.15) we know that

$$A_2 = (-1/j\kappa)\, dA_1/dz \tag{10.20}$$

So the boundary conditions could be written in the alternative form:

$$A_1 = 1 \text{ and } dA_1/dz = 0 \text{ on } z = 0 \tag{10.21}$$

It is easy to see the solution must then be

$$A_1 = \cos(\kappa z) \tag{10.22}$$

The solution for A_2 can be found by differentiation using Eq. (10.20). The result is

$$A_2 = -j \sin(\kappa z) \tag{10.23}$$

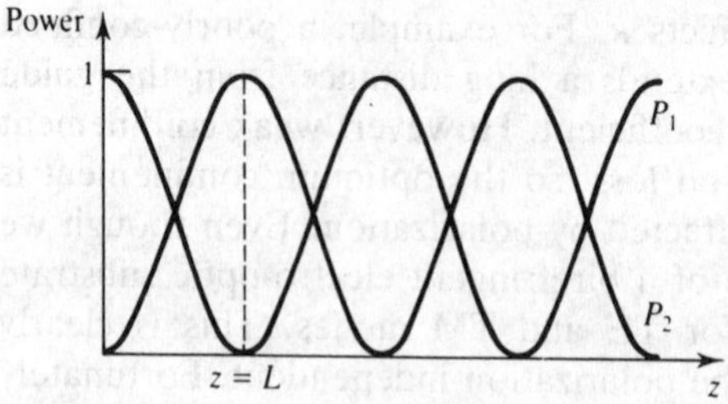

Figure 10.6 Variation of the powers P_1 and P_2 with distance z.

Note that A_2 is not real in Eq. (10.23), showing that A_2 is in quadrature with A_1. It is probably more useful to consider the powers in the two guides, rather than mode amplitudes. These can be found in the usual way via Poynting's Theorem. Normalizing relative to the input power, we get

$$P_1 = A_1 A_1^* = \cos^2(\kappa z) \qquad P_2 = A_2 A_2^* = \sin^2(\kappa z) \tag{10.24}$$

This demonstrates one of the nicer features of coupled mode theory. After a string of complicated manipulations we still end up with a physically convincing result: total power, which is of course given by $P_1 + P_2$, is conserved by the device. This is what we would expect since there is no loss. We can plot the variations of P_1 and P_2 with distance as shown in Fig. 10.6.

It can be seen that the variations are in complete agreement with our earlier qualitative discussion, and the power distribution between the guides is an oscillatory function of z. All the power is transferred from guide 1 to guide 2 at the point $z = L$ when $\kappa L = \pi/2$. Complete transfer also occurs when $\kappa L = 3\pi/2$, $5\pi/2$ and so on. Generally, however, the device parameters are chosen for the lowest of these values. We can arrange this either by altering the length L or the coupling coefficient κ. For Ti:$LiNbO_3$ channel waveguide devices, $L_{(\pi/2)}$ is typically 5–10 mm.

10.5 ASYNCHRONOUS SOLUTION

What happens if there is a mismatch in propagation constant, so that $\Delta\beta$ is non-zero? Well, we can still solve the coupled mode equations. First, it is best to make the substitutions

$$A_1 = a_1 \exp(-j\Delta\beta z/2) \text{ and } A_2 = a_2 \exp(+j\Delta\beta z/2) \tag{10.25}$$

Clearly we can obtain derivatives of these new variables in the form:

$$dA_1/dz = [da_1/dz - j\Delta\beta/2\, a_1] \exp(-j\Delta\beta z/2) \tag{10.26}$$

and so on. Substituting into Eqs (10.13) we find that all the exponentials cancel out, leaving only constant terms, so that

$$da_1/dz - j\Delta\beta/2\, a_1 + j\kappa\, a_2 = 0$$
$$da_2/dz + j\Delta\beta/2\, a_2 + j\kappa\, a_1 = 0 \tag{10.27}$$

These equations are exactly equivalent to Eqs (10.13), but much easier to solve. We won't repeat the mathematics involved, but merely quote the general solution. For the same input as before, the power output from guide 2 is given by

$$P_2 = \sin^2 [\sqrt{(\nu^2 + \xi^2)}] / (1 + \xi^2/\nu^2) \tag{10.28}$$

where we have introduced two new parameters, a *normalized coupling length* ν and a *normalized dephasing parameter* ξ, given by

$$\nu = \kappa L \qquad \xi = \Delta\beta L/2 \tag{10.29}$$

Equation (10.28) looks a bit like a sinc^2 function (remember that $\text{sinc}(x) = \sin(x)/x$). We can plot it out as a function of ξ (together with the power in guide 1, given by $P_1 = 1 - P_2$), for our design value of $\nu = \pi/2$. This gives the typical switching characteristic shown in Fig. 10.7. When $\xi = 0$ (i.e. at synchronism) all the power emerges from guide 2 as before, but as $|\xi|$ increases the power transfer is destroyed and P_2 falls with a sort of 'filter' characteristic. Similarly P_1 rises as $|\xi|$ increases, so that total power is once again conserved. The first zero in Eq. (10.28) occurs when

$$\sqrt{(\nu^2 + \xi^2)} = \pi \tag{10.30}$$

Given that we chose a normalized coupling length of $\nu = \pi/2$, this implies that

$$\xi = \pi\sqrt{3}/2 \text{ or } \Delta\beta L = \pi\sqrt{3} \tag{10.31}$$

To make a switch, we therefore need only to be able to define two states: (i) state A, when no voltage is applied to the electrodes (so $\Delta\beta L = 0$). The light then emerges from guide 2; (ii) state A′, when a voltage is applied such that $\Delta\beta L = \pi\sqrt{3}$. The light then emerges from guide 1.

Design Example

We can compare the voltages required to switch a coupler and a Mach-Zehnder interferometer, as follows. For example, we know from Chapter 9 that the output of the interferometer will fall to zero when there is π radians phase difference between the arms. The equivalent value in a coupler can be found from Eq. (10.31) as $\pi\sqrt{3}$ rads. At first sight, therefore, the coupler seems less efficient because it will need $\sqrt{3}$ times the voltage for the same electrode geometry. However, coupler electrodes provide push-pull modulation from a narrow electrode gap. This would not be practical in the Mach-Zehnder modulator because the guides are further apart (a *bipolar* drive is actually required in Fig. (9.28) for push–pull operation). A more realistic figure for the voltage ratio is therefore $\sqrt{3}/2$.

Practical Switches

The astute reader will have noticed that one switch state in Fig. 10.7 is electrically-adjustable, but the other (at $\xi = 0$) is set by the coupling length. Because κ is dependent on modal confinement, an error in fabrication may easily lead a value

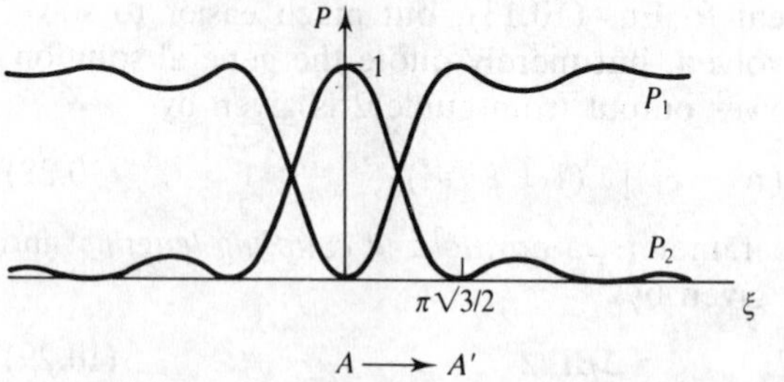

Figure 10.7 Directional coupler switch characteristic.

of ν different from $\pi/2$. It is natural to ask whether the two switch states may then still be obtained. The short answer is, no: the cross-coupled state will in general suffer crosstalk. As a result, an alternative set of electrodes, known as *stepped $\Delta\beta$ electrodes*, are often used. Each electrode is divided into two at its midpoint, giving two pairs of electrodes, both of length $L/2$. Using two separate control voltages (one for each pair) it is then possible to tune the device so that both states are always achievable. However, this can only be done if the value of ν is greater than $\pi/2$.

Directional coupler switches are often integrated together as a *switch array*. If a number of couplers are suitably arranged on a substrate, they can connect 2^N inputs to 2^N outputs very simply. Figure 10.8 shows an example of a 4×4 switch connecting four input fibres to four outputs. Although 8×8 Ti:$LiNbO_3$ switch arrays are now common, there is a limit to array size set by the dimensions of available crystal substrates.

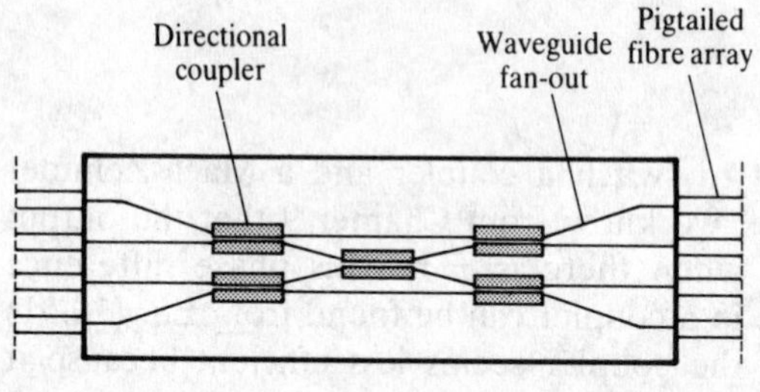

Figure 10.8 A 4×4 switching array.

10.6 FIBRE DIRECTIONAL COUPLERS

Directional couplers can also be made in single-mode optical fibre form. The fibre is first mounted in a silica or glass block which has a curved groove cut in it, and the block is then polished down until the fibre core is just exposed. Two such blocks are then placed together so that the polished areas are in contact, with a little index-matching fluid at the join. The evanescent fields in the two fibres then overlap, so power is transferred between them much as in the integrated device above. This arrangement is known as a *polished fibre coupler*, and is shown in Fig. 10.9.

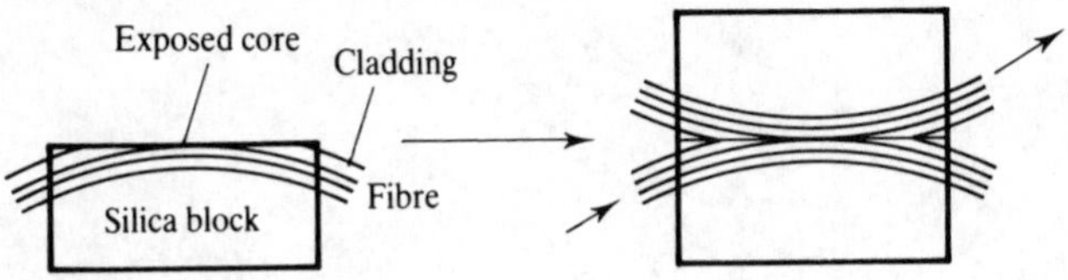

Figure 10.9 A polished fibre directional coupler.

The main differences from the integrated optic device are that the coupling is non-uniform because the spacing between the fibre cores varies with distance, and that the fibre version is not switchable. However, the coupling efficiency may easily be tuned, simply by sliding the blocks across each other as shown in Fig. 10.10(a) on page 262. This alters the core-to-core separation, and hence the coupling coefficient, allowing simple variation of the splitting ratio. Often this is adjusted for a coupling length of $\nu = \pi/4$, so the device acts as a 50:50 splitter. Figure 10.10(b) shows a tunable single-mode fibre coupler.

If ordinary circular fibre is used to fabricate a polished coupler, the resulting component will effectively scramble the input polarization because there will be coupling between orthogonal polarization modes. To make a *polarization-preserving coupler* this process must be suppressed. The easiest way to do this is to use polarization-preserving fibre, so that orthogonal polarization modes are not synchronous. D-fibre and bow-tie fibre (described in Chapter 8) are both suitable. The only complication is that the polarization axes of the fibres in the two blocks must be parallel; the usual approach is to arrange them to be parallel and perpendicular to the polished surfaces.

The Fibre Loop Mirror

Polarization-preserving couplers allow the construction of fibre circuits that operate on interferometric principles. We have already encountered the integrated optic Mach-Zehnder interferometer, and we will describe the use of fibre MZIs in Chapter 14. Here we will consider an alternative circuit, the *Sagnac interferometer*, which can be used as an all-fibre mirror. Figure 10.11 shows the circuit, which consists of a fibre coupler with two of its outputs connected together in a loop.

To show that this acts as a mirror we will consider the generation of a reflected output. There are two possible paths between input and output that lead to a reflection. In the first, light traverses the coupler without cross-coupling, travels clockwise round the loop, and is then cross-coupled on meeting the coupler again. For unity input the resulting amplitude is then

$$A_{r1} = \cos(\kappa L) \exp(-j\phi) \times -j \sin(\kappa L) \tag{10.32}$$

Here the first term is the coupler transmission *without* cross-coupling, the second is the phase shift incurred in travelling round the loop, and the third is the coupler transmission *with* cross-coupling.

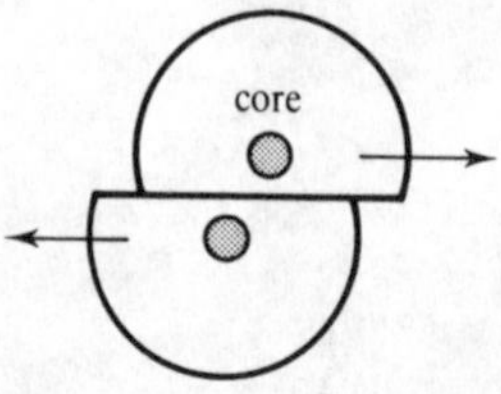

(a)

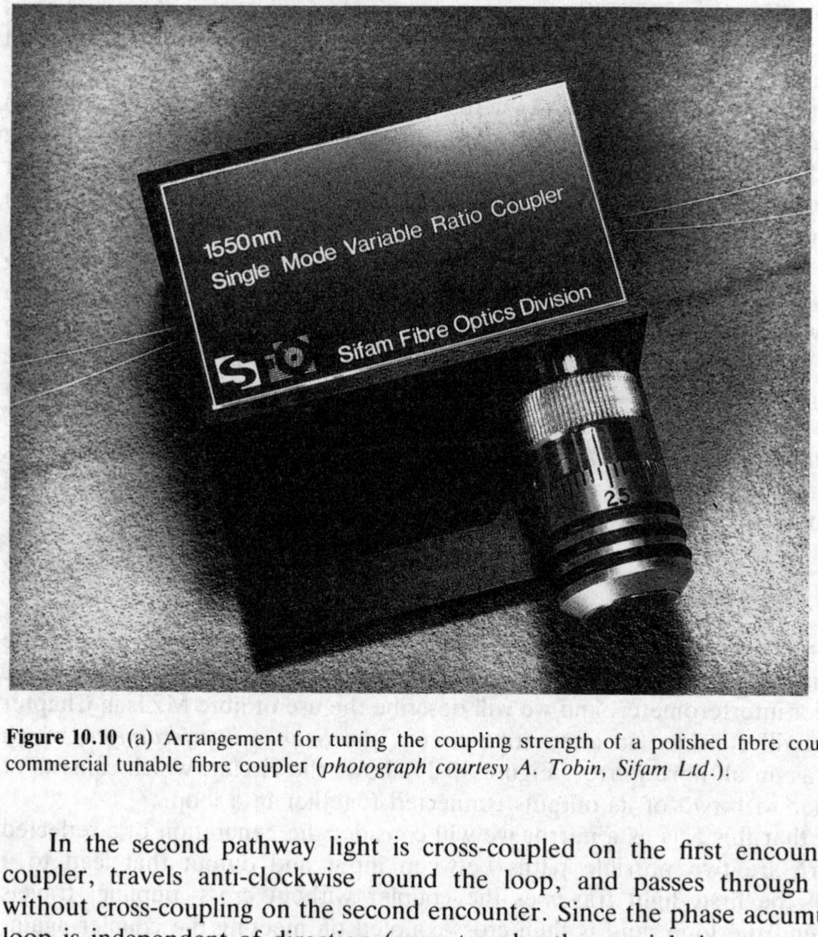

(b)

Figure 10.10 (a) Arrangement for tuning the coupling strength of a polished fibre coupler, and (b) a commercial tunable fibre coupler (*photograph courtesy A. Tobin, Sifam Ltd.*).

In the second pathway light is cross-coupled on the first encounter with the coupler, travels anti-clockwise round the loop, and passes through the coupler without cross-coupling on the second encounter. Since the phase accumulated in the loop is independent of direction (except under the special circumstances discussed in Chapter 14), the resulting amplitude is

$$A_{r2} = -j \sin(\kappa L) \exp(-j\phi) \cos(\kappa L) \tag{10.33}$$

Clearly the total reflected amplitude A_r must be the sum of these contributions, so that

$$A_r = -j \sin(2\kappa L) \exp(-j\phi) \tag{10.34}$$

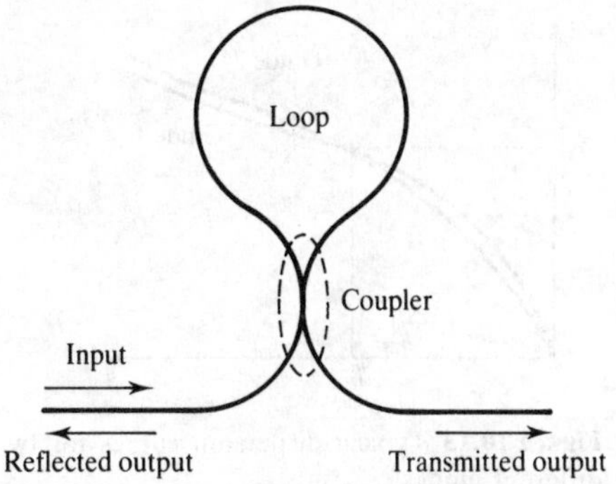

Figure 10.11 A fibre loop mirror.

The reflected power, $P_r = |A_r|^2$, is then given by

$$P_r = \sin^2(2\kappa L) \tag{10.35}$$

Equation (10.34) implies that the reflected power can reach 100 per cent if the normalized coupling length κL is suitable (for example, when $\kappa L = \pi/4$), *independent of the length of the loop*. At first sight this seems curious. To check we have not made a mistake we can evaluate the transmitted power in a similar way. The total transmitted amplitude A_t is again the sum of two components:

$$\begin{aligned} A_t &= [\cos(\kappa L) \exp(-j\phi) \cos(\kappa L)] + [-j \sin(\kappa L) \exp(-j\phi) \times -j \sin(\kappa L)] \\ &= \cos(2\kappa L) \exp(-j\phi) \end{aligned} \tag{10.36}$$

Hence the reflected power $P_t = |A_t|^2$ is given by

$$P_t = \cos^2(2\kappa L) \tag{10.37}$$

Comparing Eqs (10.34) and (10.36), we see that the result is self-consistent; the two outputs sum to unity (to be expected, since there is no loss) and the transmission is indeed zero when the reflectivity reaches 100 per cent. The physical principle in use is one of interference—the transmission falls to zero when the two components in Eq. (10.36) cancel out.

10.7 COUPLING BETWEEN DISSIMILAR WAVEGUIDES

Significant coupling can still occur between two entirely different waveguides provided they are synchronous and their modal fields overlap suitably. This principle can be exploited in a range of different devices, the most important being directional coupler filters and polarizers.

The Directional Coupler Filter

We shall begin with the directional coupler filter, which is typically formed from two quite dissimilar, parallel dielectric waveguides in close proximity. For example, these might be structurally similar, but have different core widths and refractive

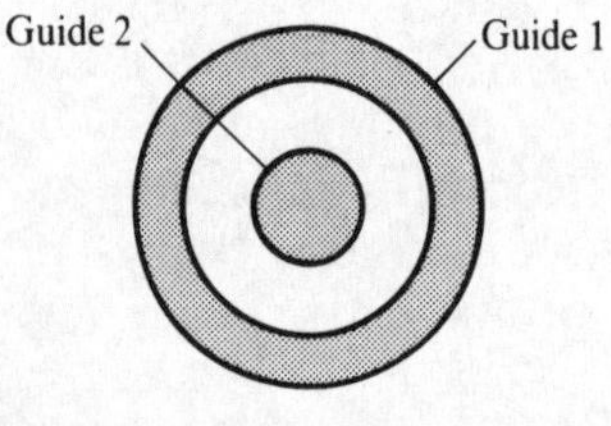

Figure 10.12 Cross-section of a coaxial fibre coupler.

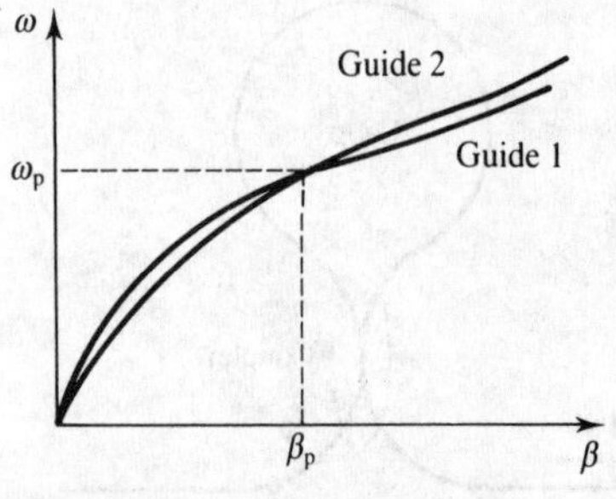

Figure 10.13 Typical dispersion curves for two different guides.

indices. Alternatively they might be quite different in shape—Fig. 10.12 shows a section through a *coaxial fibre coupler*, which operates by coupling between a central cylindrical guide and an outer annular guide.

We can understand the effect of using two different guides by considering their dispersion characteristics. For example, Fig. 10.13 shows typical ω–β diagrams for two slab guides. Note that for most of the frequency range the guides have different β-values. However, the curves have been arranged to cross (by careful choice of the waveguide parameters) so that at one frequency ω_p both propagation constants are equal to β_p. Within a narrow range of this frequency, both curves are approximately linear. Similarly the variations of β_1 and β_2 with λ_0 are roughly linear within a small range of the phase matching wavelength λ_p, as in Fig. 10.14(a).

We may approximate these curves by

$$\beta_1 \approx \beta_p + (\lambda_0 - \lambda_p)\partial\beta_1/\partial\lambda_0 \text{ and } \beta_2 \approx \beta_p + (\lambda_0 - \lambda_p)\partial\beta_2/\partial\lambda_0 \qquad (10.38)$$

where $\partial\beta_1/\partial\lambda_0$ and $\partial\beta_2/\partial\lambda_0$ represent the slopes of the dispersion characteristics near λ_p. This allows us to write the difference in propagation constant $\Delta\beta = \beta_2 - \beta_1$ between the two guides as

$$\Delta\beta \approx (\lambda_0 - \lambda_p)\,\{\partial\beta_2/\partial\lambda_0 - \partial\beta_1/\partial\lambda_0\} \qquad (10.39)$$

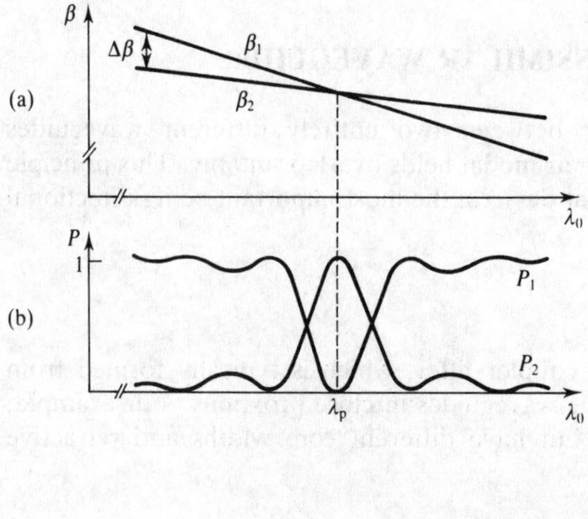

Figure 10.14 Directional coupler filter operation: (a) dispersion characteristics, and (b) filter response.

Consequently, $\Delta\beta$ depends linearly on the difference in wavelength from λ_p. Without doing any more mathematics we can deduce that the variation of the cross-coupled output P_2 with wavelength will be as previously defined for a switched coupler in Eq. (10.28), but assuming that $\Delta\beta$ is as given above. A typical response is shown plotted in Fig. 10.14(b). It is clear that the device functions as a band-pass filter, cross-coupling significant power only in the region of the synchronous wavelength λ_p. The filter bandwidth is set by the size of the proportionality constant $\{\partial\beta_2/\partial\lambda_0 - \partial\beta_1/\partial\lambda_0\}$. For maximum selectivity the difference in slopes of the two dispersion characteristics should be large. With the relatively weak guides in common use, however, there are restrictions in the possible range of slopes. This in turn limits typical filter bandwidths to around 100 Å using Ti:$LiNbO_3$ devices.

Polarizers and Polarization Splitters

The principle of codirectional coupling between two dissimilar guide modes can also be adapted to provide polarizing and polarization-splitting functions if one of the modes concerned is a surface plasma wave (which we previously introduced in Chapter 5). Surface plasmon polarizers represent a useful alternative to the birefringent devices described in Chapter 9.

We recall that a surface plasmon is a particular field structure that may be supported at the interface between a metal and a dielectric. We have already shown that it can only exist for TM-polarized light, and that it is an extremely lossy, short-range mode. Let us suppose that by appropriate choice of parameters its propagation constant may be matched to that of a mode in a nearby parallel dielectric waveguide. Synchronous coupling between these modes may then occur. However, energy transferred to the surface plasma wave will be quickly dissipated as heat, due to ohmic loss, so the coupling process will be largely one-way. In this way, the TM mode in the dielectric guide will be heavily attenuated, while the TE mode (which has no plasmon mode to couple to) is virtually unaffected. Such a device therefore acts as a TE-pass polarizer.

Figure 10.15 shows the construction of a *surface plasmon fibre polarizer*. It is based on a polished fibre block of the type previously used to fabricate a fibre directional coupler. However, there are two additional overlayers: a thin metal layer and a dielectric loading layer. In fact, the surface plasmon mode is supported by a three-layer structure, comprising the two layers just mentioned and an additional layer provided by the polished block. The form of the plasmon mode is therefore rather more complicated than our previous analysis suggests. Furthermore, detailed

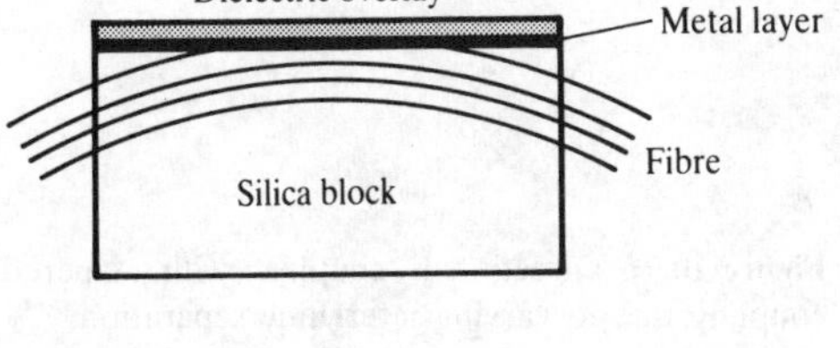

Figure 10.15 A fibre polarizer based on a polished fibre block.

analysis shows that more than one such mode can exist. The principle described above is nonetheless valid: by careful choice of the layer parameters, it is possible to phase match one of the possible surface plasma waves to the TM fibre mode, which can then be heavily attenuated. Typical results show 50–60 dB extinction ratios with excess losses for the transmitted polarization of only about 0.5 dB.

By placing another uncoated, polished fibre block on top of such a polarizer, a *polarization-splitter* can be made. This operates by using the surface plasmon as an intermediary between the two fibres. We have already seen how TM light from the original fibre can be coupled into the plasmon mode. If the additional fibre is also synchronous, the plasmon can in turn couple light into this guide, effectively transferring light between the two different fibres. Since the plasmon now acts only as a 'stepping stone' for a short period, little power is dissipated in the process. Only TM light is transferred in this way; TE light passes through without coupling.

10.8 SIDELOBE SUPPRESSION USING TAPERED COUPLING

We now consider a slight modification to the basic directional coupler design which can lead to significantly improved performance. Returning to Figs 10.7 and 10.14(b), we notice a characteristic of both responses: the existence of significant sidelobes. This can be unsatisfactory because switched states must be accurately located in the sidelobe 'nulls' for good crosstalk performance. Figure 10.16 shows an alternative coupler design which avoids this problem. The modification introduced is simply that the interguide separation is made to be slowly-varying with distance, reaching a minimum at the device centre and increasing at either end.

What is the effect of this slow tapering? We already know that the coupling coefficient κ depends on g, so it is reasonable to assume that a spatially-varying $g(z)$ will lead to a varying $\kappa(z)$. Since κ depends roughly exponentially on the gap, a parabolic change in g (due to circularly-curved guides) would lead to a Gaussian variation in κ. In this case we might expect that the coupled mode equations for the device will be as Eqs 10.13, but with the constant κ replaced by $\kappa(z)$. For simplicity we shall just consider the equation for A_2, which becomes

$$dA_2/dz + j\kappa(z)\, A_1 \exp(+j\Delta\beta z) = 0 \qquad (10.40)$$

If we knew the form of $A_1(z)$, this equation could be integrated over the device length to get

$$A_2(L) - A_2(0) = -j\int_0^L \kappa(z)\, A_1(z) \exp(+j\Delta\beta z)\, dz \qquad (10.41)$$

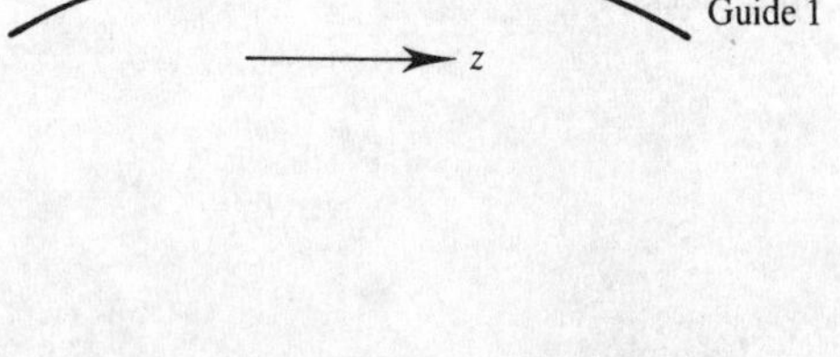

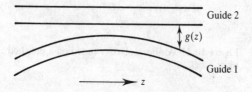

Figure 10.16 Directional coupler with tapered coupling due to varying interguide separation.

Without solving the equations we do not know $A_1(z)$. However, we can make an approximation, valid for weakly-coupled devices: if little power is transferred we may take $A_1(z)$ to be virtually constant. For unity input to guide 1 we then get

$$A_2(L) \approx -j \int_0^L \kappa(z) \exp(+j\Delta\beta z)\, dz \tag{10.42}$$

Equation (10.42) now shows that there is an integral relationship between the output $A_2(L)$ and $\kappa(z)$. For constant coupling we may evaluate this very simply to get

$$A_2(L) \approx -jkL \exp(+j\Delta\beta L/2) \operatorname{sinc}(\Delta\beta L/2) \tag{10.43}$$

This implies that the cross-coupled output power P_2 is given by

$$P_2 \approx \nu^2 \operatorname{sinc}^2(\xi) \tag{10.44}$$

where $\nu = \kappa L$ and $\xi = \Delta\beta L/2$, as before.

Equation (10.44) could of course have been found from the exact solution (Eq. (10.28)) merely by assuming a small value of ν. However, this derivation shows an interesting feature: the amplitude response (Eq. (10.42)) looks very like a Fourier transform of the coupling distribution. The sinc function obtained in Eq. (10.43) therefore arises from having uniform coupling over a finite distance, i.e. because $\kappa(z)$ is a rectangle function. To suppress the sidelobes we should choose alternative variations for $\kappa(z)$ whose Fourier transforms have lower lobes. One example is the Gaussian variation described earlier, since this transforms to another Gaussian. If the coupling is tapered in this way, the levels of the sidelobes may be greatly reduced. Most usefully, it turns out that this principle is still valid for strongly-coupled devices, even though the relation between the switch response and the coupling variation is no longer a simple Fourier transform.

10.9 THE REFLECTION GRATING FILTER—BASIC PRINCIPLES

We now move on to consider another extremely useful device, which operates this time by *contradirectional* coupling—the *reflection grating filter*. This is very similar to the planar waveguide gratings described in qualitative terms in Chapter 7. It can be made by etching a periodic corrugation on the surface of a waveguide, as shown in Fig. 10.17. The device can be used to convert a mode travelling in the forward direction into a backward-travelling mode, through the Bragg effect. Each groove acts like a weak mirror, which would individually give only a very small reflection. However, if the components all sum constructively, the result can be a very strong combined reflection. For this to occur we require the components reflected from adjacent grooves to add in-phase. The optical path between grooves is $n_{\text{eff}}\Lambda$, where n_{eff} is the effective index and Λ is the grating wavelength. If a component reflected from one groove is to sum in-phase with the contribution of its neighbour, we require that twice this path is a whole number of optical wavelengths (the factor of two arises from the reversal in direction). We can write this as

$$2n_{\text{eff}}\Lambda = m\lambda_0 \tag{10.45}$$

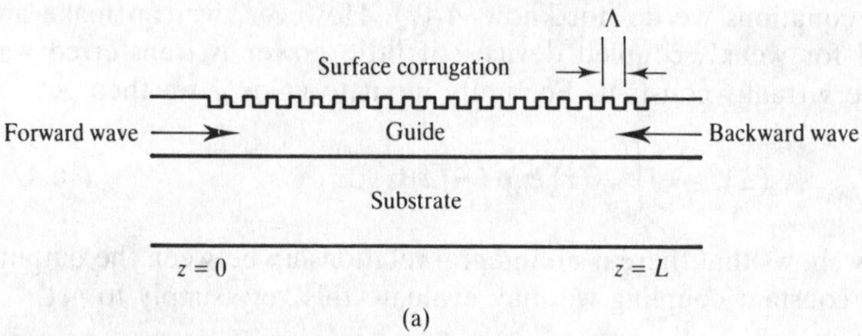

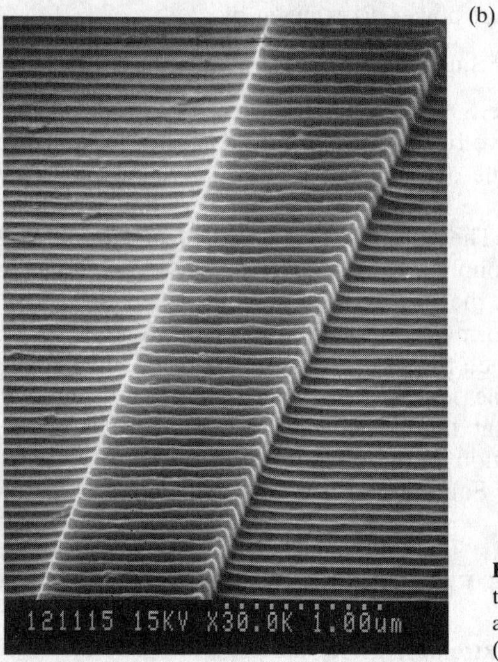

Figure 10.17 (a) Schematic of a corrugated reflection grating, and (b) electron microscope view of a reflection grating on a ridge guide formed in InP (*photograph courtesy G. Heise, Siemens AG*).

where m is an integer. This is a special case of Bragg's Law, and if we take m to be unity, the guided wave is incident at the first Bragg condition. We shall assume this is the case from now on. With this assumption we can clearly rearrange Eq. (10.45) as

$$2(\Lambda/2\pi) = (\lambda_0/2\pi n_{\text{eff}}) \tag{10.46}$$

As in Chapter 7, we now define a grating vector $\mathbf{K}$, whose direction is normal to the corrugations, and whose magnitude is given by:

$$K = 2\pi/\Lambda \tag{10.47}$$

Bearing in mind that the propagation constant β is equal to $2\pi n_{\text{eff}}/\lambda_0$, Eq. (10.46) implies that:

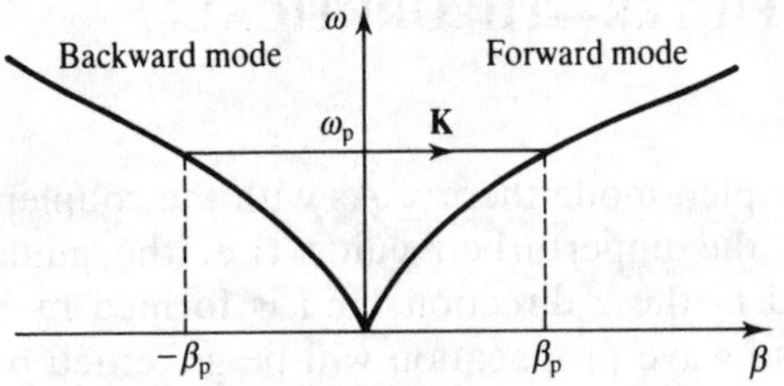

Figure 10.18 Phase matching by a grating, illustrated on ω–β plane.

$$K = 2\beta \tag{10.48}$$

Thus, when the Bragg condition is satisfied, the magnitude of the grating vector is exactly twice β. This corresponds to the K-vector closure condition discussed in Chapter 7, and could be represented by a K-vector diagram. However, an alternative viewpoint is provided in Fig. 10.18. This shows the dispersion characteristics of both a forward- and a backward-travelling mode; clearly the latter merely has a negative value of β. Here Eq. (10.48) is represented by the dashed-line construction, which shows that the forward and backward modes are phase matched by the grating vector for incidence exactly at the phase matching frequency ω_p.

Design Example

We can work out the period needed to satisfy the Bragg condition for a typical design, as follows. We take the optical wavelength to be $\lambda_0 = 1.5\ \mu m$, a common near-infrared wavelength. We assume that the guides are made in titanium-indiffused $LiNbO_3$ (whose refractive index is about 2.2). Because guides made by indiffusion are very weak, we may take $n_{eff} \approx n_s$, so $\beta \approx 2\pi n_s/\lambda_0 = 9.2 \times 10^6$. Hence $K = 2\beta = 18.4 \times 10^6$ and $\Lambda = 2\pi/K = 0.34\ \mu m$.

The small figure for Λ found above implies that gratings are hard to make. Generally a surface pattern is first prepared, either by *electron beam lithography* or by exposure to a *holographic interference pattern*. This is then etched into the guide using *ion beam milling, reactive ion etching* or *wet etching*. In semiconductor materials the refractive index is generally higher, which results in an even smaller value of Λ. The required pattern is then so hard to generate that a different grating with period 2Λ or 3Λ is often used instead, with incidence at a higher Bragg condition.

Off-Bragg Incidence

Following the argument above, we would expect a strong reflection from the grating when Eq. (10.48) holds because the scattered components will add constructively. What happens if this is not the case (if we change the optical wavelength, for example)? There will still be scattered components, but these will now sum with different phases. The further we are off-Bragg, the greater the differences involved, and there will soon come a point when they are so large that the effective sum of the components is zero. At this point there is no net reflection. The grating is therefore wavelength-selective, and can be used as a filter.

10.10 THE REFLECTION GRATING FILTER—THEORETICAL ANALYSIS

We can also analyse gratings using scalar coupled-mode theory. As with the coupler, we first briefly consider the properties of the unperturbed guide (i.e. the guide without the grating). This is again oriented in the z-direction, and is formed by a two-dimensional index distribution $n(x,y)$, so wave propagation will be governed by a scalar wave equation similar to Eq. (10.1) (ignoring the subscript i). As before, we assume that the guide is single-moded, so a forward-travelling mode $E_y(x,y,z) = E(x,y) \exp(-j\beta z)$ will satisfy a waveguide equation similar to Eq. (10.3), as does a backward-travelling mode $E_y(x,y,z) = E(x,y) \exp(+j\beta z)$. We now assume that the guide is perturbed by a periodic index change $\delta n(x,y,z)$, which represents the grating. In the simplest model, δn could be defined by a periodic function of the form:

$$\delta n(x,y,z) = \Delta n(x,y) \cos(Kz) \tag{10.49}$$

Here $\Delta n(x,y)$ describes the variation of the index perturbation in the transverse direction, while the cosine accounts for the periodicity of the grating in the z-direction. The scalar wave equation for the complete system (i.e. guide plus grating) is then

$$\nabla^2 E_y(x,y,z) + [n(x,y) + \delta n(x,y,z)]^2 k_0^2 E_y(x,y,z) = 0 \tag{10.50}$$

Given that the perturbation is likely to be small, we can approximate Eq. (10.50) by

$$\nabla^2 E_y + [n^2 + 2n\delta n] k_0^2 E_y = 0 \tag{10.51}$$

We know from the qualitative discussion above that a guided mode incident in the $+z$ direction will be reflected as a mode travelling in the $-z$ direction, provided the Bragg condition is nearly satisfied, so that $K \approx 2\beta$. We therefore assume a solution to Eq. (10.51) as the sum of a forward- and a backward-travelling mode, in the following form:

$$E_y(x,y,z) = A_F(z)\, E(x,y) \exp(-j\beta z) + A_B(z)\, E(x,y) \exp[-j(\beta - K)z] \tag{10.52}$$

Here A_F and A_B are the amplitudes of the forward- and backward-modes, respectively, and their z-dependence accounts for the gradual conversion of one mode into the other. Note that the backward mode can be recognized as varying approximately as $\exp(+j\beta z)$, if $K \approx 2\beta$.

Doing the necessary differentiation we can obtain the Laplacian as

$$\nabla^2 E_y = [A_F \nabla_{xy}^2 E + (d^2A_F/dz^2 - 2j\beta\, dA_F/dz - \beta^2 A_F)E] \exp(-j\beta z) +$$
$$[A_B \nabla_{xy}^2 E + (d^2A_B/dz^2 - 2j\{\beta - K\}\, dA_B/dz - \{\beta - K\}^2 A_B)E] \exp[-j(\beta - K)z] \tag{10.53}$$

We now substitute this and the assumed solution into the wave equation (Eq. (10.51)). This produces a very lengthy equation, which can however be simplified using much the same techniques as we have used before. We first eliminate all the

∇^2_{xy} terms using the unperturbed waveguide equation, and neglect second derivatives of A_F and A_B (on the grounds that these vary slowly). This gives

$$\{-2j\beta\, dA_F/dz \exp(-j\beta z) + [-2j(\beta - K)\, dA_B/dz - [(\beta - K)^2 - \beta^2]A_B] \exp[-j(\beta - K)z] + 2n\Delta n\, k_0^2 \cos(Kz)\,[A_F \exp(-j\beta z) + A_B \exp[-j(\beta - K)z]]\}E = 0 \tag{10.54}$$

We then note that the cosine term above can be written as

$$\cos(Kz) = (1/2)\,[\exp(jKz) + \exp(-jKz)] \tag{10.55}$$

With this substitution we can write

$$\cos(Kz)\,\{A_F \exp(-j\beta z) + A_B \exp[-j(\beta - K)z]\} = (1/2)\,\{A_F \exp[-j(\beta - K)z] + A_F \exp[-j(\beta + K)z] + A_B \exp[-j(\beta - 2K)z] + A_B \exp(-j\beta z)\} \tag{10.56}$$

Notice that terms containing exponentials like $\exp[-j(\beta + K)z]$ and $\exp[-j(\beta - 2K)z]$ have not appeared before, and are not contained in our assumed solution (Eq. 10.52)). In fact, these represent higher diffraction orders of the grating and we can safely ignore them here because they will not be phase matched. With this approximation, Eq. (10.54) reduces to:

$$\{-2j\beta\, dA_F/dz \exp(-j\beta z) - [2j(\beta - K)\, dA_B/dz + [(\beta - K)^2 - \beta^2]A_B] \exp[-j(\beta - K)z] + n\Delta n\, k_0^2\,[A_F \exp[-j(\beta - K)z] + A_B \exp(-j\beta z)]\}E = 0 \tag{10.57}$$

If Eq. (10.57) is to be true for all z, the coefficients of the terms $\exp(-j\beta z)$ and $\exp[-j(\beta - K)z]$ must vanish separately. We therefore equate them both to zero, to get the following equations

$$\{-2j\beta\, dA_F/dz + n\Delta n\, k_0^2 A_B\}E = 0$$

$$\{-2j(\beta - K)\, dA_B/dz - [(\beta - K)^2 - \beta^2]A_B + n\Delta n\, k_0^2 A_F\}E = 0 \tag{10.58}$$

Further approximations are also possible. Because the grating will only have a significant effect if we are close to the Bragg condition, we can put

$$\beta - K/2 = \Delta\beta \tag{10.59}$$

where $\Delta\beta$ is a small quantity. Consequently, we can write

$$(\beta - K)^2 - \beta^2 \approx -4\beta\Delta\beta \tag{10.60}$$

We will use this in the lower equation in Eq. (10.58). We also need an approximation for $\beta - K$; a reasonable substitution is simply to put $\beta - \mathrm{K} \approx -\beta$. If this is done, Eqs (10.58) become

$$\{dA_F/dz + j(k_0^2/2\beta)\, n\Delta n\, A_B\}\, E = 0$$

$$\{dA_B/dz - j2\Delta\beta\, A_B - j(k_0^2/2\beta)\, n\Delta n\, A_F\}\, E = 0 \tag{10.61}$$

These look very like coupled differential equations once again, apart from the x- and y-variations contained in the terms $n(x,y)$, $\Delta n(x,y)$ and $E(x,y)$. We can get rid of these by multiplying by E^* and integrating over the waveguide cross-section. This gives

$$dA_F/dz + j\kappa A_B = 0 \qquad dA_B/dz - j2\Delta\beta A_B - j\kappa A_F = 0 \tag{10.62}$$

These are now true coupled-mode equations; once again the rate of change of each mode amplitude depends on the amplitude of the other mode, through a coupling coefficient κ given by

$$\kappa = (k_0^2/2\beta)\,\langle n\Delta n\, E, E\rangle \,/\, \langle E, E\rangle \tag{10.63}$$

Notice how κ depends on an overlap between the index perturbation Δn and the transverse field E. If we put the grating in the wrong place, therefore, the field will simply not see it, and there will be little reflection. In Fig. 10.17, the corrugation is at the upper surface of the guide. Consequently, the grating will only affect the evanescent part of the field, where the amplitude is weak. This will give a low value of κ, so a long grating will be needed for high reflectivity.

10.11 SOLUTION OF THE EQUATIONS AT SYNCHRONISM

As with the directional coupler, we will start by investigating the solution of the equations at synchronism (i.e. when $\Delta\beta = 0$, so that $K = 2\beta$). The coupled-mode equations are then

$$dA_F/dz + j\kappa A_B = 0 \qquad dA_B/dz - j\kappa A_F = 0 \tag{10.64}$$

Notice how similar these are to the equations for a synchronous coupler (Eq. (10.15))—the only real difference is a sign change in the lower equation. We can therefore solve them in a very similar way. We start by differentiating the upper equation and then eliminating A_B using the lower equation. This gives a second-order equation for A_F in the form:

$$d^2A_F/dz^2 - \kappa^2 A_F = 0 \tag{10.65}$$

Because of the sign change mentioned above, solutions of this equation are not trigonometric functions; instead, they are hyperbolic. We may therefore put

$$A_F = D_1 \cosh(\kappa z) + D_2 \sinh(\kappa z) \tag{10.66}$$

Here D_1 and D_2 are constants, chosen to satisfy the boundary conditions. With a forward-going input A_F is clearly specified at $z = 0$. However, the backward wave must grow from zero at the far end of the device, which is of length L. Suitable boundary conditions are therefore

$$A_F = 1 \text{ at } z = 0 \qquad A_B = 0 \text{ at } z = L \tag{10.67}$$

It is then quite simple to show that the solutions must be:

$$A_F(z) = \cosh[\kappa(L - z)] \,/\, \cosh(\kappa L) \qquad A_B(z) = -j \sinh[\kappa(L - z)] \,/\, \cosh(\kappa L) \tag{10.68}$$

Notice that the modal amplitudes at any point inside the grating now depend on the precise value of κL. The power carried in the $+z$ direction by the forward wave is clearly $P_F = A_F A_F^*$. However, the power carried in the same direction by the backward wave is $-P_B$, where $P_B = A_B A_B^*$. Once again it can be shown that power is conserved by the system of equations, since $P_F - P_B$ = constant.

Figure 10.19 shows the variation of P_F and P_B with distance through the grating for a typical normalized thickness of $\kappa L = 1.5$. As can be seen, the incident wave decays as it travels through the grating. However, as it loses energy, the reflected wave grows. The separation of the two curves is constant, consistent with the power conservation relation above. In general we will actually be most interested in the two outputs from the grating. These are the grating transmissivity T and reflectivity R, defined as $T = A_F A_F{}^*$ at $z = L$ and $R = A_B A_B{}^*$ at $z = 0$. It is easy to show that these are given by

$$T = 1/\cosh^2(\kappa L) \quad \text{and} \quad R = \tanh^2(\kappa L) \tag{10.69}$$

Equation (10.69) implies that 100 per cent peak reflectivity is possible, provided the normalized coupling length κL is large enough. For example, if $\kappa L = 2$, we obtain $R = 0.93$; similarly for $\kappa L = 4$, we get $R = 0.999$. Figure 10.20 shows a plot of T and R versus κL. Power conservation is again demonstrated, since the sum of T and R is unity throughout.

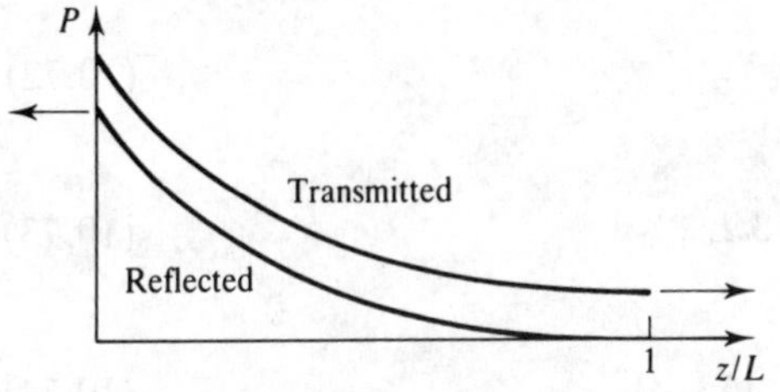

Figure 10.19 Variation of reflected and transmitted powers with distance.

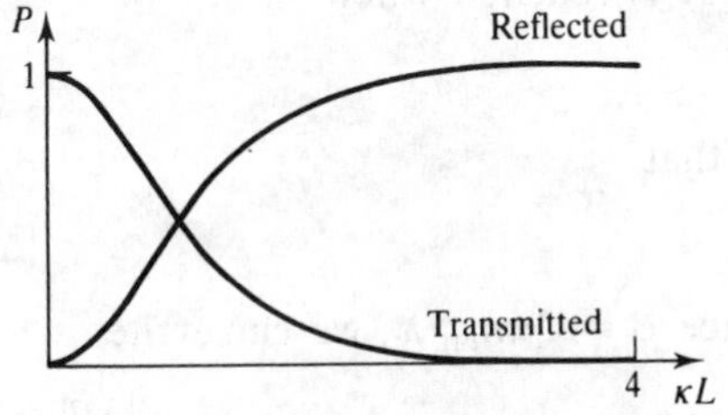

Figure 10.20 Grating reflectivity and transmissivity versus normalized length.

10.12 ASYNCHRONOUS SOLUTION

Now we consider the case when the optical wavelength is slightly incorrect, so that $\Delta\beta \neq 0$. Once again the solutions can be found by quite straightforward methods at each of the outputs as

$$T = |1/[\cosh(\Psi) - j(\xi/\Psi)\sinh(\Psi)]|^2$$
$$R = |1/[(\xi/\mu) + j(\Psi/\mu)\coth(\Psi)]|^2 \tag{10.70}$$

where μ, ξ and Ψ are defined by

$$\mu = \kappa L \qquad \xi = -\Delta\beta L \quad \text{and} \quad \Psi = \sqrt{(\mu^2 - \xi^2)} \tag{10.71}$$

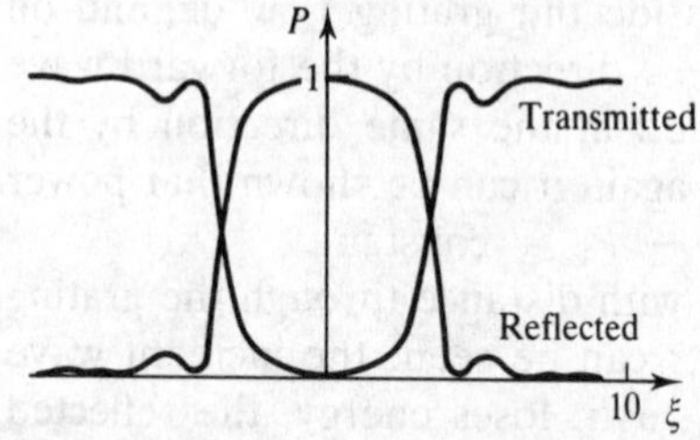

Figure 10.21 Typical filter response of a reflection grating.

These solutions can be plotted against ξ (i.e. as a function of the difference in β from the design value) for a given normalized thickness. Figure 10.21 shows the response for $\kappa L = 3$. As expected, the results are in keeping with our earlier qualitative discussion. For $\xi = 0$ there is high reflectivity. As $|\xi|$ increases, the reflected power falls, with a typical filter response. This is accompanied by a corresponding rise in transmissivity, and for sufficiently large $|\xi|$, there is no significant reflection.

Design Example

We can estimate the bandwidth of a typical reflection grating filter as follows. For a reasonable range of normalized lengths, the half-power point in curves like Fig. 10.21 is reached when

$$|\xi| \approx 3 \tag{10.72}$$

so that

$$|\Delta\beta| \approx 3/L \tag{10.73}$$

Since $\beta = 2\pi n_{\text{eff}}/\lambda_0$ we can write

$$|\Delta\beta| \approx 2\pi n_{\text{eff}}|\Delta\lambda_0|/\lambda_0^2 \tag{10.74}$$

and therefore that

$$|\Delta\lambda_0| \approx 3\lambda_0^2 \,/\, 2\pi n_{\text{eff}}L \tag{10.75}$$

For the typical parameters used before, namely $n_{\text{eff}} \approx 2.2$ and $\lambda_0 = 1.5\ \mu\text{m}$, and a grating length of $L = 1$ mm, we obtain $|\Delta\lambda_0| \approx 0.5$ nm. Grating filters are therefore highly wavelength-selective.

10.13 FIBRE GRATINGS

We have already described how gratings can be made in integrated optics. They can also be made in fibre optic form by etching a corrugation over a fibre that has been polished back to expose its core. Figure 10.22 shows how this is done, starting with one half of a polished fibre coupler. A fibre grating might be used as a frequency selective mirror in a fibre laser, or as a channel selector in a wavelength-

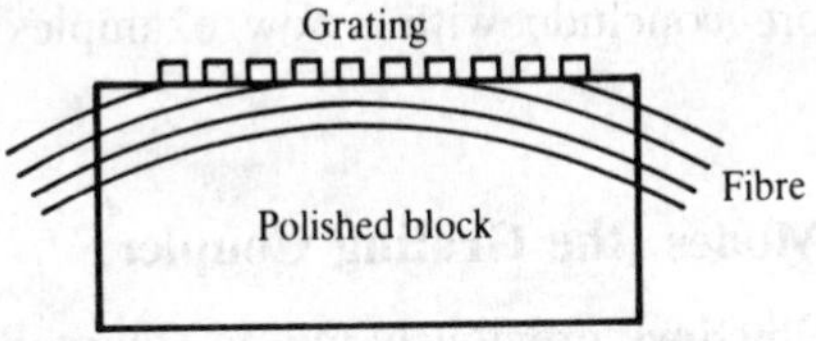

Figure 10.22 Optical fibre grating based on half a polished fibre coupler.

division multiplexed (WDM) communications system. Notice that the distance from the fibre core to the grating corrugation varies through the device due to the curvature of the fibre. Consequently, the coupling coefficient must also vary. It can be shown that a tapered coupling of this type has much the same effect as was found in the directional coupler, so we might expect the sidelobes of the filter response to be supressed.

More recently, an alternative method of grating fabrication has been investigated, based on the inherent photosensitivity of Ge-doped silica to short-wavelength radiation. This allows a periodic change in refractive index to be obtained inside the fibre core itself, rather than in the cladding. The fibre is simply exposed to an optical interference pattern, either derived from two coherent, counter-propagating guided waves or from two external beams. In the latter case, a grating period suitable for reflection of a guided infrared beam may be obtained, allowing the construction of filters for use in fibre lasers. Figure 10.23 shows a typical setup for external recording. The exposure source is normally an Ar^+ laser, operating either at $\lambda_0 = 0.488\ \mu m$ or at the frequency-doubled wavelength of $0.244\ \mu m$. Although the work is at an early stage, and the index change obtained is rather weak ($\Delta n \approx 6 \times 10^{-5}$), high-efficiency gratings have already been produced by this method.

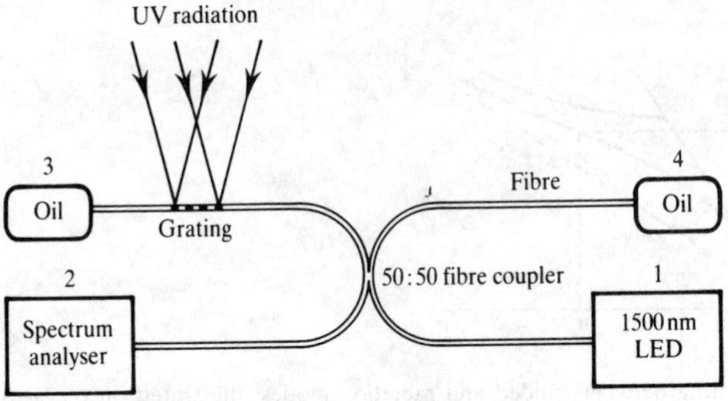

Figure 10.23 Writing and detection of a reflection grating in germanosilicate fibre (*after R. Kashyap* et al., *1990*).

10.14 OTHER COUPLED MODE INTERACTIONS

It turns out that the directional coupler and the reflection grating are merely the simplest examples of a wide range of devices that operate by some combination of co- and contra-directional coupling. We therefore conclude with a few examples illustrating further coupled mode interactions.

Conversion Between Guided and Radiation Modes (the Grating Coupler)

The first example occurs in the grating coupler, which we previously met in Chapter 7 as an input/output coupling component. There, its operation was described in terms of a K-vector diagram, but an alternative view is provided by the ω–β diagram of Fig. 10.24.

This is the same as for a simple reflection grating (Fig. 10.18), except that the frequency has been raised (or the wavelength lowered) so that the left hand end of the grating vector no longer lies in the region where backward-travelling, guided modes exist; instead, it falls in the area of radiation modes. We would therefore expect the grating exactly to phase match a forward-travelling guided mode to a backward-travelling radiation mode, and that this interaction could be described by a pair of coupled mode equations. This simple picture does not, unfortunately, represent the whole story. We have already seen that significant power can be coupled to modes that are not exactly phase matched, provided they are reasonably close to synchronism. Since the radiation modes are a continuum, the guided mode will be coupled to an entire *group* of such modes, whose propagation constants lie close to the phase matched value. The longer the grating, however, the more selective it becomes, and the more restricted the group; eventually, the description above must become a reasonable one. When this occurs the radiated output travels predominantly in a defined direction, which must of course be wavelength-dependent. This allows a grating coupler to be used as a form of spectrometer.

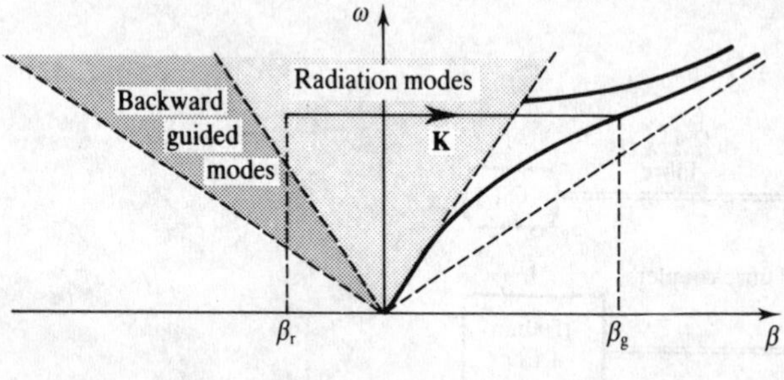

Figure 10.24 Phase matching by a grating between guided and radiation modes, illustrated on the ω–β plane.

Grating-assisted Coupling

A grating can also be used to achieve phase matching between two otherwise asynchronous, codirectional guided modes. These might be the two different polarization modes in a single-moded guide formed in a birefringent substrate (e.g.

$LiNbO_3$). Phase matching is obtained when one mode is excited as the first diffraction order of the grating, so that

$$|\mathbf{K}| = \beta_{TE} - \beta_{TM} = 2\pi\, \Delta n_{eff}/\lambda_0 \tag{10.76}$$

where Δn_{eff} is the difference in effective indices of the TE and TM modes. Since Δn_{eff} is usually only moderate, $|\mathbf{K}|$ is small, and the grating wavelength Λ is large (7 μm in Ti:$LiNbO_3$ at $\lambda_0 = 0.6\ \mu$m). To obtain coupling between orthogonal modes, a periodic perturbation to an off-diagonal element of the dielectric tensor is needed. This cannot be achieved using a simple surface corrugation, but it can be induced electrically through an off-diagonal electro-optic coefficient, by using a periodic electrode as shown in Fig. 10.25. This type of device is known as a *TE-TM mode converter*, and can be used as a polarization rotator or a filter. Filter operation depends on the difference in dispersion of the two modes, as for the dissimilar-guide directional coupler described earlier. However, greater wavelength selectivity (around 10 Å in Ti:$LiNbO_3$) can be obtained in this case because larger differences in slope of the two dispersion characteristics are possible with orthogonal modes.

It is not necessary for the two modes concerned to exist in the same guide. If a grating is combined with a directional coupler, it may be used to phase match codirectional modes of the same polarization in the two guides (if they are dissimilar) or orthogonal polarizations (if they are the same). Equally, the two modes do not have to be codirectional. Figure 10.26 shows a device which acts as a track-changing

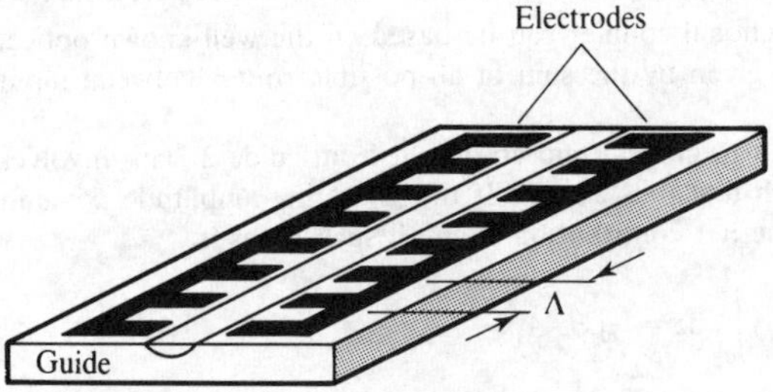

Figure 10.25 A TE-TM mode converter filter.

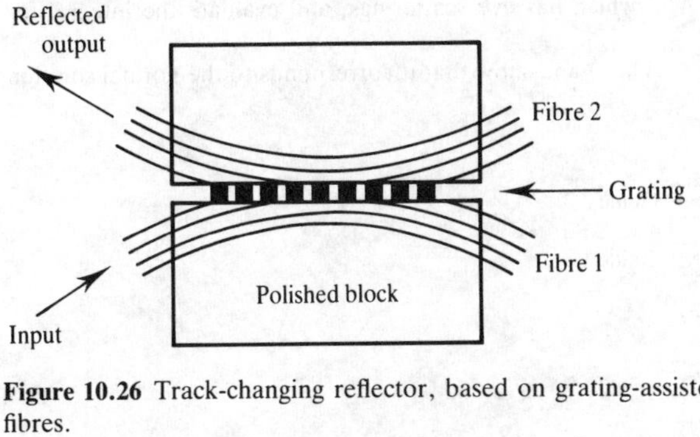

Figure 10.26 Track-changing reflector, based on grating-assisted coupling between two dissimilar optical fibres.

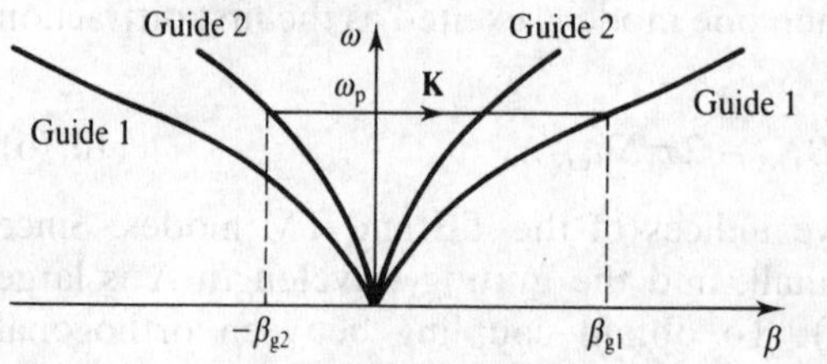

Figure 10.27 Phase matching by a grating between forward- and backward-travelling modes in different guides, illustrated on the ω–β plane.

reflector by using a reflection grating to phase match contradirectional modes in a fibre coupler formed between entirely different guides.

Without the grating, any codirectional coupling would be asynchronous, as shown in the ω–β diagram of Fig. 10.27. Similarly, the grating period is chosen so that it cannot retroflect light in either guide 1 or guide 2. However, at the correct frequency, it can phase match a forward mode in guide 1 to a backward mode in guide 2. This occurs when

$$|\mathbf{K}| = \beta_{g1} + \beta_{g2} \tag{10.77}$$

The result is that the reflected output now emerges from a different fibre from that used for the input—this is particularly convenient for channel-dropping operations.

PROBLEMS

10.1 Discuss the approximations inherent in the derivation of the coupled-mode equations for a directional coupler. Are they valid? If not, can you suggest the likely consequences for device performance?

10.2 An alternative view of a synchronous directional coupler can be based on the well-known optical principle that the output from a linear device is given by the sum of all possible routes between input and output.

Figure (i) below shows one path between an input to guide 1 and an output from guide 2. This involves a single scattering at an arbitrary position z, such that $0 \le z \le L$. If the scattering amplitude per unit length is $-\mathrm{j}\kappa$, and the input amplitude is A_{10}, the net contribution from all such paths is

$$A^1_{2L} = -\mathrm{j}\kappa A_{10} \int_0^L \mathrm{d}z = -\mathrm{j}\kappa L A_{10}$$

Figure (ii) below shows another possible route, involving three scatterings.

(a) If the sum of all such paths is denoted by A^3_{2L}, evaluate A^3_{2L}.

(b) Draw the paths contributing to A^5_{2L}, which has five scatterings, and evaluate the integral for this case.

(c) Sum the series $A_{2L} = A^1_{2L} + A^3_{2L} + A^5_{2L} \ldots$ and show that it corresponds to the normal solution for a synchronous directional coupler.

10.3 A directional coupler is constructed from two identical, symmetric slab waveguides. The guide thickness is h and the guide and substrate indices are n_g and n_s, respectively. Using the methods outlined in Chapter 6, derive a solution for the transverse field of the lowest-order, symmetric TE mode of a single such guide. Then, by direct evaluation of the overlap integral concerned, show that the coupling coefficient κ decreases exponentially with the interguide separation g in the twin-guide device.

10.4 The coupled differential equations for an asynchronous directional coupler are given by:

$$dA_1/dz + j\kappa A_2 \exp(-j\Delta\beta z) = 0 \qquad dA_2/dz + j\kappa A_1 \exp(+j\Delta\beta z) = 0$$

Assuming that total power is defined as $P_T = |A_1|^2 + |A_2|^2$, show (without solving the equations) that power is conserved by the device.

10.5 From a series of experiments with Ti:LiNbO$_3$ directional couplers made with different interwaveguide gaps g, the following data were obtained for the length required for complete coupling: $g = 4$ μm; $L_{\pi/2} = 5$ mm; $g = 6$ μm; $L_{\pi/2} = 7$ mm; $g = 7.5$ μm, $L_{\pi/2} = 9$ mm. Estimate the length required to achieve 50 per cent coupling efficiency when $g = 7$ μm.
[4.14 mm]

10.6 A fibre directional coupler is to be made by polishing two glass blocks which have fibres embedded in curved grooves.

(a) Suggest how the polishing process can be monitored, so it can be terminated when the core is reached.

(b) When the coupler is assembled it is found that the amount of cross-coupled light is very small, despite attempts to design it for 100 per cent coupling. Suggest two possible explanations, and the remedy in each case.

10.7 The output of a synchronous reflection grating can be found as the sum of all possible routes between the input and output. Figure (i) below shows one path between a forward-going input and a backward-travelling output. This involves a single scattering at an arbitrary position z, such that $0 \le z \le L$. If the scattering amplitude per unit length is $-j\kappa$, and the input amplitude is A_{F0}, the net contribution from all such paths is

$$A^1_{B0} = -j\kappa A_{F0} \int_0^L dz = -j\kappa L A_{F0}$$

Figure (ii) below shows another possible route, involving three scatterings.

(a) If the sum of all such paths is denoted by A^3_{B0}, evaluate A^3_{B0}.

(b) Draw the paths contributing to A^5_{B0}, which has five scatterings, and evaluate the integral for this case.

(c) Sum the series $A_{B0} = A^1_{B0} + A^3_{B0} + A^5_{B0}, \ldots$ and show that it corresponds to the normal solution for a reflection grating, when replayed at the Bragg wavelength.

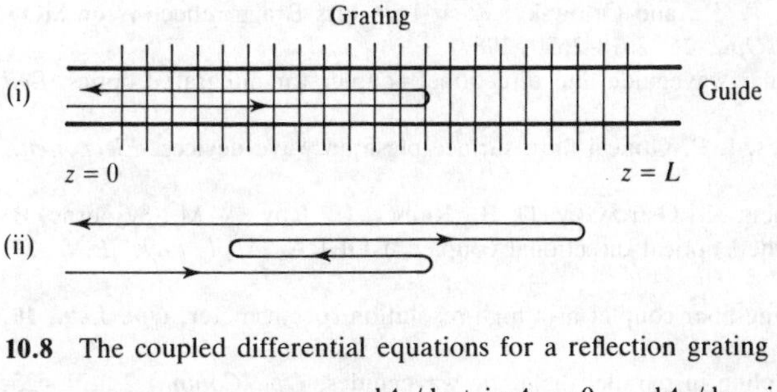

10.8 The coupled differential equations for a reflection grating at Bragg incidence are given by

$$dA_F/dz + j\kappa A_B = 0 \qquad dA_B/dz - j\kappa A_F = 0$$

Verify that the general solutions to these equations, subject to the boundary conditions $A_F = 1$ at $z = 0$, $A_B = 0$ at $z = L$, are

$$A_F(z) = \cosh[\kappa(L - z)]/\cosh(\kappa L) \qquad A_B(z) = -j \sinh[\kappa(L - z)]/\cosh(\kappa L).$$

10.9 A Bragg reflector is to be used with an integrated optic channel waveguide. Past measurements with this system suggest that the coupling coefficient is $\kappa = 200/\text{m}$ at the design wavelength. How long should the grating be to ensure reflectivity not less than 99 per cent?
[15 mm]

10.10 Compare the wavelength selectivity of waveguide filters based on (a) codirectional coupling between dissimilar waveguides, and (b) contradirectional coupling via a grating. What are the factors limiting selectivity in each case?

SUGGESTIONS FOR FURTHER READING

Alferness, R. C., and Buhl, L. L. Long-wavelength Ti:$LiNbO_3$ waveguide electro-optic TE-TM convertor, *Elect. Lett.* **19**, 40–41, 1983.

Alferness, R. C., and Cross, P. S. Filter characteristic of codirectionally coupled waveguides with weighted coupling, *IEEE J. Quant. Elect.*, **QE-14**, 843–847, 1978.

Alferness, R. C., Joyner, C. H., Divino, M. D., and Buhl, L. L. InGaAsP/InP waveguide grating filters for $\lambda = 1.5\ \mu m$, *Appl. Phys. Lett.*, **45**, 1278–1280, 1984.

Alferness, R. C., Schmidt, R. V., and Turner, E. H. Characteristics of Ti-diffused lithium niobate optical directional couplers, *Appl. Opt.*, **18**, 4012–4016, 1979.

Alferness, R. C.,and Veselka, J. J. Simultaneous modulation and wavelength multiplexing with a tunable Ti:$LiNbO_3$ directional coupler filter, *Elect. Lett.*, **21**, 466–467, 1985.

Bennion, I., Reid, D. C. J., Rowe, C. J., and Stewart, W. J. High reflectivity monomode-fibre grating filters, *Elect. Lett.*, **22**, 341–342, 1986.

Cozens, J. R., Boucouvalas, A. C., Al-Assam, A., Lee, M. J., and Morris, D. G. Optical coupling in coaxial fibres, *Elect. Lett.*, **18**, 679–681, 1982.

Digonnet, M. J. F., and Shaw, H. J. Analysis of a tunable single mode optical fiber coupler, *IEEE J. Quant. Elect.*, **QE-18**, 746–754, 1982.

Dyott, R. B., and Bello, J. Polarization-holding directional coupler made from elliptically cored fibre having a D section, *Elect. Lett.*, **19**, 601, 1983.

Flanders, D. C., Kogelnik, H., Schmidt, R. V., and Shank, C. V. Grating filters for thin-film optical waveguides, *Appl. Phys. Lett.*, **24**, 194–196, 1974.

Granestrand, P., Stolz, B., Thylen, L., Bergvall, K., Döldissen, W., Heinrich, H., and Hoffmann, D. Strictly nonblocking 8 × 8 integrated optical switch matrix, *Elect. Lett.*, **22**, 816–817, 1986.

Hill, K. O., Fujii, Y., Johnson, D. C., and Kawasaki, B. S. Photosensitivity in optical fiber waveguides: application to reflection filter fabrication, *Appl. Phys. Lett.*, **32**, 647–649, 1978.

Kogelnik, H., and Schmidt, R. V. Switched directional couplers with alternating $\Delta\beta$, *IEEE J. Quant. Elect.*, **QE-12**, 396–401, 1976.

Kondo, M., Ohta, Y., Fujiwara, M., and Sakaguchi, M. Integrated optical switch matrix for single-mode fibre networks, *IEEE J. Quant. Elect.*, **QE-18**, 1759–1765, 1982.

Lee, H. J., Henry, C. H., Kazarinov, R. F., and Orlowsky, K. J. Low loss Bragg reflectors on SiO_2-Si_3N_4-SiO_2 rib waveguides, *Appl. Opt.*, **26**, 2618–2620, 1987.

Marcatili, E. A. J. Dielectric rectangular waveguide and directional coupler for integrated optics, *Bell Syst. Tech. J.*, **48**, 2071–2102, 1969.

Markatos, S., Zervas, M. N., and Giles, I. P. Optical fibre surface plasmon wave devices, *Elect. Lett.*, **24**, 287–288, 1988.

Papuchon, M., Combemale, Y., Mathieu, X., Ostrowsky, D. B., Reiber, L., Roy, A. M., Sejourne, B. and Werner, M. Electically switched optical directional coupler: COBRA, *Appl. Phys. Lett.*, **27**, 289–291, 1975.

Russell, P. St. J., and Ulrich, R. Grating-fiber coupler as a high-resolution spectrometer, *Opt. Lett.*, **10**, 291–293, 1985.

Taylor, H. F. Frequency-selective coupling in parallel dielectric waveguides, *Opt. Comm.*, **7**, 421–425, 1973.

Veselka, J. J., and Korotky, S. K. Optimisation of Ti:$LiNbO_3$ optical waveguides and directional coupler switches for 1.56 μm wavelength, *IEEE J. Quant. Elect.*, **QE-22**, 947–951, 1986.

Whalen, M. S., Divino, M. D., and Alferness, R. C. Demonstration of a narrowband Bragg-reflection filter in a single-mode fibre directional coupler, *Elect. Lett.*, **22**, 681–682, 1986.
Whalen, M. S., and Walker, K. L. In-line optical-fibre filter for wavelength multiplexing, *Elect. Lett.*, **21**, 724–725, 1985.
Yariv, A. Coupled mode theory for guided wave optics, *IEEE J. Quant. Elect.*, **QE-9**, 919–933, 1973.

CHAPTER

ELEVEN

OPTOELECTRONIC INTERACTIONS IN SEMICONDUCTORS

11.1 WAVE–PARTICLE DUALITY

In Chapter 3 we described the interaction of light and matter in terms of electric and magnetic fields imposing forces on discrete charges. These charges—electrons—were considered to be either basically free (as in a metal) or bound (as in a dielectric). Furthermore, the dynamics of the interaction were explained in classical terms, so that the electron was regarded as a particle with charge and mass, obeying the laws of mechanics. There is considerable evidence to support this view—for example, measurements of the charge and charge-to-mass ratio of an electron—and it has led to a variety of hugely successful devices, many still of great importance (the oscilloscope, the TV tube, and high-power, high-frequency valves). Equally, the model leads to a perfectly adequate explanation of many optical phenomena—for example, the slowing of a wave in matter. This has allowed us to analyse many guided wave optical devices merely by assuming different distributions of refractive index.

Unfortunately light displays some features beyond the scope of this model. So far we have treated optical waves purely as solutions of Maxwell's equations. This provides a perfectly good explanation of the propagation of electromagnetic energy on a large scale, but it is seriously inadequate for the description of energy transfer between the electromagnetic field and matter. The need for a different analysis was first proposed by Max Planck in 1900, in connection with *black body radiation*. Planck found that the wavelength distribution of the energy radiated by a body in thermal equilibrium could only be explained if the energy exchanged between the field and the body (idealized as a collection of oscillators) took place in discrete amounts, comprising an integral number of elementary units known as *quanta*. He also showed that the magnitude of these quanta must be hf, where h is a constant, subsequently called *Planck's constant* (with the value 6.62×10^{-34} J s), and f is the frequency. The quanta of visible radiation have become known as *photons*. Further support for this idea came later through observations by P. von Lenard of the *photoelectric effect*, where the illumination of a metal surface under vacuum was found to result in the ejection of electrons. Here the critical feature was that the threshold of electron emission was determined by the frequency of the radiation and not by its power. This effect was explained completely by Albert Einstein in 1905.

Black body radiation and the photoelectric effect both relate to the interaction of radiation with matter. We might argue that a classical electromagnetic picture is still valid in the absence of absorption or emission effects. In fact, even this cannot really be justified. The first evidence came with an analysis of the scattering of X-rays by electrons, by A. H. Compton in 1923. Compton showed that there is a change in the X-ray wavelength with scattering angle; this is only consistent with a billiard-ball type of collision, in which the X-rays are represented by quanta of energy $\mathscr{E}|_{\text{photon}} = hf$, carrying momenta of $p|_{\text{photon}} = h/\lambda$. At this point, we introduce the notation $\hbar = h/2\pi$, which allows us to write

$$\mathscr{E}|_{\text{photon}} = \hbar\omega|_{\text{photon}} \text{ and } p|_{\text{photon}} = \hbar k|_{\text{photon}} \tag{11.1}$$

It is therefore clear that light actually possesses some of the attributes of both waves and particles, a property known as *wave–particle duality*. In practice, the relative importance of each facet of behaviour depends on the circumstances. As mentioned earlier, a wave analysis is often adequate, but the quantum nature of radiation should be invoked if any energy interchanges take place.

It is also a reasonable guess that if a quantity we have previously considered to be a wave may have particle-like properties, something described earlier as a particle may display wave-like behaviour. This turns out to be true: electrons also share the properties of both particles and waves, and their wave nature is important in explaining their behaviour in crystalline solids, especially the semiconductor materials used for optoelectronic devices. If we regard the electron as being similar to a photon so that in each case we are dealing with a localized, particle-like entity, we might say that the momentum of an electron ($p|_{\text{electron}}$) corresponds to $p|_{\text{photon}}$, while its energy ($\mathscr{E}|_{\text{electron}}$) is analogous to $\mathscr{E}|_{\text{photon}}$. In general the electron energy will be the sum of kinetic and potential contributions, so that

$$\mathscr{E} = 1/2\, mv^2 + \mathscr{V} \tag{11.2}$$

where m is the electron mass, v is its velocity and $\mathscr{V}$ is its potential energy. Since $p = mv$ we could write this alternatively as

$$\mathscr{E} = p^2/2m + \mathscr{V} \tag{11.3}$$

We now take one more step and say that if a photon has momentum and energy while at the same time having angular frequency ω and propagation constant (or *wavenumber*) k, it is possible that an electron—which clearly has momentum and energy—should also have a frequency and wavelength, i.e. wave-like characteristics associated with a particle-like model. If this were true we could introduce an electron wavenumber and angular frequency, following

$$\hbar k|_{\text{electron}} = p|_{\text{electron}} \quad \text{and} \quad \hbar\omega|_{\text{electron}} = \mathscr{E}|_{\text{electron}} \tag{11.4}$$

Consequently we might also write

$$\hbar\omega = p^2/2m + \mathscr{V} = \hbar^2k^2/2m + \mathscr{V} \tag{11.5}$$

Equation (11.4), first suggested by L. de Broglie in 1924, is the key to the link between the particle and wave nature of electrons. So far it is just conjecture; the most we can say is that it is plausible, and direct experimental evidence is required for confirmation. This has been provided in many ways. The earliest was the observation of *electron diffraction*—an interference phenomenon associated only with waves—by C. J. Davisson and L. H. Germer in 1927. As we saw in Chapter 7, diffraction can be observed from a periodic structure with a spacing comparable to the wavelength in question. For electrons we can easily calculate the wavelength for a given energy. When $\mathscr{V} = 0$, Eq. (11.3) implies that

$$p = \sqrt{(2m\mathscr{E})} \tag{11.6}$$

so that the equivalent wavelength may be found as

$$\lambda = h/p = h/\sqrt{(2m\mathscr{E})} \tag{11.7}$$

For electrons accelerated through V volts we have $\mathscr{E} = eV$. For $V = 1$ kV this corresponds to $\lambda = 0.04$ nm, and for 10 kV, $\lambda = 0.012$ nm. Thus for typical oscilloscope accelerating voltages we obtain wavelengths roughly corresponding to X-rays. This should result in observable diffraction from the periodic array of atoms in a crystal. In the event, Davisson and Germer showed that the scattering of electrons from a perfect crystal of nickel was not isotropic, but showed a clear dependence on angle (as expected from Bragg's law). This and other experiments established the validity of the idea of the wave nature of electrons, and the accuracy of the de Broglie relation.

11.2 PHOTONS

We shall begin our discussion of these new ideas by considering the description of radiation in terms of photons. To use a classical analogy, we might regard these localized fields as groups of plane waves, or *wavepackets*. For propagation in free space we might write

$$\mathsf{E}(z,t) = \int_{-\infty}^{\infty} E(\omega) \exp\{\mathrm{j}(\omega t - kz)\}\, \mathrm{d}\omega \tag{11.8}$$

for a one-dimensional wavepacket. This will travel at a speed given by the group velocity $v_g = \mathrm{d}\omega/\mathrm{d}k$ (which is equal to the phase velocity c in free-space).

The relation between the shape of the packet and its frequency spectrum $E(\omega)$ is determined by the laws of Fourier transformation. For example, Fig. 11.1 shows

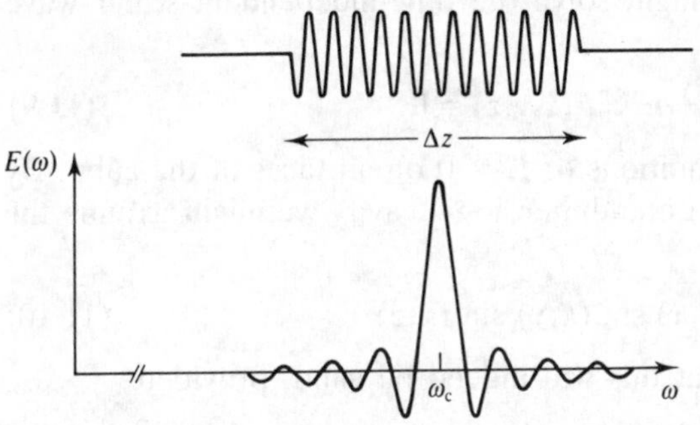

Figure 11.1 A wavepacket and its frequency spectrum.

a shapshot of a burst of radiation of uniform amplitude and length Δz. The corresponding angular frequency spectrum (which is directly analogous to one previously computed in Chapter 2) is also shown. In this case the angular frequency ω_c of the radiation determines the centre frequency of the spectrum, and the length Δz of the packet the range of frequencies present. As Δz increases the range of frequencies decreases, until in the limit we are left only with a single plane wave of frequency ω_c.

A typical radiation field will comprise many photons of random lengths separated by random intervals. The classical and the quantum pictures can then be reasonably brought together by treating the classical power density as being a measure of the photon density at any place. Except for very low power levels, the photon density is likely to be sufficiently high that the inherent 'graininess' of the radiation field is effectively smoothed out (although it will still be responsible for a level of noise).

We must be more careful in our description of a radiation field, however, if it is confined inside a volume. For example, Fig. 11.2 shows a cube of side h, consisting of material of refractive index n. This structure is effectively a three-dimensional resonant cavity, so we can guess from our earlier analysis that only particular standing-wave patterns may exist inside it. Any localized field we care to postulate—for example, a wavepacket—may then only be constructed from these allowed fields.

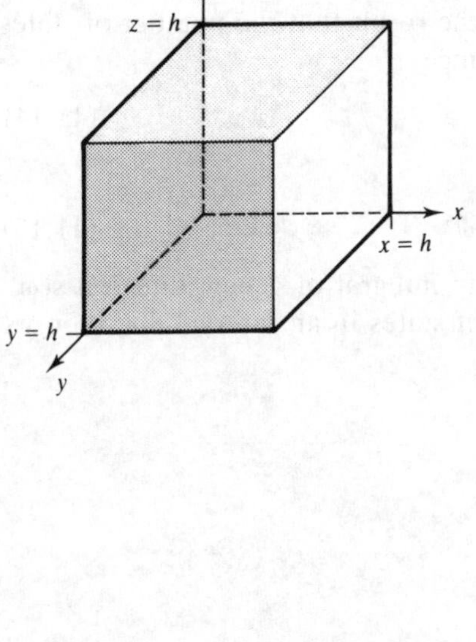

Figure 11.2 A cube of matter of side h.

To determine the allowed fields we might solve the time-independent scalar wave equation

$$\nabla^2 E(x,y,z) + n^2 k_0^2 E(x,y,z) = 0 \tag{11.9}$$

subject to the simplified boundary conditions of $E = 0$ on all faces of the cube. By analogy with our previous results for a one-dimensional cavity we might assume the solution:

$$E = E_0 \sin(k_x x) \sin(k_y y) \sin(k_z z) \tag{11.10}$$

Substitution into Eq. (11.9) shows that this will indeed be valid, provided

$$k_x^2 + k_y^2 + k_z^2 = n^2 k_0^2 \tag{11.11}$$

while the boundary conditions will be satisfied if

$$k_x = \nu_x \pi/h \qquad k_y = \nu_y \pi/h \text{ and } k_z = \nu_z \pi/h \tag{11.12}$$

where ν_x, ν_y and ν_z are integers. Equations (11.11) and (11.12) imply that the possible values of k_0 are restricted. In particular, we find that the allowed frequencies are specified by

$$\omega = ck_0 = c\pi\nu/nh \tag{11.13}$$

where $\nu^2 = \nu_x^2 + \nu_y^2 + \nu_z^2$. Any localized field may therefore only be constructed as a summation of components at these frequencies, rather than as a continuous integral. Because this is a complicated process, it is now sensible to consider the allowed solutions (or *states*) as being of significance in themselves. One question we might ask is, how many of them are there? If h is large, the minimum difference in frequency between any two states is small, so that the allowed frequencies approximate to a continuum. Note that if ν is constant, then so is ω, and if the allowed frequencies are plotted in a three-dimensional space with axes ν_x, ν_y and ν_z, then a sphere of radius ν represents a surface of constant ω. In this space the distance between neighbouring states is one unit when measured in the x- y- or z-directions, so the number of states contained in a cube of unit volume is one. The number of states with frequency less than ω might therefore be taken to be equal to the volume of the sphere, namely $4\pi\nu^3/3$. However, the indices ν_x, ν_y and ν_z must all actually be positive for the results to be meaningful, so the true figure should really be one-eighth of this value. Including the effect of polarization (which doubles the number of states), we finally obtain the result that the number of states with frequency less than ω is $\pi\nu^3/3$. However, since

$$\nu = C\,\omega \tag{11.14}$$

where $C = nh/\pi c$, it can easily be seen that

$$\pi\nu^3/3 = (\pi/3)C^3\,\omega^3 \tag{11.15}$$

If we now define this quantity to be equal to the integral of a new function $s(\omega)$, i.e. as $\int_0^\omega s(\omega)\,d\omega$, where $s(\omega)\,d\omega$ is the number of states in an interval $d\omega$, then we may extract $s(\omega)$ by differentiation, as

$$s(\omega) = \pi C^3 \omega^2 \tag{11.16}$$

We will often be more interested in a function $S(\omega)$, the number of states per unit frequency interval per unit volume (generally known as the *density of states*). This may be obtained very simply by dividing Eq. (11.16) by the volume of the cavity (h^3) to get

$$S(\omega) = n^3\omega^2/\pi^2c^3 \tag{11.17}$$

Note that this result is entirely independent of the cavity dimension h.

We may also care to enquire whether any of these possible states are indeed filled in a practical cavity, i.e. how much radiation exists at a given temperature. Using the methods of statistical thermodynamics it can be shown that the probability of a state of frequency ω being filled at absolute temperature T is

$$F(\omega) = 1/\{\exp(\hbar\omega/kT) - 1\} \tag{11.18}$$

where k is *Boltzmann's constant*. Equation (11.18) is known as the *Bose-Einstein probability function* and it was originally derived under two assumptions: (1) that radiation exists only in quanta whose energy is a multiple of $\hbar\omega$, and (2) that there is no restriction on the occupation of any state. We may use it to deduce the number of filled states lying in a frequency range between ω and $\omega + d\omega$. If we write this quantity as $n(\omega)\, d\omega$, then clearly $n(\omega)\, d(\omega) = S(\omega)\, F(\omega)\, d\omega$ or

$$n(\omega)\, d\omega = [(n^3\omega^2/\pi^2c^3)/\{\exp(\hbar\omega/kT) - 1\}]\, d\omega \tag{11.19}$$

Equally, we may use it to find the energy density of the radiation field in the same range. Since each state of frequency ω carries energy $\hbar\omega$, this is given by $\rho(\omega)\, d\omega = \hbar\omega\, S(\omega)\, F(\omega)\, d\omega$, or

$$\rho(\omega)\, d\omega = [(\hbar n^3\omega^3/\pi^2c^3)/\{\exp(\hbar\omega/kT) - 1\}]\, d\omega \tag{11.20}$$

This is the famous *black-body radiation law* mentioned earlier, and its agreement with experiment helped establish the quantum theory. As we shall see later, we can also interpret the existence of radiation in terms of the interaction of three processes for the absorption and emission of light. In equilibrium these balance, so that the radiation density is stabilized at the black-body level. However, by manipulating them away from their equilibrium rates, they may be exploited for device applications.

11.3 ELECTRONS

We now turn our attention to electrons. Having established their wave nature it would be natural to try to develop a suitable electron wave equation. In Chapter 2 the wave equation for light arose from Maxwell's equations. The problem here is that we do not have a corresponding set of 'electron laws'. However, recall that we used the electromagnetic wave equation to deduce the dispersion characteristic of light. By contrast, we already know the dispersion characteristic of electrons; it must be Eq. (11.5). An alternative route would therefore be to work backwards

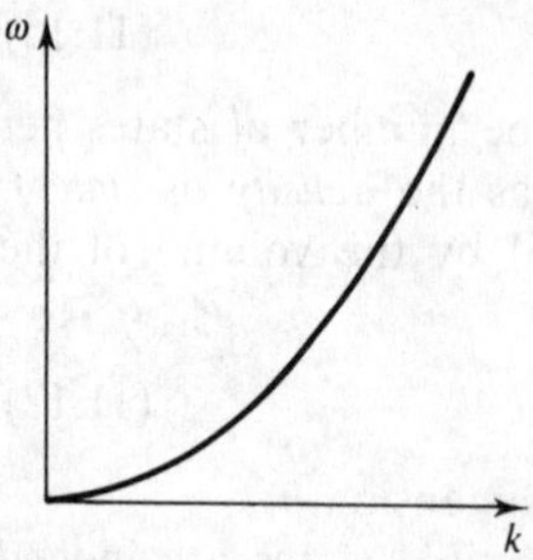

Figure 11.3 The dispersion characteristic of an electron.

from this. We start by plotting the dispersion characteristic (Fig. 11.3) for free electrons (i.e. with $\mathscr{V}$ = constant, say zero). Note that it is parabolic, and hence quite different from the linear characteristic of light waves, and that the curve might also represent the variation of electron energy with k (since $\mathscr{E} = \hbar\omega$). Remember that the dispersion characteristic provides a relation between ω and k for a travelling wave, which is itself descriptive of the electron in a uniform potential. In our earlier electromagnetic analysis, a z-propagating plane wave solution for the electric field was written in the vector form:

$$\mathbf{E}(z,t) = \mathbf{E}_0 \exp\{j(\omega t - kz)\} \tag{11.21}$$

For an electron wave we choose a similar type of solution, writing:

$$\mathbf{\Psi}(z,t) = \Psi_0 \exp\{j(\omega t - kz)\} \tag{11.22}$$

where the quantity $\mathbf{\Psi}$ (known as the *wavefunction*) describes some characteristic of electrons that is roughly analogous to the electric field, but is now a scalar. Differentiating with respect to t, we find

$$\partial\mathbf{\Psi}/\partial t = j\omega\Psi_0 \exp\{j(\omega t - kz)\} = j\omega\mathbf{\Psi} \tag{11.23}$$

while differentiating with respect to z gives

$$\partial\mathbf{\Psi}/\partial z = -jk\,\mathbf{\Psi} \quad \text{and} \quad \partial^2\mathbf{\Psi}/\partial z^2 = -k^2\mathbf{\Psi} \tag{11.24}$$

We can now form a wave equation from the dispersion relation by substituting for ω and k^2. If this is done we get

$$j\hbar\partial\mathbf{\Psi}/\partial t = \hbar^2/2m\ \partial^2\mathbf{\Psi}/\partial z^2 - \mathscr{V}\mathbf{\Psi} \tag{11.25}$$

We could easily extend this argument to three dimensions, where we would find

$$j\hbar\partial\mathbf{\Psi}/\partial t = \hbar^2/2m\ \nabla^2\mathbf{\Psi} - \mathscr{V}\mathbf{\Psi} \tag{11.26}$$

This equation was first derived by Erwin Schrödinger in 1926, and is known as the *time-dependent Schrödinger wave equation*.

One obvious question is, what does the wave function $\mathbf{\Psi}$ mean? Since the Schrödinger equation is complex, it is a reasonable guess that $\mathbf{\Psi}$ will also be complex, and therefore of no direct physical significance. In fact, the exact interpretation of $\mathbf{\Psi}$ was the cause of some controversy. After much debate it was concluded that the quantity $\Psi^*\Psi\, dv$ (or $|\Psi|^2\, dv$) could be taken to represent the probability that an electron will be found in the volume dv. This situation is not so different from that occurring in electromagnetic theory. We may feel more familiar and comfortable with the fields E and H, and hence tend to regard them as real and observable, but when we want to observe or detect an electromagnetic wave we have to extract energy from it, and the measure of energy content is proportional to $|E|^2$ (or $|H|^2$). We are therefore inevitably sampling the field intensities, much as we sample $|\Psi|^2$. In order to obtain the realistic result that the probability of finding a given electron somewhere in all space must be unity, the wavefunction must be normalized, so that

$$\iiint_V |\Psi|^2\, dv = 1 \tag{11.27}$$

Clearly the value of $|\Psi|^2\, dv$ is the same everywhere for the wave of Eq. (11.22). We therefore conclude that a single electron wave is spread throughout space; in order to model a mobile, localized electron we must again group a number of plane wave components together into a wavepacket. This will travel at the group velocity, which may be found by differentiating the $\mathscr{E}$–k characteristic to get

$$v_g = d\omega/dk = \hbar k/m = \sqrt{(2\mathscr{E}/m)} \tag{11.28}$$

A moment's thought shows that this is identical to the speed of our earlier particle-like electron, showing that our new model is consistent with the old. Note, however, that the speed of the electron wavepacket is variable, and depends on its energy. Consequently, when we increase the kinetic energy of an electron (e.g. by accelerating it in an electric field), its velocity increases monotonically.

The Schrödinger wave equation thus fills the role played by the electromagnetic wave equation that we have used so far. Although we have based the derivation on a simple travelling wave, we would expect that other solutions will describe electron behaviour under more complicated boundary conditions. For example, we would expect the 'trapping' of an electron wave to be possible since we have already seen that an optical wave may be confined in a waveguide. Here the details of the guiding structure were introduced through a variation of the refractive index; for an electron, we now have variations in the potential energy caused by some distribution of charge or electric potential.

Before examining this aspect we adopt a simplification used earlier, namely to eliminate any time-dependence by assuming a harmonic solution. We therefore write

$$\Psi(x,y,z,t) = \Psi(x,y,z) \exp(j\omega t) \tag{11.29}$$

If this is done, the Schrödinger wave equation reduces to its time-independent form

$$\nabla^2\Psi(x,y,z) + (2m/\hbar^2)\{\mathscr{E} - \mathscr{V}\}\Psi(x,y,z) = 0 \tag{11.30}$$

Electron in a Potential Well

We shall now attempt to find solutions to Eq. (11.30) for a particular situation analogous to an electromagnetic problem considered in Chapter 6, the propagation

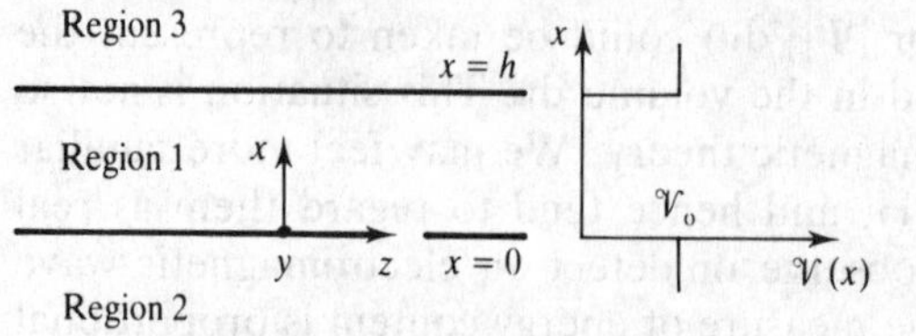

Figure 11.4 Potential energy distribution forming a one-dimensional well.

of an optical wave in a slab guide. This time, however, we will examine the behaviour of an electron wave confined to a one-dimensional *potential well*, and propagating in the z-direction. To describe the well we will assume a potential energy distribution of the form $\mathscr{V} = 0$ for $0 \leq x \leq h$; $\mathscr{V} = \mathscr{V}_0$ otherwise. The geometry is shown in Fig. 11.4; this distribution is analogous to the index variation of a symmetric slab guide if decreased potential is taken to correspond to increased refractive index.

In this geometry there will be no y-variation in the solution, so that $\partial\Psi/\partial y = 0$. The solution we assume in each of the three regions may then be taken in the separated form:

$$\Psi_i(x,z) = \psi_i(x) \exp(-j\beta z) \tag{11.31}$$

With this substitution Eq. (11.30) reduces to

$$d^2\psi_i/dx^2 + [(2m/\bar{h}^2)\{\mathscr{E} - \mathscr{V}_i\} - \beta^2]\psi_i = 0 \tag{11.32}$$

in each region. This is similar to the waveguide equation for the dielectric guide, which had the form:

$$d^2E_i/dx^2 + (n_i^2k_0^2 - \beta^2)E_i = 0 \tag{11.33}$$

In fact, the analogy is perfect if we take the term $(2m/\bar{h}^2)\{\mathscr{E} - \mathscr{V}_i\}$ to correspond to $n_i^2k_0^2$. We may therefore assume the features of the likely solutions directly from previous results. If the term $(2m/\bar{h}^2)\{\mathscr{E} - \mathscr{V}_i\}$ is greater than β^2, the solution for $\psi_i(x)$ will be in the form of sines and cosines; otherwise, it will be exponential. We would expect the former situation to occur inside the well (where $\mathscr{V} = 0$) when $(2m/\bar{h}^2)\mathscr{E} > \beta^2$. In region 1, therefore, we could put

$$\psi_1(x) = \psi_0 \cos(\kappa x - \phi) \tag{11.34}$$

with

$$\kappa = \sqrt{\{(2m/\bar{h}^2)\mathscr{E} - \beta^2\}} \tag{11.35}$$

However, when $\mathscr{V}_i$ is large (e.g. outside the well) we might find $(2m/\bar{h}^2)\{\mathscr{E} - \mathscr{V}_i\} < \beta^2$, so the solution will be an exponential instead. In regions 2 and 3 we would then expect the solutions:

$$\psi_2(x) = \psi_0' \exp(\gamma x) \qquad \psi_3(x) = \psi_0'' \exp\{-\gamma(x - h)\} \tag{11.36}$$

with

$$\gamma = \sqrt{[\beta^2 - (2m/\bar{h}^2)\{\mathscr{E} - \mathscr{V}_0\}]} \tag{11.37}$$

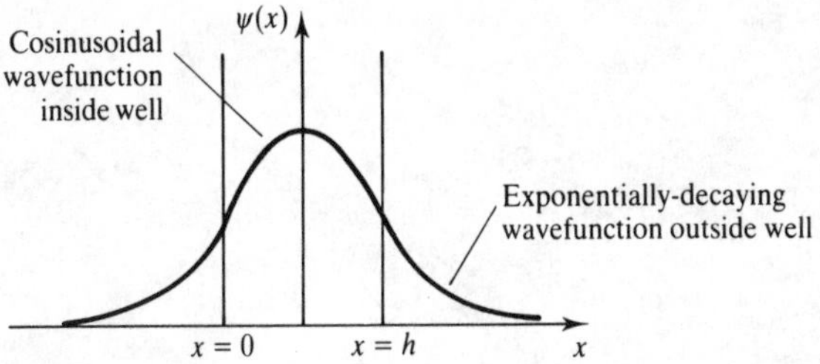

Figure 11.5 Variation of the wavefunction for the lowest-energy bound solution in a potential well.

The solutions may be matched at $x = 0$ and $x = h$ by imposing the reasonable requirement that ψ and $d\psi/dx$ must be continuous. Since the mathematics is identical to that used in Chapter 6, it should be obvious that bound solutions will exist, provided the eigenvalue equations $\tan(\kappa h/2) = \gamma/\kappa$ or $\tan(\kappa h/2) = -\kappa/\gamma$ are satisfied. Just as we found for the slab guide, only a discrete number of solutions can be found, each corresponding to a state in which the electron is 'trapped' inside the well. A typical solution—with the lowest energy— is shown in Fig. 11.5. Clearly the probability of finding the electron is highest inside the well, and diminishes rapidly outside it, although there is a finite expectation of locating it close to the well. States with a higher energy (analogous to higher-order guided modes) penetrate the region outside the well to a greater extent. In fact, if the energy is high enough, the electron is not confined at all, but is free to roam. This is analogous to the onset of cut-off in a waveguide, and the unbound states are roughly equivalent to radiation modes. We will not examine these in detail; instead, we focus our attention on the energy levels of the bound states.

In Fig. 11.5, the exponential decay of the wavefunction outside the well is defined by γ, which in turn depends on β, $\mathscr{E}$ and $\mathscr{V}_0$. If we assume additionally that $\mathscr{V}_0$ is very large (tending to infinity) γ will be so great that we can take ψ to be zero outside the well. This geometry is known as the *infinite potential well.* Since the electron is now completely confined, the boundary conditions resemble those of the metal-walled guide, and the eigenvalue equations reduce to

$$\sqrt{\{(2m/\hbar^2)\,\mathscr{E} - \beta^2\}}h = \nu\pi \tag{11.38}$$

We then obtain for the possible electron energies:

$$\mathscr{E} = \nu^2(\pi^2\hbar^2/2mh^2) + \hbar^2\beta^2/2m \tag{11.39}$$

These results are plotted in Fig. 11.6; they suggest that the energies are quantized and can take only specific, discrete values. Note that the levels are not equally separated—the energy difference between the νth and $(\nu+1)$th levels increases rapidly with ν. Furthermore, the separation of the levels depends inversely on the size of the well, and is only apparent for small values of h; as $h \to \infty$, the energies revert to those of a free electron. In the absence of motion in the z-direction (i.e. when $\beta \approx 0$) the energy is entirely potential. In this case Eq. (11.39) reduces to

$$\mathscr{E} = \nu^2(\pi^2\hbar^2/2mh^2) \tag{11.40}$$

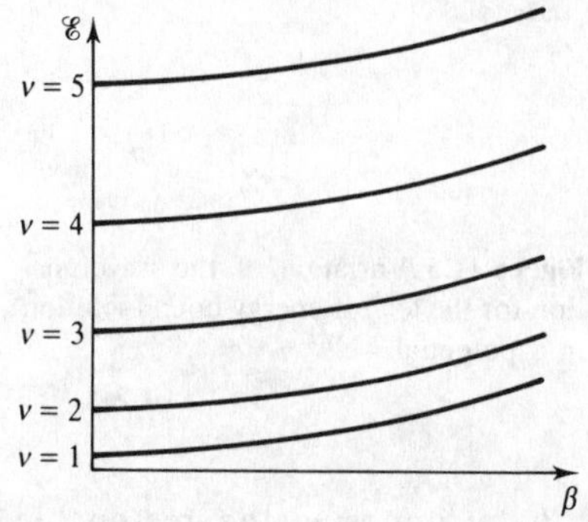

Figure 11.6 The energy levels of an electron in an infinite potential well.

However, as β increases, the total energy rises, since there are now kinetic and potential contributions.

Design Example

We can estimate the size of well for which quantization effects become significant, by comparing the value of the term $\pi^2\bar{h}^2/2mh^2$ with the thermal energy of a free electron. From the principle of equipartition of energy, the latter will be $\mathscr{E}_{th} = (3/2)kT$, where k is *Boltzmann's constant* (1.38×10^{-23} J/K). At room temperature (when $T \approx 300$ K), $\mathscr{E}_{th} = 6.21 \times 10^{-21}$. This value is reached by $\pi^2\bar{h}^2/2mh^2$ when $h = \surd\{(6.62 \times 10^{-34})^2/(8 \times 9.1 \times 10^{-31} \times 6.2 \times 10^{-21})\} = 3.1 \times 10^{-9}$ m. Quantization effects will thus be important for well sizes approaching interatomic distances.

Two- and Three-dimensional Effects

In a one-dimensional well we cannot easily restrict electrons to propagate in the z-direction alone, so that motion in the y-direction should really be allowed. The first question we might ask is, does this make any difference? Not really; the analysis above can be repeated assuming the more general solution $\Psi(x,y,z) = \psi(x) \exp\{-j(\beta_y y + \beta_z z)\}$; if this is done, we reach similar conclusions, with $\beta^2 = \beta_y{}^2 + \beta_z{}^2$. Equally, we might wish to confine the electron in more than one direction. For example, we might have an infinite well, in the form of a cube of sides h in each of the x-, y-, and z-directions (as in our earlier Fig. 11.2). This does make more of a difference, as we now show.

In this case we must satisfy the equation

$$\nabla^2\Psi(x,y,z) + \{(2m/\bar{h}^2)\,\mathscr{E}\}\,\Psi(x,y,z) = 0 \qquad (11.41)$$

inside the well, subject to the boundary conditions of $\Psi = 0$ on all faces of the cube. This situation is again analogous to the three-dimensional Fabry-Perot resonator, so we should look for standing-wave solutions in each of the three directions. We might put

$$\Psi(x,y,z) = \psi_0\, \psi_x(x)\, \psi_y(y)\, \psi_z(z) \tag{11.42}$$

where ψ_0 is a constant, and

$$\psi_x = \sin(k_x x) \qquad \psi_y = \sin(k_y y) \qquad \psi_z = \sin(k_z z) \tag{11.43}$$

Substituting into Eq. (11.41), we see that this solution will be valid, provided

$$k_x^2 + k_y^2 + k_z^2 = k^2 = (2m/\hbar^2)\,\mathscr{E} \tag{11.44}$$

Similarly, the boundary conditions will be satisfied if we choose

$$k_x = \nu_x \pi/h \qquad k_y = \nu_y \pi/h \text{ and } k_z = \nu_z \pi/h \tag{11.45}$$

This yields the final result for the allowed energies:

$$\mathscr{E} = (\nu_x^2 + \nu_y^2 + \nu_z^2)\,(\pi^2\hbar^2/2mh^2) \tag{11.46}$$

This result should be compared with Eq. (11.40); clearly confinement in each direction has contributed a similar amount to the total energy. However, there is a new feature—the spacing of the levels is different. While the single index ν^2 may only have the values 1, 4, 9 and so on, the term $\nu_x{}^2 + \nu_y{}^2 + \nu_z{}^2$ may assume the values 1, 2, 3, 4, 5, 6, ... depending on the exact choices of ν_x, ν_y and ν_z. Interestingly, some of the values of $\nu_x^2 + \nu_y^2 + \nu_z^2$ may be made up in more than one way. For example, a total of three may be obtained only if $\nu_x = 1$, $\nu_y = 1$, $\nu_z = 1$, but six may be obtained from $\nu_x = 1$, $\nu_y = 1$, $\nu_z = 2$, or from $\nu_x = 1$, $\nu_y = 2$, $\nu_z = 1$, or from $\nu_x = 2$, $\nu_y = 1$, $\nu_z = 1$. As a result, different states may now have the same energy; such states are described as *degenerate*. Furthermore, as the energy rises, the number of states with the same (or nearly the same) energy will also increase. We will return to this later when we consider the density of states for electrons.

Electronic Transitions

It might be argued that the quantization of energy levels in a potential well is of purely academic interest. However, we do not need to add many refinements to describe situations of fundamental importance. For example, the function:

$$\mathscr{V}(r) = -e^2/4\pi\varepsilon_0 r \tag{11.47}$$

describes the potential energy of an electron at a distance r from a single positive charge, e.g. a proton. This is clearly a three-dimensional well, albeit of a more complicated form. If we now substitute Eq. (11.47) into the Schrödinger equation and look for bound solutions, we will actually determine the electronic energy levels of a single hydrogen atom. As it turns out, this problem has a complete analytic solution. Most importantly, differences between the energy levels can be shown to correspond exactly to the energies of photons that are absorbed or emitted by hydrogen, demonstrating not only that the levels themselves are substantially correct, but also that the absorption and emission of radiation may be linked directly to electronic transitions. This is illustrated by Fig. 11.7, which shows the interaction

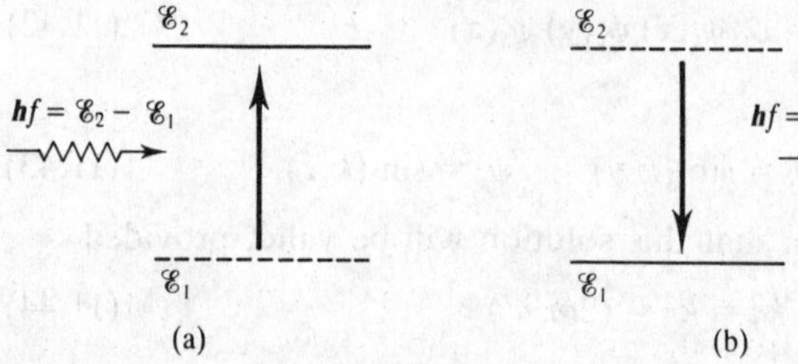

Figure 11.7 (a) Absorption and (b) emission of radiation, modelled by electronic transitions.

of a photon with a simple two-state system. If an incident photon has energy $hf = \mathscr{E}_2 - \mathscr{E}_1$, it may cause an electron to jump from the low energy state $\mathscr{E}_1$ to the high energy state $\mathscr{E}_2$, being itself absorbed in the process (Fig. 11.7(a)). Similarly, a transition in the reverse direction may cause the emission of a photon, again of energy $hf = \mathscr{E}_2 - \mathscr{E}_1$ (Fig. 11.7(b)). This is known as *spontaneous emission*.

This view of absorption and emission is clearly rather different from the mechanistic model previously given in Chapter 3. Most importantly, it was found necessary to postulate the existence of a third process in order to explain the black-body radiation law. Here an incident photon of the correct energy may induce a downward electronic transition, resulting in the emission of a photon (Fig. 11.8). Since the original photon is not absorbed, the result is that two photons actually emerge.

At first sight this is a very strange effect, since we might expect that when a photon hits an electron in the state $\mathscr{E}_2$ it would force the electron to move to a still higher state (say, $\mathscr{E}_3$). To appreciate why this does not occur we must remember that the photon is essentially a wave train of electric field with some characteristic frequency, and that the electron states $\mathscr{E}_1$ and $\mathscr{E}_2$ are simply normal modes of the unperturbed atom. If these modes are perturbed by interacting with an oscillating field, then mode coupling will occur strongly if the frequency of the field matches the difference in frequency between the two modes. This is clearly analogous to the mode coupling situations encountered in Chapter 10, where a spatially periodic perturbation was used to phase match two propagating modes.

Since the new photon has been emitted in response to the stimulation of the driving field, the phases of the incident and stimulated photons must be linked. In fact, they are identical, so that the two photons are effectively joined together in a single wavetrain. We may describe this as a 'doubly occupied photon mode'. This

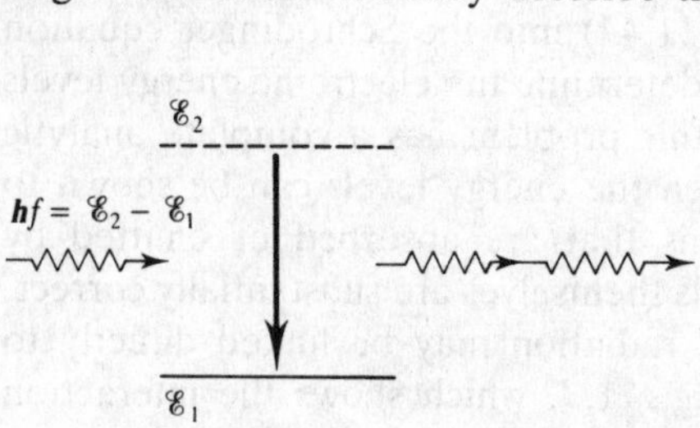

Figure 11.8 Stimulated emission.

effect, known as *stimulated emission*, is the key to the amplification of light in optoelectronic devices. Spontaneous emission can also be interpreted in terms of coupling between higher and lower states, except that in this case the stimulating perturbation is not derived from an external agency (an incident photon) but from fluctuations in the local field within the atom itself.

For the other elements, a similar model may be constructed. For example, helium has two protons and two electrons, so each electron will derive potential energy from its companion, as well as from the protons in the nucleus. Unfortunately, the relevant mathematical description turns out to be of the type known as a many-body problem, which cannot be solved analytically. In fact, the higher the atomic number of the element, the worse the difficulty. However, through suitable approximations, quantum mechanics has allowed good estimates of the energy levels of all the elements to be made.

Two additional refinements are needed to account completely for all the spectroscopic observations. First, some energy levels are found experimentally to be split into two closely-spaced levels. This feature may be explained by attributing to electrons an additional property known as *spin*, which may have one of two states, spin-up and spin-down. Originally, the physical description of spin was mechanistic—electrons were assumed to spin about their own axes—but no such analogy really exists; it is simplest to view spin as a vehicle for increasing the number of possible states to conform with experiment. Second, to account for details of the actual occupation of the levels, it is necessary to assume that there can only be one electron in any given state at any time. This idea was first postulated by W. Pauli in 1925, and is known as the *exclusion principle*.

11.4 BAND THEORY

The arguments above apply to isolated potential wells or to single atoms, however, most electronic and optoelectronic devices are based on crystalline solids. Here, the electron waves are propagating in a lattice, i.e. in a medium that contains a regular periodic array of fixed, positively-charged nuclei, each of which generates its own potential distribution. To model this geometry, we must somehow combine the results for single wells with the additional feature of periodicity.

To do so, we must in general express $\mathcal{V}$ as a periodic function of all three coordinates. Solution of the resulting Schrödinger equation is extremely difficult, forcing some simplifications. First, we shall assume a one-dimensional variation of potential energy, $\mathcal{V}(x)$. Second, we might guess that any new features in the solution will emerge from the periodic nature of $\mathcal{V}(x)$, and not from its detailed form. In this, we follow Krönig and Penney, who gave a particularly simple explanation of the behaviour of electrons in crystals—known (naturally) as the *Krönig-Penney model*—in 1931. Since any periodic variation will do, we will use the simplest consistent with the physics. For a one-dimensional lattice, the potential energy of an electron will be maximum exactly between the ions, falling to a minimum at each ion. We shall therefore adopt the potential distribution of Fig. 11.9.

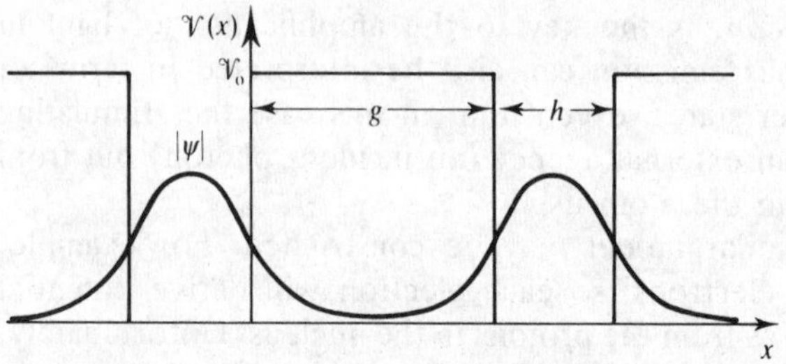

Figure 11.9 Variation of $\mathcal{V}(x)$ used in the Krönig-Penney model.

The potential energy near each ion is modelled as a well of height $\mathcal{V}_0$ and width h. The edges of neighbouring wells are g apart, so that the lattice periodicity is $L = g + h$. For simplicity, we shall ignore electron motion in the y- and z-directions, and concentrate on the effect of the lattice on the variation of the wavefunction in the x-direction. The equation that must be solved is therefore

$$d^2\psi(x)/dx^2 + [(2m/\hbar^2)\{\mathcal{E} - \mathcal{V}\}]\,\psi(x) = 0 \tag{11.48}$$

with $\mathcal{V} = 0$ inside the wells and $\mathcal{V} = \mathcal{V}_0$ outside.

We have tackled wave propagation problems in a layered structure many times before. In each case a solution was first found for each layer in turn; the solutions were then matched at the interfaces using suitable boundary conditions. The difficulty here is that we will have to solve an infinite number of simultaneous equations since there are an infinite number of layers. Clearly we must try to find some relation between the values of the wavefunction at different points along the lattice, which will reduce the number of equations. This was originally done by Bloch in 1928, who based his analysis on a mathematical theorem governing the solutions of differential equations with periodically-varying coefficients (known as *Floquet's theorem*). The physical argument runs like this: since the lattice is infinite, a displacement of the coordinate origin by exactly one period can make no difference to the boundary conditions. Consequently, if $\psi(x)$ is a solution, $\psi'(x) = \psi(x + L)$ must also be one. The two can only be related by a simple constant, so we may write

$$\psi(x + L) = C\psi(x) \tag{11.49}$$

Similarly, for a displacement of more than one period (say, N whole periods), we may write

$$\psi(x + NL) = C^N\,\psi(x) \tag{11.50}$$

This follows from repeated application of Eq. (11.49). In order for the solution to be finite everywhere we must have $|C^N| = 1$. Consequently, C can only be a complex number of modulus unity, so that $C = \exp(j\phi)$. As a result C^N must be of the form $C^N = \exp(j\phi N)$. If we now define a new quantity k as $k = \phi/L$, Eq. (11.50) can be written as

$$\psi(x + NL) = \exp(jkNL)\,\psi(x) \tag{11.51}$$

Since NL is a distance, this can be seen to be a strange type of travelling electron wave if k is interpreted as the wavenumber. Because $|\psi|$ is not constant, it is not a plane wave, but it has features that are clearly wave-like. For example, within each period, the wave-functions are all similar to $\psi(x)$, but each solution is shifted in phase by a progressive amount along the lattice, and the pattern repeats itself after a distance $\lambda = 2\pi/k$. Most importantly, Eq. (11.51) is the desired connection between the value of ψ at two different points along the lattice. We may now return to our original problem.

We begin by assuming that the energy $\mathscr{E}$ is less than $\mathscr{V}_0$, so that electrons tend to become trapped inside the potential wells. In these regions we would therefore expect the wavefunction to be a standing-wave pattern. Locating the x-origin conveniently at the right-hand edge of one of the wells, we might therefore guess a solution inside the well (i.e. for $-h < x < 0$) of

$$\psi_1(x) = \psi_0 \exp(\mathrm{j}\kappa x) + \psi_0' \exp(-\mathrm{j}\kappa x) \tag{11.52}$$

where ψ_0 and ψ_0' are constant coefficients, and

$$\kappa = \sqrt{\{(2m/\hbar^2)\mathscr{E}\}} \tag{11.53}$$

Using Eq. (11.51) we can write the wavefunction in the next well along (i.e. for $g < x < L$) in terms of this solution, as $\psi_1'(x) = \exp(\mathrm{j}kL)\ \psi_1(x - L)$, or

$$\psi_1'(x) = \exp(\mathrm{j}kL)\,[\psi_0 \exp\{\mathrm{j}\kappa(x - L)\} + \psi_0' \exp\{-\mathrm{j}\kappa(x - L)\}] \tag{11.54}$$

The solution inside other wells may be found in the same way.

Outside each well, on the other hand, we can guess that the solution will be exponentially-varying. However, since the wells are close together, we would expect that an *electronic tunelling*, analogous to the frustrated total-internal reflection occurring in similar optical geometries, might take place between them. It is then prudent to allow a solution in the form of exponentially-growing and decaying waves. In the region to the right of the well of interest (i.e. for $0 < x < g$) we might put

$$\psi_2(x) = \psi_0'' \exp(\gamma x) + \psi_0''' \exp(-\gamma x) \tag{11.55}$$

where ψ_0'' and ψ_0''' are further constants, and

$$\gamma = \sqrt{[(2m/\hbar^2)\{\mathscr{V}_0 - \mathscr{E}\}]} \tag{11.56}$$

We can now match the solutions at the boundaries. At $x = 0$, for example, we must equate ψ_1 with ψ_2, and $\mathrm{d}\psi_1/\mathrm{d}x$ with $\mathrm{d}\psi_2/\mathrm{d}x$. This yields

$$\psi_0 + \psi_0' = \psi_0'' + \psi_0'''$$
$$\mathrm{j}\kappa\{\psi_0 - \psi_0'\} = \gamma\{\psi_0'' - \psi_0'''\} \tag{11.57}$$

Similarly, at $x = g$, we must equate ψ_1' with ψ_2 and $\mathrm{d}\psi_1'/\mathrm{d}x$ with $\mathrm{d}\psi_2/\mathrm{d}x$. If this is done, we get

$$\exp(\mathrm{j}kL)\,\{\psi_0 \exp(-\mathrm{j}\kappa h) + \psi_0' \exp(+\mathrm{j}\kappa h)\} = \psi_0'' \exp(\gamma g) + \psi_0''' \exp(-\gamma g)$$
$$\mathrm{j}\kappa \exp(\mathrm{j}kL)\,\{\psi_0 \exp(-\mathrm{j}\kappa h) - \psi_0' \exp(+\mathrm{j}\kappa h)\} = \gamma\{\psi_0'' \exp(\gamma g) - \psi_0''' \exp(-\gamma g)\} \tag{11.58}$$

Equations (11.57) and (11.58) represent four equations, which may be solved for the four unknowns ψ_0, ψ_0', ψ_0'' and ψ_0'''. It can be shown that in any such system of simultaneous equations a non-trivial solution exists only if the determinant of the matrix of the equation coefficients is zero. In this case we find (after some manipulation) that this condition is satisfied when

$$\cos(kL) = \cos(\kappa h)\cosh(\gamma g) + \{(\gamma^2 - \kappa^2)/2\kappa\gamma\}\sin(\kappa h)\sinh(\gamma g) \quad (11.59)$$

Since the electron energy $\mathscr{E}$ depends on κ and γ, Eq. (11.59) is a form of dispersion equation, relating to $\mathscr{E}$ to k. Hence it should reveal any modifications to the dispersion characteristic imposed by the lattice. Assuming it is soluble, we might return to Eqs (11.57) and (11.58) to find the values of ψ_0, ψ_0', ψ_0'' and ψ_0'''. If this is done it should be no suprise that the overall solution is as shown in Fig. (11.9), where we have plotted the variation of $|\psi|$ along the lattice. This repeats periodically, as expected from Eq. (11.51). Not unnaturally, the 'strange waves' (or *Bloch waves*) peak in the neighbourhood of the ion cores, where electrons are most likely to be found.

As it turns out, Eq. (11.59) is most easily interpreted graphically. However, before doing this, we shall make one simplification. We allow $\mathscr{V}_0$ to become very large and g to tend to zero, in such a way that the product $\mathscr{V}_0 g$ remains finite. In the process, the following limiting behaviour occurs: $h \to L$ and $\gamma g \to 0$, while $\gamma^2 g$ remains constant and much greater than $\kappa^2 g$. Equation (11.59) then tends to

$$\cos(kL) = \cos(\kappa L) + P\sin(\kappa L)/(\kappa L) \quad (11.60)$$

Where P is a new parameter, given by:

$$P = \gamma^2 gL/2 = m\mathscr{V}_0 gL/\hbar^2 \quad (11.61)$$

Equation (11.60) may now be interpreted as follows. The right-hand side can take a wide range of values, depending on the exact values of P and κL. However, the left-hand side is limited to the region between -1 and $+1$ (for real k). For a given value of P, we conclude that there will only be a restricted range of κL (and hence of $\mathscr{E}$) for which a solution can be found. This is illustrated in Fig. 11.10, which shows the right-hand side of Eq. (11.60) plotted against κL, for $P = 10$.

Only the shaded areas allow solutions for real k. For large P, these are centred on the zeros of the function $\sin(\kappa L)/(\kappa L)$, i.e. on $\kappa L \approx \nu\pi$. Breaks in the solution

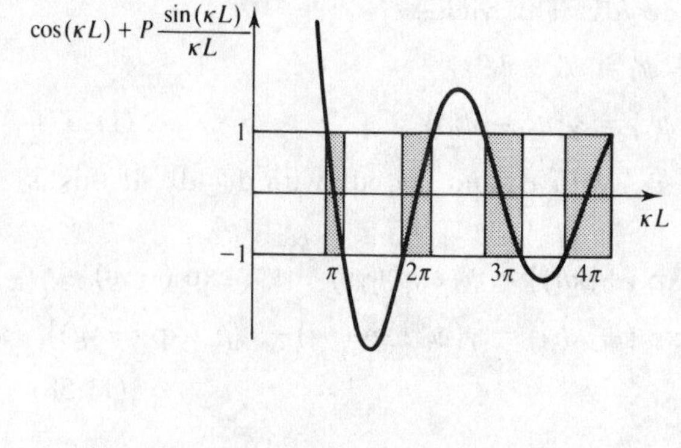

Figure 11.10 A plot of the function $\cos(\kappa L) + P\sin(\kappa L)$ against κL, for $P = 10$.

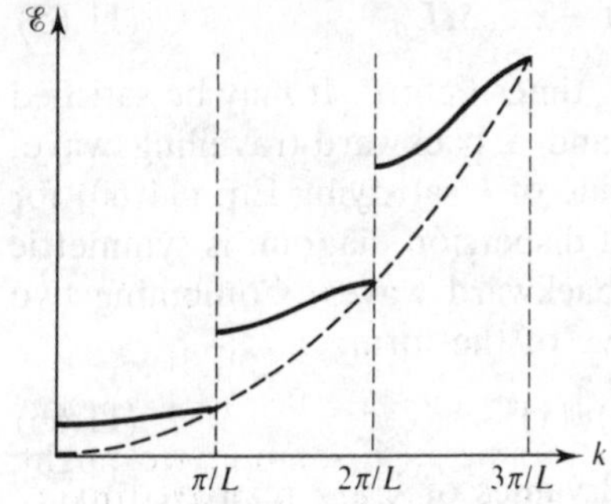

Figure 11.11 ℰ–k diagram for the Krönig-Penney potential, with $P = 10$.

occur when the right-hand side of Eq. (11.60) lies outside the limits of ± 1, i.e. when $kL = \nu\pi$. The corresponding ℰ–k diagram is shown in Fig. 11.11. As predicted, it is discontinuous, and there are ranges of energy with no solutions. The widths of these 'forbidden bands' depend on the value of P, but their positions do not; they lie at $k = \nu\pi/L$. The diagram combines some of the features of both free and bound electrons—it has an overall shape that is parabolic, but nonetheless only discrete *bands* of energy are allowed.

We can see why this must be so by considering Eq. (11.60) in the limits of small and large P. For small P (when the barriers are small) it approximates to $\cos(kL) = \cos(\kappa L)$, which has the trivial solution $\kappa = k$. The energy is then found from Eq. (11.53) as

$$\mathcal{E} = \hbar^2 k^2/2m \tag{11.62}$$

This is the parabolic ℰ–k characteristic of a free electron, and is shown as a dashed line in Fig. 11.11. Similarly, if P is very large (when the barriers are much greater), Eq. (11.60) tends to

$$P \sin(\kappa L)/(\kappa L) = 0 \tag{11.63}$$

This has the solution $\kappa = \nu\pi/L$, with the corresponding allowed energy levels:

$$\mathcal{E} = \nu^2 \hbar^2 \pi^2/(2mL^2) \tag{11.64}$$

These should be compared with the solutions previously given for an isolated well (Eq. (11.40); the two are clearly identical if the width of the well (h) is set equal to L.

The Effect of a Finite Crystal Lattice

If the crystal is infinitely large, then the sections of dispersion characteristic outside the forbidden bands are continuous lines, and there are no further restrictions on the possible values of ℰ and k. Any real crystal will have finite dimensions, however. We can model this through the introduction of additional boundary conditions. For our 1-D lattice, the length of an M-atom crystal will be ML. Since we would expect electrons to be confined inside the crystal, we might arbitrarily require

$$\psi(x) = 0 \quad \text{at} \quad x = 0 \quad \text{and at} \quad x = ML \tag{11.65}$$

We have met this type of boundary condition many times before. It may be satisfied by assuming a solution as a sum of a forward- and a backward-travelling wave, which combine into a standing-wave pattern. A value of k satisfying Eq. (11.60) for a given energy $\mathscr{E}$ is also satisfied by $-k$, so the full dispersion diagram is symmetric about $k = 0$, with the left-hand half describing backward waves. Combining two waves of the same energy will give a standing wave of the form:

$$\psi(x + NL) = \sin (kNL)\psi(x) \tag{11.66}$$

Clearly Eq. (11.65) then requires that the allowed values of k are restricted to

$$k = \nu\pi/ML \tag{11.67}$$

Consequently, the dispersion characteristic cannot be continuous, but must consist of discrete points. In each band (which has width π/L) there will be M states, each representing one standing-wave solution. Since M is likely to be large, the energy levels are likely to be so close together that this granularity will be hard to detect; nonetheless, it is significant. The effect of incorporating M atoms into a 1-D lattice is summarized in Fig. 11.12, which shows that each level of an isolated atom is split into a band of M states. Including spin, this figure rises to $2M$.

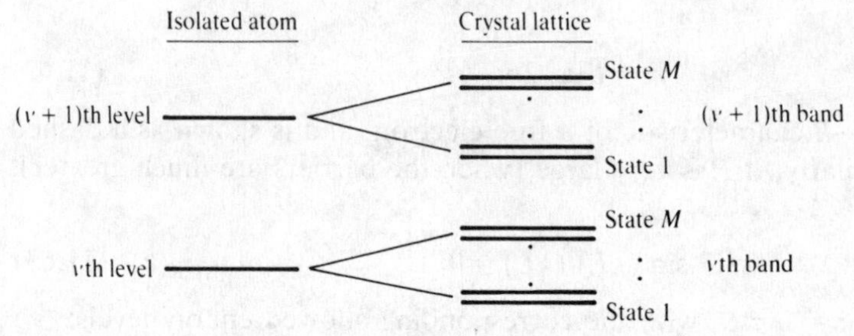

Figure 11.12 Comparison of the energy levels of an isolated atom and of a finite crystal lattice.

Reduced Dispersion Characteristics

We now consider briefly one further way of representing the dispersion characteristic. As we have seen, forward- and backward-travelling waves are of the form $\psi(x + NL) = \exp(\pm jkNL)\psi(x)$. However, since these functions are periodic in k, they are actually indistinguishable from the very similar solutions $\psi(x + NL) = \exp(+jk'NL)\,\psi(x)$, provided $k' = k + 2\nu\pi/L$. This shows that a very compact diagram may be drawn for both forward and backward waves, by taking suitable elements from the extended diagram and shifting them to the left or right by multiples of $2\pi/L$. If this is done, the *reduced dispersion characteristic* shown in Fig. 11.13 is obtained.

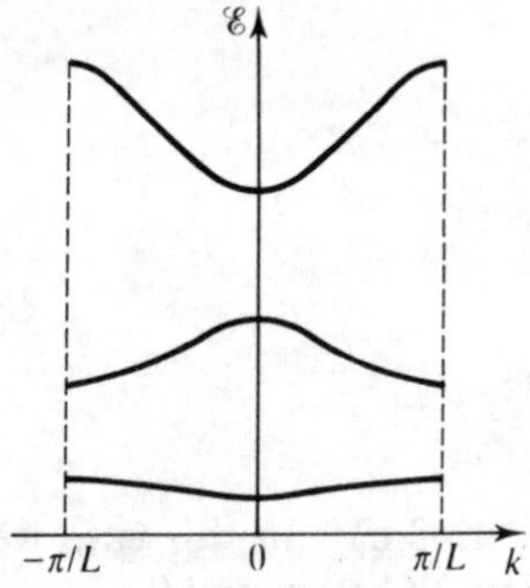

Figure 11.13 Reduced ℰ–k diagram for the Krönig-Penney potential.

Three-dimensional Crystals

Real crystals differ from this simple model in several respects. First, they are three-dimensional; to account for this we should allow a three-dimensional periodic potential, and consider the variation of ℰ with the three components of a vector $\mathbf{k}$, describing propagation in three-dimensional space. Second, the ion cores need not be located on a regular cartesian grid—their positions are determined by the particular lattice structure of the crystal. Third, the potential distribution near each ion will be more complicated than the one used above. Despite this, the broad conclusions remain qualitatively valid; the energy levels of a real crystal comprise a series of bands, the investigation of which has been a major task for solid state physicists. The results are normally represented as ℰ–k diagrams for motion in a given direction, or as surfaces of constant energy in k-space.

As we shall see in the next section, the exclusion principle dictates that most optical and electronic effects will be determined by the outer bands because low-energy bands are likely to be completely filled under normal conditions (so that no transitions may be made to them). The outermost and second outermost bands are known as the *conduction* and *valence* bands, respectively.

Direct-gap Materials

The important features are mostly determined by the shape of the ℰ–k diagram near the bottom of the conduction band and the top of the valence band. One particular result, which cannot really be predicted at all by our current one-dimensional model, is that the value of k required for maximum energy in the valence band may or may not correspond to that for minimum energy in the conduction band. This is illustrated in the two possible ℰ–k diagrams of Fig. 11.14. In Fig. 11.14(a) the two bands lie one above the other at their point of closest approach. Crystals with this type of band-structure are known as *direct-gap* materials.

We would expect the absorption and emission of radiation in crystals to involve electronic transitions. For example, in Fig. 11.14(a), absorption of a photon of energy $\hbar\omega > \mathscr{E}_g$ can provide sufficient energy to promote an electron from a position near the top of the valence band into the conduction band. The III-V semiconductor

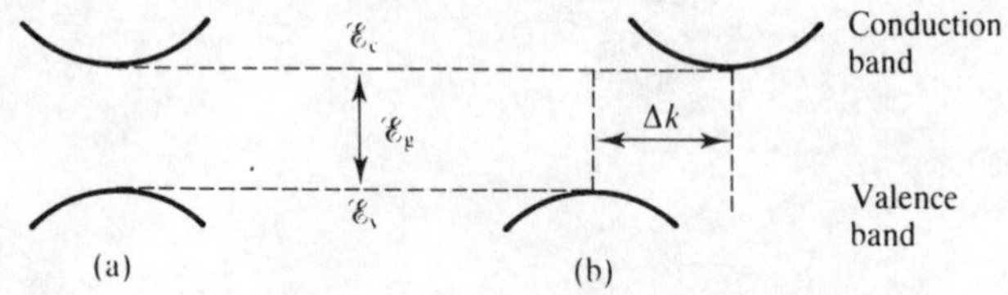

Figure 11.14 $\mathscr{E}$–k diagrams for (a) direct- and (b) indirect-gap materials.

GaAs is one example of a direct-gap material, with $\mathscr{E}_g \approx 1.42$ eV. In this case we would expect the wavelength of any absorbed or emitted light to be around 0.87 μm. Other materials of similar band structure can be made, with bandgaps chosen for the particular application. For example, the ternary alloy $GaAs_{1-x}P_x$ has an energy gap $\mathscr{E}_g \approx 1.42 + 1.24x$ eV in the range $0 \leq x \leq 0.45$ (after which the bandgap becomes indirect), so the corresponding optical wavelength may be tuned between 0.63 and 0.87 μm by changes in the alloy composition. $Ga_{1-x}Al_xAs$ has a similar linear energy gap variation up to $x = 0.35$. Using these materials, devices ranging from visible LEDs to near-infrared lasers may be constructed.

Figure 11.15 shows a contour map of the variation of $\mathscr{E}_g$ with composition for another important optoelectronic material, the quaternary alloy $In_{1-x}Ga_xAs_{1-y}P_y$. As can be seen, the variation is quite complicated. However, most devices are made from layers of InGaAsP grown on InP substrates. These layers must have compositions that are lattice-matched to InP for strain-free growth to be possible. As we will show in Chapter 13, the locus of all possible compounds meeting this condition is the approximate straight line linking InP to $In_{0.53}Ga_{0.47}As$. The overall materials system therefore allows the fabrication of devices with direct bandgaps

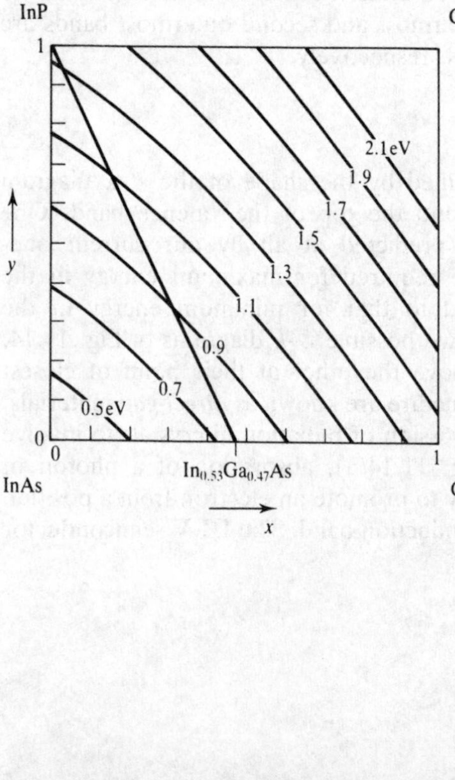

Figure 11.15 Variation of energy gap with alloy composition, for the InGaAsP materials system.

lying between $\mathscr{E}_g = 1.35$ eV (InP) and $\mathscr{E}_g = 0.74$ eV ($In_{0.53}Ga_{0.47}As$). The corresponding photon wavelengths are 0.92 μm and 1.67 μm. Sources emitting at the important wavelengths of $\lambda_0 = 1.3$ μm (where dispersion is minimum in silica fibres) and 1.55 μm (where loss is minimum) may thus be made using the InP/InGaAsP system.

Design Example

Optical absorption in a direct-gap material satisfies the principle of energy conservation, but Newton's laws also require conservation of momentum. While a photon carries a fixed momentum $p_{photon} = \hbar\omega/c$ in vacuum, an electron can have a momentum $p_{electron} = \hbar k$, whose exact value depends on the position of the electron in the $\mathscr{E}$–k diagram. What do the relative sizes of these quantities imply? We can estimate the momentum of a typical photon as follows. For a photon energy of 1 eV, p_{photon} is approximately 5×10^{-28} N s in vacuum. In a material of refractive index n, this rises by a factor of n. Taking $n \approx 4$, we have $p \approx 2 \times 10^{-27}$ N s. The variation of the electron wavelength between the bottom and top of a band is $\Delta p = \hbar\Delta k = \hbar\pi/L$. If we assume the lattice constant $L \approx 2$ Å, then $\Delta p = 1.6 \times 10^{-24}$ N s. Hence the momentum carried by a photon is approximately 1000 times less than the momentum change required when transferring an electron across a band. It follows that while visible photons have enough energy to carry electrons over the energy gap, their momenta are insufficient to translate them significantly in the k-direction. Electron transitions due to optical interactions *alone* therefore follow essentially vertical paths on the $\mathscr{E}$–k diagram.

Indirect-gap Materials

Other materials have different band structures. For example, in Fig. 11.14(b) there is an offset of Δk between the top of the valence band and the bottom of the conduction band. In this case the crystal is said to be *indirect-gap*; the Group IV elements Si and Ge are typical indirect-gap materials. Clearly a large momentum mismatch must be made up if optical absorption or emission is to occur. Where is this to come from? The answer is that it can be supplied through the agency of a phonon.

We have so far concentrated on the behaviour of electrons in the lattice, which was itself assumed to be fixed. However, the lattice is simply an arrangement of atoms held together by bonds, which are only approximately rigid. Consequently, it can vibrate if supplied with energy. Since it possible to describe any such vibrations in terms of travelling sound waves, the lattice vibrations are wave-like, and we might expect that they may also be quantized. This is indeed the case, and the elementary units of vibration are known as *phonons*. By analogy with photons, we can guess that a phonon carries an energy of $\mathscr{E}_{phonon} = \hbar\omega$, where ω is the angular frequency of the vibration, and a momentum of $p_{phonon} = \hbar k = \hbar\omega/v$, where v is the velocity of sound in the material. A typical upper value for ω might be 10^{13} rad/s and for v, 5×10^3 m/s, so that phonon energies of (say) 10^{-3} eV would have momenta of 3×10^{-26} N s. Comparing these figures with those calculated for photons, we see that phonons typically carry less energy but more momentum. A combination of photon and phonon emission or absorption can therefore allow conservation of both energy and momentum.

In Si and Ge, most interband electron transitions involving photons would also include a phonon. Since phonon absorption is relatively improbable at room temperature, such a transition does not have a high rate, so we should not expect significant light emission from indirect-gap materials. (However, in some instances, efficient emission *can* be obtained by providing impurity levels within the bandgap, in which electrons can be trapped. The impurity site is capable of absorbing any momentum imbalance. For example, in the $GaAs_{1-x}P_x$ system with $x > 0.45$, the band structure becomes indirect, but the addition of nitrogen can provide electron traps near the valence band edge, allowing effective emission down to the yellow part of the visible spectrum.) The band structures of a number of important optoelectronic materials are summarized in Table 11.1.

Table 11.1 Characteristics of a number of important semiconducting and insulating materials

Material	$\mathscr{E}_g$ (eV) @ 300 K	Band structure	Type
Ge	0.66	Indirect	Semiconductor
Si	1.12	Indirect	Semiconductor
GaP	2.2	Indirect	Semiconductor
InP	1.35	Direct	Semiconductor
GaAs	1.42	Direct	Semiconductor
C (diamond)	5.47		Insulator
SiO_2	8		Insulator

11.5 EFFECTIVE MASS

Other important characteristics of electrons may be deduced from the $\mathscr{E}$–k diagram. Recall that a free electron can be modelled as a wavepacket, travelling at the group velocity. Clearly we would like to describe motion of electrons in a crystal in a similar way, but this time we have found solutions to the wave equation as a vast array of stationary states. However, standing-wave solutions are merely sets of counterpropagating waves, and these travelling waves may also be grouped together into a wavepacket that can describe a partially localized electron. This time, motion of the packet amounts to a 'hopping' of the electron between the atoms. Most importantly, there is now a restriction on the acceleration of an electron by an electric field. As the electron gains energy, the constituents of the packet must jump to higher levels. However, the exclusion principle implies that this is possible only if the levels are unoccupied. If all are filled, there can be no conduction. Herein lies the distinction between conductors and insulators, as we will see later. Even if appropriate levels are available, the process of conduction is more complicated than might be expected, as we now show.

Returning to the reduced $\mathscr{E}$–k diagram of Fig. 11.13, we note that since the band structure is periodic, a reasonable approximation to the shape of the νth band is provided by the relation:

$$\mathscr{E}(k) = \mathscr{E}_0 - \Delta\mathscr{E} \cos(kL) \tag{11.68}$$

where $\mathscr{E}_0$ and $\Delta\mathscr{E}$ are constants. This function is shown in Fig. 11.16(a). Note that when k is small (i.e. near the bottom of the band), we can use a binomial expansion for the cosine, to get

$$\mathscr{E} - \mathscr{E}_{min} = \Delta\mathscr{E} k^2 L^2/2 \tag{11.69}$$

where we have defined $\mathscr{E}_{min}$ as the minimum energy level of the band. Consequently, the $\mathscr{E}$–k relation is approximately parabolic in this region.

The velocity of a wavepacket inside the crystal may again be found by differentiating the dispersion characteristic. In this case v_g is given by

$$v_g = (1/\hbar)\, d\mathscr{E}/dk = (L\, \Delta\mathscr{E}/\hbar) \sin(kL) \tag{11.70}$$

The resulting variation is shown in Fig. 11.16(b). Most interestingly, an increase in energy does not necessarily give an increase in v_g. We can explain this curious

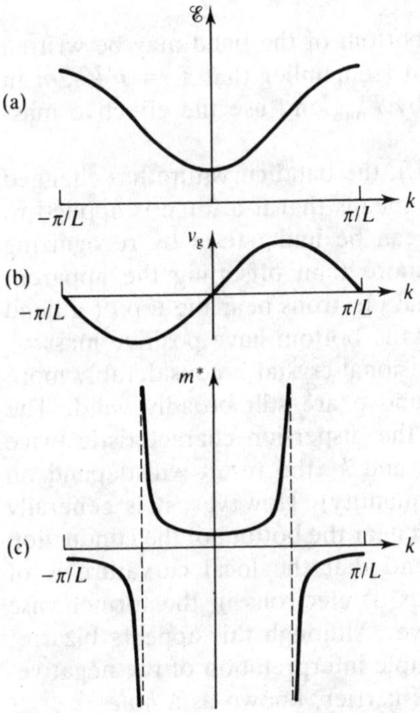

Figure 11.16 Variation of (a) electron energy, (b) group velocity and (c) effective mass through a band.

fact as follows. By differentiating Eq. (11.5) we can express the mass of a free electron as

$$m = \hbar/\{\mathrm{d}^2\omega/\mathrm{d}k^2\} \tag{11.71}$$

Thus m is simply related to the curvature of the dispersion characteristic. We can define a similar quantity m^*—known as the *effective mass*—for an electron in a crystal lattice. In this way, we can still treat the electron as a pseudo free particle, but allow for the interaction with the lattice through differences in the mass from that of a free electron. Differentiating Eq. (11.68) we obtain

$$m^* = \{\hbar^2/(L^2\Delta\mathscr{E})\} \sec(kL) \tag{11.72}$$

The effective mass clearly varies through the band in an extraordinary way, as shown in Fig. 11.16(c). Near the bottom of the band (where $k \approx 0$), the $\mathscr{E}$–k characteristic is roughly parabolic, so the electron has a positive, constant mass. This can be found by approximating Eq. (11.72) as

$$m^* \approx \hbar^2/(L^2\Delta\mathscr{E}) \tag{11.73}$$

Note that this may be entirely different from the mass of a free electron; however, by combining Eqs (11.69) and (11.73), we may still find a parabolic relation linking $\mathscr{E}$, k and m^*, given by

$$\mathscr{E} - \mathscr{E}_{\text{min}} = \hbar^2k^2/2m^* \ (\text{for } \mathscr{E} > \mathscr{E}_{\text{min}}) \tag{11.74}$$

This shows that the $\mathscr{E}$–k characteristic near the bottom of the band may be written in the same form as we used for a free electron (remember that $\mathscr{E} = \hbar^2k^2/2m$ in this case), provided we shift the energy origin by $\mathscr{E}_{\text{min}}$ and use the effective mass instead of the free mass.

Near to the top of the band (where $|k| \approx \pi/L$), the band curvature has changed sign, so the effective mass is now negative! This means that if a force is applied to an electron, it will decelerate. This odd result can be understood by recognizing that there can be a gain in potential energy, more than offsetting the apparent reduction in kinetic energy. Thus we conclude that electrons near the top of a band have a negative effective mass, while those near the bottom have positive mass.

Although the band structure of a three-dimensional crystal is considerably more complicated, it turns out that the conclusions above are still broadly valid. The effective mass may be found by differentiating the dispersion characteristic twice (although since this is now a function of k_x, k_y and k_z the result will depend on direction, so that m^* turns out to be a tensor quantity). However, it is generally true that the $\mathscr{E}$–k relation is roughly parabolic both near the bottom of the conduction band and near the top of the valence band, and that the local curvature is of opposite sign. Consequently, the effective masses of electrons in the former case are positive, while those in the latter are negative. Although this appears bizarre, we shall see in the next section that there is a simple interpretation of the negative-mass states in terms of a different type of charge carrier, known as a *hole*.

11.6 CARRIER STATISTICS

So far we have discussed solutions to the Schrödinger equation without considering the actual occupation of the states by electrons. We will now consider this point for a three-dimensional cube of crystal. We start by deriving a suitable density of states function, as we did earlier for photons.

The Density of States

To begin with, we return to the result previously obtained for a three-dimensional well, namely that for an electron confined in a cubic well of side h the allowed energy levels are given by

$$\mathscr{E} = \nu^2\pi^2\hbar^2/2mh^2 \tag{11.75}$$

where $\nu^2 = \nu_x^2 + \nu_y^2 + \nu_z^2$ and ν_x, ν_y and ν_z are integers. If h is large, the minimum difference between levels is small, so the energies again tend to a continuum. With the restriction that ν_x, ν_y and ν_z must be positive, and including the effect of spin (which doubles the number of states), the number of states with energy less than $\mathscr{E}$ can be shown by our previous method to be $\pi\nu^3/3$. However, since

$$\nu^2 = C\mathscr{E} \tag{11.76}$$

where $C = 2mh^2/\pi^2\hbar^2$, it can easily be seen that

$$\pi\nu^3/3 = (\pi/3)C^{3/2}\mathscr{E}^{3/2} \tag{11.77}$$

If we now define this quantity to be equal to the integral of a new function $s(\mathscr{E})$ (i.e. $\int_0^{\mathscr{E}} s(\mathscr{E})\, d\mathscr{E}$, by analogy with $s(\omega)$ for photons), then we may find $s(\mathscr{E})$ by differentiation to be

$$s(\mathscr{E}) = (\pi/2)\, C^{3/2}\sqrt{\mathscr{E}} \tag{11.78}$$

The density of states may then be obtained by dividing Eq. (11.78) by h^3 to get

$$S(\mathscr{E}) = (\pi/2)\, C'^{3/2}\sqrt{\mathscr{E}} \quad \text{where } C' = 2m/\pi^2\hbar^2 \tag{11.79}$$

We have obtained this result by assuming that electrons are confined inside a finite potential well. However, we really wish to find the density of states in a lattice. Most especially, we will be interested in the variation of $S(\mathscr{E})$ near the bottom of the conduction band and the top of the valence band. How should we modify Eq. (11.79) to cope with this? Arguably, the result we have obtained has followed from combining the quantization effect of the well with the parabolic relation between $\mathscr{E}$ and k for a free electron. Since we showed in Eq. (11.74) that the $\mathscr{E}$–k diagram is also parabolic near the bottom of the conduction band, we might expect a similar result to be obtainable merely by replacing $\mathscr{E}$ by $\mathscr{E} - \mathscr{E}_c$ (where $\mathscr{E}_c$ is the minimum energy of the conduction band) and m by m_e^* (where m_e^* is the value of m^* near $\mathscr{E}_c$). In this case, we get

$$S(\mathscr{E} - \mathscr{E}_c) = (\pi/2)\, C''^{3/2}\sqrt{\{\mathscr{E} - \mathscr{E}_c\}} \quad \text{for } \mathscr{E} > \mathscr{E}_c, \quad \text{where } C'' = 2m_e^*/\pi^2\hbar^2 \tag{11.80}$$

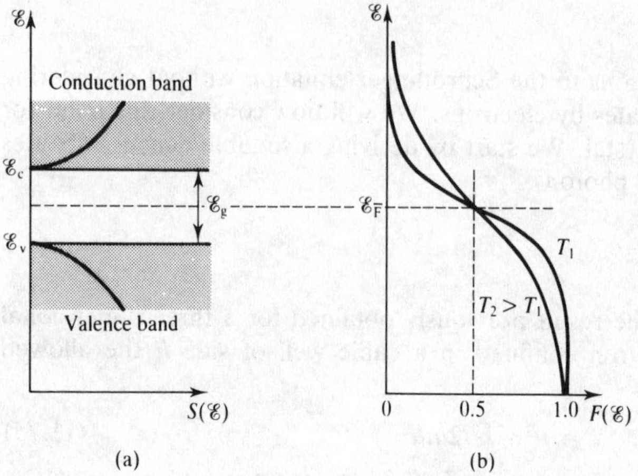

Figure 11.17 Typical variation of (a) the density of states, and (b) the Fermi-Dirac distribution function.

The $\mathscr{E} - k$ relation is also parabolic near the top of the valence band, although the parabola is now upside-down. Applying a similar argument, we can show that the density of states here is:

$$S(\mathscr{E}_v - \mathscr{E}) = (\pi/2)\, C'''^{3/2} \sqrt{\{\mathscr{E}_v - \mathscr{E}\}} \qquad \text{for } \mathscr{E} < \mathscr{E}_v, \text{ where } C''' = 2m_h^*/\pi^2\hbar \qquad (11.81)$$

where m_h^* is the modulus of m^* near $\mathscr{E}_v$. Combining Eqs (11.80) and (11.81), the variation of $S(\mathscr{E})$ throughout the conduction and valence bands is much as shown in Fig. 11.17(a).

The Fermi-Dirac Distribution Function

We now turn our attention to the actual occupation of the levels. Once again the details are determined by statistical thermodynamics. Given that the total energy of the electrons and their number are fixed, and that they satisfy the Pauli exclusion principle, it can be shown that the probability of occupation of a state of energy $\mathscr{E}$ is given by a function known as the *Fermi-Dirac probability function* (since it was originally derived independently by E. Fermi and P. A. M. Dirac in the 1920s). This represents the most likely distribution of electron energies, and has the form:

$$F(\mathscr{E}) = 1/[1 + \exp\{(\mathscr{E} - \mathscr{E}_F)/kT\}] \qquad (11.82)$$

where $\mathscr{E}_F$ (the *Fermi energy*) is the energy at which the probability of a state being filled is exactly 0.5. The Fermi-Dirac probability function is rather different from the Bose-Einstein function, and is shown in Fig. 11.17(b) for two temperatures T_1 and T_2. As can be seen, increased temperature causes a greater spread of the function, so that the probability of occupation of higher-energy states rises. A reduction in temperature causes the opposite effect, so that lower states are filled preferentially. In the limit as $T \to 0$, we find that $F(\mathscr{E}) = 1$ for $\mathscr{E} < \mathscr{E}_F$, and $F(\mathscr{E}) = 0$ for $\mathscr{E} > \mathscr{E}_F$.

The electronic properties of most materials are, in general, determined by the absolute position of the Fermi level, and by the size of the energy gap $\mathscr{E}_g$ by comparison with kT. We shall consider first what happens when $\mathscr{E}_F$ lies in the energy gap, when $\mathscr{E}_g \gg kT$ as shown in Fig. 11.17. Since the Fermi-Dirac function tails off rapidly into the conduction band, we should find only a few electrons in this range. In pure or *intrinsic* material these must have originated in the valence band, but thermal energy provides a means for them to move up into the conduction band. Together with these filled states, there are a vast array of empty ones. Electrons can easily move into these states, so they can gain energy in an electric field. Some electrical conductivity is therefore to be expected.

Because the Fermi-Dirac function is close to unity in the valence band, we would expect most states in this band to be filled. In intrinsic material, however, some empty states must be generated as electrons are promoted to the conduction band. The presence of even a few such states can allow nearby electrons to move into them, and hence contribute to conduction. In this case the process of conduction is best viewed rather differently. Since the availability of an empty state allows a neighbouring particle—a negatively-charged, negative-mass electron—to move into it, the empty state and the electron can effectively change places. Because only a small number of spaces are available in a great sea of electrons, it is simpler to treat the spaces themselves as the particles. These positively-charged pseudo-particles are known as *holes*; since they move in the opposite direction to electrons, they may be described by a positive mass. This leads to great simplifications.

To summarize; we would expect that a small number of electrons in the conduction band and holes in the valence band will be able to conduct electricity. Furthermore, an increase in temperature will extend the tail of the Fermi-Dirac distribution, and hence the number of electrons and holes available for conduction, and a shift of $\mathscr{E}_F$ from the middle of the bandgap will markedly increase the number of carriers in one band at the expense of the other. Materials with this kind of energy band diagram are classed as either semiconductors or insulators. The distinction is essentially one of degree. If $\mathscr{E}_g$ is so large that the numbers of carriers in the two bands are very small, then the material is an insulator. If the energy gap is moderate, it is a semiconductor. Table 11.1 illustrates this distinction.

Finally, we note that in most metals, the conduction band is partly filled even at low temperatures, so the conductivity is always high. Despite this advantage, the limited conductivity of a semiconductor is more than compensated for by the degree of control that can be exercised through the positioning of the Fermi level. This is achieved by *doping* the material with impurity atoms. Since the properties of the semiconductor are then largely determined by the dopants, it is called *extrinsic material*. For example, crystalline silicon is formed from Group IV atoms, each of which has four outer-shell or valence electrons and combines in a tetrahedral covalent bond with its nearest neighbours. Group V atoms (e.g. arsenic or phosphorous) are of a similar size, but have five valence electrons. If such an atom is substituted for silicon in the lattice, four of its five electrons will be used up in

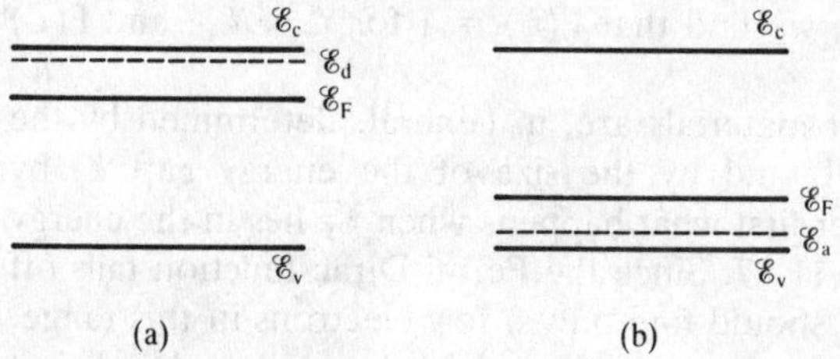

Figure 11.18 Impurity states for (a) donors and (b) acceptors.

nearest-neighbour bonding. Very little energy (typically <0.05 eV) is required to detach the remaining one from its parent atom, since it is only weakly bound. If this can be supplied, the fifth electron will be available for conduction. We can represent the process by defining additional localized impurity states in the energy diagram, located slightly below the bottom of the conduction band. Since the dopants provide electrons in this case, they are known as *donors*, and the level of the impurity states is defined as $\mathscr{E}_d$ (as in Fig. 11.18(a)). Less energy is required to promote electrons into the conduction band from these states than from the valence band, and, in the process, the number of conduction electrons is increased, without altering the number of holes. Doping with donors has the additional effect of lifting the Fermi level up towards $\mathscr{E}_c$, as we shall see later.

Group III atoms (such as gallium or boron) have only three valence electrons. Consequently, if these are substituted for silicon, they lack the correct number of electrons for complete bonding. Given a small amount of energy, an electron may be attracted from a neighbouring silicon atom, leaving a hole in the silicon bond network. We may represent this process by defining other impurity states at the *acceptor level* $\mathscr{E}_a$, just above $\mathscr{E}_v$ as shown in Fig. 11.18(b). Because holes may be created from these states very easily, the effect is to alter the hole density without changing the number of electrons. A further consequence is that the Fermi level is pulled down towards $\mathscr{E}_v$.

Dopants may also be added to III-V semiconductors, such as GaAs and InP. Clearly, a Group VI atom will provide an additional electron if it replaces a Group V atom in the lattice, so it can act as a donor. Similarly, Group II dopants can replace Group III atoms and act as acceptors. However, Group IV dopants may be donors or acceptors, depending on which atom they replace.

As we move $\mathscr{E}_F$ around in the energy gap (or even out of it) by suitable doping, we can alter the conductivity of semiconductors, and produce material with either predominantly 'normal' electron conduction (known as *n-type* material), or mainly 'abnormal' electron or hole conduction (*p-type* material). We have already seen one application for this technique in Chapters 3 and 9; a change in the number of free carriers can be used to generate a change in refractive index, via the plasma contribution to the dielectric constant. However, many other highly important properties follow from further application of the same basic principle.

11.7 INTRINSIC AND EXTRINSIC SEMICONDUCTORS

The equilibrium densities of electrons and holes can be calculated very simply for intrinsic material. To find the total density n of electrons in the conduction band, we merely multiply the density of available states (i.e. $S(\mathscr{E})$) by their probability of being filled ($F(\mathscr{E})$), and integrate the result from the bottom of the conduction band to the top. Thus we may write

$$n = \int_{\mathscr{E}_c}^{\mathscr{E}_t} F(\mathscr{E})\, S(\mathscr{E})\, d\mathscr{E} \tag{11.83}$$

where $\mathscr{E}_t$ is the level at the top of the conduction band. Returning to Fig. 11.17, however, we note that it is often true that $\mathscr{E} - \mathscr{E}_F \gg kT$ within the conduction band, so that $F(\mathscr{E})$ declines roughly exponentially with $\mathscr{E}$ in this range. We may therefore write

$$F(\mathscr{E}) \approx \exp\{-(\mathscr{E} - \mathscr{E}_F)/kT\} \tag{11.84}$$

Since this function tends to zero very quickly as $\mathscr{E}$ increases, the upper limit $\mathscr{E}_t$ of the integral above may be allowed to tend to infinity without appreciably changing the answer. Using the density of states function given in Eq. (11.80), the integral may then be approximated by

$$n \approx \int_{\mathscr{E}_c}^{\infty} \exp\{-(\mathscr{E} - \mathscr{E}_F)kT\}\, (\pi/2)\, C'^{3/2}\, \sqrt{(E - E_c)}\, d\mathscr{E} \tag{11.85}$$

This is a standard integral which may be evaluated analytically to give

$$n \approx N_c \exp\{-(\mathscr{E}_c - \mathscr{E}_F)/kT\} \tag{11.86}$$

where N_c is a term varying only slowly with temperature, given by

$$N_c = 2(2\pi m_c^* kT/\boldsymbol{h}^2)^{3/2} \tag{11.87}$$

The net result is that n depends (for the most part) exponentially on temperature, and on the position of the Fermi level relative to the base of the conduction band. The first feature is as we would expect, since an increase in temperature means that more carriers will have sufficient energy to move from the valence band into the conduction band. We will return to the second shortly.

To find the total density p of holes in the valence band, we merely multiply the density of available states by their probability of *not* being filled $(1 - F(\mathscr{E}))$, and integrate the result from the bottom of the valence band to the top. Thus we may write

$$p = \int_{\mathscr{E}_b}^{\mathscr{E}_v} \{1 - F(\mathscr{E})\}\, S(\mathscr{E})\, d\mathscr{E} \tag{11.88}$$

where $\mathscr{E}_b$ is the energy at the bottom of the valence band. A suitable approximation to the Fermi-Dirac function may again be used. Returning to Fig. 11.17, we note that $\mathscr{E} - \mathscr{E}_F \ll kT$ throughout the valence band, so that $F(\mathscr{E})$ is close to unity, and $1 - F(\mathscr{E})$ decreases roughly exponentially with $-\mathscr{E}$ in this range. We may therefore write

$$1 - F(\mathscr{E}) \approx \exp\{-(\mathscr{E}_F - \mathscr{E})/kT\} \tag{11.89}$$

Since this tends to zero very quickly as $\mathscr{E}$ decreases, the lower limit $\mathscr{E}_b$ of the integral may be allowed to tend to minus infinity, without much effect. Using the density of states function given in Eq. (11.81), and evaluating the integral by standard methods, we get

$$p \approx N_v \exp\{-(\mathscr{E}_F - \mathscr{E}_v)/kT\} \tag{11.90}$$

where the term N_v is also slowly-varying with temperature, and given by

$$N_v = 2(2\pi m_h^* kT/\mathbf{h}^2)^{3/2} \tag{11.91}$$

The result is similar to that previously obtained for n, but this time we find that the density of holes is exponentially dependent on the position of the Fermi level relative to the top of the valence band.

Carrier Generation and Recombination in Intrinsic Material

The generation of carrier pairs should yield equal electron and hole densities, so that $n = p$. We may use this to deduce the position of the Fermi level. Substituting for n and p from Eqs (11.86) and (11.90), we find after a little rearrangement that $\mathscr{E}_{Fi}$ is given by

$$\mathscr{E}_{Fi} = \{\mathscr{E}_c + \mathscr{E}_v\}/2 + (3kT/4)\log_e(m_h^*/m_e^*) \tag{11.92}$$

Since m_e^* and m_h^* are of the same order, the Fermi level lies about half way between the conduction and valence bands, as anticipated. We may also calculate the quantity $n_i = n = p$ (the *intrinsic carrier concentration*). We first form the product np using Eqs (11.86) and (11.90), as

$$np = N_c N_v \exp\{-(\mathscr{E}_c - \mathscr{E}_v)/kT\} = N_c N_v \exp\{-\mathscr{E}_g/kT\} \tag{11.93}$$

As can be seen, this result is independent of the position of the Fermi level, a point we will return to shortly. Intrinsic carrier concentration n_i may then be found as the square-root of the above, namely

$$n_i = \sqrt{(N_c N_v)} \exp\{-\mathscr{E}_g/2kT\} \tag{11.94}$$

The intrinsic carrier concentration therefore decreases exponentially with $\mathscr{E}_g$. This explains the poor conductivity of insulators, which have large energy gaps. Why, however, should n_i depend only on temperature? To answer this we must examine the dynamics of carrier-pair generation. Since this is a thermal effect, we would expect its rate to depend only on T. We might define this as $G(T)$. However, in equilibrium it must be exactly balanced by an inverse process called *recombination*, in which a conduction electron falls back into the valence band, annihilating an electron and a hole. The probability of an electron finding a hole available for recombination must be proportional to the hole density, i.e. to p. Since the electron density is itself n, the recombination rate R must be proportional to the product of n and p. In equilibrium we must have $R = G(T)$, which implies that R can only be a function of temperature. Consequently, the product $n \times p = n_i^2$ is constant at a given value of T.

Extrinsic Material

We now consider the properties of extrinsic material, starting with the electron density in n-type material. Let us assume there is a density N_d of donor atoms in an otherwise pure semiconductor. At a given temperature, we might expect a fraction N_d^+ of these to be ionized, and so have made their electrons available for conduction. As a result, the electron density will increase. The details of what actually occurs depend on the value of T. At low temperatures, N_d^+ is small, so that $n \approx n_i$. At higher temperatures, there is a range where the donors are fully ionized, so $N_d^+ \approx N_d$. Normally, the dopant density is high enough that $N_d \gg n_i$ at room temperature. In this case, the donated electrons dominate the electron density, and we may write $n \approx N_d$. Electrons are then the *majority carriers*; their density is essentially equal to that of the dopant atoms, and independent of temperature. At higher temperatures still, the rate of thermal carrier generation increases sufficiently that n tends to n_i again.

In the normal regime, where $n \approx N_d$, we may use Eq. (11.86) to find an approximate position for the Fermi level, as

$$\mathscr{E}_{Fn} \approx \mathscr{E}_c - kT \log_e (N_c/N_d) \tag{11.95}$$

Consequently, in n-type material, the Fermi level moves close to the base of the conduction band, as we have previously stated. (Note that if $\mathscr{E}_{Fn}$ is very near to the band edge, or even in the band itself, Eq. (11.86) is no longer accurate. This is the case in heavily-doped material.)

We must also consider the effect of doping on the hole density. To first-order, the generation rate G will be unchanged by the introduction of dopant atoms because carrier-pair generation is a thermal effect. Consequently, the recombination rate R must also be unaffected, so that the expression $np = n_i^2$ must still be valid. In the normal regime, where $n \approx N_d$, we would therefore expect the hole density to decrease, so that the product np is held constant. The holes are therefore *minority carriers*, since their density is reduced by doping to a value $p = n_i^2/n \approx n_i^2/N_d$, which is much lower than n_i. In n-type material, therefore, conduction occurs predominantly by electrons.

Results for p-type material (where doping is through acceptor atoms) may be found by analogy. In the normal regime the acceptors are fully ionized, and the density of holes induced in the valence band by the doping greatly outweighs the effects of thermal carrier-pair generation. In this case holes are the majority carriers, and we find that $p \approx N_a$, where N_a is the acceptor density. Using Eq. (11.90) we can show that the position of the Fermi level becomes

$$\mathscr{E}_{Fp} \approx \mathscr{E}_v + kT \log_e (N_v/N_a) \tag{11.96}$$

The Fermi level then shifts to a position near the top of the valence band, also as described earlier.

Motion of Charge Carriers

Having established the carrier densities, we must examine how they may move. Clearly an electric field will accelerate an electron, however, the acceleration will not continue indefinitely; collisions at random intervals will cause the electron to lose its kinetic energy, reducing its velocity to zero. The instantaneous speed of the electron will therefore be random, but its average value—known as the *drift velocity*, v_d—in the direction of the field will be non-zero. The motion of n electrons per unit volume results in a current density J_e given by

$$J_e = nev_d \tag{11.97}$$

Clearly v_d should increase with the applied field, so we may write

$$v_d = \mu_e E \tag{11.98}$$

where μ_e is a new parameter, the *electron mobility*. The current density J_e is therefore also proportional to the field. A similar motion of holes may occur; because holes move in the opposite direction, but carry the opposite charge, the analogous expression in this case is $J_h = pe\mu_h E$, where μ_h is the *hole mobility*. The total current density is the sum of these two components, namely

$$J = \{ne\mu_e + pe\mu_h\} E \tag{11.99}$$

Maxwell's equations imply that a one-dimensional field E will yield a current density $J = \sigma E$. From the above, we can see that this is substantially true, provided we define the conductivity σ as

$$\sigma = ne\mu_e + pe\mu_h \tag{11.100}$$

This is not the end of the story, however, since there is a further mechanism for motion of free carriers that is not included in the version of Maxwell's equations given in Chapter 2: *diffusion*.

In any system of mobile particles (e.g. a gas, or a collection of free carriers) random thermal motion occurs independently of any other effect. If concentration gradients exist, this will tend to remove them, so that the concentration everywhere tends to equalize. The necessary mathematics is common to many branches of physics, and may be derived by considering a simple field-free system involving one-dimensional diffusion. Figure 11.19 shows two adjacent layers of a mixture of mobile carriers (say, electrons); each has unit cross-sectional area and thickness dx, but the left-hand layer contains a density n of electrons, while the right-hand one has a

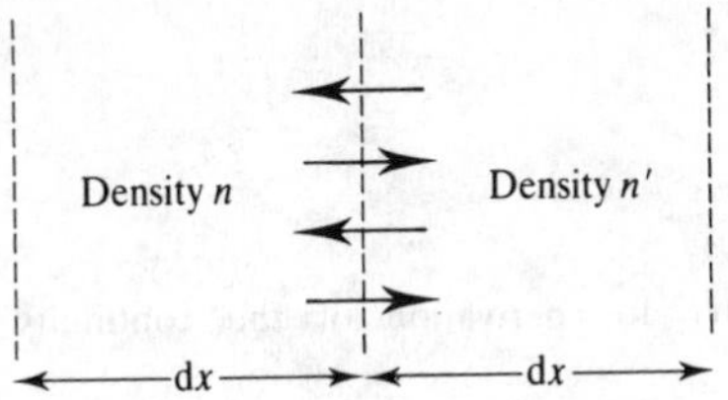

Figure 11.19 Geometry for calculation of diffusion current.

different density $n' = n + \partial n/\partial x \, dx$. Due to the random motion, we would expect a fraction of the electrons in the left-hand region to move to the right and cross the boundary between the layers. The number moving into the right-hand region per unit area and time will be proportional to n, so we may write it as $D_e n$, where D_e is a constant known as the electron *diffusion coefficient*. Simultaneously, electrons from the right-hand region will move in the opposite direction. The probability of moving in either direction must be the same, so the number of electrons crossing the boundary from right to left is $D_e\{n + \partial n/\partial x \, dx\}$. The two processes yield a net motion to the right of $D_e n - D_e\{n + \partial n/\partial x \, dx\} = -D_e \, \partial n/\partial x \, dx$ electrons. These move a distance dx in unit time, so this is equivalent to a flux density F of electrons, where

$$F = -D_e \, \partial n/\partial x \tag{11.101}$$

Because electrons carry a negative charge, this is equivalent to a *diffusion current density* of

$$J = D_e e \, \partial n/\partial x \tag{11.102}$$

Equation (11.102) shows that a non-uniform electron density will result in a flow of current in the direction of the concentration gradient. A similar equation may be written for holes; since these carry a positive charge, this time we get $J = -D_h e \, \partial p/\partial x$, where D_h is the diffusion coefficient for holes. Thus, including both drift and diffusion components, the electron and hole current densities are

$$J_e = ne\mu_e E + D_e e \, \partial n/\partial x \quad \text{and} \quad J_h = pe\mu_h E - D_h e \, \partial p/\partial x \tag{11.103}$$

In the steady-state, the partial derivatives are replaced by total ones. The total current density is then the sum of J_e and J_h. However, before we can use Eq. (11.103), we must determine the actual carrier distributions. In general the distribution of (say) electrons will depend on x and t. We can find the variation $n(x,t)$ by setting up a continuity equation, starting by neglecting the effect of recombination. Figure 11.20 shows another view of our one-dimensional system, where a layer of thickness dx has an electron flux density F flowing in from the left, and $F' = F + \partial F/\partial x \, dx$ flowing out to the right.

The effect of the difference between F and F' is a net outflow of $F - \{F + \partial F/\partial x \, dx\} = -\partial F/\partial x \, dx$ electrons per unit area per unit time. Since these are lost from a layer of thickness dx, the local electron concentration must drop at a rate $\partial n/\partial t$, where

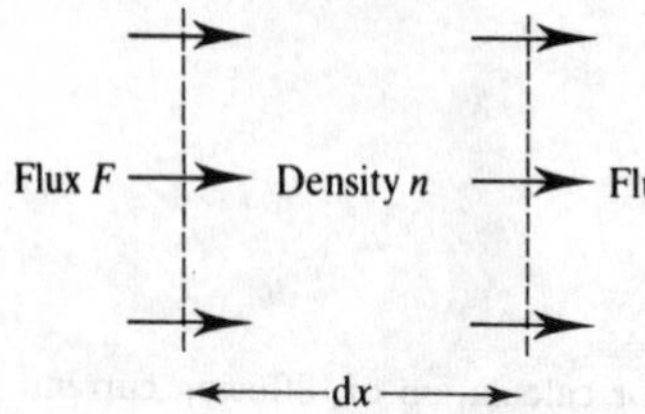

Figure 11.20 Geometry for derivation of the continuity equation.

$$\partial n/\partial t = -\partial F/\partial x \tag{11.104}$$

Equation (11.104)—the *continuity equation*—shows that a flux gradient will cause a decrease in the local electron density, which tends to equalize the value of n everywhere. Assuming that D_e is independent of n, we may combine Eqs (11.101) and (11.104) to eliminate F. We then get

$$\partial n/\partial t = D_e\, \partial^2 n/\partial x^2 \tag{11.105}$$

Equation (11.105)—known as the *diffusion equation*—may be solved to find the electron density, subject to given boundary conditions. We shall see how this is done in the next chapter.

We now consider the additional effect of recombination. Let us suppose that the electron concentration is uniform and higher than the equilibrium value n_E for some reason. Due to recombination, n will tend to decrease. As we will show later, the rate of decrease will be proportional to the excess electron density, so we may write

$$\partial n/\partial t|_{\text{recomb}} = -(n - n_E)/\tau_e \tag{11.106}$$

where τ_e is a constant known as the *electron lifetime*. This has the general solution

$$n(t) - n_E = \{n(0) - n_E\} \exp(-t/\tau_e) \tag{11.107}$$

where $n(0)$ is the initial value of n. As a result, the excess electron concentration decays exponentially with time. A synthesis of the effects of recombination and diffusion is simple; we merely modify Eq. (11.105) to include the right-hand side of Eq. (11.106).

11.8 DETAILED BALANCE

Having previously described the optical and electronic characteristics of semiconductors in isolation, we shall now consider combined optoelectronic interactions. We shall start by returning to the black body radiation law, to examine its origins in the elementary processes occuring within the material. The processes we must consider are absorption, spontaneous emission and stimulated emission; these were previously shown in Figs 11.7 and 11.8, but we shall now consider their operation in detail. For simplicity we shall consider transitions between single states in two levels of a direct-gap semiconductor. The lower state might lie in the valence band

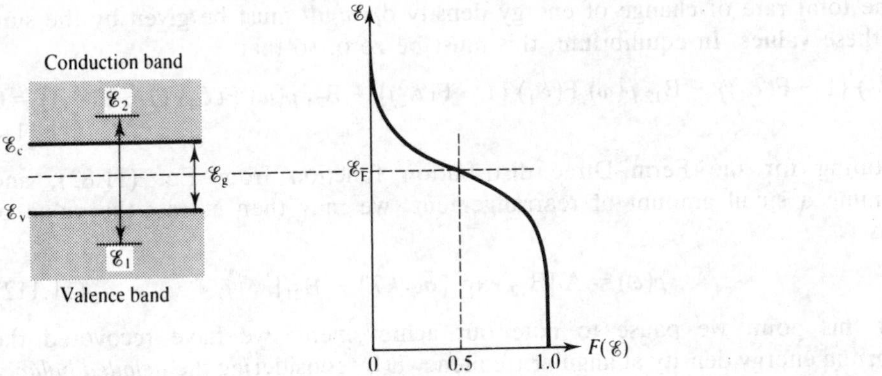

Figure 11.21 Two particular states in the valence and conduction bands of a semiconductor.

and have energy $\mathscr{E}_1$, while the upper one (with energy $\mathscr{E}_2$) might lie in the conduction band, as in Fig. 11.21. For convenience, the Fermi-Dirac distribution function is also shown in this figure.

First, we consider spontaneous emission. Here an electron from the upper state drops down to the lower state, emitting a photon of angular frequency ω (where $\hbar\omega = \mathscr{E}_2 - \mathscr{E}_1$) en route. The likelihood of this transition occurring must be proportional to the product of the probabilities of finding an electron in the upper state in the first place, i.e. $F(\mathscr{E}_2)$, and of finding the lower state unoccupied (namely $\{1 - F(\mathscr{E}_1)\}$). If it *does* actually take place, the optical energy density $\rho(\omega)$ will increase. We may therefore write the rate of increase of $\rho(\omega)$ due to this transition alone as

$$d\rho(\omega)/dt|_{spon} = A\,F(\mathscr{E}_2)\,\{1 - F(\mathscr{E}_1)\} \tag{11.108}$$

where A is a constant.

We now consider absorption. Here an electron from the lower state is promoted into the upper one, through the absorption of a photon. The likelihood of this occurring must depend on the probabilities of finding an electron in the lower state, i.e. $F(\mathscr{E}_1)$, and of finding the upper state unoccupied, namely $\{1 - F(\mathscr{E}_2)\}$, and also on the energy density $\rho(\omega)$. If the transition occurs, $\rho(\omega)$ will decrease this time, so the rate of change of energy density due to absorption can be written as

$$d\rho(\omega)/dt|_{abs} = -B_{12}\,\rho(\omega)\,F(\mathscr{E}_1)\,\{1 - F(\mathscr{E}_2)\} \tag{11.109}$$

where B_{12} is a constant.

Finally we consider stimulated emission, where the effect of an incident photon is to cause an electron to drop from the upper state to the lower one, releasing an additional photon. In this case we have

$$d\rho(\omega)/dt|_{stim} = B_{21}\,\rho(\omega)\,F(\mathscr{E}_2)\,\{1 - F(\mathscr{E}_1)\} \tag{11.110}$$

where B_{21} is a further constant.

The total rate of change of energy density $d\rho(\omega)/dt$ must be given by the sum of all these values. In equilibrium, this must be zero, so that

$$A\,F(\mathscr{E}_2)\,\{1 - F(\mathscr{E}_1)\} - B_{12}\,\rho(\omega)\,F(\mathscr{E}_1)\,\{1 - F(\mathscr{E}_2)\} + B_{21}\,\rho(\omega)\,F(\mathscr{E}_2)\,\{1 - F(\mathscr{E}_1)\} = 0 \qquad (11.111)$$

Substituting for the Fermi-Dirac distribution function from Eq. (11.82), and performing a small amount of rearrangement, we may then extract the value of $\rho(\omega)$ as

$$\rho(\omega) = A/[B_{12}\exp\{\hbar\omega/kT\} - B_{21}] \qquad (11.112)$$

At this point we pause to note our achievement: we have recovered the equilibrium energy density at angular frequency ω by considering the *detailed balance* of three processes which tend either to increase or to decrease the density of photons of this frequency. In fact, we could perform a similar analysis for transitions between any two states, and we would obtain a similar result. Why should this be? The answer must lie in the universal black body radiation law. Returning to Eq. (11.20), we have already shown that the energy density $\rho(\omega)$ must be

$$\rho(\omega) = [\hbar n^3\omega^3/\pi^2c^3)/\{\exp(\hbar\omega/kT) - 1\}] \qquad (11.113)$$

Comparison between Eqs (11.112) and (11.113) shows that this will be the case, if $B_{12} = B_{21} = B$ and $A/B = \hbar n^3\omega^3/\pi^2c^3$. We have therefore deduced two important truths: the coefficients associated with absorption and stimulated emission are identical, while those for spontaneous and stimulated emission have a definite ratio. Historically, the A and B coefficients are known as the *Einstein coefficients* (after Albert Einstein, who first suggested the possibility of stimulated emission).

11.9 RATE EQUATIONS

In optoelectronics we will normally be concerned with operation away from thermal equilibrium. For example, we might increase the photon density above its equilibrium value by shining a beam of light on a semiconductor, or we might increase the carrier density by injecting extra electrons. To analyse the results, we must make some simplifications. We should really describe the transfer of electrons between bands in terms of the probabilities of occupation of a particular pair of states, integrating over all possible states. Since this is a complicated procedure, we shall use a much simpler model, in which we regard the system as having only two states. Among other things, this implies that all photons involved have roughly the same energy. This model is so far from the truth that we might expect any of its predictions to be valueless. In fact, if the conditions are carefully chosen, the predictions are consistent with observation and are simple enough to be useful.

As before, the rate of change of the concentration ϕ of photons of energy $\hbar\omega \approx \mathscr{E}_g$ has three contributions. We begin by considering the case where there is no optical feedback, and the electron and hole densities are low. In these circumstances we can neglect stimulated emission entirely. The rate of spontaneous emission will obviously depend on the electron density in the conduction band. However, it will also depend on the density of holes in the valence band, because this band is so full that the availability of unoccupied states for electrons to move into is a rate-limiting factor. Absorption depends on the photon density only, since there are a very large number of electrons in the valence band and many spaces in the conduction band. The rate of change of photon density is therefore

$$\mathrm{d}\phi/\mathrm{d}t = anp - b\phi \tag{11.114}$$

where a and b are constants related to (but not identical to) the Einstein A and B coefficients used in the previous section. The generation of each photon consumes one electron and one hole, so we may write the rate of change of electron density as

$$\mathrm{d}n/\mathrm{d}t = -anp + b\phi \tag{11.115}$$

We could write an identical equation for holes, but we shall not bother: the essence of the problem is already fully described. The problem with these equations is that they are non-linear. However, they can be linearized through suitable approximations, as we now show. In thermal equilibrium, the electron, hole and photon densities are stabilized at their equilibrium values n_E, p_E and ϕ_E, and all derivatives are zero. We may therefore write:

$$an_Ep_E - \mathrm{b}\phi_E \approx 0 \tag{11.116}$$

We now assume that all quantities are perturbed from their equilibrium values, so that $n = n_E + \Delta n$, $p = p_E + \Delta p$ and $\phi = \phi_E + \Delta\phi$. In this case, Eq. (11.115) becomes

$$\mathrm{d}(n_E + \Delta n)/\mathrm{d}t \approx -a\{n_E + \Delta n\}\,\{p_E + \Delta p\} + b(\phi_E + \Delta\phi) \tag{11.117}$$

We may now remove a number of terms using Eq. (11.116). At this point, we also note that the requirement for charge neutrality implies that Δn must equal Δp. Assuming additionally that Δn and Δp are small, Eq. (11.117) may be reduced to

$$\mathrm{d}(\Delta n)/\mathrm{d}t \approx -a\Delta n\{n_E + p_E\} + b\Delta\phi \tag{11.118}$$

This implies that an excess photon density will increase the electron density through absorption. However, this process is counteracted by spontaneous emission, which will tend to remove any such increase by recombination. The change due to spontaneous emission alone can be written

$$\mathrm{d}(\Delta n)/\mathrm{d}t|_{\mathrm{spon}} \approx -\Delta n/\tau_{rr} \tag{11.119}$$

where τ_{rr} is a constant known as the *radiative recombination lifetime*, defined as

$$\tau_{rr} = 1/a\{n_E + p_E\} \tag{11.120}$$

Design Example

We may estimate the radiative recombination lifetime in p-type silicon with a doping level of $N_a \approx 10^{24}\ \mathrm{m}^{-3}$ as follows. For Si, the relevant recombination coefficient is

$a \approx 2 \times 10^{-21}$ m^3/s. The recombination time is $\tau_{rr} = 1/a\{n_E + p_E\}$; this may be approximated by $1/ap_E \approx 1/aN_a$ since $p_E \gg n_E$. Thus, $\tau_{rr} \approx 1/(2 \times 10^{-21} \times 10^{24}) = 0.5 \times 10^{-3}$ s. This is so long that the corresponding rate $r_{rr} = 1/\tau_{rr}$ is extremely small. By comparison, the recombination coefficient in GaAs is $a \approx 7 \times 10^{-16}$ m^3/s, a factor of $\approx 10^5$ larger.

Non-Radiative Recombination

It is possible for recombination to involve the generation of phonons rather than photons. This is termed *non-radiative recombination*, and has a characteristic lifetime τ_{nr}. Since the rates of the two processes are additive, we might therefore define an overall recombination rate as

$$d(\Delta n)/dt|_{recomb} \approx -\Delta n/\tau_e \quad (11.121)$$

where $1/\tau_e$—the overall electron lifetime used in the previous section—is given by

$$1/\tau_e = 1/\tau_{rr} + 1/\tau_{nr} \quad (11.122)$$

With this new notation, the electron *rate equation* may be written in the form:

$$d(\Delta n)/dt \approx -\Delta n/\tau_e + b\Delta\phi \quad (11.123)$$

Allowing the generation of photons to occur only through radiative recombination, a similar argument would lead to the corresponding equation for photons:

$$d(\Delta\phi)/dt \approx \Delta n/\tau_{rr} - b\,\Delta\phi \quad (11.124)$$

This shows that any excess of photons will tend to decrease through absorption, and increase through spontaneous emission, as might be expected.

Concentration Gradients

So far we have assumed uniform electron and photon densities, and we must now make allowance for concentration gradients. This is very simple—in fact, we have already covered the modifications needed for electrons in Section 11.7. We must include another term to account for the extra rate of change of Δn due to gradients in electron flux. For a one-dimensional flux in the x-direction, this term is $\partial n/\partial t = -\partial F/\partial x$. Since $F = -D_e\,\partial n/\partial x$, $\partial F/\partial x = -D_e \partial^2(\Delta n)/\partial x^2$ and Eq. (11.123) becomes

$$\partial(\Delta n)/\partial t = D_e\,\partial^2(\Delta n)/\partial x^2 - \Delta n/\tau_e + b\Delta\phi \quad (11.125)$$

This places the electron rate equation in a form similar to Eqs (11.105) and (11.106). However, since the electron diffusion current is $J = D_e e\,\partial(\Delta n)/\partial x$, we may also write Eq. (11.125) as

$$\partial(\Delta n)/\partial t = (1/e)\,\partial J/\partial x - \Delta n/\tau_e + b\Delta\phi \quad (11.126)$$

This version will prove more appropriate for the analysis of light-emitting devices in Chapter 12.

We now consider a similar modification to the photon rate equation. For photon motion in a different direction (say, the z-direction) the excess photon flux is $\Delta\Phi = v_g \Delta\phi$, where v_g is the group velocity for photons. The extra term is now $-\partial(\Delta\Phi)/\partial z = -v_g\,\partial(\Delta\phi)/\partial z$, and Eq. (11.124) becomes

$$\partial(\Delta\phi)/\partial t \approx -v_g\,\partial(\Delta\phi)/\partial z + \Delta n/\tau_{rr} - b\Delta\phi \tag{11.127}$$

Equations (11.125) and (11.127) must really be solved simultaneously to provide a full description of the interaction between electrons and photons. However, considerable insight may be gained from approximate solutions for several limiting cases.

Optical absorption We start by making an approximation to the photon rate equation, that is appropriate for describing the absorption of a travelling wave of photons. If radiative recombination can be neglected, and a steady-state exists, so that $\partial(\Delta\phi)/\partial t = 0$, Eq. (11.127) reduces to

$$\mathrm{d}(\Delta\phi)/\mathrm{d}z \approx -(b/v_g)\,\Delta\phi \tag{11.128}$$

The rate of change of $\Delta\phi$ with distance therefore depends linearly on $\Delta\phi$. As each photon carries energy $\hbar\omega \approx \mathscr{E}_g$ and travels at velocity v_g, the optical power density is $P \approx \mathscr{E}_g v_g \Delta\phi$. Since P is proportional to $\Delta\phi$, it should decay with distance as

$$\mathrm{d}P/\mathrm{d}z \approx -(b/v_g)P \tag{11.129}$$

We have already come across this behaviour before: in an absorbing medium, the power carried by an optical wave travelling in the z-direction varies with distance as $P(z) = P(0)\exp(-\alpha_p z)$, where α_p is the *power absorption coefficient* (equal to 2α, where α is the amplitude absorption coefficient of Chapter 2). In this case, $\mathrm{d}P/\mathrm{d}z = -\alpha_p P$. By comparison with Eq. (11.129), we may extract the power absorption coefficient as $\alpha_p \approx b/v_g$. We have therefore found a satisfying link between a macroscopically-observable quantity (α_p) and microscopic events (interband transitions).

Photogeneration of carriers We now make an approximation to the electron rate equation that can describe the photogeneration of carriers. If diffusion currents may be neglected, and a steady-state again exists, so that $\partial(\Delta n)/\partial t = 0$, Eq. (11.126) reduces to

$$\Delta n \approx \alpha_p v_g \tau_e\,\Delta\phi \approx \alpha_p \tau_c\,\Delta\Phi \tag{11.130}$$

This implies that absorption of photons will give rise to an increase in n (and similarly in p). We might use this mechanism in a photodetector, which works by measuring the corresponding increase in conductivity. This can be done by applying a voltage to contacts on either side of a slab of material, as shown in Fig. 11.22. Any change in the current through the circuit is then linearly proportional to the intensity of the illuminating beam. The output may be converted into a voltage by measuring the voltage dropped across a load resistor R_L. This device is called a *photoconductive detector*. While it is extremely simple, it suffers from two important

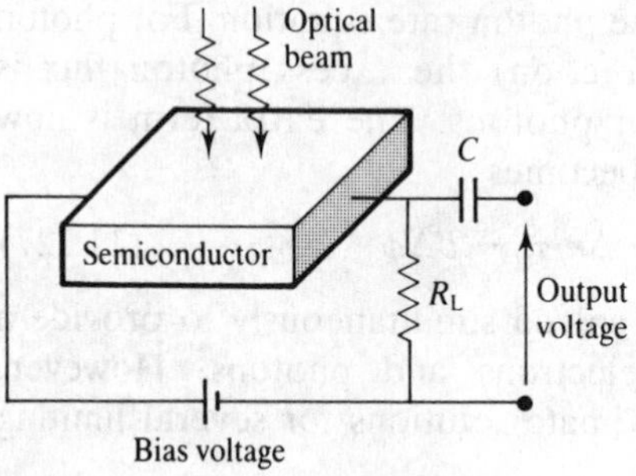

Figure 11.22 A photoconductive detector.

disadvantages. First, the conductivity will not be zero when the optical beam is switched off, due to the inevitable presence of thermally-generated carriers. Consequently, there will be a d.c. voltage across R_L. Although this may be blocked by the capacitor C, thermal carriers will also be responsible for noise in the output. Second, the value of Δn depends on τ_e, so for high sensitivity we require a long lifetime. However, a large value of τ_e implies that the photogenerated carriers will persist for some time after the beam is switched off (remember that we showed in Eq. (11.107) that the decay of Δn from an initial value follows $\Delta n(t) = \Delta n(0) \exp(-t/\tau_e)$). It is therefore difficult to combine high sensitivity with a high-speed response. As we will see later, it is possible to construct different types of detector, which do not have these limitations. These are based on *p-n junction diodes*, and are essential for the success of modern lightwave communications systems.

Electroluminescence We now make an approximation that is more appropriate to the electrical generation of light through spontaneous emission (known as *electroluminescence*), which will be useful in describing the operation of LEDs. Assuming this time that absorption is unimportant, and that photon concentration gradients are negligible, the photon rate equation may be approximated by

$$\partial(\Delta\phi)/\partial t = \Delta n/\tau_{rr} \tag{11.131}$$

Making a similar approximation to the electron rate equation we get

$$\partial(\Delta n)/\partial t = (1/e)\,\partial J/\partial x - \Delta n/\tau_e \tag{11.132}$$

These equations show that the existence of a diffusion current will lead to a rise in Δn, which will in turn result in an increase in $\Delta\phi$ through spontaneous emission.

Optical gain Finally we will extend our model to include stimulated emission; this version of the equations will be appropriate for the description of semiconductor laser operation. We will now ignore spontaneous emission, since stimulated emission will dominate this process. The rate of stimulated emission will simply depend on the product of n and p and the photon density. In this new situation, however, the rate of absorption must be reconsidered, since n and p must be very high, and we cannot assume that there will be large numbers of electrons in the valence band or empty levels in the conduction band. We can only say that there is some number N_{val} of the former and N_{con} of the latter, and the actual values of these quantities will depend on the degree of excitation of the system.

We may therefore modify the rate equation for a travelling photon wave (Eq. (11.127)) to

$$\partial\phi/\partial t \approx -v_g\ \partial\phi/\partial z - b'\phi\{N_{val}N_{con} - np\} \tag{11.133}$$

where b' is a new constant. Note that ϕ, n and p appear in the equation, rather than $\Delta\phi$, Δn and Δp, since we are no longer considering small perturbations. Normally np is much less than $N_{val}N_{con}$, so that net absorption occurs. However, Eq. (11.133) implies that *optical gain* will be possible if we can somehow increase np above $N_{val}N_{con}$. This might be achieved by taking (say) a p-type material, which already contains a large hole density, and somehow injecting a large number of electrons. Under these circumstances there will be many more electrons in the conduction band than in the valence band, so the electron population is *inverted* from its normal equilibrium distribution.

There must be some particular level of excitation at which the loss of photons by absorption is exactly balanced by gain from stimulated emission. Under these circumstances the material is said to be *transparent*. If we define the electron concentration required for transparency to be n_0, there will be a net gain of photons by stimulated emission if n is increased above n_0. In this case we may write

$$\partial\phi/\partial t \approx -v_g\ \partial\phi/\partial z + G\phi(n - n_0) \tag{11.134}$$

where G is the *gain constant* for the material. Typical values for n_0 and G are 10^{24} m^{-3} and 10^{-12} m^3/s, respectively. At this point, the amplification of a travelling optical wave will be feasible. It should be noted that the generation of photons by stimulated emission is not isotropic; the stimulated photons travel in essentially the same direction as the incident photons.

Further modification to the equations (e.g. to include all three optical processes at once) is obviously possible. This is actually required in the case of a semiconductor laser—at low injection levels, spontaneous emission is the dominant process, but stimulated emission takes over at higher levels. However, we shall postpone this complication until Chapter 12.

PROBLEMS

11.1 Show that the energy density of radiation emitted by a black body in an interval between the wavelengths λ_0 and $\lambda_0 + d\lambda$ is given by $\rho(\lambda_0)\,d\lambda = (8\pi hnc/\lambda_0^5)/\{\exp(hc/\lambda kT) - 1\}\,d\lambda$. Sketch the variation of $\rho(\lambda_0)$ with λ_0, for two different temperatures—say, 1000 K and 6000 K. Show that ρ peaks at a wavelength defined by *Wien's law* $\lambda_{max}\,T$ = constant, and calculate the value of the constant.
[2.9×10^{-3} mK]

11.2 What is the value of accelerating voltage required to obtain an electron wavelength of 0.05 nm in vacuum? What is the corresponding electron momentum?
[602 V; 1.32×10^{-23} Ns]

11.3 An electron of energy $\mathscr{E}$ is incident on the potential step shown below. By assuming a suitable solution for the wavefunction on either side of the step, calculate the probability that the electron will be reflected by the step, for the two cases of (a) $\mathscr{E} > \mathscr{V}$, and (b) $\mathscr{E} < \mathscr{V}$.

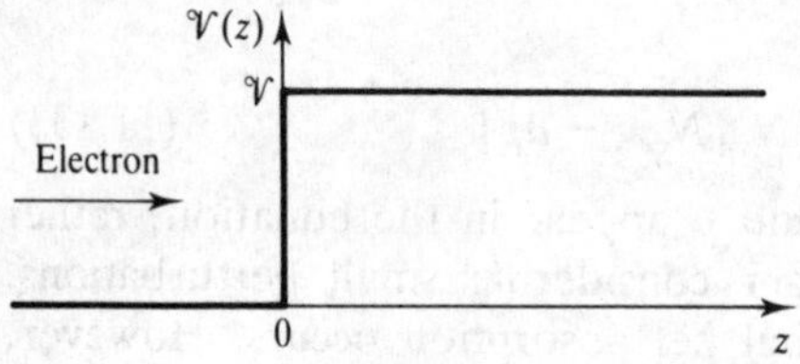

11.4 Calculate the wavelength at which the materials in Table 11.1 should first start to absorb light. Which material would you use to fabricate (a) a photoconductive detector, and (b) a transparent window, for operation at λ_0 = 1.55 μm. Compare your choice for (a) with the ternary alloy $Ga_{0.47}In_{0.53}As$ ($\mathscr{E}_g$ = 0.73 eV, direct-gap).

11.5 Calculate the position of the Fermi level in intrinsic InP, assuming that $\mathscr{E}_g$ = 1.35 eV at 300 K, and that $m_c^*/m = 0.077$, $m_h^*/m = 0.64$, where m is the mass of a free electron.
[0.716 eV above the valence band edge]

11.6 Using the data given in the table below, compare the intrinsic carrier concentration n_i in Ge, Si and GaAs at a temperature of 300 K. Assuming that N_c and N_v are roughly independent of temperature, estimate the rise in temperature needed to double n_i in Ge.

	Ge	Si	GaAs
N_c (m^{-3})	1×10^{25}	2.8×10^{25}	4.7×10^{23}
N_v (m^{-3})	6×10^{24}	1.04×10^{25}	7×10^{24}
$\mathscr{E}_g$ (eV)	0.66	1.12	1.42

[2.3×10^{19} m^{-3}; 6.95×10^{15} m^{-3}; 2.25×10^{12} m^{-3}; 16.3 K]

11.7 Which of the following can act as (a) donors and (b) acceptors in GaAs: Be, C, Cd, Mg, S, Se, Si, Sn, Te, Zn? A sample of GaAs is doped with 10^{18} Zn atoms per m^3. Is the material p-type or n-type? Using the data from Question 11.6, find the position of the Fermi level and the majority and minority carrier densities, assuming complete ionization of the dopants.
[0.408 eV above the valence band edge; 10^{18} m^{-3}; 5×10^6 m^{-3}]

11.8 Explain why the likelihood of occupation of an acceptor level may be found using the Fermi-Dirac probability function as $F(\mathscr{E}_a) = 1/[1 + \exp\{(\mathscr{E}_a - \mathscr{E}_F)/kT\}]$. Using this expression, check the validity of assuming complete ionization of the dopant atoms in Zn-doped GaAs at room temperature. You may take it that the impurity level for Zn lies 0.031 eV above the valence band edge. How would the density of ionized dopant atoms be calculated in Sn-doped GaAs?

11.9 Show that if stimulated emission is not included in the calculation, the black body radiation law modifies to $\rho(\omega)\,d\omega = (\hbar n^3/\omega^3/\pi^2c^3)\exp(-\hbar\omega/kT)\,d\omega$. When would this represent a good approximation? Derive a different approximation, valid for low photon energies.

11.10 Assuming that $n_0 \approx 10^{24}$ m^{-3}, and $G = 10^{-12}$ m^3/s, estimate the electron density required to achieve an optical power gain of 10^6 m^{-1} in a material of refractive index 3.5.
[8.7×10^{25} m^{-3}]

SUGGESTIONS FOR FURTHER READING

Carroll, J. E. "Rate Equations in Semiconductor Electronics", Cambridge University Press, Cambridge, 1985.

Kittel, C. "Introduction to Solid State Physics", 3rd ed., Ch. 9 and 10, John Wiley and Sons, New York, 1966.

Leaver, K. D. "Microelectronic Devices", Ch. 1 and 2. Longman Scientific and Technical, Harlow, 1989.

Pais, A. "Subtle is the Lord ... The Science and Life of Albert Einstein", Ch. 19, 21 and 23, Oxford University Press, Oxford, 1982.

Pulfrey, D. L., and Tarr, N. G. "Introduction to Microelectronic Devices", Ch. 2–5, Prentice-Hall International, Englewood Cliffs, 1989.

Solymar, L., and Walsh, D. "Lectures on the Electrical Properties of Materials", 4th ed., Ch. 2, 3, 4, 7 and 8, Oxford University Press, 1988.

Sze, S. M. "Semiconductor Devices: Physics and Technology", Ch. 1, 2 and 7, John Wiley and Sons, New York, 1985.

Yariv, A. "Quantum Electronics", 3rd ed., Ch. 1, 2 and 11, John Wiley and Sons, New York, 1989.

CHAPTER

TWELVE

OPTOELECTRONIC DEVICES

12.1 THE p-n JUNCTION DIODE

This chapter is devoted to the wide variety of optoelectronic devices that are made in semiconductor materials. Many (although not all) involve the absorption and emission of light, and most are based around the surprisingly versatile structure mentioned at the end of Chapter 11, the *p-n junction diode*. We shall begin by considering the electronic aspects of this device.

In its simplest form a p-n junction diode can be constructed as a *homojunction*, using an abrupt change between p- and n-type regions in a single piece of semiconductor. Figure 12.1 illustrates the formation of a diode, in a one-dimensional 'thought experiment' based on the ideas of the previous chapter. The p- and n-regions might initially be considered to be separated slabs of material. When these

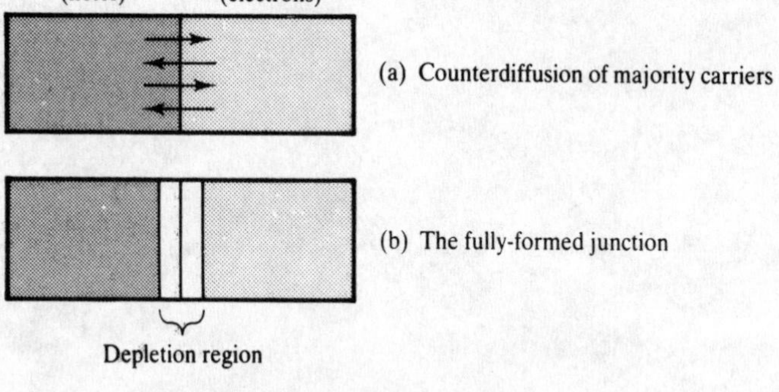

Figure 12.1 Formation of a p-n junction between p- and n-type material.

are placed in intimate contact, the difference in hole concentration between the p- and n-type regions should result in the diffusion of holes from the former into the latter (Fig. 12.1(a)). In the same way, electrons ought to diffuse in the reverse direction, from the n-type material into the p-type. If the two carrier types meet near the interface, they will recombine. The effect should then be that a layer—known as the *depletion region*—will be almost entirely denuded of carriers (Fig. 12.1(b)).

A logical question to ask is, what determines the width of the depletion region? For example, is it possible that the counter-diffusion process would continue indefinitely, until *all* the free carriers have been used up, so that the depletion region extends right through both layers of material? The short answer is that the process is self-limiting. As we shall see, carrier recombination in the depletion region sets up a space-charge; this in turn induces an electric field, which acts as an obstacle to further motion of carriers, so that the depletion region quickly stabilizes as a thin layer. However, in practice, the dynamics of junction formation are irrelevant to the operation of the device, and we need only consider the situation after equilibrium has been established.

The Distribution of Charge and Electric Field Inside the Depletion Layer

Due to the lack of free carriers, the depletion region must be locally charged, since the charges of the dopant ions embedded in the lattice are no longer compensated by those of the mobile carriers. The sign of this 'exposed charge' varies. On the p-side, for example, it is negative, due to exposure of the charge of the ionized acceptors. In contrast, the charging is positive on the n-side because the positive charges of donor ions are now exposed. To a first approximation, the charge distribution within each region is uniform (although we shall review this later). Thus, assuming total ionization of the dopant atoms, a space extending from (say) $x = -x_{dp}$ to $x = 0$ in the p-type material may be assumed to have a local charge density $\rho = -eN_a$, while in the n-type material the space between $x = 0$ and $x = x_{dn}$ has a charge density eN_d (Fig. 12.2(a)). Here x_{dn} and x_{dp} are the lengths of the p- and n-sides of the depletion layer; these are related to the total width w of the region by $w = x_{dp} + x_{dn}$.

When taken as a whole, the complete structure must be electrically neutral, so the exposed charges must balance each other. We may therefore put

$$x_{dp}N_a = x_{dn}N_d \tag{12.1}$$

This allows us to write the width of the depletion layer as

$$w = x_{dn}\{1 + N_d/N_a\} \tag{12.2}$$

Note that the 'split' of the depletion layer between the p- and n-sides may be highly unequal if $N_a \neq N_d$. For example, the p-side might be much more heavily doped than the n-side; in this case (known as a p^+-n junction) we find that $x_{dp} \ll x_{dn}$, so the depletion layer extends further into the n-side.

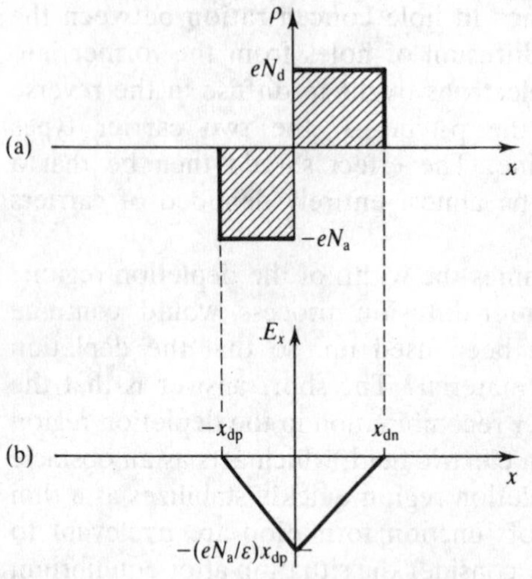

Figure 12.2 (a) Charge density and (b) electric field, for a p-n junction in equilibrium.

We can use Fig. 12.2(a) to find the electric field resulting from the charge distribution. The calculation is based on Gauss' law, which requires that the electric flux flowing out of a closed surface equals the charge contained inside it. Taking the enclosed volume to be a parallel-sided slab of unit cross-sectional area lying between two planes at $-x_{dp}$ and x, the enclosed charge is given by

$$\begin{aligned} q &= -eN_a(x + x_{dp}) \qquad (\text{for } -x_{dp} \leq x \leq 0) \\ &= -eN_d(x_{dn} - x) \qquad (\text{for } 0 \leq x \leq x_{dn}) \end{aligned} \tag{12.3}$$

This must give rise to an equal flux density which (by symmetry) has only an x-component D_x. As a result, the corresponding electric field $E_x = D_x/\varepsilon$ is given by

$$\begin{aligned} E_x &= -(eN_a/\varepsilon)\,(x + x_{dp}) \qquad (\text{for } -x_{dp} \leq x \leq 0) \\ &= -(eN_d/\varepsilon)\,(x_{dn} - x) \qquad (\text{for } 0 \leq x \leq x_{dn}) \end{aligned} \tag{12.4}$$

where ε is the dielectric constant of the semiconductor. This electric field variation is shown plotted in Fig. 12.1(b). As can be seen, the field varies linearly on each side of the depletion region. On the p-side it falls from $E_x = 0$ at $x = -x_{dp}$ to $E_x = -(eN_a/\varepsilon)x_{dp}$ at $x = 0$; similarly, on the n-side it rises from this minimum value to zero at $x = x_{dn}$. The existence of an electric field within the depletion region suggests that there will be a voltage between different points. We might therefore find the potential between (for example) the left-hand edge of the depletion region and an arbitrary point x by direct integration; i.e. as

$$V(x) = \int_{-x_{dp}}^{x} -E_x \, dx \tag{12.5}$$

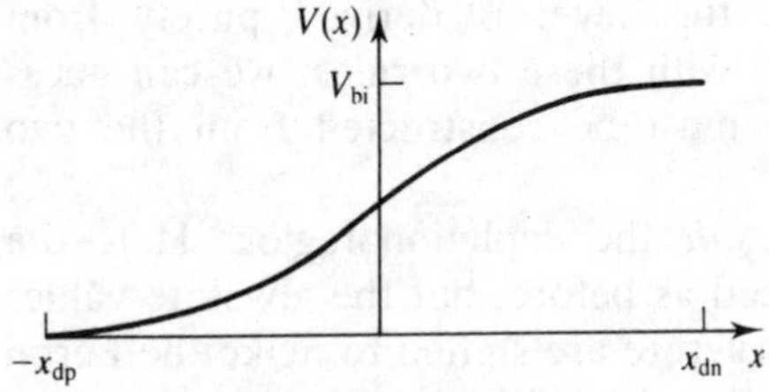

Figure 12.3 Potential distribution inside a p-n junction in equilibrium.

Substituting from Eq. (12.4) we then obtain

$$V(x) = (eN_a/\varepsilon)\{x^2/2 + xx_{dp} + x_{dp}^2/2\} \qquad (\text{for } -x_{dp} \le x \le 0)$$

$$= (eN_a/\varepsilon)\,x_{dp}^2/2 + (eN_d/\varepsilon)\{xx_{dn} - x^2/2\} \qquad (\text{for } 0 \le x \le x_{dn}) \qquad (12.6)$$

This distribution is shown schematically in Fig. 12.3. Throughout the depletion region, $V(x)$ varies parabolically, following a smooth S-shaped curve from zero at $-x_{dp}$ to a maximum (which we shall refer to as the *built-in voltage*, V_{bi}) at x_{dn}. Since the electric field is zero at either end of the depletion layer, dV/dx is also zero at these points. Clearly the built-in voltage may be found as $V_{bi} = V(x_{dn})$; substituting into Eq. (12.6) we get

$$V_{bi} = (eN_a/\varepsilon)\,x_{dp}^2/2 + (eN_d/\varepsilon)x_{dn}^2/2 \qquad (12.7)$$

Substituting for x_{dp} from Eq. (12.1), and for x_{dn} from Eq. (12.2), we can find V_{bi} in terms of the depletion layer width w as

$$V_{bi} = (ew^2/2\varepsilon)\{N_aN_d/(N_a + N_d)\} \qquad (12.8)$$

Conversely we can find the width w in terms of the built-in voltage by inverting Eq. (12.8):

$$w = \sqrt{\{(2\varepsilon V_{bi}/e)(N_a + N_d)/N_aN_d\}} \qquad (12.9)$$

At this point we have found a closed-form expression for the depletion layer width which contains just one unknown parameter, V_{bi}. We shall now consider the factors that determine V_{bi} itself.

The Band Diagram for a p-n Junction

We start by returning to the results of Chapter 11, where we drew energy band diagrams for isolated regions of n- and p-type material. How should these be modified for the composite p-n junction? We would expect some features—the relative spacings of the conduction band, the Fermi level and the valence band, for example—to be preserved at a reasonable distance from the depletion layer. Inside the depletion region the band structure will clearly be modified. However, we may make one deduction: the Fermi level (for which the occupation probability is 0.5) must be the same in all regions, because this level is derived purely from considerations of thermal equilibrium. Armed with these two rules, we can guess that the band diagram for the p-n junction must be constructed from the two diagrams of Fig. 11.18 as shown in Fig. 12.4.

We start by considering the situation *outside* the depletion region. Here the bands are flat, and the main features are spaced as before, but the absolute values of the energy levels in the two halves of the structure are shifted to make the Fermi levels coincide. After this shift the conduction band level is clearly different in the p- and n-regions. We will define the values of $\mathscr{E}_c$ in these two regions as $\mathscr{E}_{cp}$ and $\mathscr{E}_{cn}$. The position of the Fermi level in n-type material was previously defined relative to the conduction band, in terms of the density of donor atoms. Using Eq. (11.95) we may therefore define the Fermi level $\mathscr{E}_{Fn}$ as

$$\mathscr{E}_{Fn} \approx \mathscr{E}_{cn} - kT \log_e (N_c/N_d) \tag{12.10}$$

Similarly, the position of the Fermi level in p-type material can be defined relative to the valence band in terms of the acceptor density. However, it will be more convenient to relate $\mathscr{E}_{Fp}$ to $\mathscr{E}_{cp}$ instead. We can do this by combining Eqs (11.94) and (11.96), which gives

$$\mathscr{E}_{Fp} \approx \mathscr{E}_{cp} - kT \log_e (N_a N_c/n_i^2) \tag{12.11}$$

We now have two results for the Fermi level, one for each side of the junction. According to our previous argument they should be equal. Equating them allows us to find the difference between the conduction band levels on the p- and n-sides as

$$\mathscr{E}_{cp} - \mathscr{E}_{cn} = kT \log_e \{N_a N_d/n_i^2\} \tag{12.12}$$

A difference in the energy of a conduction electron on the two halves of the junction represents a potential barrier, which acts to prevent the transit of electrons across the depletion layer. There is a similar barrier to hole motion, which can be calculated from the difference between the valence band levels on the p- and n-sides. Overall, the energy barrier is equivalent to a built-in voltage across the structure of $V_{bi} = \{\mathscr{E}_{cp} - \mathscr{E}_{cn}\}/e$; V_{bi} is therefore given by

$$V_{bi} = (kT/e) \log_e \{N_a N_d/n_i^2\} \tag{12.13}$$

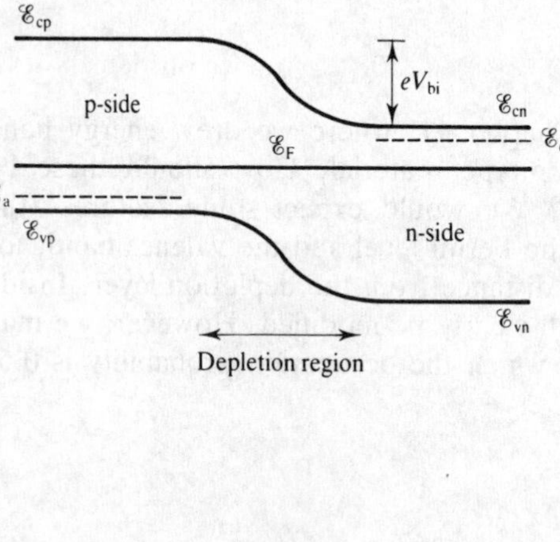

Figure 12.4 Energy band structure for a p-n junction in equilibrium.

Equation (12.13) is the missing link which will allow us to calculate the width of the depletion layer.

Design Example

We can find parameters of a typical p-n junction as follows. Assuming that the material is silicon (for which $n_i \approx 1.4 \times 10^{16}\ \text{m}^{-3}$ at $T = 300$ K), and that the donor and acceptor densities in the p- and n-regions have the typical values of $N_a = 10^{24}\ \text{m}^{-3}$, $N_d = 10^{22}\ \text{m}^{-3}$, respectively (so this is a p^+-n junction), Eq. (12.13) implies that the built-in voltage is

$$V_{bi} = \{1.38 \times 10^{-23} \times 300/1.6 \times 10^{-19}\} \log_e\{10^{24} \times 10^{22}/2 \times 10^{32}\} = 0.82\ \text{V}$$

The depletion layer width is then given by Eq. (12.9) (assuming that $\varepsilon = 11.7\ \varepsilon_0$ for Si) as

$$\text{w} = \surd\{(2 \times 11.7 \times 8.85 \times 10^{-12} \times 0.82/1.6 \times 10^{-19}) \\ (10^{24} + 10^{22})/(10^{24} \times 10^{22})\}\ \text{m} = 0.33\ \mu\text{m}$$

Most of this layer extends into the n-region since $N_a \gg N_d$. The salient features of a p-n junction are therefore a built-in voltage of the order of a volt, dropped across a layer less than a micron wide. This is equivalent to an electric field of about 2.5×10^6 V/m, a high value.

The Band Diagram Inside the Depletion Region

We may exploit the relation between voltage and the energy of a conduction electron once again to fill in the missing features of the band diagram *inside* the depletion region. We have already seen that the voltage $V(x)$ relative to the left-hand edge of the depletion layer varies parabolically. Consequently, the conduction band energy level will also be spatially-varying, following

$$\mathscr{E}_c(x) = \mathscr{E}_{cp} - eV(x) \tag{12.14}$$

The conduction band therefore bends parabolically across the depletion layer, as in Fig. 12.4. The valence band varies similarly, maintaining a constant energy gap $\mathscr{E}_g$ between the two levels.

Knowledge of this band structure may be used to calculate the free carrier distribution throughout the depletion layer. For example, if we substitute Eq. (12.14) into Eq. (11.86), we can find the variation of the conduction electron density as

$$n(x) = N_c \exp\{-[\mathscr{E}_c(x) - \mathscr{E}_F]/kT\} \tag{12.15}$$

A typical variation of $n(x)$ is shown in Fig. 12.5. Note that $n(x)$ does not abruptly fall to zero at the edge of the depletion layer as we have assumed so far; instead, the variation is smooth. However, the drop from the equilibrium value in the n-type material outside the depletion layer (labelled n_n) is sufficiently rapid for our assumed charge distribution (Fig. 12.2(a)) to be a good approximation to the truth.

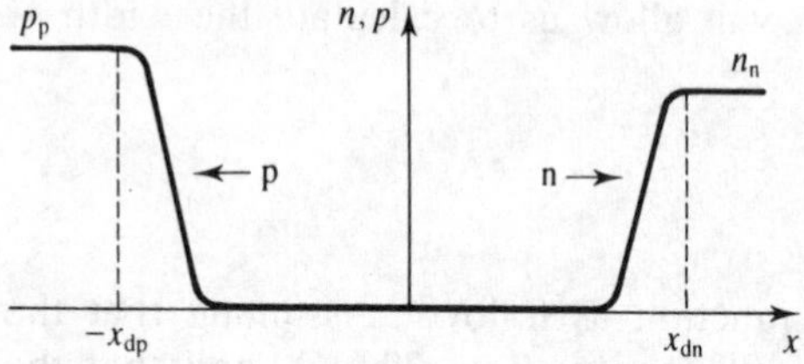

Figure 12.5 Carrier distribution for a p-n junction in equilibrium.

Our model is therefore reasonably self-consistent. The corresponding hole density variation $p(x)$ is also shown. The equilibrium hole density p_p is the p-type material may be vastly different from n_n if the junction is asymmetric (i.e. if $N_a \neq N_d$). Note also that the value of n at the depletion layer edge in the p-region (which we might call n_p) is not identically zero, although it will generally be fairly small. From Eq. (12.15) we can see that $n = n(x_{dn}) = N_c \exp\{-(\mathscr{E}_{cn} - \mathscr{E}_F)/kT\}$, so that $n_p = n(-x_{dp})$ is given by

$$n_p = n_n \exp(-eV_{bi}/kT) \tag{12.16}$$

The value of p at the depletion layer edge in the n-region (termed p_n) is found in a similar way as

$$p_n = p_p \exp(-eV_{bi}/kT) \tag{12.17}$$

where $p_p = N_v \exp\{-(\mathscr{E}_F - \mathscr{E}_{vp})/kT\}$, and $\mathscr{E}_{vp}$ is the valence band level in the p-region (Fig. 12.4). Given that V_{bi} is significantly greater than 25 mV (i.e. kT/e at 300 K), we find that $n_p, p_n \ll n_n, p_p$.

We can use these distributions to examine the details of the equilibrium inside the depletion layer, for example for electron motion. In Chapter 11 we saw that the current density for electrons is composed of two components (due to drift and diffusion, respectively). If the two contributions are calculated, it can be shown that they balance exactly at any point inside the depletion layer, and that the net current density is zero (although each component may individually be very large). A similar dynamic balance may be found for holes, so that equilibrium is established for both types of charge carrier. This contrasts with the situation outside the depletion layer, here there is no field and no concentration gradient either, so that the two current components are zero independently for both species.

The Band Diagram of a Biased p-n Junction

Suppose that electrical contacts are now attached to the two ends of the diode so that external voltages can be applied. We shall assume that these are metal, and of a particular type that forms an *ohmic contact* with the semiconductor. In this case the metal can supply or receive any number of electrons, so the carrier density at the interface will be the equilibrium value. For example, aluminium forms an ohmic contact on p- or heavily-doped n-type Si. Ohmic contacts may be also made to intrinsic or n-type Si, via an intermediate layer of n^+-Si. However, other combinations (e.g., Al on n$^-$-GaAs) form a rectifying contact called a *Schottky barrier*, which is itself a form of diode.

Since the depletion layer has a very low carrier concentration, we would expect the bulk of any applied voltage to be dropped across this highly resistive layer. If we make the n-type side more negative than the p-type by V volts—a condition known as *forward bias*—this will tend to reduce the built-in barrier from eV_{bi} to $e\{V_{bi} - V\}$. As a result, the band diagram will alter as shown in Fig. 12.6.

On the other hand, if the n-side is made more positive than the p-side—when the junction is said to be in *reverse bias*—the applied voltage will tend to increase the size of the built-in potential barrier. In this case the band diagram will modify as shown in Fig. 12.7.

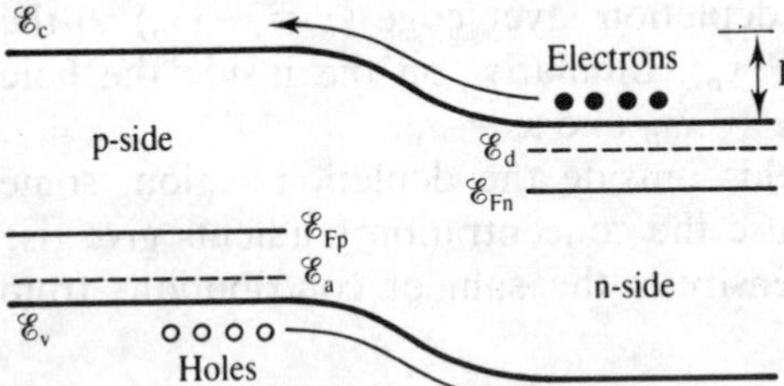

Figure 12.6 Energy band structure for a p-n junction in forward bias.

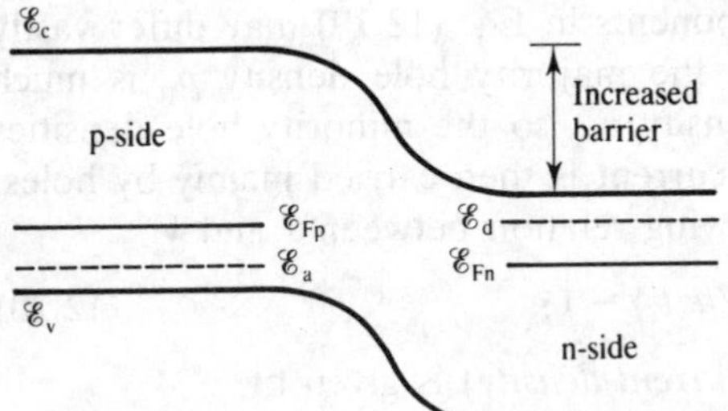

Figure 12.7 Energy band structure for a p-n junction in reverse bias.

Current—Voltage Characteristics

The reduction in the barrier height in a forward-biased junction will naturally make it easier for electrons to travel across the depletion region from the n-side to the p-side, and for holes to move in the reverse direction. The arrival of these carriers at the depletion layer edges will then cause a change in the minority carrier densities. The new values n_p' and p_n' may be found by substituting $V_{bi} - V$ for V_{bi} in Eqs (12.16) and (12.17). If this is done, it is easy to see that

$$n_p' = n_p \exp(eV/kT) \qquad \text{and} \qquad p_n' = p_n \exp(eV/kT) \tag{12.18}$$

Clearly n_p' will be much greater than the equilibrium value n_p (and similarly $p_n' \gg p_n$) for values of V greater than about 25 mV. Even a small forward bias will therefore produce a considerable increase in minority carrier density near the depletion layer edges. This is called *minority carrier injection*. Any such increase in n_p (for example) must follow from an actual transport of electrons from the n- to the p-side. However, it is important to recognize that the densities of the minority carriers are generally so much greater than those of the minority carriers that the fractional change in n_n is very small. Its value (and that of p_p) may thus be considered unchanged to a good approximation.

The fate of the injected carriers depends on the geometry. If the diode is short enough, recombination may be neglected, and the carriers simply diffuse to the contacts. In this case we must return to the results of Chapter 11, where we showed that the equations governing carrier diffusion had the form $\partial n,p/\partial t = D_{e,h}\partial^2 n,p/\partial x^2$. In the steady-state, $\partial n,p/\partial t = 0$, so the carrier densities must vary linearly with distance. The solution is then as shown in Fig. 12.8; on the p-side the electron concentration falls linearly from n_p' at the depletion layer edge ($x = -x_{dp}$) to the equilibrium value n_p at the contact ($x = -x_p$). Similarly, on the n-side the hole concentration falls from p_n' to p_n between $x = x_{dn}$ and $x = x_n$.

Despite the absence of appreciable fields outside the depletion region, some current must still flow to the contacts because the concentration gradients give rise to diffusion currents. The total current density is the sum of contributions from electron and hole diffusion, namely

$$J = D_e e\ \mathrm{d}n/\mathrm{d}x - D_h e\ \mathrm{d}p/\mathrm{d}x = D_e e\ \{n_p' - n_p\}/L_p + D_h e\ \{p_n' - p_n\}/L_n \quad (12.19)$$

where the lengths L_p and L_n are given by $L_p = x_p - x_{dp}$ and $L_n = x_n - x_{dn}$. Note that if the doping is unequal, the two components in Eq. (12.19) may differ vastly in size. For example, in a p^+-n junction, the majority hole density p_p is much greater than the corresponding electron density n_n, so the minority hole densities p_n' and p_n are greater than n_p' and n_p. The current is then carried mainly by holes. Substituting for n_p' and p_n' we find the following relation between J and V

$$J = J_s\{\exp(eV/kT) - 1\} \quad (12.20)$$

where J_s (known as the *reverse saturation current density*) is given by

$$J_s = D_e e n_p/L_p + D_h e p_n/L_n \quad (12.21)$$

This is the well-known exponential characteristic of a diode, shown in Fig. 12.9. For a device of cross-sectional area A, the current is simply $I = J \times A$, so the I–V characteristic has a similar shape.

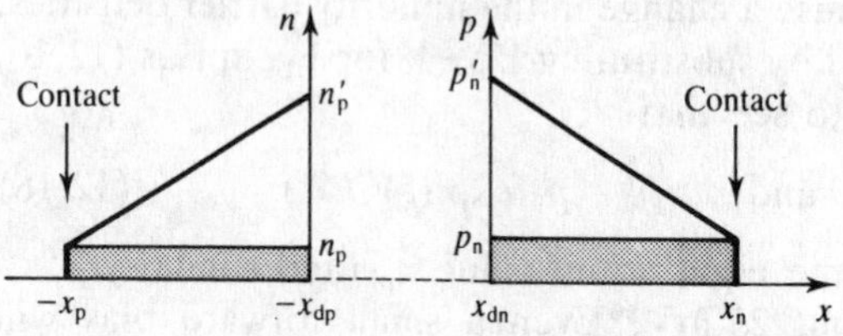

Figure 12.8 Minority carrier distribution outside the depletion layer of a forward-biased p-n junction.

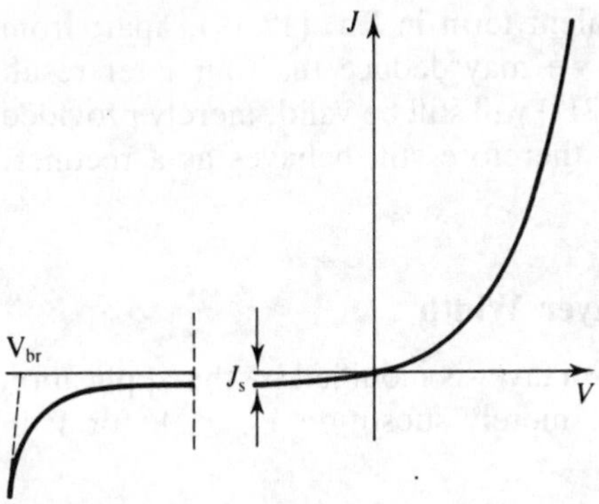

Figure 12.9 J–V characteristic for a p-n junction diode.

When V is zero, the current is also zero. For a small positive voltage, however, the forward current can be extremely high. In contrast, the reverse current rapidly saturates at $I_s = J_s \times A$ (which is typically small) when V is negative. As a result, the diode acts as a rectifier, presenting an extremely high impedance to reverse voltages and a low impedance to forward ones. Note that the behaviour of a real diode departs from the ideal at high reverse voltages, a feature we will return to shortly.

The Effect of Recombination

We now examine how this characteristic is modified if the distance to the contacts is large enough for recombination to be important (which is the case in light-emitting devices). Once again we start by returning to the results of Chapter 11. In the steady state the electron density (for example) on the p-side is governed by

$$D_e \partial^2 n/\partial x^2 - (n - n_p)/\tau_e = 0 \qquad (12.22)$$

where we have written n_p for the equilibrium level n_E. This is a standard second-order differential equation, which can be solved exactly, subject to the boundary conditions $n = n_p'$ at the edge of the depletion layer ($x = -x_{dp}$) and $n = n_p$ at the contact ($x = -x_p$). Provided x_p is sufficiently large, however, a reasonable approximation to the solution is given by

$$n(x) - n_p = \{n_p' - n_p\} \exp \{(x + x_{dp})/\sqrt{(D_e\tau_e)}\} \qquad (12.23)$$

Equation (12.23) shows that the excess electron density decays exponentially away from the depletion layer edge, falling to 1/e of its initial value in a length $L_e = \sqrt{(D_e\tau_e)}$. This is known as the *diffusion length* of electrons in the p-type material. The excess hole density on the n-side varies in a similar way, decaying to 1/e of its initial value in a corresponding length $L_h = \sqrt{(D_h\tau_h)}$.

The value of J_e at $x = -x_{dp}$ is the total injected electron current density. Performing the necessary differentiation we find that

$$J_e = D_e e\{n_p' - n_p\}/L_e \qquad (12.24)$$

This is qualitatively very similar to the equivalent term in Eq. (12.19), apart from the replacement of L_p by L_e. Consequently, we may deduce that our later result for the J–V characteristic (Eqs (12.20) and (12.21)) will still be valid, merely provided we replace L_p by L_e and L_n by L_h. A diode therefore still behaves as a rectifier, even when recombination is significant.

The Effect of Voltage on the Depletion Layer Width

We now consider how the width of the depletion layer is modified by the application of a voltage. To find the new width $w(V)$ we merely substitute $V_{bi} - V$ for V in Eq. 12.9, which gives

$$w(V) = \sqrt{\{1 - V/V_{bi}\}}\, w(0) \tag{12.25}$$

where $w(0)$ is the width of the unbiased junction. The effect of a negative value of V is therefore to increase the depletion layer width, while a positive value has the opposite effect. According to Eq. (12.25) w can become negative for voltages greater than V_{bi}; this is clearly unrealistic, and merely shows that our original approximation concerning the charge distribution inside the depletion layer has broken down at this point.

Avalanche Breakdown

At high reverse voltages our approximations will again break down (although in a slightly different manner). Consider the band diagram shown in Fig. 12.10. An electron moving to the right will travel down the sloping region of the conduction band in a series of 'hops', with each abrupt decrease in energy corresponding to a collision with the lattice. If the electric field is high enough the electron may gain a considerable amount of kinetic energy between collisions. Consequently, it is possible for the energy given up at a collision to be greater than $\mathscr{E}_g$. In this case the energy is sufficient to ionize a lattice atom by promoting a valence electron to the conduction band. If this occurs the number of carriers will increase, since there will now be two electrons moving to the right, and one hole moving to the left. After further acceleration by the field the electrons may perform further ionizations, and so on. Under high reverse bias a shower of new carriers may therefore be created. This leads to a sudden rise in the reverse current—a behaviour known as *avalanche breakdown*—at the characteristic breakdown voltage V_{br} shown in

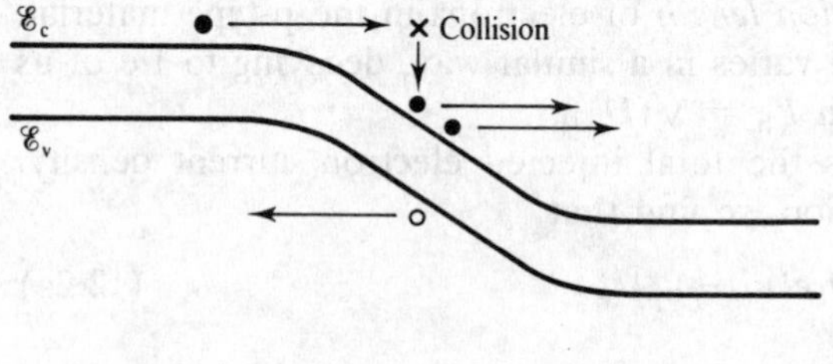

Figure 12.10 Carrier multiplication by avalanche breakdown under high reverse bias.

Fig. 12.9. This effect is exploited in the avalanche photodiode, which we will describe later.

Heterojunction Diodes

Up until now we have assumed that the materials on either side of the junction are of essentially the same type, with any differences introduced through doping. This need not be the case—for example, we have already mentioned the metal–semiconductor junction. Alternatively, different semiconductors may be used. Clearly there are certain constraints; it must be feasible to grow the layered structure as a perfect crystal, which requires the two materials to be lattice-matched. Despite this restriction, it is entirely possible to fabricate diodes with different bandgaps on either side of the junction. In this case the structure is known as a *heterojunction*.

Heterojunctions are of great importance in optoelectronics, since they allow two important new features to be built in to the basic diode structure. First, the difference in refractive index between the two materials may be used to provide optical confinement, as in the heterostructure waveguides described in Chapter 9. Second, the modification to the energy band diagram introduced by the variation in $\mathscr{E}_g$ may be used to provide different potential barriers to the motion of electrons and holes, allowing each type of carrier to be confined independently to a particular region of space. As a result, the performance of light-emitting devices may be improved enormously.

The energy band diagram of a typical heterojunction is shown in Fig. 12.11, under equilibrium conditions. Here the energy gap $\mathscr{E}_{g1}$ on the p-side is greater than the value $\mathscr{E}_{g2}$ on the n-side, so the structure is known as a P-n anisotype heterojunction (the upper-case letter denoting the material with the larger energy gap). Outside the depletion layer the diagram is constructed as before, using the continuity of the Fermi level to determine the absolute levels of the conduction and valence bands. Inside the depletion layer itself, however, the construction is slightly different.

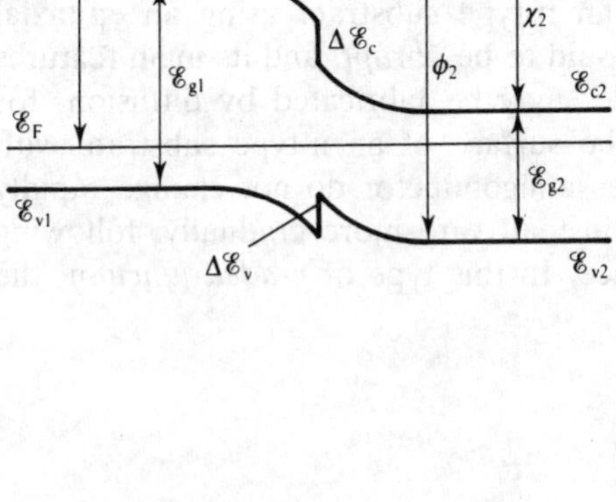

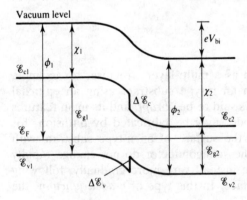

Figure 12.11 Energy band structure for a P-n heterojunction in equilibrium.

The analysis required to determine the band variation may be adapted from that used for the homojunction. The exposed charge distribution of Fig. 12.2(a) may again be used; however, the electric field variation shown in Fig. 12.2(b) must be modified to allow for the different dielectric constants ε_1 and ε_2 in the two materials—the former value must be used in the upper line in Eq. (12.4), the latter in the lower one. Integration of the electric field to find the potential variation $V(x)$ across the depletion layer may then be carried out much as before. If this is done, a slightly different proportion of the built-in voltage V_{bi} is found to be dropped across each side of the junction.

Due to the difference in energy gap between the two sides of the junction, it is no longer possible to construct the conduction and valence bands by assuming that they follow the potential variation $V(x)$. In fact, as we shall see shortly, both bands must be discontinuous. Instead, we must rely on the assumption that the vacuum level—the level above the conduction band—is continuous and follows $V(x)$. This band is generally displaced by an amount χ from the conduction band level $\mathscr{E}_c$, where χ is a quantity known as the *electron affinity* (the energy required to liberate an electron from the conduction band) and ϕ above the Fermi level, where ϕ is the *work function*.

To complete the diagram, the vacuum level is drawn in on either side of the junction. The difference in height of this level then determines the built-in voltage V_{bi}, allowing the width of the depletion layer to be found as before. Within the depletion layer the variation of the vacuum level follows $V(x)$. The conduction band is then drawn in at a height χ_1 below the vacuum level on the p-side, and χ_2 below on the n-side, where χ_1 and χ_2 are the electron affinities in the two materials. This leads to an abrupt discontinuity $\Delta\mathscr{E}_c \approx \chi_2 - \chi_1$ in the conduction band level at the boundary between the semiconductors. The valence band is then drawn in at a height $\mathscr{E}_{g1}$ below the conduction band level on the p-side, and $\mathscr{E}_{g2}$ below on the n-side, leading to a discontinuity $\Delta\mathscr{E}_v$ at the interface. The net result is that the potential barrier faced by electrons has height $eV_{bi} + \Delta\mathscr{E}_c$, while that for holes has height $eV_{bi} - \Delta\mathscr{E}_v$, making it harder for electrons to cross the junction, and easier for holes. The same state of affairs exists in the P-p isotype heterojunction, formed between two different p-type semiconductors. Other combinations may be envisaged where the reverse is true, so that holes face a higher barrier; this is the case in both p-N and n-N structures.

Practical p-n Junction Diodes

The p-n junction diode may be fabricated as a multi-layer structure, for example by growing a layer of p-type material on an n-type substrate using an epitaxial process. In this case the resulting junction is said to be *abrupt*, and its main features are as described above. Alternatively, diodes may be fabricated by diffusion, for example by heavily doping a region near the surface of an n-type substrate with acceptors. In this case the properties of the semiconductor do not change rapidly between the p- and n-type regions, but do instead vary more gradually, following the distribution of dopant beneath the surface. In this type of *graded junction*, the charge distribution in the depletion region varies more smoothly than we have assumed, and the electric field distribution must also be modified accordingly. Nonetheless, the operating principles described above are still broadly correct.

12.2 ELECTRO-OPTIC SEMICONDUCTOR DEVICES

We now return to a problem we touched on briefly in Chapters 9 and 10; that of constructing electro-optic waveguide devices in semiconducting materials. Here the difficulty is that the presence of carriers prevents the development of a high electric field across the guide. The p-n junction provides a solution to the problem. For example, Fig. 12.12(a) shows a cross-section of a typical phase modulator, built from an n-type InP homostructure strip-loaded waveguide on a <100> n^+ substrate. The guide is capped with a p^+ layer, thus forming a p^+-n homojunction at the top of the guide, with the bulk of the depletion region located inside the guide itself. One ohmic contact is placed on top of the p^+ layer, while the other is connected to the substrate. When the junction is reverse biased, the highly resistive depletion region grows downwards, so that all of the applied voltage is eventually dropped

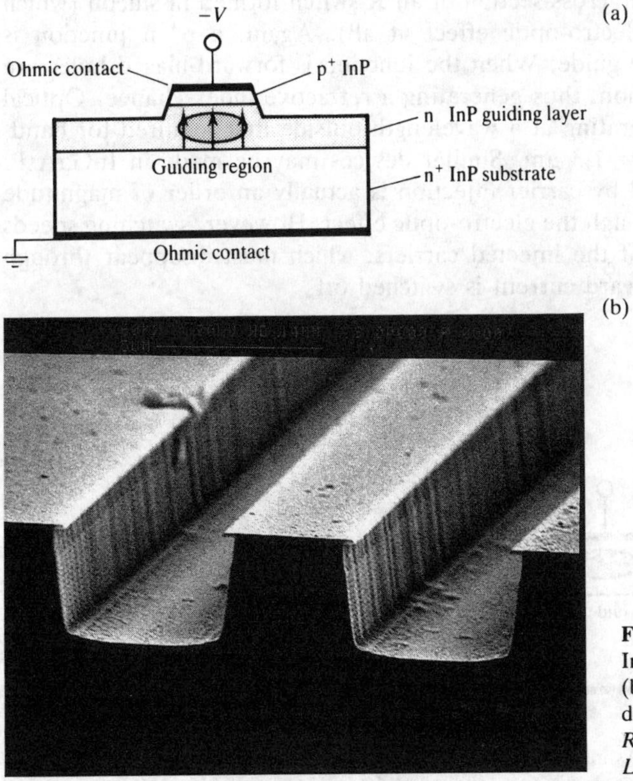

Figure 12.12 (a) Cross-section of an InP integrated optic phase modulator; (b) SEM photograph of an actual device (*photograph courtesy P.M. Rogers, British Telecom Research Laboratories*).

across the guide layer. The field is then sufficient to obtain a usable index change through the relatively weak r_{41} electro-optic coefficient, thus inducing an optical phase change. Figure 12.12(b) is an electron microscope photograph of a similar device, based on an InP/InGaAsP heterostructure.

Electro-optic directional coupler switches may be constructed in a similar way, using two closely-spaced strip-loaded guides. For example, Fig. 12.13 shows a cross-section through a typical directional coupler, this time constructed using n^--on-n^+ GaAs homostructure guides. Here a much simpler electrode structure is used, with Al electrodes deposited directly on top of the two guides. As a result, these contacts are not ohmic (although the substrate contact is), but are instead of the rectifying Schottky barrier type mentioned above. These also act as diodes, so that operation in reverse bias is much as described above. To switch the coupler, a negative voltage is applied to just one of the Schottky electrodes, thus inducing an electric field which desynchronizes the guides.

Electrically-switchable devices may even be constructed using semiconductors with weak or negligible electro-optic coefficients. We already know from Chapters 3 and 9 that the refractive index changes used in homostructure guides are obtained through variations in doping, which cause changes in the free carrier concentration. A similar index change can be created dynamically, by the electrically-controlled injection of minority carriers, this time using a forward-biased p-n junction. For example, Fig. 12.14 shows the cross-section of an X-switch formed in silicon (which does not exhibit a linear electro-optic effect at all). Again, a p^+-n junction is constructed at the top of the guide. When the junction is forward-biased holes are injected into the guiding region, thus generating a refractive index change. Optical absorption is avoided by operating at a wavelength outside that required for band-to-band transitions, e.g. $\lambda_0 = 1.3\ \mu m$. Similar devices may be made in InGaAsP, and the index change caused by carrier injection is actually an order of magnitude larger than that available through the electro-optic effect. However, switching speeds are limited by the lifetime of the injected carriers, which must disappear through recombination when the forward current is switched off.

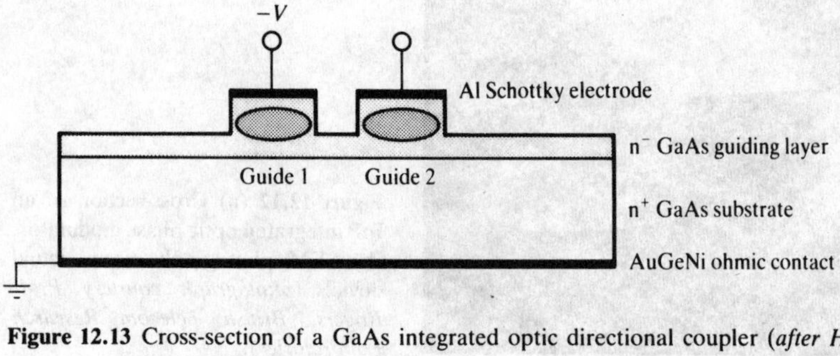

Figure 12.13 Cross-section of a GaAs integrated optic directional coupler (*after H. Inoue* et al., *1986*).

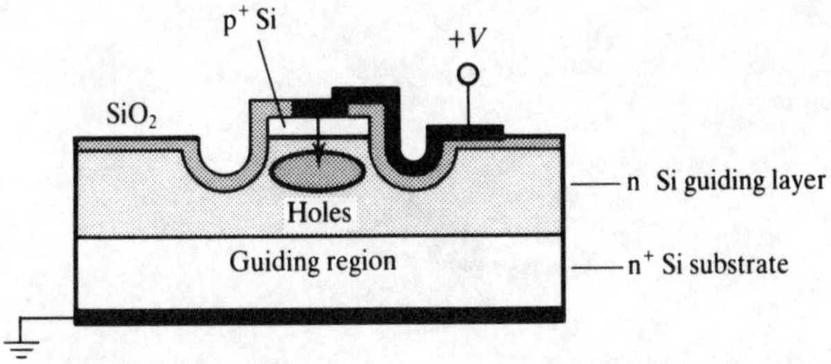

Figure 12.14 Cross-section of a Si-based carrier-injection switch (*after J. P. Lorenzo* et al., *1987*).

12.3 PHOTODIODES

We shall now consider the p-n junction in the alternative role of a light detector, or *photodiode*. In Chapter 11 we noted that absorption in a bulk semiconductor can induce band-to-band transitions, causing a measurable increase in conductivity. However, the basic weaknesses of photoconductive detection are a high intrinsic level of noise (because of the presence of thermally-excited carriers), and a trade-off between speed and sensitivity (due to recombination of carriers before they can be collected). The solution is to arrange for the absorption to occur in a region devoid of thermal carriers, and then use a high electric field to sweep the photo-generated carriers apart before they can recombine. This can be done inside the depletion layer of a p-n junction. Figure 12.15 shows the excitation of an electron from the valence band into the conduction band in this region, by a photon of energy $hf > \mathscr{E}_c - \mathscr{E}_v$. Clearly the built-in field will tend to drive the photo-generated electron out of the depletion region towards the n-side, and the corresponding hole towards the p-side.

Two types of photodiode are in common use. Most are surface-entry devices, where the light is input through (for example) the p^+ side of a p^+-n junction. Ideally, the light should pass into the depletion layer without suffering significant absorption in the p^+ region because carrier pairs generated outside the depletion layer will not be separated. Once inside this layer the absorption should be total because any light transmitted into the n-region will also be wasted. The depletion layer should therefore be as thick as possible. Side-entry devices, where the light is

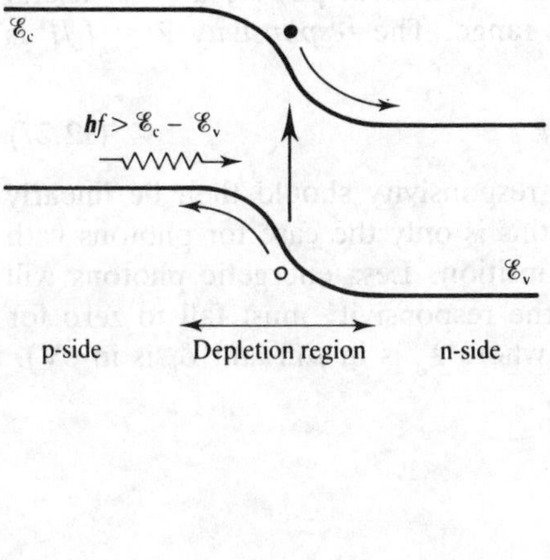

Figure 12.15 Generation of an electron-hole pair by a photon inside the depletion region of a p-n junction.

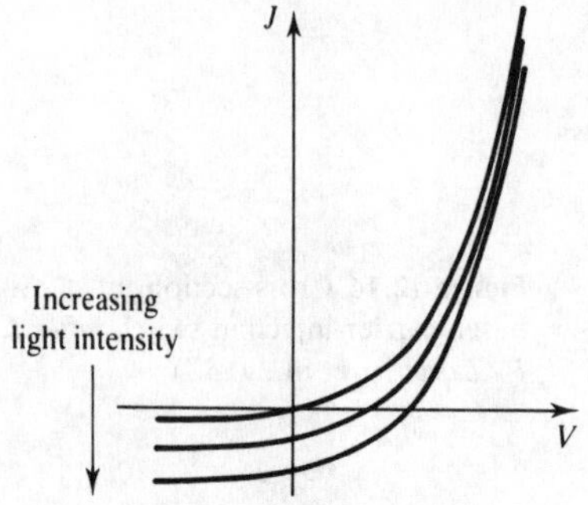

Figure 12.16 J–V characterisic for a p-n junction diode in the presence of a light flux.

passed directly into the depletion layer, are another possibility. In this case the useful absorbing region may be extremely long. However, the depletion layer should again be thick, or the light will spill out on either side.

Once out of the depletion layer the carriers diffuse to the contacts. As a result, there is an increase in the reverse current, so that the J–V characteristic is modified in the presence of light as shown in Fig. 12.16. Photodiodes are normally operated at a fixed reverse bias voltage, when they act as a current source. In this case a reverse current (known as the *dark current*) flows in the absence of illumination and contributes to noise. For a large reverse bias, this current will be I_s.

We may estimate the responsivity of a photodiode as follows. Assuming that the optical power falling on the detecting area is P, the number of photons arriving per second is $P/\mathbf{h}f = P\lambda_0/\mathbf{h}c$. However, not all of these will be absorbed, and not all the carriers generated by those that are will actually reach the contacts. We may model this by defining a *quantum efficiency* η, representing the number of useful carrier-pairs generated per photon. Clearly the maximum possible value of η is 100 per cent. In this case every photon generates one electron (which is collected on the n-side) and one hole (which arrives on the p-side). The net effect is that one unit of charge crosses the device from side-to-side. The total charge crossing in unit time, which is the *photocurrent* I_P, is then

$$I_P = \eta P e \lambda_0/\mathbf{h}c \tag{12.26}$$

so the photocurrent is linearly proportional to the optical power, a most useful result which holds good over a wide dynamic range. The *responsivity* R = I_p/P is therefore given by

$$R = \eta e \lambda_0/\mathbf{h}c \tag{12.27}$$

Assuming a uniform quantum efficiency, the responsivity should then be linearly proportional to the wavelength λ_0. However, this is only the case for photons with sufficient energy to promote an interband transition. Less energetic photons will not contribute to the photocurrent at all, so the responsivity must fall to zero for $\lambda_0 > \lambda_g$, where $\lambda_g = \mathbf{h}c/\mathscr{E}_g$ (or $\lambda_g = 1.24/\mathscr{E}_g$, where λ_g is in μm and $\mathscr{E}_g$ is in eV).

Design Example

We may calculate the theoretical responsivity of a silicon p-n junction photodiode as follows. $\mathscr{E}_g$ for Si is 1.14 eV at 300 K, so the optimum wavelength is λ_g = 1.24/1.14 = 1.09 μm. Assuming a uniform quantum efficiency of 100 per cent, R_{max} = $1.6 \times 10^{-19} \times 1.09 \times 10^{-6}/(6.62 \times 10^{-34} \times 3 \times 10^{8})$ = 0.88 A/W. The theoretical responsivity is then as shown by the dashed line in Fig. 12.17.

In practice, the quantum efficiency of a surface-entry device is variable, and peaks at about 80 per cent—at short wavelengths η is reduced because the attenuation is so strong that the light is absorbed before it reaches the depletion layer, and at long wavelengths it falls because the light is transmitted right through this layer. Furthermore, since Si has an indirect band-gap, η is reduced further near λ_g, because a simultaneous interaction with a high-energy photon is required for momentum conservation in the band-to-band transition, a relatively unlikely event. The characteristics of a real Si photodiode are therefore given by the full line in Fig. 12.17, which shows a useful responsivity only up to $\lambda_0 \approx 0.85$ μm. For high sensitivity at the near-infrared wavelengths used in second-generation fibre communication systems, materials with smaller bandgaps are required. Ge (useful up to $\lambda_0 \approx$ 1.55 μm) or GaInAs (up to 1.67 μm) are both suitable.

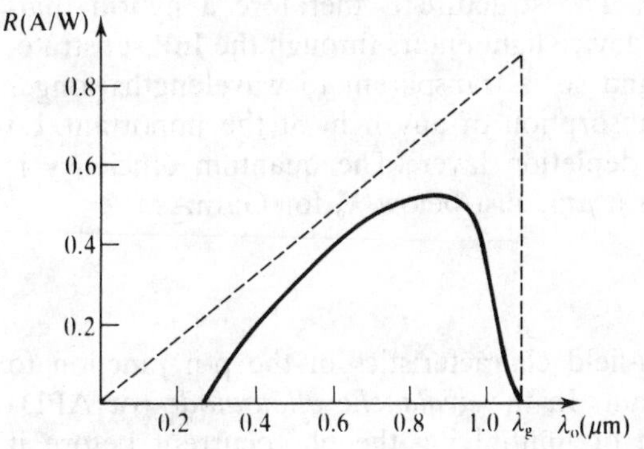

Figure 12.17 Ideal (dashed line) and actual (full line) responsivity of a silicon p-n junction photodiode.

PIN Photodiodes

The basic limitation of a p-n junction photodiode—that the depletion layer is so thin that radiation of long wavelength is only weakly absorbed—is overcome in the p-i-n structure shown in Fig. 12.18(a) (known as a *PIN photodiode*). Here a region of intrinsic or lightly-doped material is introduced between two heavily-doped p- and n-type regions. Because the doping is so low in this region, the depletion layer can be arranged to extend through it under a modest reverse bias, yielding the electric field variation of Fig. 12.18(b). The effective depletion layer width may therefore be fixed at a value far greater than the 'natural' one, approximately the width of the intrinsic region.

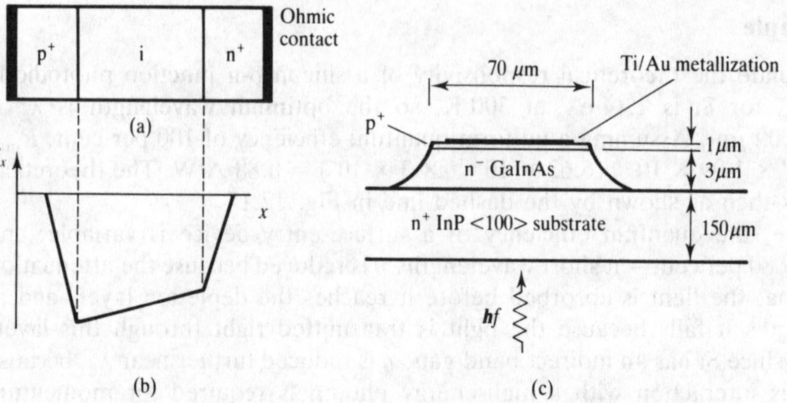

Figure 12.18 (a) A p-i-n structure, (b) the corresponding electric field variation, and (c) a substrate-entry GaInAs PIN photodiode (*after R.H. Moss* et al., *1983*).

Figure 12.18(c) shows a cross-section of a typical GaInAs PIN photodiode. The n^- region is a mesa of $Ga_{0.47}In_{0.53}As$ (a particular composition that is lattice-matched to InP). This is grown epitaxially on an n^+ InP substrate, and the p^+ region is then made by zinc diffusion. The structure is therefore a hybrid, part homo- and part hetero-junction. In this device light enters through the InP substrate, which has an energy gap of 1.35 eV and so is transparent to wavelengths longer than about 0.92 μm. This avoids the absorption of any light at the important 1.3 and 1.55 μm wavelengths outside the depletion layer. The quantum efficiency is then almost uniform between 1.0 and 1.6 μm, just below λ_g for GaInAs.

Avalanche Photodiodes

It is possible to make use of the high-field characteristics of the p-n junction to construct a photodiode with internal gain. In the *avalanche photodiode* (or APD) impact ionization is used as a method of multiplying the photocurrent before it enters the circuit of any electrical amplifier, thus providing an increase in receiver sensitivity. Figure 12.19 shows the reach-through avalanche photodiode (RAPD), which illustrates the basic principle involved.

The RAPD is a four-layer device with a p^+-i-p-n^+ structure (Fig. 12.19(a)). Under reverse bias most of the applied voltage is dropped across the p-n^+ junction, although the exact fraction depends on the doping levels. As the voltage is increased the depletion layer associated with this junctions widens, until it just reaches through to the intrinsic region. The electric field distribution is then as shown in Fig. 12.19(b). In the depletion region the field is extremely high; normally the peak field is held at roughly 10 per cent below the value at which avalanche breakdown occurs. In the long intrinsic region, however, the electric field is only moderate. Light enters the structure through the p^+ layer and travels through to the intrinsic region, where

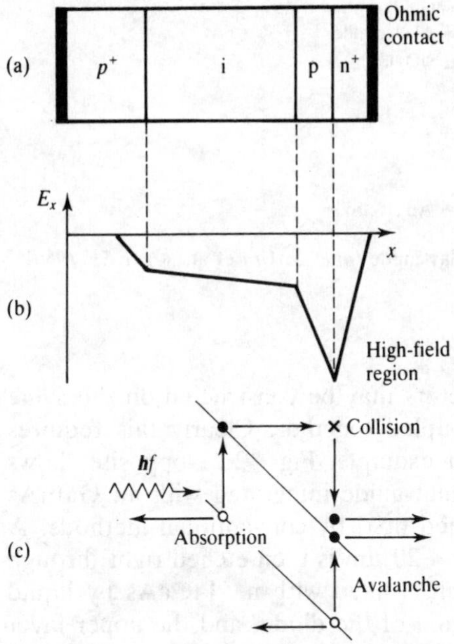

Figure 12.19 (a) Avalanche photodiode structure, (b) the corresponding electric field variation, and (c) the processes of absorption and avalanche gain.

it is absorbed. The photogenerated carriers drift in the moderate electric field, with electrons travelling towards the p-n$^+$ junction. Carrier multiplication then takes place inside the high-field region by impact ionization. As a result, there is a large increase in the photocurrent, up to 1000 times, which provides the desired improvement in detector sensitivity. However, since the avalanche process is a statistical one, there is a corresponding increase in the noise level.

Waveguide-integrated Photodiodes

The integration of photodetectors with waveguides poses an inherent compatibility problem—the former are intended to act as efficient, localized absorbers of light, while the latter should (ideally) be lossless. A number of strategies may be used to circumvent the difficulty. First, the waveguide and the detector may be constructed on different levels of a multilayer, allowing the properties of each level to be tailored individually. Light may then be transferred between the two at the appropriate point, either by directional coupling or by radiation. In the latter case, a standard component such as a grating coupler may be used, or the guide may be induced to cut off. For example, Fig. 12.20 shows the cross-section of a structure combining a passive guide made in two glassy layers with a p-n junction photodiode formed by a p-type diffusion into an n-type silicon substrate. To couple light into the detector, the height of the SiO_2 buffer layer is gradually reduced until the high-index substrate approaches the guide sufficiently to induce cut-off. The light is then radiated onto the detector.

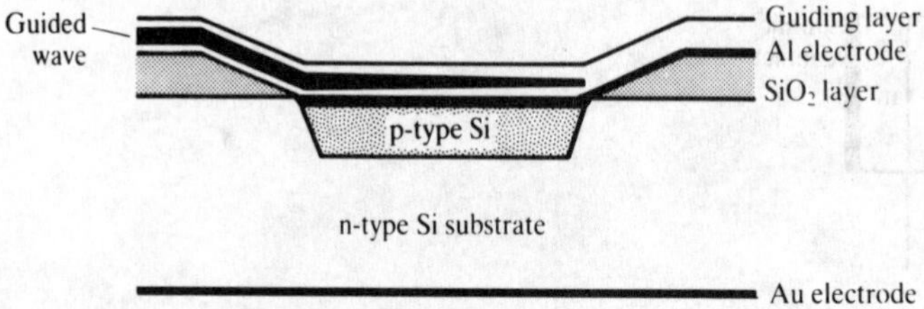

Figure 12.20 An Si p-n junction integrated with an overlay guide (*after S. Ura* et al., © *IEEE 1986*).

Alternatively, waveguides and photodetectors may be constructed on the same level of an integrated structure, and butt-coupled together. Clearly this requires material properties to be tailored *locally*. For example, Fig. 12.21 opposite shows an InP/InGaAsP double-heterostructure channel guide integrated with an GaInAs PIN photodiode. Here the waveguide is formed first, by conventional methods. A trench with dimensions approximately 10 μm × 20 μm is then etched right through the guide to the n^+-InP substrate. This trench is filled with n^- InGaAs by liquid phase epitaxy (LPE) to form the intrinsic region of the diode, and the upper layer is then Zn-doped to form the p^+ region. The net result is that the channel guide is coupled directly into the depletion region of a vertically-oriented PIN photodiode.

12.4 THE LIGHT-EMITTING DIODE

We now consider another application of the p-n junction, as a light source. In this incarnation it is known as a *light-emitting diode* (or LED). In Chapter 11 we described electroluminescence as a mechanism for generating photons in a direct-gap semiconductor. However, an isolated lump of material will not emit significant quantities of light because, in thermal equilibrium at room temperature, the number of downward electron transitions is extremely small. To improve the optical output we must move the material far from equilibrium, so that the rate of spontaneous emission is considerably increased. This might be done by taking (for example) a p-type material, which already contains a large hole density, and pouring electrons into it. This can be done in a forward-biased p-n junction. Figure 12.22 opposite illustrates the process; electrons are injected into the p-type material, where they combine with the holes already present. At the same time, holes are injected into the n-type material; however, the junction is normally highly asymmetric, so that light emission takes place mainly from one side.

The Emission Spectrum of an LED

Once again we recall that spontaneous emission generates photons that travel in random directions, so the emission is isotropic; as we will show later, this greatly

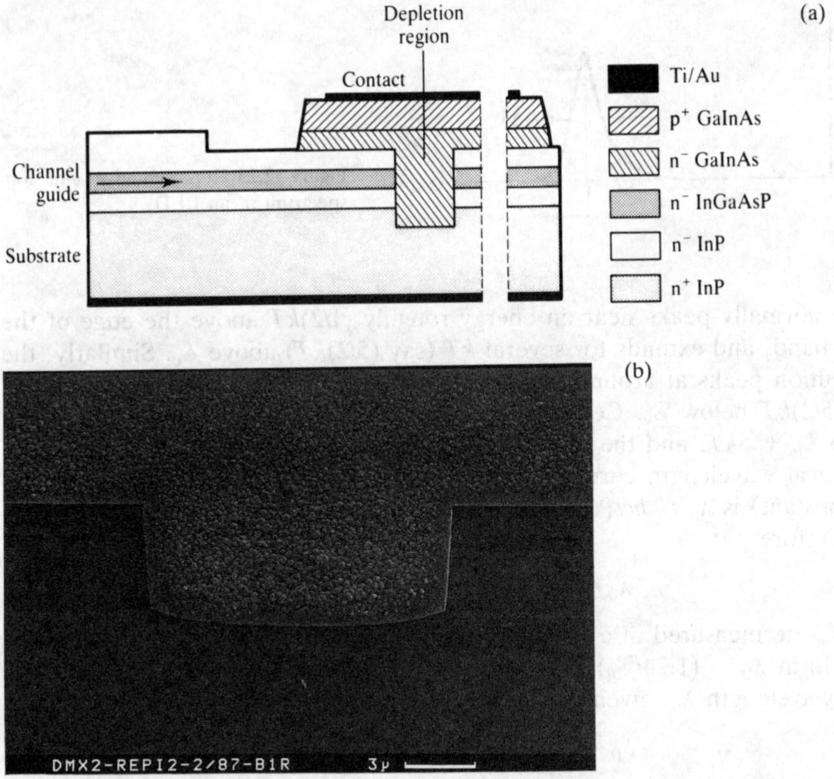

Figure 12.21 (a) Cross-section of an InGaAs PIN photodiode integrated with an InGaAsP waveguide (*after W. Döldissen* et al., *1989*), and (b) SEM view in the region of LPE overgrowth (*photograph courtesy W. Döldissen, Heinrich-Hertz Institut*).

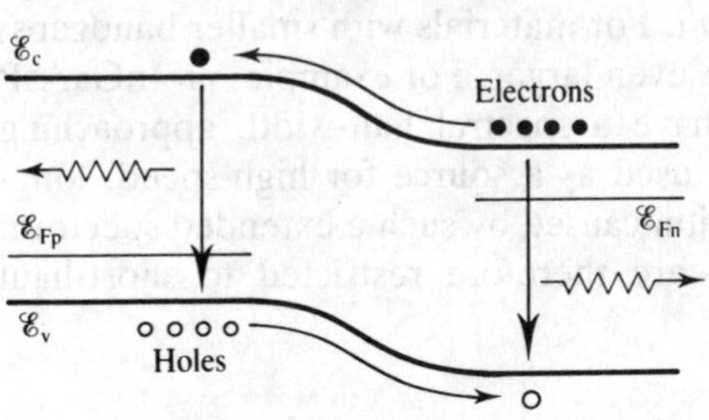

Figure 12.22 Spontaneous emission by recombination of injected minority carriers in a forward-biased p-n junction.

reduces the external efficiency of an LED. The emission is also unpolarized. Furthermore, we note that the output does not consist only of light at the wavelength $\lambda_g = hc/\mathscr{E}_g$, as a two-state model would imply. Because of the wide distribution of carrier energies within the conduction and valence band, the output is incoherent, with a spectrum consisting of a broad range of wavelengths (Fig. 12.23). The electron

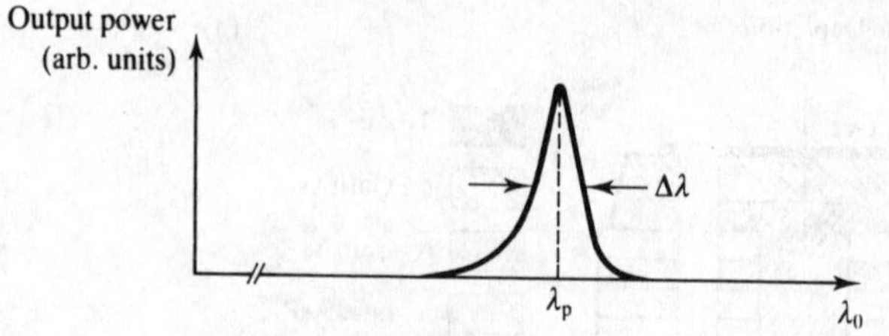

Figure 12.23 Representative emission spectrum of an LED.

distribution normally peaks near an energy roughly (1/2)kT above the edge of the conduction band, and extends for several kT (say (5/2)kT) above $\mathscr{E}_c$. Similarly, the hole distribution peaks at around (1/2)kT below the valence band edge, extending for about (5/2)kT below $\mathscr{E}_v$. Consequently, the possible photon energies lie in the range $\mathscr{E}_g$ to $\mathscr{E}_g + 5kT$, and the most likely energy is around $\mathscr{E}_g + kT$.

The optical wavelength corresponding to an energy $\mathscr{E}_g + \alpha kT$ (where α is an arbitrary constant) is $\lambda_0 = \boldsymbol{h}c/\{\mathscr{E}_g + \alpha kT\}$. Since kT is typically much less than $\mathscr{E}_g$, we may therefore put

$$\lambda_0 \approx (\boldsymbol{h}c/\mathscr{E}_g)\,\{1 - \alpha kT/\mathscr{E}_g\} \tag{12.28}$$

If kT and $\mathscr{E}_g$ are measured in eV, and λ_0 in μm, Eq. (12.28) may be written in the alternative form $\lambda_0 \approx (1.24/\mathscr{E}_g)\,\{1 - \alpha kT/\mathscr{E}_g\}$. The output from a LED therefore peaks at a wavelength λ_p, given by

$$\lambda_p \approx (1.24/\mathscr{E}_g)\,\{1 - kT/\mathscr{E}_g\} \tag{12.29}$$

Similarly, the half-power bandwidth of the emission spectrum is $\Delta\lambda$, where

$$\Delta\lambda \approx 3.1\,kT/\mathscr{E}_g^2 \tag{12.30}$$

Now, $\mathscr{E}_g$ for GaAs is 1.42 eV. For a GaAs LED at room temperature, for example, we therefore find that $\lambda_p \approx (1.24/1.42)\,\{1 - 0.0258/1.42\} = 0.857\ \mu$m. Similarly, $\Delta\lambda = 3.1 \times 0.0258/1.42^2 = 0.04\ \mu$m, or 40 nm. For materials with smaller bandgaps, designed to emit at longer wavelengths, $\Delta\lambda$ is even larger. For example, an InGaAsP LED operating near $\lambda_0 = 1.3\ \mu$m would have a spectral half-width approaching 100 nm. Consequently, the LED cannot be used as a source for high-speed, long-distance optical communications; the dispersion caused by such a extended spectrum would be far too large. LED transmitters are therefore restricted to short-haul applications.

The Frequency Response of an LED

We now consider the efficiency of the electroluminescence process inside an LED, which we might term the *internal efficiency* η. As it turns out, η is frequency-dependent. We can illustrate this by calculating the optical output of an LED for a time-varying drive current, assuming a two-state model for simplicity. We start by returning to the rate equations of Chapter 11. Considering spontaneous emission to be the dominant process, and the injected carriers to be electrons, the first relevant equation is Eq. (11.132). Since the system will be significantly perturbed, however, we must replace terms of the form Δn by n, and so on. Furthermore, for simplicity we shall consider the LED to be a lumped-element device, consisting of a uniform recombination region of cross-sectional area A and depth Δx (i.e. a finite volume $v = A\Delta x$). In this case we may rewrite Eq. (11.132) as

$$\mathrm{d}n/\mathrm{d}t + n/\tau_e = J/e\Delta x = I/ev \tag{12.31}$$

where $I = J\,A$ is the total current injected into the recombination region. In the general case, the current may consist of a d.c. bias with a superimposed a.c. modulation. We might therefore put

$$I = I' + I'' \exp(\mathrm{j}\omega t) \tag{12.32}$$

where I′ is the d.c. bias, and I'' and ω are the amplitude and angular frequency of the a.c. modulation, respectively. We assume that $I' > I''$, so that the LED is always under forward bias. We may take the solution for the electron density in a similar form, as

$$n = n' + n'' \exp(\mathrm{j}\omega t) \tag{12.33}$$

Substituting these expressions into Eq. (12.31) we get

$$n' + n''(1 + \mathrm{j}\omega\tau_e) \exp(\mathrm{j}\omega t) = \{I' + I'' \exp(\mathrm{j}\omega t)\}\, \tau_e/ev \tag{12.34}$$

Equating the d.c. and a.c. terms in Eq. (12.34), it is clear that we must have

$$n' = I'\tau_e/ev \qquad \text{and} \qquad n'' = (I''/ev)\,\{\tau_e/(1 + \mathrm{j}\omega\tau_e)\} \tag{12.35}$$

We now return to the corresponding the photon rate equation. For spontaneous emission due to electron injection, the rate of change of photons may be taken as

$$\mathrm{d}\phi/\mathrm{d}t = n/\tau_{rr} - \phi/\tau_p \tag{12.36}$$

Here we have included an additional term $-\phi/\tau_p$, which describes rate of loss of photons from the active volume (since it is now finite). This must be proportional to ϕ itself, and the proportionality constant τ_p is known as the *photon lifetime*. For an active volume whose longest dimension is L, τ_p is of the order of L/v_g, the time taken for a photon to escape from the region. For typical device dimensions of 100 μm, τ_p is $\approx 10^{-12}$ seconds. Since $1/\tau_p$ is then so large, we may neglect $\mathrm{d}\phi/\mathrm{d}t$ by comparison, and write

$$\phi/\tau_p = n/\tau_{rr} \tag{12.37}$$

For a recombination volume v, the total number of photons escaping per second is $v\phi/\tau_p$. This must represent the external photon flux Φ generated by the diode. Since each photon carries energy $\boldsymbol{h}c/\lambda_g \approx \mathscr{E}_g$, the optical power P emitted is

$$\mathrm{P} \approx \boldsymbol{h}c\Phi/\lambda_g \approx (\boldsymbol{h}c/\lambda_g)\, vn/\tau_{rr} \tag{12.38}$$

We now assume a solution for P in the form of a d.c. emission plus an a.c. modulation term; i.e. as

$$P = P' + P'' \exp(\mathrm{j}\omega t) \tag{12.39}$$

Substituting this expression into Eq. (12.38), together with our previous solution for n, we get

$$P' + P'' \exp(j\omega t) = (hc/\lambda_g)\,[(I'\tau_e/\tau_{rr}e) + (I''/\tau_{rr}e)\,\{\tau_e/(1 + j\omega\tau_e)\}\,\exp(j\omega t)] \tag{12.40}$$

Equating the d.c. and a.c. components separately in Eq. (12.39) we then obtain

$$P' = \eta' hcI'/e\lambda_g \quad \text{and} \quad P'' = \eta'' hcI''/e\lambda_g \tag{12.41}$$

where we have defined two new quantities, the d.c. and a.c. internal efficiencies η' and η'', as

$$\eta' = \tau_e/\tau_{rr} \quad \text{and} \quad \eta'' = \tau_e/\{\tau_{rr}(1 + j\omega\tau_e)\} \tag{12.42}$$

Equation (12.41) shows that the optical output from an LED is linearly proportional to current, for both d.c. and a.c. injection. However, the internal efficiency in each case is slightly different. We shall now examine the implications of this result by considering the likely values of η' and η''.

DC Internal Efficiency

Bearing in mind that $1/\tau_e = 1/\tau_{rr} + 1/\tau_{nr}$, the d.c. efficiency η' has the value

$$\eta' = r_{rr}/\{r_{rr} + r_{nr}\} \tag{12.43}$$

where $r_{rr} = 1/\tau_{rr}$ and $r_{nr} = 1/\tau_{nr}$ are the rates of radiative and non-radiative transitions, respectively. Consequently, the d.c. efficiency depends on the fraction of all downward transitions that are radiative, a physically convincing result. In simple LEDs η' may be as high as 50 per cent. Substituting for the constants, the d.c. output power may be written in the simple form $P' = 1.24\,\eta' I'/\lambda_g$, where λ_g is measured in μm. For a GaAs LED emitting at $\lambda_g \approx 0.85$ μm, with $\eta' \approx 50$ per cent, roughly 0.73 mW of optical power should thus be generated per mA of drive current. However, this figure is lowered by absorption, and by total-internal-reflection at the surfaces of the device, as we shall show later.

AC Internal Efficiency

Combining Eqs (12.42), we note that the a.c. internal efficiency may also be written as

$$\eta'' = \eta'/(1 + j\omega\eta'\tau_{rr})\} \tag{12.44}$$

Since η'' is complex, it is clear that the optical modulation cannot be in phase with the drive current modulation, but must instead lag behind. The magnitude of η'' is given by

$$|\eta''| = \eta'/\sqrt{\{1 + \omega^2\eta'^2\tau_{rr}^2\}} \tag{12.45}$$

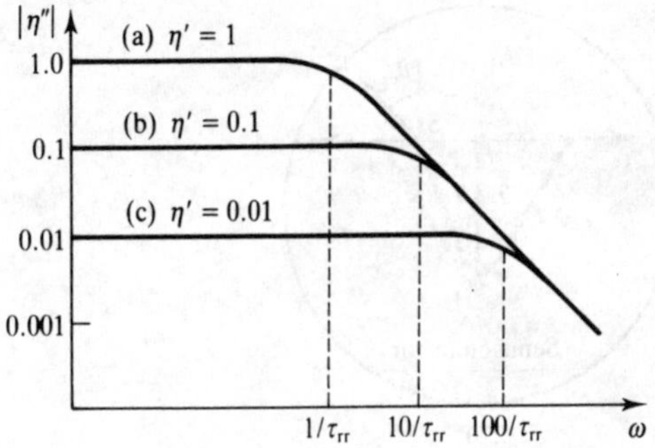

Figure 12.24 Variation of the A.C. internal efficiency η'' of an LED, for three values of D.C. internal efficiency.

At low frequencies, the a.c. internal efficiency tends to the d.c. value, but for angular frequencies greater than the break-point $\omega_c = 1/\eta'\tau_{rr}$, $|\eta''|$ is approximately

$$|\eta''| \approx 1/\omega\tau_{rr} \tag{12.46}$$

so that $|\eta''|$ is independent of η' at high frequencies.

Figure 12.24 shows a logarithmic plot of $|\eta''|$ versus ω, for three values of η'. As can be seen, there is a trade-off between the d.c. efficiency and the position of the break-point. This illustrates a basic limitation of LEDs: they may be either fast, but inefficient, or slow but efficient. However, it is not possible to obtain the best of both worlds. The d.c. efficiency may be optimized by maximizing η', by heavily doping the recombination region. The frequency response is clearly limited by τ_{rr}, and in typical devices the break-point occurs at $f_c = \omega_c/2\pi \approx$ 10–20 MHz. LEDs may therefore only be used as sources for optical links of moderate bit-rate.

The External Efficiency of an LED

Despite the reasonable internal efficiency of an LED, it is an unfortunate truth that very little of this theoretical output is available for use. To illustrate this point we show in Fig. 12.25(a) a typical surface-emitting LED; this is a layered p-n structure in which the optical output is taken from the upper surface. Because the light generation is isotropic, half of the light is emitted downwards and has no chance of escaping from this surface. Of the half that is emitted upwards, only a fraction escapes, due to the small critical angle at the semiconductor/air interface.

The proportion of the light that escapes into the air can be found as follows. We first define a source of radiation as the point L inside the LED and the point of escape from the surface of a given ray emitted by L as S, as shown in Fig. 12.25(b). If the angle made by this ray with the surface normal is θ, and the distance between L and S is r, the fraction of the radiation emitted by L lying in the solid angle between θ and $\theta + d\theta$ is $2\pi r \sin(\theta)\, r\, d\theta/4\pi r^2 = (1/2) \sin\theta\, d\theta$. The proportion of this fraction that actually crosses the interface is then given by $T(\theta) = 1 - \Gamma(\theta)^2$, where $\Gamma(\theta)$ is the Fresnel reflection coefficient for incidence at angle θ. The *external efficiency* η_e of the LED is then found as the integral of the product of the above

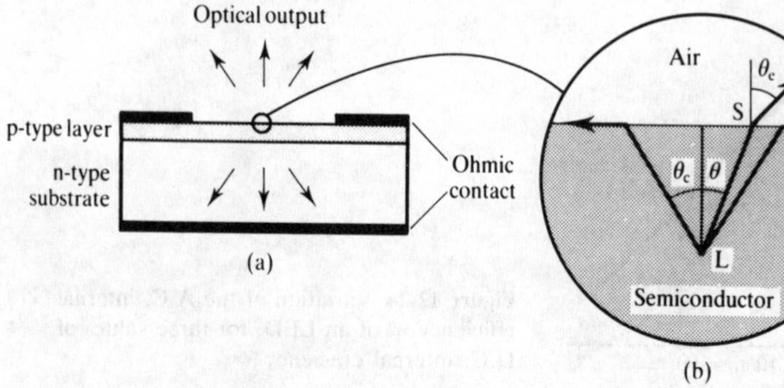

Figure 12.25 (a) A typical surface-emitting LED; (b) geometry for calculation of external efficiency.

two quantities, over the angular range for upward-travelling radiation, namely $0 \leq \theta \leq \pi/2$. We may therefore write

$$\eta_e = \int_0^{\pi/2} \{T(\theta)/2\} \sin\theta \, d\theta \tag{12.47}$$

Now, $T(\theta)$ is a slowly-varying function, falling from a maximum value $T(0) = 4n/(n+1)^2$ at normal incidence (where n is the refractive index of the semiconductor) to $T(\theta_c) = 0$ at the critical angle. A reasonable approximation to Eq. (12.47) is then:

$$\eta_e = T(0)/2 \int_0^{\theta_c} \sin\theta \, d\theta = T(0)/2 \{1 - \cos\theta_c\} \tag{12.48}$$

For typical LED materials, n is large (3.5 for GaAs), so $\theta_c = \sin^{-1}(1/n) \approx 1/n$ and $\{1 - \cos\theta_c\} \approx \theta_c^2/2 \approx 1/2n^2$. With these approximations, we may write the external efficiency as

$$\eta_e \approx 1/n(n+1)^2 \tag{12.49}$$

For GaAs, $\eta_e \approx 0.0141$, so that only about 1.4 per cent of the light generated inside the LED can actually escape from the surface. Consequently, the d.c. external output is reduced to $P_e = P'\eta_e$. For the GaAs LED discussed earlier, the usable output would be 0.73×0.0141 mW ≈ 10 μW per mA drive current.

A fraction of the light is, in fact, absorbed while travelling between L and S, reducing this figure still further. There are three partial solutions to the problem. First, a reflector can be added to the lower surface, so that some of the light emitted downwards may be exploited. Second, the upper surface of the LED may be antireflection coated, which provides an improvement in T(0). Third, the surface may be lensed (either by etching the surface into a hemisphere or by attaching an external microlens), so that a larger proportion of the light is incident on the interface at an angle less than θ_c. This solution is used extensively in display LEDs but is inappropriate for lightwave systems, where the aim is to couple the LED output into an optical fibre. In this case the power actually delivered into the fibre will be lower than the estimates obtained above by factor η_c, the *coupling efficiency*.

Calculation of η_c for the general case of an optical fibre butt-coupled to the surface of an LED is a relatively complicated problem; however, for an emission area that is small compared with the fibre core, and arranged on-axis as shown in Fig. 12.26, we can obtain a useful approximation to η_c. Light emitted by the LED travels in all directions, ranging from normal to the surface to parallel. It can be shown that the distribution of output intensity with external angle θ_e is Lambertian, following

$$I(\theta_e) \approx I(0) \cos \theta_e \tag{12.50}$$

We may use this distribution to perform a calculation very similar to the previous one. We note that the fraction of the output lying in the solid angle between θ_e and $\theta_e + d\theta_e$ is

$$dI = I(\theta_e) \sin \theta_e \, d\theta_e \Big/ \int_0^{\pi/2} I(\theta_e) \sin \theta_e \, d\theta_e \tag{12.51}$$

The proportion of this fraction that crosses the air/fibre interface is then given by $T'(\theta_e) = 1 - \Gamma'(\theta_e)^2$, where $\Gamma'(\theta_e)$ is the Fresnel reflection coefficient for incidence at an angle θ_e. Of the part that is transmitted into the fibre, only the fraction that is total-internal-reflected at the core/cladding interface will actually be guided. In Chapter 8 we showed that this condition led to a maximum acceptance angle for a multi-mode step-index fibre of

$$\theta_{emax} = \sin^{-1} \{\sqrt{(n_1^2 - n_2^2)}\} = \sin^{-1} (NA) \tag{12.52}$$

where NA is the numerical aperture of the fibre. Consequently η_c may be found by multiplying Eq. (12.50) by $T(\theta_e)$ and integrating from $\theta_e = 0$ to $\theta_e = \theta_{emax}$. Since θ_{emax} is typically small, and $T(\theta_e)$ is slowly-varying, η_c may be approximated by

$$\eta_c = T'(0) \int_0^{\theta_{emax}} I(\theta_e) \sin \theta_e \, d\theta_e \Big/ \int_0^{\pi/2} I(\theta_e) \sin \theta_e \, d\theta_e \tag{12.53}$$

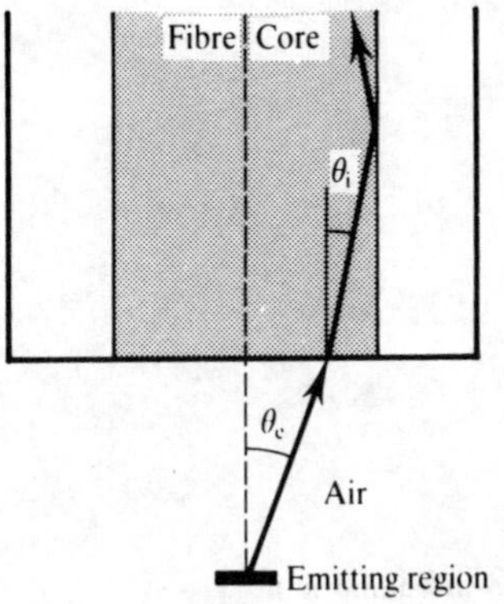

Figure 12.26 Geometry for calculation of LED: fibre coupling efficiency.

where $T'(0) = 4n'/(n' + 1)^2$, and n' is the refractive index of the fibre core. At this point we may substitute for $I(\theta_e)$ and perform the necessary integration. This yields the extremely simple result

$$\eta_c = T'(0) \sin^2 (\theta_{emax}) = T'(0)\, NA^2 \tag{12.54}$$

Thus, for a step-index fibre with $n' \approx 1.5$ and $NA \approx 0.2$ we obtain a coupling efficiency of $\eta_c \approx 0.04$. This figure may be improved by contacting the fibre directly to the LED, or by using a lens between the two (if the emitting area is smaller than the fibre core). One alternative device—the edge-emitting LED, which we will describe shortly—also offers an advantage. However, the fundamental cause of low external efficiency is the undirected nature of spontaneous emission. A major improvement may only be obtained by adopting the alternative mechanism of light production, stimulated emission.

Double Heterostructure LEDs

In place of the simple p-n junction, a more complicated structure is often used. Typically this might be a P-n-N *double heterostructure* containing three layers: a p-type layer with a wide energy gap, a narrow-gap n-type layer, and a wide-gap n-type layer. We have already mentioned the way in which a single heterojunction may be used to provide potential barriers of different heights for electrons and holes; the double heterostructure may be used to provide a high barrier at different positions for electrons and holes. In this way the recombination region may be limited to a defined region of space, resulting in an increase in the radiative recombination efficiency. We illustrate this in Fig. 12.27, which shows the energy band diagram for a P-n-N structure. The potential barrier faced by electrons is higher at the P-n interface, while that faced by holes is higher at the n-N interface. Under forward bias, electrons injected into the n-type layer from the right are prevented from diffusing away to the left by the first of these barriers. Similarly, holes injected into the n-type layer from the left are stopped from leaking away to

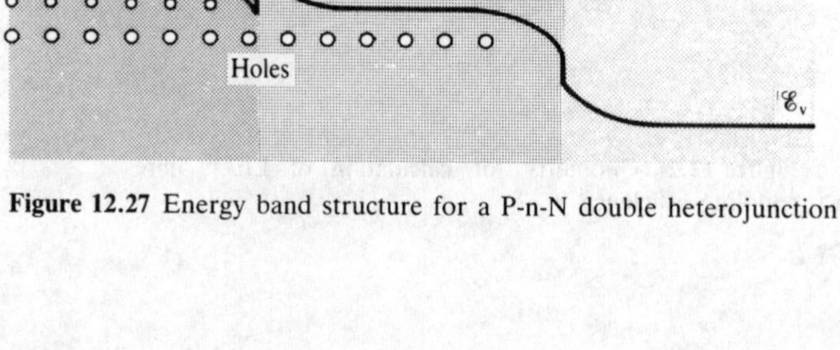

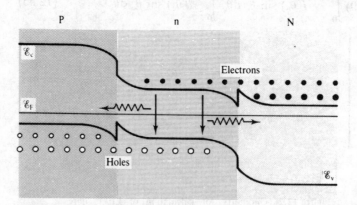

Figure 12.27 Energy band structure for a P-n-N double heterojunction in equilibrium.

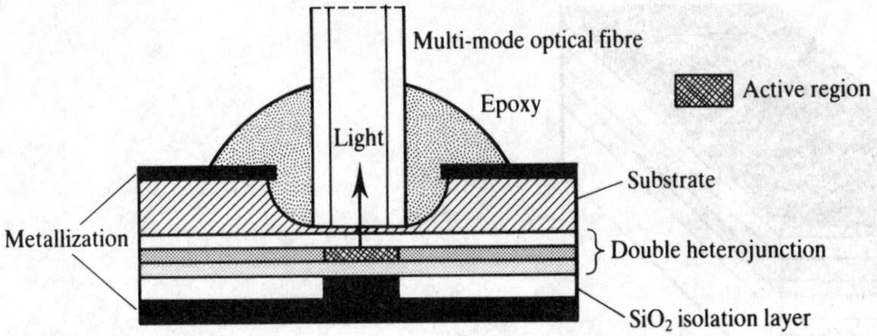

Figure 12.28 A Burrus-type double-heterojunction surface-emitting LED.

the right by the second barrier. The net effect is that carriers may be poured into the central layer, which acts as a localized recombination region. Two further advantages stem from this construction. The first is that the larger bandgaps of the P- and N-type layers can make them effectively transparent to radiation generated in the narrow-gap recombination region, thus reducing the amount of light that is absorbed before it escapes from the surface. The second is that the refractive index of the narrow-gap material at the centre of the structure may be higher than that of the material on either side. It is therefore possible to combine localization of the recombination region with confinement of an optical field. This feature is exploited extensively in the double heterostructure laser, which will be described in the next section.

Figure 12.28 shows a typical surface-emitting double-heterostructure LED, the *Burrus-type LED* (named after its inventor). This is a high-efficiency device, suitable for use with multi-mode fibre. To fabricate the device, a double heterostructure is first grown on a substrate, and an SiO_2 isolation layer is then deposited on the lower layer of the heterojunction. This layer is etched to expose the heterojunction over a small window. Metallization layers are then added to the upper and lower surfaces; clearly the lower of these makes contact with the heterojunction only in the region of the window. A deep well, aligned with this window, is then etched right through the substrate to the upper layer of the heterojunction, and a multi-mode optical fibre is epoxied into the well in contact with the LED surface. When a forward current is passed through the LED it flows mainly through the region immediately below the fibre. The recombination region (or *active region*) is therefore confined vertically by the double heterojunction and laterally by the distribution of current flow.

The Edge-emitting LED

Figure 12.29 shows an alternative geometry, the *edge-emitting* or *superluminescent LED*. This is also based on a double heterostructure, but now the emission is taken from the edge of the junction rather than its surface. Generally the current is forced to flow through a narrow strip down the device centreline by the introduction of a

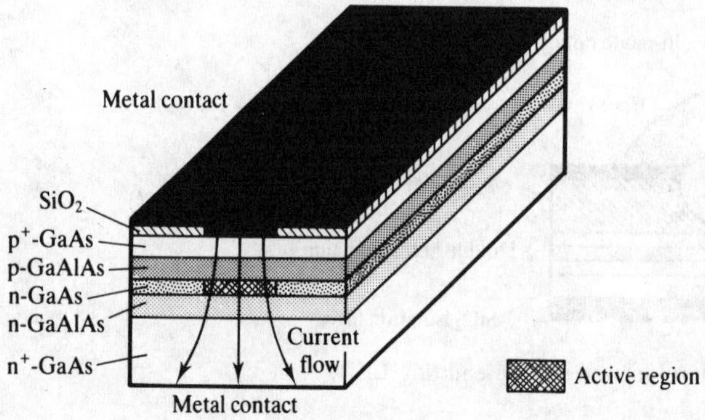

Figure 12.29 An edge-emitting LED.

current-blocking silica layer. As a result, the active volume is constrained vertically by the double heterostructure and laterally by the current flow. Most importantly, light generated by spontaneous emission is also confined vertically to a certain extent; the refractive index difference at the top and bottom of the active layer results in total internal reflection, which tends to channel a large fraction of the light towards the emission windows at relatively shallow angles. While the radiation pattern is still Lambertian in the horizontal plane (with a half-power beam width approximately 120°), it is considerably compressed in the vertical plane (into a width of about 30°).

The large distance travelled before the light escapes does result in an increased level of absorption (although this may be minimized by making the active layer thin, and forcing most of the light to travel in the wide-bandgap confining layers). Consequently, the external efficiency is lower than in a surface-emitting LED. However, this is more than compensated for by the increased directivity of the output, which allows much higher coupling efficiency to a fibre. In turn, this makes a reduction in the internal efficiency possible, thus improving the high-frequency response. Antireflection coatings may again be used to improve the transmission through the emitting facet. Edge-emitting LEDs may therefore have bandwidths around five times greater than their surface-emitting counterparts, while still coupling five times the power into a low-NA fibre. Despite all this, the performance improvement offered is only moderate, and a major difference can only be obtained using the alternative mechanism of light production; stimulated emission. Surprisingly, the structure required is very similar, but it has entirely different optical characteristics. We shall now consider this device—the *semiconductor laser*—in detail.

12.5 THE SEMICONDUCTOR LASER: BASIC OPERATION

To explain the difference between the semiconductor laser and the LED we must return to the results of Chapter 11. When absorption and stimulated emission are the dominant processes the rate equation for a travelling photon wave in our two-level model may be taken as

$$\partial\phi/\partial t = -v_g\,\partial\phi/\partial z + G\phi(n - n_0) \quad (12.55)$$

In the steady-state, $\partial\phi/\partial t = 0$, so that

$$\partial\phi/\partial z = g_P\phi \quad (12.56)$$

where the power gain is $g_P = G(n - n_0)/v_g$. In this case we can find the variation of ϕ with z as

$$\phi(z) = \phi(0)\exp(g_P z) \quad (12.57)$$

Equation (12.57) describes an exponentially-growing wave. This type of amplification ought to offer a method of light generation, if we can harness it. Clearly we require $n > n_0$ for g_P to be positive, so that net gain occurs. To achieve this we must find some means to raise the electron density far above the equilibrium level. The injection of electrons across a forward-biased diode immediately suggests itself as a suitable method. Figure 12.30 shows a three-dimensional sketch of the band-structure of a p-n$^+$ junction diode under heavy forward bias. Here the forward voltage V is so large that the relative levels of the conduction band on either side of the junction are the opposite of those in equilibrium. In this case, we would expect a veritable flood of electrons to be injected into the p-side. An optical wave travelling parallel to and sufficiently close to the depletion layer will then move through a region with greatly increased electron concentration. Our previous expression for n_p' is, unfortunately, inaccurate at the high forward bias required to invert the population. However, by a more accurate calculation, it can be shown that the net gain is obtained when the difference between the quasi Fermi-levels $\mathscr{E}_{Fn}$ and $\mathscr{E}_{Fp}$ is greater than the bandgap $\mathscr{E}_g$.

The injected electrons will diffuse away from the depletion layer edge, so the gain must fall off in the same direction. Arguably, the excess electron concentration will fall to zero within about a diffusion length. Since no amplification will occur outside this region, the optical wave should be confined near the depletion layer to take full advantage of the gain. One approach would be to force it to propagate in a waveguide that passes through the gain region. We might then ask, how should

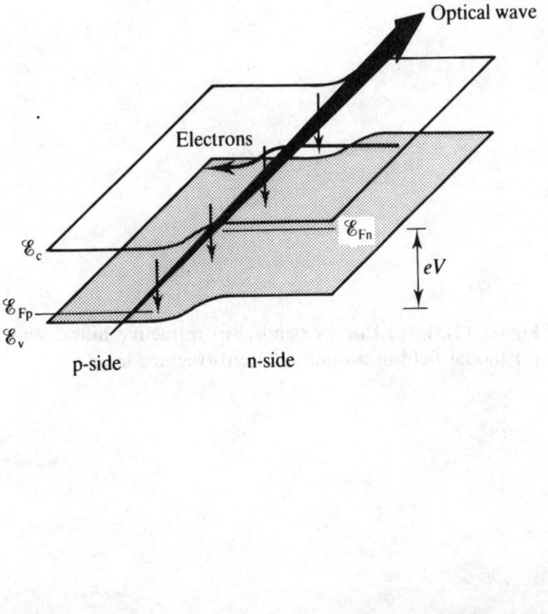

Figure 12.30 Amplification of a travelling optical wave near the depletion layer of a forward-biased p-n junction.

the guide be designed for optimum gain? To answer this we must find the effective power gain due to a distribution $g_p(x)$ that varies over the guide cross-section. This can be done using the methods of Chapter 9, where we calculated the effect of a non-uniform index perturbation Δn on a guided mode in a phase modulator. There we found that the change in effective index was found from the overlap between the perturbation and the normalized transverse field, as $\Delta n_{eff} = \langle \Delta n, |E|^2 \rangle$. It is easy to show that the effective power gain is also found as an overlap, this time between the transverse field and the gain distribution, i.e. as

$$g_{peff} = \langle g_p, |E|^2 \rangle \tag{12.58}$$

Consequently, we may deduce that the most important condition for high effective gain is a strong overlap between the modal field and the gain distribution. Early semiconductor lasers were made from p-n homojunctions. These contain an in-built planar guide, formed by the increase in refractive index that occurs naturally in the depletion region due to the absence of free carriers. However, since the index change obtained this way is very small ($\Delta n/n \approx 0.1\%$ to 1%), such a guide provides a very weak confinement of the optical field. The lack of any kind of potential barrier also allows the diffusion of injected carriers away from the junction, so that homostructure lasers have a low effective gain and are of historical interest only.

A much more successful geometry is the *double heterostructure laser*, which uses compositional variations to optimize the optical and carrier confinement. For example, Fig. 12.31(a) shows the band diagram for a P-n-N heterostructure under strong forward bias. Note that there is a potential well in the n-region in both the conduction band and the valence band. As mentioned earlier, electrons injected into the upper well from the N-side are highly localized and available for recombination with holes injected into the lower well from the P-side.

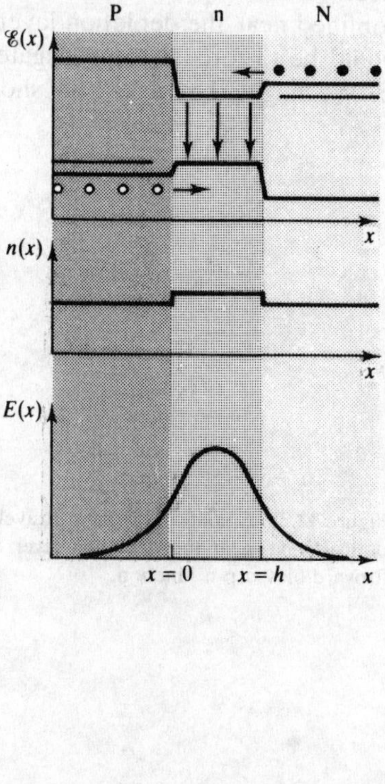

Figure 12.31 (a) Energy bands, (b) refractive index, and (c) modal field in a double heterostructure laser.

Figure 12.31(b) shows the corresponding refractive index distribution across the junction; the wide bandgap P- and N-type regions have a lower index than the n-region, so a roughly symmetric slab guide (with $\Delta n/n \approx 5$ per cent) is created exactly in the high-gain region. The modal field (Fig. 12.31(c)) is then well-confined and has a strong overlap with the gain. Furthermore, the confinement is highly stable since it is hardly altered by changes in the injection level. Modern semiconductor lasers are therefore almost exclusively built around variants of the double heterostructure. For example, in the *separate confinement heterostructure* (SCH), optical field confinement and localization of the recombination region are optimized separately, using a quadruple heterojunction. The two inner junctions limit the active region, while the two outer ones confine the optical field.

Design Example

We may estimate the effective power gain in a double heterostructure laser as follows. In the central active layer we may take the power gain to be constant, and equal to g_{p1}. However, there may also be additional sources of loss (free carrier absorption, for example), which we might define by a power absorption coefficient α_{p1}. In the unpumped outer layers there will be no gain, only loss defined by α_{p2} and α_{p3} in the left- and right-hand layers, respectively. For the coordinates of Fig. 12.31 the overlap integral in Eq. (12.58) may then be evaluated as

$$g_{peff} = \Gamma_1(g_{p1} - \alpha_{p1}) - \Gamma_2\alpha_{p2} - \Gamma_3\alpha_{p3} \tag{12.59}$$

where the overlap terms Γ_1, Γ_2 and Γ_3 are defined as

$$\Gamma_1 = \int_0^h |E|^2 \, dx \qquad \Gamma_2 = \int_{-\infty}^0 |E|^2 \, dx \qquad \Gamma_3 = \int_h^\infty |E|^2 \, dx \tag{12.60}$$

and $\Gamma_1 + \Gamma_2 + \Gamma_3 = 1$. To maximize the effective gain Γ_1 should be made much greater than Γ_2 and Γ_3 by confining as large a fraction of the optical field as possible inside the central pumped region. For a more complicated channel guide geometry we might write by analogy

$$g_{peff} = \Gamma(g_p - \alpha_{pcore}) - (1 - \Gamma)\alpha_{pclad} \tag{12.61}$$

where Γ is the overlap of the guided mode with the active region—known as the *confinement factor*, and α_{pcore} and α_{pclad} are the losses in the core and cladding, respectively.

Practical Laser Geometries

An optical amplifier may be turned into an oscillator by providing positive feedback. We have already described how this is done in Chapter 4; all that is needed is to add mirrors to the gain block, so that the structure becomes a Fabry-Perot cavity. In the semiconductor laser the mirrors may be fabricated very simply by cleavage down crystal planes orthogonal to the junction plane, as shown in Fig. 12.32.

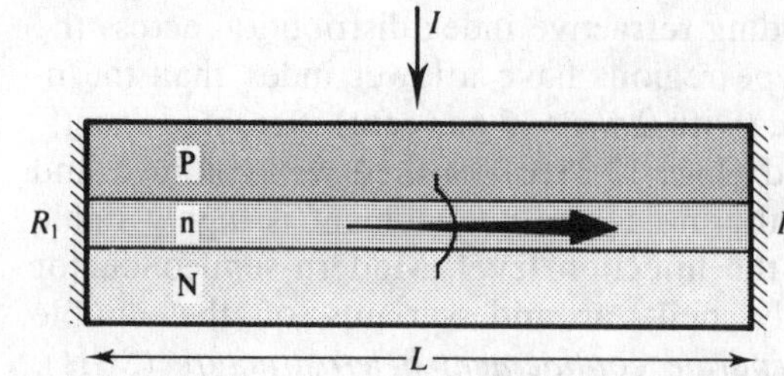

Figure 12.32 Basic geometry of a planar double heterostructure laser.

Naturally their reflectivity is fixed; for a material of refractive index 3.5 surrounded by air, the amplitude reflectivity is $R_1 = R_2 \approx (3.5 - 1)/(3.5 + 1) \approx 0.56$. This figure is high enough to obtain a suitable level of feedback, while still allowing a good fraction of the light generated to escape from the cavity. Gain is provided by the injection of current in a direction normal to the junction. Because a planar structure provides confinement of the light only in one direction, we would expect light to travel between the mirrors as a sheet beam, growing in amplitude as it propagates.

In practice, the use of a simple layered structure is undesirable. First, both the drive current and the heat generated by non-radiative transitions will be large, since they are proportional to the area of the junction. Second, the finite lateral extent of the guide implies that it will actually function as a very broad, multi-moded channel guide. The optical field inside the cavity will then consist of a random mixture of modes, which may lead to a highly non-uniform intensity distribution across the beam. Third, broad-area emission will result in very low coupling efficiency into a fibre. Extra refinements are very often used to reduce the active volume and introduce transverse confinement, so that the light propagates in a single-moded channel guide.

The primary problem with semiconductor lasers is therefore to combine two-dimensional confinement of the optical wave with localization of the recombination region. There are many possible solutions. One is to restrict the flow of current laterally, for example by using an SiO_2 isolation layer with a stripe contact window, as shown earlier in Fig. 12.29. This geometry is known as the *oxide stripe laser*. Alternatively, a full-area contact might be used, but the semiconductor might be made highly resistive in all except a narrow stripe region, for example by proton bombardment. Both methods restrict the gain region laterally. However, the lateral optical confinement is only moderate since this is introduced by a local increase in gain rather than by an index variation. Such lasers are therefore often described as *gain-guided*. Due to the weak confinement the guide may support several transverse modes, and the relative importance of each mode may be highly dependent on the level of injection. For stable operation on a single transverse mode, stronger lateral confinement must be established using refractive index changes. This requires a more complicated structure.

Figure 12.33 shows the simplest example of a suitable geometry, the *ridge waveguide laser*. In this example, the laser is designed to emit long-wavelength

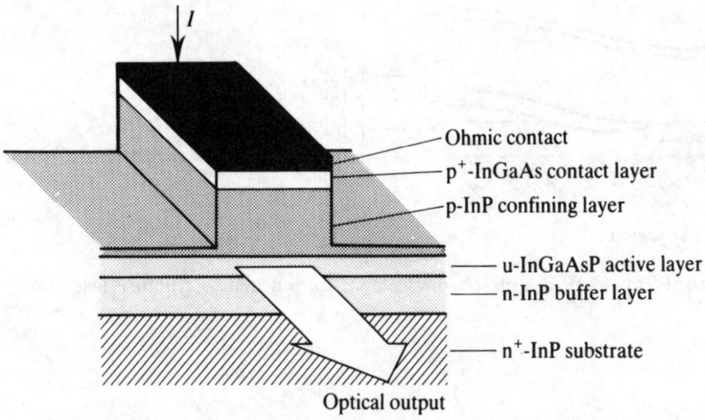

Figure 12.33 A ridge waveguide laser (*after A.W. Nelson* et al., *1986*).

infrared radiation (at $\lambda_0 \approx 1.5\ \mu$m), so the InP/InGaAsP materials system has been used. The starting point is a planar InP-InGaAsP-InP double heterostructure, but the upper InP layer is partially etched into a rib (typically 5 μm wide) to provide lateral confinement of the mode by strip-loading. Note that the current is also automatically confined by the finite width of the rib, although it may start to spread as it passes into the active layer.

Figure 12.34 shows a more complicated example, the *buried heterostructure laser*. The laser shown is designed to emit at a shorter wavelength ($\lambda_0 \approx 0.83\ \mu$m), and is therefore based on the GaAs/GaAlAs materials system. The starting point is a GaAlAs-GaAs-GaAlAs double heterostructure, which is again fashioned into a ridge. This time, however, the ridge is etched completely down to the substrate, and buried on either side by additional high-resistivity GaAlAs. Lateral optical confinement is not lost, however, since the burying layer has a lower refractive index than the active region. Contact is then made to the top of the heterostructure via an oxide stripe window, and the current is constrained to flow through the active region by the high resistivity of the surrounding material.

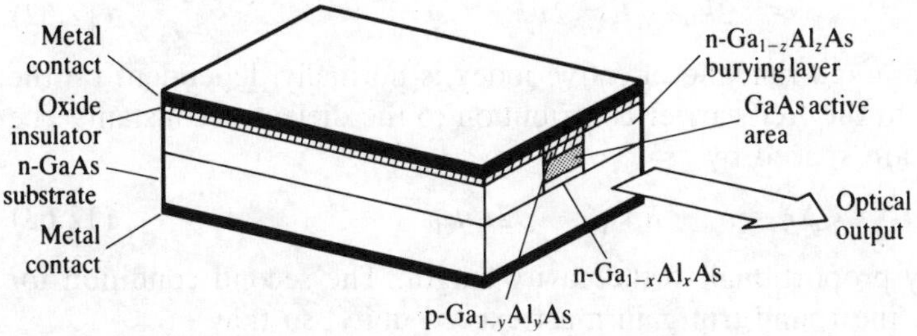

Figure 12.34 A GaAs/GaA1As buried heterostructure laser.

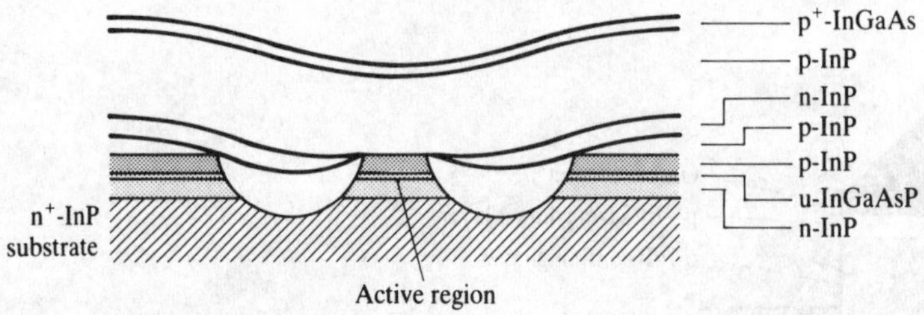

Figure 12.35 Cross-section of an InP/InGaAsP double-channel planar buried heterostructure laser (*after A. W. Nelson* et al., *1986*).

Still better current confinement is possible using more complicated structures. For example, Fig. 12.35 shows the *double-channel planar buried heterostructure laser*. This begins life as a planar InP-InGaAsP-InP double heterostructure, which has two deep channels etched through it to the substrate. Four additional layers are then added by further epitaxial growth. In order, these consist of p-InP, n-InP, p-InP and p^+-InGaAs. The important feature is that by utilizing a peculiarity of liquid-phase epitaxial growth known as *selective area epitaxy*, the n-InP layer may be constrained to grow everywhere except over the central ridge. As a result, the structure contains several in-built, reverse-biased p-n junctions, which provide far better current blocking than high-resistivity material. Not only do they force the current to flow into the top of ridge, but they also prevent it from spreading out of the sides of the active region.

Conditions for Laser Operation

We now consider the basic analysis of semiconductor lasers. From Chapter 4 we know that two conditions are required for laser operation. The first is that a longitudinal resonance is satisfied, so that the phase-change accumulated in a round-trip up and down the cavity is a whole number of multiples of 2π. Only then will the feedback be positive, so that the amplitude of the optical field inside the cavity is high. For a single-mode guide with effective index n_{eff}, this condition may be written as

$$2k_0 n_{\text{eff}} L = 2\nu\pi \tag{12.62}$$

However, it should be noted that the effective index is normally dependent on the level of injection, due to the free carrier contribution to the dielectric constant. The resonant wavelengths are spaced by

$$\Delta\lambda = \lambda_\nu - \lambda_{\nu+1} = \lambda_\nu^2 / 2Ln_{\text{eff}} \tag{12.63}$$

so that $\Delta\lambda$ is inversely proportional to the cavity length. The second condition for laser operation is that the round-trip gain must exceed unity, so that

$$R_1 R_2 \exp(2g_{\text{eff}} L) > 1 \tag{12.64}$$

The effective power gain at the threshold of lasing is therefore

$$g_{\text{pefft}} = 2g_{\text{eff}} = 1/L \log_e (1/R_1R_2) \quad (12.65)$$

Consequently, the shorter the cavity, the higher the threshold power gain. For a cavity of length 250 μm with end-mirrors of reflectivity $R_1 = R_2 = 0.56$ we obtain $g_{\text{pefft}} \approx 4{,}640\ \text{m}^{-1}$. Note that the gain that must be provided in the active region is even higher, since from Eq. (12.61) we obtain

$$g_{\text{pt}} = \alpha_{\text{pcore}} + \{(1 - \Gamma)/\Gamma\}\alpha_{\text{pclad}} + (1/\Gamma L) \log_e (1/R_1R_2) \quad (12.66)$$

The gain in the active core must therefore be sufficient to overcome all absorption and output coupling losses. The likely size of g_{pt} emphasizes the efforts that must be made to optimize Γ.

In our earlier discussion of LEDs we described how the distribution of electron and hole energies leads to a finite spectrum for spontaneous emission. For the same reason, gain can be obtained in a semiconductor laser over a finite range known as the *gain bandwidth*. Normally the variation of gain with wavelength is written in the form:

$$g_{\text{p}}(\lambda_0) = a(n - n_0) - b(\lambda_0 - \lambda_{\text{p}})^2 \quad (12.67)$$

where a and b are constants, and λ_{p} is the wavelength for peak gain. With this in mind, we would expect the output of a laser with a relatively long cavity to consist of a series of closely-spaced lines, as shown in Fig. 12.36(a). Each corresponds to a particular longitudinal mode, when Eq. (12.62) is satisfied for a given value of ν. The gain bandwidth of (for example) an InGaAsP laser has a half-width of around 30 nm. For the typical parameters of $\lambda_\nu = 1.5\ \mu$m, $L = 250\ \mu$m and $n_{\text{eff}} = 3.2$ we obtain $\Delta\lambda \approx 1.4$ nm, so many lines must lie within the gain bandwidth. We cannot comment on the relative powers in each of the lines yet (except to say that there will be more power in those nearest to the gain peak) since the details of the spectrum depend on the level of injection. However, an extended spectrum is generally undesirable in fibre communications systems. Furthermore, we note that a laser with a broader stripe that supports several transverse modes would have a more complicated spectrum. Since each transverse mode will have a different effective index, each will have its own associated spectrum of longitudinal resonances.

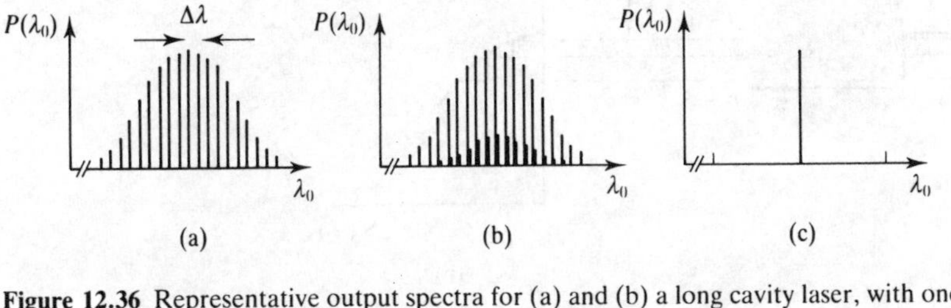

Figure 12.36 Representative output spectra for (a) and (b) a long cavity laser, with one and two transverse modes, respectively; (c) a short cavity laser supporting a single transverse mode.

However, the effective gain of higher-order modes is likely to be lower since they are more weakly confined than the fundamental mode. The spectrum of a two-moded laser might, therefore, be as shown in Fig. 12.36(b), where the smaller lines correspond to the second-order mode. Since the lowest-order mode is generally the most useful, single-transverse-mode operation is highly desirable. In this case, and with a much shorter cavity, we might expect only one longitudinal resonance to fall inside the gain bandwidth, so that the output is effectively single-line (Fig. 12.36(c)). However, this requires an extremely small value of L ($\approx 50\ \mu m$).

Quantum-well and Multi-quantum-well Lasers

Bandgap engineering offers further scope for the control of laser properties. For example, Fig. 12.37(a) shows the equilibrium band diagram of an undoped GaAlAs-GaAs-GaAlAs double heterostructure. Due to the difference in bandgap between the materials, there are potential wells in both the conduction band and the valence band. In conventional lasers these are exploited purely for carrier confinement. However, as the width h of the central layer is reduced, new effects start to occur. In Chapter 11 we analysed the confinement of an otherwise free electron in a one-dimensional well, and showed that the result was a quantization of the electron energy. For an infinite well of width h with a minimum potential energy of $\mathcal{V} = 0$, the allowed energies were

$$\mathcal{E} = \nu^2(\pi^2\bar{h}/2mh^2) + \bar{h}\beta^2/2m \qquad (\text{where } \nu = 1, 2, \ldots) \qquad (12.68)$$

Of these two contributions, the first arises from the quantization effect of the well, while the second is a kinetic energy term due to motion in the two remaining allowed directions. By analogy we might expect the energies of conduction electrons in the well of Fig. 12.37 to be of the form:

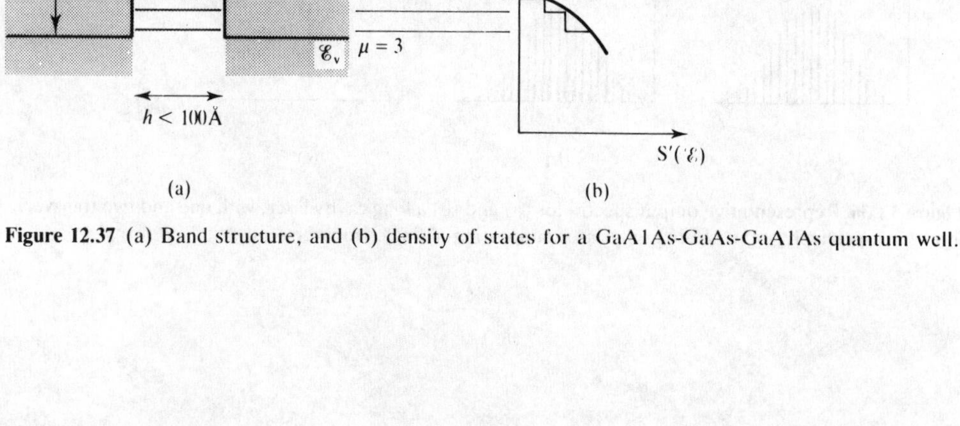

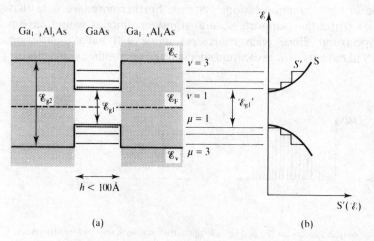

Figure 12.37 (a) Band structure, and (b) density of states for a GaAlAs-GaAs-GaAlAs quantum well.

$$\mathscr{E}'_{c1} = \mathscr{E}_{c1} + \nu^2(\pi^2\hbar^2/2m_e^*h^2) + \hbar\beta^2/2m_e^* \quad \text{(where } \nu = 1, 2, \ldots) \tag{12.69}$$

while the corresponding hole energies might be

$$\mathscr{E}'_{v1} = \mathscr{E}_{v1} - \mu^2(\pi^2\hbar/2m_h^*h^2) - \hbar\beta^2/2m_h^* \quad \text{(where } \mu = 1, 2, \ldots) \tag{12.70}$$

where $\mathscr{E}_{c1}$ and $\mathscr{E}_{v1}$ are the extremities of the conduction and valence bands in the GaAs layer, and m_e^* and m_h^* are the electron and hole effective masses. As a result, the minimum difference between conduction and valence levels is no longer $\mathscr{E}_{g1} = \mathscr{E}_{c1} - \mathscr{E}_{v1}$, but

$$\mathscr{E}'_{g1} = \mathscr{E}_{g1} + (\pi^2\hbar/2h^2)(1/m_e^* + 1/m_h^*) \tag{12.71}$$

This energy difference corresponds to a transition between stationary electrons and holes in the levels defined by $\nu = 1$, $\mu = 1$. This is equivalent to the creation of an artificial semiconductor with a bandgap different from that of either of the materials forming the well. Note that $\mathscr{E}'_{g1}$ is controlled essentially by the value of h. Consequently, the emission wavelength of a laser constructed in this way should be similarly variable. In fact, the wavelength may be tuned over almost the entire range set by the bandgaps of the two constituent materials, GaAs and $Ga_{1-x}Al_xAs$. The tuning range is therefore limited mainly by the transition of GaAlAs to an indirect bandgap material at $x \approx 0.35$. Using quantum wells made from InP and InGaAs (which are both direct-gap), the wavelength may be tuned over the staggering range 1.07–1.55 μm by varying the well width between 10 and 110 Å.

The quantum size effect offers other advantages besides wavelength tunability. For example, it is possible to repeat the calculation to find the density of states $S'(\mathscr{E})$ in the conduction band inside a single well. If this is done, we obtain

$$S'(\mathscr{E}) = \nu m_e^*/\pi h\hbar^2 \quad \text{for } \mathscr{E}_\nu \leq \mathscr{E} \leq \mathscr{E}_{\nu+1} \tag{12.72}$$

where $\mathscr{E}_\nu = \mathscr{E}_{c1} + \nu^2(\pi^2\hbar/2m_e^*h^2)$. Consequently, $S'(\mathscr{E})$ is discontinuous, as shown in Fig. 12.37. Although this is very different from our previous parabolic result for $S(\mathscr{E})$—which is also the density of states for an infinite well width—the most important point is that $S'(\mathscr{E})$ intersects $S(\mathscr{E})$ at each energy $\mathscr{E}_\nu$, but otherwise lies inside it. We may therefore deduce that there must be a smaller total density of states in the conduction band in a quantum well (actually this is obvious, since the effect of quantization is to disbar many possible states). Remember that we must invert the electron population to obtain optical gain. With fewer states to invert, the electron concentration required for transparency must be lower, so a quantum well laser must have a lower threshold than a conventional one. Furthermore, states with higher energy are filled more easily, leading to an improvement in optical gain at the short-wavelength end of the spectrum. This increase in gain bandwidth is exploited in wide-range tunable semiconductor lasers. However, it should be noted that the peak gain of a quantum well laser must be reduced, since the conduction band will be filled more easily. If the well is large enough, a double heterostructure laser can be made using a single quantum well as the active layer. However, as the well size is reduced, it becomes impossible to provide suitable optical confinement—the resulting waveguide is too narrow, so that the optical field extends considerably outside the gain region. It is then normal to construct the active layer as a multi-layer stack, consisting (for example) of many alternating layers of InP and InGaAs, and to provide the optical confinement separately. In this case the structure is known as a *multi-quantum-well* (MQW) laser.

12.6 THE SEMICONDUCTOR LASER: STEADY-STATE ANALYSIS

Assuming that the laser operates on a single transverse mode, and that the output is concentrated in a single line, we may model the operation of a laser using relatively minor modifications of the two-state, lumped-element rate equations derived earlier for LEDs. Although the optical field inside the cavity will actually consist of a standing-wave pattern, we shall ignore this and simply assume an *average* value for the photon density ϕ. This will be high at the resonant wavelength (which we take to be λ_g), and zero otherwise. For simplicity we shall ignore optical loss, and assume a confinement factor of $\Gamma = 1$. We may then take the photon rate equation for a cavity of volume v in the form:

$$d\phi/dt = \beta n/\tau_{rr} + G\phi(n - n_0) - \phi/\tau_p \tag{12.73}$$

Here we have included terms describing spontaneous emission ($\beta n/\tau_{rr}$), stimulated emission and absorption ($G\phi(n - n_0)$), and loss of photons from the cavity (ϕ/τ_p). The factor β has been introduced into the first of these to describe the fraction of the isotropically-generated spontaneous emission which is actually coupled into the guided mode; typically, this is very small ($\approx 10^{-3}$) due to the low *NA* of a single-mode guide. Similarly, the electron rate equation may be taken as

$$dn/dt = I/ev - n/\tau_e - G\phi(n - n_0) \tag{12.74}$$

Since Eqs (12.73) and (12.74) are non-linear, we will solve them by means of approximations that are valid in the two important regimes of operation. In the steady-state, the left-hand side of each of these equations is zero, so we may then combine them to eliminate $G\phi(n - n_0)$. This yields

$$\phi/\tau_p = \beta n/\tau_{rr} + (I/ev - n/\tau_e) \tag{12.75}$$

The light flux out of the cavity is $\Phi = v\phi/\tau_p$ as before. As we have ignored optical loss, this flux must all emerge from the cavity end-mirrors. If each photon carries energy hc/λ_g, the optical power output is then $P = hc\Phi/\lambda_g$, or

$$P = (hcv/\lambda_g)\{\beta n/\tau_{rr} + (I/ev - n/\tau_e)\} \tag{12.76}$$

Note that Eq. (12.76) represents the *total* output, so that $P/2$ emerges from each of the mirrors.

Operation Below Threshold

For a low photon density, spontaneous emission is the dominant process, so we may set the term $\{I/ev - n/\tau_e\}$ (which describes stimulated emission) equal to zero. In this case, we have $n = I\tau_e/ev$, so that

$$P = \eta h c I/e\lambda_g \tag{12.77}$$

where $\eta = \beta\tau_e/\tau_{rr}$. For low currents, the output is therefore proportional to current, as in the left-hand part of Fig. 12.38(a). The laser acts as an LED in this regime, with a reduced internal efficiency.

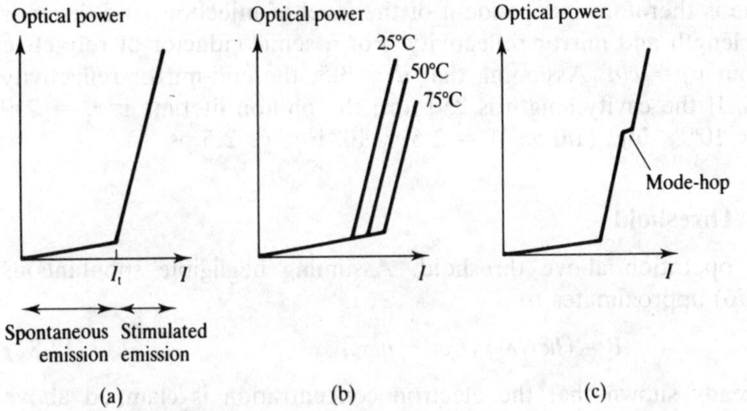

Figure 12.38 (a) Optical power–current characteristic of a semiconductor laser, (b) dependence of threshold current on temperature, and (c) kink in characteristic due to mode-hopping.

The Threshold Condition

The existence of spontaneously-emitted light in the laser cavity provides a source of optical noise from which laser oscillation may start. The noise will travel up and down the cavity, and, if there is net gain, it will be amplified as it propagates. The photon density will very quickly become high, so that stimulated emission becomes the dominant process. In the steady-state, assuming that spontaneously-emitted light is now negligible, we may approximate Eq. (12.73) by

$$G\phi(n - n_0) - \phi/\tau_p = 0 \tag{12.78}$$

We may then obtain the electron density during lasing as

$$n = n_0 + 1/G\tau_p \tag{12.79}$$

As a result, the electron concentration is entirely independent of ϕ. Equation (12.79) must also represent the electron density at threshold, n_t, and therefore describes the *threshold condition*.

Design Example

We may use the result above to estimate the photon lifetime. Since we have already established the threshold condition, we may find n_t by an alternative route. Setting $\Gamma \approx 1$ and $\alpha_{core} = \alpha_{clad} \approx 0$ in Eq. (12.66), the threshold gain is $g_{pt} \approx (1/L) \log_e (1/R_1R_2)$. Since $g_{pt} = G(n_t - n_0)/v_g$, we may find the threshold electron density as

$$n_t = n_0 + (v_g/GL) \log_e (1/R_1R_2) \qquad (12.80)$$

Comparing Eqs (12.79) and (12.80) we may then obtain the photon lifetime τ_p as

$$\tau_p = L/\{v_g \log_e (1/R_1R_2)\} \qquad (12.81)$$

The photon lifetime is therefore independent of the level of injection, and depends only on the cavity length and mirror reflectivity. For a semiconductor of refractive index n, we may put $v_g \approx c/n$. Assuming that $n = 3.5$, the end-mirror reflectivity is $R_1 = R_2 \approx 0.56$. If the cavity length is 250 μm, the photon lifetime is $\tau_p = 250 \times 10^{-6} \times 3.5/\{3 \times 10^8 \times \log_e (1/0.56^2)\} = 2.5 \times 10^{-12}$ s, or 2.5 ps.

Operation Above Threshold

We now consider operation above threshold. Assuming negligible spontaneous emission, Eq. (12.76) approximates to

$$P = (hcv/\lambda_g)(I/ev - n/\tau_e) \qquad (12.82)$$

Since we have already shown that the electron concentration is clamped above threshold, n/τ_e must be constant. We might define it to be equal to I_t/ev, where I_t is a new parameter known as the *threshold current*: I_t is therefore proportional to the active volume v and the threshold electron density n_t, and inversely proportional to τ_e. Its exact value is highly dependent on the laser type; typical values are in the range 10–100 mA. Using this definition, Eq. (12.82) may be written as

$$P = hc\{I - I_t\}/e\lambda_g \qquad (12.83)$$

The output of a semiconductor laser is therefore a discontinuous function of the drive current, being proportional to $I - I_t$ in the lasing regime as shown in the right-hand part of Fig. 12.38(a). The peak output power is normally limited either by mirror damage (caused by the high optical intensity passing through the end-facets) or by thermal breakdown (caused by the high current density passing through the junction). Output powers in the range 1–10 mW are routinely available at all the major wavelengths (near 0.85 μm for GaAs/GaAlAs lasers, and 1.3 and 1.5 μm for InP/InGaAsP).

Generally I_t rises with temperature because the spread of energies in the conduction and valence band increases, reducing the density of carriers that can generate a photon of a given wavelength. The recombination time τ_e is also decreased. The power–current characteristic of a laser at three values of T might therefore be as shown in Fig. 12.38(b). In fact, operation at a greatly reduced temperature was often used to lower the threshold current to an acceptable level in early lasers, which had poor confinement factors. Recombination time τ_e also decreases as the active region ages, so I_t rises with time.

Multi-line Operation

The two-state model we have used cannot describe multi-line operation. However, we may obtain a feel for the likely behaviour, as follows. If the spontaneous emission term is retained in Eq. (12.73), a steady-state solution for the photon density may be found as

$$\phi = (\beta n/\tau_{rr})/\{1/\tau_p - G(n - n_0)\} \tag{12.84}$$

This shows that the light in the cavity may be considered to arise from spontaneously-emitted noise, 'amplified' by the factor $F = 1/\{1/\tau_p - G(n - n_0)\}$. Since the denominator of this expression will be very small, F must be very large. In a more complicated model including several longitudinal modes, we might assign a different gain to each, following Eq. (12.67). We might then write

$$F_\nu = 1/\{1/\tau_p - [G(n - n_0) + b'(\lambda_\nu - \lambda_p)^2]\} \tag{12.85}$$

for the amplification factor for mode ν, where λ_ν is the wavelength of the mode and $b' = bv_g$. The relative sizes of each F_ν then depends on the closeness of each term $G(n - n_0) + b'(\lambda_\nu - \lambda_p)^2$ to $1/\tau_p$. For low currents (i.e. when $n - n_0$ is small), all the amplification factors are small, and roughly equal, so that multi-line operation occurs. As $n - n_0$ increases every value of F_ν will increase, but the line whose wavelength is closest to λ_p will have a proportionately large amplification factor. We can then guess that the mode spectrum will tend to alter as shown in Fig. 12.39, where the power is gradually concentrated into one dominant line as the current rises. This is highly desirable because it reduces the effect of dispersion when a

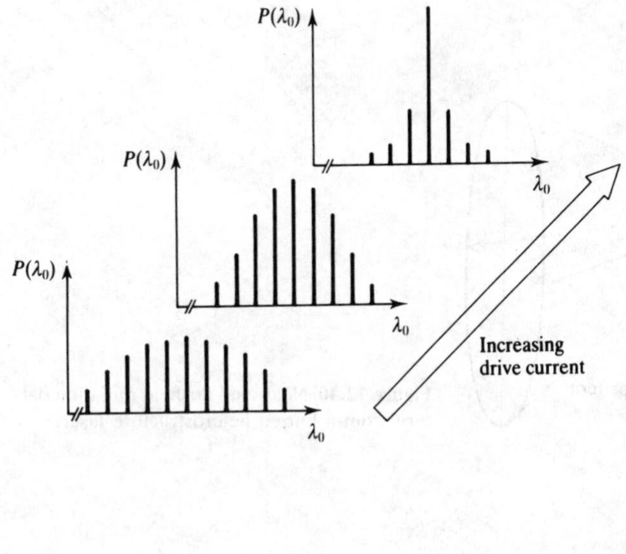

Figure 12.39 Typical variation of the mode spectrum of a semiconductor laser with current.

laser is used as a source for optical communications. Note that an increase in drive current may also make it possible for the transverse lasing mode to change, for example, from the lowest-order mode at low current to the second-lowest at higher current. This behaviour is known as *mode-hopping*. It may be observed as a kink in the laser output characteristic, as shown in Fig. 12.38(c), and is generally undesirable.

Optical Characteristics

We now consider briefly a number of other characteristics of a semiconductor laser. First, in order to obtain good overlap between the optical field and the gain distribution, the height h of the active region of a double heterostructure laser supporting a single transverse mode is normally very small, of the order 0.1–0.5 μm, while the corresponding width w is typically 1–5 μm. The core of the laser waveguide is therefore rectangular, with an aspect ratio of about 10:1. The mode supported by the guide is similarly distorted, being highly compressed perpendicular to the plane of the layering, as shown in the near-field pattern of Fig. 12.40.

To a good approximation, each facet behaves as a point emitter, so that the far-field radiation pattern is roughly that of a paraxial spherical wave. We saw in Chapter 4 that the spread of a bounded beam is inversely dependent on its width. Because of the shape of the near-field pattern, we would expect the angular spread to be greatest in the plane perpendicular to the junction (also as shown in Fig. 12.40). A typical value for the full-angle, half-maximum spread in intensity in this plane is $\theta_\perp \approx$ 30–50°; the corresponding spread $\theta_\parallel$ parallel to the junction is much smaller, in the range 10–15°, but both figures depend strongly on the construction of the laser. The output may be collimated using a spherical lens simply by placing the emitting area at the focal point. The resulting beam will be roughly parallel, but (due to the differences in beam divergence) its cross-section will be elliptical. Approximate circular symmetry may be restored by passing the beam through an anamorphic expansion telescope, which magnifies in one direction only. However, the presence of gain in the active region causes a slight astigmatism of the modal

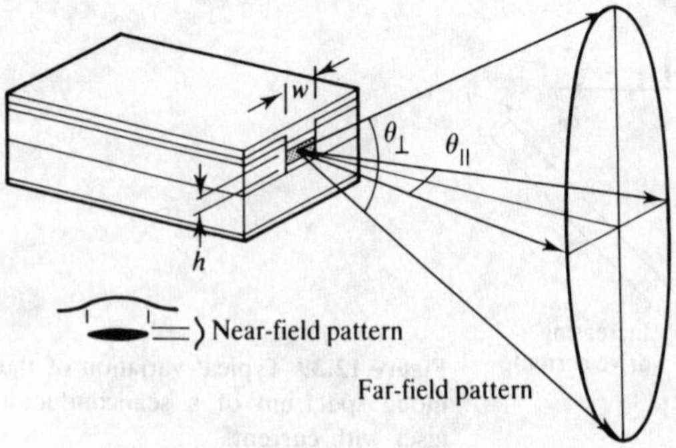

Figure 12.40 Near- and far-field radiation pattern from a buried heterostructure laser.

phase-front. If accurate collimation is required, this must be corrected by passing the beam through an additional cylindrical lens.

Because of the increased directivity and coherence of the light generated by stimulated emission, much higher coupling efficiency to a low-NA single-mode fibre can be obtained with a laser than with an LED. The coupling efficiency may be found by evaluating the overlap between the modal fields of the laser and fibre, much as was done in Chapter 8. Since the laser mode is typically rather small, an improvement in efficiency may be obtained by using a lens. This magnifies the laser mode slightly, so that it makes a better match to the fibre mode.

Finally we note that a nominally single-moded guide may support two orthogonal polarization modes. However, due to differences in end-mirror reflectivity, the mode polarized parallel to the junction plane experiences greater round-trip gain, and will lase preferentially, so that the stimulated emission is linearly polarized in this direction. Spontaneous emission will, however, add unpolarized light to the output, which will also have a component perpendicular to the junction plane. This lowers the degree of polarization, i.e. the ratio of orthogonally-polarized components. This is limited and will vary with the output power. A typical value for a laser driven well past threshold is greater than 50:1.

12.7 THE SEMICONDUCTOR LASER: MODULATION

We now consider the behaviour of a semiconductor laser when it is modulated by a sinusoidal time-varying drive current, to demonstrate the improvement in its high-frequency response over that of an LED. It can be shown that the time taken to build up the photon density in the cavity from zero is relatively long, due to the slow response of the device at low current (when it operates as an LED). For high-speed operation, a laser is therefore normally biased past threshold, using a d.c. bias current. We shall therefore assume that the drive current is of the form $I = I' + I'' \exp(j\omega t)$, where I' is the d.c. bias, and I'' and ω are the amplitude and angular frequency of the superimposed a.c. modulation, respectively. Note that the bias is chosen so that $I' > I_t$ and $I' \gg I''$. We shall also define the other quantities of interest in a similar way, as the sum of a d.c. bias and an a.c. modulation term. We therefore put $\phi = \phi' + \phi'' \exp(j\omega t)$, $n = n' + n'' \exp(j\omega t)$ and $P = P' + P'' \exp(j\omega t)$.

We now substitute these expressions into Eqs (12.73) and (12.74). To simplify the results we perform the following operations: (1) we neglect spontaneous emission as small by comparison with stimulated emission, and (2) we eliminate as many terms as possible, using the relations between the d.c. terms ϕ', n', I' and P' derived in the previous section. The photon rate equation then reduces to

$$j\omega\phi'' = G\phi' n'' \tag{12.86}$$

Similarly, the electron rate equation becomes

$$j\omega n'' = I''/ev - n''/\tau_c - \phi''/\tau_p - G\phi' n'' \tag{12.87}$$

Eqs (12.86) and (12.87) may then be combined to yield the small-signal optical modulation amplitude P'' in terms of the corresponding current modulation term I'', as

$$P''(\omega) = (\boldsymbol{h}cI''/e\lambda_g)\,(G\phi'/\tau_p)/\{-\omega^2 + j\omega[1/\tau_e + G\phi'] + G\phi'/\tau_p\} \qquad (12.88)$$

Because this expression is rather complicated, we now introduce a number of new parameters. First we define the modulation amplitude at zero frequency as

$$P''(0) = \boldsymbol{h}cI''/e\lambda_g \qquad (12.89)$$

Using this, we define the complex modulation ratio as $m(\omega) = P''(\omega)/P''(0)$. Finally we introduce a resonant frequency ω_r and a damping factor ζ, defined as

$$\omega_r = \sqrt{(G\phi'/\tau_p)} \qquad \text{and} \qquad \zeta = \{1/\tau_e + \tau_p\omega_r^2\}/2\omega_r \qquad (12.90)$$

With this new notation, the modulation ratio may be written as

$$m(\omega) = \omega_r^2/\{(\omega_r^2 - \omega^2) + 2j\zeta\omega\omega_r\} \qquad (12.91)$$

The small-signal frequency response of a laser is therefore that of a damped simple harmonic oscillator. This result might well have been anticipated; since the rate equations amount to two coupled first-order differential equations, the transient response is effectively described by two uncoupled second-order equations, each in the form of the governing equation of an oscillator. Since $m(\omega)$ is complex, the optical modulation must lag behind the current once again. The modulus of $m(\omega)$ is of most interest since this describes the depth of the optical modulation. This is given by

$$|m(\omega)| = \omega_r^2/\sqrt{\{(\omega_r^2 - \omega^2)^2 + (2\zeta\omega\omega_r)^2\}} \qquad (12.92)$$

Figure 12.41 shows a logarithmic plot of $|m(\omega)|$ versus ω, for typical values of ω_r and ζ. At low angular frequencies, $|m(\omega)|$ is unity. As the frequency rises, there is a peak in the response due to a resonance between the electron and photon populations. This is known as a *relaxation resonance*; for low damping, it occurs at a frequency close to ω_r. Finally, at high frequencies, the response falls off roughly as $|m(\omega)| = \omega_r^2/\omega^2$. The bandwidth is therefore set almost entirely by the resonant frequency ω_r. Using some simple manipulations it is possible to define ω_r alternatively as

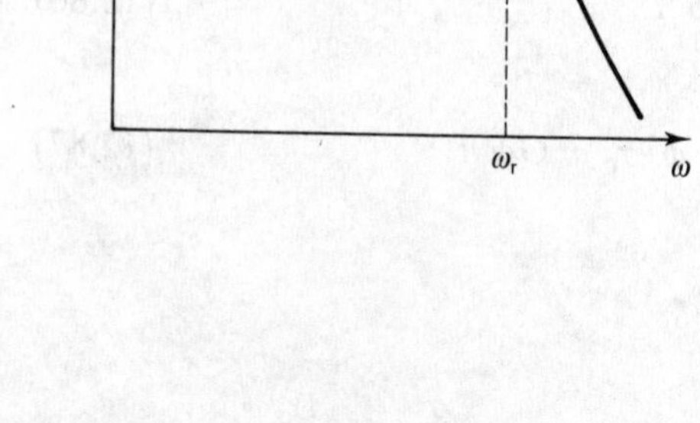

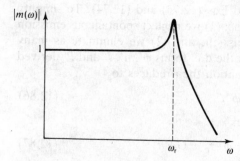

Figure 12.41 Small-signal frequency response of a semiconductor laser.

$$\omega_r = \surd(G\lambda_g P'/hcv) \tag{12.93}$$

The resonant frequency is therefore variable, rising as the square root of the d.c. power level. In order to obtain the greatest modulation bandwidth, the laser should therefore be driven as hard as possible.

Design Example

We may estimate ω_r and ζ for GaAs laser as follows. Assuming a cavity length of 250 μm, and an active area cross-section of 0.1 μm $\times$ 3 μm, a gain constant of 10^{-12} m^3/s, and an output power level of 3 mW at λ_g = 820 nm, we obtain $\omega_r = \surd\{10^{-12} \times 0.82 \times 10^{-6} \times 3 \times 10^{-3}/(6.62 \times 10^{-34} \times 3 \times 10^8 \times 250 \times 0.1 \times 3 \times 10^{-18})\} = 1.29 \times 10^{10}$ rad/s, corresponding to a bandwidth of roughly 2 GHz. This is entirely typical; the modulation bandwidth of a laser is typically 100 times greater than that of an LED, making the laser the preferred source for high bit-rate communications systems. For modulation at still higher speeds, an external modulator must be used. We have previously estimated the photon lifetime for a 250 μm-long GaAs laser as $\tau_p \approx 2.5$ ps. Assuming a recombination time of $\tau_e \approx 1$ ns, the damping factor is $\zeta = \{10^9 + 2.5 \times 10^{-12} \times (1.29 \times 10^{10})^2\}/(2 \times 1.29 \times 10^{10}) \approx 0.05$. This is also typical; the relaxation oscillations are underdamped at normal powers, leading to the sharp peak in the response near ω_r in Fig. 12.41.

12.8 DBR AND DFB LASERS

Although it is blessed with the important advantage of simplicity, the Fabry-Perot semiconductor laser suffers from several undesirable characteristics. The first is that its output is often multi-line, a major drawback for optical communication systems. In this section we shall consider a number of solutions to the problem. The two most successful approaches involve reflection gratings of the type described in Chapter 10; these are known as the *distributed Bragg reflector* (or DBR) and the *distributed feedback* (or DFB) laser, respectively. Figure 12.42 shows a comparison between the two structures.

In the DBR laser, either or both of the cleaved end-mirrors are replaced by a corrugated reflection grating. As a result, the DBR laser has several sections: a central active region, which provides the gain; and passive grating reflectors, which provide feedback. We would expect high reflectivity from the grating only for longitudinal modes whose wavelengths approximately satisfy the Bragg condition, $2n_{\text{eff}\nu}\Lambda = \lambda_\nu$. When this is not satisfied, the reflectivity is quickly reduced to zero, drastically reducing the photon lifetime in the process. We have already shown that the relative importance of different longitudinal modes is governed by amplification factors F_ν. Including a mode-dependent photon lifetime $\tau_{p\nu}$, these might be written as $F_\nu = 1/\{1/\tau_{p\nu} - [G(n - n_0) + b'(\lambda_\nu - \lambda_p)^2]\}$. Since they depend inversely on the difference between two comparable quantities, even a small increase in $1/\tau_{p\nu}$ can have a significant effect. Consequently, the introduction of grating feedback

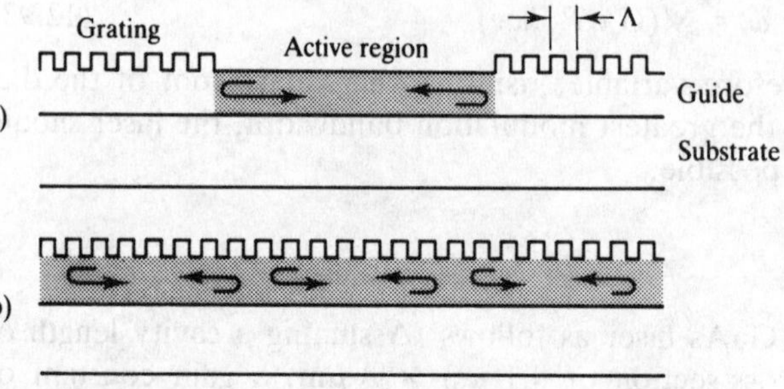

Figure 12.42 (a) DBR and (b) DFB lasers.

causes a vastly greater purification of the spectrum than the grating filter response might suggest. In fact, the mode selectivity is governed almost entirely by the curvature of the response near the Bragg wavelength, and the output of a DBR laser may consist of only a single line.

The exact wavelength of each longitudinal mode is governed by the optical path length between the reflectors. Since the effective index in the active region will depend on the level of pumping, the positions of the longitudinal modes relative to the peak in grating reflectivity must alter with current. To obtain optimum reflectivity for one desired mode, independent of variations in drive current, a slightly more complicated structure is required. For example, Fig. 12.43 shows a three-section DBR laser. This has a multi-quantum-well active section; feedback is provided by a cleaved mirror at the left-hand end and a grating reflector at the right. However, in this case the Bragg wavelength may be altered by a tuning current, which varies the effective index beneath the grating through minority carrier injection. A central passive phase-shifting section (operating by the same mechanism) then allows the postion of one longitudinal mode to be matched to the peak in grating reflectivity. Through an interplay between the three currents, tunable single-line operation is therefore possible.

DFB lasers (Fig. 12.42(b)) achieve a similar end result, but operate by an entirely different principle. In this case the grating and the active region coincide, and it is no longer possible to consider the feedback as being provided by reflectors at either end of a cavity; instead, it must be distributed throughout the active region. To show that this may give rise to laser operation, we need a simple model. Below threshold, linear coupled mode theory is suitable. All that is required is the insertion of an effective gain parameter g into the coupled equations previously derived for a reflection grating. If this modification is made to Eqs (10.62), for example, we obtain

$$\mathrm{d}A_{\mathrm{F}}/\mathrm{d}z - gA_{\mathrm{F}} + \mathrm{j}\kappa A_{\mathrm{B}} = 0$$

$$\mathrm{d}A_{\mathrm{B}}/\mathrm{d}z + \{g - j2\Delta\beta\}A_{\mathrm{B}} - \mathrm{j}\kappa A_{\mathrm{F}} = 0 \tag{12.94}$$

We can check that the extra terms are realistic by examining the solutions for the forward and backward mode amplitudes A_{F} and A_{B} when the coupling coefficient is zero. They grow exponentially in the $\pm z$ directions, as $\exp(\pm gz)$, which seems reasonable. When κ is non-zero, the equations may still be solved analytically.

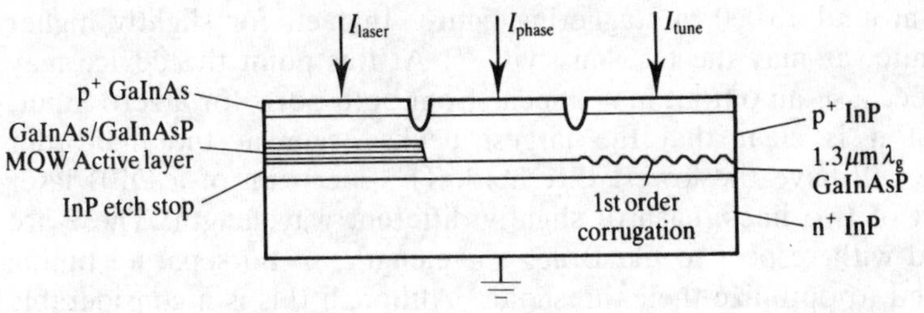

Figure 12.43 A three-section, continuously-tunable MQW-DBR laser (*after T.L. Koch* et al., *1988*).

However, the solutions are hard to interpret, as they contain many complex quantities. It is therefore simpler to discuss a typical numerical solution for the boundary conditions of unity input to one end of the device (i.e. for $A_F = 1$ on $z = 0$ and $A_B = 0$ on $z = L$). For example, Fig. 12.44 shows logarithmic plots of reflectivity $R = A_B A_B^*$ on $z = 0$ versus detuning $\Delta\beta L$, for a coupling length of $\kappa L = 0.4$ and different values of normalized gain gL.

For $gL = 0$, the response is that of a passive grating, reaching a peak when the Bragg condition is satisfied and falling gradually to zero with a standard filter response as $|\Delta\beta L|$ increases. As gL rises, however, the shape of the response alters dramatically. The reflectivity improves overall, which is only to be expected. More interestingly, sharp peaks start to appear in the curves. These are displaced symmetrically on either side of the Bragg condition, and their characteristics are broadly analogous to those of the longitudinal resonances occuring in a Fabry-Perot cavity. They arise when the contributions that are scattered by the grating and then amplified by the gain mechanism combine with a suitable relative phase to ensure high net reflectivity. For $gL = 3$, the first-order resonances are extremely pronounced,

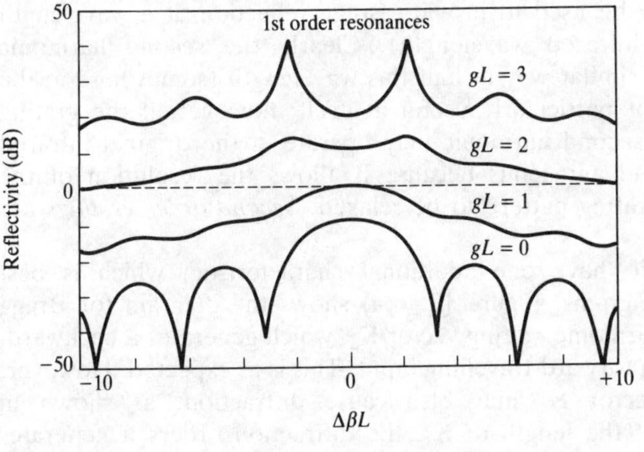

Figure 12.44 Reflectivity as a function of detuning $\Delta\beta L$, for a reflection grating with gain.

reaching a peak of around 13000, a staggering figure. In fact, for slightly higher gain, R may be infinite, as may the transmissivity T. At this point the device may be said to be lasing because an output may appear from both ports for a zero input.

From Fig. 12.44 it is clear that the largest peaks occur at the first-order resonances, so these will have the lowest threshold. The spectrum of a DFB laser will therefore consist of two lines, each of slightly different wavelength. These are automatically aligned with respect to the Bragg wavelength, so no separate tuning mechanism is required to optimize their thresholds. Although this is a considerable improvement over a Fabry-Perot laser, further advances are possible. All that is required is to generate some form of asymmetry in the grating reflectivity. This may be done by inserting a step in the phase of the grating at a suitable point along the cavity, a simple process for gratings fabricated by electron-beam writing. Since the required shift is one quarter of a period, such devices are known as *Λ/4-shifted DFB lasers*. The net result is that the threshold gains of the two first-order resonances are displaced from one another, so that one longitudinal mode lases preferentially and the output is single-line.

Figure 12.45(a) shows a schematic of a typical long-wavelength ridge waveguide DFB laser. This is very similar in construction to the Fabry-Perot laser shown in Fig. 12.33, but an extra layer of material has been inserted above the active layer. The grating is fabricated by etching a pattern of electron-beam-defined grooves into this layer, using a preferential wet chemical etch that forms self-stopping V-shaped grooves with a precise depth (giving a well-controlled coupling coefficient). Note that etching of the active layer itself is inadvisable because most etch processes create surface damage sites that degrade the recombination efficiency, increasing the threshold.

Because it is possible to fabricate complicated grating profiles, the perturbation of the guide need not be sinusoidally-varying (although it must be periodic in Λ). In general the perturbation may be written as a Fourier series, with components of period Λ, $\Lambda/2$, $\Lambda/3$, etc. These have associated grating vectors $K_1 = 2\pi/\Lambda$, $K_2 = 2K_1$ and $K_3 = 3K_1$, respectively. We have already seen that the fundamental component of a grating may be used to provide Bragg reflection at a wavelength $\lambda_0 = 2n_{eff}\Lambda$ (say, a near-infrared wavelength). Clearly the second harmonic component will operate in a similar way at half this wavelength (somewhere in the visible spectrum). This is not particularly useful in itself; however, if the grating period is now doubled, the second harmonic may operate at the original infrared wavelength. This is highly advantageous because it allows the resolution of the process used to define the grating pattern to be relaxed. *Second-order gratings* are therefore common.

Second-order gratings do have one additional characteristic, which is best illustrated using K-vector diagrams. Figure 12.46(a) shows the diagram for Bragg reflection using the second-harmonic grating vector $\mathbf{K}_2$, which generates a backward-going diffraction order from a forward-travelling input. This is as expected. However, the fundamental grating vector $\mathbf{K}_1$ may also cause diffraction, as shown in Fig. 12.46(b). Since it is half the length of $\mathbf{K}_1$, the diffraction orders it generates propagate vertically upwards into the guide cover layer and downwards into the

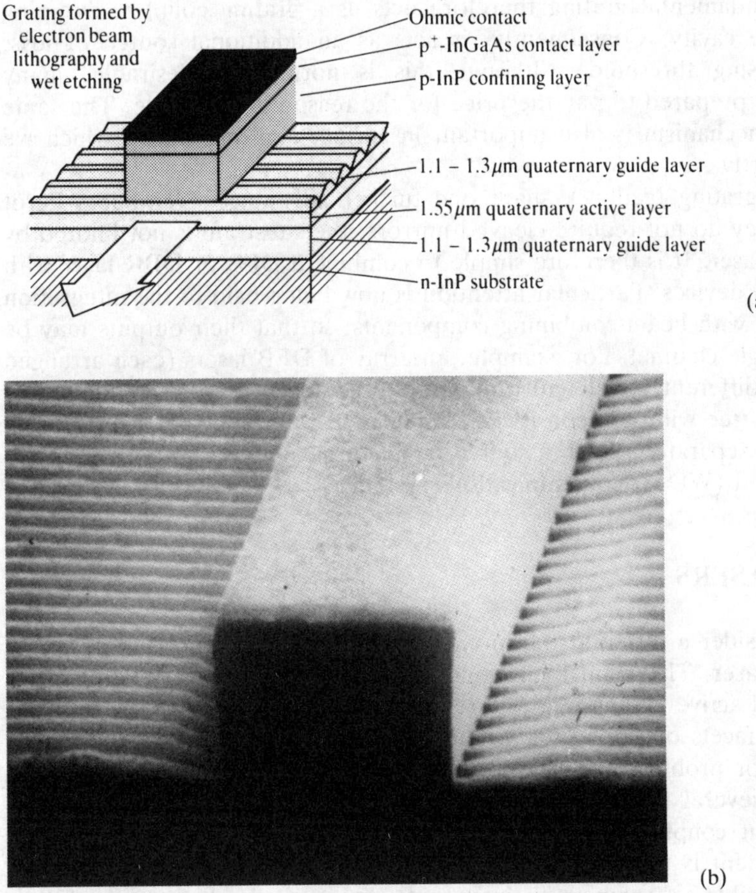

Figure 12.45 (a) Schematic of a typical InP/InGaAsP ridge waveguide DFB laser, and (b) SEM view of a similar device (*photograph courtesy C.J. Armistead, S.T.C. Technology Ltd.*)

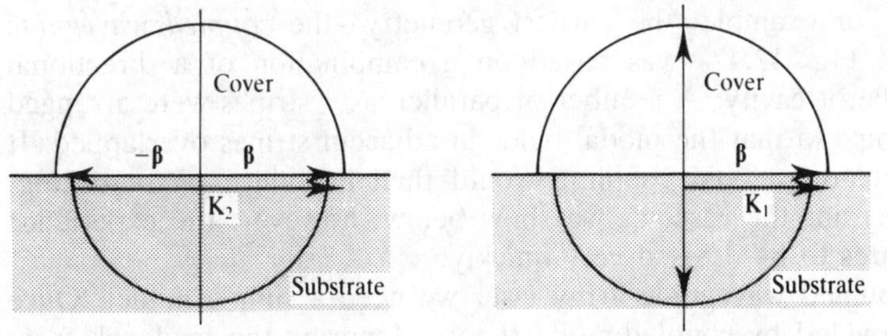

Figure 12.46 K-vector diagrams for diffraction from (a) the second harmonic, and (b) the fundamental of a second-order grating.

substrate. The fundamental grating therefore acts as a grating coupler, dumping energy out of the cavity. Consequently, it acts as an additional source of loss, increasing the lasing threshold. Although this is normally undesirable, many manufacturers are prepared to pay the price for the reason given above. The same output coupling mechanism is also important in surface-emitting lasers, which we shall describe shortly.

All forms of grating feedback share one further advantage over Fabry-Perot lasers. Because they do not require cleaved mirrors, the substrate is not limited by the length of the laser. It is therefore simple to combine a DFB or DBR laser with other guided wave devices. Particular attention is now being paid to the integration of arrays of lasers with beam-combining components, so that their outputs may be combined in a single channel. For example, an array of DFB lasers (each arranged to emit a slightly different wavelength) might be integrated with a Y-junction tree. The output of the tree will then consist of a number of wavelengths, each of which can be modulated separately. Such a source might form the basis of a wavelength-division multiplexed (WDM) communications system.

12.9 ARRAY LASERS

We shall now consider a different question, namely how to increase the power of a semiconductor laser. The main limitations of stripe-geometry Fabry-Perot lasers are their restricted active volume and the requirement that the optical output must pass through end-facets of small cross-sectional area—indeed, catastrophic mirror failure was a major problem in early devices. One obvious way to improve total output is to use several lasers, placed side-by-side on a chip, so that the active volume and output coupling window are increased simultaneously. The problem with this arrangement is that each of the lasers will oscillate independently. As a result, it is impossible to combine their outputs into a single, high-power beam, because they must sum together with random phases.

To obtain a useful increase in power, the laser must be forced to emit light of a fixed spatial pattern. The outputs of all emitters should, preferably, be of equal amplitude and phase, and a variety of possible structures have been investigated in pursuit of this goal. For example, the earliest geometry—the *coupled waveguide array laser*, shown in Fig. 12.47—was based on a combination of a directional coupler and a Fabry-Perot cavity. A number of parallel laser stripes were arranged close together on a chip, so that the modal fields in adjacent stripes overlapped. It was anticipated that co-directional coupling would then introduce *phase-locking*. Unfortunately this was not the case, as we show below; however, the experience allowed better structures to be devised very quickly.

To analyse the coupled waveguide array laser we need a simple model. Once again this may be provided by coupled-mode theory. Ignoring the feedback from the mirrors for the time being, we may start by considering the related problem of wave propagation in an N-element coupled waveguide array. We shall assume there is equal gain in each stripe (defined by an effective gain g) and nearest-neighbour

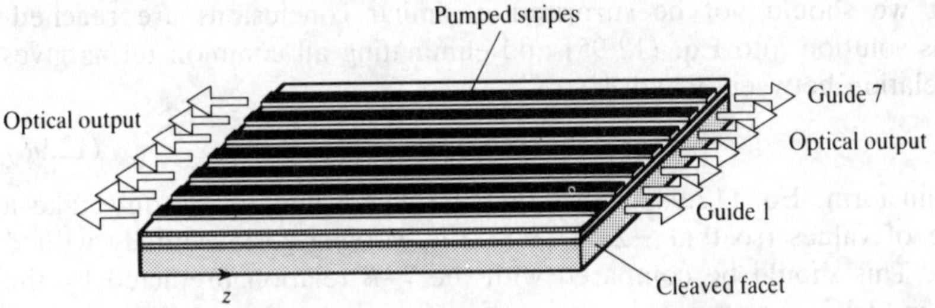

Figure 12.47 A seven-element co-directionally coupled ridge waveguide array laser.

coupling between stripes (defined by a coupling coefficient κ). In this case the coupled-mode equations governing the change in amplitude of the mode in each stripe with distance along the array may be obtained by a simple modification of our earlier two-guide equations, as

$$\mathrm{d}A_1/\mathrm{d}z - gA_1 + \mathrm{j}\kappa A_2 = 0$$

$$\mathrm{d}A_n/\mathrm{d}z - gA_n + \mathrm{j}\kappa \{A_{n+1} + A_{n-1}\} = 0 \qquad \text{for } n = 2, 3, \ldots, N-1$$

$$\mathrm{d}A_N/\mathrm{d}z - gA_N + \mathrm{j}\kappa A_{N-1} = 0 \tag{12.95}$$

Equation (12.95) shows that the two outermost guides are coupled to a single neighbour, while all the other guides are coupled to two. In Chapter 10 we solved two similar equations for fixed boundary conditions. This time we shall follow a different course and look for eigenmode solutions. These describe patterns of mode amplitude that may propagate along the array with a change in overall amplitude or phase, but not of shape. Each such pattern may therefore be written in the form:

$$A_n = a_n \exp(-\mathrm{j}\gamma z) \tag{12.96}$$

Here the coefficients a_n describe the fixed shape of the pattern, while the eigenvalue γ (which may be complex) accounts for any change in gain or phase additional to the average propagation constant β_0 implicit in the equations. Initially we shall assume that the array is infinite, so that the central equation in Eq. (12.95) is valid everywhere. Because the array now represents an infinite 'medium' (albeit with rather complicated properties), it is natural to look for wave solutions. We therefore put

$$a_n = A \exp(-\mathrm{j}\phi n) \tag{12.97}$$

where ϕ is a constant. Equation (12.97) describes a form of travelling wave, moving across the array and changing in phase by discrete amounts between guides. There is a strong analogy between this solution and the Bloch waves of Chapter 11; indeed, the behaviour of light in a periodic array of guides is very similar to that of electrons in an array of potential wells. Here we have approached the mathematics slightly differently, but we should not be surprised if similar conclusions are reached. Substituting this solution into Eq. (12.95) and eliminating all common terms gives the following relation between γ and ϕ:

$$\gamma = \mathrm{j}g + 2\kappa \cos(\phi) \tag{12.98}$$

Ignoring the gain term, Eq. (12.98) shows that the eigenvalue γ may only take a restricted range of values (so that $-2\kappa \leq \gamma \leq 2\kappa$), varying cosinusoidally with ϕ over the range. This should be compared with the $\mathscr{E}$–k relation predicted by the Krönig-Penney model in Chapter 11.

We now reintroduce the finite lateral extent of the array by assuming that the upper and lower equations Eq. (12.95) will effectively be satisfied if we assume coupling to two ficticious guides (at $n = 0$ and $n = N + 1$), which support modes of zero amplitude. This may be done by taking $a_n = A' \sin(\phi n)$, i.e. by assuming a standing-wave solution. Applying the boundary conditions $a_0 = a_{N+1} = 0$, we then find that ϕ is restricted to the following set of discrete values:

$$\phi_\nu = \nu\pi/(N+1) \tag{12.99}$$

where $\nu = 1, 2, \ldots, N$ is the mode number. The complete solution to Eq. (12.96) is therefore defined by mode amplitudes $a_{\nu n}$ and propagation terms γ_ν, given by

$$a_{\nu n} = A \sin[\nu n\pi/(N+1)] \qquad \gamma_\nu = \mathrm{j}g + 2\kappa \cos[\nu\pi/(N+1)] \tag{12.100}$$

The terms $a_{\nu n}$ are the amplitudes of the modes in each guide. If each individual mode is defined by a transverse field $E_i(x,y)$, the total field pattern of the νth eigenmode must therefore be

$$E_\nu(x,y) = \sum_{i=1}^{N} a_{\nu i}\, E_i(x,y) \tag{12.101}$$

Figure 12.48(a) shows a one-dimensional plot of the lowest-order transverse mode ($\nu = 1$) for a seven-guide array. Within each guide the pattern is like that of the individual guide mode, but there is an overall sinusoidal envelope defined by Eq. (12.100). Similarly, Fig. 12.48(b) shows the second-order mode ($\nu = 2$). This is similar, but the spatial frequency of the envelope is now doubled. The trend

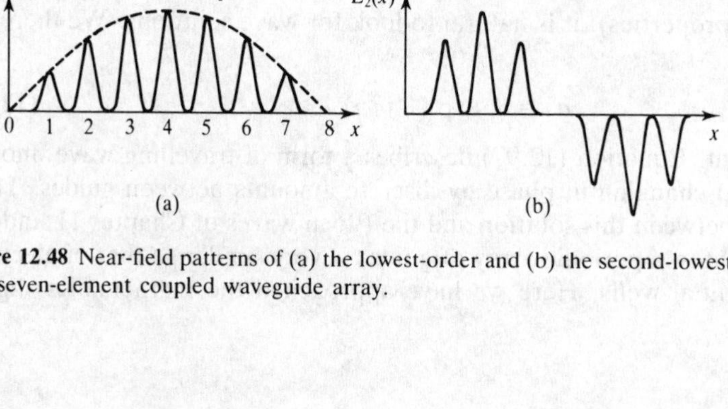

Figure 12.48 Near-field patterns of (a) the lowest-order and (b) the second-lowest-order transverse modes of a seven-element coupled waveguide array.

continues as ν increases, so that the envelope of the highest-order mode has the greatest spatial frequency, with adjacent mode amplitudes of opposite sign.

We now return to the provision of feedback by the end-facets. If their reflectivity is uniform across the array, each eigenmode will be reflected without any change in shape, i.e. without mode conversion. The eigenmodes found above are therefore identical to the transverse modes of the waveguide array laser itself, since they may travel a round-trip through the cavity unaltered. If the laser can be forced to oscillate on just one of these modes (the lowest-order is the most uniform, and hence the most useful), the required phase-locking will have been introduced, because the output from each stripe will then have a fixed phase relationship with its neighbours.

To investigate the mode selectivity of the structure we must examine the conditions for lasing. The total propagation constant of the νth transverse mode is $\beta_0 + \gamma_\nu$, including the average term β_0. Only Re (γ_ν) will affect phase variations, so the longitudinal resonance condition is therefore

$$2L\{\beta_0 + \text{Re}\ (\gamma_\nu)\} = 2\mu\pi \tag{12.102}$$

where μ is an integer and L is the cavity length. Since Re (γ_ν) is different for each transverse mode, we may deduce that each will have a different spectrum of longitudinal modes. Similarly, only Im (γ_ν) governs the overall gain of each mode, so the condition for unity round-trip gain is

$$R_1 R_2 \exp\ \{2L\ \text{Im}\ (\gamma_\nu)\} = 1 \tag{12.103}$$

where R_1 and R_2 are the amplitude reflectivities of the mirrors. Examining Eq. (12.100), however, we can see that Im $(\gamma_\nu) = g$ for all ν. Consequently, the threshold gain must be the same for each transverse mode, so co-directional coupling has actually introduced no selectivity at all. We would therefore expect the laser to oscillate on a random mixture of all the possible transverse modes.

In practice, early array lasers had a tendency to lase only on the highest-order mode. The explanation lay in the fact that only the active region of each stripe is pumped, so that there is very little gain between the stripes. An accurate calculation shows that the coupling coefficient is then complex, so that Im (γ_ν) is actually different for each mode, being greatest for the highest-order one. This is physically realistic; this particular mode has a field that changes in sign from guide-to-guide, falling to zero in between, so it is unaffected by interstripe loss. However, being the least uniform mode, it also is the least usable. Despite some attempts to introduce selectivity by separately pumping each stripe or using non-uniform stripe spacings, directionally-coupled arrays have largely been abandoned. Nevertheless, high outputs were achieved; for example, 2.6 W CW was obtained from a 40-stripe array.

A more effective mechanism is required for single-transverse-mode operation. The obvious solution is to introduce a mode-dependent loss, which reduces the effective gain of undesired modes while allowing desired ones to travel up and down the cavity unaffected. In Chapter 9 it was shown that a single-mode Y-junction will radiate any antisymmetric component that is injected at its forked end. Consequently,

Y-junctions can be used to construct a filter that discriminates against transverse field patterns of high spatial frequency (which have fields of opposite sign in adjacent stripes). Figure 12.49 shows a device based on such a filter, the *Y-junction-coupled array laser*. At either end of the device there is an array of uncoupled single-moded laser stripes. In this example there are seven stripes in the left-hand array and eight in the right. The two arrays are linked together in the central section by a network of back-to-back Y-junctions, leaving two unconnected stubs on the left. All sections of guide are equally pumped and the structure is contained in a Fabry-Perot cavity.

To analyse the structure we assume for generality that there are N stripes at the left-hand end of the array, and $N + 1$ at the right. Once more we assume that each guide has a total length L, an amplitude gain g, a propagation constant β_0, and end-mirror amplitude reflectivities R_1 and R_2. The Y-junctions act as linear beamsplitters or combiners, as described in Chapter 9. We proceed by defining the amplitudes of the modes in the individual stripes at the left-hand end of the cavity as $a_1, \ldots, a_N$ at the start of a round trip. It is then simple to show that the amplitudes $b_1, \ldots, b_{N+1}$ before reflection at the right-hand facet are

$$b_n = 1/2\{a_n + a_{n-1}\} \exp(-j\beta_0 L) \exp(gL) \tag{12.104}$$

where once again we have introduced two ficticious amplitudes a_0 and a_{N+1} (those of the modes in the stubs), which are both zero. After reflection at the right-hand facet, propagation back to the left-hand end of the cavity, and further reflection, the new mode amplitudes $a'_1, \ldots, a'_N$ are

$$a'_n = 1/4\{a_{n+1} + 2a_n + a_{n-1}\} R_1 R_2 \exp(-j2\beta_0 L) \exp(2gL) \tag{12.105}$$

If the original amplitudes $a_1, \ldots, a_N$ represent an eigenmode, they should be related to the new amplitudes $a'_1, \ldots, a'_N$ by a constant factor (otherwise the pattern will have changed in a round-trip). We may thus put $a'_n = \delta a_n$, where δ is a constant. This allows us to solve Eq. (12.105) using our previous methods. Initially we ignore the finite extent of the array, and assume a solution as a kind of travelling wave, writing $a_n = A \exp(-j\phi n)$. We then obtain

$$\delta = 1/2\{1 + \cos(\phi)\} R_1 R_2 \exp(-j2\beta_0 L) \exp(2gL) \tag{12.106}$$

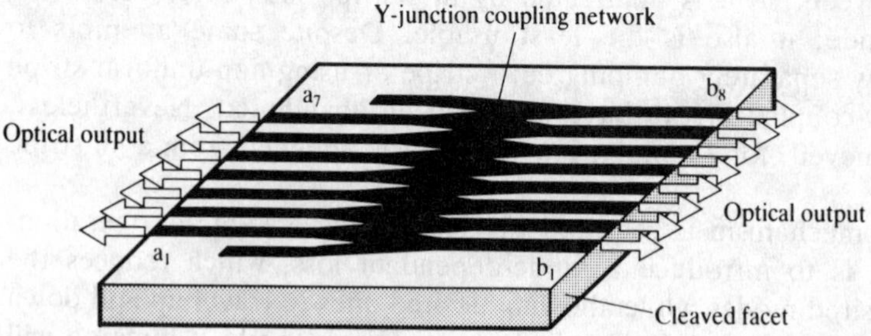

Figure 12.49 A Y-junction-coupled semiconductor array laser.

However, since the array is actually finite, we must satisfy the boundary conditions $a_0 = a_{N+1} = 0$. The solution must therefore again be a standing-wave pattern, so that $a_n = A' \sin(\phi n)$. Since it is clear that the analysis will give very similar results to those obtained for the co-directionally-coupled array, we may immediately deduce that ϕ is again restricted to the set of values in Eq. (12.99), and that the overall shapes of the modes at the left-hand-end of the array are as specified in Eq. (12.100).

Each mode will lase when its round-trip gain is unity, i.e. when $\delta = 1$. From Eq. (12.106), we can see that the longitudinal resonance condition for all transverse modes is now the same, namely $2\beta_0 L = 2\mu\pi$. More interestingly, the threshold gain g_ν is now different for each, being given by

$$g_\nu = 1/2L \log_e (1/f_\nu R_1 R_2) \tag{12.107}$$

where the factor f_ν—the transmission efficiency of the Y-junction network for the νth mode—is

$$f_\nu = 1/2\{1 + \cos[\nu\pi/(N+1)]\} \tag{12.108}$$

From this, we can see that the highest transmission efficiency (slightly less than unity) is obtained for the lowest-order transverse mode, which must therefore have the lowest lasing threshold. In contrast, the transmission of the highest-order mode is close to zero, so its threshold will be very high. The Y-junction network has therefore introduced the desired selectivity in favour of a single transverse mode with a useful spatial distribution. Practical Y-junction-coupled lasers have excellent stability, showing little change in their near- or far-field patterns over the entire range of operation, and their output powers are typically of the order of 400 mW CW at present.

12.10 SURFACE-EMITTING LASERS

In the future it may be desirable to arrange for the emission from a laser chip to be taken from its surface, rather than an end-facet. This would allow the fabrication of two-dimensional arrays of lasers, which might have applications in optical processors. At the very least it would enable an entire wafer of lasers to be tested for failures before individual devices are separated by cleaving. A variety of *surface-emitting semiconductor laser* geometries are currently being investigated. These are all still based on single-mode waveguides, but there are two major distinctions between types.

First, the laser cavity may be in the conventional orientation, parallel to the surface. Feedback may be obtained without cleaving, by using distributed Bragg reflectors. If these are based on second-order gratings, diffraction from the grating fundamental can be used to obtain output coupling normal to the surface, as described earlier (although unnecessary cladding layers over the grating are generally removed, to minimise absorption). For example, Fig. 12.50 shows a *Y-junction-coupled, surface-emitting DBR array laser*. A laser of this type has already been

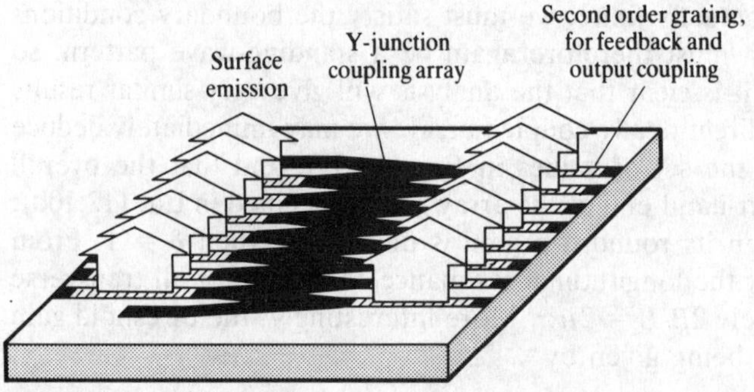

Figure 12.50 A Y-junction-coupled, surface-emitting DBR array laser.

demonstrated with a 10 × 10 array of emitters occupying an area of 60 μm × 5 mm, with a total output power of more than 1 W.

Alternatively, the cavity may be normal to the substrate surface. In this case, the device is known as a *vertical cavity surface-emitting laser*. Figure 12.51 shows a typical example, a circular buried heterostructure GaAs/GaAlAs laser. Although the double heterostructure is still formed from layers parallel to the surface, the cavity is a roughly cylindrical pillar approximately 6 μm in diameter and 7 μm long with its axis normal to the surface. The current is injected from a ring electrode into the lower end of the pillar, which is surrounded by lateral current-blocking layers. Feedback is provided by a gold/silica mirror at the bottom of the cavity, and by a silica/titania multi-layer stack deposited in a deep well etched right through the substrate at the top. The optical output is taken from the upper mirror. More recent vertical cavity lasers have used distributed feedback or distributed Bragg reflection, based on epitaxial multi-layers, thus eliminating the requirement for

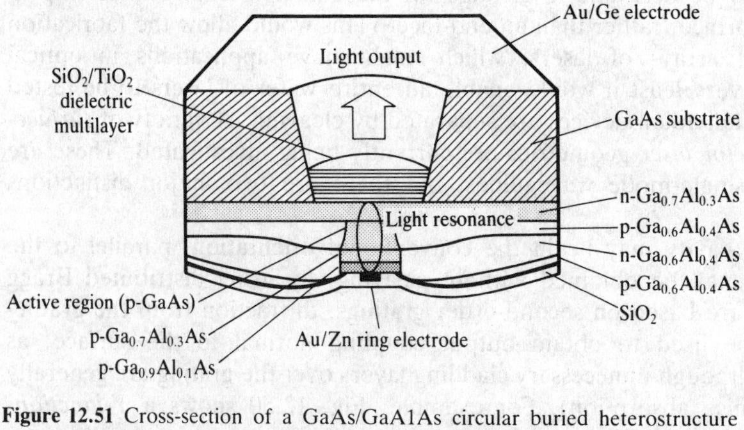

Figure 12.51 Cross-section of a GaAs/GaA1As circular buried heterostructure vertical cavity surface-emitting laser (*after S. Kinoshita* et al. © *IEEE 1987*).

external mirrors. Due to the short length of the cavity, the output consists only of a single longitudinal mode, a feature of all such lasers. The particular laser shown had a threshold current of 6 mA and an output in excess of 1 mW. Although all surface-emitting lasers are at very early stages of development, they represent a promising new technology, which may have many applications outside communications.

12.11 TRAVELLING-WAVE LASER AMPLIFIERS

Besides acting as a versatile source, the semiconductor laser can provide other functions. For example, if the end-facets are suitably coated so that they are anti-reflecting, laser operation will be suppressed due to the lack of feedback. However, optical gain will still be available in the cavity, provided the electron density is driven above threshold. In this case the device may act as a *travelling-wave laser amplifier* (TWLA). Figure 12.52 shows the principle, and a typical application.

The TWLA is connected between two fibres as an in-line amplifier. A relatively weak optical input will grow exponentially as it passes through the TWLA, emerging as a much stronger amplified signal. The maximum possible input is limited because the gain saturates at high optical power, but this level is quite sufficient for use in typical systems. Recent devices based on double-channel planar buried heterostructure lasers show little polarization sensitivity (1 dB variation in gain between orthogonal polarizations), gains of up to 30 dB (although this figure is reduced by coupling losses), and bandwidths of around 75 GHz. Unfortunately, the amplification is not noiseless, due to the presence of spontaneous emission. This may be minimized by passing the output through a narrow-band filter, which blocks the majority of the wide-band spontaneous emission noise, while passing the desired signal. Figure 12.53 shows a fully packaged travelling wave laser amplifier with fibre pigtails.

Travelling wave laser amplifiers can have applications other than as simple repeaters. For example, Fig. 12.54 shows an amplifying switch based on a passive coupler combined with two TWLAs. The coupler acts as a splitter which directs the input to the two amplifiers. Since the output of each TWLA may be switched off

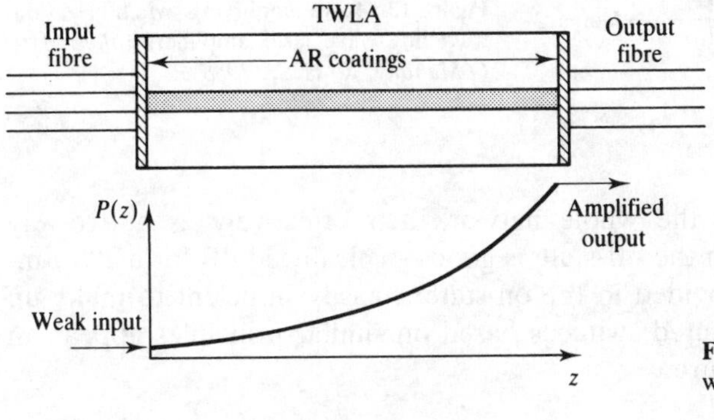

Figure 12.52 Schematic of a travelling-wave semiconductor laser amplifier.

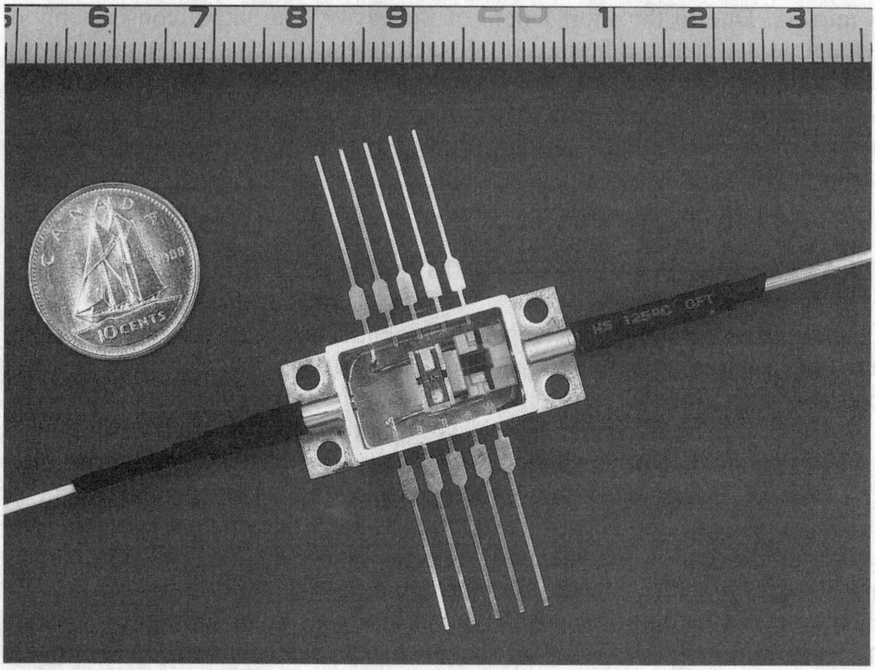

Figure 12.53 A travelling-wave laser amplifier (*photograph courtesy Telesis, a publication, of Bell Northern Research Ltd*).

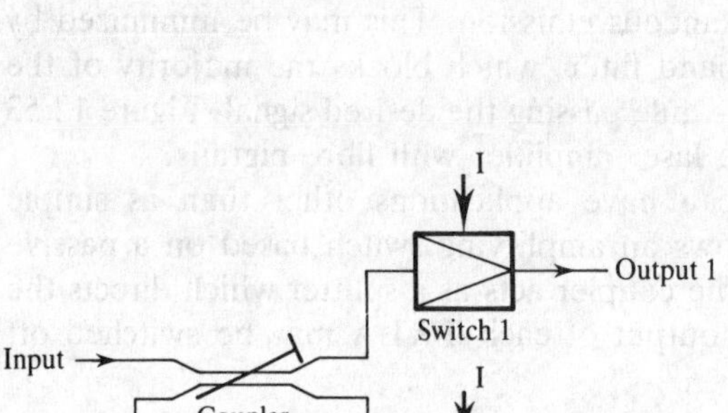

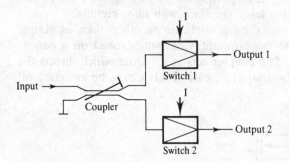

Figure 12.54 An amplifying switch based on travelling wave laser amplifiers (*after M.J. O'Mahony, © IEEE 1988*).

by removing the bias current, the whole network acts effectively as a two-way switch. The isolation achieved in the off-state is good—typically 30 dB for a 200 μm-long device—while the gain provided in the on-state is easily sufficient to make up for coupling losses. Fully integrated switches based on similar principles appear an attractive possibility for the future.

PROBLEMS

12.1 A silicon p-n junction diode has doping levels of $N_a = 10^{24}$ m^{-3} and $N_d = 10^{23}$ m^{-3}. Estimate the electric field across the depletion layer, under 5 V reverse bias. Assuming that the mobilities of electrons and holes in Si are $\mu_e = 0.135$ m^2/Vs and $\mu_h = 0.048$ m^2/Vs, respectively, estimate their transit times across the depletion layer. Discuss the implications for photodiodes.
[20.4×10^6 V/m; 0.1 ps; 0.3 ps]

12.2 A silicon p-n$^+$ diode has a cross-sectional area of 1 mm^2, and doping levels of $N_a = 10^{21}$ m^{-3} and $N_d = 10^{24}$ m^{-3}. Assuming that the distance L_p from the depletion layer to the p-side contact is 0.1 μm, and that the corresponding distance L_n is $\gg L_p$, calculate the current through the diode under a forward voltage of 0.5 V. The hole and electron diffusion coefficients may be found using the *Einstein relation* $D_{e,h} = \mu_{e,h}kT/e$. If the minority carrier lifetimes are ≈ 1 μs, are your assumptions valid?
[284 mA]

12.3 Compare the performance of InP/InGaAsP phase modulators based on (a) the electro-optic effect and (b) minority carrier injection. For the former you may assume that $r_{41} = -1.4 \times 10^{-12}$ m/V is the relevant electro-optic coefficient, that the overlap factor Γ is 0.5, and that the breakdown field is 3×10^7 V/m. For the latter you may assume that $m_e^*/m = 0.077$, $m_h^*/m = 0.64$, and that an upper limit on the injected carrier density is $\Delta N = 10^{24}$ m^{-3}. The refractive index of InP is 3.2, and the operating wavelength is $\lambda_0 = 1.5$ μm.

12.4 A $Ga_{0.47}In_{0.53}As$ PIN photodiode with a cross-sectional area of 0.1 mm^2 produces a photocurrent of 0.996 μA when illuminated by a beam of 1.3 μm wavelength and 1 mW/cm^2 intensity. Calculate (a) the quantum efficiency, and (b) the photocurrent at 1.5 μm and 1.7 μm wavelength.
[0.95; 1.15 μA; 0]

12.5 A surface-emitting GaAs LED emitting at approximately 820 nm launches 50 μW of optical power into a step-index fibre when driven by a current of 50 mA. Assuming that the fibre has an NA of 0.3, estimate the internal efficiency of the LED. If the minority carrier recombination time is 10 ns, what is the radiative recombination time? Estimate the 3 dB modulation bandwidth of the LED.
[54 %; 18.4 ns; 27.6 MHz]

12.6 Show that the effective power gain experienced by a guided mode with normalized transverse electric field E due to the presence of a non-uniform power gain distribution g_p may be written as $g_{peff} = \langle g_p, |E|^2\rangle$. Discuss the way in which the double heterostructure maximises g_{peff}.

12.7 Show that, if a non-unity confinement factor and non-zero losses are included in the calculation, the power obtained from a semiconductor laser in the lasing regime alters to

$$P = \eta hc\{I - I_t\}/e\lambda_g,$$

where

$$\eta = \{1/L \log_e (1/R_1R_2)\}/\{\Gamma\alpha_{pcore} + (1 - \Gamma)\alpha_{pclad} + (1/L) \log_e (1/R_1R_2)\} .$$

The table below shows measurements of the power out of one facet of a GaAs/GaAlAs laser at different currents. The active volume is $250 \times 0.1 \times 3$ μm, and emission is at 820 nm. Estimate the threshold current density and the average power absorption coefficient in the cavity.

I	P
20 mA	15 μW
30 mA	500 μW
35 mA	2 mW
37.5 mA	2.75 mW

[3.8×10^7 A/m^2; 7065 m^{-1}]

12.8 Assuming that the drive current of a semiconductor laser is in the form of a d.c. bias current I' plus a small time-varying modulation $I''(t)$, and that the optical output power is the sum of a constant term P' and a similar time-varying term $P''(t)$, show that the differential equation governing $P''(t)$ is

$$d^2P''/dt^2 + 2\zeta\omega_r \, dP''/dt + \omega_r^2 P'' = \{\mathbf{h}c\omega_r^2/e\lambda_g\}I''$$

where the resonant frequency and damping factor are

$$\omega_r = \sqrt{(G\lambda_g P'/\mathbf{h}cv)} \text{ and } \zeta = \{1/\tau_c + \tau_p\omega_r^2\}/2\omega_r$$

12.9 Solve the differential equation in Question 12.8 for the case of a step change in the modulating current from $I'' = 0$ to $I'' = i''$ at $t = 0$, assuming ζ is small. Sketch the response for $\zeta = 0.05$.

12.10 A Y-junction-coupled GaAs/GaAlAs array laser has nine lasing stripes at one end of the cavity, and ten at the other. Calculate the lasing threshold for all the transverse modes, assuming a cavity length of 500 μm. Sketch the near-field patterns of the lowest- and highest-order transverse modes at either end of the cavity. How would you calculate the far-field patterns?

SUGGESTIONS FOR FURTHER READING

Botez, D., and Ettenberg, M. Comparison of surface- and edge-emitting LEDs for use in fiber-optical communications, *IEEE Trans. Electron Devices,* **ED-26**, 1230–1238, 1979.

Burrus, C. A., and Miller, B. I. Small-area double-heterostructure aluminium-gallium arsenide electroluminescent diode sources for optical fiber transmission lines, *Opt. Comm.*, **4**, 307–369, 1971.

Carenco, A., Menigaux, L., and Linh, N. T. InP electro-optic directional coupler, *Appl. Phys. Lett.*, **40**, 653–655, 1982.

Casey, H. C., and Panish, M. B. "Heterostructure Lasers", Academic Press, New York, 1978.

Casey, H. C., Somekh, S., and Ilegems, M. Room-temperature operation of low-threshold separate-confinement heterostructure injection laser with distributed feedback, *Appl. Phys. Lett.*, **27**, 142–144, 1975.

Chen, K.-L., and Wang, S. Single-lobe symmetric coupled laser arrays, *Elect. Lett.*, **21**, 347–349, 1985.

Gillessen, K., and Schairer, W. "Light Emitting Diodes—An Introduction", Prentice-Hall International, Englewood Cliffs, 1987.

Dutta, N. K., Cella, T., Piccirilli, A. B., and Brown, R. L. Tunable single wavelength lasers, *Proc. SPIE,* **800**, 56–62, 1987.

Evans, G. A., Carlson, N. W., Hammer, J. M., Lurie, M., Butler, J. K., Palfrey, S. L., Amantea, R., Carr, L. A., Hawrylo, F. Z., James, E. A., Kaiser, C. J., Kirk, J. B., Reichert, W. F., Chinn, S. R., Shealy, J. R., Zory, P. S. Coherent, monolithic two-dimensional (10 × 10) laser arrays using grating surface emission, *Appl. Phys. Lett.*, **53**, 2123–2125, 1988.

Holonyak, N., Kolbas, R. M., Dupuis, R. D. and Dapkus, P. D. Quantum well heterostructure lasers, *IEEE J. Quant. Elect.*, **QE-16**, 170–185, 1980.

Inoue, H., Hiruma, K., Ishida, K., Sato, H., and Matsumura, H. Switching characteristics of GaAs directional coupler optical switches, *Appl. Opt.*, **25**, 1484–1489, 1986.

Kapon, E., Katz, J., and Yariv, A. Supermode analysis of phase-locked arrays of semiconductor lasers, *Opt. Lett.*, **10**, 125–127, 1984.

Kinoshita, S., and Iga, K. Circular buried heterostructure (CBH) GaAlAs/GaAs surface emitting lasers, *IEEE J. Quant. Elect.*, **QE-23**, 882–888, 1987.

Mikami, O., and Nakagome, H. InGaAsP/InP optical waveguide switch operated by carrier-induced change in the refractive index, *Opt. Quant. Elect.*, **17**, 449–455, 1985.

Moss, R. H. and Ritchie, S. The development of gallium indium arsenide for photodetectors, *Br. Telecom J.*, **1**, 7–22, 1983.

Nelson, A. W., Spurdens, P. C., Close, S., Walling, R. H., Moss, R. H., Wong, S., Harding, M. J., Cooper, D. M., Devlin, W. J., and Robertson, M. J. The role of MOVPE in the manufacture of high performance InP based optoelectronic devices, *J. Crystal Growth,* **93**, 792–802, 1988.

Okuda, H., Hirayama, Y., Furuyama, H., Kinoshita, J.-I., and Nakamura, M. Five-wavelength integrated DFB laser arrays with quarter-wave shifted structures, *IEEE J. Quant. Elect.*, **QE-23**, 843–847, 1987.

O'Mahony, M. J. Semiconductor laser optical amplifiers for use in future fiber systems, *IEEE J. Lightwave Tech.*, **LT-6**, 531–544, 1988.

Scifres, D. R., Lindstrom, C., Burnham, R. D., Streifer, W., and Paoli, T. L. Phase locked (GaAl)As laser diode emitting 2.6 W CW from a single mirror, *Elect. Lett.*, **19**, 169–171, 1983.

Stillman, G. E., Cook, L. W., Tabatabaie, N., Bulman, G. E., and Robbins, V. M. InGaAsP photodiodes, *IEEE Trans. Electron Devices,* **ED-30**, 364–381, 1983.

Streifer, W., Cross, P. S., Welch, D. F., and Scifres, D. R. Analysis of a Y-junction semiconductor laser array, *Appl. Phys. Lett.*, **49**, 58–60, 1986.

Thompson, G. H. B. "Physics of Semiconductor Laser Devices", John Wiley and Sons, New York, 1980.

Wada, O., Yamakoshi, S., Hamaguchi, H., Sanada, T., Nishitani, Y., and Sakurai, T. Performance and reliability of high-radiance InGaAsP/InP DH LEDs operating in the 1.15–1.5 μm wavelength region, *IEEE J. Quant. Elect.*, **QE-18**, 368–373, 1982.

Welch, D. F., Cross, P., Scifres, D., Streifer, W., and Burnham, R. D. In-phase emission from index-guided laser array up to 400 mW, *Elect. Lett.*, **22**, 293–294, 1986.

Yariv, A. "Optical Electronics", Ch. 11, 15, Holt, Rinehart and Winston, New York, 1985.

CHAPTER

THIRTEEN

OPTICAL DEVICE FABRICATION

13.1 OVERVIEW

We will now consider some of the methods used to make the guided wave devices discussed in the preceding chapters. Because the range of these devices is itself so great, it is hardly surprising that many different techniques are needed. However, these can be broken down reasonably conveniently into two groups, which we will describe separately.

First we note that planar and channel guide integrated optics are both analogous to very large scale integrated (VLSI) microelectronics, in that the aim in each case is to fabricate a number of small, interconnected devices on a single substrate. LEDs, semiconductor lasers and detectors are also mass-produced on common substrates, which are subsequently separated into individual elements by dicing. The planar processing strategy of VLSI can therefore be used for optoelectronics, and many of the individual steps involved can be carried out using similar equipment.

The contrasting aim of optical fibre manufacture is to produce a waveguiding medium in extremely long lengths. Since this lacks any planar characteristics or localized features, the fabrication procedure is entirely different. In fact, the whole operation is much more akin to wire-drawing—optical fibre is also manufactured by a drawing process, but in a unique construction known as a fibre-pulling tower. While it can be pulled directly from melts, a preform is generally used in modern processes. In effect, this is a large size 'copy' of the fibre, which is drawn down to the final cross-section. Such preforms are often built up using deposition methods analogous to those developed for planar substrates.

13.2 PLANAR PROCESSING

The basic steps involved in planar processing are as follows. The starting point is always a substrate, which must be specially prepared so that it is correctly orientated, polished flat, and cleaned. Thin layers of material can then be added to the substrate surface. If the additional material is amorphous, it may be deposited directly onto any type of substrate; crystalline layers, on the other hand, may only be added to crystalline substrates, and then only by strain-free epitaxial growth. However, both of these techniques may be carried out in a number of different ways, each offering a specific advantage. Alternatively, the properties of the substrate can be modified near its surface, for example, by diffusion or implantation of different material species. Finally, material can be removed from the surface by etching. All of these techniques can be applied to the entire area of the substrate, but it is often necessary to delineate small areas for local processing. This is carried out using a pattern transfer process called lithography, which itself involves deposition and removal of material.

Because the dimensions involved in most optoelectronic devices are very small (of the order 1 μm—waveguide tracks are usually several microns wide, while grating structures normally have sub-micron periodicity), nearly all the processing stages are carried out in a special-purpose room known as a *clean room*. This is fed with filtered air to remove any dust particles that might otherwise introduce flaws during pattern transfer, and maintained at constant temperature and humidity to ensure process repeatability. Similarly, the services to the room (water, and process gases) are also highly purified, and clean-room technicians wear special clothing, designed to minimize the contamination they introduce. These precautions greatly increase the yield of working devices, and improve their reliability and lifetime (extremely important in devices destined for undersea use, where replacement of failed components is difficult and costly).

After the majority of the process operations are completed, the device must be packaged and sealed. At this point, any required electrical connections must be made by wire-bonding between electrode contacts on the substrate and terminals on the package. Similarly, any optical connections—e.g. optical fibre pigtails—must be attached by precision mechanical alignment.

13.3 SUBSTRATE GROWTH AND PREPARATION

We begin by considering typical substrate materials. For passive integrated optic devices, the substrate could be a sheet of amorphous material (say, glass), alternatively, a wafer of crystalline material such as Si might be required as a substrate beneath a passive silica guide for its mechanical or electronic properties. For electro-optic devices, insulating crystalline substrates (e.g. $LiNbO_3$) are used, while light-emitting devices require semiconducting materials (InP or GaAs). Substrate shapes and dimensions vary widely, but usually the thickness is of the order 0.5 mm, while the lateral dimensions are typically several centimetres.

Crystalline substrates originate from a *boule*, which is itself a large, perfect crystal of the material concerned. In general the process of crystallization can involve a phase change from any of the following phases—vapour, liquid, or amorphous solid—into a crystalline solid phase. However, the liquid-to-crystalline solid transition is used most often. In this case the crystal is grown by the slow solidification of material at the interface between a liquid (a highly purified melt of the desired material composition) and a solid (which itself originates as a small, perfect seed crystal), so that each successive plane of atoms attaches to the previous one in the correct relative position. For example, silicon is often grown by the *Czochralski process*, using a crystal-pulling rig. Here, a large charge—normally, tens of kg—of previously-prepared electronic-grade Si is placed in a crucible and melted by an RF induction heater or a resistive heating coil. Ideally the crucible should be extremely unreactive with molten Si to avoid contamination by impurities. The melting point of silicon is 1412°C, which eliminates most potential crucible materials except fused silica (SiO_2). Since this is rather fragile, the crucible requires further mechanical support by a graphite susceptor, and is normally destroyed at the end of the run by stress fracture on cooling. To avoid contamination, melting is performed in either an inert gas atmosphere (e.g. argon) or vacuum.

The seed crystal is held by a chuck on the end of a long screw thread, so that the desired growth plane is accurately parallel to the surface of the melt. At the start of the growth the seed is lowered into the melt, and it is then gradually withdrawn by a slow rotation of the screw. The crucible is generally also rotated and slowly raised to keep the surface level at a constant height as the melt is consumed. Careful control of a number of parameters—e.g. temperature, crucible rotation rate and pulling rate—is required for defect-free growth, and to maintain a uniform boule diameter. However, commercial processes are now highly automated, and extremely large boules are grown routinely. These may be several metres in length, and yield wafers of 8″ diameter after grinding and sawing.

Gallium arsenide (melting point 1238°C) may also be prepared by Czochralski growth, using a variant known as the *liquid encapsulated* (LEC) process. The basic difference is that since GaAs decomposes when heated, the As vapour pressure above the melt must be controlled or the stoichiometry of the melt will alter as the As evaporates. This is done by covering the melt and the hottest part of the growing crystal in a thick layer of molten boric oxide (B_2O_3—a glassy viscous liquid insoluble to both Ga and As), while pressurizing the growth chamber to 300–500 psi using an inert gas (e.g. N_2). Other differences are the use of a boron nitride crucible, and the application of a magnetic field to the melt to damp down thermal convection through the 'magnetic viscosity' effect; this greatly reduces the number of dislocations in the crystal. GaAs growth technology is less highly developed than that used for silicon, but wafers are now available up to four inches in diameter. Indium phosphide (melting point 1062°C) also suffers from the problem of decomposition on heating, and so must be grown by the LEC process. There are two additional problems, however: the dissociation pressure of InP is higher than that of GaAs, and the increased viscosity of boric oxide at 1062°C results in an increase in stacking faults and a smaller yield of perfect crystals.

After recovery from the crystal puller, the boule is reshaped. The ends are first removed by sawing, together with any flawed regions, and the remainder is ground into a cylinder on a lathe. The crystal axes are then located by x-ray diffraction, and orientational flats are ground on the cylinder so that the axes may be identified later on. The boule is then sawn into wafers using a diamond-impregnated saw blade. To minimize wastage (or *kerf loss*) this must be as thin as possible. For small boules a disc-shaped blade may be used, with the cutting edge at its periphery. This type of blade cannot be used for large boules, however, because of the likelihood of buckling as the blade diameter increases. Instead, an annular blade is used; this is supported and tensioned at its outer rim, while the inner rim provides the cutting edge. After separation the wafers are lapped flat, and their edges are rounded (to minimize edge-chipping in later mechanical handling). The wafers are then chemically etched to a depth of about 10 μm (more for GaAs), to remove crystallographic defects induced by the cutting and shaping operations. Finally, one or both surfaces are polished to a flatness of better than 2 μm.

13.4 DEPOSITION AND GROWTH OF MATERIAL

Deposition

There are several methods of depositing material in the form of dielectric or metallic films, either for insulating layers or for contacts. All are limited to about 1 μm coating thickness. The simplest technique is *vacuum evaporation*. The principle involved is trivial; the substrate and the coating material are both placed in an evacuated enclosure, some distance apart. The coating material is then heated to its melting point so that it evaporates. Enough thermal energy is supplied to enable individual atoms to escape from the surface of the molten material, and (since there are no intervening gas molecules to collide with) these travel in a straight line towards the substrate, where they adhere. Several methods may be used to melt the evaporant. For example, Fig. 13.1(a) shows *resistive heating*; here, the evaporant is simply loaded into a coiled wire (typically a tungsten filament), through which a heavy d.c. current is passed. Alternatively, the heating element may be boat-shaped and made of sheet metal (e.g. Mo) or graphite.

Many metals (e.g. Ag, Au and Al) may be evaporated very successfully using resistive heating. However, some metal evaporants (e.g. Ti) have high melting points, and are extremely reactive when molten. This may lead to dissolution and failure of the filament, and contamination of the film by impurities. Similarly, many dielectrics have even higher melting points, and also very poor thermal conductivity. Consequently, alternatives to resistive heating are often required.

One suitable alternative is *electron-beam heating*, shown in Fig. 13.1(b). Here a hot-wire filament is again used, but this time simply as a thermionic electron emitter. A charged aperture above the filament then accelerates the electrons in the vertical direction (through, say 5 keV) while simultaneously limiting them to a roughly collimated beam. A magnetic field is arranged to lie perpendicular to the direction of the electron beam, so that it is bent into an arc and falls on a water-cooled crucible containing the evaporant. This is now heated by the kinetic energy of the electrons, rather than by direct contact with a hot body, eliminating the problems described above.

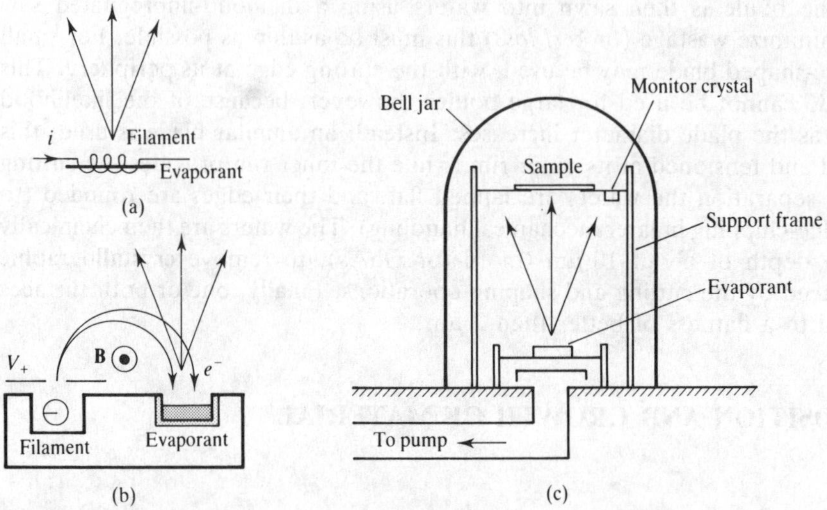

Figure 13.1 (a) Resistive heating, (b) electron-beam heating, and (c) schematic of an evaporator.

Figure 13.1(c) shows a complete evaporator. The evacuated enclosure is a glass bell-jar, held at a pressure of about 10^{-6} Torr; higher pressures result in oxide formation during the evaporation of metal films. The sample is held face-down over the evaporant by a support frame, and evaporation takes place upwards. The thickness of the film is monitored by allowing the simultaneous coating of a small quartz crystal, arranged as a part of an oscillator circuit. Changes in the resonant frequency of the circuit may then be related to the mass-loading of the crystal, and hence to the thickness of the coating. The uniformity of the deposition may be improved by mounting the substrate on a planetary stage, so that it follows a rotating, spinning orbit during coating. It should be noted that, due to the direct line-of-flight of atoms from the evaporant to the substrate, evaporation cannot be used to coat any features of the surface topology that lie 'in shadow'.

An entirely different method of vacuum coating is *RF sputter-deposition*, which can provide conformal coatings of both metals and dielectrics, albeit at the expense of requiring a more complicated apparatus. Here the substrate is placed much closer to the coating material, which is normally arranged in a large sheet known as the *target* (Fig. 13.2). This time the process does not take place under high vacuum; instead, an inert gas (e.g. Ar) is bled into the chamber at a pressure of 10^{-3}–10^{-2} Torr. The gas is then excited into a plasma—a gaseous mixture of ions and free electrons—by an a.c. radio-frequency electric field (usually at 13.56 MHz, a standard scientific frequency) in the space between target and substrate. However, the former is arranged as the cathode, and the latter as the anode, so that positively charged Ar^+ ions strike the target preferentially, and eject atoms from it. This process—known as *sputtering*—is a physical one, which operates by transfer of momentum between the incident, fast-moving ion and the static atom. Although

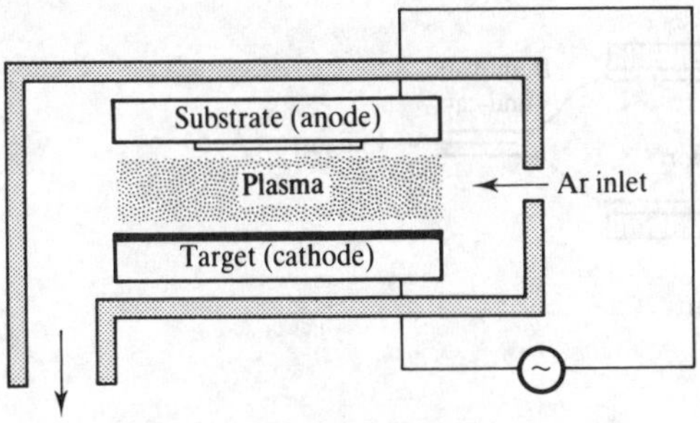

Figure 13.2 Apparatus for planar RF sputtering.

the sputtered atoms are ejected in random directions, many move towards the nearby substrate where they adhere to the surface. Because of further collisions with heavy ions inside the plasma, the sputtered atoms strike the substrate from a wide range of angles and so the technique can be used to coat non-planar substrates. However, since argon atoms are chemically inert, no additional reactions take place. Sputtering may also be induced by a d.c. field, but with a dielectric target the arrival of positively-charged ions leads to a build-up of surface charge, which very quickly halts the process. Consequently, d.c. sputter–deposition is only appropriate for metals. Naturally, just as the target is bombarded by positive ions, the substrate is continually struck by electrons. Since these are much lighter they do not eject material, but they can raise the substrate temperature considerably (by several hundred degrees), unless water-cooling is used. Alternatively, a permanent magnetic field may be used to force the electrons to follow cycloidal tracks, which avoid the substrate for the most part. The sputtering rate is highly dependent on the target material, and is generally low ($\approx 0.01\ \mu m$/min for glassy materials). The time taken to deposit material to the thickness required for a guiding layer (microns) may therefore be large. Despite this, sputtering was one of the first techniques used successfully for the fabrication of step-index planar waveguides, using SiO_2 and Al_2O_3 films.

We now consider a further method of deposition, *chemical vapour deposition* (or CVD). This is not a vacuum process. Instead, a hot gas mixture (often at atmospheric pressure) is passed over the substrate in a furnace, so that a film of the correct composition is gradually deposited following chemical reactions which take place at the substrate surface. The physical layout of the equipment used can take a variety of different forms, depending on the direction of the gas flow (which might be horizontal or vertical) and the form of heating used. For example, Fig. 13.3 shows a horizontal–tube hot-wall reactor, consisting of a quartz tube with an external resistive heater. The substrate is placed on a quartz support and loaded into the tube through a removable end-cap. The process gases are injected from one end and heated prior to arrival at the substrate by contact with the walls. They flow over the substrate as a laminar boundary layer and are adsorbed on the surface where they react to form the desired film. By-products of the reaction and unreacted process gas are removed to exhaust at the downstream end of the tube.

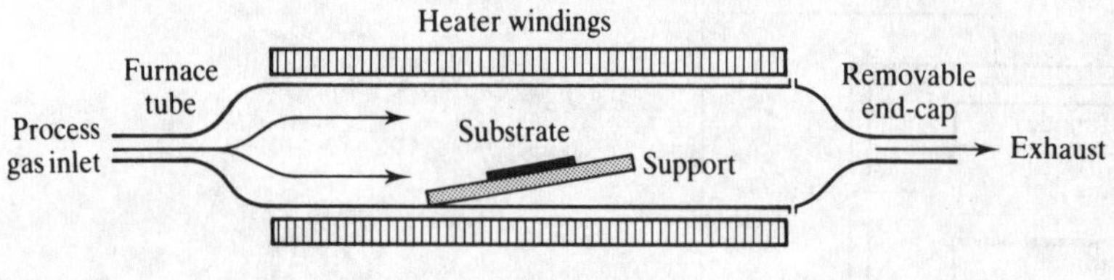

Figure 13.3 A horizontal-tube hot-wall CVD reactor.

The temperature and pressure are both highly dependent on the process used. One major branch of CVD involves the thermally-activated oxidation of hydrogen compounds, such as silane (SiH_4), phosphine (PH_3) and arsine (AsH_3). For example, silica layers may be deposited at atmospheric pressure by a mixture of silane and oxygen, at a temperature of 400–450°C, following the reaction

$$SiH_4 + O_2 \rightarrow SiO_2 + 2H_2 \qquad (13.1)$$

Similarly, silicon nitride might be produced from dichlorosilane and ammonia at about 750°C, or from silane and ammonia at around 900°C, following

$$3SiCl_2H_2 + 4NH_3 \rightarrow Si_3N_4 + 6HCl + 6H_2 \qquad (13.2)$$

or

$$3SiH_4 + 4NH_3 \rightarrow Si_3N_4 + 12H_2$$

As a final example of the versatility of the basic CVD process, we mention that it is possible to fabricate doped glassy layers easily. For example, phosphosilicate glass (SiO_2 doped with P_2O_5) may be produced following the two simultaneous reactions:

$$SiH_4 + 2O_2 \rightarrow SiO_2 + 2H_2O$$
$$2PH_3 + 4O_2 \rightarrow P_2O_5 + 3H_2O \qquad (13.3)$$

The index changes needed in planar silica-based waveguides (for example) may therefore be obtained merely by alterations in the process gas flows.

A number of important variations on the standard CVD process exist. For example, SiO_2 may also be produced by decomposing the metal organic compound TEOS (tetraethylorthosilicate, $Si(OC_2H_5)_4$) at a reduced pressure and a higher temperature (650–750°C). A profusion of acronyms is used to distinguish these process variants; low-pressure processes come under the heading of LPCVD, while the initials MOCVD describe processes based on metal organic compounds. The latter are rapidly increasing in importance for the growth of thin crystalline films, as we will describe shortly.

Though CVD is conceptually a simple process, in reality it is significantly complicated by two aspects. The first is the difficulty of ensuring uniform film deposition—in terms of both thickness and stoichiometry—since the temperature, concentration, chemical composition and velocity of the gas mixture can vary considerably along the length of the tube and across its diameter. The second is the need to handle gases that are often toxic, explosive or corrosive (or a combination of all three). Some gases are highly dangerous; in particular, we mention silane,

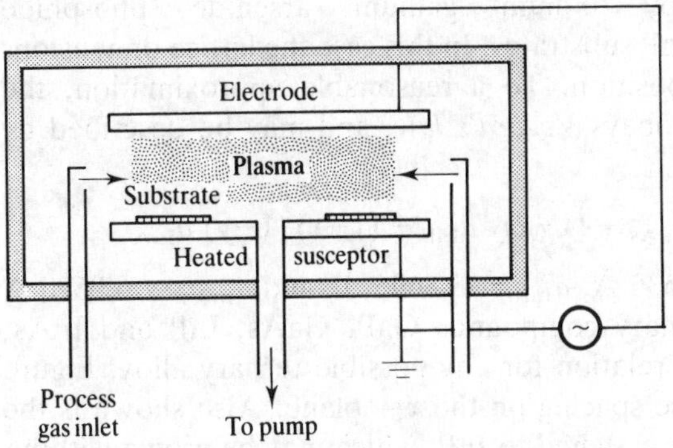

Figure 13.4 Basic PECVD apparatus.

which is both toxic and explodes in contact with air. Double-layer stainless steel pipework is therefore a standard safety precaution.

Using a modification of the CVD process, it is possible to deposit films at lower substrate temperatures (200–350°C) by supplying the necessary energy from an electrically-excited gas plasma rather than as heat. This variant is known as *plasma-enhanced chemical vapour deposition*, or PECVD. Figure 13.4 shows typical apparatus. The substrate is placed on a heated susceptor, which is arranged to act as one of a pair of RF electrodes. A plasma is then established between the electrodes, and the process gases are then bled into this region (which may have a very high electron temperature). The plasma might be derived from an inert gas; for example, silicon dioxide films may be deposited by reacting silane and nitrous oxide in an argon plasma. Alternatively, it might itself form one of the reactants. Considerable interest is currently being shown in PECVD processing for the fabrication of waveguides on silicon substrates. These require a thick dielectric spacer (at least 12 μm of SiO_2) to separate the waveguide (doped silica) from the substrate.

Epitaxial growth

All of the above processes deposit amorphous films with a random rather than a regular crystalline structure. To add material while maintaining a regular arrangement of atoms (as required in III–V optoelectronic devices), the layer must be grown epitaxially. This can be done by matching the lattice parameters of the film to those of the substrate so that the latter acts as a template for ordered growth. If the grown material has the same chemical composition as the substrate (e.g. as when a GaAs layer is grown on a GaAs substrate), the process is called *homoepitaxy*. Often, however, the two materials have different compositions. In this case the process is known as *heteroepitaxy*. For guided wave optical devices there are currently two important materials systems. First, it is possible to grow the ternary alloy gallium aluminium arsenide ($Ga_{1-x}Al_xAs$) on GaAs substrates over the complete range of alloy composition, since the lattice dimensions of the former are almost entirely independent of the mole fraction x.

Second, the quaternary alloy indium gallium arsenide phosphide ($In_{1-x}Ga_xAs_{1-y}P_y$) can be grown on InP substrates. In this case the lattice dimensions are a strong function of alloy composition. To a reasonable approximation, the lattice parameter $a(x,y)$ of the alloy obeys *Vegard's law*, and may be described by the polynomial function

$$a(x,y) = xy\ a_{GaP} + x(1-y)\ a_{GaAs} + (1-x)y\ a_{InP} + (1-x)(1-y)\ a_{InAs} \quad (13.4)$$

where a_{GaP} = 5.4512 Å, a_{GaAs} = 5.6536 Å, a_{InP} = 5.8696 Å, and a_{InAs} = 6.0590 Å are the lattice parameters of the binary compounds GaP, GaAs, InP and InAs, respectively. This reduces to a linear relation for any possible ternary alloy. Figure 13.5 shows contours of constant lattice spacing on the x–y plane. Also shown is the locus of all compounds that are lattice matched to InP, which may be grown without strain on InP substrates. This is almost exactly a straight line, obeying the relation $y \approx 1$–$2.14x$. When $y = 0$ the locus intercepts the x-axis at $x \approx 1/2.14$, i.e. at the alloy $In_{0.53}Ga_{0.47}As$.

Just as in crystal growth, epitaxial processes may be based on vapours or liquids, but epitaxial growth may be performed at temperatures well below the melting point of the substrate. Processes based on liquids are known as *liquid-phase epitaxy* (LPE); these exploit the principle that mixtures may have lower melting points than those of their stoichiometric counterparts. For example, a mixture of 5 per cent of arsenic with 95 per cent gallium melts at 880°C (considerably lower than the melting point of GaAs, 1238°C). However, when the mix is slowly cooled, some of the arsenic must precipitate out of solution. Most importantly, it does to stoichiometrically, with each atom of As combining with one of Ga. The result is GaAs, which grows

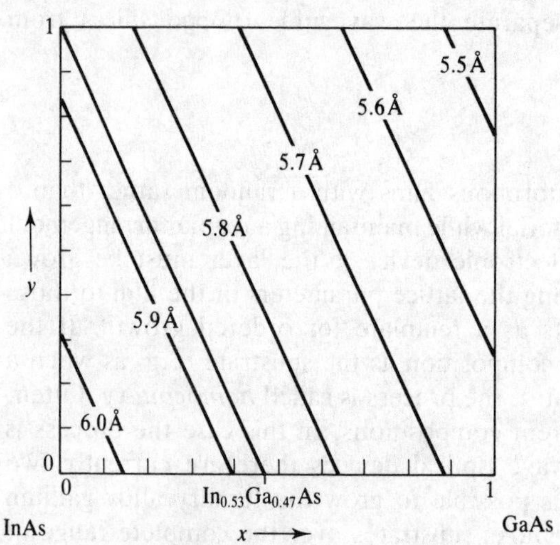

Figure 13.5 Variation of the lattice parameter with composition, for InGaAsP.

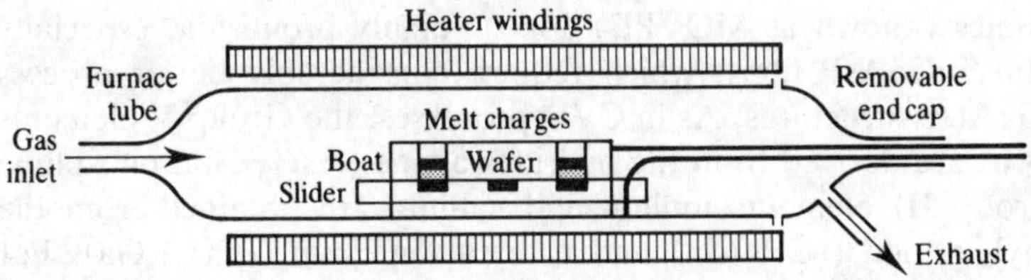

Figure 13.6 The horizontal slider method for growth by liquid phase epitaxy.

epitaxially on the substrate surface. Clearly the ratio of gallium to arsenic in the melt will change in the process, but if the ratio is high to begin with, the effect will be negligible. As a result, continuous growth of GaAs films on GaAs substrates can be carried out, without melting the latter. $Ga_{1-x}Al_xAs$ films may be grown in a similar way from melts containing Ga, Al and As, and $In_{1-x}Ga_xAs_{1-y}P_y$ can be grown from four-component melts. However, it is difficult to maintain a specified composition throughout the thickness of a layer.

Figure 13.6 shows one LPE technique, the *horizontal slider method*, for the fabrication of GaAlAs/GaAs epitaxial films. This has been extremely successful in its time. The key component is a carefully-machined graphite boat, fitted with a horizontal graphite slider. Normally, several substrates are placed in shallow wells machined into the top of the slider; one of these is the wafer destined for overgrowth, while the others are sacrificial dummies. Similarly, charges of the growth material are placed in deeper wells in the boat, which are sealed with further wafers capped with graphite plugs. The whole assembly is placed inside a resistively-heated furnace tube in an inert gas atmosphere, so that the growth material is melted. Using push rods, the relative positions of the boat and slider may be manipulated so that the wafer is covered in the required melt. This will grow an epitaxial layer on the wafer surface when the temperature is suitably lowered. The dummy wafers are used to seal all unused melts. When it is time to change the composition of the epitaxial film, the slider is simply moved under a new melt. Due to the close fit of slider and boat, the old melt is wiped cleanly off the wafer surface, while the new melt is wiped on. Further growth then takes place over a prescribed program.

Although a simple process, LPE suffers some disadvantages. First, some crystal faults are inevitably introduced between the layers as melts are exchanged, reducing the perfection of interfaces. Second, because of the high temperature, some interlayer diffusion also occurs; this renders the junctions between layers of different composition less abrupt, which can be undesirable for many device applications. Third, since the process can operate relatively near to equilibrium, slight changes in temperature can cause dissolution of the substrate instead of growth. Finally, it is hard to maintain the composition of quaternary layers with the precision required for accurate lattice matching. As a result, LPE is rapidly being overtaken by methods based on *vapour-phase epitaxy* (VPE).

Vapour-phase epitaxy simply involves the use of chemical vapour deposition for epitaxial rather than disordered growth. In particular, non-equilibrium techniques based on metal organics (known as MOVPE) appear highly promising, especially for the growth of InP/InGaAsP films, which require considerably better process control than GaAs/GaAlAs structures. As in CVD processes, the Group V elements arsenic and phosphorus are derived from the hydrides arsine (AsH_3) and phosphine (PH_3), while the Group III elements indium and gallium are obtained from the metal alkyls trimethyl indium ($((CH_3)_3In)$) and trimethyl gallium ($((CH_3)_3Ga)$). For example, GaAs may be deposited at 600–800°C, following the reaction:

$$(CH_3)_3Ga + AsH_3 \rightarrow GaAs + 3CH_4 \tag{13.5}$$

Deposition of InP follows a similar reaction, using trimethyl indium and phosphine, while InGaAsP requires all four constituents. Hydrides of sulphur or selenium (H_2S and H_2Se) are used as sources of n-type dopants for both InP and InGaAsP; p-type dopant sources include dimethyl zinc and dimethyl cadmium. Figure 13.7 shows a schematic of an MOVPE process for the growth of InGaAsP on InP. Here the hydrides and dopants are added directly to the vapour stream, while the metal organics (contained in temperature controlled baths) are transported by a carrier gas such as H_2 or He.

The main defect of earlier CVD processes—lack of uniformity—appears to be well under control in modern MOVPE rigs, which offer a number of advantages. First, the stochiometry of the gas mixture may be set very accurately using mass-flow controllers. Second, any transitions between layers of different composition may be made merely by switching from one gas mixture to another, and gas switching may be extremely fast with electrically-controlled solenoid valves. Third, the process temperature is lowered to around 650°C, which reduces interlayer diffusion. As a result, the extent of any interlayer lattice imperfections is greatly reduced, and heterostructure interfaces may be extremely abrupt (a few atomic layers in thickness). This allows the fabrication of very thin, pure films. More importantly, multilayer

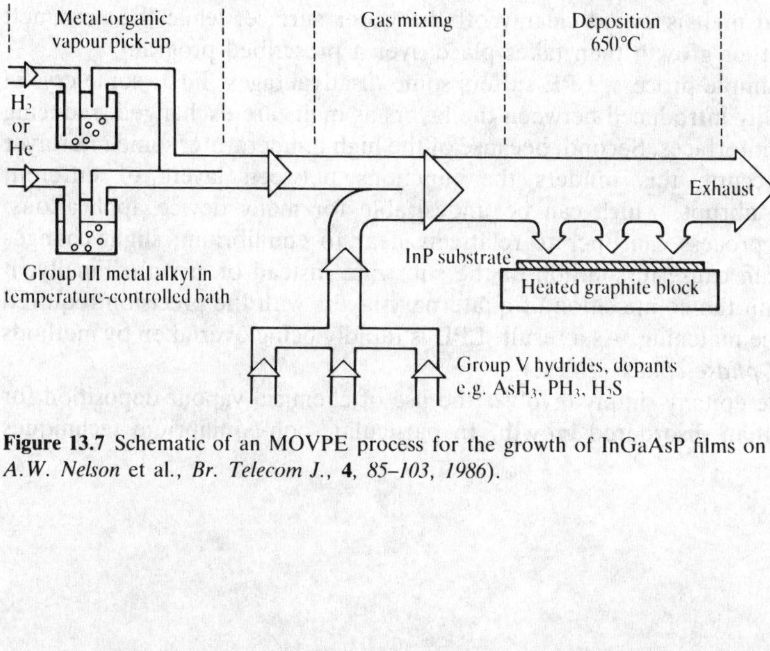

Figure 13.7 Schematic of an MOVPE process for the growth of InGaAsP films on InP substrates (*after A.W. Nelson* et al., *Br. Telecom J.*, **4**, *85–103, 1986*).

films may be built up with very sharp boundaries between the layers. These may be used as *multiple quantum well* (MQW) structures.

A final growth method is *molecular beam epitaxy*, or MBE. This is a low-temperature process, based on evaporation under ultrahigh vacuum (10^{-8}–10^{-10} Torr). Essentially, several crucibles containing the constituents of the growth material are heated (normally, by electron-beam heating) so that their contents evaporate simultaneously. The departing atoms combine in flight to form a molecular beam, which is then adsorbed on the substrate. Precise control of the evaporation rates of the separate materials is required to establish the correct stoichiometry of the growing layer. However, the advantage is that growth occurs at such a low temperature that hardly any diffusion takes place between the layer and the substrate, so that MBE may also be used to grow MQW structures—in fact, it was the original MQW process. Its main disadvantages as a production tool is that the growth rate is extremely small compared with other epitaxial methods, typically 0.01–0.03 μm/hr.

13.5 MATERIAL MODIFICATION

There are several methods of modifying the properties of a substrate near its surface. The simplest is *diffusion*, where the substrate is placed in intimate contact with the dopant material—which may be a solid, a liquid or a gas—and both are heated to 800–1000°C. Due to their greatly increased thermal energy, the dopant and substrate atoms are all considerably more mobile than at room temperature. However, the latter are often anchored reasonably well to fixed lattice sites, about which they vibrate. As they do so, some sites may be vacated, into which migrant dopant atoms may then move (as shown in Fig. 13.8(a)). Alternatively, dopant atoms may diffuse into the substrate without the necessity for such vacancies, by moving through the spaces between the lattice sites, known as *interstitials* (Fig. 13.8(b)). The overall process is known as *in-diffusion*. It is often possible to avoid any large-scale transport of material in the opposite direction; however, this can and does occur, and is known as *out-diffusion*; it may involve either host atoms or impurities.

Mathematically, diffusion is governed by two rate equations whose derivation is based on the assumption that the behaviour of any system of mobile particles—for example, a gas—is a random process driven by concentration gradients. Since we have already derived the necessary mathematics in Chapter 11 for the analogous

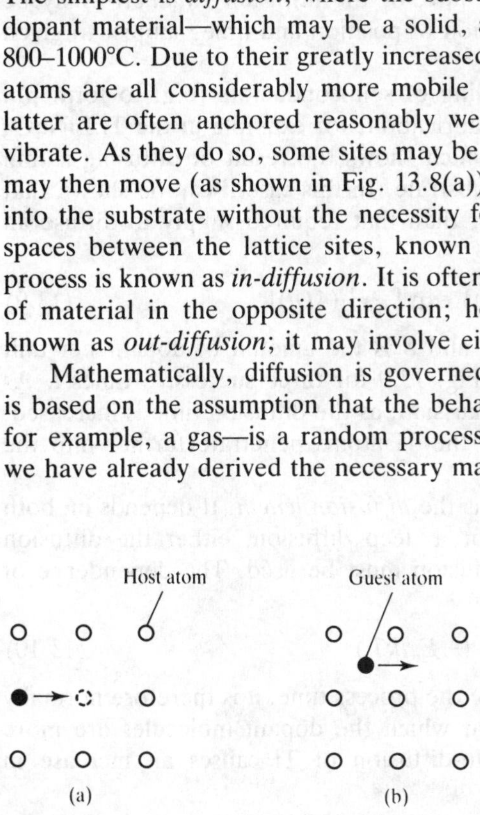

Figure 13.8 Diffusion by (a) occupation of vacancies, (b) occupation of interstitials.

phenomenon of carrier diffusion, we will not repeat it here. Suffice it to say that a one-dimensional gradient in dopant concentration C leads to a flux of material of

$$J = -D\partial C/\partial x \tag{13.6}$$

where D is the *diffusion coefficient*. Equation (13.6) is known as *Fick's first law*; it shows that a non-uniform concentration will result in a flux of material in the opposite direction to any concentration gradient. At the start of an in-diffusion process there is a large difference in the value of C inside and outside of the substrate. Consequently, dopant atoms will tend to diffuse inwards, altering the local concentration in the process. To analyse this effect we require a continuity equation. Once again a suitable equation has already been derived for carrier diffusion in Chapter 11. Restated in terms of the dopant concentration, this becomes

$$\partial C/\partial t = -\partial J/\partial x \tag{13.7}$$

Equation (13.7) shows that a flux gradient will result in a decrease in the local concentration, which tends to equalize the concentration everywhere. Assuming the diffusion coefficient is independent of concentration, we may combine Eqs (13.6) and (13.7) to obtain a single equation in C alone:

$$\partial C/\partial t = D\partial^2 C/\partial x^2 \tag{13.8}$$

Equation (13.8)—known as *Fick's second law*, or the *diffusion equation*—may be solved to find the concentration as a function of position and time, subject to given boundary conditions.

In-diffusion is often used for the doping of semiconductors (e.g. to form p-n junctions). It is also used for waveguide fabrication, for example in the Ti:$LiNbO_3$ process. This is illustrated in Fig. 13.9, which shows the result of diffusing a thin surface layer of Ti metal into a $LiNbO_3$ substrate. In this case it can be shown that the solution to Eq. (13.8) for times larger than that required simply to diffuse all the metal into the crystal is

$$C(x,t) = \{S/\surd(\pi Dt)\} \exp[-x^2/(4Dt)] \tag{13.9}$$

where x is the distance from the surface, and S is the amount of dopant per unit area at $t = 0$. This solution is plotted in Fig. 13.9 for three successive times t. As can be seen, the concentration profile is Gaussian; as the diffusion time is increased, the surface concentration is lowered, but the Ti atoms penetrate further into the crystal.

The quantity $L_D = 2\surd(Dt)$ is known as the *diffusion length*. It depends on both the diffusion coefficient and the time; for a deep diffusion, either the diffusion coefficient must be large or a lengthy diffusion must be used. The dependence of D on temperature is normally of the form

$$D = D_0 \exp(-E_a/kT) \tag{13.10}$$

where E_a is an activation energy. To reduce the process time, it is therefore normally carried out at an elevated temperature, at which the dopant molecules are more energetic (giving a large value of D). In-diffusion of Ti causes an increase in

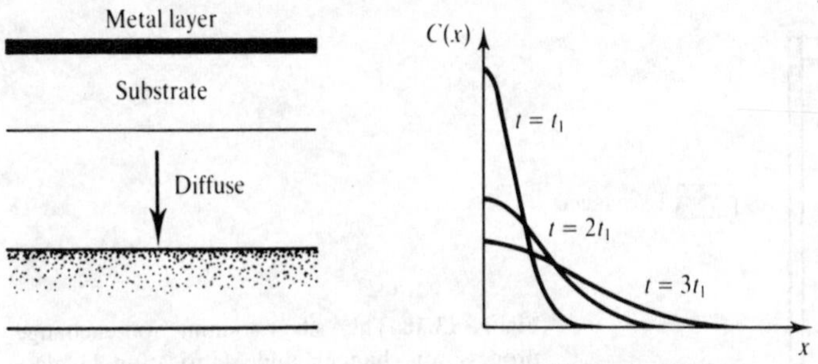

Figure 13.9 Concentration profiles resulting from the diffusion of Ti metal into $LiNbO_3$.

refractive index that is proportional to the local concentration of Ti metal (at least for one polarization). Consequently, a planar guide with the index variation

$$n(x) = n_0 + \Delta n \exp\{-(x/L_D)^2\} \tag{13.11}$$

may be obtained very simply by coating a substrate with a layer of metal and placing it in a furnace for a number of hours (e.g. for 8 hours, at 1050°C). More generally, diffusion is a three-dimensional process. In this case we may replace the one-dimensional flux J by a vector $\mathbf{J}$ indicating the flow direction as well as its strength, and Eqs (13.6) and (13.7) modify to

$$\mathbf{J} = -D\,\nabla C \quad \text{and} \quad \partial C/\partial t = -\nabla \cdot \mathbf{J} \tag{13.12}$$

Similarly, the diffusion equation becomes

$$\partial C/\partial t = D\,\nabla^2 C \tag{13.13}$$

Equation (13.13) may be used to model the effect of, say, the diffusion of a narrow strip of Ti metal into a $LiNbO_3$ crystal. In this case the concentration profile varies in two dimensions, so it is not hard to see that a two-dimensional index change (and hence a buried channel guide) will be formed.

An alternative to diffusion is *ion exchange*. This time the substrate is placed in a liquid melt, usually at a much lower temperature (200–400°C). Normally there is a higher concentration of one type of mobile ion inside the substrate than in the melt, and a higher concentration of another inside the melt than in the substrate. As a result, these two species counter-diffuse, so that material is exchanged between substrate and melt. Ion exchange may also be used for waveguide fabrication, since differences in the polarizability of the two species cause a change in refractive index. For example, Fig. 13.10 shows a process for the fabrication of channel guides in soda-lime glass (a mixture of SiO_2, Na_2O and other metal oxides). Here the glass substrate has been coated in a layer of mask material in which narrow stripe openings have been made, and placed in a bath of molten $AgNO_3$. The mask acts as an effective barrier to the ions, so that ionic motion only occurs through the openings. In these regions Na^+ ions diffuse out of the substrate and into the melt, to be replaced by Ag^+ ions. The exchange of relatively light Na^+ ions with heavier Ag^+

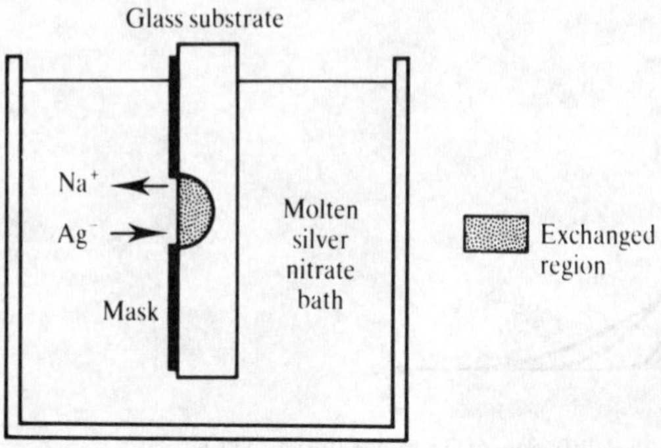

Figure 13.10 The silver-sodium ion-exchange process for channel guide fabrication in glass (*after C.D.W. Wilkinson* et al., *1978*).

ions causes an increase in refractive index in the region of the stripe opening, and hence a channel guide. The motion of the Ag^+ ions into the substrate may be enhanced by the application of an electric field, which drastically reduces the process time and allows deep, uniform diffusion profiles.

Waveguides may be fabricated in $LiNbO_3$ using a similar process known as *proton exchange*, which involves the replacement of Li^+ ions by H^+ ions from a melt of benzoic acid. However, owing to the crystallinity of the substrate, the refractive index change that results is strongly dependent on the crystal orientation and the optical polarization. For example, Fig. 13.11 shows typical refractive index profiles for X-cut, MgO-doped $LiNbO_3$, exchanged for three hours in pure benzoic acid. The change in n_e near the surface is large, positive, and almost constant, so a near-step-index guide will be formed. However, the change in n_o is actually negative, the opposite sign to that required. A waveguide is therefore formed for one polarization only.

As a final example of a technique for material modification, we mention *ion implantation*, which is mainly used for doping semiconductors. Here the change also follows from the insertion of ions into the substrate, but in an entirely different manner. The process is carried out under high vacuum, and the equipment required is both sophisticated and expensive. The source of ions is a crucible of molten

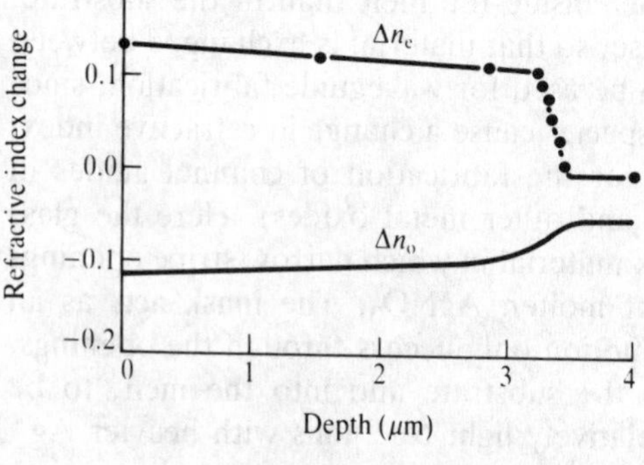

Figure 13.11 Change in refractive index with depth, for a typical proton-exchange process (*data taken from M. Digonnet* et al., *1985*).

material, from which a beam of ions is extracted electrostatically. This may contain several different types of ion—singly- and doubly-charged ions of the desired element, impurity ions, and so on—so it is passed through a filter sensitive to charge and atomic weight (known as a *mass filter*, or a *Wien filter*) to select a single ion species. The emerging beam is then accelerated by a high voltage (typically 100–1000 kV) so that ions strike the substrate at speed. Instead of sputtering atoms from the surface, they now have enough energy to penetrate some distance before stopping. As they slow, the ions lose energy by undergoing several successive collisions with the atoms in the substrate, as in Fig. 13.12(a). Since the substrate atoms may be displaced considerably from their normal positions as a result, a thermal annealing step is almost always required after implantation.

One attractive feature is that the distribution of implanted ions is very well defined, and controllable by varying the acceleration voltage. To a reasonable approximation, the variation of the concentration of implanted ions with depth follows the Gaussian distribution

$$N(x) = \{\phi/\sqrt{(2\pi)}\sigma\} \exp\{-(x-r_p)^2/2\sigma^2\} \qquad (13.14)$$

where ϕ is the density of implanted ions per unit surface area, defined by $\phi = \int_0^\infty N(x)\,dx$, r_p is the projected range of the ions, and σ is the standard deviation of the projected range. The peak concentration occurs at $x = r_p$, and is given by $N_p = N(r_p) = \phi/\sqrt{(2\pi)}\sigma \approx 0.4\phi/\sigma$. Projected range r_p depends on several parameters: the mass and charge of the ionic species, the substrate material, and the acceleration voltage. However, r_p is generally limited by the maximum practical voltage ($\approx$1 MV) to about 1 μm. When plotted on a logarithmic scale, the concentration distribution becomes a parabola; Fig. 13.12(b) shows typical profiles obtained for three different acceleration voltages, such that $V_1 < V_2 < V_3$. The larger the voltage, the deeper the implantation, but the broader and weaker the distribution. Other, more complicated

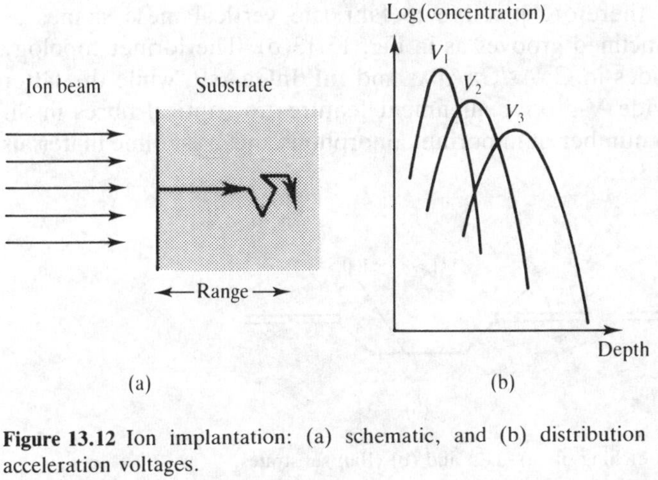

Figure 13.12 Ion implantation: (a) schematic, and (b) distribution of implanted ions, for different acceleration voltages.

profiles may be synthesized from the cumulative effect of a succession of implants, each with a different acceleration voltage and implantation dose, or from a combination of an initial implantation with a drive-in diffusion step.

Ion implantation is potentially attractive for the direct fabrication of waveguides, since it suffers less from the lateral spreading inherent in comparable processes like diffusion. Channel waveguide formation has been investigated in a number of crystalline materials (e.g. $LiNbO_3$, by implantation of He^+ ions). However, the lattice damage causes high optical propagation loss and a significant reduction in the electro-optic coefficient, which is only partially restored by annealing.

13.6 ETCHING

Material removal or etching methods can generally be subdivided into classes which are based either entirely on physical or chemical principles, or on some combination of the two. Additional divisions may also be made between those techniques which do not require the use of vacuum and those which do, and between masked and maskless processes.

Wet chemical etching is commonly required to pattern thin metal and dielectric films on the substrate, either to fabricate specific features (e.g. electrodes), or to open windows in surface mask layers for subsequent local processing. These materials normally etch isotropically, so that the fidelity of the pattern transfer is limited by undercut of the etch mask. However, etches are also often used to alter the surface topology of crystalline substrates. In this case the etching may follow a particular direction. *Anisotropic etching* can be performed in crystals with diamond and zincblende lattices (e.g. Si, GaAs and InP), because the ⟨111⟩ crystal plane is more closely packed than the ⟨100⟩ plane, and so is normally etched more slowly. For example, using the water: ethylene diamine pyrocatechol (EDP) etch system, the etch rates for ⟨100⟩, ⟨110⟩ and ⟨111⟩ planes in Si are 50, 30 and 3 μm/hr at 100°C. The ⟨111⟩ etch rate is clearly considerably slower. By choosing the substrate orientation correctly, it is therefore possible to fabricate vertical mesa shapes as shown in Fig. 13.13(a) or inclined grooves as in Fig. 13.13(b). The former topology is appropriate for ridge guides in GaAs/GaAlAs and InP/InGaAsP, while the latter is most often used to provide V-groove alignment features for optical fibres in Si. Suitable etch mixtures for a number of important amorphous and crystalline materials are summarized in Table 13.1.

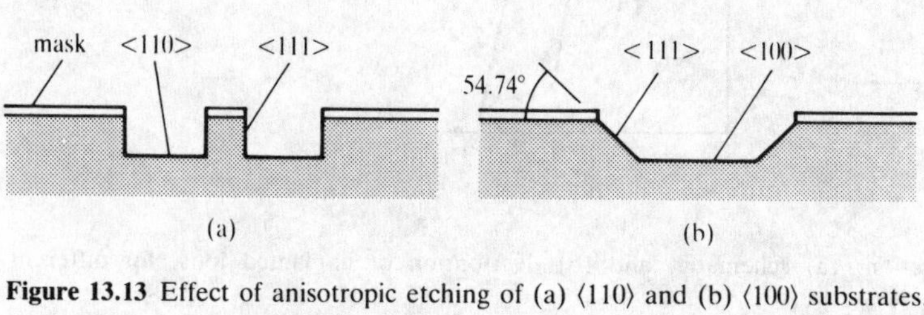

Figure 13.13 Effect of anisotropic etching of (a) ⟨110⟩ and (b) ⟨100⟩ substrates.

Table 13.1 Wet etch mixtures for a variety of materials used in optoelectronic device fabrication

Material	Etch system	Comments
Al	$HNO_3/CH_3COOH/H_3PO_4/H_2O$	Used for electrode patterning
Au	$KI/I_2/H_2O$	Used for electrode patterning
Ti	$HF/HNO_3/H_2O$	Used to pattern Ti prior to diffusion into $LiNbO_3$
SiO_2	$HF/H_2O/NH_4F$	Used to open windows in SiO_2 surface mask
GaAs	$H_2SO_4/H_2O_2/H_2O$	Used to etch rib guides on ⟨100⟩ substrates
InP	HCl/H_3PO_4	Has negligible effect on InGaAsP
InGaAsP	$H_2SO_4/H_2O_2/H_2O$	Has negligible effect on InP
Si	$KOH/H_2O/C_3H_7OH$	Used to etch V-grooves in ⟨100⟩ substrates

Dry Etching

Although wet etching is still used to a considerable extent, many modern microfabrication procedures involve one of a battery of rather different vacuum etching techniques, all loosely known as *dry etches*. Although these require complicated equipment, they have the advantage of greatly improved process control and a high degree of selectivity and anisotropy in the etching.

In dry etching, either purely physical or combined physical/chemical methods are used. The simplest physical method is sputtering, in which atoms are ejected from a surface by ion bombardment. The most obvious implementation is *RF sputter etching*, a variant of RF sputter deposition (previously shown in Fig. 13.2). The electrode connections are simply reversed, so the substrate is bombarded by positive ions, rather than the target (which may be a simple electrode). This technique is often used to clean a substrate prior to deposition, but it may also be used as a method of etching. However, since the ions strike the substrate from a range of angles, etching is roughly isotropic.

Another physical method is *ion beam milling*, shown in Fig. 13.14. This is similar to d.c. sputter etching (performed by reversing the electrode connections in a d.c. sputter coater); however, the plasma is generated rather differently. Here an inert gas (e.g. argon) is bled into the far end of an evacuated chamber. A thermionic emitter is then used to generate electrons, which ionize the argon. Generally the interaction between the electrons and the plasma is increased by using a magnetic field, which makes the electrons follow a long spiral trajectory. The Ar^+ ions are extracted from the plasma and accelerated towards the target by a biased grid, and atoms are then sputtered off the substrate by this beam in the normal way. To prevent the accumulation of positive charge at the substrate, a *neutralizer filament* is used. This is a further thermionic emitter, which injects electrons into the beam. In contrast to RF sputter etching, the ions are incident from a defined direction, so that the etching is anisotropic. However, this advantage is limited by ion 'bounce', which causes ions to recoil in random directions, eroding the walls of any deep

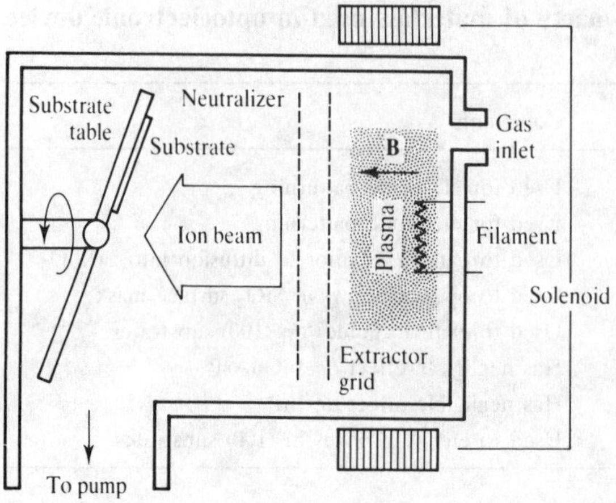

Figure 13.14 An argon ion beam mill.

features. The substrate can be rotated to ensure even etching, and enhanced etch rates in most materials result from machining at a tilt.

Most often it is necessary to apply particular processes locally; for example, we might wish to define a small area of a substrate for sputter etching or ion milling. This can be done by depositing a layer of a mask material over the substrate with openings where the etching is to be performed. Clearly the property required of the mask material is that it should have a much lower sputter yield than the substrate, so that it erodes more slowly. Unfortunately, most purely physical etching techniques offer little differentiation between materials. As a result, the majority of the mask is etched at roughly the same rate as the substrate, and its edges (which define the periphery of the etched area) are often attacked even faster, since they present tilted faces to the beam. This limits the depth of etched features, and prevents the fabrication of anything other than sloping sidewalls. The fabrication of deep etched features, with walls that are accurately orthogonal to the substrate (as might be required in a waveguide mirror), is therefore extremely difficult.

Design Example

We may estimate the maximum groove depth achievable in GaAs by argon ion milling through a mask layer, as follows. Etch rates for a variety of materials milled with Ar^+ ions at normal incidence, and 1 keV beam energy, 1 mA/cm^2 beam current are given in Table 13.2. We might select photoresist as the masking material, spin-coated to a thickness of 0.3 μm. With the etch rate given, the mask layer protecting areas of GaAs that are not to be milled will erode completely in $0.3/600 \times 10^{-4} = 5$ minutes. In this time, exposed GaAs areas will be etched to a depth of $5 \times 2{,}600 \times 10^{-4} = 1.3$ μm.

Maskless Etching

One possible solution to the problem of etch selectivity is to avoid the use of a mask altogether. This is the rationale behind *focused ion beam* (FIB) *micromachining,*

Table 13.2 Etch rates for different materials by Ar^+ ions

Material	Etch rate (Å/min)
Si	360
SiO_2	420
GaAs	2,600
Cr	200
Ti	100
Al	440
Shipley AZ 1350 photoresist	600

which achieves local sputtering simply by confining the ion beam to a small area. Figure 13.15 shows the required apparatus, which is effectively a modified scanning ion microscope. Here the source of the ions is not a plasma, but a high-brightness, near-point emitter known as a *liquid metal ion source* (LMIS). The ion source is a capillary tube filled with a metal of low melting point (e.g. In or Ga), surrounded by a resistive heater. A small aperture, raised to a negative potential, is placed just beyond the tip of the capillary. On heating, the metal is liquified and the molten surface at the very end of the capillary is then able to respond to external forces. Due to the large electric field in the neighbourhood of the aperture, this surface is distorted electrostatically into a shape known as a *Taylor cone*. At the apex of the cone the field is strong enough to overcome surface forces and extract large numbers of ions, so that the tip of the cone acts as a near-point ion source. After emission,

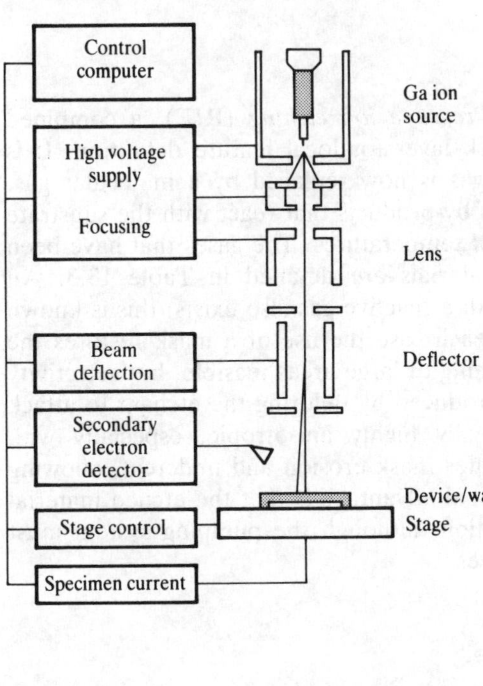

Figure 13.15 A focused ion beam micromachining system, based on a gallium liquid metal ion source (*after Harriot L.R.* et al., *1986*).

the ions pass through the aperture as an expanding beam, which may be refocused by electrostatic lenses to a point further down the column. Any portion of the substrate placed here will be subjected to an intense, highly directed flux of ions, which will eject atoms from the surface by sputtering. The charging of insulating substrates may be prevented by machining through a thin metal layer, grounded to leak any charge away.

By confining the area of the substrate that is exposed to the ions to a small spot, the sputtering is localized. The beam may also be deflected and blanked under computer control. It is thus possible to program the area that is machined so that a pattern is etched into the substrate. This may be aligned to other features on the substrate prior to etching by using the equipment in an alternative *ion microscope* mode. In this case a smaller ion flux is used, and the beam is raster scanned across the entire substrate. Variations in the numbers of secondary electrons (also generated by the sputtering process) emitted during the scan are then used to build up a picture of the surface, so that the substrate can be moved to a desired coordinate origin by an x–y stage. The substrate is also eroded in this alignment stage, albeit to a very small extent. Normally the machining is performed in a number of passes, using a raster scan of the addressed area. Due to the absence of an erodable mask, the etched features may be extremely deep (tens of microns), with near-vertical walls.

Focused ion beam micromachining may be used to cut very deep features, and has been used in a number of niche applications in optoelectronics, such as mirror fabrication. However, it suffers from two major defects. First, it is a serial process, which limits throughput. It is therefore inappropriate for the definition of, say, ridge guides, where most of the substrate area must be etched. Second, it shares the disadvantage of other physical processes; there is no distinction between the substrate and the sputtered material (which may be redeposited elsewhere on the surface).

Reactive Etching

Both of these problems are avoided by *reactive ion etching* (RIE), a combined physical/chemical process requiring a mask layer for local feature definition. It is similar to sputter etching, but the inert gas is now replaced by a molecular gas, whose decomposition by the plasma yields by-products that react with the substrate material to form volatile compounds at low temperatures. The gases that have been used to etch a number of important materials are detailed in Table 13.3. An analogous process based on ion milling with a reactive gas also exists; this is known as *reactive ion beam etching* (RIBE). In each case the use of a mask restores the parallelism of the process, making the etching of large areas feasible, but selectivity between the mask and the substrate is introduced by tailoring the etchant to attack mainly the latter. Etching is also normally highly anisotropic, especially with crystalline substrates; this virtually eliminates mask erosion and undercut, allowing deep features to be etched with ease. A final advantage is that the etched material may be pumped away, avoiding redeposition, although the pumping system must be specially protected against the etch gases.

Table 13.3 Dry etch mixtures for some crystalline materials used in optoelectronics

Material	Etch system
GaAs	CCl_2F_2
InP	$CCl_2F_2/Ar/O_2$
$LiNbO_3$	CCl_2F_2 or CF_4

13.7 LITHOGRAPHY

The process of pattern transfer is known as *lithography*. This is normally a two-step process; in the first, a *mask*—normally a glass plate, containing an opaque pattern in chromium metal—is made. In the second, the features on this master are copied onto the substrate by a form of printing.

Mask Fabrication

We begin by considering mask fabrication using *electron beam lithography*. Initially, the design dimensions of the device are specified, and its salient features are assigned to one of a number of mask levels. For example, Fig. 13.16 shows the plan view of a directional coupler. To fabricate the device using the Ti:$LiNbO_3$ process, two lithographic steps will be needed: patterning of a layer of Ti metal to define the guides prior to diffusion, and patterning of a layer of Al metal to outline the electrodes. All waveguide features—generally in the form of thin, curving tracks—are therefore assigned to level 1 (unshaded), while the rectangular electrode blocks form level 2 (shaded). A separate mask will be made for each level, and often each will contain additional alignment features so that the two patterns may be registered together with high accuracy. *Computer-aided design* (CAD) tools are then used to specify the pattern still further. As a result of direct interaction with the designer, a set of special purpose code is generated. This is written in a high-level computer language, and each line typically contains a description of the vertices of an elementary shape, corresponding to the whole or part of one feature in the design.

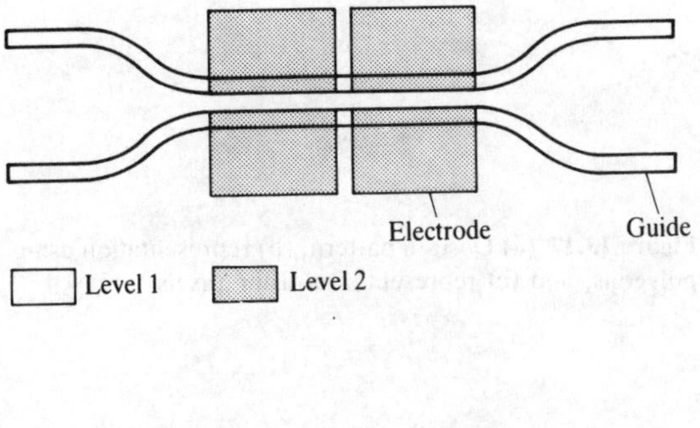

Figure 13.16 Mask levels required for a directional coupler switch in the Ti:$LiNbO_3$ process.

The code can be interpreted by plotting packages, allowing the pattern to be verified on a computer screen. When the design is finalized, a compiler is used to generate a new set of low-level code capable of driving the mask-making machine directly. One problem for integrated optics is that while the features contained in waveguide devices are normally simple, some of these (curved guide sections) are inappropriate for conventional CAD systems, which have historically been developed to support integrated electronics, and are therefore based around rectangular shapes, tracks or polygons. Consequently, using high-level code, a typical curved guide is normally approximated by subdivision into a set of polygonal shapes. It is then further approximated by the low-level code as a pixellated pattern, each element of which is an elementary square corresponding to the minimum feature size of the mask-making system.

Figure 13.17 illustrates this gradual degradation. Figure 13.17(a) shows the desired pattern—a section of curved guide. In Fig. 13.17(b) this is approximated as a track, formed by concatenating a number of basic parallelogram shapes. Figure 13.17(c) shows the result of breaking the shapes into individual pixels. Clearly some distortion of the original pattern has been introduced, together with considerable edge roughness. The latter can be a major problem since waveguide bends are typically shallow—the schematic of Fig. 13.16 is artificially compressed in the horizontal direction—and the result can be high scattering loss, unless the pixel size is made small.

The low-level code is then used to write the masks directly by computer-controlled deflection of an electron beam over a sensitized plate under high vacuum. The equipment required is basically a modified scanning electron microscope, fitted with additional beam-deflection and beam-blanking electronics and a precision motorized positioning table. Figure 13.18 shows a typical *electron-beam mask-making machine*. The source of electrons—known as an *electron gun*—is a thermionic emitter located at the upper end of the evacuated column on the left. Its output is focused by a set of fixed electrostatic lenses to a spot (with a diameter of 0.5–0.01 μm) in the writing chamber below the column. The sensitized plate is placed on an x–y positioning table at the focal plane, and writing is carried out by a combination of beam deflection and table motion.

By deflecting the beam electrostatically, a *field* of about 0.1 mm^2 may be addressed without moving the table. The field is broken up into a number of pixels

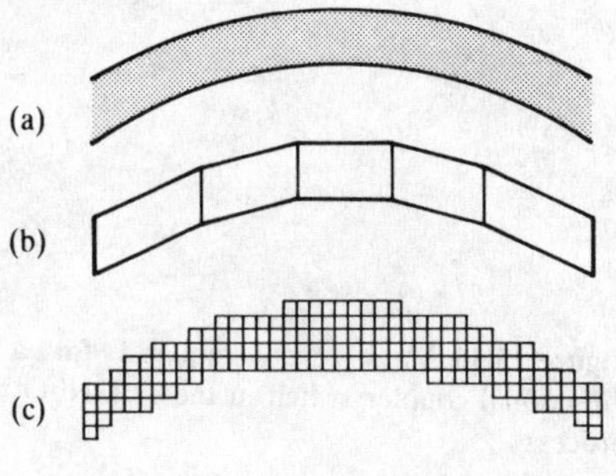

Figure 13.17 (a) Desired pattern, (b) representation using polygons, and (c) representation using pixels.

Figure 13.18 An electron-beam mask-making machine (*photograph courtesy R.A. Lawes, SERC Rutherford Appleton Laboratory*).

on an x–y grid; typically, each is of the order of 0.1 μm × 0.1 μm and corresponds to one pixel generated by the low-level code. Shapes that pass outside the field boundaries are then composed of a number of fields 'stitched' together, with the entire plate being moved by the motorized table to each new field position. Interferometric position sensors are used to ensure that the stitching is accurate. A modern electron-beam mask-making machine writes at about 10^7 pixels per second. Given that each is about 10^{-14} m^2 in area, about 10^{-7} m^2 can be written per second. Since a moderate-sized mask is about 10 cm × 10 cm, the maximum required writing time is of the order of 10^5 seconds, or a few hours. However in practice, this figure is greatly reduced, since only those areas that contain features (a small percentage of the total in waveguide devices) need be exposed to the electron beam.

Two different methods are used in the actual writing of the pattern. Both are based on organic polymers known as *electron resists*. In negative resist, cross-linking is induced by irradiation with electrons, making the irradiated areas insoluble. Unirradiated areas are then dissolved away during a later development step. In positive resist, on the other hand, cross-links are broken by electron irradiation, making the exposed areas soluble. Each type therefore forms a relief image after exposure and processing. The most common negative resist is poly (glycidylmethacrylate-co-ethyl acrylate) or COP, while its positive counterpart is polymethyl-methacrylate, or PMMA, developed using methyl isobutyl ketone, or MIBK. PMMA is considerably less sensitive than COP, but offers higher resolution. However, one of the main factors limiting resolution in electron beam lithography is backscatter of electrons that have passed through the resist into the substrate.

Further processing may then be carried out to transfer the written features to a chromium metal layer. Additive or subtractive methods may be used; for example, Fig. 13.19(a) shows a subtractive process for chrome-on-glass masks, using negative electron-beam resist.

The mask plate—a parallel-sided, highly-polished glass substrate—is first coated with Cr metal (e.g. by evaporation) to a thickness of about 1000 Å (1). A layer of negative resist is then added; typically, this is spin-coated in liquid form, and then baked to drive off any solvent (2). Next, it is exposed by the scanned electron beam (3), and developed to remove resist from the unexposed areas (4). The remaining resist is then used as a mask in an etching step in which any unprotected areas of chrome are dissolved away (5). Finally, the resist is also removed, leaving a metal pattern on the glass (6).

Figure 13.20 shows a typical set of curved S-bend guide patterns in chromium metal, on a mask fabricated using the processes described here. The guide widths are 7 μm.

Figure 13.19(b) shows the corresponding additive process, which requires one less step. To begin with, the plate is coated only in resist (1). It is then exposed to the electron beam, but this time in different areas (2). After development, the areas of glass that are to be metallized are therefore unprotected (3). Next, the whole plate is coated with chromium (4). The remaining resist is then dissolved away, simultaneously removing the unwanted metal, which simply breaks up in the resist solvent. This is known as *lift-off*; it offers some advantages in the fabrication of fine

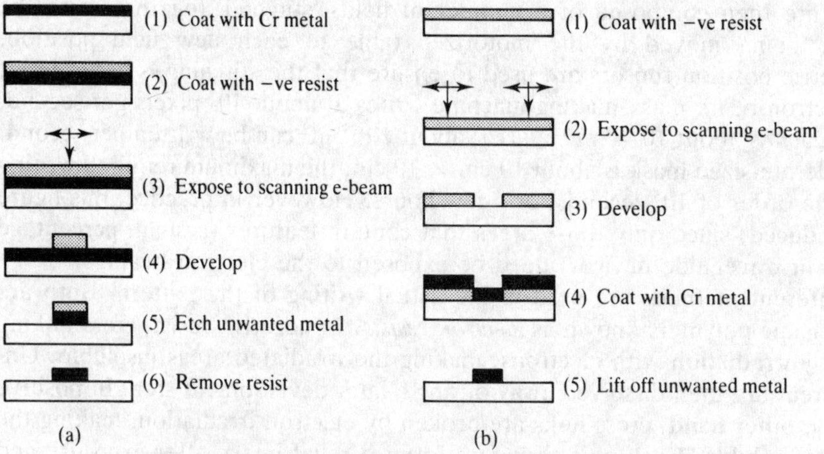
(1) Coat with Cr metal
(2) Coat with −ve resist
(3) Expose to scanning e-beam
(4) Develop
(5) Etch unwanted metal
(6) Remove resist
(a)
(1) Coat with −ve resist
(2) Expose to scanning e-beam
(3) Develop
(4) Coat with Cr metal
(5) Lift off unwanted metal
(b)

Figure 13.19 Mask fabrication using negative resist and (a) subtractive, (b) additive processing.

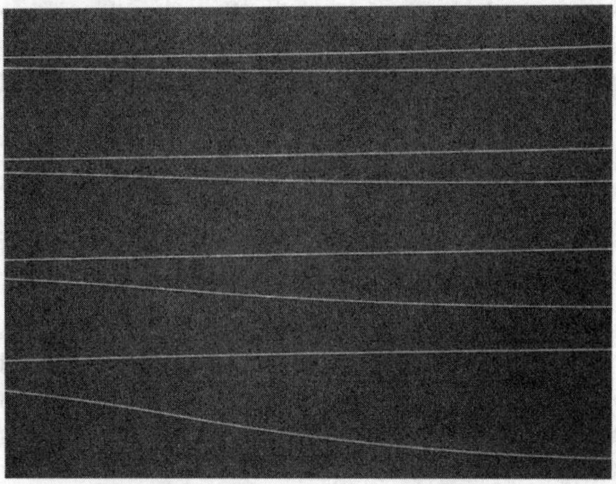

Figure 13.20 S-bend pattern on a chromium metal mask.

features, but special processing is needed to give the edge of the resist an undercut profile, so there is a clean break of the metal layer during lift-off.

Pattern Transfer

In the second stage the pattern is transferred from the mask to the substrate using an optical printing step known as *photolithography*. It is based on the properties of light-sensitive organic materials or *photoresists*, whose behaviour is qualitatively similar to the electron resists described above. Negative and positive optical resists both exist; in the former, exposure to ultraviolet light polymerizes the material, and in the latter, cross-links are broken by exposure. Figure 13.21 shows a typical process for subtractive pattern transfer using positive resist, with the mask of Fig. 13.19. It is assumed here that the substrate has already been coated with the film layer to be patterned, together with a layer of photoresist. Exposure is carried out in a machine known as a *mask aligner*; this contains accurate translation stages, a high-power microscope and a UV exposure source. First, the relative positions of mask

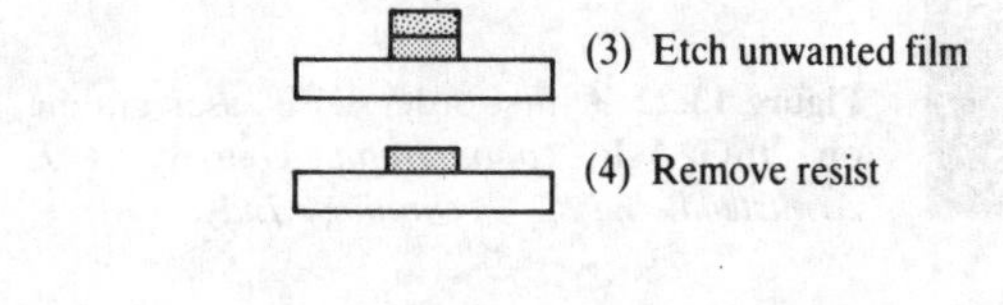

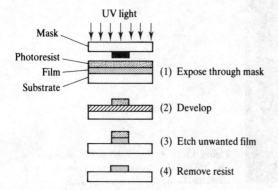

Figure 13.21 Optical photolithography: a typical subtractive pattern transfer process.

and substrate are carefully adjusted, so that the mask pattern is aligned with existing features on the substrate. The two are then clamped together by evacuating the space between them, and the resist is exposed by a flood beam (1). After removal from the mask aligner, the resist is developed; negative resists generally require organic solvents, while alkaline aqueous solutions are used for positive resist (2). The result is a pattern of resist on the film layer; this may be transferred to the film itself by etching (3), followed by removal of the remaining photoresist (4).

Grating Fabrication

The two-step procedure described above suffices for most patterns in integrated optics. However, the resolution of the copy process is limited by diffraction to approximately the wavelength of the incoherent ultraviolet exposure source. This is inadequate for making reflecting gratings, especially in semiconductor materials. For these, single-step processes are generally employed. There are two common techniques. In the first, the grating pattern is written directly by e-beam lithography onto electron resist, which is coated on the substrate itself. After development of the resist, the pattern may be transferred to the substrate by (for example) anisotropic wet etching. Figure 13.22 shows a first-order grating for a DFB laser (period $\approx 0.23\ \mu m$) etched into InGaAsP in this way.

In the second method, an ultraviolet laser is used as a maskless exposure source for photoresist. We have previously seen in Chapter 5 that the local distribution of irradiance resulting from two coherent plane waves that intersect at an angle is a periodic fringe pattern. This may be used to record a grating pattern directly in

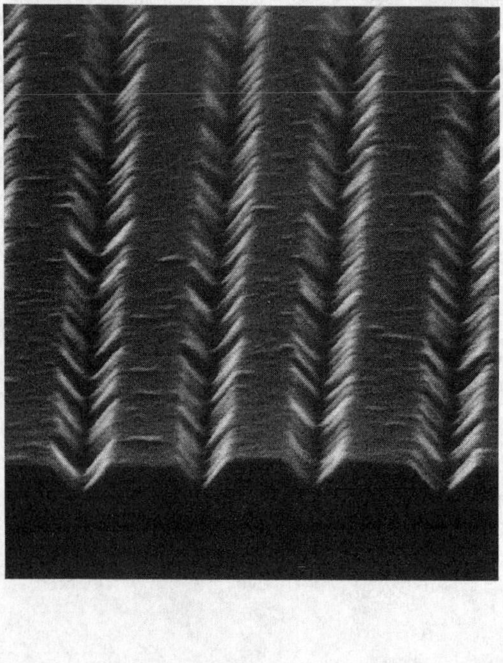

Figure 13.22 A first-order DFB laser grating on InGaAsP (*photograph courtesy C.J. Armistead, S.T.C. Technology Ltd*).

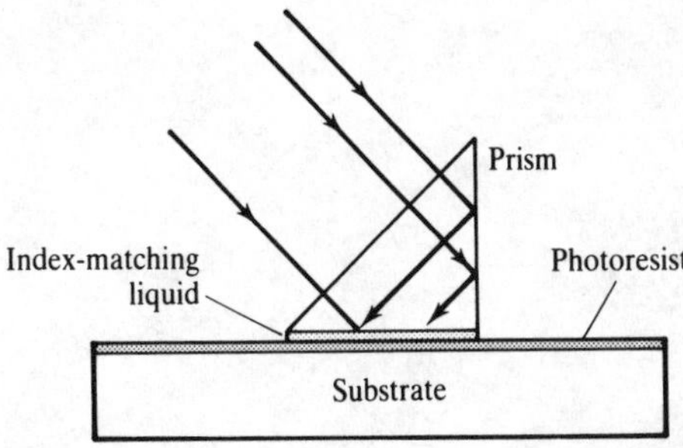

Figure 13.23 Simple experimental configuration for grating fabrication by holographic exposure.

photoresist, in a technique known as *holographic exposure*. Figure 13.23 shows a particularly simple set-up, which requires only a single recording beam. This uses a prism, and the geometry is arranged so that the second beam necessary for the exposure is generated by total internal reflection of the input wave at the end-face of the prism. Both beams then pass directly into the photoresist layer without reflection, since the prism is contacted onto the resist-coated substrate using index-matching liquid.

Design Example

We can estimate the parameters of the holographic setup needed to record a second-order grating for a GaAs/GaAlAs DFB laser, as follows. Assuming that the laser wavelength is 850 nm, and that $n \approx 3.6$ for GaAs, the propagation constant of the laser mode will be $\beta \approx 2\pi \times 3.6/0.85 \times 10^{-6} = 26.61 \times 10^{6}\ \text{m}^{-1}$. For Bragg reflection from a first-order grating, a grating vector of magnitude 2β is needed. However, with a second-order grating, the spatial frequency is halved, so here we require $K = \beta$. This corresponds to a grating period of $\Lambda = 2\pi/K = 0.236\ \mu\text{m}$. We shall assume that a He-Cd laser ($\lambda_0 = 0.442\ \mu\text{m}$) will be used for recording, and that the two recording beams are generated inside a glass prism (for which $n \approx 1.5$) and intersect symmetrically with an interbeam angle of 2θ. In this case the propagation constant inside the glass is $\beta_r = 2\pi \times 1.5/0.442 \times 10^{-6} = 21.3\ \text{m}^{-1}$, and the spatial frequency generated by the recording is $K = 2\beta_r \sin\theta$. To obtain the value of K above, the recording angle should be $\theta = \sin^{-1}(K/2\beta_r) = \sin^{-1}(26.61/2 \times 21.3) = 38.65°$. This is sufficiently close to 45° for the setup shown in Fig. 13.23 to be used.

13.8 OPTICAL FIBRE FABRICATION

We now turn our attention to the methods of optical fibre fabrication. Broadly speaking, these may be divided into two groups: those which do not require a preform, and those which do. However, processes based on preforms allow so much more control over the characteristics of the fibre that they are now entirely dominant. We will therefore limit our discussion of preformless methods to a brief description of one historically interesting example, the *double crucible method*.

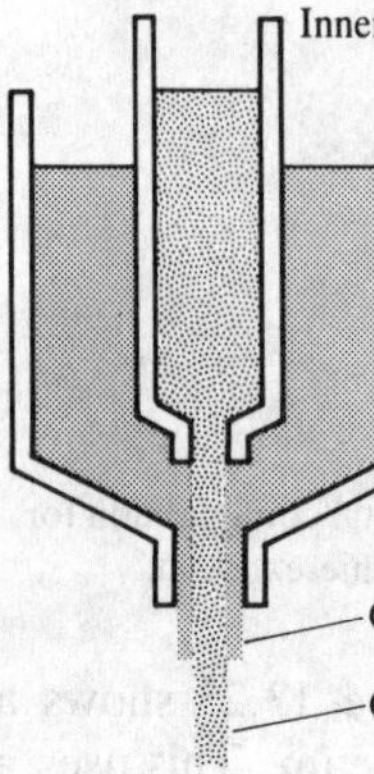

Figure 13.24 The double crucible method, for fibre pulling directly from melts.

As its name suggests, this involves the use of two concentric platinum crucibles, placed one inside and slightly above the other and heated by external windings (Fig. 13.24). The inner crucible is loaded with the core glass (supplied in the form of rods), the outer one with the cladding glass. When the assembly is raised to a sufficient temperature, both glasses melt. The core material is then allowed to flow through a nozzle at the base of the inner crucible to mingle with the cladding material, which can flow through a larger nozzle at the bottom of the outer crucible. Because of the concentricity of the arrangement, little mixing of the two glasses occurs; the flow stream of the core material is cylindrical, while that of the cladding is annular. The result is that the core glass emerges inside the cladding, in the geometry of a step-index fibre. After leaving the heated region the glass quickly solidifies, and can be coated in a protective thermosetting plastic jacket and wound on a drum. Correct choice of nozzle dimensions and pulling speed then yields a fibre of the desired core and cladding diameters.

Although simple, the double crucible method suffers from the disadvantage that the refractive index profile obtained is entirely dictated by the process. If little interdiffusion of the two glasses occurs, the profile is almost exactly a step-index one, but the diffusion of a mobile, highly polarizable dopant (e.g. thallium) out of the core may be used as a mechanism for obtaining graded-index fibres with near-parabolic profiles. Fibre bundles (consisting of a number of parallel fibre cores) may even be drawn, using a multiple nozzle arrangement. However, it is not possible to exercise the degree of control required to fabricate low-dispersion or polarization-preserving fibres, both of which require complicated and well-specified index profiles. This can only be achieved by using a preform.

Preform Fabrication

A *preform* is, in effect, a large-scale model of the fibre cross-section, which is subsequently reduced to the final diameter by drawing. As an example of an early method of making step-index preforms, we mention the *rod-in-tube method*. Here the preform was simply fabricated from a rod of core glass, fused inside an annular cylinder of cladding glass. Although direct, and adaptable to more complicated geometries (e.g. non-circular cross-sections), the resulting fibre suffered severely from scattering loss, caused by scratches and flaws at the interfaces between the glasses.

Figure 13.25 shows the alternative *built-in casting method*, used to fabricate preforms in multi-component fluoride glasses (e.g. a ZrF_4-BaF_2-GdF_3 mixture). In Fig. 13.25(a) molten cladding glass is poured into a cylindrical brass mould, heated to around the glass transition temperature. The mould is then upset (Fig. 13.25(b)) so that the central portion of the casting, which is still liquid, runs out, and the core glass is then poured into the resulting cylindrical hollow (Fig. 13.25(c)). After cooling, the casting is removed from the brass mould and its ends are trimmed to yield a cylindrical preform of step-index profile (Fig. 13.25(d)).

Neither of these techniques allows the construction of graded-index preforms of arbitrary refractive index profile. To overcome this problem, a range of entirely different fabrication methods based on chemical vapour deposition were developed. There are three main variants. The first is the *outside vapour phase oxidation* (OVPO) or *soot* process. This was developed by Corning Glassworks in 1970, and used to fabricate the first fibre with losses as low as 20 dB/km. In this case the preform is built up by depositing layers of silica glass on a ceramic mandrel or 'bait'. The silica is produced by hydrolysing a mixture of $SiCl_4$ and O_2 in a glassburner, following the reaction:

$$SiCl_4 + 2H_2O \rightarrow SiO_2 + 4HCl \qquad (13.15)$$

The result is a stream of small silica particles, or 'soot', which are blown onto the bait by the oxyhydrogen flame, where they stick. During deposition the bait is rotated in a glassblowing lathe, and the soot stream is traversed up and down the axis to ensure uniform coverage. In early experiments titania was used as a dopant to raise the refractive index of the inner glass layers, which form the fibre core. However, this caused high propagation loss (due to the presence of Ti_3^+ ions), necessitating an additional annealing step. Considerably better results were obtained when germania was used instead, deposited following the similar reaction:

$$GeCl_4 + 2H_2O \rightarrow GeO_2 + 4HCl \qquad (13.16)$$

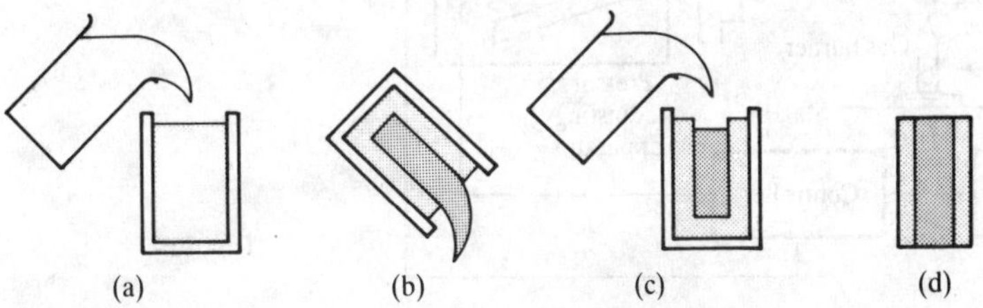

Figure 13.25 The built-in casting method for fibre preform fabrication (*after T. Katsuyama and Matsumusa, 1989*).

Because the composition of each layer may be varied merely by adjusting the flow of the relevant gas, graded-index preforms of arbitrary radial profile may be built up by OVPO. After the deposition of sufficient material, the bait is removed, and the rather porous soot is compacted or 'sintered' in a furnace at around 1500°C to produce glass of optical quality. Chlorine gas is used as a drying agent during sintering; the effect is to reduce the concentration of hydroxyl (OH^-) ions, which would otherwise greatly increase propagation loss in the final fibre.

The OVPO process suffers from two main disadvantages. The first is that the removal of the bait from the preform is moderately difficult, and may cause cracking. It also leaves a large hole in the centre of the preform; this is closed up during the fibre drawing process, but a central depression in the refractive index profile normally remains. The second is that the process is a batch one, which limits its applicability to the production of fibre in very long lengths.

The second process variant is the *modified chemical vapour deposition* (MCVD) process, developed by Bell Telephone Laboratories and others in 1974. This avoids one of the difficulties above; the requirement for a central bait. In effect, the process is one of inside vapour phase oxidation (IVPO), with deposition now taking place inside a silica tube (Fig. 13.26). One major difference from OVPO is that the chemical reaction does not take place in a flame. Instead, unreacted gases are mixed together and injected into one end of the tube. This is heated externally to 1500–1650°C by a traversing oxyhydrogen burner, and the reaction is induced at the growing surface by contact with the heated region. Since this process takes place in an enclosure, an improvement in purity of the deposited material might be expected over that obtained by OVPO.

Once again, the preform is built up slowly in layers, with changes in refractive index obtained by fine control of the gas composition. This time, however, it grows inwards, and the soot is simultaneously sintered by the oxyhydrogen flame. In early

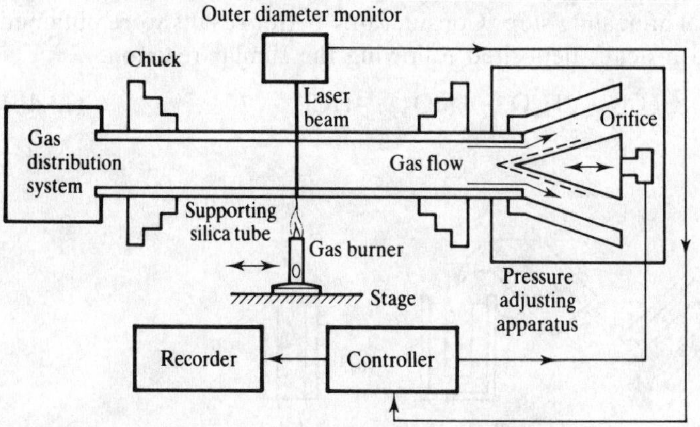

Figure 13.26 Improved CVD system with control of the internal pressure of the supporting tube (*after M. Okada* et al., *1978*).

processes there was a tendency for the silica tube to shrink at high temperatures as the deposition progressed, due to surface tension. This resulted in an increase in the thickness of the tube, which made it difficult to maintain the correct internal temperature. The effect was suppressed using slight positive pressure inside the tube, under feedback control from a laser-based measurement of the tube diameter.

Originally the deposition reaction was based on the oxidation of hydrides (e.g. SiH_4) in an oxygen gas stream, much as in the planar CVD techniques described earlier. However, slow deposition rates (coupled with relatively high contamination by hydroxyl ions) induced a switch to alternative chloride-based processes. For example, the two materials most commonly required, silica and germania, may be deposited from $SiCl_4$ and $GeCl_4$, respectively, following

$$SiCl_4 + O_2 \rightarrow SiO_2 + 2Cl_2$$
$$GeCl_4 + O_2 \rightarrow GeO_2 + 2Cl_2 \qquad (13.17)$$

This particular MCVD process has allowed the fabrication of graded-index fibres with extremely low loss (0.2 dB/km at 1.55 μm wavelength). A further process variant exists, in which the heating is performed by a microwave plasma inside the tube instead of an external flame; this is known as *plasma chemical vapour deposition* (PCVD) and yields similar results.

One advantage of MCVD is the ease with which more complicated preforms may be made. For example, Fig. 13.27 shows the process for making *bow-tie* high-birefringence fibres (previously described in Chapter 8), developed at Southampton University. These require the insertion of stress-producing sectors into the preform, so the fibre is not circularly symmetric. Figure 13.27(a) shows the first step, which involves the deposition of a uniform layer of fluorophosphorus silica glass inside a silica tube, followed by a similar layer of borosilicate glass (which will supply the stress-inducing sectors). In Fig. 13.27(b), the rotation of the lathe is stopped, and two traversing burners are used to provide localized hot zones on either side of the

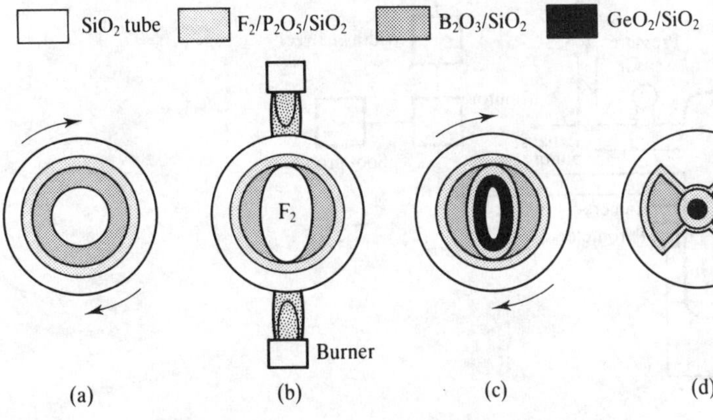

Figure 13.27 Fabrication of bow-tie fibre (*after R.D. Birch* et al., *1982).*

tube. The borosilicate layer is then etched away from two diametrically-opposed strips by passing a fluorine-liberating gas (SF_6 mixed with N_2) down the tube. In Fig. 13.27(c), two more uniform layers (of fluorophosphorus silica glass, which acts as a buffer, and germania-doped silica, which provides the core) are then deposited. Finally, in Fig. 13.27(d), the tube is collapsed, to yield the characteristic bow-tie geometry.

Despite this flexibility, MCVD still suffers from being inherently a batch process. So far, the only method suitable for continuous preform fabrication is the *vapour-phase axial deposition* (VAD) process, developed by NTT Ibaraki Laboratories. Figure 13.28 shows a schematic of a VAD preform plant. Once again, the process involves the production of soot from chlorides, which are hydrolysed in a flame exactly as in the OVPO process. The reactions followed in the deposition of silica and germania are therefore as in Eqs. (13.15) and (13.16), respectively. The main difference is that the preform is grown axially, starting from one end. A silica rod held in a rotating chuck is used as a bait, and soot is blown onto the end of this rod by a number of fixed torches. As the deposition procedes, the preform is slowly raised, to keep the growing surface at a fixed height. The porous soot preform is consolidated into a transparent glassy mass by passing it directly from the growth chamber into a sintering furance (at 1500–1600°C), where it is dried in a chlorine or thionyl chloride ($SOCl_2$) atmosphere to reduce hydroxyl ion contamination. In this stage isolated OH^- ions are replaced by Cl^- ions, generating Si-Cl bonds; however, these do not affect the fibre propagation loss, since their fundamental

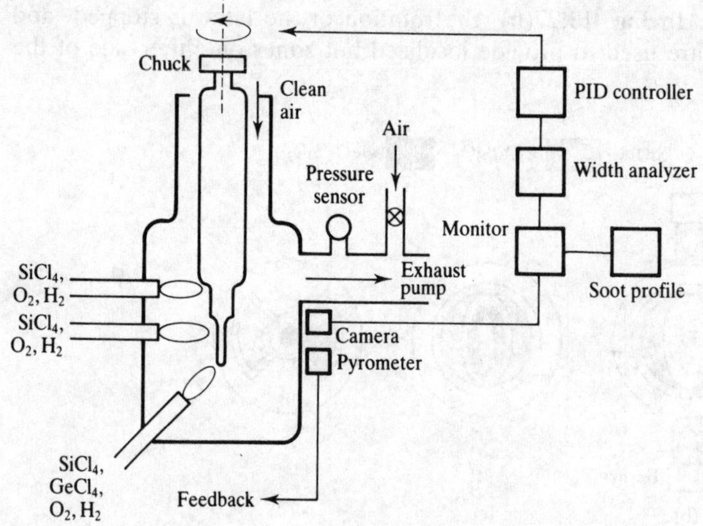

Figure 13.28 Schematic of a VAD preform plant (*after H. Murata, 1989*).

Table 13.4 Comparison of fibre preform fabrication processes (after H. Murata, 1989)

	OVPO	MCVD	PCVD	VAD
Chemical reaction	Hydrolysis	Oxidation	Oxidation	Hydrolysis
Dehydration	Yes	No	No	Yes
Heat source	Oxyhydrogen flame	Oxyhydrogen flame	Microwave plasma	Oxyhydrogen flame
Deposition rate (g/min)	9.0	0.5–1.0	0.5	5–10
Production maximum	>10	2	2.5	>10
Profile control	Easy	Easy	Very easy	Easy for SM fibre
Typical fibre length (km)	90	15–20	5–10	100
SM loss @ 1.3/1.55 μm (dB/km)	0.35/0.2	0.35/0.2		0.35/0.2

absorption peak lies at around $\lambda_0 = 25$ μm, far beyond the operating wavelength. A typical dehydration reaction follows

$$\text{Si-OH} + SOCl_2 \rightarrow \text{SO-Cl} + SO_2 + HCl \quad (13.18)$$

Figure 13.28 shows a multiple-torch plant, where a set of three fixed torches is used to fabricate a germanium-doped silica core surrounded by a thick silica cladding layer in a single deposition run. By careful design of the torch nozzles, the spatial distribution of the individual gas flows may be controlled. This allows the local dopant concentration, deposition rate and surface temperature to be specified accurately, so that graded-index guides with the required profile may be fabricated.

Table 13.4 shows a comparison of the methods described above—the OVPO, MCVD, PCVD and VAD processes. As can be seen, all can be used to make very low-loss optical fibre in long lengths, but vapour-phase axial deposition probably allows the most rapid construction of the largest preforms.

Fibre Pulling

After consolidation and drying, the preform is inserted into a pulling tower for the final *drawing* stage, shown in Fig. 13.29. The preform is fed at constant speed into the electrically-heated carbon furnace at the top of the tower, which is maintained at a precisely-controlled temperature (≈2000°C) under a clean atmopshere. As the end of the preform melts, it is drawn out of the furnace as a thin liquid stream, which solidifies rapidly on cooling. For single-mode fibre, the outer diameter is

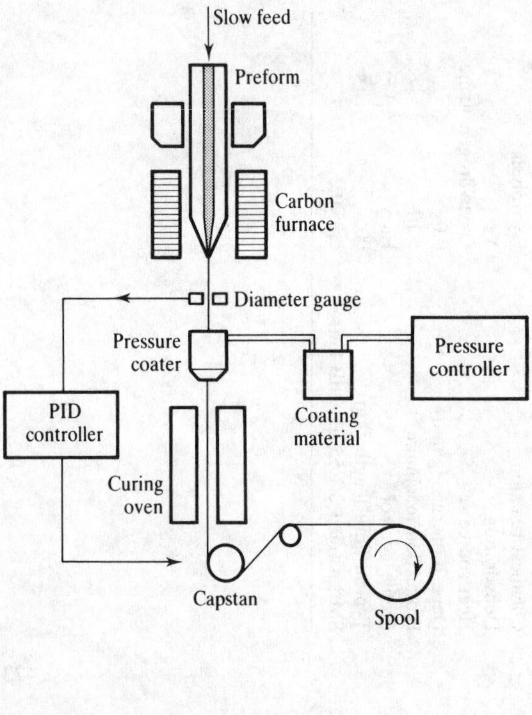

Figure 13.29 Fabrication of optical fibre in a fibre pulling tower.

typically 125 μm. At this point the fibre is extremely fragile, and contact with any surface can induce flaws and scratches which will subsequently cause mechanical failure. It is therefore passed directly into a coating die; this applies a protective plastic jacket, typically in the form of a 100 μm-thick layer of silicone resin, which is hardened in a curing oven further downstream. In order to achieve high coating speeds, the coater is pressurized by a high-pressure gas source; without this, the fibre has been found to slip through the resin without being coated. The fibre is pulled by a capstan whose rotational speed is controlled in a feedback loop by a non-contact outer-diameter monitor situated just outside the furnace, and then wound onto a take-up spool.

PROBLEMS

13.1 Why are binary compounds used as substrate materials for III–V optoelectronic devices, rather than ternary or quarternary compounds? Distinguish between homoepitaxy and heteroepitaxy. In what optoelectronic applications might the latter growth technique be required? What are the advantages of MOVPE over LPE for the overgrowth of $In_{1-x}Ga_xAs_{1-y}P_y$?

13.2 Assuming that the lattice parameter $a(x,y)$ of the quaternary alloy $In_{1-x}Ga_xAs_{1-y}P_y$ may be described by the function $a(x,y) = xy\, a_{GaP} + x(1-y)\, a_{GaAs} + (1-x)y\, a_{InP} + (1-x)(1-y)\, a_{InAs}$ (where $a_{GaP} = 5.4512$ Å, $a_{GaAs} = 5.6536$ Å, $a_{InP} = 5.8696$ Å, and $a_{InAs} = 6.0590$ Å are the lattice parameters of the binary compounds GaP, GaAs, InP and InAs, respectively), show that the locus of compounds lattice matched to InP is almost exactly a straight line, obeying the relation $y \approx 1-2.14x$. What is the stoichiometry of lattice-matched indium gallium arsenide?

13.3 Show by direct substitution that the concentration variation $C(x,t) = \{S/\surd(\pi Dt)\} \exp(-x^2/4Dt)$ is a solution of the one-dimensional diffusion equation $\partial C/\partial t = D\, \partial^2 C/\partial x^2$ for $t > 0$. Show also that the constant S is given by $S = \int_0^\infty C(x,t)\, dx$. What does S represent?

13.4 A particular doping process requires the diffusion of an impurity into silicon for eight hours. Assuming that the process activation energy is 3 eV, and that the diffusion coefficient tends to 10^{-1} cm^2 s^{-1} as $T \rightarrow \infty$, calculate the diffusion length for (a) $T = 900$°C and (b) $T = 1000$°C.
[0.39 μm; 1.25 μm]

13.5 (a) Show that the concentration variation $C(x,y,t) = (S/4\pi Dt) \exp\{-(x^2 + y^2)/4Dt\}$ represents a solution of the two-dimensional diffusion equation $\partial C/\partial t = D\, \nabla^2 C$ for $t > 0$. What boundary conditions would this solution correspond to?

(b) A channel waveguide is to be formed by in-diffusion of a strip of Ti metal into a $LiNbO_3$ crystal. Assuming that the strip occupies the range $-w/2 \leq y \leq w/2$ on the crystal surface prior to diffusion, show that the resulting refractive index distribution is given by:

$$n(x,y) = n_s + \Delta n \exp(-x^2/L_D^2)\, [\text{erf}\{(w+2y)/2L_D\} + \text{erf}\{(w-2y)/2L_D\}]$$

where n_s is the substrate index prior to diffusion, x is the depth below the crystal surface, $L_D = 2\surd(Dt)$ and Δn is a constant. Sketch the variation in refractive index along lines of (a) constant x and (b) constant y.

13.6 A single-mode fibre is to be clamped between two silicon wafers which carry V-shaped alignment grooves. The fibre will be ⟨100⟩ Si, etched anisotropically down the ⟨111⟩ crystal planes, which are oriented at 54.74° to the wafer surface. Assuming that the fibre has a diameter of 125 μm, that 5 μm clearance is required between the wafers, and that the etch selectivity is perfect, what should be the groove width? What will be the groove depth?
[149.55 μm; 105.76 μm]

13.7 The following lines of code are a short section of a CAD design for the mask set used to fabricate an integrated optical device, written in the now-obselete GAELIC design language. The first line sets the design grid to 0.1 μm intervals on each axis, while the last line terminates execution. Each of the lines RECT(*mask*)x,y:Δx,Δy; generates a filled rectangular shape on level *mask* of the mask-set. The bottom left-hand corner of the rectangle is located at the point (x,y), while the sides of the rectangle are of length Δx and Δy in the x- and y-directions. Sketch the device, indicating the main design parameters. Assuming that the device is to be fabricated by the Ti:$LiNbO_3$ process, what crystal orientation should be used?

```
UNITS=MICRONS,GRID=0.1;
RECT(1)-25000,35:50000,70;
RECT(1)-25000,-105:50000,70;
RECT(2)-25000,35:50000,500;
RECT(2)-25000,-535:50000,500;
FINISH;
```

13.8 (a) The device design in Question 13.7 is to be completed by the addition of symmetric, circular-arc S-bend sections at either end, so that the centre-to-centre separation of the guides at the input and output is increased to 300 μm, for connection to optical fibres. Assuming that the minimum bend radius that gives acceptable loss in the Ti:$LiNbO_3$ process used is 30 mm, estimate the minimum length of each transition.

(b) The GAELIC command POLY(*mask*)L,x,y:Δx_1,Δy_1,Δx_2,Δy_2,. . ., Δx_n,Δy_n; generates an n-sided polygonal shape on level *mask* of the mask-set. The point (x,y) specifies the first vertex of the polygon, while the increments $(\Delta x_i,\Delta y_i)$ define the position of the $(i+1)$ th vertex relative to the i th vertex. The vertex positions are listed in anti-clockwise order along the command line. By breaking the curved waveguide sections into a number of elementary polygonal shapes, use the POLY command to add S-bends to the design of Question 13.7.
[4.14 mm]

13.9 The GAELIC code of Question 13.7 is to be used as input to an electron beam mask-making machine. Assuming that negative electron resist and subtractive processing are used to fabricate the resulting chrome-on-glass mask set, outline the process steps by which the complete Ti:$LiNbO_3$ device might be fabricated. You should assume that the waveguide patterns are to be defined by wet etching, and the electrodes by lift-off.

13.10 (a) A fibre preform is to be fabricated by MCVD inside a silica tube of length 1 m and inside diameter 20 mm. Assuming that the material is mainly pure silica (density 2.6 g/cm^3) deposited at a rate of 2 g/min, estimate the mass of the preform and the time taken to form it.

(b) The preform is to be pulled down to fibre in a single step. Assuming that the preform feed and fibre pulling rates are 3.5 mm/min and 1.5 m/s respectively, calculate the length and diameter of the resulting fibre.
[0.817 kg; 6.81 hours; 25.6 km; 125 μm]

SUGGESTIONS FOR FURTHER READING

Beales, K. J., Day, C. R., Duncan, W. J., Dunn, A. G., Dunn, P. L., Newns, G. R., and Wright, J. V. Low loss graded index fiber by the double crucible technique, *Proc. 5th European Conf. on Optical Fiber Communication*, Amsterdam, Paper 3.2, 1979.

Bean, K. E. Anisotropic etching of silicon, *IEEE Trans. Electron. Devices,* **ED-25**, 1185–1193, 1981.

Birch, R. D., Payne, D. N., and Varnham, M. P. Fabrication of polarisation-maintaining fibres using gas-phase etching, *Elect. Lett.*, **18**, 1036–1038, 1982.

Brodie, I., and Muray, J. J. "The Physics of Microfabrication", Plenum Press, New York, 1982.

Chartier, G. H., Jaussaud, P., de Olivera, A. D., and Parriaux, O. Optical waveguides fabricated by electric-field controlled ion exchange in glass, *Elect. Lett.*, **14**, 132–134, 1978.

Flavin, P. G. Fabrication of curved structures by electron-beam lithography, *Elect. Lett.*, **18**, 865–867, 1982.

Garvin, H. L., Garmire, E., Somekh, S., Stoll, H., and Yariv, A. Ion beam micromachining of integrated optics components, *Appl. Opt.,* **12**, 455–459, 1973.

Harriott, L. R., Scotti, R. E., Cummings, K. D., and Ambrose, A. F. Micromachining of integrated optical structures, *Appl. Phys. Lett.,* **48**, 1704–1706, 1986.

Hu, E. L., and Howard, R. E. Reactive-ion etching of GaAs and InP using $CCl_2F_2/Ar/O_2$, *Appl. Phys. Lett.,* **37**, 1022–1024, 1980.

Izawa, T., and Sudo, S. "Optical Fibers: Materials and Fabrication", KTK Scientific Publishers, Tokyo, 1987.

Kapron, F. P., Keck, D. B., and Maurer, R. D. Radiation losses in optical waveguides, *Appl. Phys. Lett.,* **10**, 423–425, 1970.

Katsuyama, T., and Matsumura, H. "Infrared Optical Fibers", Adam Hilger, 1989.

Mai, X., Moshrefzadeh, R., Gibson, U., Stegeman, G. I., and Seaton, C. I. Simple versatile method for fabricating guided wave gratings, *Appl. Opt.,* **24**, 3155–3161, 1985.

McIlroy, P. W., Rogers, P. M., Singh, J. S., Spurdens, P. C., and Henning, I. D., Low loss single mode InP/GaInAsP waveguides grown by MOCVD, *Electr. Lett.,* **23**, 701–703, 1987.

Mitachi, S., and Miyashita, T. Preparation of low-loss fluoride glass fibre, *Elect. Lett.,* **18**, 170–171, 1982.

Murata, H. Recent developments in vapor phase axial deposition, *IEEE J. Lightwave Tech.,* **LT-4**, 1026–1033, 1986.

Murata, H. "Development of Optical Fibers in Japan", Gordon and Breach Science Publishers, New York, 1989.

Naden, J. M., and Weiss, B. L. Optical properties of planar waveguides formed by He^+ implantation in $LiNbO_3$, *IEEE J. Lightwave Tech.,* **LT-3**, 855–859, 1985.

Nagel, S. R., MacChesney, J. B., and Walker, K. L. An overview of the modified chemical vapour deposition (MCVD) process and performance, *IEEE J. Quant. Elect.,* **QE-18**, 459–477, 1982.

Nahory, R. E., Pollack, M. A., and Johnston, W. D. Band gap versus composition and demonstration of Vegard's law for $In_{1-x}Ga_xAs_yP_{1-y}$ lattice matched to InP, *Appl. Phys. Lett.,* **33**, 659–661, 1978.

Nelson, A. W., Spurdens, P. C., Close, S., Walling, R. H., Moss, R. H., Wong, S., Harding, M. J., Cooper, D. M., Devlin, W. J., and Robertson, M. J. The role of MOVPE in the manufacture of high performance InP based optoelectronic devices, *J. Crystal. Growth,* **93**, 792–802, 1988.

Okada, M., Kawachi, S., and Kawana, A. Improved chemical vapour deposition method for long-length optical fibre, *Elect. Lett.,* **14**, 89–90, 1978.

Panish, M. G., and Cho, A. Y. Molecular beam epitaxy, *IEEE Spectrum*, April Issue, 18–23, 1980.

Ritchie, S., Rogers, P. M. Optical waveguides in III–V semiconductors, *J.I.E.R.E.,* **57**, S44–S50, 1987.

Schultz, P. C. Fabrication of optical waveguides by the outside vapor deposition process, *Proc. IEEE,* **68**, 1187–1190, 1980.

Stulz, L. W. Titanium in-diffused $LiNbO_3$ optical waveguide fabrication, *Appl. Opt.,* **18**, 2041–2044, 1979.

Sze, S. M. "VLSI Technology", McGraw-Hill International Book Company, London, 1983.

CHAPTER

FOURTEEN

SYSTEMS AND APPLICATIONS

14.1 INTRODUCTION

We have encountered a wide range of guided wave components, and it is now time to show how a number of these can be combined together in a complete optical circuit. We have chosen to illustrate the process of integration with concrete examples from the literature. Some of these are old applications which have unfortunately not stood the test of time. Either the performance of the system turned out to be too poor to be usable in the end, or it was simply apparent that there were better ways of solving the problem which did not involve optics at all. Others are so new that they have yet to reveal their limitations, or indeed their potential. Only a small subset can be described as 'successful'. We make no apology for this; new technology obeys the law of the jungle, just like everything else, and there will naturally be few eventual winners. Our criteria for selecting the examples was that they illustrate the process of integration itself, rather than the judgement of history.

14.2 THE PLANAR INTEGRATED OPTIC RF SPECTRUM ANALYSER

We begin with applications that use one single waveguiding system on its own. The first such example is the planar integrated optic *radio-frequency spectrum analyser*. The function of this particular circuit is to perform a real-time, parallel, spectral analysis of RF signals. Such a package might form a key component in an aircraft-based electronic countermeasures system; used to monitor enemy radio or radar signals. Its operation requires a Fourier transform of the incoming radio signal to be carried out. Although electronics could be used, optics immediately suggests itself because of the ability of a lens to perform a parallel Fourier transform on a spatial distribution of light, as mentioned in Chapter 4. Having chosen an optical implementation, the advantages of integration are light weight, and a small, rugged

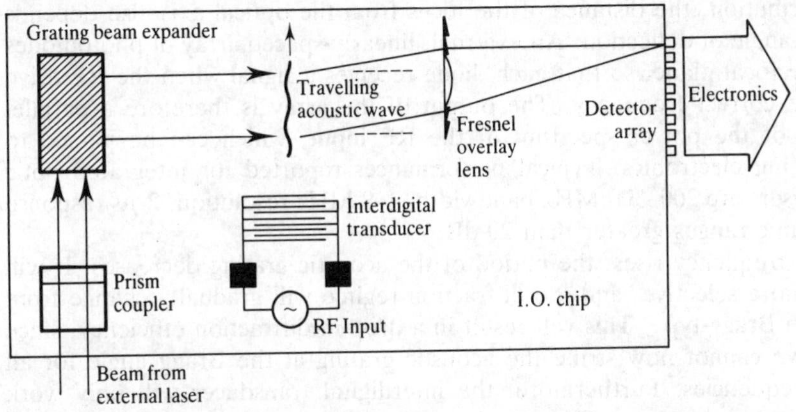

Figure 14.1 The planar integrated optic RF spectrum analyser.

package which will be immune to misalignment through shocks and vibration (e.g. during takeoff and landing).

Figure 14.1 shows one possible realization of the circuit, which consists of a planar integrated optic chip containing a number of the components described in Chapter 7. This particular example is conceptual, being an amalgam of several real prototypes demonstrated using different technologies. It is based on a waveguide fabricated on a piezoelectric, acousto-optic substrate. In this case we may assume that $LiNbO_3$ has been used, and that the guide has been formed by Ti indiffusion; however, other systems have used silicon as the substrate material.

The circuit works as follows. First, a prism coupler is used to launch light directly from an external laser (typically a HeNe laser) into the guide. This results in a relatively narrow beam of guided light, which is then passed to a corrugated Bragg grating; this has a rectangular boundary, with fringes oriented at 45° to the input beam direction. It is used to expand the beam cross-section, and also to deflect it through a right-angle. The expanded beam then interacts with a travelling surface acoustic wave, which is itself excited by the radio-frequency signal through the piezoelectric effect, using an interdigital transducer. As a result there are now two waves propagating near the substrate surface: the guided optical wave, and the SAW. Since the substrate is acousto-optic, the surface acoustic wave creates a travelling phase grating, which may diffract the optical beam. Because the SAW has a relatively small period at low frequencies, the grating wavelength is rather large, so diffraction takes place in the Raman-Nath regime. This results in the production by the grating of many diffraction orders, which may be a nuisance. However, the angle of deflection of the most important order—the first—depends linearly on the RF frequency to a good approximation. For simplicity we shall ignore all higher diffraction orders from now on, and simply consider the principal order as 'the diffracted beam'.

An overlay Fresnel lens is used to focus this beam to a point somewhere in the focal plane. In fact, because the lens performs a Fourier transform of the optical amplitude distribution, the distance of the focus from the optical axis also depends linearly on the angle of deflection. An external, linearly-spaced array of photodiodes is placed at the focal plane, so that each diode receives a signal when the RF drive is tuned to the correct frequency. The output of the array is therefore a parallel representation of the power spectrum of the RF input, which can be passed to further processing electronics. Typical performances reported for integrated optic spectrum analysers are 200–500 MHz bandwidth, 4–8 MHz resolution, 2 μs response time and dynamic ranges greater than 20 dB.

As the RF frequency rises, the period of the acoustic grating decreases. It will then become more selective, and the diffraction regime will gradually change from Raman-Nath to Bragg-type. This will result in a drop in diffraction efficiency, since the optical wave cannot now strike the acoustic grating at the Bragg angle for all possible RF frequencies. Furthermore, the interdigital transducer will only work over a limited frequency range. There are several ways to overcome these problems, so that the overall RF bandwidth is increased. The most common is to use several transducers, each with a different 'finger' period and angle of orientation, driven in parallel from the same RF source. At any given frequency, only the transducer with approximately the correct period will launch a surface acoustic wave, which will then travel at the correct angle. Alternatively, the acoustic wave may be continuously 'steered' with frequency, using a single transducer arranged as a *phased array*.

The circuit shown above could have been implemented in a variety of different ways—a telescope could have been used as the beam expander, the Fresnel lens could be replaced by a geodesic lens, and so on. The important constraint is that all the components must be made using a common technology, and should therefore be compatible. It does not necessarily follow that the component with the best performance in isolation should be used.

Design Example

We can estimate the parameters of a typical spectrum analyser using results derived in Chapters 4 and 7. Figure 14.2 shows the geometry; the input optical beam is assumed to be roughly collimated, with a Gaussian amplitude distribution of characteristic width $2w_{10}$.

Even though the results of Chapter 4 were for three-dimensional waves, we can guess that the effect of focusing the beam by a two-dimensional lens of focal length

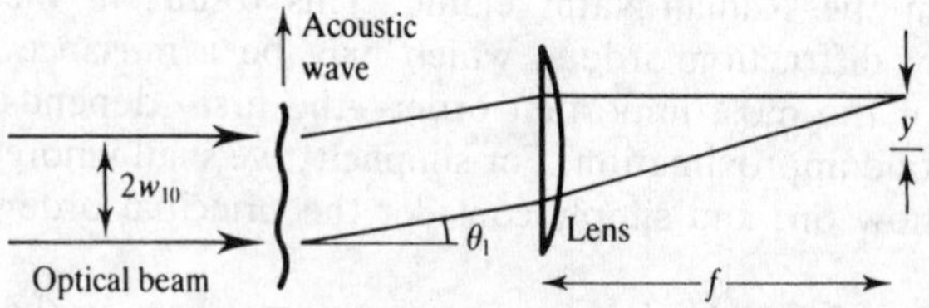

Figure 14.2 Geometry for calculation of RF spectrum analyser design.

f will be to produce a Gaussian focal spot of characteristic width $2w_{20}$, found by substituting β for k_0 in Eq. (4.72) to get

$$w_{20} = 2f/\beta w_{10} = f\lambda_0/\pi n_{\text{eff}} w_{10} \tag{14.1}$$

For a beam of width $2w_{10}$ approximately 2 mm and wavelength 0.633 μm propagating in a guide of effective index 2.2 (typical for Ti:$LiNbO_3$) and focused by a lens of 2 cm focal length, the width of the focal spot is

$$2w_{20} \approx 4 \times 2 \times 10^{-2} \times 0.633/(\pi \times 2.2 \times 2 \times 10^{-3}) \approx 3.7\ \mu\text{m}$$

Since it will be impossible to resolve finer details using such a lens, the width of each detector should be of this order.

In Chapter 7 we showed that the directions of the diffraction orders are defined by the grating equation, $\sin \theta_L = \sin \theta_0 + L\lambda_0/n\Lambda$. For acousto-optic diffraction, the grating period Λ is F/v, where v is the acoustic velocity and F is the frequency. Assuming a small angle of deflection and normal incidence, the direction of the first order is then approximately

$$\theta_1 \approx \lambda_0 F/n_{\text{eff}} v \tag{14.2}$$

Assuming that $v = 6.57 \times 10^3$ m/s in $LiNbO_3$, and that the transducer centre frequency is $F = 500$ MHz, we get $\theta_1 \approx 0.022$ rad (or 1.25°). Neglecting the distance between the acoustic beam and the lens, the centre of the detector array should then be located at a distance $y = f\theta_1$ away from the optical axis. From the data above, we obtain $y \approx 2 \times 10^{-2} \times 0.022$ m, or 440 μm.

Differentiating Eq. (14.2), we can find the displacement Δy of the focal spot that accompanies a change in frequency ΔF, as

$$\Delta y = f\lambda_0 \Delta F/n_{\text{eff}} v \tag{14.3}$$

For a transducer of (say) 200 MHz bandwidth we obtain

$$\Delta y \approx 2 \times 10^{-2} \times 0.633 \times 2 \times 10^8/(2.2 \times 6.57 \times 10^3)\ \mu\text{m} = 175.2\ \mu\text{m}$$

The detector array should therefore be of this length to cover the full RF bandwidth. The number of output channels may then be found as the number of resolvable spots in this distance, i.e. as $N \approx \Delta y/2w_{20}$. Using the data above we obtain $N \approx 175.2/3.7 \approx 48$. The channels will then be located at intervals of $\approx 200/48 = 4.2$ MHz in frequency-space.

Since the number of channels is so low, we might enquire as to the limitations on N. Combining Eqs (14.3) and (14.1), we obtain the surprisingly simple result

$$N = (\pi/4)\,\Delta F \tau \tag{14.4}$$

where τ is the time-of-flight of the acoustic wave across the optical beam, defined as $\tau = 2w_{10}/v$. To increase N while maintaining the same acoustic bandwidth, we must use either a wider optical beam or a material with a smaller acoustic velocity.

14.3 THE PLANAR INTEGRATED OPTIC DISC READ HEAD

The second application is more recent, a *read head* for an *optical disc player*. Here the aim is to use a focused laser beam to read the information contained in tiny pits, a few microns in diameter, which are etched in the surface of a rotating optical disc. Such a device might be used in a compact disc player, or in an optical data storage system (known as a CD ROM). The advantage to be gained from using optics in either case is the high potential storage capacity. We may estimate this by calculating the number of pits of diameter d which may be stored on a disc of diameter D. The value of d depends mainly on the size of focused spot we can create. This is limited by diffraction, and a typical (conservative) value might be $d \approx 2\ \mu m$. Taking $D \approx 10$ cm, we obtain a figure of $\pi(D/2)^2/\pi(d/2)^2 = 2.5$ Gbits, a capacity much larger than that of current magnetic discs. Optical data storage must clearly involve some form of optical reading or writing, however it is configured. The benefits of integration are the cost-reductions made possible by using a mass-production technology. This particular application shows that guided-wave optics may even penetrate the consumer goods market, a considerable achievement for such an embryonic technology.

Figure 14.3 shows a prototype read head, which is again based on a planar integrated optic chip, with an external laser diode and processing electronics. The waveguide is fabricated on a silicon substrate, using sputtered Corning 7059 glass as the guiding layer, with an Si-N cladding layer and a SiO_2 buffer layer. The circuit works as follows. The laser diode is butt-coupled to the chip edge, and so produces a diverging beam in the guide. This beam then encounters a Bragg grating device, a *twin-grating focusing beamsplitter* (or TGFBS), which is effectively a split grating lens. The grating is fabricated by writing a curved, periodic pattern in a resist layer (using an electron beam deflected under computer control), and then transferring the pattern to the Si-N layer by reactive ion etching. Most of the beam passes through the TGFBS without diffraction to another grating component known as a *focusing grating coupler* (or FGC). This is used to couple light out of the guide and create a focal spot in a plane above the chip, where the optical disc is rotating.

Light reflected from the pit pattern on the disc is then coupled back into the guide, also by the FGC. The return signal is then split into two converging beams by the TGFBS, and each focused beam falls on a split detector. These are made from p-n junction photodiodes fabricated directly in the Si substrate, and the SiO_2 buffer layer is tapered from the guiding region to the detecting area so that the light is gradually led onto the diodes. The four detectors give different signals, depending on the precise positions of the two return foci. Using combinations of these signals, external electronics can provide not only the desired readout signal, but also tracking and focusing error signals, which are used to ensure the correct position of the read head relative to the disc, through a feedback control loop. The tracking error signal is particularly important because the track width is approximately one pit diameter (a few μm). Because likely errors in the concentricity of the central hole locating the disc are vastly greater than this figure, it would be impossible to follow the track correctly without accurate feedback control.

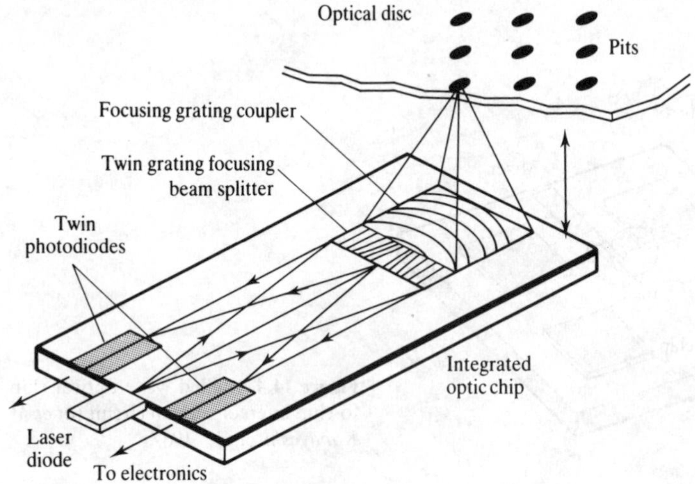

(a)

(b)

Figure 14.3 Integrated optic read head for an optical disc player: (a) schematic (*after S. Ura* et al., © *IEEE 1986*, and (b) the actual device (*photograph courtesy T. Suhara, Osaka University*).

14.4 GUIDED-WAVE OPTICAL CHIP-TO-CHIP INTERCONNECTS

We shall now present some different examples of optical integration using the channel guide components of Chapters 9 and 10. The first is an application designed for use with VLSI electronics, which is based on the following premise. It is becoming accepted that electrical connections among integrated circuits and between circuit boards will encounter serious problems if either the speed of the circuitry or its complexity rises much further. This is mainly due to two factors. The first is the limitation on signal transmission bandwidth caused by RC time constants in the circuit. Contrary to expectations, these do not scale down with decreasing circuit dimensions, so that propagation delays eventually become more significant than gate

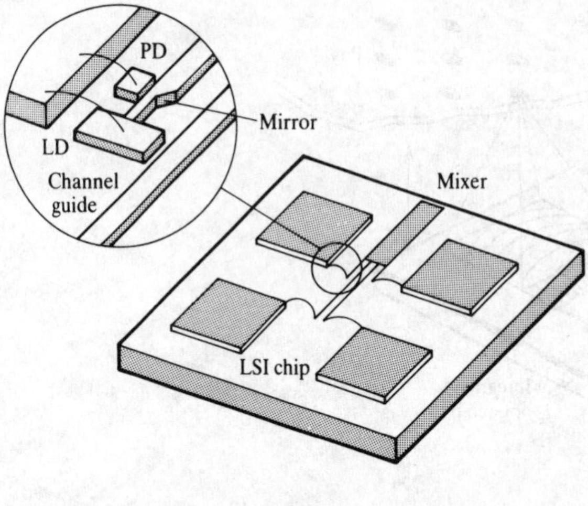

Figure 14.4 Guided wave optical chip-to-chip interconnection circuit (*after M. Kobayashi* et al., *1987*).

delays. The second is interference caused by stray capacitances, which also become more significant as components are packed closer together.

Optical interconnects have therefore been proposed as an alternative to electronic links. These could involve either free-space or guided-wave optics, and both techniques are currently the subject of research (although it is likely that guided wave devices will prove more successful in the long run, because they are smaller, more rugged and optically more efficient). Figure 14.4 shows an early prototype circuit based on channel guide technology, which is designed to connect four LSI chips together on a common substrate. This particular circuit is a hybrid owing to the difficulty of fully integrating the different technologies involved. The optical part is composed of semiconductor laser diodes, a multi-mode ridge waveguide circuit, and photodiodes. The waveguide circuit has four guides for transmitting optical signals, and a reflective mixing device to broadcast the resulting signal uniformly back to receivers at each LSI chip.

Each guide is used for both transmitting and receiving. The laser diodes (LD), which are butt-coupled to the end of the guide, act as the transmitters; these are InGaAsP diodes operating at $\lambda_0 = 1.3\ \mu$m, which are bonded onto the substrate upside-down. Angled reflection mirrors tap a fraction of the return signal from the mixer and pass it to the photo-detectors (PD), which act as the receivers. These are PIN photodiodes, also bonded to the substrate in a hybrid fashion. The waveguides are fabricated in a high index material, an SiO_2-TiO_2 mixture, and a thick SiO_2 buffer layer is used to separate the guides from the high index Si substrate. In this example the measured signal transmission capacity was 1 Gbit/s. Though an early result, this clearly indicates considerable potential for the future.

14.5 THE CHANNEL WAVEGUIDE INTEGRATED OPTIC A-TO-D CONVERTER

We now consider an alternative channel guide device, with an application in signal processing. It turns out that *analogue-to-digital conversion* is one of the more difficult tasks to perform quickly using electronics because many of the most common algorithms (e.g. successive approximation) are rather slow. What is really required is a device that compares a signal with a number of different reference levels at once, and then outputs a suitable binary code. Such devices do exist in electronics, and are known as parallel or 'flash' A-to-D convertors. However, they are expensive, and it was thought at one time that optical A-to-D conversion offered an advantage at the high-speed end of the market.

The operation of the optical A-to-D converter is based on the periodic response of the electro-optic Mach-Zehnder interferometer. We have already encountered this device in Chapter 9, in the guise of a modulator. It consists of two back-to-back Y-junctions linked by straight waveguides. The first Y-junction splits the incoming beam into two components which pass along the straight guides and are recombined at the second Y-junction. Either or both of the interferometer arms may have surface phase-shifting electrodes to allow the relative phase of the recombining components to be altered. If the two are in-phase the guided output is high, and if they are out-of-phase, it is low. More generally, the output is periodic in the phase imbalance, as shown in Fig. 14.5(a). If an additional static phase shifter is used to bias the interferometer, the sinusoidal response is shifted by a quarter of a period. If the optical output is now detected and then thresholded by an electronic comparator at a level of half the maximum possible detector output, the comparator output will vary as a square-wave with voltage (Fig. 14.5(b)). This corresponds to a one-bit binary representation of the applied voltage.

Because the electro-optic effect is linear, the output of a Mach-Zehnder interferometer must depend on both the drive voltage and the electrode length. For example, a short device will be less sensitive to voltage than a long device, and so on. Figure 14.6 shows the output characteristics that might be expected from three similar interferometers with electrode lengths of L, $L/2$ and $L/4$. Clearly all the responses are still periodic with voltage, but the longest device has twice the sensitivity of the intermediate one, and four times that of the shortest. If the additional processing described above—static phase-shifting, detection and thresholding—is applied to each interferometer, the comparator outputs will all vary

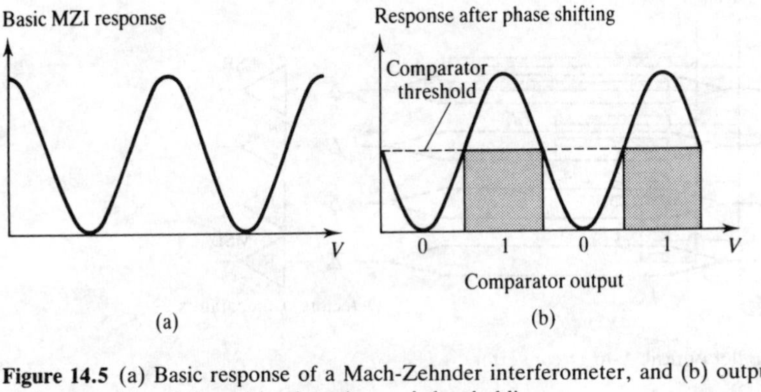

Figure 14.5 (a) Basic response of a Mach-Zehnder interferometer, and (b) output after application of a static phase bias, followed by detection and thresholding.

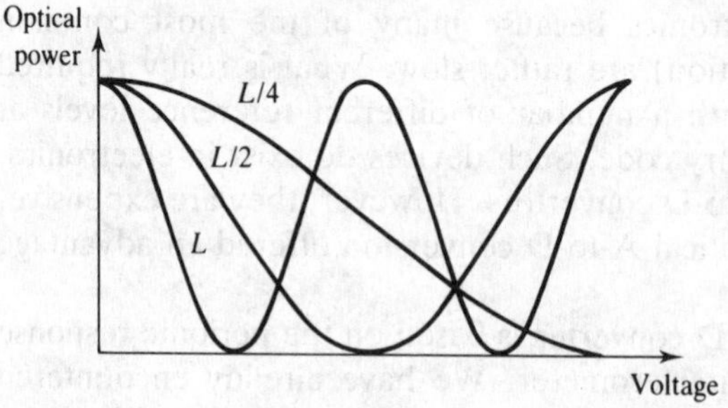

Figure 14.6 Power/voltage characteristics for Mach-Zehnder interferometers with different electrode lengths.

as square waves, but with periods differing by factors of two. Taken together, they yield a three-bit binary representation of the drive voltage.

Figure 14.7 shows a prototype high-speed, four-bit parallel analogue-to-digital converter for radio-frequency signals, operating on the principles described above. The optical circuit consists of four Ti:$LiNbO_3$ Mach-Zehnder interferometers, which are arranged in parallel on a common substrate and fed by a common optical input through a Y-junction 'tree'. Each interferometer has the same RF input, and is equipped with travelling-wave electrodes for high-speed response, but the electrode lengths differ by factors of two, so that the longest has electrodes of length L and the shortest of $L/8$. The circuit therefore performs a four-bit analogue-to-digital conversion of the input voltage. In this example the optical input is not continuous-wave; instead, it consists of a train of short pulses, so that the RF signal is sampled optically at a very high rate. Suitable sources include pulsed semiconductor lasers and mode-locked YAG lasers. Conversion speeds of almost 1 Gbit/s have been reached with prototype Ti:$LiNbO_3$ devices, but difficulties with the comparators and other readout electronics have prevented the anticipated market success. Despite this, the electro-optic A-to-D converter remains an ingenious and novel application of guided wave technology.

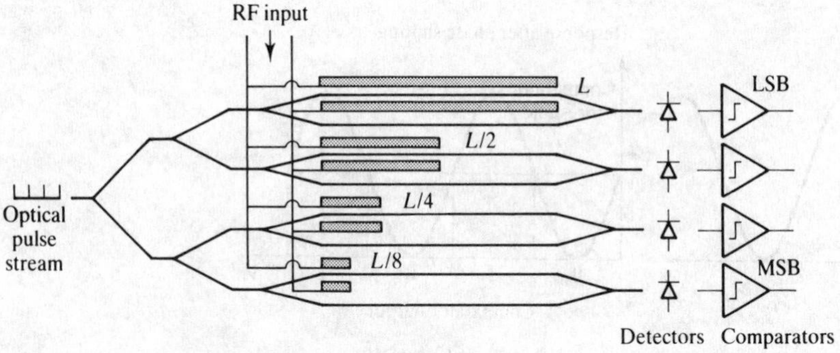

Figure 14.7 A 4-bit parallel optical A-to-D converter.

14.6 OPTICAL FIBRE SENSORS

We now move on to consider systems based around optical fibre and the all-fibre components of Chapters 8 and 10, beginning with *fibre sensors*. This is not an obvious application because fibre is normally designed to provide a low-loss transmission medium for optical waves, which are confined near the fibre core and are therefore isolated from their environment by a substantial thickness of cladding. However, it is also possible to arrange for a guided wave to be affected quite strongly by external perturbations. For example, the core might be relocated much nearer the edge of the cladding, so that the evanescent field extends outside it. The guided wave is then directly affected by any external changes, and the fibre can act as a sensor.

Once this is accepted, there are many convincing arguments to support the use of fibre as a sensing medium. Because it is possible to arrange a considerable length of fibre in a coil, high sensitivity to a particular parameter can be achieved within a small sample volume. Furthermore, it has already been shown that fibre can be sensitized to a wide variety of measurands, including temperature, humidity, strain, electric and magnetic fields, pressure, acoustic waves, and the presence of different inorganic and organic chemicals. Optical sensors are non-electrical, which is important in many hazardous environments (e.g. petroleum plants), and can be read out remotely. It is even possible to build multiplexed sensor highways, which contain a number of sensors in series that may be interrogated individually. Optical fibre sensors are therefore a promising future technology.

It is obviously impossible to do justice to a topic of this importance in a small section. All we can really do is discuss the major issues involved. Broadly speaking, fibre sensors are divided into two classes, operating on non-interferometric and interferometric principles, respectively. In the former, an external parameter is arranged directly to affect the transmission through the fibre by altering the propagation loss. One simple example is the *microbend sensor*, which works by inducing a periodic bend in the fibre by compression between two grooved plates (Fig. 14.8). The amount of bending and the consequent radiation loss are proportional to the depth of compression or the pressure applied. This allows very cheap, simple sensors to be constructed, in which a change in the sensed parameter is directly reflected as a variation in optical transmission. However, all such sensors suffer from the following major disadvantage: it is hard to differentiate this particular loss mechanism from any other—for example, from the loss which would arise if the

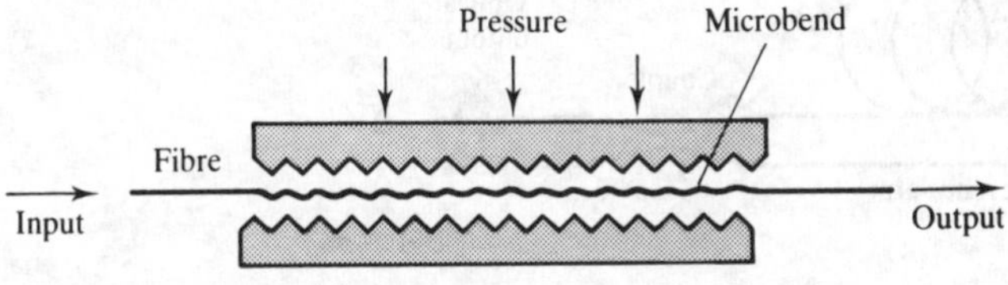

Figure 14.8 A microbend sensor.

fibre links to and from the sensor section were accidently bent. This type of sensor is therefore 'lead sensitive'.

In *interferometric sensors* the external parameter is arranged to affect the phase of the light instead. Unwanted perturbations generally have little affect on phase, so this minimizes the lead sensitivity problem. However, it is still possible to recover the sensor output in the form of intensity variations, by using interferometry. Figure 14.9 shows an example of a generic architecture for interferometric fibre sensors, based once again on the Mach-Zehnder interferometer.

The optical circuit is slightly different this time, since it contains two 3 dB fibre couplers (which act as 50:50 beam splitters) instead of Y-junctions. The couplers are linked together by a short *reference arm* and a much longer *sensing arm*, which is arranged as a coil. As before, the first splitter divides the optical input into two components. The upper part passes through the sensor coil, and so is affected by changes in the measurand. For example, if the device is to be used as a *hydrophone*, it is arranged for pressure waves passing through the coil to dilate it, thus altering the phase of the sensing beam. The beam passing through the reference arm is not exposed to the measurand, so its phase is unaffected by such changes. The second coupler is used to recombine the beams. These will add together coherently, provided that (i) a laser source is used, with a coherence length greater than the path imbalance of the two arms, and (ii) the two components have the same polarization. The latter condition generally requires the entire circuit to be constructed from polarization-preserving fibre, with the input adjusted so that only a single polarization component is launched.

Unlike the Y-junction Mach-Zehnder interferometer, the coupler-based MZI has two possible optical outputs. We can calculate these quite simply, as follows. First, we recall that the transmitted and cross-coupled amplitudes A_1 and A_2 of a synchronous coupler with coupling coefficient κ and length L can be written in the form:

$$A_1 = \cos(\kappa L) \quad \text{and} \quad A_2 = -\mathrm{j}\sin(\kappa L) \tag{14.5}$$

In a 3 dB coupler (for which $\kappa L = \pi/4$) these reduce to

$$A_1 = 1/\sqrt{2} \quad \text{and} \quad A_2 = -\mathrm{j}/\sqrt{2} \tag{14.6}$$

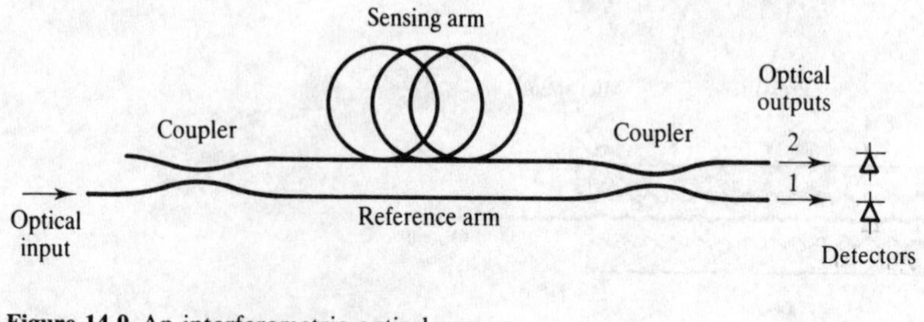

Figure 14.9 An interferometric optical sensor.

An optical input of amplitude A_0 will therefore be split by the first coupler into two components, of amplitude $A_0/\sqrt{2}$ in the reference arm and $-jA_0/\sqrt{2}$ in the sensing arm. We now assume that the sensing beam will incur an additional phase delay Φ relative to the reference arm, due to the intrinsic difference in their lengths and to the effect of the measurand. Ignoring any common phase delay, the amplitudes of the two outputs are

$$B_1 = A_0(1/\sqrt{2})^2 + A_0(-j/\sqrt{2})^2 \exp(-j\Phi) = (A_0/2)\{1 - \exp(-j\Phi)\} \quad (14.7)$$

and

$$B_2 = A_0(1/\sqrt{2}) \times (-j/\sqrt{2}) + A_0(-j/\sqrt{2}) \times (1/\sqrt{2}) \exp(-j\Phi)$$
$$= -j(A_0/2)\{1 + \exp(-j\Phi)\} \quad (14.8)$$

The output powers are then proportional to

$$P_1 = B_1B_1^* = (A_0^2/2)\{1 - \cos\Phi\} = P_0 \sin^2(\Phi/2) \quad (14.9)$$

and

$$P_2 = B_2B_2^* = (A_0^2/2)\{1 + \cos\Phi\} = P_0 \cos^2(\Phi/2) \quad (14.10)$$

The two outputs are therefore complementary, in that their sum is equal to the input power P_0. They are shown plotted as a function of Φ in Fig. 14.10. Normally the phase imbalance is biased so that the deviations imposed by the measurand occur only in a small range about the typical bias point Φ_0. This can be accomplished using additional static phase-shifters. Near the bias point, small changes in P_1 and P_2 are both linearly dependent on small changes in Φ, and hence on variations in the measurand, which makes assessment of the state of the sensor by processing electronics particularly simple. Larger excursions of the measurand (e.g. of more than one period) may be tracked by a variety of techniques, including 'fringe-counting'.

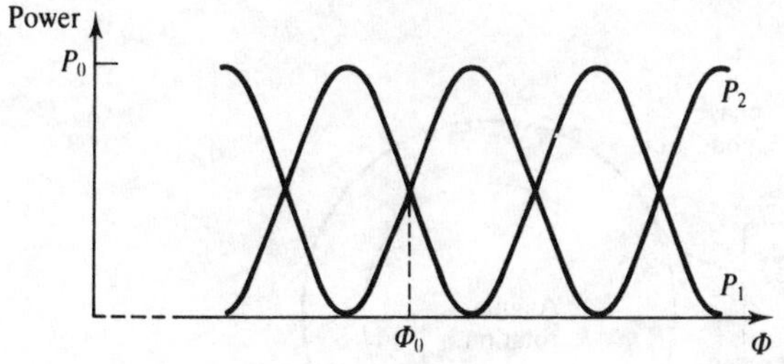

Figure 14.10 Variation of the power P_1 and P_2 with the phase imbalance ϕ.

14.7 THE INTEGRATED OPTIC FIBRE GYROSCOPE

We now consider mixed systems employing both fibre and integrated optics. We begin with another type of interferometer, the *Sagnac interferometer* (which we have already encountered as the fibre loop mirror in Chapter 10). This has also been investigated extensively for sensor applications. It shows little sensitivity to any parameter except rotation, and is therefore of great potential use, since it can act as a *gyroscope*. In fact, the fibre gyro has considerable advantages over its mechanical equivalent, since it is lightweight, requires no 'run-up' time, and needs little maintenance. Three such gyros (sensing rotation about three orthogonal axes), combined with three linear accelerometers, could form the basis of future solid-state inertial navigation systems, and they are actively being developed for applications as diverse as aircraft navigation, missile guidance and oil-well logging.

Since we have already described the static behaviour of a Sagnac interferometer, we will concentrate here on the situation when the interferometer is rotating. In this case it can be shown that there is a phase shift between the two counterpropagating beams in the loop, a phenomenon known as the *Sagnac effect*. Although an accurate description actually requires the invocation of general relativity, we shall base the discussion on approximate theory, which provides the correct result to first-order.

Figure 14.11 shows an experimental gyroscope based on a Sagnac interferometer. It is constructed from a Ti:$LiNbO_3$ channel waveguide integrated optic chip (which contains a number of signal routeing and processing components), an optical fibre sensing coil, an external laser diode source, and detector (which will typically be followed by futher processing electronics). The optical circuit works as follows. The N-turn fibre coil, which is of radius R, is arranged as the sensing loop. Light from the laser is butt-coupled into the chip, split into two by Y-junction #1, and launched in opposite directions round the fibre coil. After passing round the coil, which has length $2\pi NR$, the two components are recombined by Y-junction #1. A second Y-junction, #2, is then used to pick off a proportion of the power, which is passed to a detector. According to the usual principle of interferometry, the detected signal is high if the two recombining components are in-phase, and low otherwise.

We can calculate the phase change accompanying rotation using the following simple argument. Normally the time-of-flight round the loop is τ given by

$$\tau = 2\pi NR/c \tag{14.11}$$

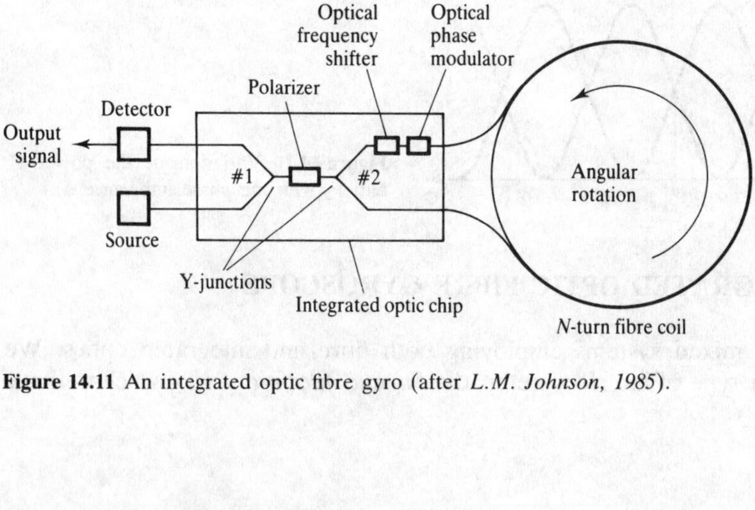

Figure 14.11 An integrated optic fibre gyro (after *L.M. Johnson, 1985*).

We now assume that the coil rotates at an angular speed Ω, so the recombination point will continually move away from one beam and towards the other. In a time τ it moves an angular distance $\Omega\tau$, so one beam must travel a distance $2\pi NR + \Omega\tau R$ to reach it, the other a distance $2\pi NR - \Omega\tau R$. The beams therefore take different times to exit the loop. The time-of-flight difference is

$$\Delta\tau = (2\pi N + \Omega\tau)R/c - (2\pi N - \Omega\tau)R/c \qquad (14.12)$$

This corresponds to a phase shift (known as the *Sagnac phase shift*) of

$$\Delta\Phi = \omega\Delta\tau = 4\pi NR^2\omega\Omega/c^2 \qquad (14.13)$$

where ω is the optical frequency. Consequently, the relative phase of the recombining beams is linearly dependent on the angular rotation rate. The sensitivity is controlled by the area enclosed by the coil, and the number of turns; this illustrates the main advantage of fibre optics which has turned the Sagnac interferometer into a practical proposition—it is easy to make N large without incurring high loss. The phase shift can also be written in terms of the length L of the fibre, the coil diameter D, and the wavelength of the light used. If this is done, we obtain

$$\Delta\Phi = 2\pi LD\Omega/\lambda_0 c \qquad (14.14)$$

The phase shift is therefore linearly dependent on the length of fibre used and the coil diameter.

Design Example

We can calculate the typical sensitivity of a gyroscope as follows. Using a 1 km length of fibre arranged in a coil of 10 cm diameter, and a wavelength of $\lambda_0 \approx 1\ \mu$m, we obtain from Eq. (14.14)

$$\Delta\Phi \approx 2\Omega \qquad (14.15)$$

The earth's rotation rate is $\Omega_e = 2\pi/(24 \times 3600)$ rad/s $= 7.3 \times 10^{-5}$ rad/s. To register this motion, we must therefore be able to measure a phase-shift of $\Delta\Phi \approx 1.5 \times 10^{-4}$ rad with the design above. For an inertial navigation system we need to do considerably better than this—say, at least a thousand times better. A typical phase shift will thus be about 10^{-7} rad, an incredibly small figure. This presents a considerable measurement problem, which is only partially solved at the moment.

The Minimum Configuration

To achieve this high sensitivity, the optical paths travelled by the two beams must be identical when the gyro is not rotating, so the system must be reciprocal. Figure 14.11 is an example of the so-called *'minimum configuration'* needed to achieve reciprocity in a gyro. The output is taken from the loop at exactly the point where the input is injected, and a polarizer is used to ensure that only a single polarization mode is launched and detected. Generally a number of further optical components must also be integrated on the substrate to improve performance. Here phase- and frequency-shifters have been added. These may be used to increase the signal-to-noise ratio of the detected signal through various signal processing tricks, or to allow closed-loop tracking of the Sagnac phase shift.

14.8 HIGH SPEED, GUIDED WAVE OPTICAL COMMUNICATIONS

A major application of guided wave optics is in high-speed optical communications. In fact, point-to-point links have now reached such an advanced stage that data can be transmitted at very high bit rates over staggering distances. Such progress represents a handsome confirmation of the original promise of fibre optics, and shows just how well the considerable investment required to develop low-loss, low-dispersion fibre and the other vital components of a communications system—lasers of high spectral purity, high-speed modulators, and fast, low-noise detectors—has paid off.

A High-speed Point-to-point Link

To illustrate this, Fig. 14.12 shows a recent experiment, in which data were transmitted at 4 Gbit/s and received after travelling the phenomenal distance of 117 km without regeneration. The importance of the elimination of repeaters cannot be overstressed—not only does this lessen the need for fleets of maintenance engineers for land-based systems, but it drastically improves the reliability of undersea links, which can only be repaired at quite considerable expense (or not at all!). The source used was a high-power InGaAsP/InP buried heterostructure laser, operating at 1.53 μm wavelength and externally modulated by a travelling-wave Ti:$LiNbO_3$ directional coupler. The detector was an InGaAs avalanche photodiode with a GaAs FET preamplifier, coupled to the fibre using a GRIN-rod lens. This type of system might well act as the basic trunk route of a future long-distance telecommunications network.

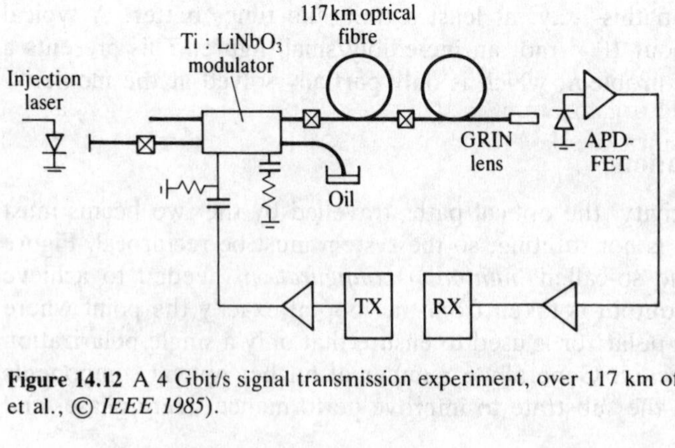

Figure 14.12 A 4 Gbit/s signal transmission experiment, over 117 km of optical fibre (*after S.K. Korotky* et al., © *IEEE 1985*).

Bidirectional links are clearly feasible using similar technology, and a variety of demonstration land-lines have now been installed for assessment and characterization. However, no single user is likely to require access to the entire bandwidth of an optical fibre, which is vast. Instead, this will be shared between a number of users, or multiplexed. The two major strategies used are *wavelength-division multiplexing* (WDM), whereby each user is assigned a specific wavelength interval with which to communicate, and *time-division multiplexing* (TDM), where each user is instead offered a set fraction of the available time. Once again we can do little in the space available here to describe the advantages and disadvantages of each method in depth—the reader is best advised to consult a specialized optical communications text for details. Instead, we offer examples of the implementation of each method, using some of the hardware we have encountered so far.

Optical Wavelength-division Multiplexing Devices

We begin with wavelength-division multiplexing. Figure 14.13 shows the main features of an optical WDM system, which operates as follows. A set of N users are first assigned wavelengths, $\lambda_1, \lambda_2, \ldots, \lambda_N$ with which to communicate; these might be derived from a set of lasers, emitting light of different wavelengths. Each user may modulate his source independently, so that wavelength λ_n acts as a carrier of information for Channel #*n*. Data from all N channels is combined together by the multiplexer (MUX), and then passed to a common channel (a single optical fibre) for transmission. At the far end, the signals are separated by the demultiplexer (DEMUX), and forwarded to their final destinations.

The MUX component may consist of a set of discrete optical filters arranged in cascade, as shown in Fig. 14.14(a). For example, a set of asymmetric directional coupler filters might be used as the basic combiner element; alternatives include mode-conversion filters and Bragg gratings. Each filter is arranged to couple maximum power to the common channel at a different wavelength, so that Filter #*n* has a response that peaks at wavelength λ_n (as in Fig. 14.14(b)). In this way, all N channels may be combined into a single output without loss of power. The DEMUX component may be a similar network, operating backwards.

Alternatively, the multiplexer may be a single dispersive component, which effectively provides all N filters in parallel. This is normally based on a grating of some kind. For example, Fig. 14.15 shows a prototype four-channel optical WDM chip fabricated by the LETI organization, France. The chip is formed from a silicon substrate, carrying three silica overlayers fabricated by plasma-enhanced chemical

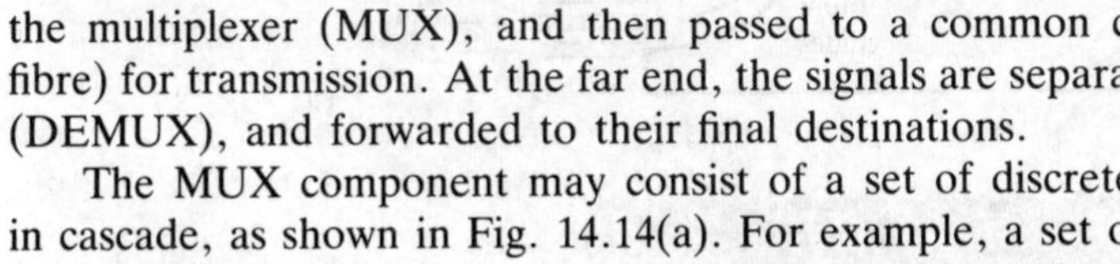

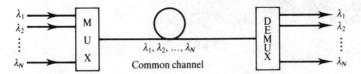

Figure 14.13 The principle of an optical WDM communications system.

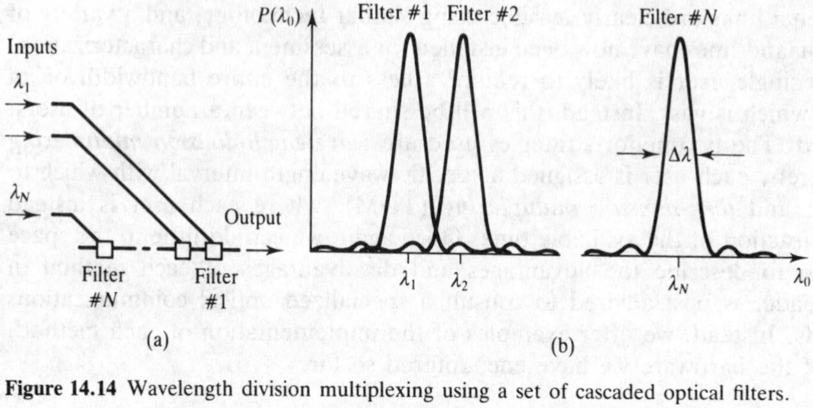

Figure 14.14 Wavelength division multiplexing using a set of cascaded optical filters.

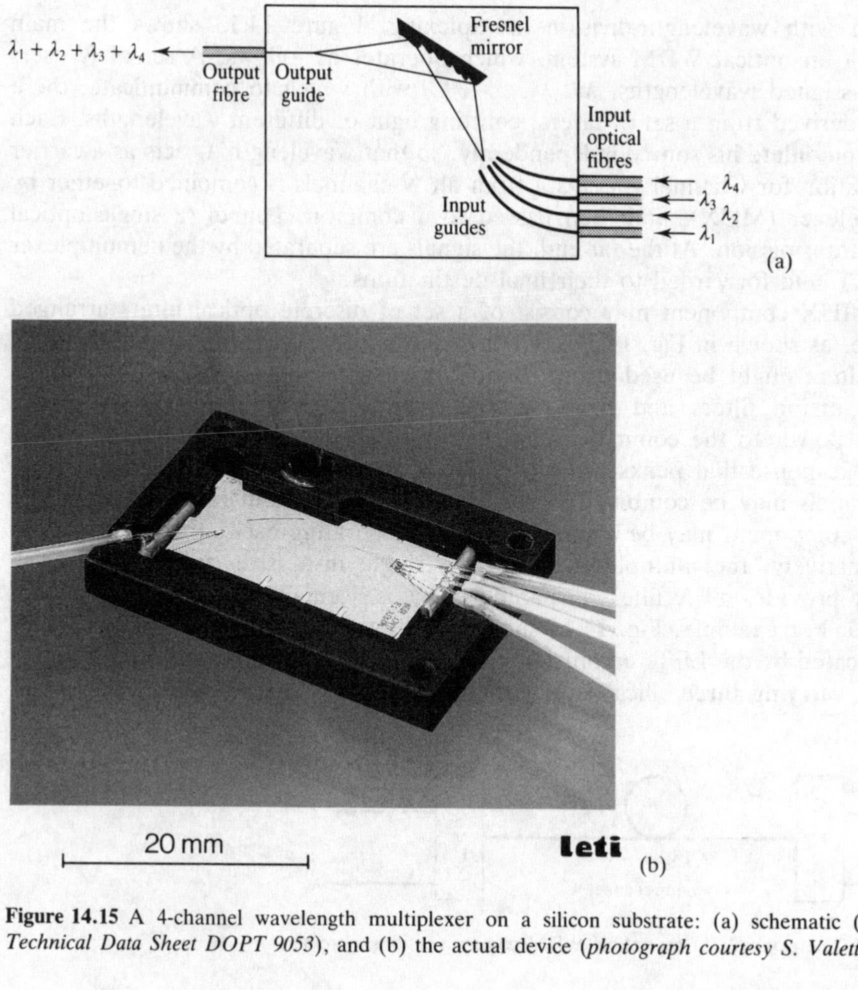

Figure 14.15 A 4-channel wavelength multiplexer on a silicon substrate: (a) schematic (*after LETI Technical Data Sheet DOPT 9053*), and (b) the actual device (*photograph courtesy S. Valette, LETI*).

vapour deposition. The lowest layer acts as a low-index buffer layer, separating the guiding layer from the high-index substrate. The middle (guiding) layer is phosphorus-doped, to increase its refractive index by $\Delta n \approx 7 \times 10^{-3}$ over that of undoped SiO_2. This layer is etched into ridges near the chip edges before the deposition of the upper layer, which acts as a cladding. The input and output channels therefore consist of buried ridge guides, which interface to a planar guiding region near the chip centre and to optical fibres at the chip edge. The fibres are automatically aligned in the correct position by grooves anisotropically etched directly into the silicon substrate.

After leaving the input channel guides, the input beams propagate as diverging cylindrical waves, which are combined into a single cylindrical beam converging on the output channel guide by a dispersive imaging device. This is a *Fresnel mirror* (the reflective equivalent of a Fresnel lens), which provides a wavelength-dependent, reflective focusing action. The mirror is formed by etching a blazed pattern right through the silica overlayers to the silicon substrate, so that total internal reflection is obtained at the silica/air interface. The chip combines four optical channels into a single output; the centre wavelength is 1.545 μm, and the channel separation is 20 nm. Operating in the reverse direction, the chip can separate an input containing all four wavelengths into individual outputs. The insertion loss is less than 6 dB, while the crosstalk between channels is better than 20 dB.

The advantage of wavelength-division multiplexing is that it is asynchronous, so that each user may have access to the common channel instantaneously. It is an efficient process, in that the total bandwidth required is approximately N times that of the individual channels. The disadvantage is the need to fabricate the MUX and DEMUX components. This may be an extremely demanding task if the number of channels is large and the centre wavelengths are closely-spaced.

Optical Time-division Multiplexing Devices

We now consider the alternative of a time-division multiplexed communications system, the main features of which are shown in Fig. 14.16. This time the N users all communicate on the same carrier wavelength. However, they may no longer have access to the shared channel at will; instead, they are allocated specific time-slots. In Fig. 14.16, Channel #1 is assigned to the slot centred on time t_1, Channel #2 to that on time t_2, and so on. Data from all N users is combined together by the multiplexer, which interleaves the time-slots of the individual channels so that the overall data rate is uniform, and then passed to the common channel for transmission. At the far end the signals are separated out again by the demultiplexer. This is done by sampling and redirecting the data at exactly the rate and sequence of the original interleaving process. Consequently, synchronization information must also be passed between the MUX and DEMUX components.

Optical time-division multiplexers and demultiplexers may be constructed from a set of channel waveguide switches, arranged to connect a set of inputs to a single output in turn. For example, Fig. 14.17 shows a 1 $\times$ 4 demultiplexer fabricated on a $LiNbO_3$ substrate using three electro-optic Mach-Zehnder interferometric (MZI)

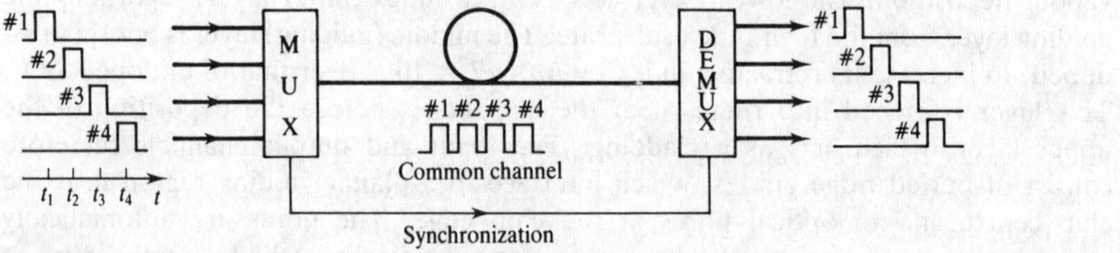

Figure 14.16 The principle of an optical TDM communications system.

switches. These are modified versions of the devices described in Chapter 9; at their output ends a hybrid X-junction has been used instead of a Y-junction. This works like a 3 dB coupler, and allows each interferometer to function as a two-way switch, rather than as a simple modulator. In fact, the precise device used is unimportant—the MZI could equally well be replaced by a directional coupler. The important feature is that the switch can act as a high-speed 'gate' when driven by a sinusoidal signal. This is clearly true for the Mach-Zehnder switch, which will give an output from one arm when the driving voltage is close to zero, and from the other when it is close to that required for $\pm\ \pi$ radians phase shift between the arms. It is also the case for a directional coupler, which will give a cross-coupled output only when the voltage is close to zero.

The optical input is shown at the left, and is assumed to consist of a train of pulses at regular intervals. These are to be distributed cyclically to each of the outputs in turn, following a clocking sequence. The switches are arranged as a binary tree, and are driven from a common signal generator synchronized to the

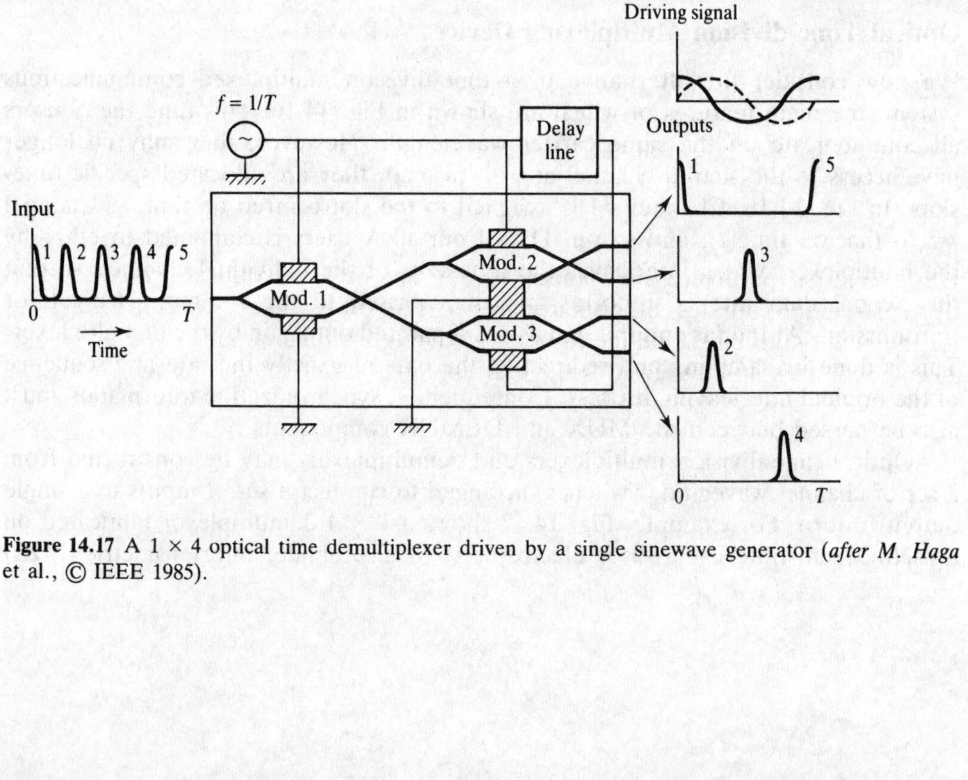

Figure 14.17 A 1 × 4 optical time demultiplexer driven by a single sinewave generator (*after M. Haga et al.*, © IEEE 1985).

input data stream and having a period of T seconds, equivalent to the time frame occupied by four input pulses (this driving signal is shown as the full line in Fig. 14.17).

There are slight, but important, differences between the switches. Modulator #1 has no optical bias, and its electrode length is chosen so that the peak driving voltage causes a $\pm\ \pi$ radians shift. An output is therefore obtained from its upper arm near the times $t = 0$, $T/2$, T and so on. Similarly, light passes to the lower arm at $t = T/4$, $3T/4$, etc. These outputs are used to feed the second-stage devices in the tree. Modulator #2, on the other hand, is arranged to have a static phase bias of $\pi/2$ radians between its arms, and its electrodes are half as long as those of modulator #1. As a result, its sensitivity to voltage is halved, and the effect of the voltage can only be an additional $\pm\ \pi/2$ phase shift. At $t = 0$ and $t = T$, the total shift is then π radians, so an output is obtained from the uppermost guide. At $t = T/2$, though, the net phase shift is zero, so the output emerges from the second uppermost guide. The operation of modulator #3 is similar, but an electrical delay line is used to shift the phase of the driving signal (shown dashed in Fig. 14.17) through 90°. Consequently, outputs are obtained from the second-lowest and lowest guides at $t = T/4$ and $t = 3T/4$, respectively.

The net effect is to distribute the pulsed input among the four outputs in turn. Experimentally, the chip was driven by a 1 GHz signal, and was able to demultiplex an input optical signal from 4 Gbit/s to 1 Gbit/s. Alternatively, the device could have been run backwards, and used to combine four inputs as a single output data stream. This demonstrates the impressive capability of integrated guided wave components for high-speed data handling.

14.9 FIBRE LASERS AND AMPLIFIERS

Our final example is rather different. In Chapter 12 we described the operation of semiconductor lasers and travelling-wave amplifiers. Here we will discuss more briefly some similar devices based on rare-earth-doped optical fibre, which appear very promising for use in all-fibre systems.

Remember that optical gain is obtained in semiconductors by inverting the electron population between the conduction and valence bands. In electrically-pumped devices, this is achieved by minority carrier injection using a p-n junction. However, this particular mechanism is not unique, and optical gain has been demonstrated in countless other materials systems. More generally, it is not necessary for the material to be crystalline (or even solid), or for the pumping to be performed electrically; all that is required is the existence of a set of suitable energy levels and a method for inverting the population between two particular levels. Electrically-excited gas lasers and optically-pumped solid-state lasers (which operate by transitions in guest atoms embedded in crystalline or amorphous hosts) have therefore been the dominant laser types of the previous two decades. Of the former, He-Ne and Ar^+ lasers are the most common, while ruby, Nd:YAG and Nd:glass lasers typify the latter.

The inversion mechanism normally involves three or four energy levels. For example, in a three-level system, optical pumping might involve the excitation of an electron from the lowest state (level #1) to the highest (level #2) through the absorption of a photon of energy $\hbar\omega = \mathscr{E}_2 - \mathscr{E}_1$, followed by a rapid decay to a slightly lower state (level #3). If electrons have a long lifetime in level #3, it is possible to build up the population in this state by continuous pumping until it exceeds that in level #1. Gain is then available at the wavelength corresponding to the transition between levels #3 and #1.

It has recently been found possible to incorporate lasing species in the core of silica glass fibres, thus forming the fibre equivalent of a solid-state laser medium. For the active material, attention has been focused on the rare-earth elements erbium, neodymium, praesodymium and thulium. Of these, the most important are praesodymium, which may be used to provide gain at 1.3 μm wavelength, and erbium, which operates at 1.55 μm. For erbium, windows for optical pumping exist at 0.8, 0.98 and 1.48 μm wavelength, allowing the use of powerful and potentially inexpensive laser diode pump sources (especially the 0.98 μm window).

Figure 14.18 shows a typical travelling-wave *fibre amplifier* experiment involving Er^{3+}-doped fibre. Two 0.82 μm GaAs/GaAlAs lasers are used as pump sources; these inject light into opposite ends of the Er^{3+}-doped fibre, so that population inversion is maintained along its entire length. Each pump beam is coupled into the amplifier using a wavelength-selective fibre coupler; this allows the signal beam (derived from a 1.535 μm laser diode) to be combined efficiently with the pump at the input, and the amplified signal to be separated from any residual pump radiation at the output. A similar arrangement can be used with a 1.48 μm pump, although a high-performance WDM coupler is required because of the close proximity of the pump and signal wavelengths. In this example the gain medium was 90 m of GeO_2-SiO_2 fibre doped with 34 ppm of erbium. The pump threshold for transparency was 15 mW, and 22 dB of gain was obtained from a total pump power of 54 mW.

A travelling-wave fibre amplifier may be converted into a *fibre laser* through the provision of suitable feedback. To construct a linear cavity, fibre reflectors are required. Several suitable devices exist. For example, the *fibre loop mirror* may be

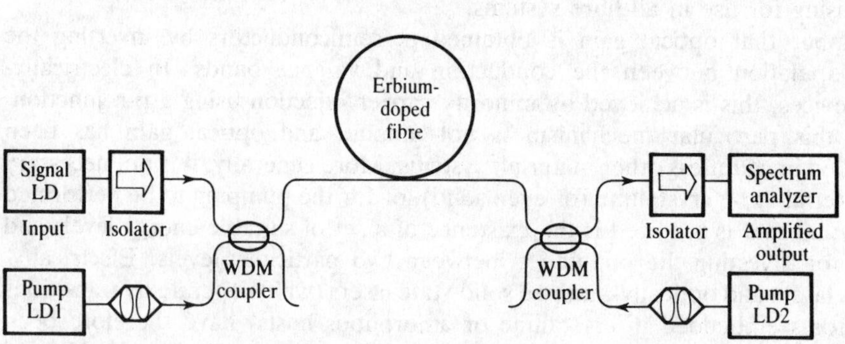

Figure 14.18 Set-up for in-line amplification using an Er^{3+}-doped fibre travelling wave amplifier (*after K. Suzuki* et al., 1990).

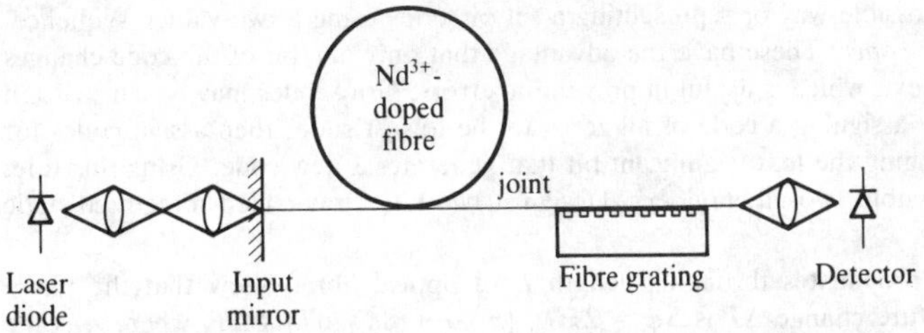

Figure 14.19 A DBR fibre laser (*after I.M. Jauncey* et al., *1986).*

used as a broad-band reflector in a Fabry-Perot cavity. Alternatively, a *fibre grating* (either based on half of a polished coupler, or on an intracore grating recorded holographically in a photosensitive fibre) might be used as a distributed Bragg reflector, providing wavelength-selective feedback and hence narrow-linewidth operation.

Figure 14.19 shows a typical example of a DBR fibre laser based on a polished fibre coupler grating. This time the active medium is a 6-metre length of silica fibre doped with 330 ppm of neodymium, and the pump laser is a 0.83 μm GaAs/GaAlAs laser diode coupled into the fibre through a wavelength-selective mirror. The threshold for laser operation was 2 mW, and approximately 1.5 mW of output power was obtained for a pump power of 10 mW at the fibre laser wavelength of 1.084 μm.

PROBLEMS

14.1 An integrated optic RF spectrum analyser is to be constructed on a $LiNbO_3$ substrate, with a tilted interdigital transducer. Assuming that the optical wavelength is 0.633 μm, calculate the angle of transducer tilt required for optimum operation in the Bragg diffraction regime at the transducer centre frequency of 500 MHz. If the transducer bandwidth is 200 MHz, how wide may the acoustic beam be before Bragg selectivity limits the diffraction efficiency? The effective index of the Ti:$LiNbO_3$ waveguide is 2.2, and the acoustic velocity is 6570 m/s.
[0.627°; 7.5 mm]

14.2 The figure below shows a small portion of a conducting path on a VLSI electronic circuit. The conductor has cross-sectional area $A = w \times t$ and length L, and is separated from the ground plane by an insulator of thickness h. Show that, if all the linear dimensions of the circuit are scaled by α, the RC time constant associated with the path remains unchanged. Why is this a potential problem? Describe several advantages of guided wave optical interconnects over this type of circuit.

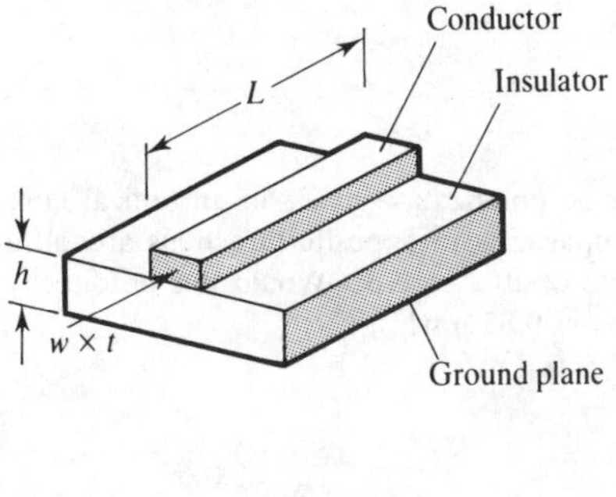

14.3 Binary codes are just one possible way of representing a set of states using a two-valued sequence. An alternative is offered by *Gray codes*. These have the advantage that only one bit of the code changes in moving from one state to the next, which is useful in preventing errors. Gray codes may be constructed using the following rule. Start by assigning a code of all zeros to the lowest state; then assign codes for successively higher states by changing the least significant bit that generates a new code. Using this rule, write down the codes for all possible two-bit numbers. Design a two-bit, Gray-coded integrated optic A-to-D converter.

14.4 A beam of wavelength λ_0 propagates through a length L of optical fibre. Show that the phase change accompanying a temperature change ΔT is $\Delta\phi = 2\pi/\lambda_0 \{n_{eff}\alpha + dn_{eff}/dT\} L\Delta T$, where α is the thermal expansion coefficient of the fibre and n_{eff} is its effective index. Assuming that $\alpha = 0.004 \times 10^{-4}/°C$ and $dn_{eff}/dT = 1.1 \times 10^{-5}/°C$ for silica fibre, and that $n = 1.458$ for bulk silica, estimate the phase change per °C for a 10 cm length of fibre at $\lambda_0 = 0.85\ \mu m$. Which effect contributes most to the phase change, thermal expansion or refractive index variation?
[8.56 rad/°C]

14.5 The figure below shows an interferometric sensor contructed from two 3 dB couplers linked by fibres. The additional time delay experienced by a wave propagating along the sensing arm over that experienced along the reference arm is τ. Show that, for a single-frequency optical input, the output is ambiguous in τ, i.e. that the same output will be obtained for a variety of different values of τ. Show further that the sensitivity of the device to variations in τ will go to zero whenever $\omega\tau = \nu\pi$, where ω is the angular frequency of the input. When is the sensitivity maximum? Discuss the implications of these features for a practical sensing system.

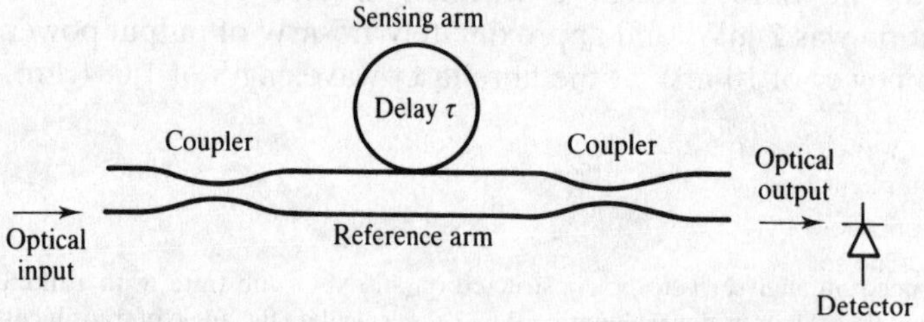

14.6 One solution to the ambiguity problem identified in Question 14.5 is to increase the bandwidth of the optical input. The figure below shows the input used in the *linear FM* method. Here, the input is frequency modulated by a sawtooth wave of period T. For $0 \leq t \leq T$, for example, the frequency variation is linear with time, following $f(t) = f_1 + (f_2 - f_1)t/T$. Assuming that $\tau < T$, calculate and sketch the response of the sensor of Question 14.5 to this input.

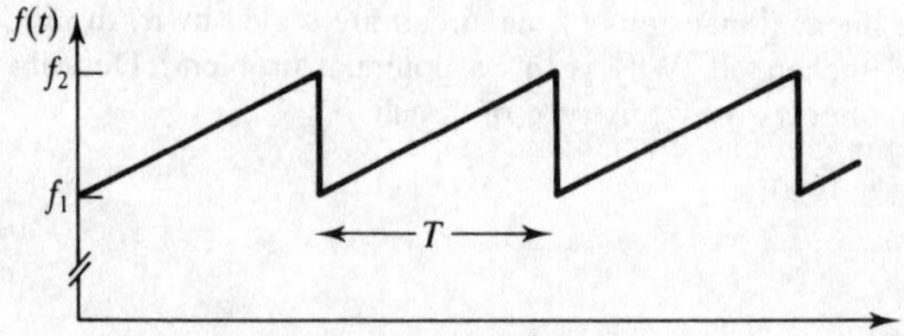

14.7 Discuss the importance of launching and detecting a single polarization mode in an optical fibre gyro. Calculate the time-of-flight difference for two beams propagating in opposite directions around a 20 cm diameter coil containing 1 km of fibre, at a rotation rate of 10^{-5} rad/sec. Would this be directly measurable? What is the corresponding phase difference, at $\lambda_0 = 0.85\ \mu m$?
[2.22×10^{-20} s; 4.93×10^{-5} rad]

14.8 (a) The transfer function $P_n(\lambda_0) = \text{sinc}^2\{\pi(\lambda_0 - \lambda_n)/\Delta\lambda\}$ is a good approximation to the response of many real filters. Sketch the variation of $P_n(\lambda_0)$ with λ_0. What does the term $\Delta\lambda$ represent?

(b) The figure below shows a demultiplexer used in a WDM system. The channels are regularly spaced at wavelengths $\lambda_1, \lambda_2, \ldots, \lambda_N$, such that $\lambda_{n+1} - \lambda_n = \Delta\Lambda$, and the peak in the passband of the n th filter is centred on λ_n. However, due to the finite extent of the filter response, a fraction of the power at wavelength λ_{n+1} will be accepted by the n th filter. Use the response above to estimate the worst-case interchannel crosstalk, assuming that $\Delta\Lambda \geq \Delta\lambda$.
[−13.46 dB]

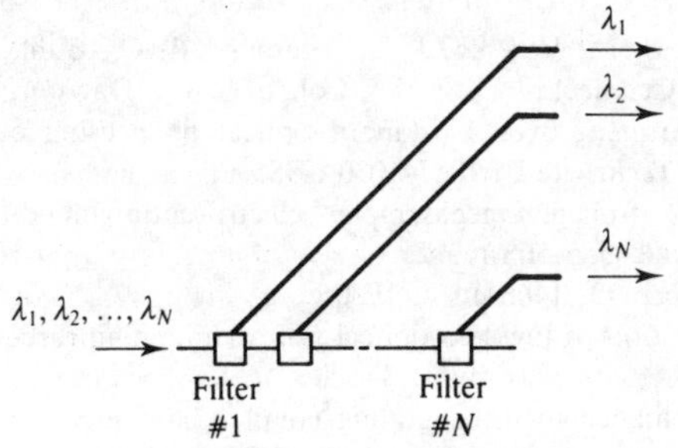

14.9 A directional coupler (with a coupling length of $\kappa L = \pi/2$) is driven by a 200 V peak-to-peak sinusoidal voltage of frequency 1 GHz. The voltage required to switch the coupler from the cross-state to the bar-state is 10 V. Draw a dimensioned sketch of the cross-coupled output with time, for a CW optical input. What application might this have in a TDM system?

14.10 A point-to-point communications link operating at $\lambda_0 = 1.55\ \mu m$ consists of 50 km of single-mode fibre with a loss of 0.2 dB/km. The source is a 10 mW laser diode, coupled to the fibre via a GRIN-rod lens. The coupling efficiency is 25 per cent. There are 10 fusion splices in the link, each contributing 0.1 dB loss. Assuming that the receiver requires a signal of 1 mW to operate, calculate the length of Er^{3+}-doped fibre required in a travelling-wave amplifier arranged as a pre-detector signal booster. When pumped at 0.82 μm wavelength the Er^{3+}-doped fibre has a gain of 0.25 dB/m.
[28 m]

SUGGESTIONS FOR FURTHER READING

Ball, G. A., Morey, W. W., and Waters, J. P. Nd^{3+} fibre laser utilising intra-core Bragg reflectors, *Elect. Lett.*, **26**, 1829–1830, 1990.

Barnes, W. L., and Townsend, J. E. Highly tunable and efficient diode pumped operation of Tm^{3+} doped fibre lasers, *Elect. Lett.*, **26**, 746–747, 1990.

Barnoski, M. K., Chen, B., Joseph, T. R., Lee, J., and Ramer, O. G. Integrated optic spectrum analyser, *IEEE Trans. Circuits and Systems*, **CAS-26**, 1113–1124, 1979.

Bergh, R. A., Lefevre, H. C., and Shaw, H. J. An overview of fiber-optic gyroscopes, *IEEE J. Lightwave Tech.*, **LT-2**, 91–107, 1984.

Bucholtz, F., and Yurek, A. M. Fiber interferometric sensors: technology and applications, *Optics News*, November Issue, 20–27, 1989.

Culshaw, B. "Optical Fibre Sensing and Signal Processing", Peter Peregrinus Ltd, London, 1984.

De La Rue, R. M., Stewart, C., Wilkinson, C. D. W., and Williamson, I. R. Frequency-controlled beam steering of surface acoustic waves using a stepped transducer array, *Elect. Lett.*, **9**, 326–327, 1973.

Giallorenzi, T. G., Bucaro, J. A., Dandridge, A., Sigel, G. H., Cole, J. H., Rashleigh, S. C., and Priest, R. G. Optical fiber sensor technology, *IEEE J. Quant. Elect.*, **QE-18**, 626–665, 1982.

Goodman, J. W., Leonberger, F. J., Kung, S. Y., and Athale, R. A. Optical interconnections for VLSI systems, *Proc IEEE*, **72**, 850–866, 1984.

Haga, M., Izutsu, M., and Sueta, T. An integrated 1 × 4 high-speed optical switch and its applications to time demultiplexer, *IEEE J. Lightwave Tech.*, **LT-3**, 116–120, 1985.

Johnson, L. M. Integrated-optical components for fiber gyroscopes, *Proc. SPIE*, **566**, 96–98, 1985.

Kersey, A. D., Dandridge, A. and Burns, W. K. Fiber optic gyroscope technology, *Optics News*, November Issue, 12–19, 1989.

Kobayashi, M., Yamada, M., Yamada, Y., Himeno, H., and Terui, H. Guided-wave optical chip-to-chip interconnections, *Elect. Lett.*, **23**, 143–144, 1987.

Korotky, S. K., Eisenstein, G., Gnauck, A. H., Kasper, B. L., Veselka, J. J., Alferness, R. C., Buhl, L. L., Burrus, C. A., Huo, T. C. D., Stulz, L. W., Ciemiecki-Nelson, K., Cohen, L. G., Dawson, R. W., Campbell, J. C. 4-Gb/s transmission experiment over 117 km of optical fiber using a Ti:$LiNbO_3$ external modulator, *IEEE J. Lightwave Tech.*, **LT-3**, 1027–1030, 1985.

Leonberger, F. J., Woodward, C. E., and Becker, R. A. 4-bit 828-megasample/s electro-optic guided-wave analog-to-digital converter, *Appl. Phys. Lett.*, **40**, 565–568, 1982.

Marcatili, E. A. J. Optical subpicosecond gate, *Appl. Opt.*, **19**, 1468–1476, 1980.

Poole, S. B., Payne, D. N., and Fermann, M. E. Fabrication of low-loss optical fibres containing rare-earth ions, *Elect. Lett.*, **21**, 737–738, 1985.

Suhara, T., Nishihara, H., and Koyama, J. High-performance focusing grating coupler fabricated by electron-beam writing, Presented at Topical Meet. on Integrated and Guided-wave Optics, Kissimme, FL, Apr. 24–27, paper ThD4, 1984.

Suzuki, K., Kimura, Y., and Nakazawa, M. High gain Er^{3+}-doped fibre amplifier pumped by 820 nm GaAlAs laser diodes, *Elect. Lett.*, **26**, 949–949, 1990.

Taylor, H. F. An optical analog-to-digital converter—design and analysis, *IEEE J. Quant. Elect.*, **QE-15**, 210–216, 1979.

Tucker, R. S., Korotky, S. K., Eisenstein, G., Koren, U., Raybon, G., Veselka, J. J., Buhl, L. L., Kasper, B. L., Alferness, R. C. 4 Gb/s optical time-division multiplexed system experiments using Ti:$LiNbO_3$ switch modulators, *Proc. OSA Top. Meet. on Photonic Switching*, Incline Village, Nevada, March 18–20, Paper FD-3, 1987.

Ura, S., Suhara, T., Nishihara, H., and Koyama, J. An integrated-optic disc pickup device, *IEEE J. Lightwave Tech.*, **LT-4**, 913–918, 1986.

Valette, S., Lizet, J., Mottier, P., Jadot, J. P., Renard, S., Fournier, A., Grouillet, A. M., Gidon, P., Denis, H. Integrated optical spectrum analyser using planar technology on oxidised silicon substrate, *Elect. Lett.*, **19**, 883–885, 1983.

Valette, S., Gidon, P., and Jadot, J. P. New integrated optical multiplexer/demultiplexer realized on Si substrate, *Proc. 4th European Conf. on Integrated Optics*, Glasgow, May 11–13, 1987.

Wagoner, R. E., and Clark, T. E. Overview of multiplexing techniques for all-fiber interferometer sensor arrays, *Proc. SPIE*, **718**, 80–91, 1986.

ANSWERS TO SELECTED PROBLEMS

Chapter 2

2.1 Start with Maxwell's equations

$$\nabla \times \mathbf{E} = -\partial\mathbf{B}/\partial t \text{ and } \nabla \times \mathbf{H} = \mathbf{J} + \partial\mathbf{D}/\partial t$$

Then assume that in a dielectric medium $\mathbf{J} = \mathbf{0}$, $\mathbf{B} = \mu_0\mathbf{H}$ and $\mathbf{D} = \varepsilon\mathbf{E}$. Hence we can say that

$$\nabla \times \mathbf{E} = -\mu_0\, \partial\mathbf{H}/\partial t \qquad (1)$$

$$\nabla \times \mathbf{H} = \varepsilon\partial\mathbf{E}/\partial t \qquad (2)$$

Now take the curl of Eq. (2), to get

$$\nabla \times (\nabla \times \mathbf{H}) = \nabla \times (\varepsilon\partial\mathbf{E}/\partial t)$$

Furthermore, if we assume that ε is constant, we have

$$\nabla \times (\nabla \times \mathbf{H}) = \varepsilon\{\nabla \times (\partial\mathbf{E}/\partial t)\} \qquad (3)$$

Reversing the order of differentiation in Eq. (3), we get

$$\nabla \times (\nabla \times \mathbf{H}) = \varepsilon\{\partial(\nabla \times \mathbf{E})\partial t\} \qquad (4)$$

Substituting for $\nabla \times \mathbf{E}$ from Eq. (1) then gives

$$\nabla \times (\nabla \times \mathbf{H}) = -\mu_0\varepsilon\partial^2\mathbf{H}/\partial t^2$$

Now, the following standard identity exists

$$\nabla \times (\nabla \times \mathbf{H}) = \nabla \cdot \nabla\mathbf{H} - \nabla^2\mathbf{H}$$

Since $\nabla\mathbf{B} = 0$, it follows that $\nabla(\mu_0\mathbf{H}) = 0$, so $\nabla\mathbf{H} = 0$ and $\nabla \times (\nabla \times \mathbf{H}) = -\nabla^2\mathbf{H}$. Hence, the required vector wave equation in $\mathbf{H}$ alone is

$$\nabla^2\mathbf{H} = \mu_0\varepsilon\, \partial^2\mathbf{H}/\partial t^2.$$

2.2 Referring to the figure below:

(a) $v_{ph} = c = f\lambda_0$ so that $\lambda_0 = (3 \times 10^8 \times 2\pi)/(2.978 \times 10^{15})$ m $= 0.633 \times 10^{-6}$ m.

(b) $\lambda = \lambda_0/n = \lambda_0/\sqrt{\varepsilon} = 0.633 \times 10^{-6}/\sqrt{2.25}$ m $= 0.422 \times 10^{-6}$ m.

2.3 $\mathbf{E} = E_{y0} \exp\{-jk_0(ax + bz)\}\, \mathbf{j}$ so that $\mathbf{E} = E_{y0} \exp\{-jk_0 n(x \sin\theta + z \cos\theta)\}\, \mathbf{j}$ where

(a) E_{y0} is the wave amplitude.

(b) The E-vector points in the **j**-direction, so the wave is polarized in the **j**-direction.

(c) There is no y-variation, so the wave is travelling in the x–z plane; the angle of propagation relative to the x-axis is $\theta = \tan^{-1}(a/b) = \tan^{-1}(1/\sqrt{3}) = 30°$.

(d) $n = a/\sin\theta = 1/0.5 = 2$.

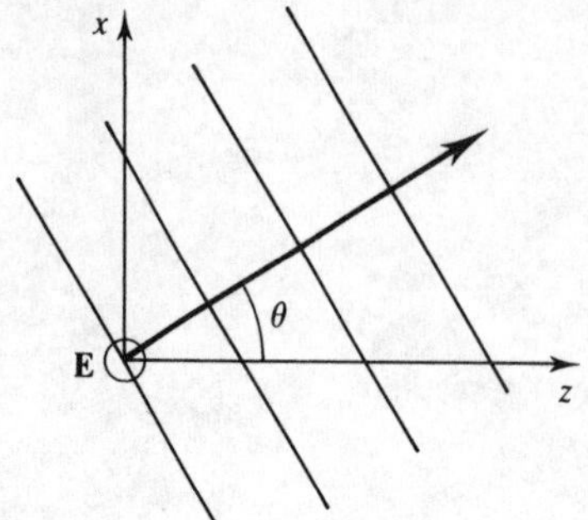

2.4 $k^2 = \omega^2\mu_0\varepsilon$ so that $k = \omega\surd(\mu_0\varepsilon)$, where k is the propagation constant. If

$$E_y = E_{y0} \exp(\gamma x) \exp(-\mathrm{j}\beta z)$$

then

$$\partial E_y/\partial x = \gamma E_{y0} \exp(\gamma x) \exp(-\mathrm{j}\beta z) \text{ and } \partial^2 E_y/\partial x^2 = \gamma^2 E_{y0} \exp(\gamma x) \exp(-\mathrm{j}\beta z)$$

and

$$\partial E_y/\partial z = -\mathrm{j}\beta E_{y0} \exp(\gamma x) \exp(-\mathrm{j}\beta z) \text{ and } \partial^2 E_y/\partial x^2 = -\beta^2 E_{y0} \exp(\gamma x) \exp(-\mathrm{j}\beta z)$$

If:

$$\nabla^2 E_y + k^2 E_y = 0$$

then

$$\gamma^2 - \beta^2 + k^2 = 0, \text{ so } \gamma = \surd(\beta^2 - k^2)$$

2.5 The wave is travelling in the z-direction; its amplitude varies exponentially in the x-direction. If γ is to be real, we must have $\beta > k$. Hence $v_{ph} = \omega/\beta < \omega/k$, so the wave travels slower than a plane wave travelling in the z-direction. This type of wave is an evanescent or boundary wave.

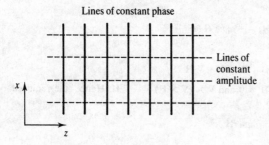

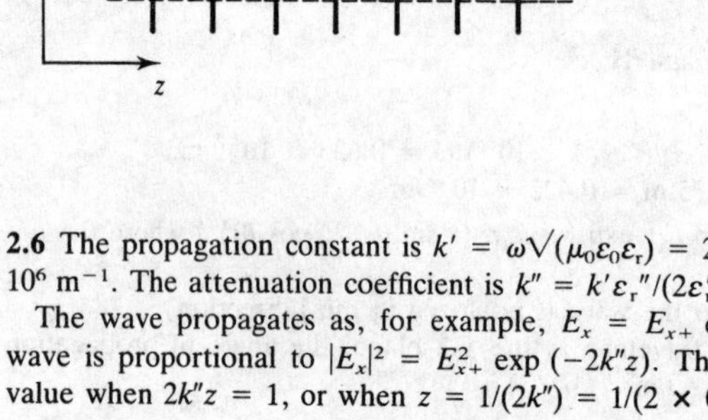

2.6 The propagation constant is $k' = \omega\surd(\mu_0\varepsilon_0\varepsilon_r) = 2\pi\surd\varepsilon_r'/\lambda_0 = 2\pi\ \surd(2.25)/0.633 \times 10^{-6} = 14.89 \times 10^6\ \mathrm{m}^{-1}$. The attenuation coefficient is $k'' = k'\varepsilon_r''/(2\varepsilon_r') = 14.89 \times 10^6 \times 10^8/(2 \times 2.25) = 0.0331\ \mathrm{m}^{-1}$.

The wave propagates as, for example, $E_x = E_{x+} \exp(k''z) \exp(-\mathrm{j}k'z)$. The power carried by the wave is proportional to $|E_x|^2 = E_{x+}^2 \exp(-2k''z)$. The power will therefore decay to 1/e of an initial value when $2k''z = 1$, or when $z = 1/(2k'') = 1/(2 \times 0.0331) = 15.11$ m.

2.7 The time averaged Poynting vector is $\mathbf{S} = (1/T)\int_0^T (\mathbf{E}\times\mathbf{H})\,dt$. Put

$$\mathbf{E} = \text{Re}\,\{\mathbf{E}\exp(j\omega t)\} \text{ and } \mathbf{H} = \text{Re}\,\{\mathbf{H}\exp(j\omega t)\}$$

Now

$$\text{Re}\,(z) = 1/2\,\{z + z^*\}$$

hence

$$\mathbf{S} = (1/4T)\int_0^T \{\mathbf{E}\exp(j\omega t) + \mathbf{E}^*\exp(-j\omega t)\}\times\{\mathbf{H}\exp(j\omega t) + \mathbf{H}^*\exp(-j\omega t)\}\,dt$$

so

$$\mathbf{S} = (1/4T)\int_0^T \{(\mathbf{E}\times\mathbf{H})\exp(2j\omega t) + (\mathbf{E}^*\times\mathbf{H}^*)\exp(-2j\omega t) + (\mathbf{E}\times\mathbf{H}^*) + (\mathbf{E}^*\times\mathbf{H})\}\,dt$$

or

$$\mathbf{S} = (1/2T)\,\text{Re}[\int_0^T \{(\mathbf{E}\times\mathbf{H})\cos(2\omega t) + (\mathbf{E}\times\mathbf{H}^*)\}\,dt]$$

The first term in the integral averages to zero, if T is long enough, leaving $\mathbf{S} = (1/2)\,\text{Re}\,(\mathbf{E}\times\mathbf{H}^*)$.

2.8 Assume that

$$\mathbf{E} = E_{yo}\exp\{-jk_0 n(x\sin\theta + z\cos\theta)\}\,\mathbf{j} = E_y\,\mathbf{j}$$

Then

$$\nabla\times\mathbf{E} = -\partial E_y/\partial z\,\mathbf{i} + \partial E_y/\partial x\,\mathbf{k} = -j\omega\mu_0\mathbf{H}$$

hence

$$H_x = -(k_0 n\cos(\theta)\,\omega\mu_0)\,E_y;\; H_y = 0;\; H_z = (k_0 n\sin(\theta)\,\omega\mu_0)\,E_y$$

so

$$\mathbf{S} = (1/2)\,\text{Re}\,(\mathbf{E}\times\mathbf{H}^*) = (1/2)\,\text{Re}\,\{(E_y\mathbf{H}_z^*)\,\mathbf{i} - (E_y H_x^*)\,\mathbf{k}\}$$

Substituting the relevant components of **E** and **H** then gives

$$\mathbf{S} = (k_0 n E_{yo}^2/2\omega\mu_0)\,\{\mathbf{i}\sin\theta + \mathbf{k}\cos\theta)$$

or

$$\mathbf{S} = 1/2\,\surd(\varepsilon/\mu_0)\,E_{yo}^2\,\{\mathbf{i}\sin\theta + \mathbf{k}\cos\theta)$$

Hence

$$S_z = 1/2\,\surd(\varepsilon/\mu_0)\,E_{yo}^2\cos\theta$$

Given that $\surd(\varepsilon_0/\mu_0) = 377\Omega$, $n = \surd\varepsilon_r = 2$ and $\theta = 30°$, we find that $S_z = 2.3\times 10^{-3}\,E_{yo}^2$ W/m².

The direction of maximum irradiance is parallel to the vector $\mathbf{i}\sin\theta + \mathbf{k}\cos\theta$; the maximum irradiance is given by

$$|\mathbf{S}| = 1/2\,\surd(\varepsilon/\mu_0)\,E_{yo}^2 = 2.65\times 10^{-3}\,E_{yo}^2 \text{ W/m}^2.$$

2.9 $|\mathbf{S}| = 1/2\,\surd(\varepsilon_0/\mu_0)\,E^2 = 1000$ W/m². Hence $E = \surd(2000\times 377)$ V/m $= 868.3$ V/m.

2.10 If the dispersion characteristic is $\omega = \surd[\omega_p^2 + c^2k^2]$, then $v_{ph} = \omega/k = \surd[(\omega_p^2/k^2) + c^2]$ and $v_g = d\omega/dk = c^2/\surd[(\omega_p^2/k^2) + c^2]$. When $\omega\to\omega_p$, $k\to 0$, so $v_g\to 0$. This implies that the velocity of propagation of information will tend to zero as $\omega\to\omega_p$.

Rearranging the dispersion characteristic, we get

$$k = (1/c)\,\surd(\omega^2 - \omega_p^2).$$

When $\omega < \omega_p$, k must be purely imaginary, so we can write: $k = -jk''$. Hence a z-propagating wave will have the form

$$E = E_0 \exp[j(\omega t - kz)] = E_0 \exp(j\omega t) \exp(-k''z).$$

This solution does not represent a travelling wave, but one that decays exponentially with distance.

Chapter 3

3.1 At $\lambda_0 = 1.3\ \mu$m:

$$n'^2 = 1 + 0.696\,166\,3 \times 1.3^2/(1.3^2 - 0.068\,404\,3^2) + 0.407\,942\,6 \times 1.3^2/(1.3^2 - 0.116\,241\,4^2) + 0.897\,479\,4 \times 1.3^2/(1.3^2 - 9.896\,161\,0^2) \Rightarrow n' = 1.449\,691\,75$$

At $\lambda_0 = 1.55\ \mu$m:

$$n'^2 = 1 + 0.696\,166\,3 \times 1.55^2/(1.55^2 - 0.068\,404\,3^2) + 0.407\,942\,6 \times 1.55^2/(1.55^2 - 0.116\,241\,4^2) + 0.897\,479\,4 \times 1.55^2/(1.55^2 - 9.896\,161\,0^2) \Rightarrow n' = 1.444\,023\,6.$$

3.2 For a single resonance, Sellmeier's equation reduces to

$$n'^2 = 1 + A_1/\{1 - (\lambda_1/\lambda_0)^2\}$$

Assuming that $\lambda_0 \gg \lambda_1$ (λ_0 a visible wavelength, λ_1 in the UV), this may be expanded as a power series:

$$n'^2 = 1 + A_1\{1 + (\lambda_1/\lambda_0)^2 + (\lambda_1/\lambda_0)^4 \ldots\}.$$

For small A_1, we may then approximate this as

$$n' \approx 1 + (A_1/2)\{1 + (\lambda_1/\lambda_0)^2 + (\lambda_1/\lambda_0)^4 \ldots\}.$$

This has the same general form as Cauchy's equation $n' \approx A + (B/\lambda_0^2) + (C/\lambda_0^4)$, with $A = 1 + A_1/2$, $B = A_1\lambda_1^2/2$ and $C = A_1\lambda_1^4/2$. Since Cauchy's equation contains only even powers of λ_0, it cannot describe the whole of the dispersion curve, e.g. the region of anomalous dispersion.

3.3 $N = \omega_p^2 m\varepsilon_0/e^2 = (12.2 \times 10^{15})^2 \times 9.1 \times 10^{-31} \times 8.85 \times 10^{-12}/(1.6 \times 10^{-19})^2 = 4.68 \times 10^{28}/\text{m}^3$. The real part of the relative dielectric constant of a metal is given by

$$\varepsilon_r' = 1 - \omega_p^2/(\omega^2 + \omega_\tau^2)$$

For $\varepsilon_r' < 0$, we require

$$\omega^2 < \omega_p^2 - \omega_\tau^2, \text{ or } \omega < \sqrt{\{(12.2 \times 10^{15})^2 - (0.09 \times 10^{15})^2\}} = 12.1996 \times 10^{15}\text{rad/s}.$$

Hence $\lambda_0 > 2\pi c/\omega = 1.545 \times 10^{-7}$ m.

3.4 The behaviour of an electron in the ionized gas is essentially similar to that in a metal, except that collision damping may be neglected (since there is no lattice, and the gas is low-pressure). We may therefore deduce from previous results that $\varepsilon_r' = 1 - \omega_p^2/\omega^2$. Since $k^2 = \omega^2\mu_0\varepsilon$, we then get $k^2 = \omega^2\mu_0\varepsilon_0\{1 - \omega_p^2/\omega^2\}$. Now $c = 1/\sqrt{(\mu_0\varepsilon_0)}$, so this may be rearranged to give the required result, $\omega = \sqrt{\{\omega_p^2 + c^2k^2\}}$.

3.5 Start by assuming that if contributions from bound and free electrons are both included, the relative dielectric constant can be written as $\varepsilon_r \approx \varepsilon_{r\infty} - j\sigma/\varepsilon_0\omega$, where $\sigma = Ne^2/(r + jm^*\omega)$. Hence

$$\varepsilon_r \approx \varepsilon_{r\infty} - jNe^2/\{m^*\varepsilon_0\omega(\omega_\tau + j\omega)\}$$

where $\omega_\tau = r/m^*$. Separating this into real and imaginary parts, we get

$$\varepsilon_r \approx \varepsilon_{r\infty} - Ne^2/\{m^*\varepsilon_0(\omega_\tau^2 + \omega^2)\} - jNe^2\omega_\tau/\{m^*\varepsilon_0(\omega_\tau^2 + \omega^2)\omega\} = \varepsilon_r' - j\varepsilon_r''$$

The introduction of free carriers therefore results in a non-zero value of ε_r''; consequently, if an attempt is made to modify the refractive index of a semiconductor by free carrier injection, there will also be significant optical loss.

3.6 The relative retardation of the two polarization components by the birefringent wave plate is

$$\phi_x - \phi_y = 2\pi(n_o - n_e)d/\lambda_0 = 2\pi(2.286 - 2.2) \times 3.684 \times 10^{-3}/0.663 \times 10^{-6} = 2\pi(m + 0.5)$$

where $m = 500$. Hence, the birefringent plate is a half-wave plate, whose action is to rotate the input polarization by 2θ, where θ is the rotation of plate relative to the x-axis. If transmission is initially maximum, the input polarization and the axis of polarizer must be aligned parallel to x-axis. Hence (a) if $\theta = 22.5°$, the polarization will be rotated by 45°, so the resulting field will have equal components parallel and perpendicular to the polarizer axis. The perpendicular component will be blocked, giving a transmission of 50 per cent. (b) if $\theta = 45°$, the polarization will be rotated by 90°, and so will be entirely perpendicular to the axis of the polarizer.

3.7 If the coordinates are represented by a radius vector **r**, and the B-coefficients are assembled into a symmetric, second-rank tensor [B], we may write:

$$[x\ y\ z]\begin{bmatrix} B_{xx} & B_{xy} & B_{xz} \\ B_{xy} & B_{yy} & B_{yz} \\ B_{xz} & B_{yz} & B_{zz} \end{bmatrix}\begin{bmatrix} x \\ y \\ z \end{bmatrix} = [x\ y\ z]\begin{bmatrix} B_{xx}x & + B_{xy}y & + B_{xz}z \\ B_{xy}x & + B_{yy}y & + B_{yz}z \\ B_{xz}x & + B_{yz}y & + B_{zz}z \end{bmatrix}$$

$$= B_{xx}x^2 + B_{yy}y^2 + B_{zz}z^2 + 2B_{xy}xy + 2B_{yz}yz + 2B_{zx}zx$$

Hence, if $B_{xx}x^2 + B_{yy}y^2 + B_{zz}z^2 + 2B_{xy}xy + 2B_{yz}yz + 2B_{zx}zx = 1$, then $\mathbf{r}^T\ [B]\ \mathbf{r} = 1$.

3.8 The individual eigenvalues and eigenvectors of the second-rank tensor $[B]$ are defined by three matrix/vector equations of the form $[B]\ \mathbf{v}_i = \lambda_i\mathbf{v}_i$. These may all be combined into a single matrix equation $[B]\ [V] = [V]\ [\Lambda]$, where $[\Lambda]$ is a diagonal matrix whose elements are λ_i, and $[V]$ contains the vectors $\mathbf{v}_i$ arranged in columns.

Post-multiplying by $[V]^{-1}$, we then obtain $[B]\ [V]\ [V]^{-1} = [B] = [V]\ [\Lambda]\ [V]^{-1}$. However, since the eigenvectors $\mathbf{v}_i$ are orthonormal, we note that $[V]^{-1} = [V]^T$, so the matrix $[B]$ can be written $[B] = [V]\ [\Lambda]\ [V]^T$. Consequently, the equation for the index ellipsoid may be written in the new quadratic form $\mathbf{r}^T\ [B]\ \mathbf{r} = \mathbf{r}^T\ [V]\ [\Lambda]\ [V]^T\ \mathbf{r} = 1$. This may be written as $\mathbf{r}'^T\ [\Lambda]\ \mathbf{r}' = 1$, if a new set of coordinates are defined according to $\mathbf{r}' = [V]^T\ \mathbf{r}$. The equation of the index indicatrix is thus $\lambda_1\ x'^2 + \lambda_2\ y'^2 + \lambda_3 z'^2 = 1$. This is the equation of an ellipsoid aligned to the x', y' and z' axes, and with semi-major axes of lengths $1/\sqrt{\lambda_1}$, $1/\sqrt{\lambda_2}$ and $1/\sqrt{\lambda_3}$.

3.9 KDP is a tetragonal crystal, of point group $\bar{4}2m$. Consequently, the only significant electro-optic coefficients are r_{41} and r_{63}. For an electric field applied in the z-direction, the changes to the B-coefficients are given by

$$\begin{bmatrix} \Delta B_1 \\ \Delta B_2 \\ \Delta B_3 \\ \Delta B_4 \\ \Delta B_5 \\ \Delta B_6 \end{bmatrix} = \begin{bmatrix} 0 & 0 & 0 \\ 0 & 0 & 0 \\ 0 & 0 & 0 \\ r_{41} & 0 & 0 \\ 0 & r_{41} & 0 \\ 0 & 0 & r_{63} \end{bmatrix}\begin{bmatrix} 0 \\ 0 \\ E_z \end{bmatrix}$$

Only the coefficient B_6 (which is the same as B_{xy}) changes, by an amount $r_{63}E_z$. Assuming that the equation of the unperturbed index ellipsoid was $B_{xx}x^2 + B_{yy}y^2 + B_{zz}z^2 = 1$, the new equation is $B_{xx}x^2 + B_{yy}y^2 + B_{zz}z^2 + 2r_{63}E_zxy = 1$.

3.10 The equation of the unperturbed index ellipsoid may be written in the quadratic form of Questions 3.7 and 3.8, with the second-rank tensor $[B]$ given by

$$[B] = \begin{bmatrix} B_{xx} & 0 & 0 \\ 0 & B_{yy} & 0 \\ 0 & 0 & B_{zz} \end{bmatrix}$$

After application of the electric field, the new tensor $[B']$ has the form:

$$[B'] = \begin{bmatrix} B_{xx} & 2r_{63}E_z & 0 \\ 2r_{63}E_z & B_{yy} & 0 \\ 0 & 0 & B_{zz} \end{bmatrix}$$

Now, the eigenvalues of $[B']$ are defined by the determinantal equation:

$$\begin{vmatrix} B_{xx} - \lambda & 2r_{63}E_z & 0 \\ 2r_{63}E_z & B_{yy} - \lambda & 0 \\ 0 & 0 & B_{zz} - \lambda \end{vmatrix} = 0$$

which may be expanded to give:

$$(B_{xx} - \lambda)(B_{yy} - \lambda)(B_{zz} - \lambda) - 4r_{63}^2 E_z^2 (B_{zz} - \lambda) = 0$$

or

$$(B_{zz} - \lambda)\{\lambda^2 - (B_{xx} + B_{yy})\lambda + (B_{xx}B_{yy} - 4r_{63}^2 E_z^2)\} = 0$$

To first-order in E_z, this has the solutions $\lambda = B_{xx}, B_{yy}, B_{zz}$. To a reasonable approximation, the eigenvalues are therefore unaltered by the application of the field, so the values $1/\sqrt{\lambda_1}$, $1/\sqrt{\lambda_2}$ and $1/\sqrt{\lambda_3}$ (the lengths of the semi-major axes of the index ellipsoid) are also unchanged.

Chapter 4

4.1 Use the lens maker's formula $1/f = (n - 1)(1/r_1 + 1/r_2)$ with $n = 1.5$.
(a) $1/f = 0.5\,(1/200 + 1/200) = 1/200 \Rightarrow f = 200$ mm.
(b) $1/f = 0.5\,(1/400 + 1/\infty) = 1/800 \Rightarrow f = 800$ mm.

4.2 Use the imaging formula $1/u + 1/v = 1/f$ so that $1/v = 1/f - 1/u$.
(a) $1/v = 1/200 - 1/400 = 1/400 \Rightarrow v = 400$ mm (real image).
(b) $1/v = 1/800 - 1/400 = -1/800 \Rightarrow v = -800$ mm (virtual image).

4.3 First define two coordinate systems (x,y,z) and (x',y',z') as shown below. The former are centred axially, the latter on the off-axis image point.

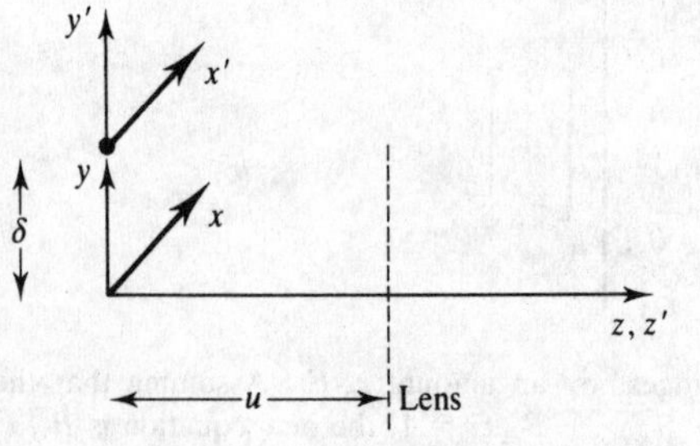

From previous results, a paraxial spherical wave is defined in (x',y',z') coordinates by

$$E(x',y',z') = A(z')\exp\{-jk_0(x'^2 + y'^2)/2z'\}$$

However, $x' = x$, $y' = y - \delta$ and $z' = z$, so we can also write

$$E(x,y,z) = A(z) \exp\{-jk_0[x^2 + (y-\delta)^2]/2z\}$$

If δ is small, we may approximate this by

$$E(x,y,z) \approx A(z) \exp\{-jk_0(x^2 + y^2 - 2y\delta)/2z\}$$

$$\approx A(z) \exp\{-jk_0(x^2 + y^2)/2z\} \exp(jk_0 y\delta/z)$$

Hence the input to the lens is

$$E(x,y,u) \approx A(u) \exp\{-jk_0(x^2 + y^2)/2u\} \exp\{jk_0 y\delta/u\}$$

The lens transform is $\tau_L = \tau_s \exp\{jk_0(x^2 + y^2)/2f\}$, so the output from the lens can be found as

$$E(x,y,u)\,\tau_L \approx A(u)\,\tau_s \exp\{jk_0(x^2 + y^2)(1/f - 1/u)/2\} \exp\{jk_0 y\delta/u\}$$

If we now define v using $1/v = 1/f - 1/u$, this can be written as

$$E(x,y,u)\,\tau_L \approx A'(u) \exp\{jk_0(x^2 + y^2)/2v\} \exp\{jk_0 y\delta'/v\}$$

where $\delta' = \delta v/u$ and $A'(u) = A(u)\tau_s$. Comparison between the expressions for the input to and the output from the lens shows that the output is a converging spherical wave, focused on a point v away from the lens, δ' below the axis. Since the result is independent of δ, the magnification of the set-up can be found for a general image as $M = \delta'/\delta = v/u$.

4.4 Lens waveguide geometry must be as shown below. Hence $d_1 = 2u$, $d_2 = 2v$. The imaging formula requires that $1/u + 1/v = 1/f$ so, the relation that must be satisfied by the lens separations is $1/d_1 + 1/d_2 = 1/(2f)$.

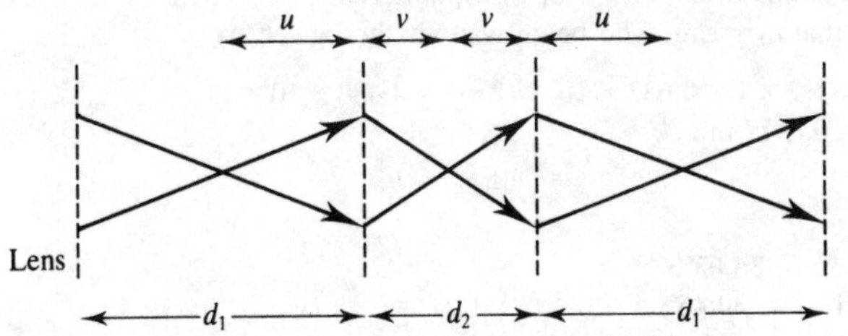

4.5 In the far-field the intensity distribution in the (ξ,η) plane can be related to the amplitude in the (x,y) plane by

$$|A(\xi,\eta)|^2 = (1/\lambda_0^2 z^2)\left|\iint_A A(x,y) \exp[+j(\omega_x x + \omega_y y)]\,dx\,dy\right|^2$$

where ω_x and ω_y are spatial frequencies given by $\omega_x = k_0\xi/z$ and $\omega_y = k_0\eta/z$.
Hence, assuming an input wave of amplitude A, the intensity distribution for the aperture is:

$$|A(\xi,\eta)|^2 = (A^2/\lambda_0^2 z^2)\left|\int_{-b/2}^{b/2}\int_{-a/2}^{a/2} \exp[+j(\omega_x x + \omega_y y)]\,dx\,dy\right|^2$$

$$= (a^2 b^2 A^2/16\lambda_0^2 z^2)\,\text{sinc}^2(\omega_x a/2)\,\text{sinc}^2(\omega_y b/2)$$

4.6 Assuming an input wave amplitude A_0 and a transparency function $\tau = \tau_0 + \Delta\tau \cos(Ky)$ for the screen, we may put

$$A(x,y) = A_0\{\tau_0 + \Delta\tau \cos(Ky)\} = A_0\{\tau_0 + (\Delta\tau/2)[\exp(jKy) + \exp(-jKy)]\}$$

The far-field diffraction integral will then average to zero, unless $\omega_x = 0$ and $\omega_y = -K$, 0 or $+K$. Consequently, the output will consist of a fan of three beams, travelling in directions defined by $\eta/z = LK/k_0$ with $L = -1$, 0 or $+1$, as shown in the construction below. These are known as *diffraction orders*.

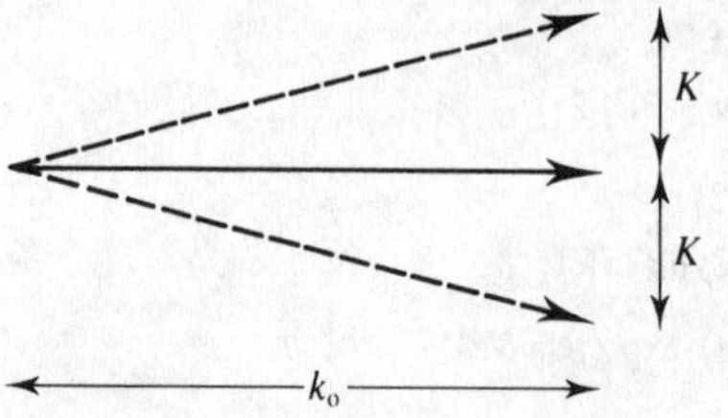

4.7 See, for example, Ramo S., Whinnery J. R., van Duzer T. "Fields and Waves in Communication Electronics", 2nd ed., John Wiley and Sons, New York, 1984; pp. 781–783.

4.8 (a) $\theta = 2/k_0w_0 = 2/\{(2\pi/0.633 \times 10^{-6}) \times (10^{-3}/2)\} = 4.0298 \times 10^{-4}$ rad.
(b) $z_0 = k_0w_0^2/2 = (2\pi/0.633 \times 10^{-6}) \times (10^{-3}/2)^2/2 = 1.2407$ m.

4.9 The lasing threshold is reached when the round-trip gain $R_1R_2 \exp(2gL)$ equals unity. Hence

$$g_{th} = (1/2L) \log_e \{1/(R_1R_2)\} = 1/0.5 \log_e(1/0.9) = 0.2107 \text{ m}^{-1}.$$

Assuming that the refractive index of the low-pressure gas plasma forming the gain medium is approximately unity, the mode separation is $\Delta f = c/2L = 3 \times 10^8/0.5$ Hz $= 600$ MHz.

4.10 The Gaussian mode is just one of the possible transverse modes of this type of cavity; the others are Hermite-Gaussian modes, which generally have most of their energy distributed further from the optical axis. Consequently, if a circular aperture is inserted into the cavity, these modes will be heavily attenuated and so will have a high lasing threshold. The Gaussian mode, on the other hand, has most of its energy concentrated on-axis, and will be affected only slightly.

The radius of curvature of a Gaussian beam at a distance z from the beam waist is given by $r(z) = \{z^2 + z_0^2\}/z$. This must match the radius of curvature of the spherical mirror, when z is equal to the cavity length. Hence $1 + z_0^2 = 2$, so that $z_0 = 1$ m. The beam waist is then given by

$$w_0 = \sqrt{(2z_0/k_0)} \Rightarrow w_0 = \sqrt{\{2 \times 1 \times 0.633 \times 10^{-6}/(2\pi)\}} = 4.488 \times 10^{-4} \text{ m}$$

and the beam diameter d is then $2w_0 = 0.8977$ mm.

Chapter 5

5.1 (a) $\theta_c = \sin^{-1}(n_2/n_1) = \sin^{-1}(1/1.7) = 36.03°$.
(b) $\theta_B = \tan^{-1}(n_2/n_1) = \tan^{-1}(1/1.7) = 30.46°$.

5.2 $\theta_c = \sin^{-1}(n_2/n_1) = 30° \Rightarrow n_2/n_1 = 0.5$. At normal incidence, $T_E = T_M = 2n_1/(n_1 + n_2) = 2/(1 + n_2/n_1) = 2/1.5 = 1./333$.

5.3 $\theta_1 = 30° \Rightarrow \theta_2 = \sin^{-1}\{(n_1/n_2)\sin\theta_1\} = 19.47°$.

$$\Gamma_E = [n_1 \cos\theta_1 - n_2 \cos\theta_2]/[n_1 \cos\theta_1 + n_2 \cos\theta_2]$$
$$= [0.866 - 1.5 \times 0.9428]/[0.866 + 1.5 \times 0.9428] = -0.2404 \Rightarrow |\Gamma_E|^2 = 0.05779$$
$$T_E = 2n_1 \cos\theta_1/[n_1 \cos\theta_1 + n_2 \cos\theta_2]$$
$$= 2 \times 0.866/[0.866 + 1.5 \times 0.9428] = 0.7596 \Rightarrow |T_E|^2 = 0.576\,99$$

5.4 If light rays are reversible, it should be possible to make the following situation arise:

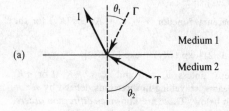

(a)

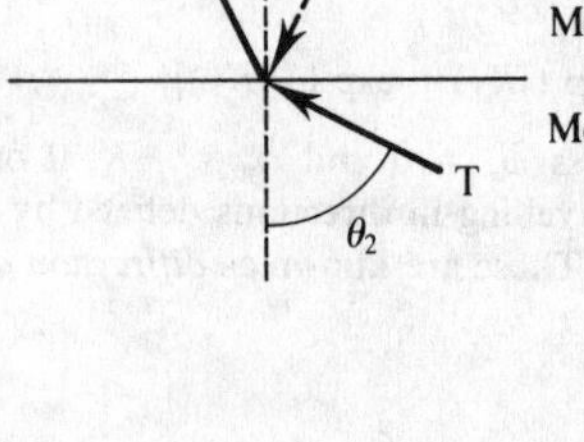

We can construct this by superposition of the sets of fields that make up the following diagrams, which define the transmission and reflection coefficients for incidence from media 1 and 2:

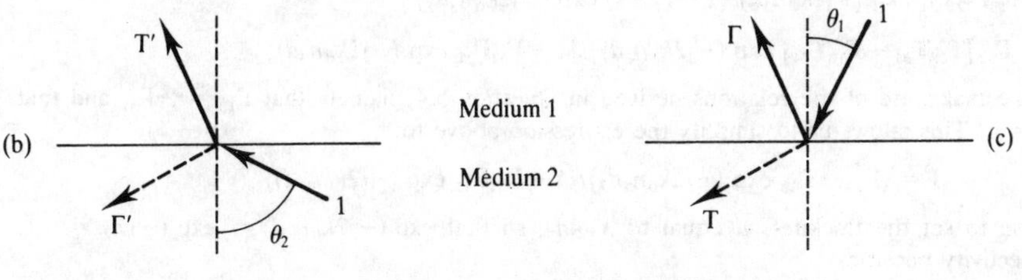

Multiplying all wave amplitudes in diagram (b) by T, and all waves in (c) by Γ then gives two new diagrams:

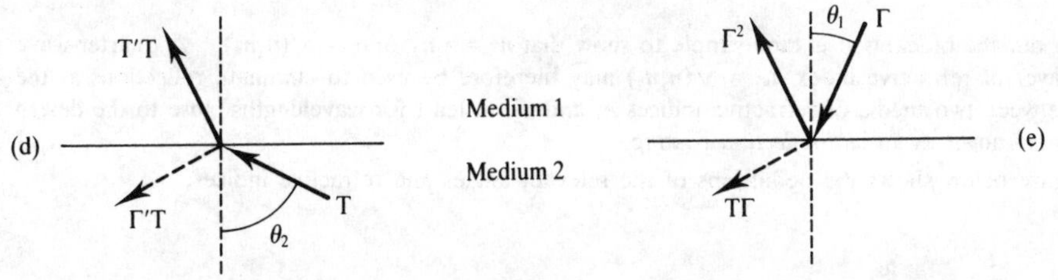

By superposition, we may add the waves in diagrams (d) and (e) together, to get:

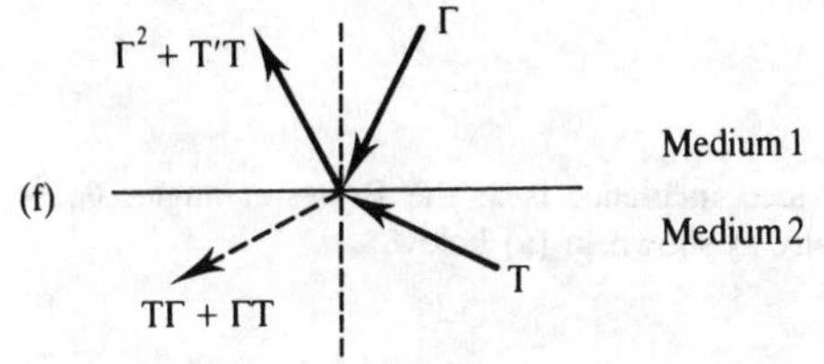

Since diagram (f) now has the same waves incident on the interface from the right as diagram (a), the waves leaving on the left must also be the same. Comparing amplitudes, we then get: $\Gamma^2 + \mathrm{T'T} = 1$ (the first required result) and $\mathrm{T}\Gamma + \Gamma'\mathrm{T} = 0$, or $\Gamma + \Gamma' = 0$ (the second).

5.5 Proof of the results of Question 5.4: for TE incidence, the relevant coefficients are given by:

$$\Gamma = [n_1 \cos\theta_1 - n_2 \cos\theta_2]/[n_1 \cos\theta_1 + n_2 \cos\theta_2]$$

$$\Gamma' = [n_2 \cos\theta_2 - n_1 \cos\theta_1]/[n_1 \cos\theta_1 + n_2 \cos\theta_2] \Rightarrow \Gamma + \Gamma' = 0 \text{ by inspection.}$$

$$\mathrm{T} = 2n_1 \cos\theta_1/[n_1 \cos\theta_1 + n_2 \cos\theta_2]$$

$$\mathrm{T}' = 2n_2 \cos\theta_2/[n_1 \cos\theta_1 + n_2 \cos\theta_2]$$

$$\Rightarrow \Gamma^2 + \mathrm{T'T} = \{[\mathrm{n}_1 \cos\theta_1 - n_2 \cos\theta_2]^2 + [2n_1 \cos\theta_1 \cdot 2n_2 \cos\theta_2]\}/[n_1 \cos\theta_1 + n_2 \cos\theta_2]^2 = 1$$

5.6 The amplitude for Path 3 is $\mathrm{T}_{13}\Gamma_{32}\Gamma_{31}\Gamma_{32}\mathrm{T}_{31} \exp(-\mathrm{j}4k_0 n_3 d)$. Hence the total reflectivity may be written as an infinite series, in the form:

$$\Gamma = \Gamma_{13} + T_{13}\Gamma_{32}T_{31}\exp(-j2k_0n_3d) + T_{13}\Gamma_{32}\Gamma_{31}\Gamma_{32}T_{31}\exp(-j4k_0n_3d) + \text{higher order paths}$$
$$= \Gamma_{13} + T_{13}\Gamma_{32}T_{31}\exp(-j2k_0n_3d)\{1 + \Gamma_{31}\Gamma_{32}\exp(-j2k_0n_3d) + [\Gamma_{31}\Gamma_{32}\exp(-j2k_0n_3d)]^2\ldots\}$$
$$= \Gamma_{13} + T_{13}\Gamma_{32}T_{31}\exp(-j2k_0n_3d)/\{1 - \Gamma_{31}\Gamma_{32}\exp(-j2k_0n_3d)\}$$
$$= \{\Gamma_{13} + \Gamma_{32}[T_{13}T_{31} - \Gamma_{13}\Gamma_{31}]\exp(-j2k_0n_3d)\}/\{1 - \Gamma_{31}\Gamma_{32}\exp(-j2k_0n_3d)\}$$

At this point we make use of the relations derived in Question 5.4, namely that $\Gamma_{13} = -\Gamma_{31}$ and that $T_{13}T_{31} + \Gamma_{13}^2 = 1$. This allows us to simplify the expression above to:

$$\Gamma = \{\Gamma_{13} + \Gamma_{32}\exp(-j2k_0n_3d)\}/\{1 - \Gamma_{31}\Gamma_{32}\exp(-j2k_0n_3d)\}$$

We now choose to set the thickness d equal to $\lambda_0/4n_3$, so that $\exp(-j2k_0n_3d) = \exp(-j\pi) = -1$. Hence the reflectivity becomes

$$\Gamma = \{\Gamma_{13} - \Gamma_{32}\}/\{1 + \Gamma_{31}\Gamma_{32}\}$$

Clearly the reflectivity can be zero if $\Gamma_{13} = \Gamma_{32}$. Using the Fresnel coefficients this requires:

$$\{n_1 - n_3\}/\{n_1 + n_3\} = \{n_3 - n_2\}/\{n_3 + n_2\}, \text{ or } \{n_1 - n_3\}\{n_3 + n_2) = \{n_3 - n_2\}\{n_1 + n_3\}$$

Multiplying out the brackets it is then simple to show that $n_3^2 = n_1n_2$, or $n_3 = \sqrt{(n_1n_2)}$. A quarter-wave dielectric layer of refractive index $n_3 = \sqrt{(n_1n_2)}$ may therefore be used to eliminate reflections at the interface between two media of refractive indices n_1 and n_2, at least for wavelengths close to the design value. This is known as an *anti-reflection coating*.

5.8 The figure below shows the definitions of the relevant angles and refractive indices.

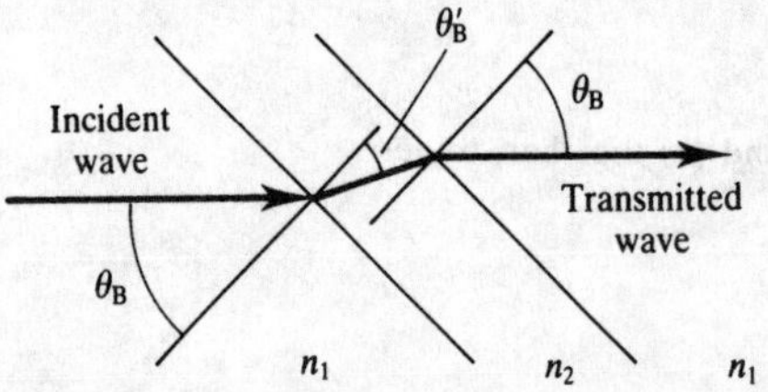

For the first interface, there will be no reflection provided incidence is at the Brewster angle, $\theta_B = \tan^{-1}(n_2/n_1)$. This angle can be represented geometrically as shown in (a) below.

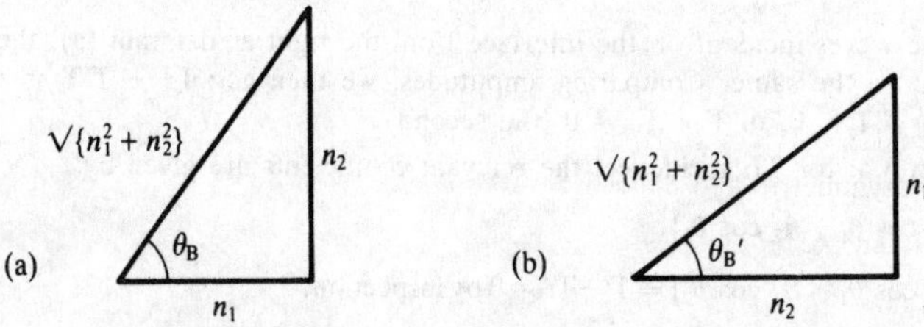

The wave is refracted as it enters medium 2. According to Snell's law, θ'_B is given by $n_1 \sin\theta_B = n_2 \sin\theta'_B$. Hence

$$n_1n_2/\sqrt{\{n_1^2 + n_2^2\}} = n_2 \sin\theta'_B \Rightarrow \sin\theta'_B = n_1/\sqrt{\{n_1^2 + n_2^2\}}$$

and θ'_B can therefore be represented geometrically as in (b) above. Therefore, $\tan \theta'_B = n_1/n_2$, so the wave is incident on the second interface at the Brewster angle for that interface. Consequently, there will be no reflection at either interface.

5.9 (a) For TE incidence

$$\Gamma^2 = \{n_1^2 \cos^2 \theta_1 - 2n_1n_2 \cos \theta_1 \cos \theta_2 + n_2^2 \cos^2 \theta_2\}/\{n_1^2 \cos^2 \theta_1 + 2n_1n_2 \cos \theta_1 \cos \theta_2 + n_2^2 \cos^2 \theta_2\}$$

$$T^2 = 4n_1^2 \cos^2 \theta_1/\{n_1^2 \cos^2 \theta_1 + 2n_1n_2 \cos \theta_1 \cos \theta_2 + n_2^2 \cos^2 \theta_2\}$$

$$\Rightarrow T^2 n_2 \cos \theta_2/\{n_1 \cos \theta_1\} = 4n_1 \cos^2 \theta_1 n_2 \cos^2 \theta_2/\{n_1^2 \cos^2 \theta_1 + 2n_1n_2 \cos \theta_1 \cos \theta_2 + n_2^2 \cos^2 \theta_2\}$$

$$\Rightarrow \Gamma^2 + T^2 n_2 \cos \theta_2/\{n_1 \cos \theta_1\} = 1 \text{ by inspection}$$

(b) Substitute previous data into (a): $0.057\,79 + 0.576\,99 \times 1.5 \cos(19.47°)/\cos(30°) = 1$.

5.10 The time-averaged power per unit area carried by a plane wave is $P = (1/2)nE^2/Z_0 \Rightarrow$ Power carried by the input wave through a length S of the boundary is $P_I = (1/2)n_1E_I^2 S \cos \theta_1/Z_0$. Similarly, the reflected power is $P_R = (1/2)n_1E_R^2 S \cos \theta_1/Z_0$ and the transmitted power is $P_T = (1/2)n_2E_T^2 S \cos \theta_2/Z_0$. Since there is no loss, we must have: $P_I = P_R + P_T$. Hence

$$E_I^2 = E_R^2 + E_T^2 n_2 \cos \theta_2/\{n_1 \cos \theta_1\} \quad \text{or} \quad \Gamma^2 + T^2 n_2 \cos \theta_2/\{n_1 \cos \theta_1\} = 1$$

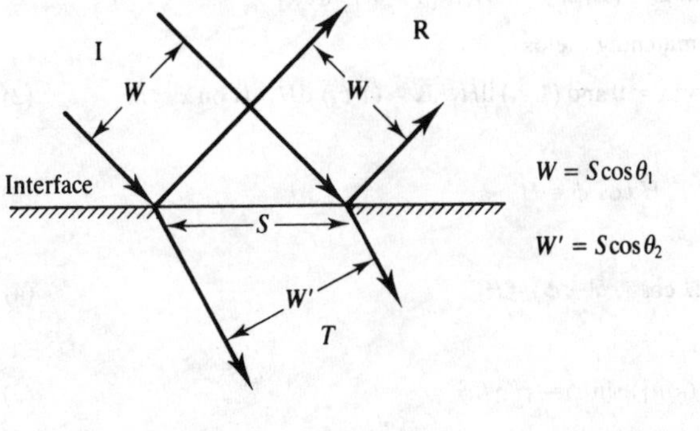

Chapter 6

6.1 From previous results we know that the propagation constant is given by $\beta = k_0 \sin \theta$, where θ is the ray angle. Similarly the eigenvalue equation requires that $2k_0h \cos \theta = 2\nu\pi$. The dispersion relation can therefore be found as $\beta^2 = k_0^2\{1 - (\nu\pi/k_0h)^2\} = (\omega/c)^2\{1 - (\nu\pi c/h\omega)^2\}$. For large β, $\omega \to c\beta$ for all modes. For small β the individual modes cut off when $\omega = \nu\pi c/h$.

6.2 For dielectric-filled metal-walled guide, the eigenvalue equation modifies to $2kh \cos \theta = 2\nu\pi$. Cut-off occurs when $\theta \to 0$, i.e. when $kh = \nu\pi$. Hence all modes are cut-off when $k < \pi/h$ or when $2\pi nf/c < \pi/h$, so $f < c/2nh = 3 \times 10^8/(2 \times 1.5 \times 0.5 \times 10^{-6}) = 2 \times 10^{14}$ Hz.

6.3 The eigenvalue equation for an asymmetric slab dielectric waveguide is given by

$$\tan(\kappa h) = \kappa(\gamma + \delta)/(\kappa^2 - \gamma\delta)$$

If the guide is symmetric, $\gamma = \delta$, so that

$$\tan(\kappa h) = 2\kappa\gamma/(\kappa^2 - \gamma^2) \Rightarrow \tan(\kappa h) = 2\gamma/\kappa/\{1 - (\gamma/\kappa)^2\}$$

Now

$$\tan(A + B) = \{\tan A + \tan B\}\{1 - \tan A \tan B\} \Rightarrow \tan(\kappa h) = 2\tan(\kappa h/2)/(1 - \tan^2(\kappa h/2))$$

Comparison with the result above shows that we must then have $\tan(\kappa h/2) = \gamma/\kappa$.

6.4 $V = (k_0h/2)\sqrt{\{n_1^2 - n_2^2\}} \approx (\pi h/\lambda_0)\sqrt{(2n\Delta n)}$; hence,

(a) $V = (6\pi/1.5)\sqrt{(2 \times 1.5 \times 0.05)} = 4.8669$

(b) $V = (6\pi/1.5)\sqrt{(2 \times 1.5 \times 0.005)} = 1.5390$

The general cut-off condition is that $V = \nu\pi/2$, where $\nu = 0, 1, 2, \ldots$ In (a), $3\pi/2 < V < 2\pi$ so the guide is 4-moded; in (b) $V < \pi/2$, but there is no cut-off for the lowest-order mode so the guide is single-moded.

6.5 Assume the geometry of Fig. 6.3. For the TM case it is simpler to work with **H** so we must solve $\nabla^2\mathbf{H} + n_i^2k_0^2\mathbf{H} = 0$ in each of the three layers. Assume $\mathbf{H} = H_i(x)\exp(-j\beta z)\mathbf{j} \Rightarrow$ waveguide equation is $\mathrm{d}^2H_i/\mathrm{d}x^2 + \{n_i^2k_0^2 - \beta^2\}H_i = 0$ in each layer. Now guess the solutions:

$$H_1 = H\cos(\kappa x - \phi);\ H_2 = H'\exp(\gamma x);\ H_3 = H''\exp[-\delta(x-h)]$$

where: $\kappa = \sqrt{[n_1^2k_0^2 - \beta^2]}$, $\gamma = [\beta^2 - n_2^2k_0^2]$ and $\delta = [\beta^2 - n_3^2k_0^2]$.

Boundary matching:

(1) tangential components of **H** must match, so that

$$H_1 = H_2 \text{ on } x = 0 \text{ and } H_1 = H_3 \text{ on } x = h \tag{1}$$

(2) tangential components of **E** must match. Now, **E** is given by

$$\mathbf{E} = (1j\omega\varepsilon)\nabla\times\mathbf{H}, \text{ or } \mathbf{E} = (1/j\omega\varepsilon)\{-\partial H_y/\partial z\,\mathbf{i} + \partial H_y/\partial x\,\mathbf{k}\}$$

E_z is the only tangential component, and matching yields

$$(1/\varepsilon_1)\,\mathrm{d}H_1/\mathrm{d}x = (1/\varepsilon_2)\,\mathrm{d}H_2/\mathrm{d}x \text{ on } x = 0 \text{ and } (1/\varepsilon_1)\,\mathrm{d}H_1/\mathrm{d}x = (1/\varepsilon_3)\,\mathrm{d}H_3/\mathrm{d}x \text{ on } x = h \tag{2}$$

From Eq. (1) we have

$$H\cos\phi = H' \tag{3}$$

and

$$H\cos(\kappa h - \phi) = H'' \tag{4}$$

From Eq. (2) we have

$$(H\kappa/n_1^2)\sin\phi = H'\gamma/n_2^2 \tag{5}$$

and

$$(H\kappa/n_1^2)\sin(\kappa h - \phi) = H'\delta/n_3^2 \tag{6}$$

From Eqs (3) and (5) we then get

$$\tan\phi = (n_1^2/n_2^2)\,\gamma/\kappa = \gamma'/\kappa \tag{7}$$

From Eqs (4) and (6) we then get

$$\tan(\kappa h - \phi) = (n_1^2/n_3^2)\,\delta/\kappa = \delta'/\kappa \tag{8}$$

Reduction of Eqs (7) and (8) to the standard form $\tan(\kappa h) = \kappa(\gamma' + \delta')/(\kappa^2 - \gamma'\delta')$ then follows the identical procedure to that used in TE case.

6.6 From the results of Question 6.5 we know that the transverse magnetic field must match at each boundary, but the *slope* of the transverse magnetic field does not, since, for a symmetric guide, $(1/n_1^2)\,\mathrm{d}H_1/\mathrm{d}x = (1/n_2^2)\,\mathrm{d}H_2/\mathrm{d}x$ on $x = 0$ and $x = h$. Because $n_1 > n_2$, $\mathrm{d}H_1/\mathrm{d}x > \mathrm{d}H_2/\mathrm{d}x$ at each boundary. Both fields then have the general form shown below, with greater penetration into the substrate by the evanescent tails in case (b).

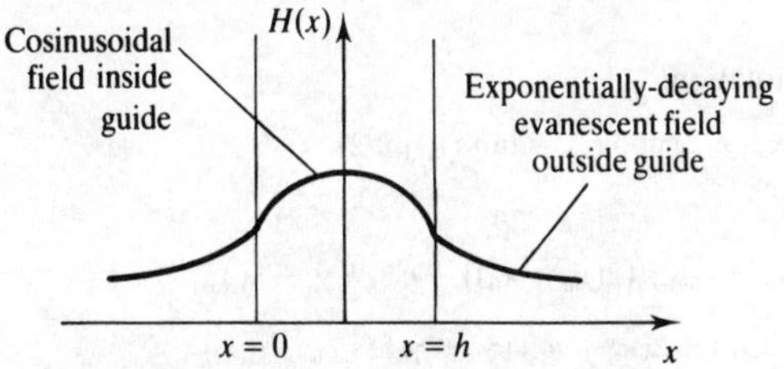

6.7 The eigenvalue equation is $\tan(\kappa h/2) = \gamma/\kappa$ for symmetric modes. Hence, when $\beta = n_{eff}k_0$ has the correct value

$$\tan\{(h/2)\surd[n_1^2k_0^2 - \beta^2]\} - \surd\{[\beta^2 - n_2^2k_0^2]/[n_1^2k_0^2 - \beta^2]\} = 0$$

For any other value of β, we have:

$$\tan\{(h/2)\surd[n_1^2k_0^2 - \beta^2]\} - \surd\{[\beta^2 - n_2^2k_0^2]/[n_1^2k_0^2 - \beta^2]\} = \alpha \neq 0$$

Solve by iteration, assuming that $n_2 < n_{eff} < n_1$.

When:

$$n_{eff} = n_2, \qquad \alpha = 0.682$$

$$n_{eff} = (n_2 + n_1)/2, \qquad \alpha = -0.54$$

$$n_{eff} = n_1, \qquad \alpha = -72$$

The sign changes between the first two values of n_{eff} so the solution lies between these points; try half-way.

When:

$$n_{eff} = (3n_2 + n_1)/4, \qquad \alpha = 4.747 \times 10^{-3}$$

This is very close to zero, so terminate the iteration $\Rightarrow n_{eff} \approx (3n_2 + n_1)/4 = 1.50125$.

6.8 Assume that the modes concerned are defined by the transverse fields $E_a(x)$ and $E_b(x)$, where

$$E_a = E\cos(\kappa_a x) \qquad \text{for } 0 < |x| < h/2$$

$$E\cos(\kappa_a h/2)\exp\{-\gamma_a[|x| - h/2]\} \qquad \text{for } |x| > h/2$$

$$E_b = E\cos(\kappa_b x) \qquad \text{for } 0 < |x| < h/2$$

$$E\cos(\kappa_b h/2)\exp\{-\gamma_b[|x| - h/2]\} \qquad \text{for } |x| > h/2$$

where $\tan(\kappa_a h/2) = \gamma_a/\kappa_a$ and $\tan(\kappa_b h/2) = \gamma_b/\kappa_b$. (NB These expressions are obtained from the standard results, merely by shifting the coordinate origin).

For orthogonality we require that $I = \int_{-\infty}^{\infty} E_a E_b \, dx = 0$. The integral can be divided into two components, I_1 and I_2, such that

$$I_1 = 2\int_0^{h/2} E\cos(\kappa_a x)\, E\cos(\kappa_b x)\, dx = 2E^2 \int_0^{h/2} [\cos\{(\kappa_a + \kappa_b)x\}/2 + \cos\{(\kappa_a - \kappa_b)x\}/2]\, dx$$

$$= 2E^2 \sin\{(\kappa_a + \kappa_b)h/2\}/\{2(\kappa_a + \kappa_b)\} + \sin\{(\kappa_a - \kappa_b)h/2\}/\{2(\kappa_a - \kappa_b)\}$$

$$I_2 = 2\int_{h/2}^{\infty} E\cos(\kappa_a h/2)\exp\{-\gamma_a[x - h/2]\}\, E\cos(\kappa_b h/2)\exp\{-\gamma_b[x - h/2]\}\, dx$$

$$= 2E^2\cos(\kappa_a h/2)\cos(\kappa_b h/2)\int_0^{\infty} \exp\{-(\gamma_a + \gamma_b)x\}\, dx$$

$$= 2E^2 \cos(\kappa_a h/2) \cos(\kappa_b h/2)/(\gamma_a + \gamma_b)$$

For simplicity we shall define the following new notation:

$$s_{a,b} = \sin(\kappa_{a,b} h/2),\ c_{a,b} = \cos(\kappa_{a,b} h/2) \text{ and } t_{a,b} = \tan(\kappa_{a,b} h/2)$$

Hence:

$$I = 2E^2 \{(s_a c_b + c_a s_b)/\{2(\kappa_a + \kappa_b)\} + (s_a c_b - c_a s_b)/\{2(\kappa_a - \kappa_b)\} + c_a s_b/(\gamma_a + \gamma_b)\}$$

$$= (2E^2/f)\{2(\kappa_a - \kappa_b)(\gamma_a + \gamma_b)(s_a c_b + c_a s_b) + 2(\kappa_a + \kappa_b)(\gamma_a + \gamma_b)(s_a c_b - c_a s_b) + 4(\kappa_a^2 - \kappa_b^2)\, c_a c_b\}$$

where $f = 4(\kappa_a^2 - \kappa_b^2)(\gamma_a + \gamma_b)$.
Recombining terms in the above, we then get

$$I = (8E^2/f)\{\kappa_a(\gamma_a + \gamma_b)s_a c_b - \kappa_b(\gamma_a + \gamma_b)c_a s_b + (\kappa_a^2 - \kappa_b^2)c_a s_b\}$$

$$= (8E^2/f c_a c_b)\{(\kappa_a t_a - \kappa_b t_b)(\gamma_a + \gamma_b) + (\kappa_a^2 - \kappa_b^2)\}$$

The individual eigenvalue equations require that $t_{a,b} = \gamma_{a,b}/\kappa_{a,b}$ so that

$$I = (8E^2/f c_a c_b)[\{(\kappa_a t_a)^2 - (\kappa_b t_b)^2\} + (\kappa_a^2 - \kappa_b^2)] = (8E^2/f c_a c_b)\{(\kappa_a/c_a)^2 - (\kappa_b/c_b)^2\}$$

Manipulation of the eigenvalue equation then shows that $c_{a,b} = \kappa_{a,b}/\sqrt{\{\kappa_{a,b}^2 + \gamma_{a,b}^2\}}$, so that

$$I = (8E^2/f c_a c_b)\{(\kappa_a^2 + \gamma_a^2) - (\kappa_b^2 + \gamma_b^2)\}$$

Recall the definition of $\kappa_{a,b}$ and $\gamma_{a,b}$:

$$\kappa_{a,b} = \sqrt{\{n_1^2 k_0^2 - \beta_{a,b}^2\}},\quad \gamma_{a,b} = \sqrt{\{\beta_{a,b}^2 - n_2^2 k_0^2\}}$$

Hence

$$(\kappa_a^2 + \gamma_a^2) = (\kappa_b^2 + \gamma_b^2) = (n_1^2 - n_2^2)\, k_0^2$$

and $I = 0 \Rightarrow$ the two modes are orthogonal. Note that this result is not valid if the two modes are identical, since $f = 0$ in this case; the integral then gives a finite, non-zero value.

6.9 Define the transverse electric fields in guides 1 and 2 as $E_y(x) = E_0 \exp(-\alpha x^2)$ and $E_y'(x) = E_0 \exp\{-\alpha(x - \delta)^2\}$, respectively. Choose E_0 so that the fields are normalized, i.e. so that:

$$\langle E_y, E_y \rangle = \langle E_y', E_y' \rangle = 1$$

This implies that we must set

$$\int_{-\infty}^{\infty} E_0^2 \exp\{-\alpha(\sqrt{2}x)^2\}\, dx = 1.$$

The coupling efficiency is then

$$\eta = |\langle E_y, E_y' \rangle|^2/\{\langle E_y, E_y \rangle \langle E_y', E_y' \rangle\} = |\langle E_y, E_y' \rangle|^2$$

so that

$$\eta = \left\{\int_{-\infty}^{\infty} E_0 \exp(-\alpha x^2) \cdot E_0 \exp\{-\alpha(x - \delta)^2\}\, dx\right\}^2$$

$$= \left\{\int_{-\infty}^{\infty} E_0^2 \exp\{-\alpha(2x^2 - 2x\delta + \delta^2)\}\, dx\right\}^2$$

$$= \left\{\exp(-\alpha\delta^2/2) \int_{-\infty}^{\infty} E_0^2 \exp\{-\alpha(\sqrt{2}x - \delta/\sqrt{2})^2\}\, dx\right\}^2$$

The integral above must yield unity—it is merely a shifted version of the normalization integral defined previously. Hence we obtain $\eta = \exp(-\alpha\delta^2)$. Clearly the coupling efficiency is maximum (and 100 per cent) when there is no lateral offset, and the variation with δ is also Gaussian.

6.10 (a) (i) If the discontinuity is vanishingly small, only the lowest-order mode of the right-hand guide will be excited. (ii) For larger discontinuities there will also be conversion to other symmetric modes (e.g. higher-order guided modes and radiation modes). Backward modes will also be excited. However, since the whole geometry is symmetric, there can be no conversion to anti-symmetric modes.

(b) (i) as (ii) above; (ii) as (i) above.

Chapter 7

7.1 In layer 1 the solution consists of incident and reflected plane waves, so that

$$E_{y1} = E_I \exp\{-jk_0n_1(z\sin\theta_1 - x\cos\theta_1)\} + E_R\exp\{-jk_0n_1(z\sin\theta_1 + x\cos\theta_1)\}$$

Define $\beta = k_0n_1\sin\theta_1$ and $\kappa = \sqrt{\{n_1^2k_0^2 - \beta^2\}}$, so that the solution can be written as:

$$E_{y1} = \{E_I\exp(+j\kappa x) + E_R\exp(-j\kappa x)\}\exp(-j\beta z)$$

If incidence is at an angle beyond θ_c, $|E_R/E_I| = 1$, so that the solution can be written as:

$$E_{y1} = E\cos(\kappa x - \phi)\exp(-j\beta z) \qquad (1)$$

In layer 2 the solution consists of growing and decaying evanescent waves, so

$$E_{y2} = E_G\exp\{-jk_0n_2(z\sin\theta_2 - x\cos\theta_2)\} + E_D\exp\{-jk_0n_2(z\sin\theta_2 + x\cos\theta_2)\}$$

where θ_2 is found from Snell's law, $n_1\sin\theta_1 = n_2\sin\theta_2$.

Define

$$j\gamma = k_0n_2\cos\theta_2 = k_0n_2\sqrt{\{1-(n_1/n_2)^2\sin^2\theta_1)\}} = j\sqrt{\{\beta^2 - n_2^2k_0^2\}}$$

$$\Rightarrow E_{y2} = \{E_G\exp(-\gamma x) + E_D\exp(+\gamma x)\}\exp(-j\beta z) \qquad (2)$$

In layer 3 the solution consists of a single transmitted wave:

$$E_{y3} = E_T\exp\{-jk_0n_1(z\sin\theta_1 - x\cos\theta_1)\} = E_T\exp(+j\kappa x)\exp(-j\beta z) \qquad (3)$$

Boundary conditions:

Match tangential electrical field at both interfaces $\Rightarrow$ match E_y at $x = 0$ and $x = h$
Match tangential magnetic field at both interfaces $\Rightarrow$ match $\partial E_y/\partial x$ at $x = 0$ and $x = h$

7.2 Write the solution in each layer in the form $E_{yi} = E_i(x)\exp(-j\beta z)$, i.e. as in Eqs (1), (2) and (3) in the solution to Question 7.1. Boundary conditions then modify to the requirement that E_i and dE_i/dx match at both interfaces.

Matching E_i: $\quad E\cos(\kappa h - \phi) = E_G\exp(-\gamma h) + E_D\exp(+\gamma h) \qquad (1)$

$$E_T = E_G + E_D \qquad (2)$$

Matching dE_i/dx: $\quad -\kappa E\sin(\kappa h - \phi) = \gamma\{E_D\exp(+\gamma h) - E_G\exp(-\gamma h)\} \qquad (3)$

$$j\kappa E_T = \gamma\{E_D - E_G\} \qquad (4)$$

Assuming that E is known, Eqs (1)–(4) are four simultaneous equations that must be solved to find the four unknowns E_G, E_D, E_T and ϕ. Combining Eqs (2) and (4) gives

$$E_G = E_T\{\gamma - j\kappa\}/2\gamma \text{ and } E_D = E_T\{\gamma + j\kappa\}/2\gamma.$$

Substitution into Eqs (1) and (3) then yields

$$E\cos(\kappa h - \phi) = E_T[(\{\gamma - j\kappa\}/2\gamma)\exp(-\gamma h) + (\{\gamma + j\kappa\}/2\gamma)\exp(+\gamma h)] \qquad (5)$$

$$-\kappa E \sin(kh - \phi) = \gamma E_T[(\{\gamma + j\kappa\}/2\gamma) \exp(+\gamma h) - (\{\gamma - j\kappa\}/2\gamma) \exp(-\gamma h)] \quad (6)$$

Finally, dividing Eqs (5) and (6) gives

$$-\kappa \tan(kh - \phi) = \gamma[(\{\gamma + j\kappa\}/2\gamma) \exp(+\gamma h) - (\{\gamma - j\kappa\}/2\gamma) \exp(-\gamma h)]/$$
$$[(\{\gamma - j\kappa\}/2\gamma) \exp(-\gamma h) + (\{\gamma + j\kappa\}/2\gamma) \exp(+\gamma h)]$$

In principle, this equation could now be solved analytically for ϕ. However, it is simpler to make some approximations. If γh is reasonably large, $\exp(-\gamma h) \approx 0$, so that $\tan(kh - \phi) \approx -\gamma/\kappa$. In this case ϕ is a constant, independent of h.

From Eq. (5) above, we get

$$E_T = E \cos(\kappa h - \phi)/[(\{\gamma - j\kappa\}/2\gamma) \exp(-\gamma h) + (\{\gamma + j\kappa\}/2\gamma) \exp(+\gamma h)]$$

Using the same approximation, we get

$$E_T \approx E \cos(\kappa h - \phi)(2\gamma/\{\gamma + j\kappa\}) \exp(-\gamma h)$$

Consequently, E_T falls off exponentially as the thickness h increases.

7.3 Phase matching occurs when $k \sin\theta_1 = \beta$, or when $k_0 n_p \sin\theta_1 = k_0 n_{eff} \Rightarrow \theta_1 = \sin^{-1}(n_{eff}/n_p)$. There will be three phase matching angles, at $\theta_A = \sin^{-1}(1.515/1.7) = 63.021°$, $\theta_B = 64.158°$ and $\theta_C = 65.343°$. This will give rise to three dips in the angular variation in reflectivity. Corresponding propagation constants are: $\beta_A = 2\pi n_{effA}/\lambda_0 = 6.346\ 10^6$/m. $\beta_B = 6.408 \times 10^6$/m and $\beta_C = 6.471 \times 10^6$/m.

7.4 Phase matching diagrams:

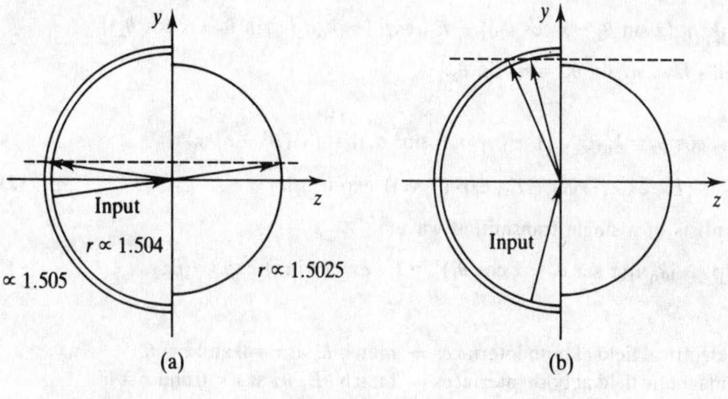

(a) Reflection and refraction take place. Transmitted beam will be the strongest; both reflected beams will be fairly weak, so little mode conversion occurs.

(b) $\theta_1 > \sin^{-1}(1.5025/1.504)$, so total internal reflection occurs. There is then no transmitted beam, and strong mode conversion probably occurs.

7.5 The lens maker's formula implies that $1/f = \{n_{Leff} - n_{Meff})/n_{Meff}\} \{1/r_1 + 1/r_2\}$, so that f is smallest when r_1 and r_2 are equal, and as small as possible, so that $2r$ = beam width in each case $\Rightarrow$ the most compact arrangement uses circular overlay lenses in the geometry shown below:

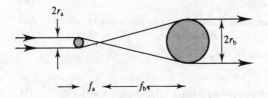

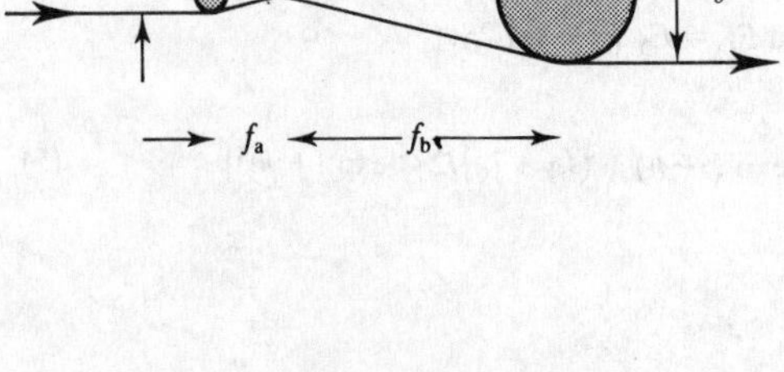

For the input lens, $f_a = \{n_{\mathrm{Meff}}/(n_{\mathrm{Leff}} - n_{\mathrm{Meff}})\}r/2 = (1.6/0.02) \times (0.5/4) = 10$ mm. For $\times$ 5 beam expansion, the focal length of the output lens is $f_b = 50$ mm so the lenses are 60 mm apart.

7.6 $2\Lambda \sin \theta_0 = \lambda_0/n \Rightarrow \theta_0 = \sin^{-1}\{1.5/(2 \times 1 \times 1.6)\} = 27.953°$. $\Delta\lambda = (\Lambda\lambda_0/d) \cot \theta_0 = (1.5 \times 10^{-6} \times 1/50) \times \cot(27.953°) = 5.653 \times 10^{-8}$ m $= 56.533$ nm. For an input angle of θ_0, $y = d \tan \theta_0 = 50 \tan(27.953°)$ μm $= 26.533$ μm. Since $\Lambda = 1$ μm, approximately 27 fringe planes are crossed.

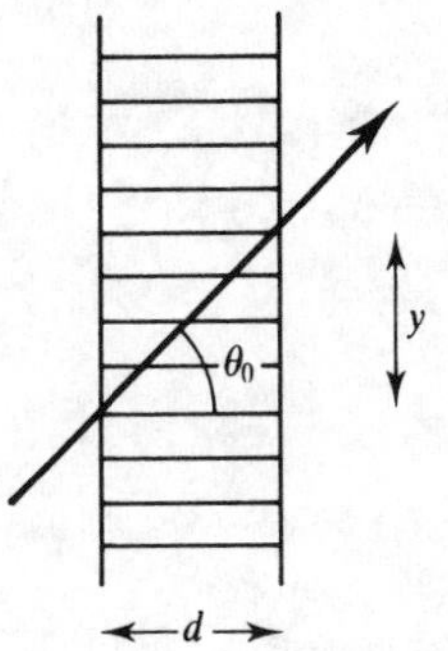

7.7 Start with Bragg's law, $2\Lambda \sin \theta_0 = \lambda_0/n$. Differentiating both sides gives $2\Lambda \cos \theta_0 \, \Delta\theta_0 = \Delta\lambda_0/n$. The wavelength bandwidth is defined by $\Delta\lambda_0 = (\Lambda\lambda_0/d) \cot \theta_0$ so substituting into the above gives $\Delta\theta_0 = \lambda_0/(2nd \sin\theta_0) = \Lambda/d$ for the angular bandwidth.

$$\Lambda = \lambda_0/(2n \sin \theta_0) = 0.633/\{2 \times 1.6 \times \sin(30°)\}\ \mu\text{m} = 0.3956\ \mu\text{m}$$

$$\Rightarrow \Delta\theta_0 = 0.3956/20 \text{ rad} = 0.01978 \text{ rad or } 1.133°.$$

7.8 Bragg's law is $2\Lambda \sin(\theta_0 - \phi) = \lambda_0/n \Rightarrow 2(2\pi n/\lambda_0) \sin(\theta_0 - \phi) = 2\pi/\Lambda \Rightarrow 2k \sin(\theta_0 - \phi) = |\mathbf{K}|$. This relationship is a reflection of the geometrical construction shown in the k-vector diagram below, where $k \sin(\theta_0 - \phi) = |\mathbf{K}|/2$.

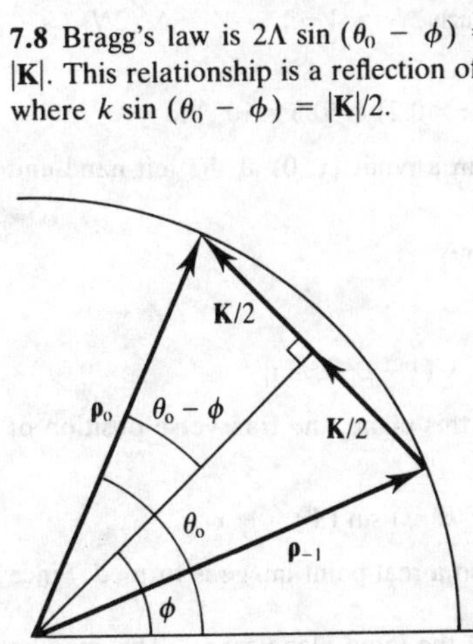

7.9 (a) Transmission grating, in this case we have:

$\Lambda = 5$ μm so $|\mathbf{K}| = 2\pi/\Lambda = 2\pi/5 \times 10^{-6} = 1.256 \times 10^6$/m and
$k = 2\pi n/\lambda_0 = 2\pi \times 1.6/0.633 \times 10^{-6} = 15.88 \times 10^6$/m.

Hence, $|\mathbf{K}| \ll k$, and $d \approx \Lambda$. Therefore diffraction will take place in the Raman-Nath regime, with many diffraction orders being produced. Deflection angle of the first diffraction order is approximately $\theta = \tan^{-1}(1.256/15.88) = 4.522°$.

(b) Reflection grating, in this case we have:

$$\Lambda = 0.5\ \mu\text{m} \Rightarrow |\mathbf{K}| = 2\pi/\Lambda = 2\pi/5 \times 10^{-7} = 1.256 \times 10^{7}/\text{m}$$

Deflection angle is given by

$$(\theta_0 - \phi) = \sin^{-1}(|\mathbf{K}|/2k) = \sin^{-1}\{12.566/(2 \times 15.88)\} = 23.3°.$$

K-vector diagrams:

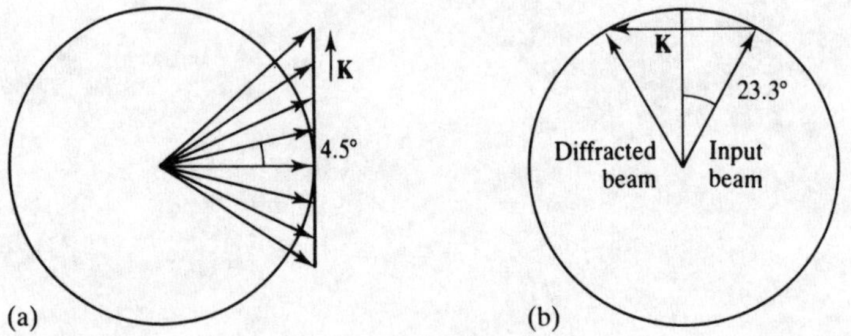

Chapter 8

8.1 If the attenuation is 4 dB/km, loss in 10 km of fibre is 4 × 10 = 40 dB ⇒ 40 = 10 $\log_{10}(P_{out}/P_{in})$ so $P_{out}/P_{in} = 10^{-4}$ or 0.01%. If loss is caused by Rayleigh scattering, it falls off as $1/\lambda_0^4$. Hence, at λ_0 = 1.3 μm, loss will have reduced by $(0.65/1.3)^4 = 1/16$ so the attenuation will be 0.25 dB/km. Fraction of power transmitted is now $P_{out}/P_{in} = 10^{-0.25} = 0.562$ or 56.2%.

8.2 $NA = \sqrt{(n_1^2 - n_2^2)}$; $V = k_0 a \sqrt{(n_1^2 - n_2^2)} \Rightarrow a = V/(k_0 NA)$. Similarly, $N \approx V^2/2 \Rightarrow V \approx \sqrt{(2N)}$. Hence

$$d = 2a = 2\sqrt{(2N)}/(k_0 NA) = 2\sqrt{(2000)} \times 0.85 \times 10^{-6}/(2\pi \times 0.2) = 60.5 \times 10^{-6}\ \text{m}$$

8.3 (i) Consider the propagation of a group of rays emanating from a point $(x, 0)$ at the left-hand end of the lens. The general solution for ray paths is

$$x = x_0\sqrt{(1 - \beta^2/k_0^2 n_0^2)}\ \sin(z/x_0 + \Psi)$$

Hence at $z = 0$, we have

$$x = x_0\sqrt{(1 - \beta^2/k_0^2 n_0^2)}\ \sin(\Psi) \Rightarrow \Psi = \sin^{-1}[x/\{x_0\sqrt{(1 - \beta^2/k_0^2 n_0^2)}]$$

After propagation through half a pitch length, $z/x_0 = \pi$. Hence at this plane, the transverse position of a general ray is

$$x' = x_0\sqrt{(1 - \beta^2/k_0^2 n_0^2)}\ \sin(\pi + \Psi) = -x_0\sqrt{(1 - \beta^2/k_0^2 n_0^2)}\ \sin(\Psi) = -x$$

All rays emanating from $(x, 0)$ therefore intersect at a single point so a real point image is formed. Since $x'' = -x$, the image is inverted.

(ii) Consider a general forward-travelling field $E_{in}(x)$, specified at the input plane $z = 0$. This may be expanded in terms of guided modes as $E_{in}(x) = \Sigma_\nu a_\nu E_\nu(x)$, where a_ν is the amplitude of the ν th mode, and $E_\nu(x)$ is its transverse field. After propagating a distance z, this field will be modified to

$$E_{out}(x) = \Sigma_\nu a_\nu E_\nu(x) \exp(-j\beta_\nu z),\ \text{where}\ \beta_\nu \approx k_0 n_0 - (2\nu + 1)/2x_0$$

Inserting these values, we get

$$E_{out}(x) = \exp\{-j(k_0 n_0 - 1/2x_0)z\}\ \Sigma_\nu a_\nu E_\nu(x) \exp(j\nu z/x_0)$$

If $z = \pi x_0$ (i.e. if $z = P/2$), this reduces to

$$E_{out}(x) = \exp(-j\psi)\ \Sigma_\nu a_\nu E_\nu(x) \exp(j\nu\pi)$$

where the phase term $\psi = (k_0 n_0 - 1/2x_0)P/2$. At this point E_{out} is very similar to E_{in} multiplied by an unimportant exponential phase factor. However, even-numbered modes have a phase shift of π with respect to odd modes. This corresponds to a sign change in all antisymmetric components in the image so the image is inverted.

8.4 TE reflection coefficient is

$$\Gamma_E = \{n_1 \cos\theta_1 - n_2 \cos\theta_2\}/\{n_1 \cos\theta_1 + n_2 \cos\theta_2\}$$

for a dielectric interface. This may be written as

$$\Gamma_E = \{1 - \alpha\}/\{1 + \alpha\}$$

where $\alpha = n_2 \cos\theta_2/\{n_1 \cos\theta_1\}$, or

$$\alpha = \sqrt{\{(n_2^2 - n_1^2 \sin^2\theta_1)/(n_1^2 - n_1^2 \sin^2\theta_1)\}}$$

This function is discontinuous, tending to $\alpha = 1$ and $\alpha = -\mathrm{j}$ as $\theta_1 \to 90°$ and $n_1 \to n_2$, for $n_1 < n_2$ and $n_1 > n_2$ respectively. In the latter situation (which corresponds to that of a turning point)

$$\Gamma_E = \{1 + \mathrm{j}\}/\{1 - \mathrm{j}\} = \exp(\mathrm{j}\pi/2)$$

8.5 $a = \sqrt{\{2r_0/n_0k_0\}} = \sqrt{\{2 \times 25 \times 10^{-6} \times 0.85 \times 10^{-6}/(1.5 \times 2\pi)\}} = 2.123 \times 10^{-6}$ m. $\theta = 2/(k_0 a) = 2 \times 0.85/(2\pi \times 2.123) = 0.1274$ rad or 7.3°.

8.6 Differentiating the assumed solution for the transverse field, we have:

$$\partial E_{\mu,\nu}/\partial x = \{(\sqrt{2}/a)\, H'_\mu(\sqrt{2}x/a) - (2x/a^2)\, H_\mu(\sqrt{2}x/a)\}\, H_\nu(\sqrt{2}y/a) \exp\{-(x^2 + y^2)/a^2\}$$

$$\partial^2 E_{\mu,\nu}/\partial x^2 = \{(2/a^2)\, H''_\mu(\sqrt{2}x/a) - (4\sqrt{2}x/a^3)\, H'_\mu(\sqrt{2}x/a) + [(4x^2/a^4) - (2/a^2)]\, H_\mu(\sqrt{2}x/a)\} \cdot H_\nu(\sqrt{2}y/a) \exp\{-(x^2 + y^2)/a^2\}$$

$\partial E_{\mu,\nu}/\partial_y$ and $\partial^2 E_{\mu,\nu}/\partial y^2$ may be found in a similar way. Substituting into the waveguide equation, eliminating all common terms, and defining $\xi = \sqrt{2}x/a$ and $\eta = \sqrt{2}y/a$, we then get:

$$\{[H''_\mu(\xi)/H_\mu(\xi)] - 2\xi[H'_\mu(\xi)/H_\mu(\xi)] - 1\} + \{[H''_\nu(\eta)/H_\nu(\eta)] - 2\eta[H'_\nu(\eta)/H_\nu(\eta)] - 1\} + (a^2/2)\{n_0^2k_0^2 - \beta_{\mu,\nu}^2\} = 0$$

The contents of the first curly bracket in the above are a function of ξ only, while those of the second are a function only of η. The third is a constant. For the equation to be satisfied for all ξ and η, we must therefore have

$$\{[H''_\mu(\xi)/H_\mu(\xi)] - 2\xi[H'_\mu(\xi)/H_\mu(\xi)] - 1\} = \text{constant} = A, \text{ say}$$

so that

$$H''_\mu(\xi) - 2\xi H'_\mu(\xi) - (1 + A)\, H_\mu(\xi) = 0$$

Similarly, we may write

$$H''_\nu(\eta) - 2\eta H'_\nu(\eta) - (1 + B)\, H_\nu(\eta) = 0$$

The propagation constant is then defined by

$$(a^2/2)\{n_0^2k_0^2 - \beta_{\mu,\nu}^2\} + A + B = 0 \text{ or } \beta_{\mu,\nu}^2 = n_0^2k_0^2 + (2/a^2)(A + B).$$

The two differential equations above are each of the standard form satisfied by Hermite polynomials, provided $A = -(2\mu + 1)$ and $B = -(2\nu + 1)$. This leads to the required definition of the propagation constant, namely

$$\beta_{\mu,\nu} = n_0 k_0 \sqrt{\{1 - 2(\mu + \nu + 1)/(n_0 k_0 r_0)\}}$$

8.9 Define the new parameter $A = w_0/a$. The end-fire coupling efficiency is then

$$\eta(A) = 4A^2/\{A^2 + 1\}^2 = 4(1/A^2)/\{1 + (1/A^2)\}^2$$

Hence, $\eta(A) = \eta(1/A)$, so the variation will be symmetric in $\alpha = \log(A)$. Now, $d\eta/dA = 8A/\{A^2 + 1\}^2 - 16A^3/\{A^2 + 1\}^3$ so $d\eta/dA = 0$ when $8A\{A^2 + 1\} - 16A^3 = 0$. Hence, η is maximum when $A = 1$, or when $a = w_0$.

Variation of η with α:

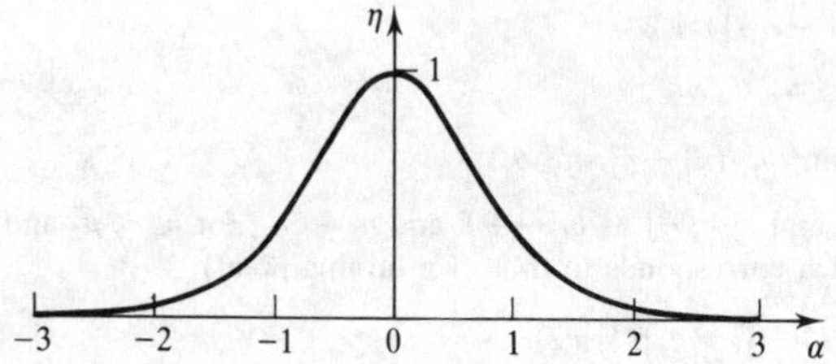

8.10 Use two 8 × 8 stars and 8 extra fused tapered couplers, as shown below.

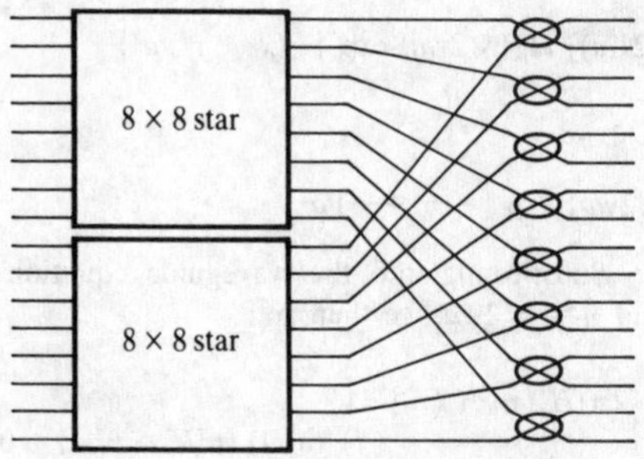

Chapter 9

9.1 For $Ga_{1-x}Al_xAs$, $n \approx 3.57$ when $x = 0$, and $n \approx 3.35$ when $x = 0.35$ so $n \approx 3.57 - 0.6286x$. Hence, for $x = 0.15$, $n = 3.57 - 0.6286 \times 0.15 = 3.476$.

From Chapter 6 a symmetric slab guide will be single-moded provided $h \leq \lambda_0/2\sqrt{(n_1^2 - n_2^2)}$. Assuming that $n_1 = 3.57$, $n_2 = 3.476$ we get $h \leq 0.9/2\sqrt{(3.57^2 - 3.476^2)}\ \mu m = 0.553\ \mu m$.

9.2 For bends of radius r formed in a given channel guide system, the optical power P_z at a distance z round the bend is related to the input power P_0 by $P_z = P_0 \exp(-\alpha z)$, where $\alpha = c_1 \exp(-c_2 r)$ and c_1 and c_2 are constants independent of r.

For 90° bends made using bend radii of r_1 and r_2, define the distance travelled as $z_1 = \pi r_1/2$ and $z_2 = \pi r_2/2$ and the transmission as $P_{z1}/P_0 = T_1$ and $P_{z2}/P_0 = T_2$. Using the formula above

$$T_1 = \exp\{(-c_1 \pi r_1/2) \exp(-c_2 r_1)\} \text{ and}$$

$$T_2 = \exp\{(-c_1 \pi r_2/2) \exp(-c_2 r_2)\}$$

Hence

$$\log_e(T_1)/\log_e(T_2) = (r_1/r_2)\exp\{-c_2(r_1 - r_2)\} \Rightarrow c_2 = \log_e\{r_2 \log_e(T_1)/r_1 \log_e(T_2)\}/\{r_2 - r_1\}$$

For $r_1 = 1$ mm, the radiation loss is 10 dB $\Rightarrow T_1 = 0.1$; for $r_2 = 10$ mm, the radiation loss is 3 dB $\Rightarrow T_2 = 0.5$. Hence, $c_2 = \log_e\{10 \log_e(0.1)/\log_e(0.5)\}/\{10 - 1\}$/mm $= 0.389$/mm.

Similarly, $c_1 = -2 \log_e (T_1)/\{\pi r_1 \exp(-c_2 r_1)\} = -2 \log_e (0.1)/\{\pi \times \exp(-0.389)\}/\text{mm} = 2.163/\text{mm}$. Hence, the transmission through a 45° bend of radius $r_3 = 5$ mm is

$$T_3 = \exp\{(-c_1 \pi r_3/4) \exp(-c_2 r_3)\} = \exp\{-2.163\pi \times (5/4) \exp(-0.389 \times 5)\} = 0.2968$$

corresponding to a loss of 5.27 dB.

9.3 Transmission through the reflective bend may be analysed by considering propagation through an equivalent image guide, the reflection of the actual output guide in the line M–M′ as shown below. Ignoring the diffraction that must occur in the short distances when the light is not fully confined, we may then consider the problem as one of butt-coupling between the two sections of guide shown. If the normalized transverse field in guide 1 is E and that in guide 2 is E', the transmission efficiency is then $\eta_r = |\langle E', E\rangle|^2$ as usual.

Even if $\delta = 0$, the transmission will not reach exactly 100 per cent: since the mirror does not extend over all space, it cannot reflect the entire transverse field. The evanescent tails extending outside the guide will not be reflected, and consequently some power must be lost to radiation.

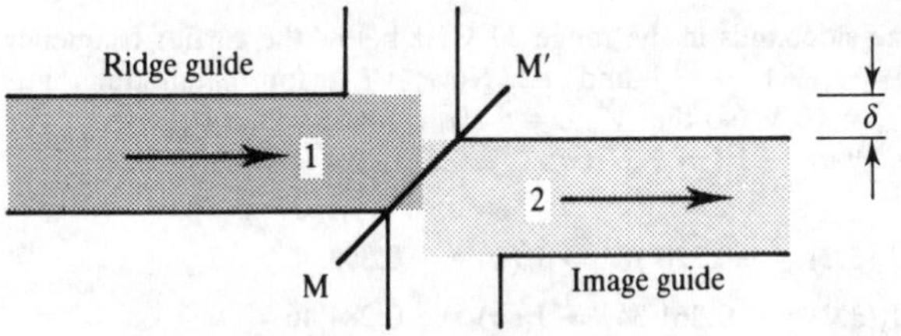

9.4 The input power is proportional to

$$P_{in} = \Sigma_i |A_i|^2 = 1 + 4 + 9 + 0 + 1 + 4 + 16 + 1 = 36$$

The Y-junction tree is an amplitude-summing device, giving the output

$$A_{out} = (1/\sqrt{2})^3 \Sigma_i A_i = (1/\sqrt{2})^3 \{1 - 2 + 3 + 0 - 1 + 2 + 4 - 1\} = 3/\sqrt{2}$$

The output power is proportional to $P_{out} = |A_{out}|^2 = 4.5$. The efficiency of the summing process is then $\eta = P_{out}/P_{in} = 4.5/36 = 0.125$, corresponding to 12.5%.

If the inputs are all arranged to be in-phase, the output amplitude rises to

$$A_{out} = (1/\sqrt{2})^3 \{1 + 2 + 3 + 0 + 1 + 2 + 4 + 1\} = 7/\sqrt{2}$$

Hence $|A_{out}|^2 = 24.5$ and $\eta = P_{out}/P_{in} = 24.5/36 = 0.681$, or 68.1%.

9.5 (i) Uses the horizontal component of the electric field, (ii) the vertical component. Field lines and waveguide positions are as shown below.

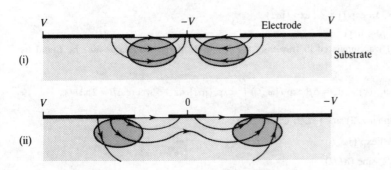

9.6 Cut-off will occur when the electrically-induced index change Δn_{eff} reduces the effective index of the guide to a value equal to the index of the substrate. Hence we require:

$$|\Delta n_{eff}| = n_e^3 r_{33} |V| \Gamma/2g = n_{eff} - n_e$$

or

$$|V| = 2g(n_{eff} - n_e)/n_e^3 r_{33}\Gamma = 2 \times 7 \times 10^{-6} \times 0.001/\{2.2^3 \times 30.8 \times 10^{-12} \times 0.6\} \text{V} = 71.1 \text{ V}$$

As cut-off is approached, the confinement of the optical field will be greatly reduced. Hence the overlap factor Γ will also reduce, and a larger voltage will actually be required to obtain reasonable extinction of the guided beam.

9.7 For a sinusoidal time-varying phase shift $\Phi(t) = \Phi_{max} \sin(\omega_m t)$, the output electric field from the modulator is

$$E_{out}(x,y,t) = E_{out}(x,y)\, \Sigma_n (-1)^n J_n(\Phi_{max}) \exp(jn\omega_m t)$$

leading to an infinite number of sidebands at angular frequencies $\omega + n\omega_m$, where ω is the carrier frequency.

For a 10 MHz modulation frequency, the sidebands in the range 20 MHz below the carrier frequency to 20 MHz above it are those with $n = -2, -1, 0, +1$ and $+2$. Now $(VL)_\pi$ for the modulator is 50 V mm; given that $L = 10$ mm and $V_{p-p} = 10$ V (so that $V_{max} = 5$ V), it follows that $\Phi_{max} = \pi$. The fractional power in the range is then $P = J_0^2(\pi) + 2J_1^2(\pi) + 2J_2^2(\pi)$.

From standard tables:

$$J_0(3.1) = -0.292\,06 \quad J_0(3.2) = -0.320\,18 \Rightarrow J_0(\pi) \approx -0.303\,76$$

$$J_1(3.1) = 0.300\,92 \quad J_1(3.2) = 0.261\,34 \Rightarrow J_1(\pi) \approx 0.284\,46$$

$$J_2(3.1) = 0.486\,20 \quad J_2(3.2) = 0.483\,52 \Rightarrow J_2(\pi) \approx 0.485\,08$$

$\Rightarrow P = (-0.303\,76)^2 + 2 \times (0.284\,46)^2 + 2 \times (0.485\,08)^2 = 0.7247$, corresponding to 72.5%.

9.8 The modulating signal is a travelling wave of the form $V(z,t) = V_0 \exp[j\{\omega_m t - \beta_m z\}]$, where V_0 is the peak voltage across the electrodes and β_m is the propagation constant of the stripline. This wave travels at a phase velocity $v_{pm} = \omega_m/\beta_m$, and gives rise to a moving variation in effective index in the optical waveguide. From our previous analysis of d.c. phase modulators, we may write the travelling variation in effective index as

$$\Delta n_{eff}(z,t) = -\{n_o^3 r\Gamma/2g\}\, V(z,t) = -\{n_o^3 r V_0 \Gamma/2g\} \exp[j\{\omega_m t - \beta_m z\}]$$

where r, Γ and g are the electro-optic coefficient, overlap factor and electrode gap, respectively.

Assume that the optical wave is at $z = 0$ at $t = t_0$. Since it travels at a phase velocity v_{po}, it will reach a point $z = v_{po}(t - t_0)$ at time t. In this case, the effective index change it encounters at any point along the modulator may be written as

$$\Delta n_{eff}(z,t) = \Delta n \exp[j\{\omega_m(z/v_{po} + t_0) - \beta_m z\}] \qquad (\text{where } \Delta n = -\{n_o^3 r V_o \Gamma/2g\})$$

$$= \Delta n \exp[j\{\omega_m t_0 + (v_{pm}/v_{po} - 1)\beta_m z\}]$$

$$= \Delta n \exp(j\omega_m t_0) \exp\{j\alpha z\}$$

where $\alpha = (\omega_m n_m/c)(n_o/n_m - 1)$.

The phase change $\Phi(t_0)$ accumulated in travelling the length L of the modulator may be found by integration, as

$$\Phi(t_0) = \int_0^L 2\pi\Delta n_{eff}(z)/\lambda_0 \, dz = \Delta\beta \exp(j\omega_m t_0) \int_0^L \exp\{j\alpha z\}\, dz \qquad \text{where } \Delta\beta = 2\pi\Delta n/\lambda_0$$

$$= \Delta\beta L \exp(j\alpha L/2) \operatorname{sinc}(\alpha L/2) \exp(j\omega_m t_0)$$

$$= \Delta\beta L\, \phi(\alpha) \exp(j\omega_m t_0)$$

where $\phi(\alpha) = \exp(j\alpha L/2) \operatorname{sinc}(\alpha L/2)$.

Therefore $\Phi(t_0)$ varies sinusoidally with t_0, as might be expected. The term $\phi(\alpha)$ accounts for the frequency-dependence of the induced phase shift, and falls off as the modulation frequency is increased. The effectiveness of the modulator is determined by $|\phi|$. In terms of angular frequency, this can be written $|\phi(\omega_m)| = |\text{sinc}\,(\omega_m n_m L/2c)\,(n_o/n_m - 1)\}|$. This falls to zero for the first time when $(\omega_m n_m L/2c)\,|(n_0/n_m - 1)| = \pi$, or when $f_m = (c/n_m L)/|(n_o/n_m - 1)|$.

Consequently, the effectiveness of a travelling-wave modulator is reduced at high modulation frequencies, if there is any difference between n_m and n_o (and hence if there is any difference between the velocities of the electrical and optical waves). The larger the velocity difference, the lower the frequency at which the induced phase modulation is reduced significantly.

9.9 The normalized output from a dual-modulator Mach-Zehnder interferometer is $P = \cos^2 \Phi$, where Φ is the phase shift applied by one of the modulators. If $(VL)_\pi = 50$ V mm, and $L = 10$ mm, then $V_\pi = 5$ V. Consequently, for $V = 1.25$ V, $\Phi = \pi/4$, and for V $\doteq$ 6.25 V, $\Phi = 5\pi/4$. The interferometer therefore has a static phase bias of $\pi/4$ radians, with a superimposed dynamic variation over a further range of π radians. The output is therefore as shown below.

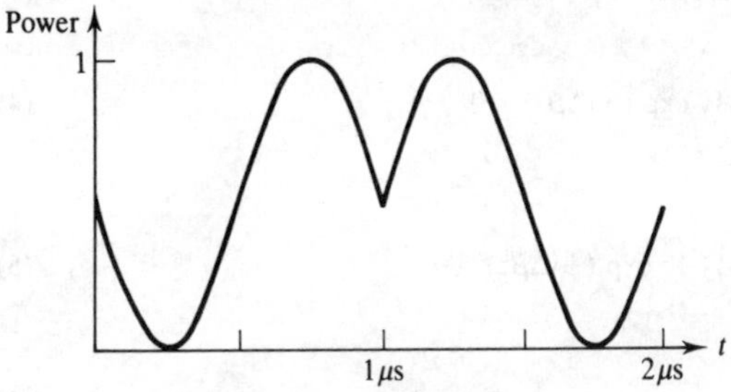

9.10 The device is a reflective variant of the Mach-Zehnder interferometer known as a *Michelson interferometer*. The normalized reflected output is $P = \cos^2 \Phi$, where Φ (the phase imbalance) is given by $\Phi = \beta L$. Given that both L and the effective index are likely to vary with temperature, the device could be used as an integrated optic temperature sensor.

Chapter 10

10.1 The main questionable approximation is that the overlap term $\langle E_2, E_1\rangle$ is negligible, while assuming that the coupling term $\langle k_0^2\,(n_T^2 - n_2^2)\,E_2, E_1\rangle$ is not. This is inconsistent—if the coupling term is to have a finite value, the two modes must overlap to a reasonable extent, so that $\langle E_2, E_1\rangle \neq 0$. In fact, numerical calculations show typically that $\langle E_2, E_1\rangle$ lies in the range 0.1–0.2 for practical devices, compared with unity for $\langle E_1, E_1\rangle$ and $\langle E_2, E_2\rangle$. As a result, it cannot logically be possible to obtain exactly 100 per cent efficiency from a synchronous coupler, or zero crosstalk in a switch—there must always be some light in the "wrong" guide.

10.2 (a) The contribution from all paths involving three scatterings is

$$A_{2L}^3 = (-\text{j}\kappa)^3 A_{10} \int_0^L \int_{z_1}^L \int_{z_2}^L \text{d}z_3\,\text{d}z_2\,\text{d}z_1 = (-\text{j}\kappa L)^3/3!\,A_{10}$$

(b) Paths contributing to A_{2L}^5:

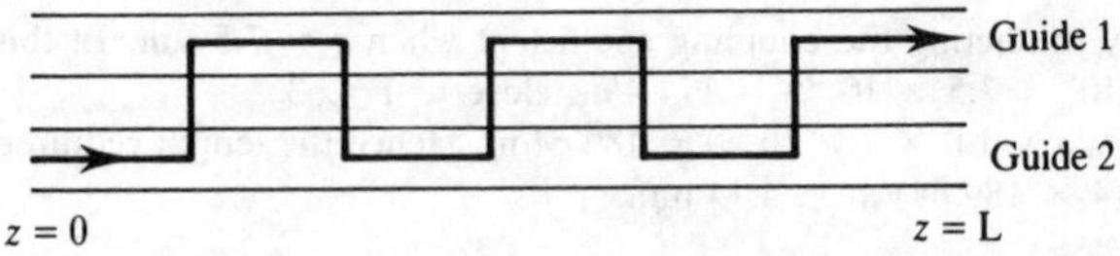

$$A_{2L}^5 = (-j\kappa)^5 A_{10} \int_0^L \int_{z_1}^L \int_{z_2}^L \int_{z_3}^L \int_{z_4}^L dz_5\, dz_4\, dz_3\, dz_2\, dz_1 = (-j\kappa L)^5/5!\, A_{10}$$

Hence

$$A_{2L} = -j\, A_{10}\, \{\kappa L - (\kappa L)^3/3! + (\kappa L)^5/5! \ldots\} = -j\, A_{10} \sin(\kappa L)$$

10.4. Start with

$$dA_1/dz + j\kappa A_2 \exp(-j\Delta\beta z) = 0 \qquad (1)$$

Multiply Eq. (1) by A_1^* to get

$$A_1^*\, dA_1/dz + j\kappa A_1^* A_2 \exp(-j\Delta\beta z) = 0 \qquad (2)$$

Then take the complex conjugate of Eq. (2)

$$\Rightarrow A_1\, dA_1^*/dz - j\kappa A_1 A_1^* \exp(+j\Delta\beta z) = 0 \qquad (3)$$

Now perform a similar process with:

$$dA_2/dz + j\kappa A_1 \exp(+j\Delta\beta z) = 0 \qquad (4)$$

Multiply Eq. (4) by A_2^* to get

$$A_2^* dA_2/dz + j\kappa A_2^* A_1 \exp(+j\Delta\beta z) = 0 \qquad (5)$$

Then take the complex conjugate of Eq. (5)

$$\Rightarrow A_2\, dA_2^*/dz - j\kappa A_2 A_1^* \exp(-j\Delta\beta z) = 0 \qquad (6)$$

Now sum Eqs (2), (3), (5), (6) to get

$$A_1^*\, dA_1/dz + A_1\, dA_1^*/dz + A_2^* dA_2/dz + A_2\, dA_2^*/dz = 0$$

Or

$$d(A_1 A_1^*)/dz + d(A_2\, dA_2^*)/dz = d\,|A_1|^2/dz + d\,|A_2|^2/dz = 0$$

10.5 Assume that the coupling coefficient depends roughly exponentially on the interwaveguide gap, following, for example $\kappa = \kappa_0 \exp(-\alpha g)$. Now $\kappa = \pi/(2L_{\pi/2})$. From the data given we obtain

$$g = 4\ \mu\text{m},\ \kappa = 1.5708/5 \times 10^{-3} = 314.16/\text{m} = \kappa_0 \exp(-\alpha \times 4 \times 10^{-6}) \qquad (1)$$

$$g = 6\ \mu\text{m},\ \kappa = 1.5708/7 \times 10^{-3} = 224.39/\text{m} = \kappa_0 \exp(-\alpha \times 6 \times 10^{-6}) \qquad (2)$$

$$g = 7.5\ \mu\text{m},\ \kappa = 1.5708/9 \times 10^{-3} = 174.53/\text{m} = \kappa_0 \exp(-\alpha \times 7.5 \times 10^{-6}) \qquad (3)$$

From results (1) and (2) we get

$$\exp(-\alpha \times 2 \times 10^{-6}) = 224.39/314.16 = 0.714\,25$$

$$\Rightarrow \alpha = -(1/2 \times 10^{-6}) \log_e (0.714\,25) = 1.6826 \times 10^5/\text{m}$$

Hence

$$\kappa_0 = 314.16/\exp(-1.6826 \times 10^5 \times 4 \times 10^{-6}) = 615.81/\text{m}$$

We can check that this is consistent, by predicting the coupling coefficient when $g = 7.5\ \mu$m. In this case we get $\kappa = 615.81 \exp(-1.6826 \times 10^5 \times 7.5 \times 10^{-6}) = 174.4$/m, close to 174.53.

When $g = 7\ \mu$m, $\kappa = 615.81 \exp(-1.6826 \times 10^5 \times 7 \times 10^{-6}) = 189.64$/m. Hence the length required for 50% coupling efficiency is $L_{\pi/4} = \pi/(4 \times 189.64)$ m $= 4.14$ mm.

10.6 (a) The fabrication process may be controlled by monitoring the optical transmission through the fibre. Typically, the grinding and polishing will take place in a liquid slurry. When sufficient material has been removed, the evanescent field of the fibre mode will extend into this slurry, and there will be considerable attenuation.

(b) The theoretical cross-coupled power is $P_2 = \sin^2(\kappa L)$. This will be very small if $\kappa L \approx \nu\pi$. If the coupler is assembled, and the two blocks are translated across one another, κ will vary, reaching a maximum when the distance between the fibres is minimum. Consequently, if $\kappa L \approx 0$, there will be only a single maximum in P_2 at this point as the blocks are shifted, and this will be very small. The remedy is to remove more material from the fibre by further polishing. If $\kappa L \approx \pi$, on the other hand, there will be two maxima, displaced on either side of the previous point. In each case, the cross-coupling efficiency will be roughly 100 per cent. The solution in this case is to operate the coupler at one of these points. Higher values of κL are similar.

10.7 (a) The contribution from all paths involving three scatterings is

$$A_{B0}^3 = (-j\kappa)^3 A_{F0} \int_0^L \int_0^{z_1} \int_{z_2}^L dz_3\, dz_2\, dz_1 = (-j\kappa L)^3/3\, A_{10}$$

(b) Paths contributing to A_{B0}^5:

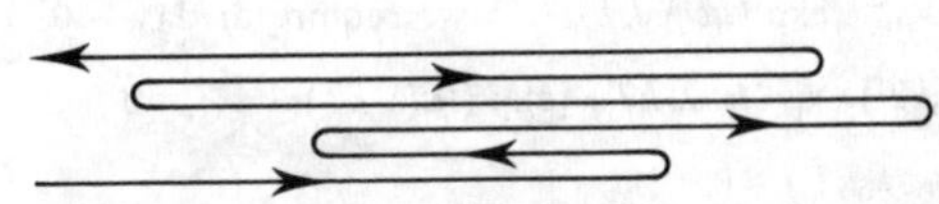

$$A_{B0}^5 = (-j\kappa)^5 A_{10} \int_0^L \int_0^{z_1} \int_{z_2}^L \int_0^{z_3} \int_{z_4}^L dz_5\, dz_4\, dz_3\, dz_2\, dz_1 = (-j\kappa L)^5 (2/15)\, A_{10}$$

Hence

$$A_{B0} = -j A_{10} \{\kappa L - (\kappa L)^3/3 + (\kappa L)^5(2/15) \ldots\} = -j A_{10} \tanh(\kappa L)$$

10.8 Start with

$$dA_F/dz + j\kappa A_B = 0 \Rightarrow d^2 A_F/dz + j\kappa\, dA_B/dz = 0 \qquad (1)$$

Now

$$dA_B/dz - j\kappa A_F = 0 \qquad \Rightarrow dA_B/dz = j\kappa A_F \qquad (2)$$

Substitute in Eq. (1) to get

$$d^2 A_F/dz - \kappa^2 A_F = 0 \qquad (3)$$

Assume the general solution:

$$A_F = A \cosh(\kappa z) + B \sinh(\kappa z) \qquad (4)$$

The boundary conditions are

$$A_F = 1 \text{ at } z = 0 \Rightarrow A = 1 \qquad (5)$$

and

$$A_B = 0 \text{ at } z = L \Rightarrow dA_B/dz = 0 \text{ at } z = L \text{ (from Eq. (2))}$$

Hence

$$\kappa \sinh(\kappa L) + \kappa B \cosh(\kappa L) = 0 \Rightarrow B = -\tanh(\kappa L) \qquad (6)$$

so that

$$A_F = \cosh(\kappa z) - \tanh(\kappa L)\sinh(\kappa z)$$
$$= \{\cosh(\kappa L)\cosh(\kappa z) - \sinh(\kappa L)\sinh(\kappa z)\}/\cosh(\kappa L)$$
$$= \cosh\{\kappa(L - z)\}/\cosh(\kappa L) \quad (7)$$

From Eq. (1) we have

$$A_B = j/\kappa \, dA_F/dz \quad (8)$$
$$= -j\sinh\{\kappa(L - z)\}/\cosh(\kappa L) \quad (9)$$

Equations (7) and (9) then represent the desired solutions.

10.9 Reflection efficiency is

$$R = \tanh^2(\kappa L) \Rightarrow L = 1/\kappa \tanh^{-1}(\sqrt{R}) = 1/200 \tanh^{-1}(\sqrt{0.99})\ \text{m} = 15\ \text{mm}$$

Chapter 11

11.1 For a maximum in the function $\rho(\lambda_0) = (8\pi hnc/\lambda_0^5)/\{\exp(hc/\lambda_0 kT) - 1\}$ we require $d\rho/d\lambda_0 = 0$.

$$\Rightarrow -(5/\lambda_0^6)/\{\exp(hc/\lambda_0 kT) - 1\} + (1/\lambda_0^5)(hc/\lambda_0^2 kT)\exp(hc/\lambda_0 kT)/\{\exp(hc/\lambda_0 kT) - 1\}^2 = 0$$

$$\Rightarrow -5\{\exp(hc/\lambda_0 kT) - 1\} + (hc/\lambda_0 kT)\exp(hc/\lambda_0 kT) = 0.$$

If we now define $\beta = hc/\lambda_0 kT$, this equation can be written as $\beta/5 + \exp(-\beta) - 1 = 0$. This is a transcendental equation, with a single solution for β. Consequently, $\lambda_0 T = hc/k\beta$ = constant. By successive approximation, it may be shown that the solution is $\beta = 4.965\,11$, which yields

$$\lambda_0 T = 6.62 \times 10^{-34} \times 3 \times 10^8/(1.38 \times 10^{-23} \times 4.965\,11) = 2.898 \times 10^{-3}\ \text{m K}$$

11.2 $\lambda = h/\sqrt{(2meV)}$ so that $V = h^2/(2me\lambda^2)$. Hence

$$V = (6.62 \times 10^{-34})^2/\{2 \times 9.1 \times 10^{-31} \times 1.6 \times 10^{-19} \times (0.05 \times 10^{-9})^2\} = 601.98\ \text{V}$$

$$p = \sqrt{(2m\mathscr{E})} = \sqrt{(2 \times 9.1 \times 10^{-31} \times 1.6 \times 10^{-19} \times 601.98)} = 1.3239 \times 10^{-23}\ \text{Ns}$$

11.3 For $z \le 0$ the wavefunction comprises incident and reflected waves, so assume the solution:

$$\boldsymbol{\Psi} = \Psi_0 \exp(j\mathscr{E}t/\hbar)\{\exp(-jk_1 z) + R\exp(+jk_1 z)\} \qquad \text{where } k_1 = \sqrt{(2m\mathscr{E}/\hbar^2)}$$

For $z \ge 0$ the wavefunction comprises a transmitted wave only, so assume the solution:

$$\boldsymbol{\Psi} = \Psi_0 \exp(j\mathscr{E}t/\hbar)\, T\exp(-jk_2 z) \quad \text{where } k_2 = \sqrt{\{2m(\mathscr{E} - \mathscr{V})/\hbar^2\}}$$

Boundary conditions:

$$\text{Match } \boldsymbol{\Psi} \text{ at } z = 0 \Rightarrow 1 + R = T$$

$$\text{Match } d\boldsymbol{\Psi}/dz \text{ at } z = 0 \Rightarrow k_1\{1 - R\} = k_2 T$$

Solve the two simultaneous equations above to determine R and T, giving:

$$R = \{k_1 - k_2\}/\{k_1 + k_2\} \text{ and } T = 2k_1/\{k_1 + k_2\}$$

The probability of reflection is $|R|^2 = |\{k_1 - k_2\}/\{k_1 + k_2\}|^2$. If $\mathscr{E} < \mathscr{V}$, k_2 is complex, so $k_2 = -jk_2'$ where $k_2' = \sqrt{\{2m(\mathscr{V} - \mathscr{E})/\hbar^2\}}$. In this case the probability of reflection is $|R|^2 = |\{k_1 + jk_2'\}/\{k_1 - jk_2'\}|^2 = 1$.

11.4 $\lambda_g = hc/(e\mathscr{E}_g) = 6.62 \times 10^{-34} \times 3 \times 10^8/(1.6 \times 10^{-19} \times \mathscr{E}_g) = 1.24 \times 10^{-6}/\mathscr{E}_g\ \text{m} = 1.24/\mathscr{E}_g\ \mu\text{m}$. So, for Ge: $\lambda_g = 2.12\ \mu$m; for Si: $\lambda_g = 1.107\ \mu$m; for GaP: $\lambda_g = 0.563\ \mu$m; for InP: $\lambda_g = 0.918\ \mu$m; for GaAs: $\lambda_g = 0.837\ \mu$m; for C: $\lambda_g = 0.227\ \mu$m; for SiO_2: $\lambda_g = 0.155\ \mu$m.

Choose Ge for the detector, and SiO_2 for the transparent window at $\lambda_0 = 1.55\ \mu m$. For $Ga_{0.47}In_{0.53}As$, $\lambda_g = 1.698\ \mu m$, so this would also be suitable as a detector material. The advantage over Ge would be improved sensitivity near λ_g, as it is a direct-gap material.

11.5

$$\mathscr{E}_{Fi} = (\mathscr{E}_c + \mathscr{E}_v)/2 + (3kT/4)\log_e(m_h^*/m_e^*) = \mathscr{E}_v + \mathscr{E}_g/2 + (3kT/4)\log_e(m_h^*/m_e^*)$$

Hence, for intrinsic InP, $\mathscr{E}_{Fi} = \mathscr{E}_v + 1.35/2 + 3 \times 0.0259/4 \log_e(0.64/0.077) = \mathscr{E}_v + 0.716$ eV.

11.6 At a temperature T

$$n_i = \sqrt{(N_c N_v)} \exp(-\mathscr{E}_g/2kT)$$

Hence

$$\text{for Ge: } n_i = \sqrt{(1.04 \times 10^{25} \times 6 \times 10^{24})} \exp\{-0.66/(2 \times 0.0259)\} = 2.31 \times 10^{19}\ \text{m}^{-3}$$

$$\text{for Si: } n_i = \sqrt{(2.8 \times 10^{25} \times 1.04 \times 10^{25})} \exp\{-1.12/(2 \times 0.0259)\} = 6.95 \times 10^{15}\ \text{m}^{-3}$$

$$\text{for GaAs: } n_i = \sqrt{(4.7 \times 10^{23} \times 7 \times 10^{24})} \exp\{-1.42/(2 \times 0.0259)\} = 2.25 \times 10^{12}\ \text{m}^{-3}$$

At a temperature $T + \Delta T$

$$n_i' = \sqrt{(N_c N_v)} \exp(-\mathscr{E}_g/2k(T + \Delta T))$$

If ΔT is small

$$n_i' \approx \sqrt{(N_c N_v)} \exp\{(-\mathscr{E}_g/2kT)(1 - \Delta T/T)\} \approx n_i \exp(+\mathscr{E}_g \Delta T/2kT^2)$$

Hence if $n_i' = 2n_i$

$$\Delta T = 2kT^2/\mathscr{E}_g \log_e(2) = (2 \times 0.0259 \times 300/0.66)\log_e(2) = 16.32\ \text{K}$$

11.7 Be: Gp IIA ⇒ acceptor; C: Gp IVB, replaces As atom ⇒ acceptor; Cd: Gp IIB ⇒ acceptor; Mg: Gp IIA ⇒ acceptor; S: Gp VIB ⇒ donor; Se: Gp VIB ⇒ donor; Sn: Gp IVB, replaces Ga atom ⇒ donor; Te: Gp VIB ⇒ donor: Zn: Gp IIB ⇒ acceptor.

For Zn-doping, the material is p-type.

$$\mathscr{E}_{Fp} = \mathscr{E}_v + kT\log_e(N_v/N_a) = \mathscr{E}_v + 0.0259\log_e(7 \times 10^{-24}/10^{18})\ \text{eV} = \mathscr{E}_v + 0.408\ \text{eV}$$

For complete ionization of the dopant atoms, the majority carrier density is $p \approx N_a^- \approx N_a = 10^{18}\ \text{m}^{-3}$. The minority carrier density is then $n = n_i^2/p = (2.25 \times 10^{12})^2/10^{18} = 5 \times 10^6\ \text{m}^{-3}$.

11.8 For acceptors, the probability of ionization is same as the probability of finding an electron at the acceptor level to perform the ionization, i.e. $F(\mathscr{E}_a) = 1/[1 + \exp\{(\mathscr{E}_a - \mathscr{E}_F)/kT)\}]$. Now

$$\mathscr{E}_a - \mathscr{E}_F = (\mathscr{E}_a - \mathscr{E}_v) - (\mathscr{E}_F - \mathscr{E}_v) = 0.031 - 0.408 = -0.377\ \text{eV}$$

Hence

$$F(\mathscr{E}_a) = 1/[1 + \exp\{-0.377/0.0259\}] = 0.99\,999\,5$$

This is so close to unity that the assumption of complete ionization is a good one.

Sn is an n-type dopant. For donors, the probability of ionization is the same as the probability of *not* finding an electron at the donor level, i.e. $\{1 - F(\mathscr{E}_d)\}$. Hence, $N_d^+ = N_d\{1 - F(\mathscr{E}_d)\}$.

11.9 The relationship between the Einstein A and B coefficients may be derived by comparing the expression

$$\rho(\omega)\,d\omega = A/[B_{12}\exp(\hbar\omega/kT) - B_{21}]\,d\omega$$

with the black-body radiation law

$$\rho(\omega)\,d\omega = (\hbar n^3\omega^3/\pi^2 c^3)/[\exp(\hbar\omega/kT) - 1]\,d\omega.$$

Clearly, the term $-B_{21}$ (which is associated with stimulated emission) in the former translates to -1 in the latter. If this process is neglected, we would expect the black-body radiation law to modify to

$$\rho(\omega)\,d\omega = [(\hbar n^3\omega^3/\pi^2c^3)/\exp(\hbar\omega/kT)]\,d\omega = (\hbar n^3\omega^3/\pi^2c^3)\exp(-\hbar\omega/kT)\,d\omega$$

This relation was originally proposed by Wien in 1896 before the idea of stimulated emission was suggested; it is a reasonable approximation to the correct result when $\exp(\hbar\omega/kT) \gg 1$ (i.e. for large values of ω or low temperatures).

If $\hbar\omega/kT \ll 1$, on the other hand, the black-body radiation law may be approximated as

$$\rho(\omega)\,d\omega \approx (\hbar n^3\omega^3/\pi^2c^3)/[1 + (\hbar\omega/kT) - 1]\,d\omega \approx n^3\omega^2kT/\pi^2c^3\,d\omega$$

This is the *Rayleigh-Jeans law*, originally derived using classical arguments. Its failure as $\omega \to \infty$ (a difficulty referred to as the *ultraviolet catastrophe*, since it implies the existence of infinite energy at short wavelengths) was one of the main spurs to the development of quantum theory.

11.10 In the steady-state, $\partial\phi/\partial z = \{G(n - n_0)/v_g\}\phi = g_p\phi$, where g_p is the optical power gain coefficient. Hence, $g_p = G(n - n_0)/v_g$. For a given power gain g_p, we require an electron density of $n \approx n_0 + g_pv_g/G$. Assuming a refractive index of 3.5, we may approximate v_g by the phase velocity, $c/3.5$. For $g_p = 10^6\ \mathrm{m}^{-1}$, $n_0 = 10^{24}\ \mathrm{m}^{-3}$ and $G = 10^{-12}\ \mathrm{m}^3/\mathrm{s}$ we therefore obtain $n \approx 10^{24} + 10^6 \times 3 \times 10^8/(3.5 \times 10^{-12}) \approx 8.7 \times 10^{25}\ \mathrm{m}^{-3}$.

Chapter 12

12.1 The built-in voltage across a p–n junction is

$$V_{bi} = (kT/e)\log_e\{N_aN_d/n_i^2\}$$

Assuming that the intrinsic carrier concentration is $n_i = 1.4 \times 10^{16}\ \mathrm{m}^{-3}$ we get

$$V_{bi} = 0.0258\log_e\{10^{24} \times 10^{23}/2 \times 10^{32}\} = 0.873\ \mathrm{V}$$

Similarly, the depletion layer width is

$$w(0) = \sqrt{\{(2\epsilon V_{bi}/e)(N_a + N_d)/N_aN_d\}}$$

with no bias voltage. Assuming that $\varepsilon = 11.7\varepsilon_0$ for Si, we obtain

$$w(0) = \sqrt{(2 \times 11.7 \times 8.85 \times 10^{-12} \times 0.873/1.6 \times 10^{-19})(10^{24} + 10^{23})/(10^{24} \times 10^{23}))}$$
$$= 1.11 \times 10^{-7}\ \mathrm{m, or}\ 0.111\ \mu\mathrm{m}.$$

Under a forward-bias of V volts, the depletion layer width becomes $w(V) = w(0)\sqrt{\{1 - V/V_{bi}\}}$; hence under a reverse bias of 5 V, $w = 0.111\sqrt{\{1 + 5/0.873\}} = 0.288\ \mu\mathrm{m}$. Assuming that all the voltage is dropped across this highly resistive region, the average electric field across the depletion layer is then $E = (5 + 0.873)/0.288 \times 10^{-6} = 20.39 \times 10^6\ \mathrm{V/m}$.

The drift velocity resulting from a given electric field is $v_d = \mu E$. Hence for electrons, $v_d = 0.135 \times 20.39 \times 10^6 = 2.75 \times 10^6\ \mathrm{m/s}$. The transit time across the depletion layer is therefore $\tau = w/v_d = 0.288 \times 10^{-6}/2.75 \times 10^6 = 0.105 \times 10^{-12}$ s, or 0.105 ps; for holes, $v_d = 0.978 \times 10^6$ m/s, and $\tau = 0.294$ ps. The implication for photodiodes is that any photo-generated carriers will be swept out of the depletion region before they can recombine.

12.2 The current through the diode is $I = I_s\{\exp(eV/kT) - 1\}$, where $I_s = A \times J_s$ and $J_s = \{D_een_p/L_p + D_hep_n/L_n\}$. Since $N_d \gg N_a$, the current will be carried almost entirely by electrons, so we may write $J_s \approx D_een_p/L_p$. The diffusion coefficient may be calculated using the Einstein relation, $D_e = \mu_ekT/e$. Since $\mu_e = 0.135\ \mathrm{m^2/V\,s}$, $D_e = 0.135 \times 0.0258 = 3.48 \times 10^{-3}\ \mathrm{m^2/s}$. The minority carrier concentration on the p-side is $n_p = n_i^2/N_a$; since the intrinsic carrier concentration in Si is $1.4 \times 10^{16}\ \mathrm{m}^{-3}$, $n_p = 1.96 \times 10^{11}\ \mathrm{m}^{-3}$. Hence $J_s = 3.48 \times 10^{-3} \times 1.6 \times 10^{-19} \times 1.96 \times 10^{11}/1 \times 10^{-7} = 1.09 \times 10^{-3}\ \mathrm{A/m^2}$. For a cross-sectional area of $10^{-6}\ \mathrm{m}^2$, $I_s = 1.09$ nA. For a forward voltage of 0.5 V we then obtain $I = 1.09 \times 10^{-9} \times \{\exp(0.5/0.0258) - 1\} = 0.284$ A, or 284 mA. For a lifetime $\tau_e = 1\ \mu$s, the diffusion length is $L_e = \sqrt{(D_e\tau_e)} = \sqrt{\{3.48 \times 10^{-3} \times 10^{-6}\}} = 5.9 \times 10^{-5}$ m, or 59 μm. Since this is much greater than L_p, the thin-diode approximation is valid.

12.3 From Chapter 9 the index change in an electro-optic modulator is $\Delta n_{eff} \approx -n^3 rVT/2g$. Taking the breakdown electric field to be $V/g = 3 \times 10^7$ V/m, assuming that $n = 3.2$, that the electro-optic coefficient is $r_{41} = -1.4 \times 10^{-12}$ and that the overlap factor is $\Gamma \approx 0.5$, we obtain $\Delta n_{eff} \approx 3.2^3 \times 1.4 \times 10^{-12} \times 3 \times 10^7 \times 0.5/2 \approx 3.5 \times 10^{-4}$.

From Chapter 3 the index change obtained by carrier injection is $\Delta n = -\Delta N e^2 \lambda_0^2 / 8\pi^2 c^2 m^* \varepsilon_0 n$, where ΔN is the density of injected carriers and m^* is their effective mass. Since $m_h^* \gg m_e^*$, only electron injection will be useful. Assuming that m_e^*/m is 0.077, and that $\lambda_0 = 1.5$ μm, we obtain

$$|\Delta n| = \Delta N \times (1.6 \times 10^{-19} \times 1.5 \times 10^{-6})^2 / \{8\pi^2 \times (3 \times 10^8)^2 \times 0.077 \times 9.1 \times 10^{-31} \times 8.85 \times 10^{-12} \times 3.2\} \approx \Delta N \times 4 \times 10^{-27}$$

Taking the upper limit of the injected electron density to be $\Delta N \approx 10^{24}$ m^{-3}, we then get $|\Delta n| \approx 4 \times 10^{-3}$. Carrier injection can therefore yield an index change an order of magnitude larger than the value obtained by the electro-optic effect.

12.4 Since the photocurrent generated in the PIN diode is defined as $I_P = \eta P e \lambda_0 / \mathbf{h}c$, the quantum efficiency is $\eta = (\mathbf{h}c/e)(I_P/P\lambda_0)$. This may be written as $\eta = 1.24\ (I_P/P\lambda_0)$ when λ_0 is measured in μm. For a diode of cross-section 0.1 mm^2 and an illuminating beam of intensity 1 mW/cm^2, $P = 0.1 \times 10^{-6} \times 10^{-3} \times 10^4 = 10^{-6}$ W. Consequently, the quantum efficiency is $\eta = 1.24 \times 0.996 \times 10^{-6}/\{1.3 \times 10^{-6}\} = 0.95$.

Assuming constant quantum efficiency and illumination intensity, I_P is proportional to λ_0. At $\lambda_0 = 1.5$ μm, we would then expect $I_P = 0.996 \times 1.5/1.3 = 1.15$ μA. At $\lambda_0 = 1.7$ μm, however, we would expect $\eta = 0$ and $I_P = 0$, since $\lambda_0 > \lambda_g$ for $Ga_{0.47}In_{0.53}As$.

12.5 The power launched into the optical fibre is $P_L = \eta_e \eta_c P'$, where P' is the internally-generated power (given by $P' = \eta' \mathbf{h}cI'/e\lambda_g$), η_e is the external efficiency, and η_c is the coupling efficiency into the optical fibre. The internal efficiency is then found as $\eta' = P_L e \lambda_g / \mathbf{h}cI' \eta_e \eta_c$, or as $\eta' = P_L \lambda_g / 1.24 I' \eta_e \eta_c$, if λ_g is measured in μm.

Now $\eta_e = 1/n(n+1)^2$, so for GaAs (for which $n = 3.5$) we obtain $\eta_e = 0.0141$. Similarly $\eta_c = \{4n'/(n'+1)^2\}\ NA^2$, so for an optical fibre (for which $n' \approx 1.5$) with an NA of 0.3, we obtain $\eta_c = 0.96 \times 0.3^2 = 0.0864$. We therefore obtain $\eta' = 50 \times 10^{-6} \times 0.82/\{1.24 \times 50 \times 10^{-3} \times 0.0141 \times 0.0864\} = 0.543$. Since $\eta' = \tau_e/\tau_{rr}$, the radiative recombination time is $\tau_{rr} = \tau_e/\eta' = 10/0.543 = 18.42$ ns.

When the LED is modulated, the ratio of a.c. to d.c. optical power generated is $|P''/P'| = 1/\sqrt{\{1 + \omega^2\tau_e^2\}}$. When this ratio has fallen to 3 dB below its d.c. value, $|P''/P'| = 1/2$, so that $\omega = \sqrt{3}/\tau_e$. Hence, the 3 dB modulation bandwidth is $\sqrt{3}/\{2\pi \times 10 \times 10^{-9}\}$ Hz, or 27.6 MHz.

12.6 The results of Chapter 4 suggest that a medium with complex relative dielectric constant $\varepsilon_r = \varepsilon_r' + j\varepsilon_r''$ will have a propagation constant $k = k' + jk''$. When $\varepsilon_r'' \ll \varepsilon_r'$, these quantities may be written in terms of a complex refractive index $n = n' + jn''$ as $k' = 2\pi n'/\lambda_0$ and $k'' = k'n''/n'$. Now, k'' is the amplitude gain coefficient g, equal to $g_p/2$. We therefore deduce that a power gain g_p will follow from the existence of an imaginary index n'' equal to $g_p n'/2k'$ (or gn'/k').

The results of Chapter 9 also show that the change in the real part of the effective index of a guided mode due to an index perturbation $\Delta n'$ can be found as an overlap between the index perturbation and the normalized transverse modal field, i.e. as $\Delta n'_{eff} = \langle \Delta n', |E|^2 \rangle$. We would therefore expect that an imaginary index perturbation n'' would alter the imaginary part of the effective index by $\Delta n''_{eff}$, found by analogy as $\Delta n''_{eff} = \langle n'', |E|^2 \rangle = n'/k' \langle g, |E|^2 \rangle$. As a result of this change in effective index, the propagation constant β' should alter to $\beta' + j\beta''$, where $\beta'' = 2\pi \Delta n''_{eff}/\lambda_0 = \langle g, |E|^2 \rangle$. Clearly β'' is the effective amplitude gain g_{eff}, and is found as an overlap between the normalized transverse field and the gain distribution. Similarly, the effective power gain is $g_{peff} = 2g_{eff} = \langle g_p, |E|^2 \rangle$.

12.7 Including a non-zero Γ, the rate equations modify to $d\phi/dt = \beta n/\tau_{rr} + \Gamma G\phi(n - n_0) - \phi/\tau_p$ and $dn/dt = I/ev - n/\tau_e - \Gamma G\phi(n - n_0)$. Solving these gives the conventional answer, namely that the power out of the cavity is $P = \mathbf{h}c\{I - I_t\}/e\lambda_g$ in the lasing regime. This time, however, not all the power is coupled out of the end mirrors; some is lost by absorption. Including loss in both core and cladding, the threshold gain is $\Gamma g_p = \Gamma \alpha_{pcore} + (1 - \Gamma)\alpha_{pclad} + 1/L \log_e (1/R_1R_2)$. In fact, the gain is clamped to this value throughout the lasing regime; this shows that the gain goes to make up absorption and output coupling losses, in the proportions shown. The power lost from the cavity is split into power absorbed and power coupled out through the mirrors, in a similar way. The useful fraction therefore defined by an efficiency

$$\eta = \{1/L \log_e (1/R_1R_2)\}/\{\Gamma\alpha_{pcore} + (1-\Gamma)\alpha_{pclad} + 1/L \log_e (1/R_1R_2)\}$$

and the usable output is $P = \eta hc\{I - I_t\}/e\lambda_g$.

A sketch of the data suggests that the threshold is between 20 and 30 mA. Using the second and third points, we get $dP/dI = (2 - 0.5)/(35 - 30) = 0.3$ A/W. When lasing, $P = dP/dI\ \{I - I_t\} \Rightarrow I_t = I - P/(dP/dI) = 30 - 0.5/0.3$ mA, or 28.33 mA. The threshold current density is then $J_t = I_t/A$, where A is the area through which the current flows. Here, we get $J_t = 28.33 \times 10^{-3}/(250 \times 3 \times 10^{-12}) \approx 3.777 \times 10^7$ A/m^2. Similarly, the output from one mirror is $P = \eta hc\{I - I_t\}/2e\lambda_g$ ($= 0.62\eta\ \{I - I_t\}/\lambda_g$ if λ_g is in μm). Differentiating this, we obtain $dP/dI = 0.62\eta\lambda_g$, so that $\eta = (\lambda_g/0.62)\ dP/dI$. Hence, the efficiency is $\eta = 0.82 \times 0.3/0.62 = 0.4$. Defining the average loss as

$$\alpha_{pav} = \Gamma\alpha_{pcore} + (1-\Gamma)\alpha_{pclad}$$

We get

$$\eta = \{1/L \log_e (1/R_1R_2)\}/\{\alpha_{pav} + 1/L \log_e (1/R_1R_2)\}$$

Hence

$$\alpha_{pav} = (1/\eta - 1)\ \{1/L \log_e (1/R_1R_2)\}$$

Taking $R_1 = R_2 \approx 0.555$ (normal incidence on a GaAs/air interface), $1/L \log_e (1/R_1R_2) = 4710$ m^{-1}, so $\alpha_{pav} = (1/0.4 - 1) \times 4170 = 7065$ m^{-1}.

12.8 Start with the photon rate equation

$$d\phi/dt = \beta n/\tau_{rr} + G\phi(n - n_0) - \phi/\tau_p \qquad (1)$$

Now assume that $\phi(t) = \phi' + \phi''(t)$, and $n(t) = n' + n''(t)$ where ϕ' and n' are d.c. values. Ignoring all second-order terms, Eq. (1) becomes

$$d\phi''/dt = \beta\{n' + n''\}/\tau_{rr} + G\{\phi' + \phi''\}(n - n_0) + G\phi' n'' - \{\phi' + \phi''\}/\tau_p \qquad (2)$$

However, Eq. (1) implies that the d.c. terms satisfy

$$0 = \beta n'/\tau_{rr} + G\phi'(n - n_0) - \phi'/\tau_p \qquad (3)$$

so that Eq. (2) reduces to

$$d\phi''/dt = \beta n''/\tau_{rr} + G\phi''(n - n_0) + G\phi' n'' - \phi''/\tau_p \qquad (4)$$

Now, the photon lifetime τ_p is defined by $1/\tau_p = G(n - n_0)$. Using this, and ignoring the contribution of spontaneous emission, Eq. (4) reduces to

$$d\phi''/dt = G\phi' n'' \qquad (5)$$

Now turn to the electron rate equation

$$dn/dt = I/ev - n/\tau_e - G\phi(n - n_0) \qquad (6)$$

and assume that $I(t) = I' + I''(t)$, where I' is a d.c. value. Ignoring second-order terms, Eq. (6) becomes

$$dn''/dt = \{I' + I''\}/ev - \{n' + n''\}/\tau_e - G\{\phi' + \phi''\}(n - n_0) - G\phi' n'' \qquad (7)$$

However, Eq. (6) implies that the d.c. terms also satisfy

$$0 = I'/ev - n'/\tau_e - G\phi'(n - n_0) \qquad (8)$$

Using this, Eq. (7) reduces to

$$dn''/dt = I''/ev - n''/\tau_e - G\phi''(n - n_0) - G\phi' n'' \qquad (9)$$

Using the definition of τ_p once again, Eq. (9) may be written as

$$dn''/dt = I''/ev - n/\tau_e - \phi/\tau_p - G\phi' n'' \qquad (10)$$

Equations (5) and (10) may then be rearranged to yield a single second-order equation in ϕ'' alone:

$$d^2\phi''/dt^2 + \{1/\tau_e + G\phi'\}d\phi''/dt + (G\phi'/\tau_p)\phi'' = G\phi' I''/ev \tag{11}$$

Defining a resonant frequency and damping factor as

$$\omega_r = \sqrt{(G\phi'/\tau_p)} \text{ and } \zeta = \{1/\tau_e + \tau_p\omega_r^2\}/2\omega_r \tag{12}$$

Equation 11 may be written as

$$d^2\phi''/dt^2 + 2\zeta\omega_r d\phi''/dt + \omega_r^2\phi'' = \tau_p\omega_r^2 I''/ev \tag{13}$$

The photon flux out of the cavity is $\Phi(t) = v\phi(t)/\tau_p$, and the optical power output is $P(t) = \mathbf{h}cv\Phi(t)/\lambda_g$. Defining $P(t) = P' + P''(t)$, where P' is the D.C. output, then $P' = \mathbf{h}cv\phi'\lambda_g\tau_p$ and $P''(t) = \mathbf{h}cv\phi''(t)\lambda_g\tau_p$. Substituting the former into Eq. (12) we can rewrite the resonant frequency as $\omega_r = \sqrt{(G\lambda_g P'/\mathbf{h}cv)}$. Similarly, substituting the latter into Eq. (13), we obtain

$$d^2P''/dt^2 + 2\zeta\omega_r dP''/dt + \omega_r^2 P'' = \{\mathbf{h}c\omega_r^2/e\lambda_g\}I''$$

12.9 Assume that the current step is described by $I''(t) = 0$ for $t \leq 0$, $I''(t) = i''$ for $t > 0$. Begin by finding the complementary function (CF), the solution to

$$d^2P''/dt^2 + 2\zeta\omega_r dP''/dt + \omega_r^2 P'' = 0$$

This can be found by assuming that $P''(t) = p \exp(\gamma t)$, where γ is the solution of the auxiliary equation $\gamma^2 + 2\zeta\omega_r\gamma + \omega_r^2 = 0$, i.e. $\gamma = -\zeta\omega_r \pm \omega_r\sqrt{\{\zeta^2 - 1\}}$. For light damping, $\gamma \approx -\zeta\omega_r \pm j\omega_r$, so the CF may be written as $P''(t) = p'' \exp(-\zeta\omega_r t)\cos(\omega_r t + \psi)$, where p'' and ψ are constants.

Now find the particular integral (PI), the solution to the differential equation

$$d^2P''/dt^2 + 2\zeta\omega_r dP''/dt + \omega_r^2 P'' = \{\mathbf{h}c\omega_r^2/e\lambda_g\}i''$$

Clearly, the PI is $P'' = \{\mathbf{h}c/e\lambda_g\}i''$.

The total solution is then found as the sum of the PI and the CF, with any unknown constants chosen to satisfy the boundary conditions (BCs). We may therefore write

$$P''(t) = \{\mathbf{h}c/e\lambda_g\}i'' + p'' \exp(-\zeta\omega_r t)\cos(\omega_r t + \psi)$$

For the BCs, we may assume that $\phi'' = 0$ and $n'' = 0$ at $t = 0$. Note that Eq. (5) in the solution to Question 12.8 implies that the latter condition reduces to $d\phi''/dt = 0$. Hence $P'' = dP''/dt = 0$ at $t = 0$, so that

$$P''(t) = p''\{1 - \exp(-\zeta\omega_r t)\cos(\omega_r t)\}$$

where $p'' = (\mathbf{h}cI''e\lambda_g)$. A plot of the solution is shown below, for $\zeta = 0.05$. The output settles to its final value p'' only after a number of oscillations. There is also considerable overshoot.

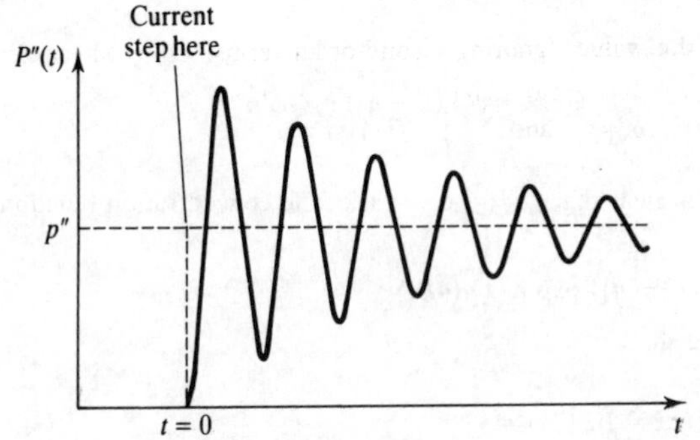

12.10 For an N-element Y-junction-coupled array laser, the threshold gains of the transverse modes are given by $g_\nu = 1/2L \log_e (1/f_\nu LR_1R_2)$, where ν is the mode number, R_1 and R_2 are the end-facet reflectivities, L is the cavity length, and $f_\nu = 1/2\{1 + \cos[\nu\pi/(N + 1)]\}$ is the transmission of the νth mode through the Y-junction network. For $N = 9$, $L = 500 \times 10^{-6}$ m, and $R_1 = R_2 \approx (3.5 - 1)/(3.5 + 1) = 0.555$ (assuming normal reflection at a GaAs/air interface), we obtain

ν	f_ν	g_ν	ν	f_ν	g_ν
1	0.9755	1200/m	6	0.3455	2238/m
2	0.9045	1276/m	7	0.2061	2755/m
3	0.7939	1406/m	8	0.0955	3524/m
4	0.6545	1599/m	9	0.0245	4886/m
5	0.5000	1868/m			

Chapter 13

13.2 The polynomial equation

$$a(x,y) = xy\, a_{GaP} + x(1-y)\, a_{GaAs} + (1-x)y\, a_{InP} + (1-x)(1-y)\, a_{InAs}$$

may be rewritten as

$$a(x,y) = \alpha + \beta x + \gamma y + \delta xy$$

where $\alpha = a_{InAs}$, $\beta = \{a_{GaAs} - a_{InAs}\}$, $\gamma = \{a_{InP} - a_{InAs}\}$ and $\delta = \{a_{Gap} - a_{GaAs} + a_{InAs} - a_{InP}\}$. It is then easy to show that compounds lattice-matched to InP are defined by the locus

$$y = \{1 - (\beta/\gamma)x\}/\{1 + (\delta/\gamma)x\}$$

Since $\beta = -0.4054$ Å, $\gamma = -0.1894$ Å and $\delta = -0.013$ Å, $\delta/\gamma \ll \beta/\gamma$. We may therefore put

$$y \approx \{1 - (\beta/\gamma)x\}$$

$$\text{or } y \approx 1 - 2.14x$$

When $y = 0$, $x = 1/2.14 = 0.47$, so lattice-matched indium gallium arsenide has the composition

$$In_{0.53}Ga_{0.47}As$$

13.3 Evaluating the relevant derivatives of the solution given, we obtain

$$\partial C/\partial x = \{S/\surd(\pi Dt)\}\,(-x/2Dt)\exp(-x^2/4Dt)$$

$$\partial^2 C/\partial x^2 = \{S/\surd(\pi Dt)\}]\,\{x^2/4D^2t^2 - 1/2Dt\}\exp(-x^2/4Dt)$$

$$\partial C/\partial t = \{S/\surd(\pi D)\}\,\{x^2/4Dt^{5/2} - 1/2t^{3/2}\}\exp(-x^2/4Dt)$$

Hence, by inspection, $\partial C/\partial t = D\,\partial^2 C/\partial x^2$.

A normalized Gaussian distribution has

$$f(x) = 1/\surd(2\pi\sigma^2)\exp[-x^2/2\sigma^2] \quad \text{and} \quad \int_{-\infty}^{\infty} f(x)\,dx = 1$$

Since the function is symmetric about $x = 0$, it is clear that $\int_0^\infty f(x)\,dx = 1/2$. The concentration function $C(x,t)$ may be written as

$$C(x,t) = 2S\,\{1/\surd(4\pi Dt)\}\exp[-x^2/(4Dt)]$$

By comparison with the previous result, we see that

$$\int_0^\infty C(x,t)\,dx = 1/2\,(2S) = S$$

The quantity S represents the surface density of dopant at $t = 0$.

13.4 The diffusion length is $L_D = 2\sqrt{(Dt)}$, where $D = D_0 \exp(-E_a/kT)$. As given, $D_0 = 10^{-1}$ cm^2 s^{-1} $= 10^{-5}$ m^2 s^{-1}. At $T = 900$°C $= 1173$ K, we therefore get

$$D = 10^{-5} \exp\{-3 \times 1.6 \times 10^{-19}/1.38 \times 10^{-23} \times 1173\} = 1.32 \times 10^{-18} \text{ m}^2 \text{ s}^{-1}$$

Hence

$$L_D = 2\sqrt{\{1.32 \times 10^{-18} \times 8 \times 3600\}} = 3.9 \times 10^{-7} \text{ m, or } 0.39\ \mu\text{m}$$

Similarly, at $T = 300$°C $= 1273$ K, $D = 1.36 \times 10^{-17}$ m^2 s^{-1} and $L_D = 1.25$ μm.

13.5 (i) Evaluating the relevant derivatives of the solution given, we obtain

$$\partial C/\partial x = (S/4\pi Dt)(-x/2Dt) \exp[-(x^2 + y^2)/4Dt]$$

$$\partial^2 C/\partial x^2 = (S/4\pi Dt)\{x^2/4D^2t^2 - 1/2Dt\} \exp[-(x^2 + y^2)/4Dt]$$

$$\nabla^2 C = (S/4\pi Dt)\{(x^2 + y^2)/4D^2t^2 - 1/Dt\} \exp[-(x^2 + y^2)/4Dt]$$

$$\partial C/\partial t = (S/4\pi D)\{(x^2 + y^2)/4Dt^3 - 1/t^2\} \exp[-(x^2 + y^2)/4Dt]$$

Hence by inspection, $\partial C/\partial t = D\,\nabla^2 C$ as required. The solution represents the concentration distribution arising from a line source parallel to the z-axis, for times much greater than that required to diffuse the source material into the substrate.

(ii) Diffusion from a strip of Ti metal may be modelled by the superposition of a distribution of line sources spaced along the line $x = 0$ from $y = -w/2$ to $y = +w/2$. Using the solution above for a line source, we therefore obtain

$$C(x,y,t) = (S/4\pi Dt) \int_{-w/2}^{w/2} \exp\{-[x^2 + (y - y')^2]/4Dt\}\, dy'$$

$$= (S/4\pi Dt) \exp(-x^2/4Dt) \int_{-w/2}^{w/2} \exp\{-(y - y')^2/4Dt\}\, dy'$$

The *error function* is defined by

$$\text{erf}(u) = 2/\sqrt{\pi} \int_0^u \exp(-u^2)\, du$$

Putting $u = (y' - y)/L_D$, where $L_D = 2\sqrt{(Dt)}$, we obtain

$$C(x,y,t) = C \exp(-x^2/L_D^2)\,[\text{erf}\{(w + 2y)/2L_D\} + \text{erf}\{(w - 2y)/2L_D\}]$$

where C is a constant. Assuming now that the index change resulting from diffusion is proportional to the local Ti concentration, we obtain the index distribution of the diffused channel guide as

$$n(x,y) = n_s + \Delta n \exp(-x^2/L_D^2)\,[\text{erf}\{(w + 2y)/2L_D\} + \text{erf}\{(w - 2y)/2L_D\}]$$

13.6 Referring to the figure below, by simple trigonometry, $w_1 = r/\cos\phi$, and $w_2 = w_1 - s\tan\phi = r/\sin\theta - s/\tan\theta$. For the ⟨111⟩ Si planes, $\theta = 54.74°$; for a fibre diameter of 125 μm, $r = 62.5$ μm; for a clearance of 5 μm, $s = 2.5$ μm. Hence, $w_2 = 74.77$ μm. The groove width is then $2w_2 = 149.55$ μm. The groove depth is $h = r/\sin\phi - s = r/\cos\theta - s = 105.76$ μm.

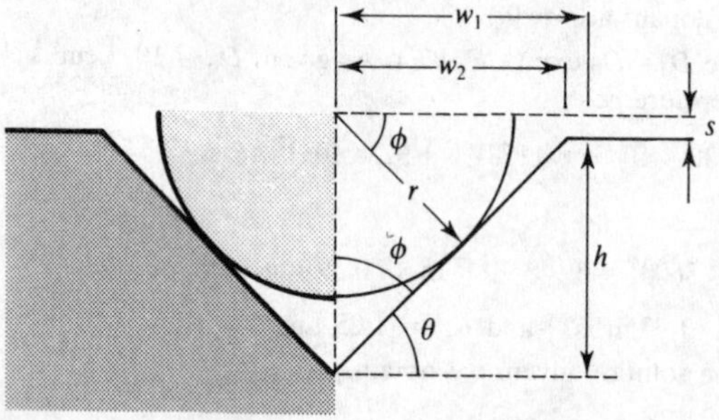

13.7 The code given will generate the central section of a 5 mm-long directional coupler, as shown below. The guides are specified on mask level #1; the guide widths are both 7 μm, and the interwaveguide gap is also 7 μm. The electrodes are specified on mask level #2 and are of width 50 μm. Since the electrodes overlap the guides, the vertical electric field component can be used to control the device, so that z-cut, y-propagating $LiNbO_3$ would be appropriate.

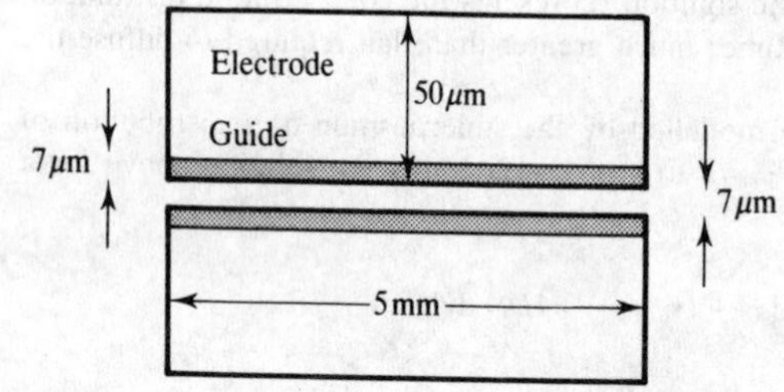

13.8 The figure below shows part of one end of the S-bend geometry, which is symmetric about the device centreline. To satisfy the design constraints, we must have

$$R\{1-\cos\theta\} \approx R\theta^2/2 = (150-7)\times 10^{-6}/2 = 71.5\times 10^{-6}\ \text{m}$$

Hence the bend angle is

$$\theta \approx \sqrt{\{71.5\times 10^{-6}\times 2/30\times 10^{-3}\}} = 0.069\,041\ \text{rad.}$$

The transition length is then

$$L = 2R\sin\theta \approx 2R\theta = 2\times 30\times 10^{-3}\times 0.069\,041 = 4.14\times 10^{-3}\ \text{m, or } 4.14\ \text{mm}$$

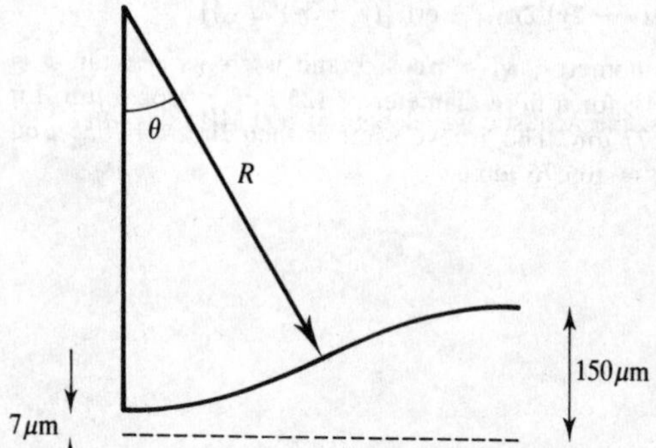

13.9 Assuming that negative electron resist and subtractive processing are used to fabricate the mask set, the shapes defined by the code will emerge as chrome blocks on the glass mask. A typical Ti:$LiNbO_3$ process involving subtractive definition of the waveguide patterns and additive definition of the electrodes might then follow steps 1–9 below for guide fabrication, 10–15 for electrode fabrication and 16–19 for packaging.

1. Acid clean $LiNbO_3$ substrate
2. Coat with e.g. 800 Å Ti metal by evaporation
3. Spin-coat with negative photoresist
4. Bake to drive off resist solvent
5. Expose to mask #1 in mask aligner
6. Develop resist
7. Wet etch to remove unwanted Ti metal
8. Remove remaining photoresist
9. Diffuse Ti metal into $LiNbO_3$ e.g. for 8 hours at 1050°C in furnace
10. Spin-coat with positive photoresist
11. Bake to drive off resist solvent
12. Expose to mask #2 in mask aligner
13. Develop resist
14. Coat with e.g. 1000 Å Al metal
15. Lift off resist, together with unwanted Al metal
16. Cut and polish device ends
17. Package device
18. Bond electrode connections
19. Fibre pigtail

13.10 (i) Preform volume $= 1 \times \pi \times (20 \times 10^{-3}/2)^2$ m^3 $= \pi \times 10^{-4}$ m^3.
Preform mass $= \pi \times 10^{-4} \times 2.6 \times 10^3$ kg $= 0.817$ kg.
Deposition time $= 0.817/2 \times 10^{-3}$ minutes $= 6.81$ hours.
(ii) Conservation of mass in the fibre drawing process suggests that the preform and fibre velocities and diameters must be related by

$$v_p \times \pi \times (d_p/2)^2 = v_f \times \pi \times (d_f/2)^2$$

so that $d_f = d_p \times \sqrt{(v_p/v_f)}$. Hence

$$d_f = 20 \times 10^{-3} \times \sqrt{(3.5 \times 10^{-3}/60 \times 1.5)} \text{ m} \approx 125\ \mu\text{m}$$

The fibre length is then found as

$$\pi \times 10^{-4}/\{\pi \times (125 \times 10^{-6}/2)^2\} = 25600 \text{ m, or } 25.6 \text{ km}$$

Chapter 14

14.1 Assuming an acoustic velocity of 6.57×10^3 m/s, the acoustic wavelength at 500 MHz is

$$\Lambda = 6.57 \times 10^3/500 \times 10^6 = 1.314 \times 10^{-5} \text{ m, or } 13.14\ \mu\text{m}$$

The K-vector associated with the acoustic grating then has magnitude $K = 2\pi/\Lambda = 0.4782 \times 10^6$ m^{-1}. For an effective index of 2.2 and an optical wavelength of 0.633 μm, the propagation constant is $\beta = 2\pi \times 2.2/0.633 \times 10^{-6} = 21.837 \times 10^6$ m^{-1}. Referring to the figure below; for Bragg diffraction of the 0th order beam into the +1th diffraction order, the transducer tilt angle should be chosen so that $2\beta \sin(\phi) = K$, i.e. so that $\phi = \sin^{-1}(K/2\beta) = \sin^{-1}\{0.4782/(2 \times 21.837)\} = 0.627°$.

For a transducer of 200 MHz bandwidth, the extreme operating frequencies are 400 and 600 MHz. Performing a similar calculation at (say) 400 MHz, we obtain

$$\Lambda' = 16.425\ \mu\text{m},\ K' = 0.3825 \times 10^6 \text{ m}^{-1}$$

The optimal transducer angle is now $\phi' = \sin^{-1}\{0.3825/(2 \times 21.837)\} = 0.502°$. If the transducer is fixed at the angle ϕ found above, the optical beam will now be off-Bragg by an angle $\Delta\theta = 0.627 - 0.502 = 0.125°$ or 2.18×10^{-3} rads. The angular bandwidth of a transmission grating of thickness d is Λ/d, so the maximum allowable acoustic beamwidth is $d = \Lambda/\Delta\theta = 16.425 \times 10^{-6}/2.18 \times 10^{-3} = 7.5 \times 10^{-3}$ m, or 7.5 mm.

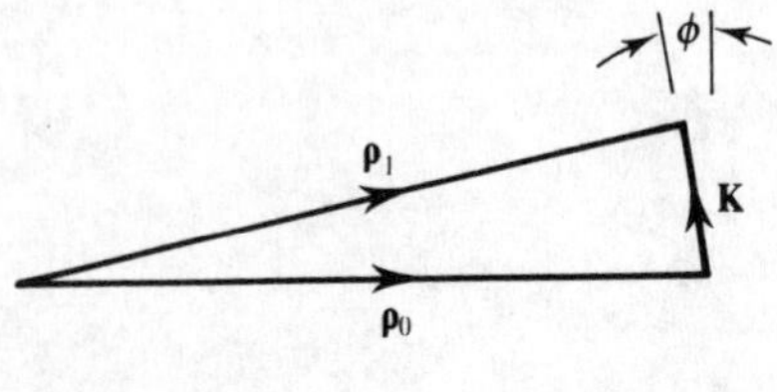

14.2 The resistance of the conductor before scaling is $R = \rho L(w \times t)$, where ρ is the resistivity. After scaling down all dimensions by α, the new resistance is $R' = \rho(L/\alpha)/\{(w/\alpha) \times (t/\alpha)\} = \alpha R$. Regarding the conductor as one plate of a parallel-plate capacitor, the capacitance to the ground-plane before scaling is approximately $C = \varepsilon Lw/h$, where ε is the dielectric constant of the insulator. After scaling, the new capacitance is $C'' = \varepsilon(L/\alpha) \times (w/\alpha)/(h/\alpha) = C/\alpha$. Hence, $R'C' = RC$, so the time constant is unchanged by scaling. This is disadvantageous, because gate operating delays are expected to reduce with scaling, leaving interconnect delays as the limit on performance. Guided wave optical devices do not suffer from RC time constants, as they are transmission-line devices with real impedance. They are also immune to mutual interference.

14.3 The Gray codes for all two-bit numbers are

0	00
1	01
2	11
3	10

A two-bit integrated optic A-to-D converter with a Gray-coded output may be constructed using two Mach-Zehnder interferometers. In contrast to a binary-coded device, each MZI should have the *same* electrode length, and be fitted with additional electrodes designed to apply a static phase shift of π (for the LSB MZI) and $\pi/2$ (for the MSB). The MZI outputs should then appear as the full lines in the figure below. After detection and thresholding, the outputs should be as shown by the shaded blocks.

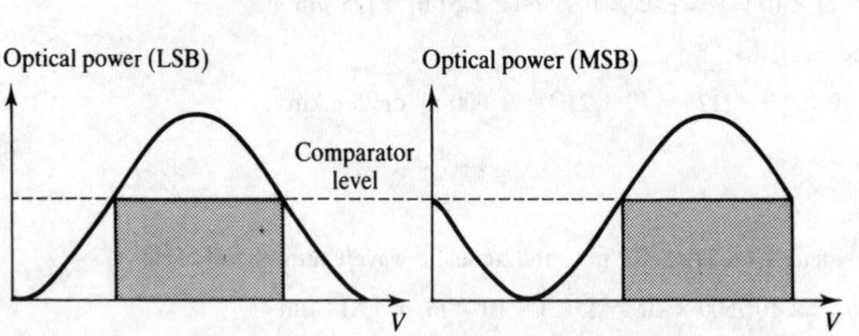

14.4 The phase change accumulated in propagating along the fibre is $\phi = 2\pi n_{\text{eff}}L/\lambda_0$. Differentiating this, we obtair

$$d\phi/dT = 2\pi/\lambda_0 \{n_{\text{eff}}\, dL/dT + L\, dn_{\text{eff}}/dT\}$$

Hence

$$\Delta\phi = 2\pi/\lambda_0 \{n_{\text{eff}}\,(1/L)\, dL/dT + dn_{\text{eff}}/dT\}\, L\Delta T$$

Since

$$\alpha = (1/L)\, dL/dT,\ \Delta\phi = 2\pi/\lambda_0\, \{n_{eff}\,\alpha + dn_{eff}/dT\}\, L\Delta T$$

Assuming that L = 10 cm, λ_0 = 0.85 μm, $n_{eff} \approx 1.458$, $\alpha = 0.004 \times 10^{-4}$/°C and $dn_{eff}/dT = 1.1 \times 10^{-5}$/°C we get

$$d\phi/dT = (2\pi/0.85 \times 10^{-6})\, \{1.458 \times 0.004 \times 10^{-4} + 1.1 \times 10^{-5}\} \times 10^{-1} = 8.562 \text{ rad/°C}$$

Thus dn_{eff}/dT is the dominant effect, contributing roughly 95 per cent of the induced phase shift.

14.5 From the analysis of Section 14.6, we may express the output power from the lower arm of the interferometer as $P = P_0 \sin^2(\Phi/2)$, where P_0 is the input power and Φ is the phase delay between the two arms. For a time delay τ, $\Phi = \omega\tau$, where ω is the angular frequency, so that $P = P_0 \sin^2(\omega\tau/2)$. Identical outputs are thus obtained for time delays $\tau' = 2\nu\pi/\omega \pm \tau$. The normalized sensitivity of the device to changes in τ is $(1/P_0)\, dP/d\tau = \omega/2 \sin(\omega\tau) \Rightarrow$ sensitivity is zero when $\omega\tau = \nu\pi$. The sensitivity is maximum when $\omega\tau = \pi/2 + \nu\pi$.

14.6 Assuming for simplicity that the input signal is described by a continuous linear ramp in frequency, and ignoring any common phase delay, the output amplitude obtained from a Mach-Zehnder sensor in a linear FM system may be written as $A(t) = 1/2\{s(t) - s(t - \tau)\}$, where $s(t) = A_0 \exp\{j2\pi f(t)t\}$ and $f(t) = \{f_1 + (f_2 - f_1)t/T\}$. $A(t)$ is therefore found as the difference between the linear FM input (shown full-line in (a) below) and a delayed copy of the input (shown dashed). Now

$$s(t - \tau) = s(t) \exp(-j2\pi\Delta f_1 t)$$

where the beat frequency Δf_1 is defined by $\Delta f_1 = (f_2 - f_1)\tau/T$, so that

$$A(t) = (1/2)\, s(t)\, \{1 - \exp(-j2\pi\Delta f_1 t)\}$$

The output *power* is then

$$AA^* = A_0^2/4\, \{1 - \exp(-j2\pi\Delta f_1 t)\}\, \{1 - \exp(+j2\pi\Delta f_1 t)\} = P_0/2\, \{1 - \cos(2\pi\Delta f_1 t)\}$$

This implies that the output is now time-varying, at a frequency Δf_1 that is linearly dependent on the imbalance τ. Measurement of Δf_1 may therefore yield τ without ambiguity, and with a constant sensitivity. For a periodic sawtooth frequency modulation, the two combining signals are as shown in (b) below, and the beat frequency oscillates between the two values Δf_1 and $\Delta f_2 = (f_2 - f_1)\,(T - \tau)/T$ as shown in (c) below.

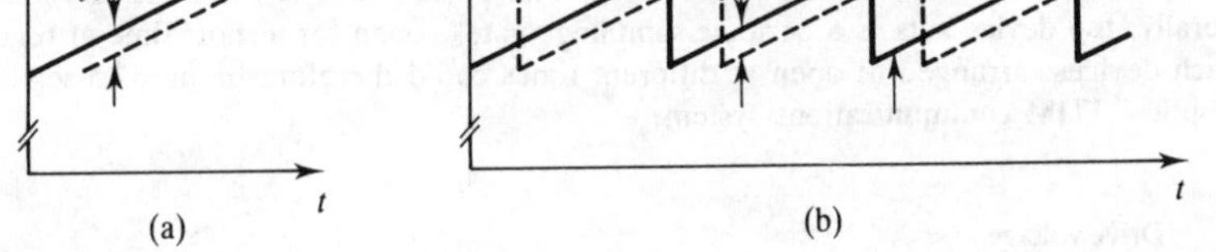

(a) (b)

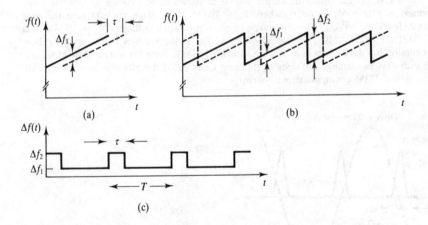

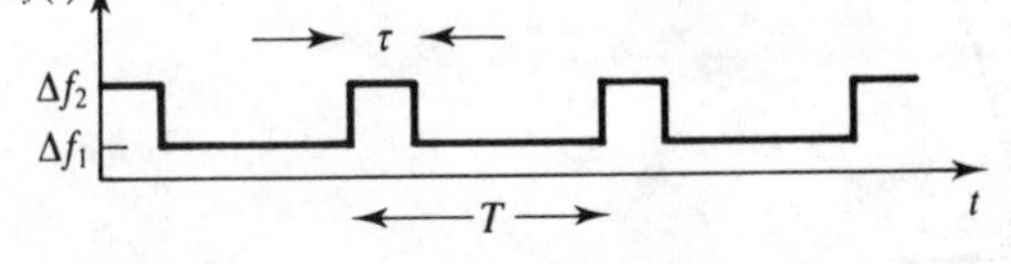

(c)

14.7 From the analysis of Section 14.7, the time-of-flight difference is $\Delta\tau = \Omega\tau D/c$, where $\tau = L/c$, i.e. $\Delta\tau = \Omega LD/c^2$. Assuming $\Omega = 10^{-5}$ rad/s, $L = 10^3$ m and $D = 20 \times 10^{-2}$ m, we obtain

$$\Delta\tau = 10^{-5} \times 10^3 \times 20 \times 10^{-2}/(3 \times 10^8)^2 = 2.222 \times 10^{-20}\ \text{s}$$

This would not be directly measurable. The corresponding phase difference is $\Delta\Phi = \omega\Delta\tau = 2\pi c\Delta\tau/\lambda_0$. Assuming that $\lambda_0 = 0.85\ \mu$m, we obtain

$$\Delta\Phi = 2\pi \times 3 \times 10^8 \times 2.222 \times 10^{-20}/0.85 \times 10^{-6} = 4.927 \times 10^{-5}\ \text{rad}$$

14.8 (i) The response $P_n(\lambda_0) = \text{sinc}^2\ \{\pi(\lambda_0 - \lambda_n)/\Delta\lambda\}$ is shown in (a) below, where $\Delta\lambda$ represents the approximate bandwidth of the filter, or, alternatively, $|\lambda_0 - \lambda_n| = \Delta\lambda$ defines the positions of the first zeros in the filter response.

(ii) For $\Delta\Lambda \geq \Delta\lambda$, the largest interchannel crosstalk will occur when the channel separation is such that the $(n + 1)$ th channel coincides with the first sidelobe in the response of the n th filter, as shown in (b) below. At this point, $\Delta\Lambda = 3\Delta\lambda/2$, and the likely crosstalk is

$$\text{sinc}^2\ (3\pi/2) = (2/3\pi)^2 = 0.045\ 03, \text{ or } -13.46\ \text{dB}$$

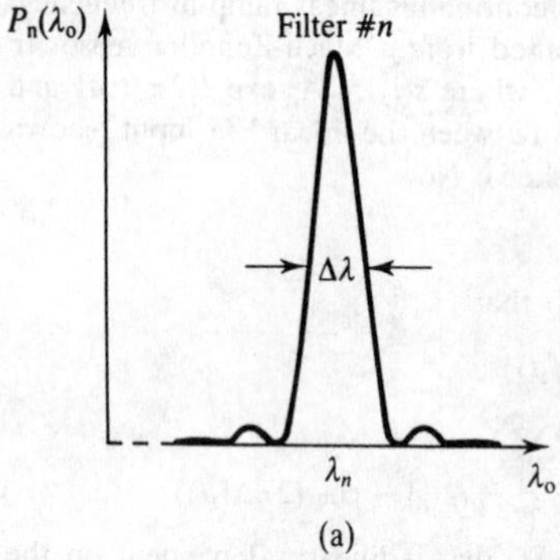

(a)

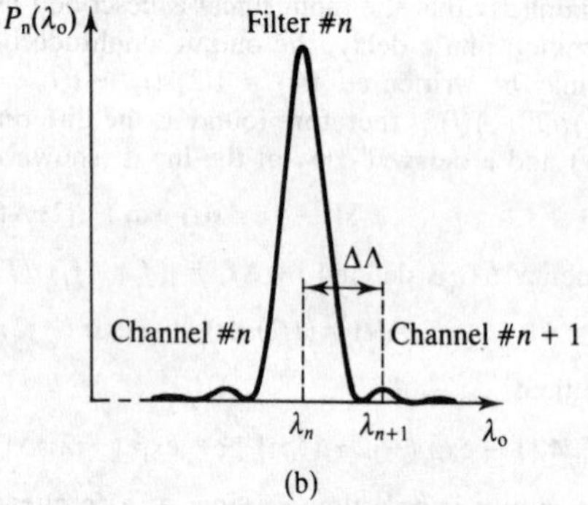

(b)

14.9 The directional coupler will be in the cross-state only near zero volts, when the two guides are synchronous. For a frequency of 1 GHz, the time between zero-crossings is $T = 1/(2 \times 10^9) = 0.5 \times 10^{-9}$ s, or 0.5 ns. Near zero volts the drive voltage varies linearly with time. In this region the cross-coupled output will therefore vary following the directional coupler switch characteristic, and will appear as a short pulse. For a CW optical input, the output will be as shown below. Assuming that the drive signal may be written as $V(t) = V_0 \cos(\omega_m t)$, where $V_0 = 100$ V, the voltage gradient near the zero crossings has magnitude $dV/dt = V_0\omega_m$. The time taken to reach the switch voltage of $V_s = 10$ V is thus $t = V_s/\{dV/dt\} = 10/\{100 \times 2\pi \times 10^9\} = 1.592 \times 10^{-11}$ s. The pulse width is therefore $\Delta t = 2t = 31.83$ ps. More generally, the device acts as a periodic sampling "gate", open for a short time at regular intervals. Several such devices, arranged to open at different times could therefore be used to separate the channels in an optical TDM communications system.

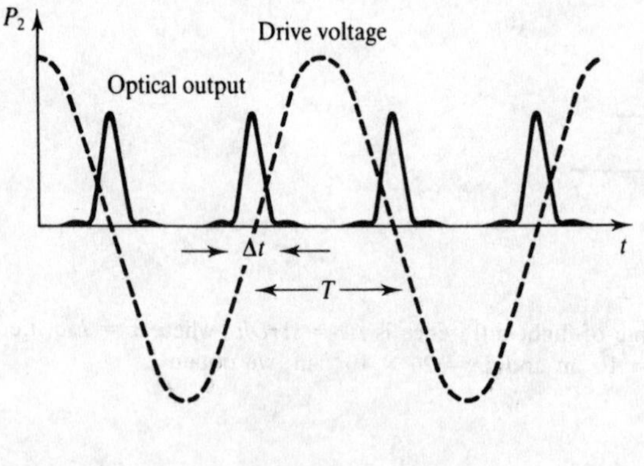

14.10 The requirement for a 1 mW signal at the detector places a constraint of 10 dB maximum loss in the system, assuming a source power of 10 mW. The power budget is

25% launch efficiency	6 dB loss
10 splices, each contributing 0.1 dB loss	1 dB loss
Propagation through 50 km of 0.2 dB/km SM fibre	10 dB loss
Total	17 dB loss

The optical amplifier must therefore have a gain of not less than 7 dB. Assuming a gain of 0.25 dB/m from the erbium-doped fibre, a 28 m fibre length is required.

INDEX